SEMICONDUCTOR-RELATED ACRONYMS

Acronym	Meaning
eV	Electron volt
FAMOS	Floating gate MOS memory element
FET	Field effect transistor
FIT	One device failure in 10^9 device hours
FPD	Focal plane deviation
FTIR	Fourier transform infrared spectroscopy
FZ	Float zone
GEW	Gram equivalent weight
GGG	Gadolinium gallium garnet
HEPA	High-efficiency particulate air (normally written as HEPA filter)
HMDS	Hexamethyldisilizane (adhesion promoter)
IC	Integrated circuit
IDLH	Immediately dangerous to life or health
IGFET	Insulated gate field effect transistor
IR	Infrared
JFET	Junction field effect transistor
KMER	Kodak metal etch resist
KTFR	Kodak thin-film resist
LCD	Liquid crystal display
LEC	Liquid encapsulation
LED	Light-emitting diode
LEED	Low-energy electron diffraction
LFPD	Local focal plane deviation
LOCOS	Local oxidation of silicon
LPCVD	Low-pressure chemical vapor deposition
LPE	Liquid phase epitaxy
LSI	Large-scale integration
MBE	Molecular beam epitaxy
MCZ	A magnetic Czochralski crystal puller
MESFET	Metal semiconductor field effect transistor
mips	Million instructions per second
MISFET	Metal insulator semiconductor field effect transistor
MNOS	Metal–nitride–oxide–semiconductor
MOCVD	Metalorganic chemical vapor deposition
MOMBE	Metalorganic molecular beam epitaxy
MOS	Metal oxide semiconductor
MOSFET	Metal oxide semiconductor field effect transistor
MOST	Metal oxide semiconductor transistor
MOVPE	Metalorganic vapor phase epitaxy

Note: Many of these acronyms are not used in the text; some that are related to devices are included here for the sake of completeness.

SEMICONDUCTOR INTEGRATED CIRCUIT PROCESSING TECHNOLOGY

W. R. RUNYAN
Retired, formerly with Texas Instruments

K. E. BEAN
Texas Instruments

Addison-Wesley Publishing Company

Reading, Massachusetts • Menlo Park, California • New York
Don Mills, Ontario • Wokingham, England • Amsterdam • Bonn
Sydney • Singapore • Tokyo • Madrid • San Juan

This book is in the Addison-Wesley Series in Electrical and Computer Engineering

Cover photos (top to bottom): Courtesy of Bell Laboratories; Courtesy of Texas Instruments; Courtesy of Intel Corporation (®Registered trademark of Fairchild Semiconductor Corporation); Courtesy of Gordon Teal; Courtesy of Texas Instruments Incorporated.

Library of Congress Cataloging-in-Publication Data

Runyan, W. R.
Semiconductor integrated circuit processing technology / Walter R. Runyan, Kenneth E. Bean.
p. cm.
Bibliography: p.
Includes index.
ISBN 0-201-10831-3
1. Integrated circuits—Design and construction.
2. Semiconductors—Design and construction. I. Bean, Kenneth E.
II. Title.
TK7874.R86 1990
621.381'73—dc19 89-320
CIP

Reprinted with corrections May, 1994

7 8 9 10 DOC 9796

Foreword

Forty years ago the transistor was hailed as a model of the scientific process of invention. Although Shockley's work provided a basic understanding of junction transistor structure, little was really known about the fabrication processes used for semiconductors. Terms such as "deathnium" and "forming" were used to describe effects that were not understood. Most of the techniques used were crude adaptations of technology developed for other purposes, and much of the learning was empirical.

Today, after many thousands of man-years of effort, the situation has changed. New processes have been developed specifically for semiconductor use, and others have been carefully optimized. By the standards of the 1950s, all are now well understood.

The authors of this book, Walt Runyan and Ken Bean, have been key contributors to this new understanding. Both have been actively involved with semiconductors since the beginning; both, in the development of new processes and the application of processes to high-volume production.

Today, semiconductor manufacture is truly a science based industry. This book provides an excellent summary of the state-of-the-art in modern semiconductor processing, both in terms of the scientific knowledge of the field and in its practical application to devices.

J. S. Kilby

Preface

The field of semiconductor process engineering is now so broad that it is virtually impossible for a single individual to provide process engineering support or process development for the complete semiconductor manufacturing operation. Thus, subdivision and specialization are required. However, since each subdivision is intricately related to all others, it is absolutely necessary that each technology specialist (engineer) have a good understanding of all phases of the manufacturing process. This book is intended to provide a unified treatment of that portion of the semiconductor manufacturing process commonly referred to as "wafer fabrication."

The book is to be used as a text by university students and as a reference by practicing engineers. It provides background, theory, and practical discussions related to the wafer fabrication process. It is not, however, a compendium of all of the latest research activities, which are constantly changing and continually covered in various technical journals and review book series. The serious student, whether in industry or university, must keep up-to-date by following the current technical journals most closely allied with his or her chosen specialty.

The level of presentation in the book assumes that the student has had introductory physics, introductory inorganic chemistry, mathematics through differential equations, a first course in transistor device physics, and a course covering electronic material properties including crystal structure. All of these courses are generally available by the senior year so the text is suitable for either senior- or graduate-level courses.

The book is organized into eleven chapters and three appendices. All chapters have an extensive list of references, and all but the first two chapters have problems and a list of key ideas. Since silicon remains the semiconductor used for 99 percent of the integrated circuits, it is naturally emphasized. However, where GaAs processing differs substantially from silicon, it is discussed separately.

The first chapter is an historical discussion of the development of the integrated circuit. The second chapter gives an overview of the processing steps required to manufacture an integrated circuit. Chapters 3 through 10 discuss the various technologies required for the steps, and each chapter has a section on safety. The chapter subjects are, respectively, silicon thermal oxidation, deposition of thin inorganic films, photolithography, etching, epitaxial growth, solid state diffusion, ion implantation, and contacts and interconnects.

Chapter 11 is devoted to a discussion of yields, yield analysis, and yield economics. The latter is included because all too often the student (and in some cases, practicing engineers as well) lose sight of the fact that to be practical for industrial use, semiconductor processes must be economical as well as scientifically feasible. Emphasis is on defect density measurement and interpretation, on the use of I–V plots to determine the cause of faulty integrated circuit elements, and on how yield and an economical manufacturing operation are related.

The first of three appendices is a review of crystallography as it relates to silicon and gallium arsenide. The second is a review of phase diagrams, and the third is a compilation of numerical constants and conversions. Finally, an extensive list of semiconductor-related acronyms is given inside the front and back covers of the book.

Acknowledgments: Over the course of time spent in preparing this book, many people have been very helpful, and in particular, we would like to acknowledge Hettie Smith, who typed the manuscript and took care of many of the details involved in its preparation, as well as Marion Johnson and John Powell, who provided many of the photographs. We also want to acknowledge the following people who read and commented on various portions of the manuscript: Neal Akridge, ATEQ Corp.; Frederick Strieter, Honeywell Corp.; Bruce Deal, Advantage Production Technology, Inc.; Howard Huff, Sematech; George Brown, Dwayne Carter, Rinn Cleavelin, Monte Douglas, P. B. Ghate, John Hatcher, William Keenan, Charlotte Tipton, Michael Tipton, Rick Wise, and Richard Yeakley, all of Texas Instruments Incorporated; and Isaac Trachtenberg, Chemical Engineering Department at the University of Texas. In addition, we want to thank the following people for their helpful reviews of the manuscript: D. K. Schroder, Arizona State University; John D. Shott, Stanford University; Dean P. Neikirk, University of Texas, Austin; Arthur B. Glaser, AT&T Bell Laboratories; and Richard C. Jaeger, Auburn University.

We would also like to thank our editor, Tom Robbins, and his staff for guidance during the various stages of manuscript preparation. Lastly, we wish to express our appreciation for the patience, encouragement, and indulgence of our wives, Delma Runyan and Helen Bean.

W. R. Runyan
K. E. Bean

Contents in Brief

Detailed Contents

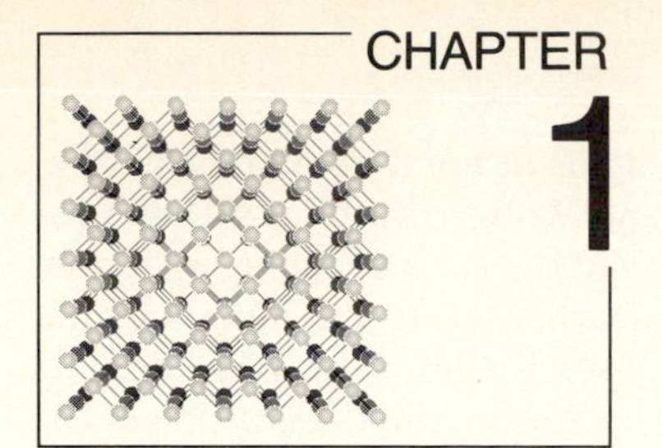

CHAPTER

1

Historical Overview

1.1 BACKGROUND

The semiconductor technology that ultimately led to the integrated circuit (IC) was over 75 years in the making.[1] Around 1875, it was observed that selenium exhibited rectification and photoconductivity. A silicon diode of sorts for detecting radio waves was described in 1906 (1). By 1935, selenium rectifiers and photodetectors, silicon carbide varistors, and galena (naturally occurring lead sulfide) and silicon point contact diodes for radio detection were on the market (2). Also in 1935, a British patent was issued, but was never reduced to practice, for a thin-film field effect transistor (3).

The development of radar just prior to and during World War II placed a great deal of emphasis on the study of silicon (Si) and germanium (Ge) since they were singled out as the most appropriate materials for the mixer and detector diodes required in the radar detection circuitry. Because of this activity, commercial sources of high-purity silicon and germanium were developed. Professor Lark-Horovitz of Purdue University prevailed on Eagle Pitcher Mining and Smelting Company to develop a high-purity germanium supply, which they did, based on extraction from the lead and zinc ores mined in the tri-state area of Kansas, Missouri, and Oklahoma. (This was the only known significant source of germanium ore in the Western Hemisphere.) During the same period, Professor Seitz of the University of Pennsylvania initiated a program at DuPont to produce high-purity silicon (4).

---

[1] The few events and names mentioned in this section should neither be construed as representing the total semiconductor activity in that period nor as even necessarily being the most significant. They were chosen to give some idea of progress as a function of time, and in fact, a large number of people, both in industry and in the universities, contributed to the development. To single out a few names and places may seem unfair, but to do otherwise would turn a processing technology text into a volume of semiconductor history. (For more definitive information, see reference 2 and the first 85 references contained therein.)

1.2 THE TRANSISTOR

In December of 1947, the first transistor (a point contact structure) was constructed by John Bardeen and Walter Brattain of Bell Telephone Laboratories. For a discussion of the events leading up to that occasion and of the subsequent work that led to the junction transistor (and to the Nobel Prize in physics for Bardeen, Brattain, and Shockley), see Shockley's paper "The Path to the Conception of the Junction Transistor," *IEEE Trans. on Electron Dev. ED-23*.

William Shockley (seated at the microscope), John Bardeen (at left), and Walter Brattain, all of Bell Telephone Laboratories, are pictured here about the time of their discovery of the point contact transistor in 1947. This work resulted in their receiving the Nobel Prize for Physics in 1956. (*Source:* Photograph courtesy of AT&T Archives.)

The first point contact transistor used polycrystalline germanium for the semiconducting material. However, when the first paper describing the transistor was submitted for publication (June 25, 1948), the transistor effect had been demonstrated in silicon as well (5). By the end of 1949, single-crystalline rather than polycrystalline material was being used (2). This conversion from polycrystalline to single-crystal source material (spearheaded by Gordon Teal of Bell Telephone Laboratories) has been one of the most significant advances in semiconductor technology since the invention of the transistor itself. Without a source of large single crystals with uniform properties, the high-volume production of small devices would have been difficult and development of large-area integrated circuits impossible.

Efforts to build useful structures using polycrystalline or amorphous material thus far have had only spotty success. Amorphous silicon solar cells are now commercially feasible, and diodes utilized as fuses for programming are sometimes built into polycrystalline silicon integrated circuit leads. However, in all cases, single-crystal devices outperform comparable devices made in amorphous or polycrystalline material.

John Bardeen, co-inventor of the point contact transistor, received a second Nobel Prize in physics in 1972 for his work on the theory of superconductivity. (*Source:* Photograph courtesy of AT&T Archives.)

Although the point contact transistors were expensive and unreliable, they were, overall, superior in many respects to vacuum tubes.[2] Consequently, they went into production at Western Electric's Allentown plant in 1951 as replacements for vacuum tubes in some telephone exchange applications. However, by the time the transistor process was licensed to other manufacturers in 1952, William Shockley had invented the grown junction transistor, and Gordon Teal and Morgan Sparks had reduced it to practice (6). This transistor became the industry workhorse for several years. It differed from all subsequent transistor structures in that junction for-

[2] Early guided missiles often used magnetic amplifiers instead of vacuum tubes in their control systems because of reliability considerations. The point contact transistor, which in retrospect appears to have been a very fragile and unreliable structure, was in fact good enough to replace the magnetic amplifiers and greatly improve the frequency response of the control systems.

FIGURE 1.1

Construction details of a silicon grown junction transistor. (Germanium pnp transistors used doped gold wires for the base contact.)

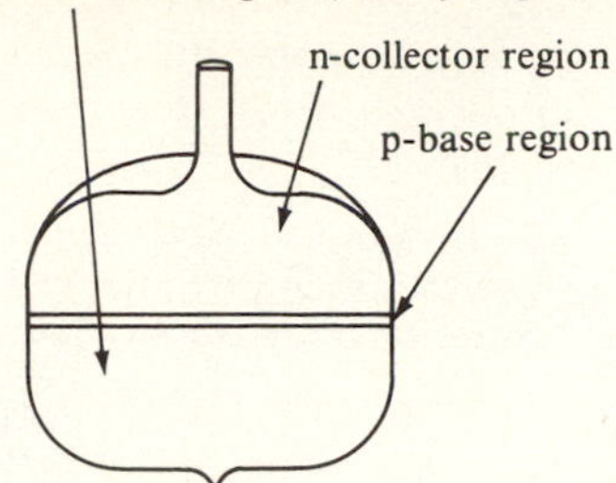

(a) Crystal cut in half for evaluation

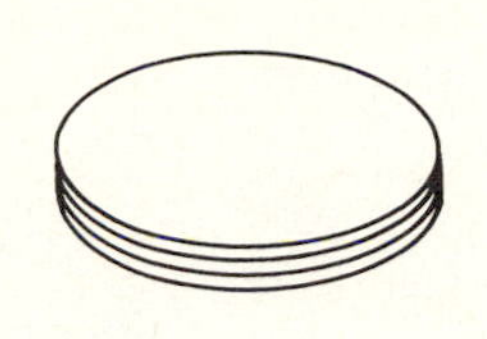

(b) Section cut from mid-region of crystal that contains base layer and small section each of collector and emitter regions

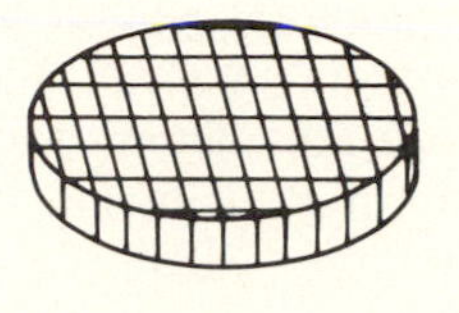

(c) Section cut into "bars" (size, varying with power rating of transistor, but typically 30× 30×150 mils)

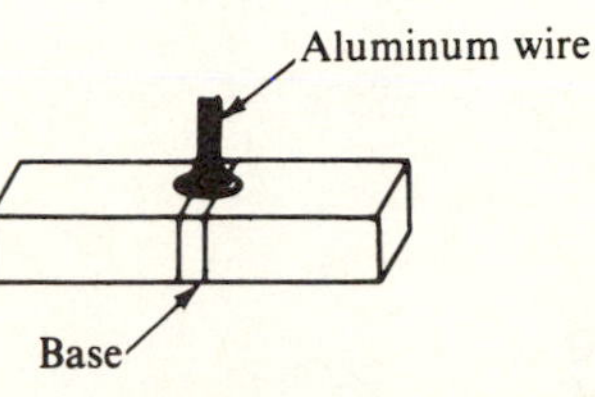

(d) Base contact formed by fusing (alloying) aluminum wire to silicon so that wire made ohmic contact to base region and rectifying contacts to collector and emitter

Walter Brattain was a co-inventor of the point contact transistor. (*Source:* Photograph courtesy of AT&T Archives.)

William Shockley, co-inventor of the point contact transistor, was also the inventor of the junction transistor. (*Source:* Photograph courtesy of AT&T Archives.)

mation was completed during the crystal-growing operation. Fig. 1.1 shows the construction details of a silicon grown junction transistor. The remainder of the manufacturing steps were to affix leads, package, and test. Fig. 1.2 shows a transistor bar[3] mounted in a header. All that remained to be done here was to add a cover and final-test.

Even though both silicon and germanium were demonstrated at the beginning to be usable for transistors, germanium was much more tractable, and silicon transistors (desirable for their higher-temperature capability) did not become commercially available until

[3] As can be seen from Fig. 1.2, the bit of silicon used in each transistor really did look like a bar. For people and companies that worked extensively with grown junctions, the term *bar* became so deeply rooted that over 30 years later integrated circuits 1 cm on a side and 0.5 mm thick may still be called bars. Those whose experience started with alloyed or diffused units used the terms *chip* or *die*, which today are still common expressions.

FIGURE 1.2

Grown junction transistor bar mounted in a header. (The collector and emitter connections support the bar; the fine wire is the base connection.)

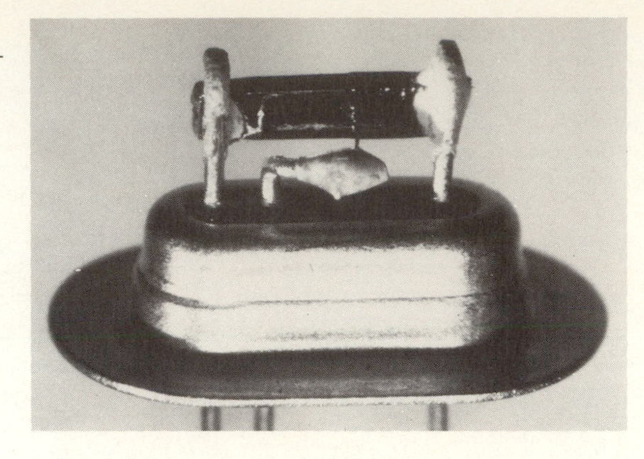

1954 when grown junction silicon transistors were introduced by Texas Instruments Incorporated (7).

Alloyed junction germanium transistors were reported by Saby of General Electric in 1951 during the same meeting at which the grown junction transistor was first openly discussed (8). These transistors were made by alloying a dot of metal, usually indium, into each side of a chip of germanium, as shown in Fig. 1.3. A pn junction formed at each alloy–germanium interface and provided the collector–base and emitter–base junctions, respectively. Germanium alloyed transistors were very successful and became an alternative to the previously introduced germanium grown junction transistors. However, silicon alloyed transistors were difficult to construct and never commanded a very large market share.

Gordon Teal pioneered the development of single-crystal growth of germanium and silicon and the use of high-purity single-crystal material for semiconductor applications. (*Source:* Photograph courtesy of Gordon K. Teal and AT&T Archives.)

The next important development was the use of gaseous diffusion at Bell Laboratories to make the necessary junctions from one side of a semiconductor slice (9). This activity gave rise to the diffused mesa transistors (10) that were commercially available in germanium in 1957 and in silicon in 1958. Fig. 1.4 shows the steps in their fabrication. There were actually several versions, only one of which is shown here. These mesa transistors had an advantage over either the grown junction or alloy transistors in that the diffusion process could give much narrower and more controlled base widths and thus allow higher-frequency operation. Since many transistors could be made at one time on each silicon or germanium slice, and since many slices could be cut from a single crystal, mesa transistors were less expensive to manufacture than grown junction transistors were. In the latter case, only one slice—or, if the crystal were rate grown (11), perhaps five or six—containing the grown junctions was available from each crystal. Diffused transistors were also less expensive than alloy transistors were since the alloy process required that each chip be processed individually rather than in slice form.

FIGURE 1.3

Construction details of a germanium alloyed transistor.

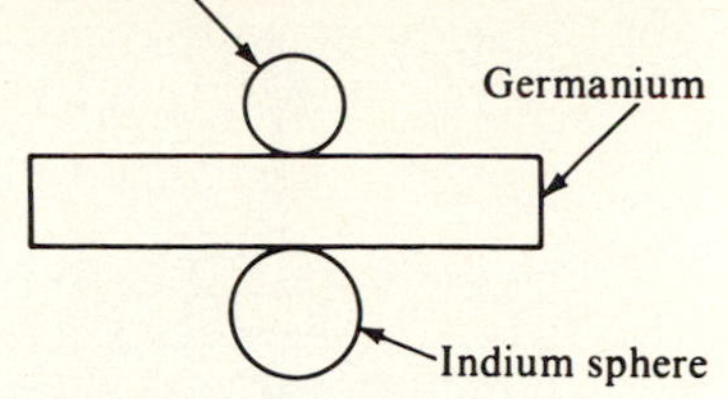

(a) Germanium chip at room temperature

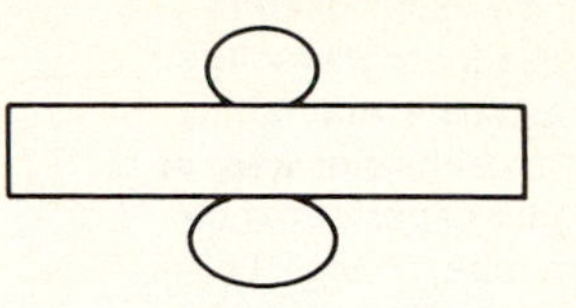

(b) At 156° C, when indium begins to melt

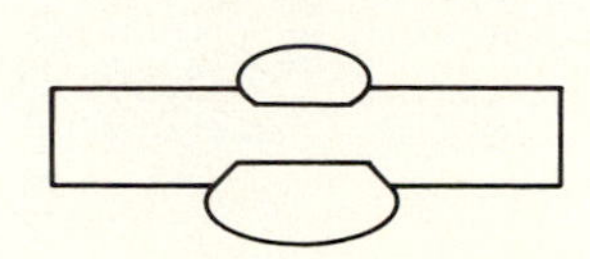

(c) At 550° C, when indium dissolves germanium

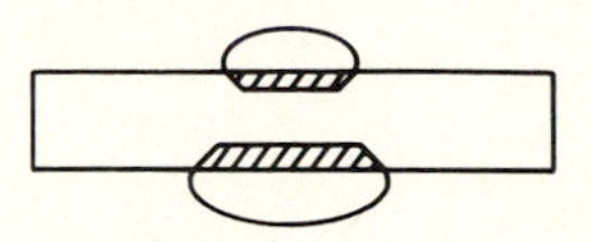

(d) At cooling stage, when germanium doped with indium regrows, providing collector–base and emitter–base junctions

FIGURE 1.4

Fabrication steps for a double diffused mesa transistor.

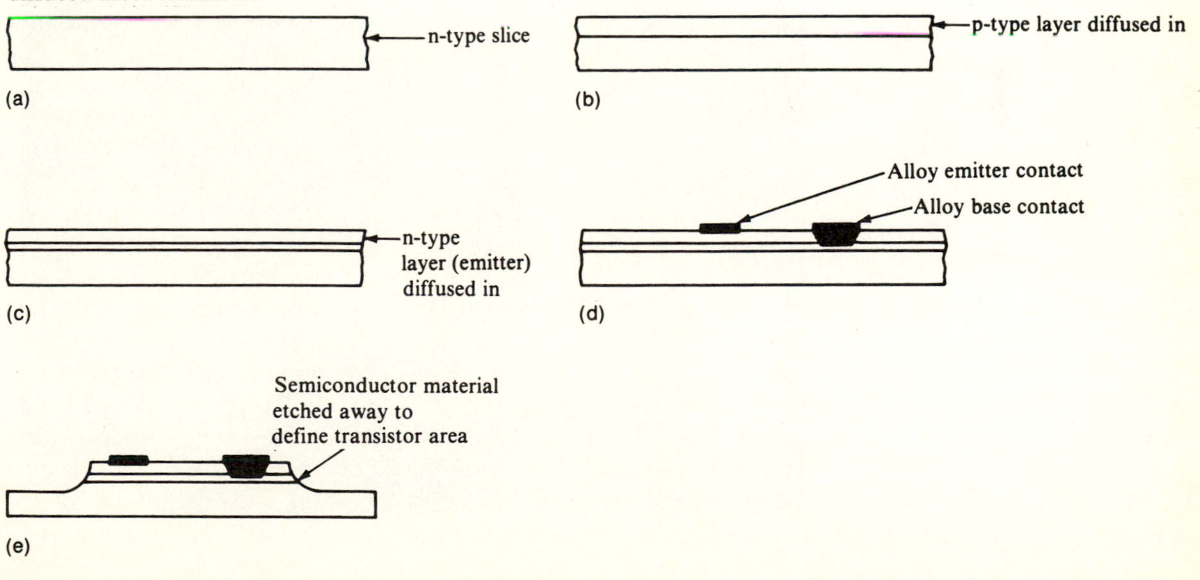

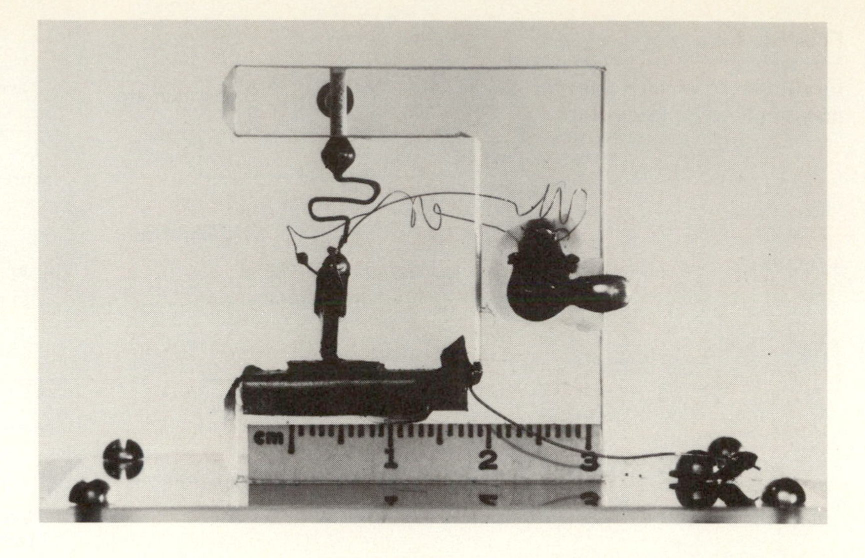

The point contact transistor went into production at Western Electric in 1951 and was licensed to other companies for manufacture. Although it was somewhat temperamental, this transistor was, in general, superior in performance to the vacuum tube. (*Source:* Photograph courtesy of AT&T Archives.)

Gordon Teal and Morgan Sparks made the first junction transistor, the construction of which eliminated many of the reliability problems of the point contact transistor. (*Source:* Photograph courtesy of AT & T Archives.)

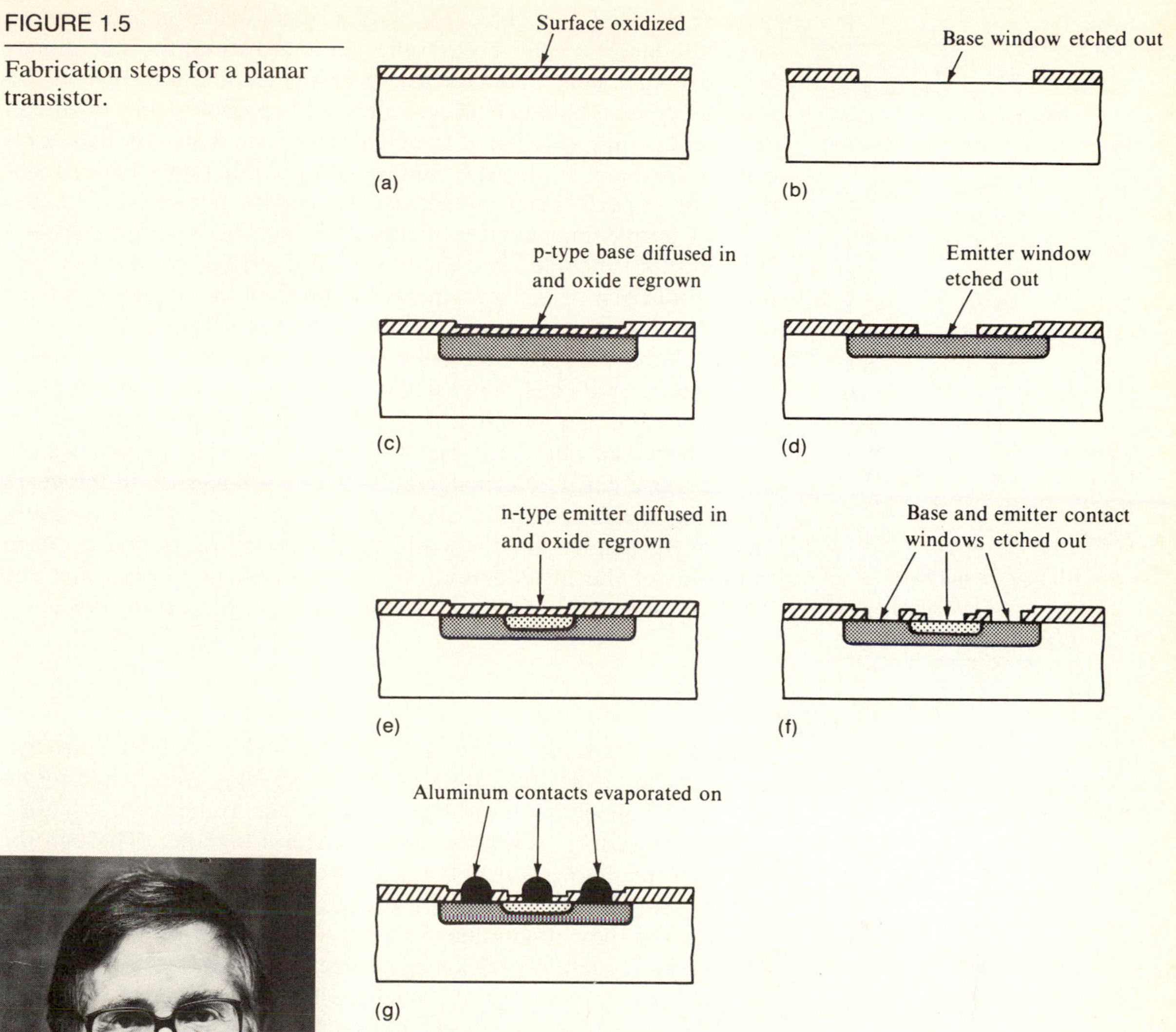

FIGURE 1.5

Fabrication steps for a planar transistor.

Jean Hoerni invented the planar process, which dramatically improved the reliability of silicon diodes and transistors. (*Source:* © Carolyn Caddes, 1986, from *Portraits of Success*.)

Unfortunately, all three processes (grown junction, alloy, and mesa) continued to have one serious flaw, and that was the exposed junctions at the semiconductor surface. Various coatings were developed to minimize electrical leakage across those junctions, but they were never completely successful.

The silicon planar transistor, invented by Jean Hoerni of Fairchild Semiconductor, followed shortly thereafter (12) and was in production by 1961. As Fig. 1.5 shows, it very effectively solved the

TABLE 1.1

Transistor Sales 1954–1966

Year	Germanium Units	Silicon Units
1954	1.3	0.02
1955	3.6	0.09
1956	12.4	0.42
1957	27.7	1.0
1958	45.0	2.1
1959	77.5	4.8
1960	119	8.8
1961	178	13.0
1962	214	26.6
1963	249	50.1
1964	289	117
1965	334	273
1966	369	481

Note: All units in millions.

Source: Electronic Industries Association Yearbook, 1967.

junction protection problem for silicon by allowing the junctions to terminate under a silicon oxide thermally grown on the silicon surface. No other semiconductor has an oxide that can be grown in situ and that possesses the near ideal electrical characteristics of silicon dioxide. The introduction of the planar process, which increased reliability and improved yields, along with the inherently better high-temperature performance of silicon, marked the beginning of the decline of germanium devices. Thus, the four-year period between 1957 and 1961 included first the move to large-scale production with the introduction of the mesa process and then the beginning of the conversion from germanium to silicon with the advent of planar processing. Table 1.1 shows the growth of transistor production through those years, and while the yearly growth rate was no larger in those four years than in any other, it was because of those advances that the phenomenal rate could be sustained. This table also shows the dramatic manner in which silicon overtook germanium in the years following the introduction of the planar process.[4] For a business view of the semiconductor world in this transition period between the times of the introduction of mesa and planar processing, see "Report on Semiconductors," *Business Week*, March 26, 1960.

1.3 THE INTEGRATED CIRCUIT

Until the invention of the integrated circuit, complete circuits, whether they used transistors or vacuum tubes, were made by individually connecting the various components (tubes or transistors, diodes, capacitors, resistors, and inductors) together. Through the years, many improvements were introduced to increase packing density, simplify fabrication, and increase reliability of electronic circuits, but they all continued to depend on the wiring together of discrete and separately packaged devices. In an early effort to attack the problem from a fresh approach, G.W.A. Dummer of the Royal Radar Establishment of the United Kingdom suggested during the May 1952 Electronics Components Symposium in Washington, D.C., an integrated approach using a monolithic block comprising "layers of insulating, conducting, rectifying and amplifying mate-

[4] It should be appreciated that "production by" and "introduced in" dates are a bit difficult to pinpoint since these dates depend on the first time anyone made and reported such a device, the first time samples were shown to a potential customer, or the time at which reasonable quantities were available to customers. Similar problems arise in other areas, such as the year of introduction of larger crystal diameters.

Jack Kilby is the inventor of the Solid Circuit® integrated circuit and a co-inventor of the pocket calculator. (*Source:* Photograph courtesy of Jack Kilby. ®Registered trademark of Texas Instruments Incorporated.)

rials, the electrical functions being connected directly by cutting out areas of the various layers." A metal model of how such a structure might be made was shown in 1957.

The figures included in a patent filed in May of 1953 by Harwick Johnson (U.S. Patent 2,816,228, issued December 10, 1957) bear a superficial resemblance to an integrated circuit. However, as expressed in the first sentence of the patent, both the discussion and claims relate only to a transistor phase-shift oscillator: "This invention pertains to semiconductor devices and particularly to semiconductor phase-shift oscillators and devices." Component isolation was not considered so that even if the concepts of the patent were extended to devices other than the phase-shift oscillator, the class of devices that could be made would be very limited.[5]

In February of 1959, Jack Kilby of Texas Instruments Incorporated filed a patent application describing a concept that allowed, using relatively simple steps, the fabrication of all of the necessary components of the desired circuit, both active and passive, in a single piece of semiconductor and their interconnection in situ. The patent states:

> In contrast to the approaches to miniaturization that have been made in the past, the present invention has resulted from a new and totally different concept for miniaturization. . . . In accordance with the principles of the invention, the ultimate in circuit miniaturization is attained using only one material for all circuit elements and a limited number of compatible processing steps for the production thereof. . . .

[5] Considerable space is being devoted to a discussion of patents and subsequent patent litigation during the development of the integrated circuit. This discussion is provided because of the historical significance and to demonstrate several practical points of the patent world quite germane to the reader. One is that the fame and fortune of an inventor may depend on the precise wording of patent claims. Unfortunately, that wording is often considered to be the domain of the patent attorney and sometimes does not exactly reflect the inventor's intent. (See, for example, the discussion of the lack of a particularly specific figure in Kilby's patent.) Another is that in any new and developing technology, the key problems will be apparent to those active in the field, and they will be offering their individual solutions at about the same time. (Note the close grouping of the events of Table 1.2, which appears later in the chapter.) It is thus very important for the engineer or scientist to keep good records. Yet another point, perhaps not quite as obvious, is that the history of invention as depicted in the popular press, in the scientific literature, and in the issued patents may not be in agreement. To decide on the "true" inventor requires a careful study of all of the literature and also may depend on the particular definition of "inventor" that is chosen.

Up to this point, the goals are perhaps not much different from those expressed in 1952 by Dummer. However, to continue with the Kilby patent:

> In a more specific conception of the invention, all components of an electric circuit are formed in or near one surface of a relatively thin semiconductor wafer characterized by a diffused p–n junction or junctions . . . .
>
> It is a primary object of the invention to provide a miniaturized electronic circuit wherein the active and passive circuit components are integrated within a body of semiconductor material, the junctions of such components being near and/or extending to one face of the body, with components spaced or electrically separated from one another as necessary in the circuit . . . .
>
> Figures 1–5a illustrate schematically various circuit components fabricated in accordance with the principles of the present invention in order that they may be integrated into, or as they constitute parts of, a single body of semiconductor material; . . .

The figures and text describe bulk resistors, diffused resistors, pn junction capacitors, MOS capacitors, transistors, and diodes. In the press coverage of the March 1959 announcement of the Kilby concept, this set of standard components was stressed (13). The patent text continues:

> Because all of the circuit designs described above can be formed from a single material, a semiconductor, it is possible by physical and electrical shaping to integrate all of them into a single crystal semiconductor wafer containing a diffused p–n junction, or junctions, and to process the wafer to provide the proper circuit and the correct component values. . . .

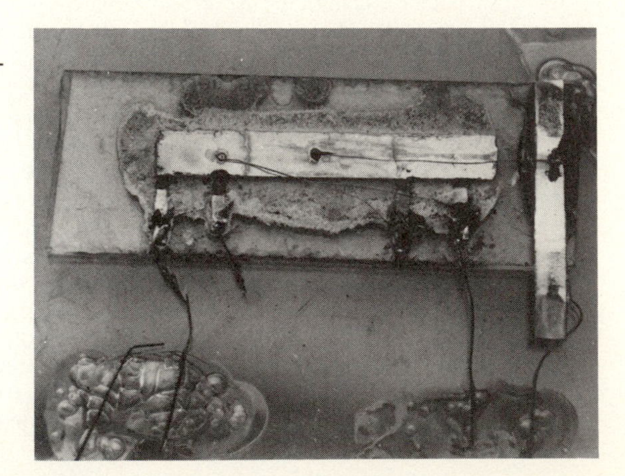

FIGURE 1.6

One of the first working germanium integrated circuits constructed by Jack Kilby in the summer of 1958. (*Source:* Courtesy of Texas Instruments Incorporated.)

FIGURE 1.7

Advertisement for the first commercially available silicon integrated circuits appeared in the March 26, 1960 issue of *Business Week*. (*Source:* Courtesy of Texas Instruments Incorporated.)

AVAILABLE NOW FROM TI

100:1 miniaturization with SOLID CIRCUIT* semiconductor networks

Now—3 years ahead of industry's expectations—*Solid Circuit* semiconductor networks from Texas Instruments for many of your high-reliability miniaturized systems!

Solid Circuit networks are a major departure from conventional components because they integrate resistor, capacitor, diode, and transistor functions into a single high-purity semiconductor wafer. Protection and packaging of discrete elements is eliminated, and contacts between dissimilar materials are minimized, reducing element interconnections as much as 80%. Fabrication steps have been reduced to one-tenth those required for the same circuit function using conventional components.

Only a few process steps and time-proved TI mesa production techniques permit a high degree of process control in *Solid Circuit* network fabrication. The result of these facts: *reliability is built into* each *Solid Circuit* network.

If you need to reduce equipment size and weight—or to design a more complex system in the same size—investigate *Solid Circuit* networks for your missile, satellite, space vehicle, and other microelectronic programs. TI engineers are ready to custom design this concept to your requirements. Contact your nearest TI Sales Engineer today. The TI Type 502 *Solid Circuit* network is immediately available for your evaluation.

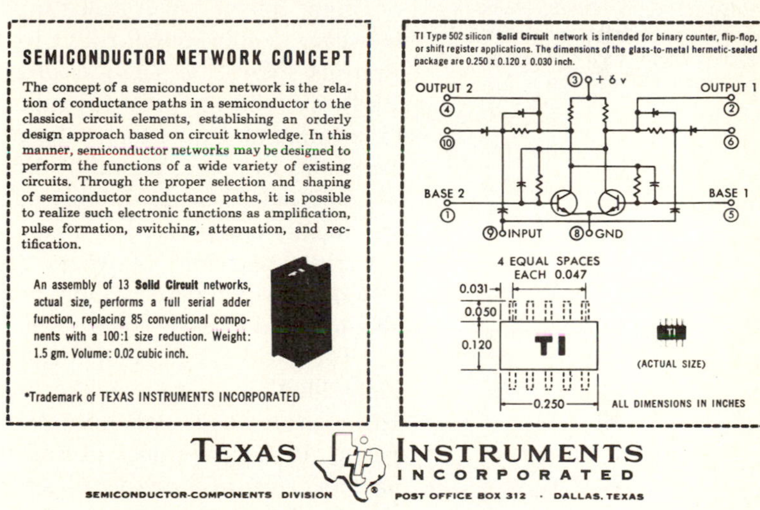

The first working integrated circuit was built by Kilby in September of 1958 and was, for expediency, fabricated of germanium (Fig. 1.6). The first commercially available integrated circuits, announced in March of 1960 with the advertisement shown in Fig. 1.7, were made of silicon.

It was after the mesa process and before the planar process that Kilby's patent application was filed, and, because of the manner in

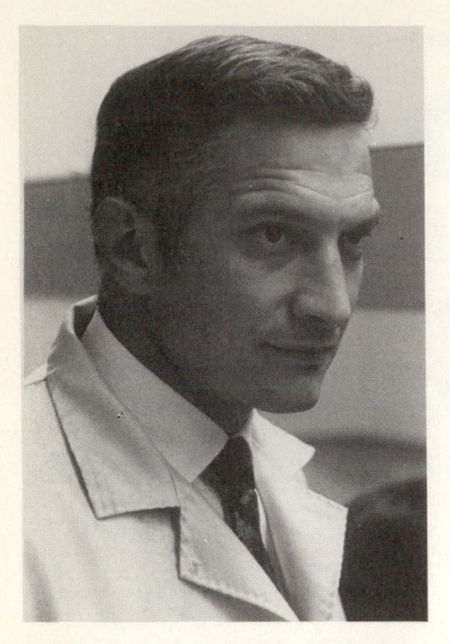

Robert Noyce is the inventor of the Microchip® integrated circuit and a co-founder of both Fairchild Semiconductor Corporation and Intel Corporation. (*Source:* Photograph courtesy of Intel Corporation.) ®Registered trademark of Fairchild Semiconductor Corporation.

which the IC construction was described in the Kilby patent (14), when the planar process was introduced, a separate patent on an IC using planar technology and adherent leads was awarded to Robert Noyce of Fairchild Semiconductor (15). That award was the subject of a heated legal battle that arose over whether Kilby's patent statement as follows:

> Instead of using the gold wires 70 in making electrical connections, connections may be provided in other ways. For example, an insulating and inert material such as silicon oxide may be evaporated onto the semiconductor circuit wafer through a mask either to cover the wafer completely except at the points where electrical contact is to be made thereto, or to cover only selected portions joining the points to be electrically connected. Electrically conducting material such as gold may then be laid down on the insulating material to make the necessary electrical connections.

covered the portion of the Noyce claims that described ". . . an electrical connection to one of said contacts comprising a conductor adherent to said layer . . . ." The disagreement centered around whether "laid down" was equivalent to "adherent to." The Board of Patent Interference, ruling in Kilby's favor, asserted that it was. However, a subsequent ruling by the Court of Customs and Patent Appeals (*Noyce* v. *Kilby*; *Kilby* v. *Noyce*, decided November 6, 1969) reversed the previous rulings and allowed the Noyce claims. (The Supreme Court then refused to review the case.) Contrary to assertions by some, the ruling did not depend on whether gold could or could not be made suitably adherent to silicon oxide. The Court specifically commented on that aspect. The ruling depended on the Court's assessment of whether or not someone reading Kilby's statement would be inevitably drawn to the conclusion that the lead should be adherent.

The Noyce patent application[6] was filed a few months after Kilby filed his, as well as after the Kilby concept was made public. However, it was issued April 25, 1961, three years before the Kilby patent. Perhaps because of this anomaly of issuing dates, and perhaps because of a mixing of the idea of a planar silicon integrated circuit with the already announced integrated circuit concept of Kilby, there has been considerable controversy in some circles over the true inventor of the integrated circuit. The sequence of events as portrayed in the patent and technical literature is shown in Table

[6] His first notebook date was in January of 1959, before the public announcement of the integrated circuit concept by Texas Instruments, but well after Kilby's original notebook entry and after the Kilby approach had been described to various government agencies.

1.2. Comments by some of the participants may be found in references 16–18. Additional comments are in references 19–20. Probably the most balanced assessment of Kilby's and Noyce's relative contributions is contained in the citations of the Franklin Institute's 1966 Ballantine Medal award, which they shared. Kilby was credited for "conceiving and constructing the first working monolithic circuit in 1958," and Noyce for "his sophistication of the monolithic circuit for more specialized use, particularly in industry."

One interesting aspect of the sequence shown in Table 1.2 is the issuance of the K. Lehovec patent for the use of a plurality of pn junctions for isolating integrated circuit components. It has been considered by some as being key to integrated circuit development in that it provided the necessary component isolation (21). Histori-

TABLE 1.2

Key Events in Development of the Integrated Circuit

Date	*Event*	*Responsible Party*
February 6, 1959	File date of U.S. Patent 3,138,743, which stated: "This invention relates to miniature electronic circuits, and more particularly to unique integrated electronic circuits fabricated from semiconductor material."	J.S. Kilby
March 6, 1959	Public announcement of concept described in Kilby patent application.	Texas Instruments Incorporated
April 22, 1959	File date of U.S. Patent 3,029,366, which covered electrical isolation through the use of at least two pn junctions between each component and described conductive ink interconnections over deposited oxide.	K. Lehovec
May 1, 1959	File date of U.S. Patent 3,025,580, which covered the planar process for making planar transistors.	J.A. Hoerni
July 3, 1959	File date of U.S. Patent 2,981,877, which stated: "Its principal objects are these: to provide improved device-and-lead structures for making electrical connections to various semiconductor regions; to make unitary circuit structures more compact and more easily fabricated in small sizes than has heretofore been feasible; and to facilitate the inclusion of numerous semiconductor devices within a single body of material." This patent covered the use of planar transistors and metallization running over the oxide.	R.N. Noyce
August 1959	Public announcement of transistors built by planar process.	Fairchild Semiconductor Corporation
March 26, 1960	Public announcement of availability of solid circuit silicon multivibrator.	Texas Instruments Incorporated
October 1960	Technical meeting at which planar process was described.	J.A. Hoerni

cally, pn junction isolation of a single discrete device was mentioned in a 1954 Shockley patent describing the use of ion implantation in transistor fabrication (22). Fig. 1.8 is a circuit diagram taken from that patent. Shockley's discussion relating to the pn junction was as follows: "It will be noted that the junction between the *N* body and the *P* layer is biased in the reverse direction. Hence the body serves essentially as a passive support for the layer." Then, before the Lehovec patent, the Kilby patent had already addressed the general problem of isolation when several components were present in one wafer and suggested various ways of providing it. One was by the use of high-resistivity material (used in gallium arsenide circuits), and one was by the use of pn junction isolation. The pertinent Kilby patent text is as follows:

> Of importance to this invention is the concept of shaping. This shaping concept makes it possible in a circuit to obtain the necessary isolation between components and to define the components or, stated differently, to limit the area which is utilized for a given component. Shaping may be accomplished in a given circuit in one or more of several different ways. These various ways include actual removal of portions of the semiconductor material, specialized configurations of the semiconductor material such as long and narrow, L-shaped, U-shaped, etc., selective conversion of intrinsic semiconductor material by diffusion of impurities thereinto to provide low resistivity paths for current flow, *and selective conversion of semiconductor material of one conductivity type to conductivity of the opposite type wherein the p–n junction thereby formed acts as a barrier to current flow* [italics added].

The Lehovec patent was filed after the filing of the Kilby patent and later was the subject of an interference proceeding by Kilby. Since the Kilby discussion supports pn junction isolation, if one of his figures had included *two* diffused resistor components instead of just one, the Lehovec patent probably would never have been is-

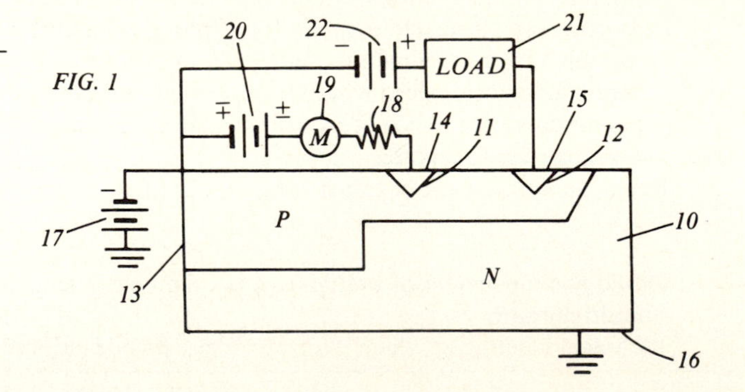

FIGURE 1.8

Fig. 1 from U.S. Patent 2,666,814 (filed April 27, 1949; issued January 19, 1954) showing use of a reverse-biased pn junction for electrical isolation.

FIGURE 1.9

(a) Use of an epitaxially grown thin layer of n-type silicon on a p-type slice to provide part of the required electrical isolation. (b) Use of that layer combined with localized n-type diffusions to improve performance of transistors built in the layer over the diffused regions.

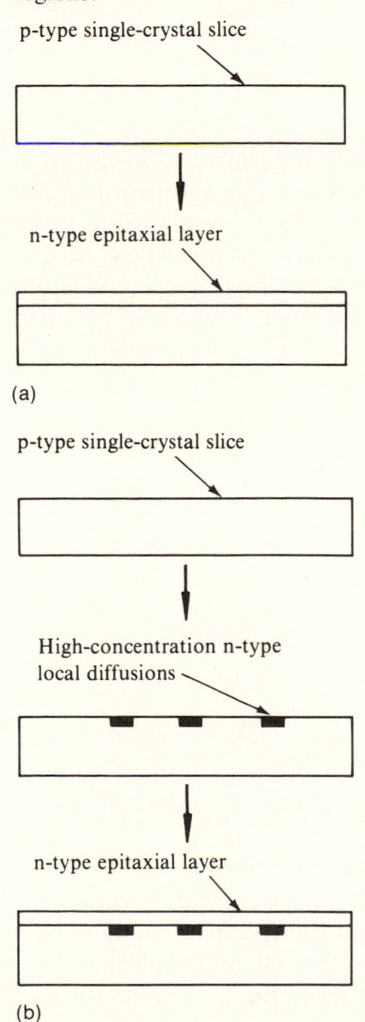

sued. The Board of Patent Interference ruled that Kilby's patent did not cover the Lehovec claims. It did not, however, rule on whether the Lehovec patent covered the methods typically used for IC isolation. In fact, the claims do not appear to support the case of ordinary bipolar IC isolation, although that aspect has not been addressed by the courts. Lehovec's text describes the use of two or more junctions for isolation that extend from one face of the semiconductor body to the other but does not mention the typical IC isolation case in which all junctions terminate on the same side of the wafer. His claims typically state:

> A multiple semiconductor assembly comprising a semiconductor slice having a plurality of regions of alternating p and n conductivity types to thereby provide a plurality of p–n junctions, two of said regions being separated by at least two of said p–n junctions, . . .

The "two of said regions being separated by at least two of said p–n junctions, . . ." would apply to an IC, but "a semiconductor slice having a plurality of regions of alternating p and n conductivity types to thereby provide a plurality of p–n junctions . . ." does not appear to apply since isolation for a whole chip is normally provided by just one p-type region and multiple n-type regions.

Despite the fact that there were no interconnection claims, Lehovec described in the body of his patent an interconnection method that used conductive ink over an evaporated quartz insulating layer. Since conductive inks are adherent and since his patent was filed before Noyce's, one can speculate that had a related claim been included in his patent application, he would now have the patent on IC interconnections.

Two later developments, while not directly related to IC development, contributed greatly to bipolar IC manufacturing ability and performance. One was the use of epitaxial overgrowth (23), which in IC applications allowed a thin n-type layer to be overgrown on a high-resistivity p-type substrate, as shown in Fig. 1.9a. The other was the use of localized high-concentration diffusions before the growth of the epitaxial layer, as shown in Fig. 1.9b, in order to reduce transistor collector resistance (24).

While early IC work was all bipolar, in 1960, a practical MOS transistor was announced (25, 26), and by 1962, a MOS IC consisting of 16 silicon n-channel transistors had been constructed (27). Efforts to use germanium, either by itself or overgrown onto semi-insulating gallium arsenide (GaAs), as a material for ICs continued for several years but never proved successful. GaAs was, however, by the mid-1980s being used for niche applications requiring very-high-frequency ICs.

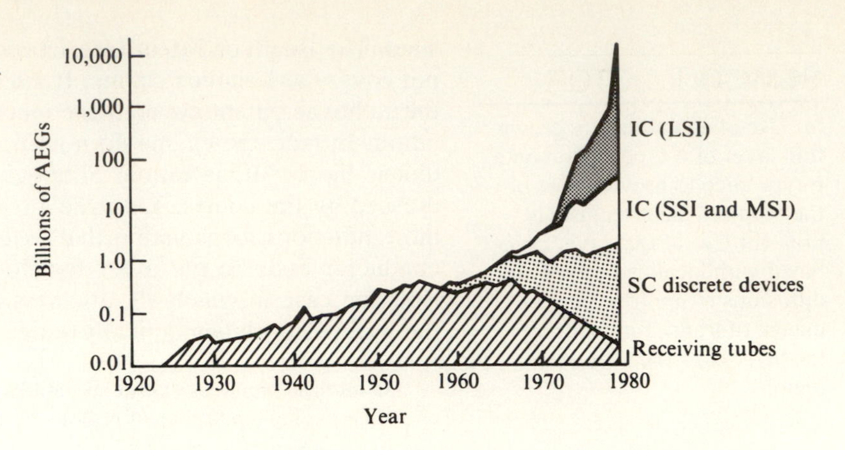

FIGURE 1.10

U.S. usage of small-signal AEGs and their distribution among various categories. (*Source:* Adapted from J. Fred Bucy, Keynote Address to IEEE Solid-State Circuits Conference, San Francisco, 1980.)

Silicon thus remains the mainstream material, and as shown in Fig. 1.10, the U.S. usage of silicon ICs had reached astronomical proportions by 1980. As represented in Fig. 1.10, this usage is given in terms of AEGs, or active element groups, which might, for example, consist of a single-stage amplifier, a single gate, or a single bit of memory. The term was apparently first applied in a limited manner to vacuum tube circuits and then by Patrick Haggerty (*IEEE Spectrum 1,* pp. 80–82, June 1964) to transistor circuits. Each AEG is made up of various combinations of the following:

1. Bipolar transistors, with or without multiple emitters or collectors
2. pn junction diodes, which can also be used as capacitors
3. Conventional metal-plate capacitors
4. MOS transistors
5. MOS capacitors
6. MIS transistors
7. Schottky diodes
8. Diffused resistors
9. Thin-film resistors
10. Interconnecting leads

Conspicuously missing from this list of available components are inductors. It is true that small inductors can be made by spiraling a conductor and that at some frequencies there can be enough phase shift to make a transistor behave as though inductance is present. However, for all practical purposes, the IC world remains inductorless.

1.4 SEMICONDUCTOR PROCESSES

The major semiconductor wafer fabrication steps in the mesa process of 1958 were diffusion, metallization, silicon etching, and various cleanups. It was with these steps that the first ICs were made. The planar process of 1960 introduced three more steps: thermal oxidation of silicon (as a step separate from diffusion), lithographic printing of a pattern in a photoresist layer on top of the oxide, and the etching of that pattern in the oxide. Since 1960, there have been only four processing innovations that allow new kinds of structures or circuits to be fabricated, and they were all developed in the 1960s. They were as follows:

1. The use of epitaxial layers in bipolar circuits to provide an easy method of device isolation. (Later, epitaxial layers were deposited over previously diffused areas of the circuit to reduce collector saturation resistance and time, thus providing high-speed circuitry.)
2. The use (in 1963) of epitaxial overgrowth of silicon onto insulating substrates such as sapphire.
3. The use of chemically vapor-deposited (CVD) materials, such as silicon nitride, first (in 1965) for protective overcoats and later as diffusion masking material.
4. The use of ion implantation (first practical application in 1968 for MOS threshold adjusting) as an adjunct to diffusion.

This is not to imply that the industry has been stagnant, however. Tremendous strides have been taken in increasing the complexity of IC circuitry available and in decreasing the cost of each AEG. However, these advances have been accomplished primarily through the use of more sophisticated equipment and through process variants such as the use of plasma etching instead of wet etching and optical steppers and the use of electron beam writers or X-ray printing instead of contact printers or the original handpainted black wax for patterning. Table 1.3 illustrates the growth in capability and cost of printing technology during the past 30 years. The increasing complexity of patterning equipment is typical of the other processes as well.

Besides the advances made in processing equipment, other parts of the industry have changed as well. The increase in circuit complexity can be judged by comparing the first commercially available ICs (like the one in the advertisement shown in Fig. 1.7, which consisted of a single flip-flop with two transistors, four diodes, four capacitors, and six resistors) with a large MOS memory such as the one megabit DRAM chip shown in Fig. 1.11, which contains a mil-

TABLE 1.3

The Increasing Complexity of Patterning Equipment

Year	*Size of Feature Patterned*	*Steps in Operation*	*Equipment Cost per Machine in Year of Purchase*	*Comments*
1957	10 mils*	Handpaint black wax on area to be protected; etch mesas	$0.10	Camel's hair brush
1960	2 mils	Apply photosensitive layer ("resist"); use contact printer and mask to expose desired area; develop resist; etch feature	$2,000	In-house constructed contact printer
1965	0.6 mils	Contact printer	$5,000	Commercial contact printer
1970	0.4 mils	Contact printer	$15,000	Commercial contact printer
1975	0.15	Optical projection printing	$80,000	Optical projection printing
1980	0.08	Optical projection printing	$120,000–500,000	Optical projection printing
1985	0.06 mils	Optical projection or electron beam printing	$500,000–3,000,000	Optical projection or electron beam printing
1990†	0.024	Optical projection, electron beam, or X-ray printing	$500,000–3,000,000	Optical projection or electron beam printing

*The semiconductor industry began by using mils, and that terminology is used for comparison throughout this table.

†Projected.

Note: For a description of patterning equipment, see Chapters 2 and 5.

lion memory cells plus thousands of additional transistors for various other operations. In ICs of the 1960s, chip area was perhaps 0.01 cm^2; it is now approaching 1 cm^2. Thus, while chip size has increased, it has by no means followed commensurately with circuit complexity. The 100× increase in area has been accompanied by a millionfold increase in circuit elements.

The slice diameter has also steadily increased since the introduction of the IC, as shown in Table 1.4. In some cases, the impetus has been to improve yields; in some, to reduce equipment and facility costs; and in others, to reduce labor costs. Not shown by the table is the small dip in the slice diameter that occurred in 1960–1962 because of the difficulty in producing uniform epitaxial depositions on slices of diameters larger than 18–20 mm.

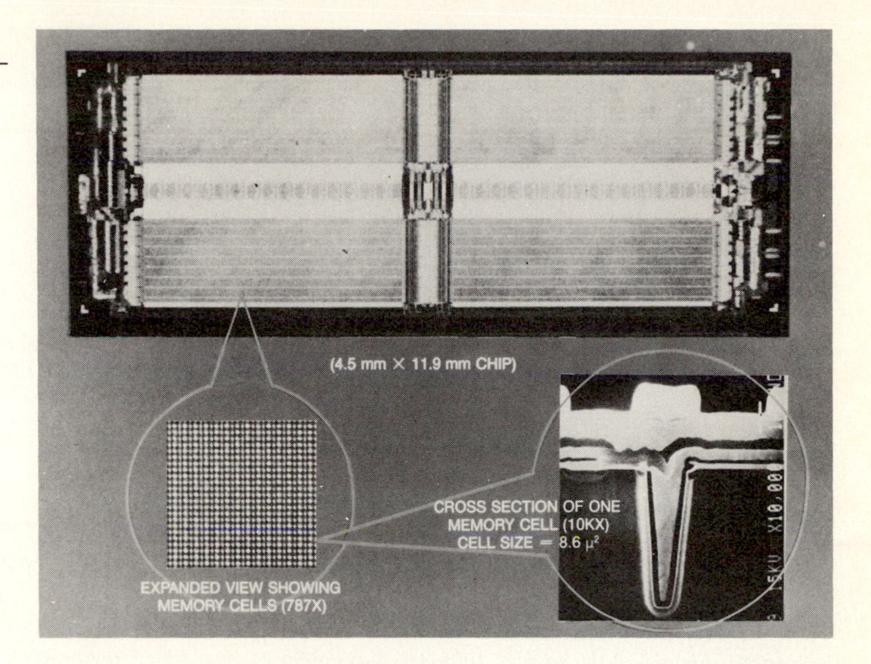

FIGURE 1.11

One megabit CMOS DRAM chip. (*Source:* Courtesy of Texas Instruments Incorporated.)

TABLE 1.4

Silicon Crystal Size Versus Year

Year	*Diameter (mm)*
1955	25*
1960	25
1965	37
1970	50
1975	75
1980	100
1985	150

*Grown junction.

Circuit design has become much more sophisticated and attuned to space-saving features since the first flip-flops were introduced. Two or three layers of interconnections between circuit elements are now used, and the number of active devices required to perform a given function has decreased. For example, some of the early DRAM memory cells required six transistors, whereas now one transistor and one capacitor suffice. Also, much more attention is given to designs that are more tolerant of the expected process variations. Simultaneously, process control has been continually tightened through better understanding of the processes and the introduction of microprocessor control of most of the equipment. With the recognition that most yield loss is associated with particulates on the slice, extensive effort has been expended to provide better cleanups, cleaner and purer chemicals, and a cleaner manufacturing environment.

Probably the most dramatic way of displaying the effect of all of the improvements on semiconductor manufacturing during the past three decades is to compare silicon usage with semiconductor output, as is done in Table 1.5. To dramatize the effect of the original inventions and the subsequent improvements on our daily lives can be done by mentally eliminating all of the things around us that depend on transistors for their successful operation.

TABLE 1.5

Silicon Usage Compared with Semiconductor Output

Year	*Polycrystalline Silicon (tons)*	*Silicon Transistors (millions)*	*IC AEGs (billions)*
1955	5	0.09	—
1960	30	8.8	—
1965	50	273	—
1970	500	—	2
1975	1500	—	200
1980	3000	—	10,000

Sources: Data on polycrystalline silicon from reference 2 and industry estimates. Data on silicon transistors from *Electronic Industries Association Yearbook*, 1967.

CHAPTER 1 REFERENCES

1. G.W. Pickard, "Thermo-Electric Wave Detectors," *Electrical World 4*, p. 1003 (L), November 24, 1906.
2. G.K. Teal et al., Chap. 6, "Semiconductor Materials," in James F. Young and Robert Shane, eds., *Material Processes*, Marcel Dekker, New York, 1985.
3. Oskar Heil, "Improvement in or Relating to Electrical Amplifiers and Other Control Arrangements and Devices," British Patent 439,457, September. 26, 1939.
4. Henry C. Torrey and Charles A. Whitmer, *Crystal Rectifiers*, McGraw-Hill Book Co., New York, 1948.
5. J. Bardeen and W.H. Brattain, "The Transistor, A Semi-Conductor Triode," *Phys. Rev. 74*, pp. 130–231, 1948.
6. W. Shockley, M. Sparks, and G.K. Teal, "p–n Junction Transistors," *Phys. Rev. 83*, pp. 151–162, 1951.
7. G.K. Teal, "Some Recent Developments in Silicon and Germanium Materials and Devices," paper presented at National IRE Conference, Dayton, May 10, 1954.
8. J.S. Saby, "Fused Impurity p–n–p Junction Transistors," *Proc. IRE 40*, pp. 1358–1360, 1952.
9. G.L. Pearson and C.S. Fuller, "Silicon p–n Junction Power Rectifiers and Lightning Protectors," *Proc. IRE 42*, p. 760 (L), 1954.
10. M. Tanenbaum and D.E. Thomas, "Diffused Emitter and Base Silicon Transistors," *Bell Syst. Tech. J. 35*, pp. 1–22, 1956. Charles A. Lee, "A High-Frequency Diffused Base Germanium Transistor," *Bell Syst. Tech. J. 35*, pp. 23–24, 1956.
11. R.N. Hall, "p–n Junctions Produced by Rate Growth Variations," *Phys. Rev. 88*, p. 139, 1952.
12. Jean A. Hoerni, "Planar Silicon Transistors and Diodes," paper presented at IRE Electron Devices meeting, Washington, D.C., October 1960. J.A. Hoerni, "Method of Manufacturing Semiconductor Devices," U.S. Patent 3,025,589, March 20, 1962 (filed May 1, 1959).
13. "Ultramicroscopic Circuits," *Electronic Design*, April 29, 1959.
14. J.S. Kilby, "Miniaturized Electronic Circuits," U.S. Patent 3,138,743, June 23, 1964 (filed February 6, 1959).
15. R.N. Noyce, "Semiconductor Device-and-Lead Structure," U.S. Patent 2,918,877, April 25, 1961 (filed July 30, 1959).
16. Jack S. Kilby, "Invention of the Integrated Circuit," *IEEE Trans. on Electron Dev. ED-23*, pp. 648–654, 1976.
17. Robert Noyce, "Microelectronics," *Scientific American 237*, pp. 63–69, 1977.
18. Kurt Lehovec, "Invention of p–n Junction Iso-

lation in Integrated Circuits," *IEEE Trans. on Electron Dev. ED-25*, pp. 495–496, 1978.

19. M.F. Wolff, "The Genesis of the Integrated Circuit," *IEEE Spectrum 13*, pp. 45–53, August 1976.
20. C. Lester Hogan, "Reflections on the Past and Thoughts about the Future of Semiconductor Technology," *Interface Age 2*, pp. 24–36, March 1977.
21. Sorab K. Ghandhi, *The Theory and Practice of Microelectronics*, John Wiley & Sons, New York, 1968.
22. W. Shockley, "Semiconductor Translating Device," U.S. Patent 2,666,814, January 19, 1954 (filed April 27, 1949).
23. H.C. Theuerer et al., "Epitaxial Diffused Transistors," *Proc. IRE 48*, pp. 1642–1643, 1960.
24. B.T. Murphy, "Monolithic Semiconductor Devices," U.S. Patent 3,237,062, February 22, 1966 (filed October 20, 1961).
25. D. Kahng and M.M. Atalla, "Silicon-Silicon Dioxide Field Induced Surface Devices," paper presented at IRE-AIEE Solid-State Device Conference, Pittsburg, 1960.
26. Dawon Kahng, "A Historical Perspective on the Development of MOS Transistors and Related Devices," *IEEE Trans. on Electron Dev. ED-23*, pp. 655–657, 1976.
27. S.R. Hofstein and F.P. Heiman, "The Silicon Insulated-Gate Field-Effect Transistor," *Proc. IEEE 51*, pp. 1190–1202, 1963.

CHAPTER 2

Processing Overview

2.1 INTRODUCTION

In Chapter 1, several processing operations were referenced, and some steps were described to illustrate differences in construction of various early devices. In this chapter, the individual processes required to produce an IC will be briefly discussed, and then the manner in which they are combined to make the circuit elements required for a complete IC will be described.

The steps required to produce ICs, beginning with the smelting of either metallurgical-grade Si or metallurgical-grade Ga and As, are listed in Table 2.1. As indicated in the table, several operations are not considered to be part of the IC chip fabrication process.[1] These steps will be considered only briefly, and only in this chapter.

Typically, separate companies specialize in various aspects of silicon material preparation. Some manufacture semiconductor-grade Si, and some grow the single crystals and supply the slices polished and ready for the first oxidation. Similarly, GaAs compounding, crystal growth, and slice cutting and polishing are generally performed by a separate materials company.

While there must be a considerable amount of interaction between the slice supplier and the wafer-fab process engineers, substantial differences exist in the engineering disciplines required, and the division is a natural one. Tables 2.2 and 2.3 give typical silicon and gallium arsenide slice specifications and illustrate the multitude of slice parameters that are important to the performance and yield of ICs. The gross mechanical specifications relate to the need for the slice to fit standardized handling equipment. The primary and secondary flats (ground on the crystal before it is sliced) are used to indicate the orientation of the slice surface and to provide an azi-

[1]The manufacturing facility for IC fabrication is usually referred to as a "front-end" or a "wafer-fab" (wafer fabrication) area.

TABLE 2.1

IC Manufacturing Operations

Manufacturing Operation	*Discussion*
Material Preparation	
Semiconductor-grade Si and GaAs manufacturing	This chapter
Crystal growing	This chapter
Slice cutting and polishing	This chapter
Photomask manufacturing	Chapter 5
Wafer Fabrication	
Cleaning of surfaces	Various chapters
Growth of epitaxial layer	Chapter 7
Thermal oxidation of silicon	Chapter 3
Patterning of the various layers (lithography)	Chapter 5
Diffusion of impurities into silicon	Chapter 8
Ion implantation of impurities	Chapter 9
Chemical vapor deposition of polycrystalline silicon	Chapter 7
Etching of silicon and GaAs	Chapter 6
Deposition of insulating layers (silicon oxide or nitride)	Chapter 4
Etching of insulating layers (silicon oxide or nitride)	Chapter 6
Deposition of conductive layer (usually metal)	Chapter 10
Etching of conductive layers (metal, polysilicon, other)	Chapter 6
Alloying (sintering) to form metal–silicon electrical contact	Chapter 10
Backgrinding (thinning of wafer by grinding)	This chapter
Multiprobing (DC electrical testing of each IC on wafer)	Chapter 11, briefly
Assembly/Test	
Cutting or breaking of wafers into individual chips	Not covered
Packaging of individual chips	Not covered
Full AC and DC electrical testing of packaged ICs	Not covered

muthal orientation reference. The SEMI Standards for the orientation of these flats (which manufacturers usually adhere to) are shown in Fig. 2.1. Flatness affects the amount of pattern distortion produced during the printing operation. The (flatness terms used in the specification sheets are defined in the section on flatness in Chapter 5.) The oxygen content of silicon affects the slice resistance to thermal shock and the concentration of some kinds of crystal defects that form during processing. Resistivity is dictated by the elec-

TABLE 2.2

Silicon Slice Specifications:
(100) and (111) Orientation

Item	100 mm Diameter		125 mm Diameter		150 mm Diameter	
	n-type	*p-type*	*n-type*	*p-type*	*n-type*	*p-type*
Resistivity (ohm-cm)						
Max.	20.0	40.0	20.0	40.0	20.0	40.0
Min.	0.008	0.010	0.008	0.010	0.008	0.010
Resistivity gradient (%)	10		10		10	
Carbon conc. (ppma)						
Standard	< 2.0		< 2.0		< 2.0	
Low	< 0.3		< 0.3		< 0.3	
Oxygen conc. (ppma)						
High	30–40		30–40		30–40	
Medium	26–33		26–33		26–33	
Low	24–30		24–30		24–30	
Dislocation density/cm^2	<100		<100		<100	
Diameter tolerance	±0.5 mm		±1.0 mm		±1.0 mm	
Primary flat length	30–35 mm		40–45 mm		55–60 mm	
Secondary flat length	16–20 mm		25–30 mm		35–40 mm	
Slice thickness						
Min. available	375 ± 25 μm		625 ± 25 μm		675 ± 25 μm	
Max. available	710 ± 25 μm					
Typical FPD* (microns)	3.5 over 100% usable area		4.0 over 100% usable area		4.0 over 100% usable area	
Max. FPD (microns)	5.0 over 95% usable area		5.5 over 95% usable area		5.5 over 95% usable area	
Max. taper (μm)	10.0		10.0		15.0	
Max. bow (μm)	35.0		40.0		50.0	

*Local flatness screening if requested.

trical properties desired of the circuit elements. Crystal orientation affects the semiconductor surface properties and thus device performance.

The material processing flows prior to entry into wafer fabrication are shown in Fig. 2.2 and Fig. 2.3. Metallurgical-grade Si is made in large quantities by reducing quartz sand with coke in an electric furnace. It is then used in the manufacture of some grades of steel and as an alloying material in high-strength aluminum. The reaction of anhydrous HCl with metallurgical-grade Si in the 800°C range gives compounds such as $SiCl_4$ (silicon tetrachloride) and $SiHCl_3$ (trichlorsilane). $SiHCl_3$ is used as raw material for the silicon industry and for semiconductor-grade Si manufacturing. Actually,

TABLE 2.3

Gallium Arsenide Slice Specifications

Parameter	*Item**	
	50.8 mm Diameter	*76.2 mm Diameter*
Doping	Chromium	Chromium
Resistivity	$> 5 \times 10^7$ Ω-cm	$> 5 \times 10^7$ Ω-cm
Mobility	> 4000 cm²/V·s	> 4000 cm²/V·s
Etch pit density per cm²	2×10^4–5×10^4	2×10^4–5×10^4
Diameter tolerance	0.4 mm	0.6 mm
Primary flat length	14.5–17.5 mm	19–25 mm
Secondary flat length	6.5–9.5 mm	10–13 mm
Minimum slice thickness available	400 ± 25 μm	500 ± 25 μm
Maximum slice thickness available	625 ± 25 μm	750 ± 25 μm
Flatness	< 0.8 μm/cm	< 0.8 μm/cm
Bow	> 20 μm	> 30 μm

*Round and grown by liquid-encapsulated Czochralski method.

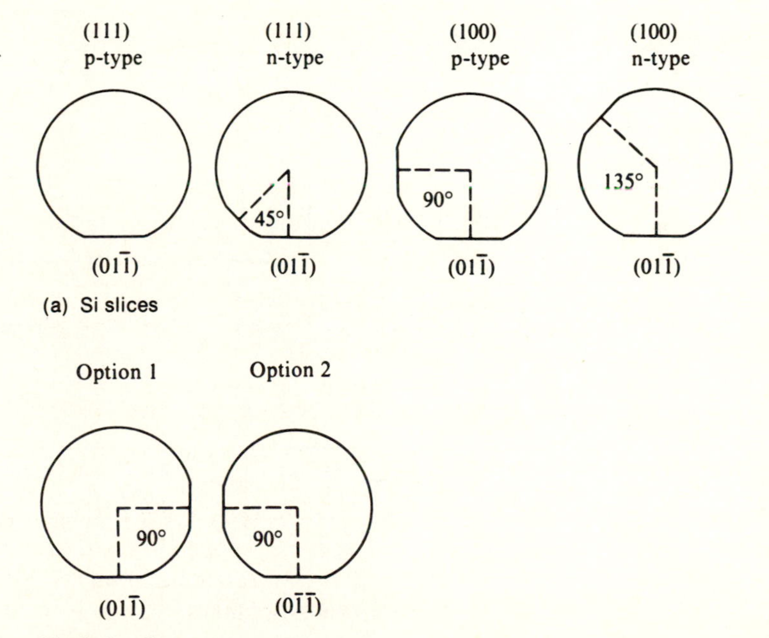

FIGURE 2.1

Location of major (primary) and minor (secondary) flats on Si and GaAs slices less than 200 mm in diameter when viewed from polished side. Two-hundred mm diameter Si slices may have either a flat or a notch to mark the (01$\bar{1}$) face. The other pertinent information is laser-scribed near the periphery on the polished side. (*Source:* Based on SEMI Specifications, Spring 1989.)

FIGURE 2.2

Material processing flow from metallurgical-grade Si to the beginning of wafer fabrication.

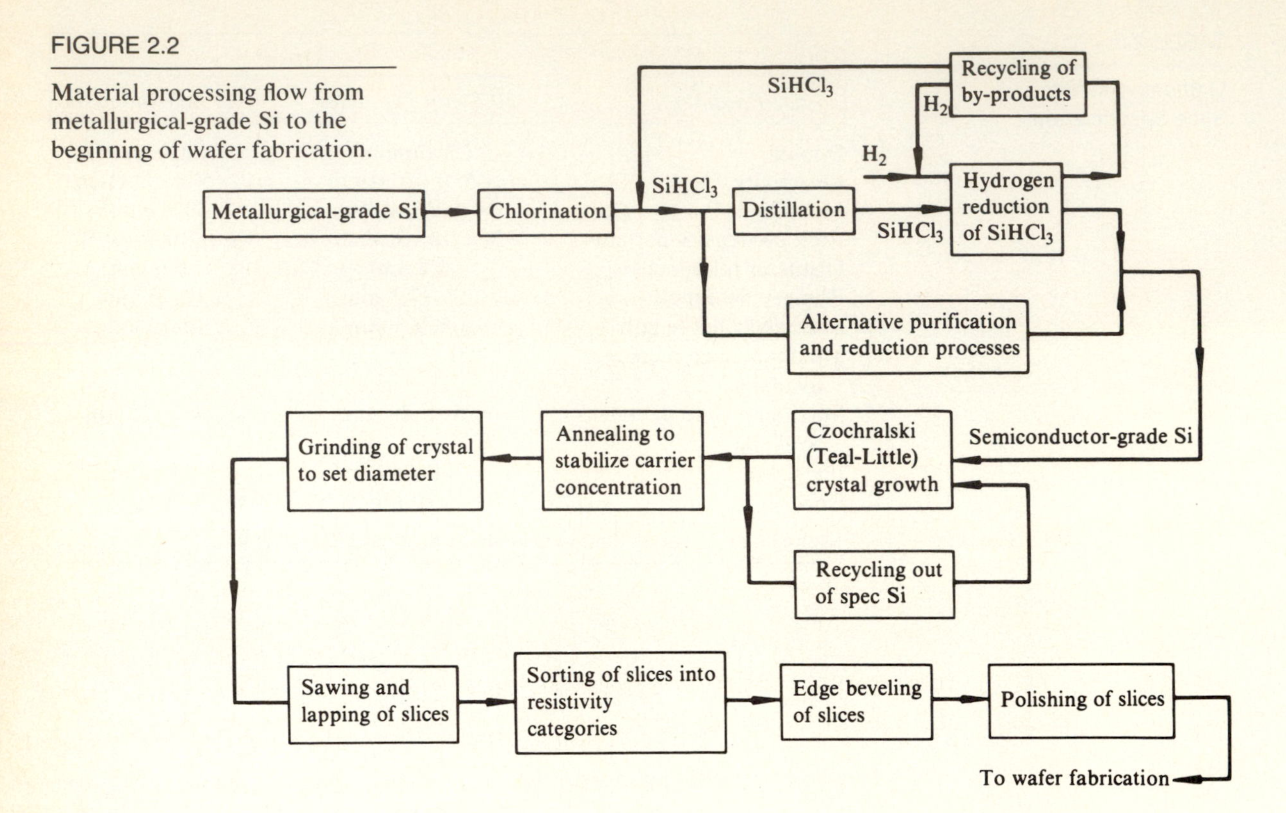

FIGURE 2.3

Material processing flow of GaAs prior to the beginning of wafer fabrication.

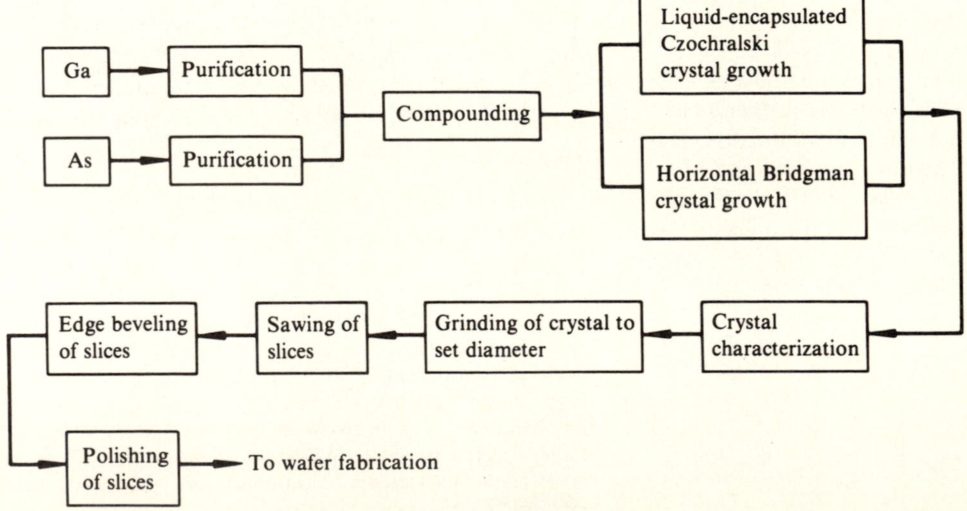

either $SiCl_4$ or $SiHCl_3$ can be purified by distillation and then reduced by hydrogen to give high-purity Si. In addition, other silicon compounds, such as SiH_4 (silane) can be used. Silicon manufacture is characterized by high power usage and poor reduction efficiency—thus the emphasis on recycling of feed stock and the continuing search for more efficient reaction chambers. In the conventional Siemens process,[2] silicon rods of up to 2 m in length are electrically heated in an atmosphere of $SiHCl_3$ and hydrogen and grow in diameter from less than 10 mm to over 125 mm. Some of the newer, more efficient processes use fluid bed reactors, where small silicon particles increase to about pea size in the reactor and are then withdrawn.

Gallium and arsenic are both produced primarily as by-products of other refining operations, with gallium coming from aluminum smelting and arsenic from copper smelting. The commercial-grade product (element or compound) is then converted to an easy-to-purify compound, is purified, is reduced, and, in the case of arsenic, is often rerefined by vacuum sublimation. The two elements are then combined (compounded) to give GaAs. The arsenic vapor pressure over gallium arsenide is high enough that compounding must be done under pressure. To directly combine the two elements at the same temperature, pressures of about 60 atm are required. Alternatively, the gallium/GaAs can be held slightly above the GaAs melting point of 1238°C, and the arsenic at about 600°C in the same closed tube. The reaction will then proceed slowly at about 1 atm as the arsenic vaporizes and diffuses to the liquid gallium/GaAs, where it reacts with the liquid gallium.

Silicon crystals are grown from the melt by either the pulling (Czochralski or Teal–Little)[3] or the float-zone (FZ) process. However, almost no float-zone crystals are used in IC manufacture. Gallium arsenide crystals are grown either by a variation of the Czochralski (CZ) method or by the horizontal Bridgman method. While it is in principle possible to grow semiconductor-material crystals

---

[2]So-called because Siemens and Halske, AG, of Germany introduced the process.

[3]Czochralski, in 1918, introduced a method for growing single-crystal metal wires that involved dipping a glass capillary tube into the metal melt and then withdrawing it and the attached thin metal thread. He is said to have gotten the idea by having mistakenly dipped his ink pen into a pot of molten tin on his laboratory workbench and then noting the attached frozen tin strand when he pulled the pen from the tin. In 1950, Teal and Little devised "crystal pulling" to grow the germanium single crystals used for early transistor development. Conceptually, their puller was the same as machines used today. The only new functions added in the intervening years have been automatic diameter control and, in some machines, magnetic damping of the melt circulation.

from solutions—for example, silicon from a molten aluminum solvent—all commercial crystals are produced from an essentially pure semiconductor melt.

In crystal pulling, shown schematically in Fig. 2.4, a single-crystal seed is dipped into the melt and slowly rotated. By controlling the temperature of the melt and the amount of heat withdrawn from the seed, freezing onto the seed (crystal growth) is possible. If too much heat is extracted, the whole surface of the melt will freeze over. If the melt is too hot, the seed will melt off above the surface. When conditions are perfectly balanced, controlled growth will occur as the seed and attached crystal are slowly withdrawn (pulled) from the melt. Stirring of the melt via crystal rotation reduces unsymmetrical growth caused by uneven heating of the melt. To minimize the number of dislocations in the crystal, the seeds must be no more than a few millimeters in diameter. To provide for economical IC fabrication, the crystals (Si) should be 100–200 mm in diameter. The growing-crystal diameter must thus be changed from 2 or 3 mm to 100–200 mm. This change and the subsequent diameter control required for the rest of the growth are accomplished by changing the melt temperature, spin (rotation) rate, and pull rate. Increasing the melt temperature decreases the crystal diameter; increasing the pull rate decreases the diameter; and increasing the crystal rotation rate increases the diameter.

Since the melt must be protected from oxidation, the crucible and crystal, as shown in Fig. 2.4, are surrounded by a chamber containing an inert or reducing atmosphere. For silicon, either atmo-

FIGURE 2.4

Various stages of crystal pulling from the melt.

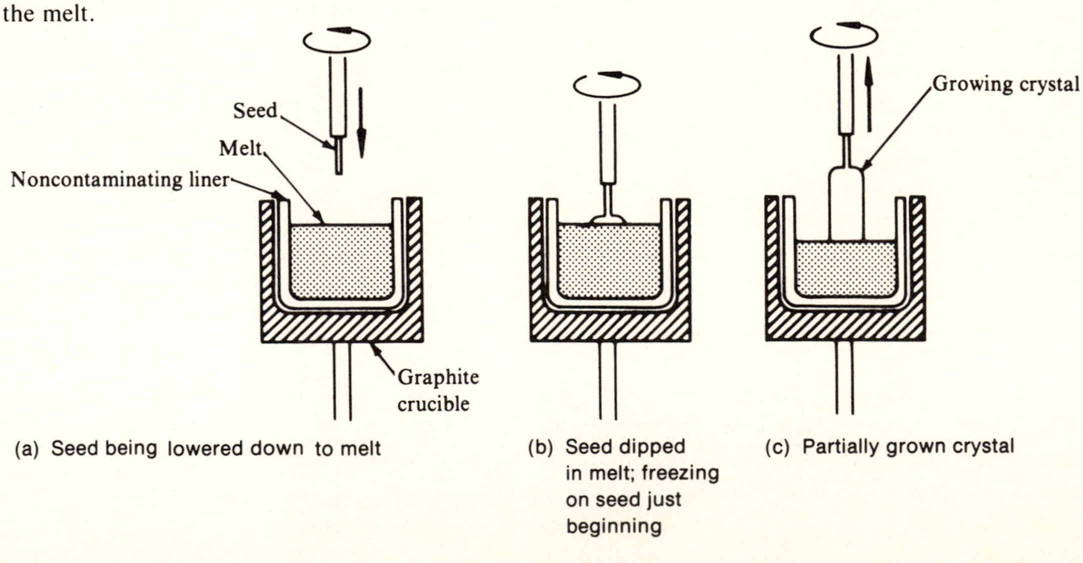

spheric or reduced pressure can be used. Gallium arsenide requires higher pressures, usually in the 5–60 atm range, and in addition, a molten layer of boric oxide on top of the melt to keep arsenic evaporation at a tolerable level. The latter procedure is referred to as liquid encapsulation. Silicon is particularly reactive but can be contained in fused silica (SiO_2) with only a small amount of the silica being dissolved. This small amount, however, is the source of a substantial amount of oxygen that becomes incorporated into silicon crystals. Silica is also used as the container for GaAs crystals and, in this case, is the source of a small amount of silicon doping.

The horizontal Bridgman method of crystal growing, shown schematically in Fig. 2.5a, consists of loading the material to be grown into a long, narrow container, usually of semicircular cross section, and moving the container from a temperature (T_1) slightly above the material's melting point to a temperature (T_2) slightly below the melting point. In order to provide for single-crystal growth, a seed crystal is placed in one end of the container and partially kept in the cool zone so that only a portion of it melts. Thus, when the rest of the container is slowly moved into the cool zone, the melt freezes onto the seed as a single crystal. This growth procedure can be used only when a container material is available that will not bond to the crystal when it freezes. In the case of silicon, none are available. For gallium arsenide, fused silica is satisfactory.

Float-zone crystal growing, as shown in Fig. 2.5b, is a method in which the melt avoids contact with any container material. Here, a rod of polycrystalline material with a short, single-crystal seed at one end is held vertically by being clamped at each end. A narrow

FIGURE 2.5

(a) Horizontal Bridgman method of crystal growing, which is used when the crystal does not stick to the container material. (b) Float-zone method of crystal growing, where the molten region is touched only by the ambient gas.

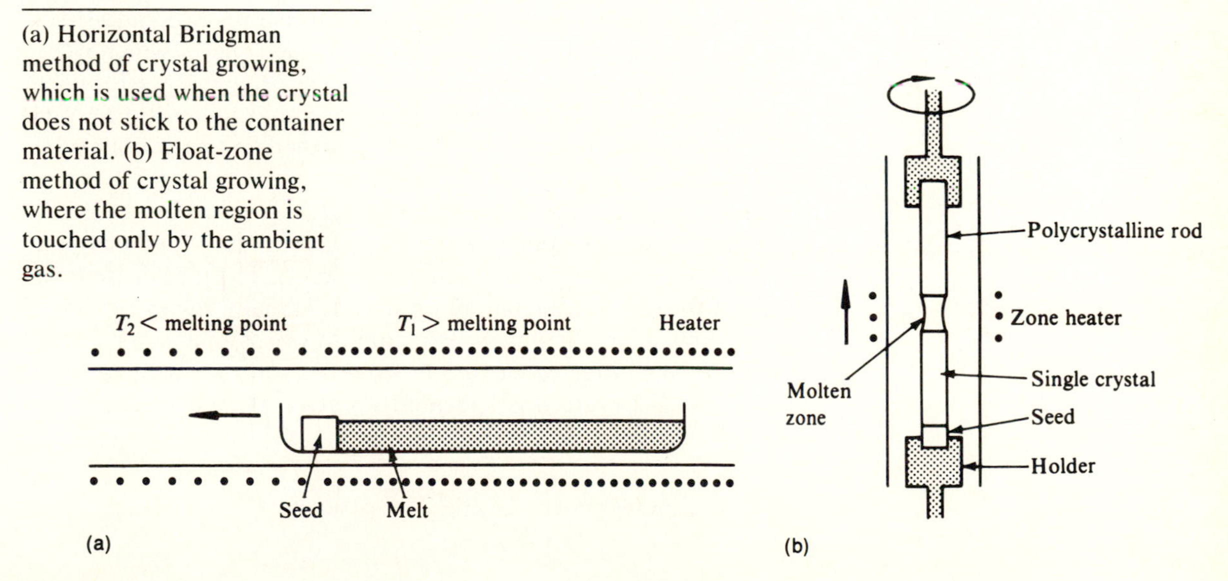

region, originally partially in the seed crystal and partially in the polycrystalline rod, is inductively melted and slowly moved the length of the rod by moving the induction coil relative to the rod. Unlike crystal pulling, in which it appears that the only limit to crystal diameter is the strength of the supporting seed, float-zone crystal diameter is limited by a combination of surface tension and levitation from the heating coil. Float-zoned silicon is used when the highest purity possible is required and when (as of 1989) the slice diameter need not be greater than 150 mm.

Silicon and gallium arsenide are brittle materials and have hardnesses that are a little less than that of quartz. They can be sawed and polished by the same techniques used for quartz. Both have pronounced cleavage planes, (111) for Si and (110) for GaAs, and in thin sections can be easily broken after scribing along traces of these planes on the wafer surface.[4] For sawing and grinding operations, diamond-bonded wheels are used. Lapping is usually done with either alumina or silicon carbide grit slurries. Polishing can be done mechanically, using, for example, diamond dust, alumina, or zirconia. It can also be done electrolytically or chemically or by a combination of chemical etching and mechanical abrasion. The process actually used is dictated by the flatness required and the need for a minimum of residual mechanical damage. In the case of silicon, a chemical–mechanical combination using fine silica powder in a hydroxide solution is commonly employed. The analogous process for gallium arsenide is silica in a sodium hypochlorite solution.

Fig. 2.6 shows the mechanical shaping operations performed on a silicon crystal. First, the undulations on the grown crystal are ground away so that all slices will have the same diameter. Next, orienting flats are ground on the crystal before it is sawed into slices. Then, in order to minimize the chipping of slice edges (which would produce yield-reducing particulates), the slice edges are rounded. Finally, each slice is lapped with progressively smaller grit in order to remove fracture damage and then is polished on one or both sides, inspected, and sent to the wafer-fab area.

Mask making, even though technologically quite closely akin to wafer fabrication, is done in external "mask shops." The manufacturing steps are briefly discussed in Chapter 5. The masks are made of an ultraviolet, transparent material (usually glass or fused quartz), with a metallic-layer overcoat into which a pattern has been etched. Masks differ conceptually from the starting slices in that a particular mask pattern is specific to a given IC. Further, not just

---

[4]See Appendix A for a discussion of planes and traces.

FIGURE 2.6

Mechanical shaping operations from as-grown crystal to polished slices.

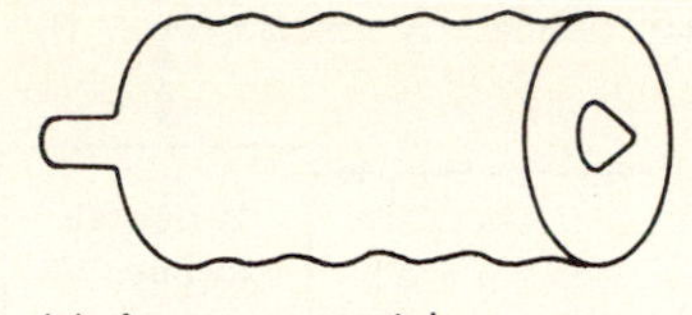

(a) As-grown crystal

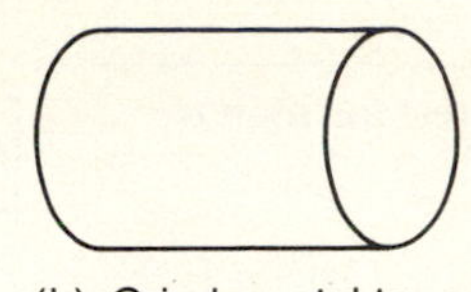

(b) Grind crystal to remove undulations and saw to remove portions not in resistivity range

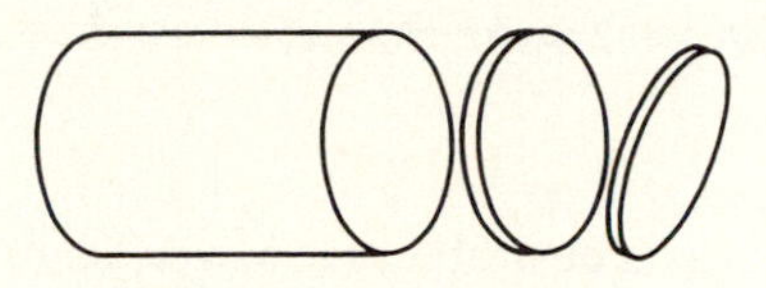

(c) Saw into slices (with orienting flats ground before sawing)

(d) Round edges of slice by grinding

(e) Lap and polish slice

one mask, but a set of masks, each with a different but related pattern, is required for each circuit.

2.2 WAFER FABRICATION FACILITY

As shown in Fig. 2.7, the slices, masks, and required chemicals are brought together in a wafer fabrication facility for conversion into IC chips. Much of the facility is specifically designed to meet the requirements necessary for high-yield IC processing.

The wafer-fab manufacturing space, besides housing the equipment required for semiconductor processing, must also supply the proper environment for the process and for the human operators. Perhaps foremost is the provision for exhausting toxic fumes. These fumes range from relatively mild organic vapors such as benzene to extremely toxic gases such as phosphine, diborane, and anhydrous HF (hydrogen fluoride). In addition, many organic fumes are highly flammable, and hydrogen and silane are quite explosive. High-capacity exhausts, coupled with specific gas detectors and great care in maintaining leakproof plumbing, are required for safety. Rooms used for epitaxial depositions and some CVD operations are laid out for quick evacuation and very rapid removal of toxic fumes in case of leaks or explosions.

In rooms where photoresist is applied, the humidity must be controlled. To improve resist adhesion to the slice surface, low hu-

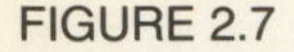

FIGURE 2.7

Elements required for manufacturing IC chips.

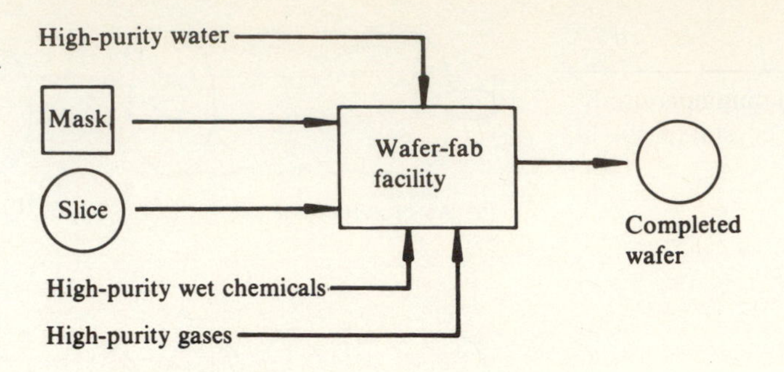

midity is desirable. However, as humidity decreases, problems with static discharge and operator discomfort emerge. As a compromise, a nominal 45% is often used. Ultraviolet (UV) resist is insensitive to longer wavelength light; consequently, yellow light is used in rooms where resist is applied, exposed, and developed to provide visibility and yet not expose the resist.

A substantial portion of the wafer-fab facility must be "cleanroom"—that is, have areas where the total airborne particles larger than, say, 0.1 μm will be 1 to 10 per cubic foot.[5] The need for such a cleanroom can be appreciated from the data of Fig. 2.8, which shows the size of some common airborne particulates. The usual criterion is that particles on critical areas of the wafer must be no larger than one-fourth of a line width. Thus, for a one Mbit DRAM with minimum geometries of 0.8 μm, even a large virus has the potential of causing a yield loss! The air in cleanrooms is cleaned by passing it through HEPA (high-efficiency particulate air) filters. To ensure that particulates generated in the room are swept out and not swirled around in eddys and deposited on wafer surfaces, the air flow must be slow and laminar. To accomplish this, the air usually comes in from the ceiling, moves down, and exits through a perforated floor.

High-purity deionized water (DI water) is essential to the semiconductor industry's slice-cleaning procedures. The purity of such water is generally specified in terms of its electrical conductivity or resistivity. Absolutely pure water, like intrinsic semiconductors, still has some conductivity, and for a similar reason—the partial dissassociation of HOH into H and OH ions, which then conduct. Its resistivity at room temperature is slightly above 18 $M\Omega$-cm, and

[5]Cleanrooms are classified by the number of particles per cubic foot larger than some specified size. A class 10 room would be one with 10 or less particles greater than some specified size per cubic foot of air. (Office space typically has about 200,000 particles greater than 0.5μm per cubic foot.)

FIGURE 2.8

Size of various common airborne particles.

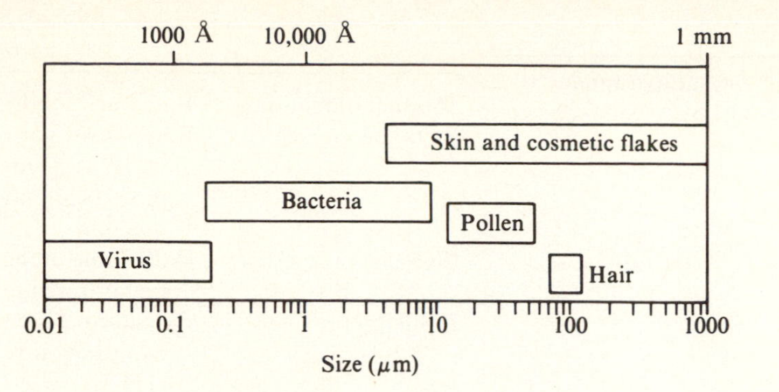

semiconductor-grade water is normally specified as 18 MΩ-cm. In addition to resistivity, the quantity of particulates and bacteria is also closely controlled. The water is irradiated with ultraviolet light to kill the bacteria, and several levels of filtration are used to remove the particulates and dead bacteria. The quantity of DI water needed is substantial, ranging from 100 to 200 gallons per completed wafer. Thus, a site producing a total of 50,000 wafers per month could use up to half a million gallons of water a day. A large and rather elaborate water treatment plant is required to produce the necessary quantity and quality from municipal water, and it is usually located in a separate building close to the wafer fabrication area.

2.3 WAFER FABRICATION OPERATIONS

Each of the wafer fabrication operations listed earlier in Table 2.1 have many steps and usually many variations, depending on exactly the kind of structure to be built. Furthermore, the order in which the steps are performed is by no means fixed, and a wafer may circulate back through some of them several times. Silicon IC wafer fabrication starts with the careful cleaning of a bare silicon slice of appropriate crystallographic orientation and the growing of a layer of thermal oxide on each surface. (Thermal oxide is SiO_2 formed by oxidizing the silicon surface.) Gallium arsenide, not being amenable to a native-grown protective oxide, is coated with a CVD silicon dioxide layer. The crystallographic orientation of the slice is important to the finished IC and is specified by the circuit designer. However, as Table 2.4 shows, the orientation also affects nearly all of the wafer-fab processing steps. As fabrication proceeds, other layers covering only one side of the wafer,[6] such as CVD oxide, poly-

[6]While the terms *slice* and *wafer* are often used interchangeably, the recommended terminology is *slice* for a structure prior to the wafer-fab operations of Table 2.1 and *wafer* for the slice and the various layers added to it.

TABLE 2.4

Effects of Crystallographic Orientation of Slice Surface on Processing

Operation	*Effect of Orientation*
Thermal oxidation	Rate varies with orientation; (100) < (110) < (111).
Epitaxial growth	Rate varies with orientation; is very low exactly on (111). Depending on orientation, depressions in surface may change shape and slightly shift position.
Etching	With some etchants, both aqueous and molten metal, rate is very low in [111] direction.
Diffusion	Orientation is dependent in silicon when done in conjunction with thermal oxidation; (100) > (111).
Ion implantation	Range is greatest in [110] direction; also is greater directly in other low-indices directions as compared with slightly off-orientation directions.

crystalline silicon, silicon nitride, and various metals will be added. Intermixed with these steps will be patterning (lithography), etching, diffusion, and ion implantation.

2.3.1 Surface Cleaning

Complete cleaning of semiconductor surfaces requires that particulates, organic films, and adsorbed metal ions be removed. Most cleaning procedures are based on immersion in liquid baths or liquid sprays. In addition, ultrasonic agitation or brush scrubbing may be required. In some cases, high-temperature vapor etching or low-pressure sputter etching may be used. A good cleanup is complicated by the fact that unless great care is taken, the materials used for cleaning may contain (and leave behind) more particulates and metals than were on the surface in the beginning. Consequently, semiconductor-grade chemicals that are specially filtered and purified are used by the semiconductor industry. Organic solvents are widely used, but since they sometimes leave residues themselves, high-purity water is ordinarily the last stage of a cleanup. This last stage is often immediately preceded by an acid to oxidize and remove any remaining organics.

2.3.2 Epitaxy

Epitaxy (epitaxial growth or epi) is the overgrowth of a thin layer of single-crystal material onto a single-crystal substrate. In the case of ordinary epitaxy, it is the growth of a doped silicon layer onto a silicon slice of different resistivity and perhaps type. The epi growth process is unique in that it is the only way in which a layer can be produced with *fewer* total impurities than the underlying slice. Depending on the specific device application, the layer may be from 1 or 2 μm to 50 μm in thickness.

Vapor phase growth is the most common silicon epitaxial method and is done in a temperature range of 1000°C–1200°C by the reduction of some silicon-bearing compound at the slice surface. Common reactions are the hydrogen reduction of $SiCl_4$, $SiHCl_3$, or SiH_2Cl_2 (dichlorosilane). For lower-temperature growth, the thermal reduction of SiH_4 is often used. Doping of the layer is accomplished by the co-deposition of the appropriate impurity—for example, by the thermal reduction of phosphine. Gallium arsenide epi for IC applications is also generally grown from the vapor, although the task is complicated by requiring sources for two elements instead of only one. For discrete light-emitting devices, growth is most often from the liquid phase (LPE). In this process, the gallium arsenide wafer surface is bathed in a gallium–arsenic melt, and the temperature is varied so that a thin gallium arsenide layer freezes (grows) onto the surface.

2.3.3 Thermal Oxidation of Silicon

The operation of oxidizing the silicon surface is performed in a temperature range of 800°C–1250°C and in an atmosphere containing oxygen or steam in an inert carrier gas. The oxide thickness increases, although not linearly, with oxidation time. Silicon is consumed during the oxidation process at a rate such that for every micron of oxide grown, the silicon surface recedes by 0.45 μm. The oxidation rate increases with increasing temperature and pressure. To reduce the oxidation time, or keep a reasonable time at a lower temperature, oxidations are sometimes made at several atmospheres pressure.

2.3.4 Diffusion

If a source of dopant is provided, the oxidation operation becomes a diffusion. The mechanics of transferring an impurity atom directly from a gas ambient into the wafer is rather inefficient. Therefore, a process is generally chosen that lets the dopant react with silicon oxide on the wafer surface to give a silicate glass that then acts as the source. In the case of gallium arsenide, a deposited silicon oxide already containing the dopant is used for the source. To achieve better control of surface concentration and diffusion depth, the overall diffusion operation is often broken into two steps. The first, a dopant deposition, or "dep," is a very thin, high-concentration diffusion, with the surface concentration set by the solubility limit of the dopant at the dep temperature. The second, the "drive," is at a different temperature and is used to adjust the overall diffusion to give both the desired surface concentration and diffusion depth. The wafers are held at their edges, separated by 1–2 mm, in a quartz boat and heated in a long, quartz or polysilicon tube that has the necessary gases flowing through it. Temperature control is required to about ± 1°C, and diffusion times range from minutes to hours.

2.3.5 Ion Implantation

An alternative to the deposition step is to accelerate the dopant atoms to a high velocity with an accelerator so that after striking the surface, they will continue on into the body of the semiconductor. Since acceleration is by electric field, ions, and not atoms, must be used—hence, the name *ion implanter*. The acceleration voltages used range from a few thousand to a few million volts. The penetration depth of the ions increases with increasing voltage and is typically a few hundred angstroms. When depths greater than those conveniently obtainable by implantation are required, implanting followed by an ordinary thermal diffusion is often used. Even if no further depth is desired, some additional heat treatment is still needed because the implanting causes substantial crystal damage, which must be annealed out. In very high dose implants, damage is so severe that amorphous layers are formed, but under appropriate heat treatment, such layers will regrow epitaxially to form high-quality single-crystal layers.

2.3.6 Lithography

Lithography, also referred to as photolithography, and sometimes as pattern printing, is the means by which a pattern of masking material (usually organic) is applied to the surface of a wafer. That pattern then provides protection to the desired portions of the wafer and allows material to be removed from the remaining area by a suitable etchant. The steps in patterning oxide on a wafer are shown in Fig. 2.9. The photosensitive resist is liquid and is applied by dropping a metered amount onto the middle of a slowly rotated wafer. The speed is then increased to a few thousand revolutions per minute so that centrifugal force can force the resist to flow out over the wafer surface and uniformly coat it. The rotational speed and resist viscosity are adjusted to give the required thickness (usually about 1 μm). The pattern in the resist is produced by exposing the resist to a similar pattern of high-intensity ultraviolet light and then developing it. The light pattern is defined by the pattern on the mask mentioned earlier. The mask is thus the counterpart to the conventional negative of photography. Either contact printing, with the mask in direct contact with the wafer, or a projection lens system can be used to transfer the mask image to the resist on the wafer. Both of these procedures are very similar to ordinary photographic processing, except that in this case, the mask pattern is either the same size or larger than the image desired on the slice. In the developing process, the unwanted resist is removed, thus leaving a pattern of bare oxide or whatever other material was under the resist.

FIGURE 2.9

Steps in patterning oxide on a wafer.

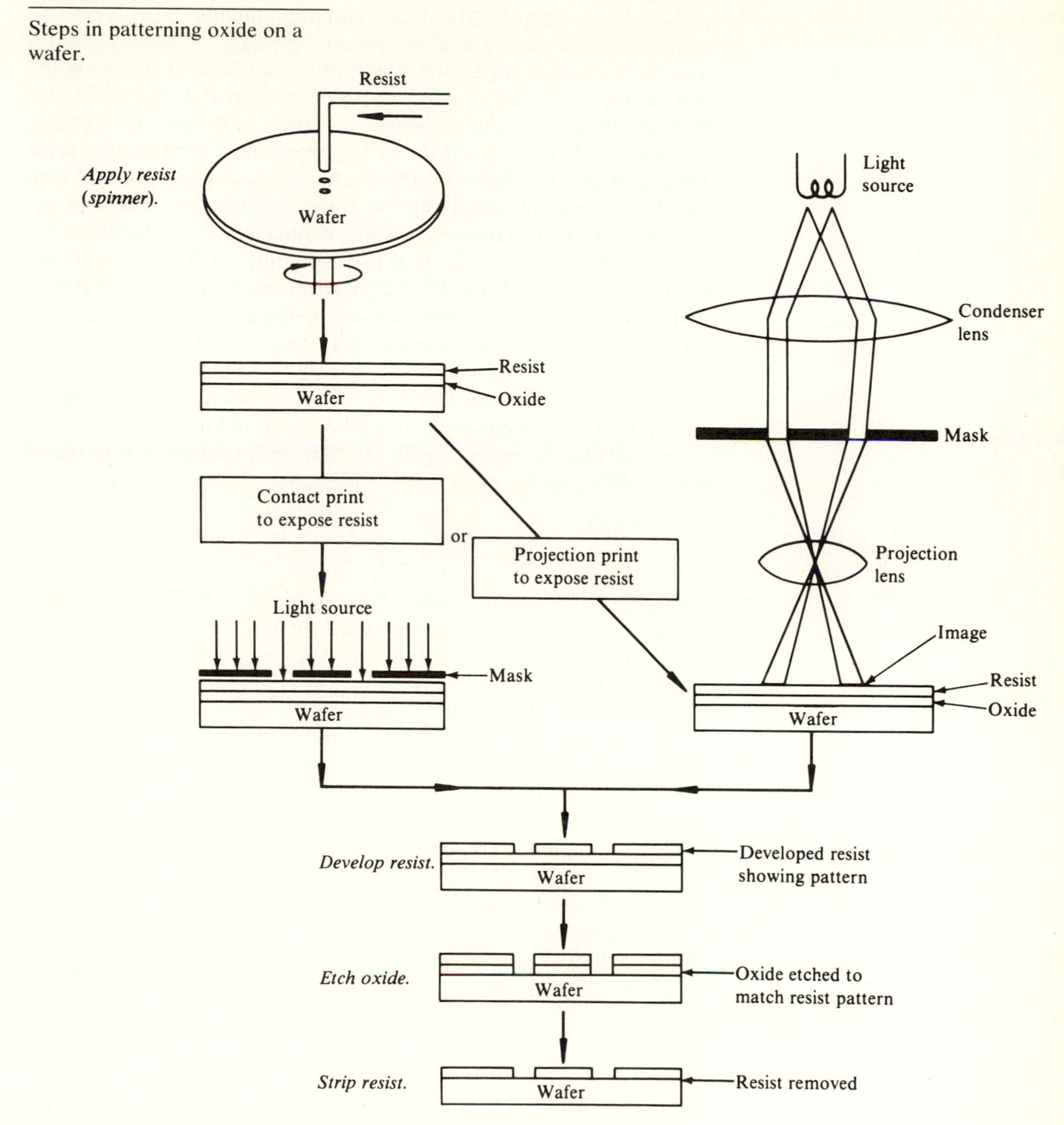

2.3.7 Etching

After the resist pattern is defined, the underlying material must be removed by etching. The most common methods are by using aqueous-based etches and by plasma etching. In either case, a means of stopping the etch at the bottom of the layer being etched is necessary. It is not satisfactory to depend on etch rate and time. Rather, the etch must be chosen so that the rate differential between the layer being etched and the layer immediately beneath it is great enough so that the desired layer can always be completely removed and an inconsequential amount of the one below it will also be removed. For fine geometrics, plasma etching is preferred and is considerably more anisotropic[7] than is wet etching. Typical wet etchants are HF–HNO_3 for silicon, HF for SiO_2, hot H_3PO_4 for silicon nitride, and cold H_3PO_4 for aluminum. Plasma etching gases usually contain fluorine or chlorine—for example, CF_4.

2.3.8 Chemical Vapor Deposition (CVD)

CVD is used to deposit layers of materials such as silicon oxide, silicon nitride, aluminum oxide, and polycrystalline silicon onto the wafers. CVD processes usually operate in a temperature range of 300°C–900°C and proceed by reactions such as follows:

$$SiH_4 + 2O_2 \rightarrow SiO_2 + 2H_2O$$
$$SiH_4 + \text{heat} \rightarrow Si + 2H_2$$

Oxides deposited in this manner do not bond to the silicon surface in the same manner as a thermal oxide would and, in addition, may have a lower density. Consequently, they are not an alternative to thermal oxides but rather an adjunct to be used when thermal oxidation temperatures cannot be tolerated or when an insulating layer is required and there is no silicon to be oxidized. By using a plasma to supply a portion of the energy required, CVD reactions such as those just listed can be done at substantially lower temperatures. In particular, usable silicon nitride films can be deposited at 300°C, whereas straight CVD requires over 800°C.

2.3.9 Metal Deposition

Metal depositions are accomplished by evaporation, sputtering, CVD, or electroplating. Evaporation is easy and inexpensive. Sputtering, with its higher pressures, is used to provide better step coverage. When refractory metals are used, either sputtering or CVD is

---

[7]Anisotropic etching has two meanings. In reference to single-crystal silicon, it means crystallographic orientation dependent etch rates. In most other contexts, it refers to etching straight down through the pattern window without any lateral undercutting.

required in order to get reasonable deposition rates. Electroplating is sometimes used to build up thin deposited layers and eliminate the need for etching through very thick metal. Aluminum is the most commonly used metal, but gold, platinum, titanium, molybdenum, tungsten, tungsten–titanium alloys, and palladium are also used. In addition, various refractory-metal silicides are alternatives to polysilicon as a gate and interconnect metallization. In their case, either co-sputtering of the components or sputtering from a silicide target is used.

2.3.10 Metal (Contact) Sintering

In order to achieve good metal-to-semiconductor contact after deposition, heating the combination to a high enough temperature to allow some interdiffusion and alloying to take place is usually necessary. For aluminum, the temperature is usually between 450°C and 500°C. The operation is performed in a diffusion-like tube in either an inert or a hydrogen atmosphere.

2.3.11 Backgrinding

For wafers to have enough strength to survive processing without any breakage, their thickness is usually in the 500–600 μm range. However, the chip separation operation works better with thinner wafers. Furthermore, thinner dies allow better heat transfer from the junctions to the lead frames on which the dies are mounted. Consequently, a thinning operation (backgrinding) is often used. This operation is done on a milling-type machine with a diamond-cup grinding wheel. Less modern equipment, based on optical industry processes, uses a large rotating table, on which many slices are mounted by hand, and a grinding wheel that may be 18–24 inches in diameter. Newer equipment is much smaller, grinds the slices individually, and loads automatically

2.3.12 Multiprobing

Multiprobing is the testing operation used to sort good electrical dies from bad ones while they are still in wafer form. A series of sharp-pointed probes is mounted so that the probes can be brought down on the wafer to simultaneously contact all of the metallized bonding pads[8] of an individual chip. A series of electrical tests, microprocesser controlled, is then sent to the chip to determine whether it functions properly. An ink dot is placed on each bad chip so that it can be sorted out later.

---

[8]When an IC is assembled, connecting wires from the lead frame to the chip attach to the chip at the bonding pads.

2.4 INTEGRATED CIRCUIT FABRICATION

2.4.1 Components and Component Construction

The wafer fabrication operations just described are used to construct the various active and passive devices (components) that may be found in an IC. The components available are as follows:

- Bipolar transistors
- Junction field effect transistors (JFETs)
- Metal oxide semiconductor field effect transistors (MOSFETs)
- Metal semiconductor field effect transistors (MESFETs)
- pn junction diodes
- Schottky barrier diodes
- pn junction capacitors
- MOS capacitors
- Diffused resistors
- Thin-film resistors
- Interconnecting leads
- Interconnecting leads crossovers

However, that a single IC will contain all possible devices is exceedingly unlikely.

Transistors are the key components in ICs and require the most sophisticated processing. Fig. 2.10 shows simple bipolar, junction field effect, and MOS transistors in cross section. Multiple bipolar (and/or junction field effect) transistors on the same chip can be electrically isolated from one another by pn junctions or by a dielectric layer completely surrounding the transistor. The most common method is pn junction isolation, and it starts with a p-type slice and then adds an epitaxial n-layer on top of it. The transistors are then made in the n-layer and isolated from one another by diffused p-regions, as shown in Fig. 2.11. Transistors designed for high-speed operation may combine thermal oxide sidewall and pn junction isolation, as depicted in Fig. 2.12. Dielectrically isolated (DI) bipolar transistors are used in some high-voltage and radiation-hardened applications. As shown in Fig. 2.13, the oxide completely surrounds the collector region and thus provides a higher collector–substrate breakdown voltage than is easily obtainable with a junction. Such oxide also prevents the possibility of a large current flow between collector and substrate that could occur with junction isolation in the event of high-intensity photon irradiation.

In some bipolar circuits, having both npn and pnp transistors on the same chip is helpful. By building lateral pnp transistors as shown in Fig. 2.14, rather than having the main portion of the junction parallel to the wafer surface and the current flow normal to the surface, current flow is parallel to the wafer surface. Because of the difficulty in spacing the two diffusions close together, gain will be quite low, but the process has the advantage of allowing the pnp emitter and collector to be made during an npn base diffusion, thus requiring no

FIGURE 2.10

Cross sections of bipolar, junction field effect, and MOS transistors. (Not to scale. Also, in these kinds of drawings, the vertical magnification is much greater than the horizontal magnification.)

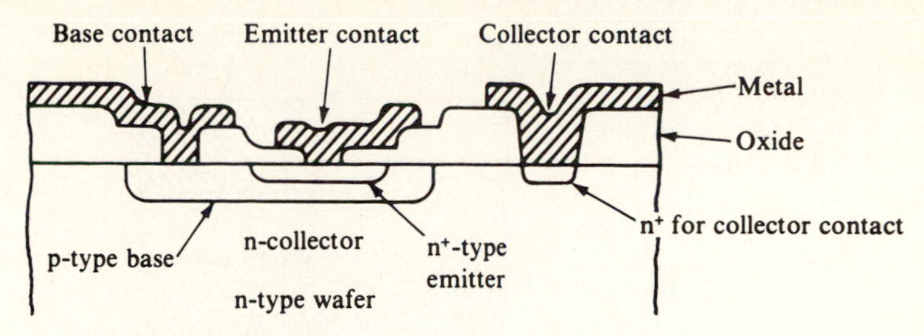

(a) npn bipolar transistor

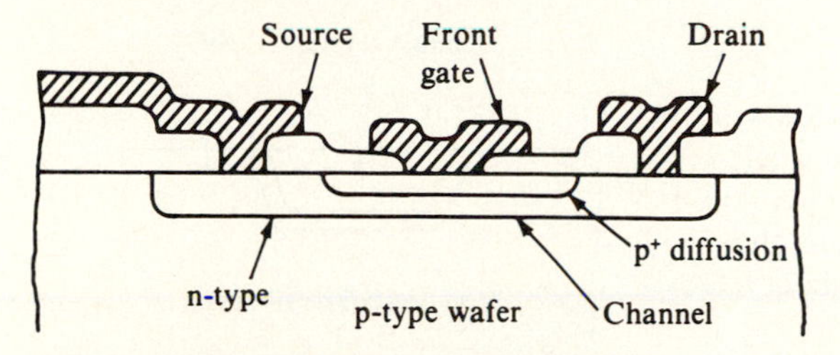

(b) n-channel junction field effect transistor

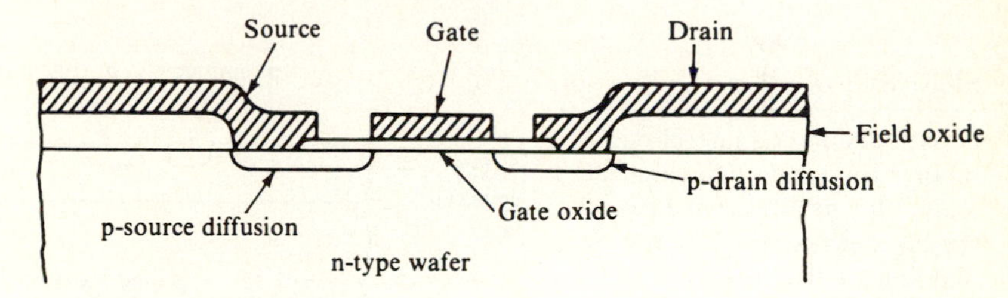

(c) p-channel MOS transistor

FIGURE 2.11

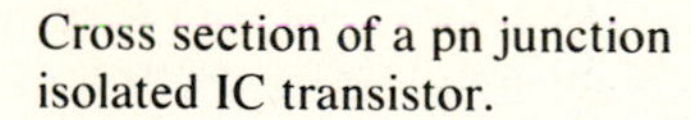
Cross section of a pn junction isolated IC transistor.

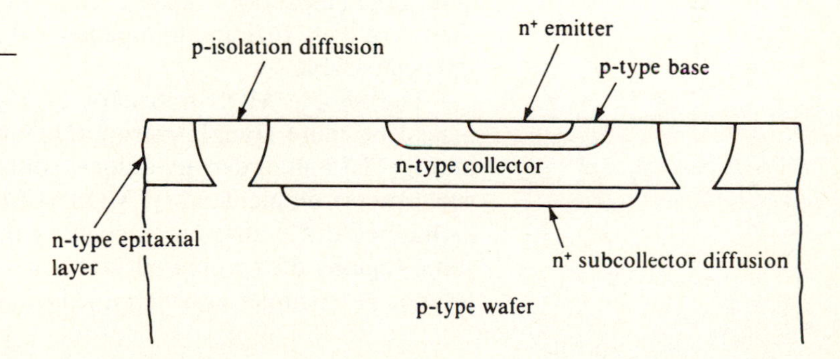

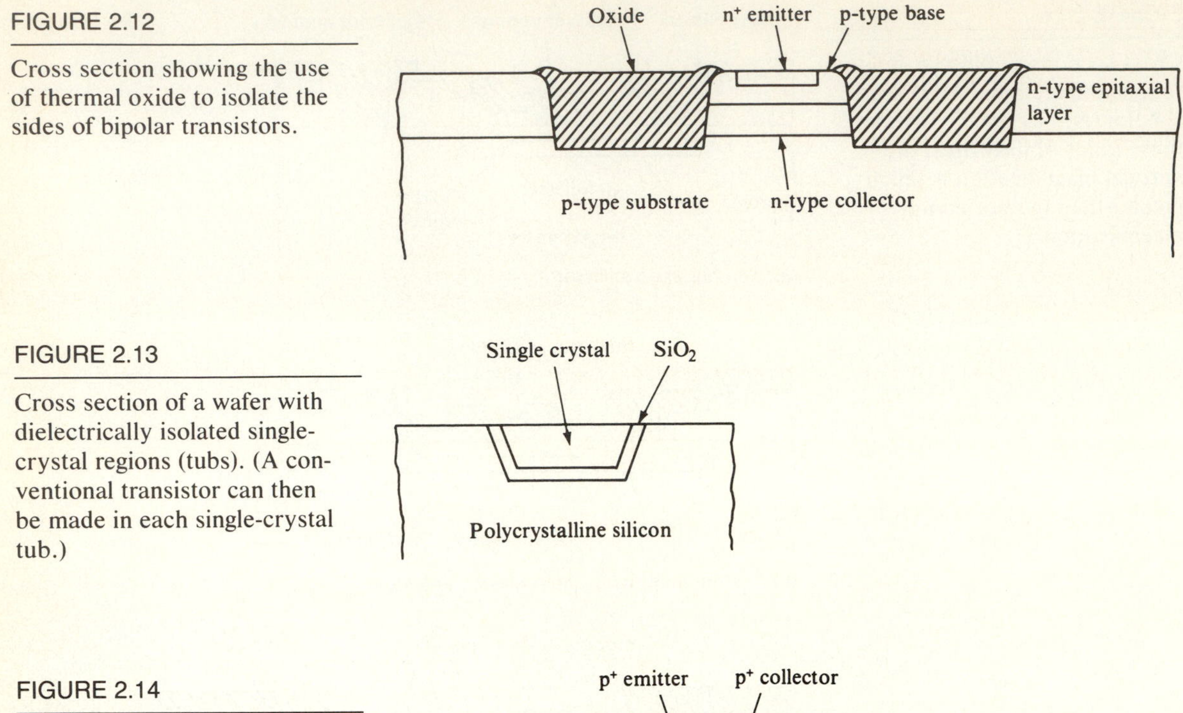

FIGURE 2.12

Cross section showing the use of thermal oxide to isolate the sides of bipolar transistors.

FIGURE 2.13

Cross section of a wafer with dielectrically isolated single-crystal regions (tubs). (A conventional transistor can then be made in each single-crystal tub.)

FIGURE 2.14

Cross section of a lateral pnp bipolar transistor. (The emitter and collector are formed by the base diffusion of the normal npn vertical transistors.)

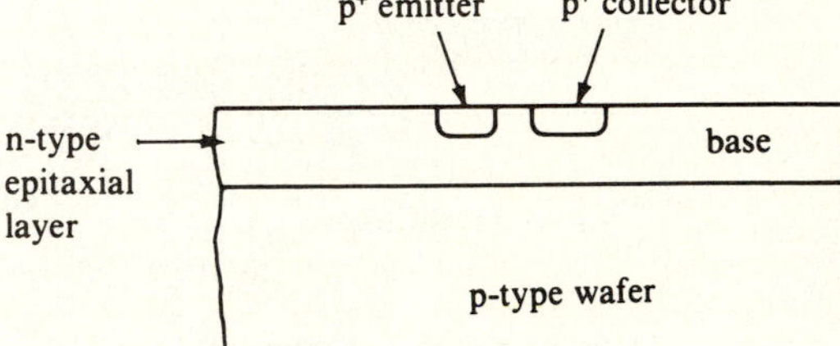

additional steps.[9] JFETs are used primarily in analog circuits where their relatively high input impedance and low noise figure are sometimes desirable.

The simple MOS transistor of Fig. 2.10 has largely been replaced by more complex structures with oxide sidewalls as shown in Fig. 2.15 in order to reduce source–drain capacitance to the substrate. Complementary MOS (CMOS) circuitry requires both n-channel and p-channel transistors in the same chip. To do this, some regions of the chip surface must be n-type and other regions p-type. The manner in which this is done leads to the n-well, p-well,

[9]If a complementary transistor circuit is being built, the npn and pnp transistors must be more closely matched than is possible with this approach.

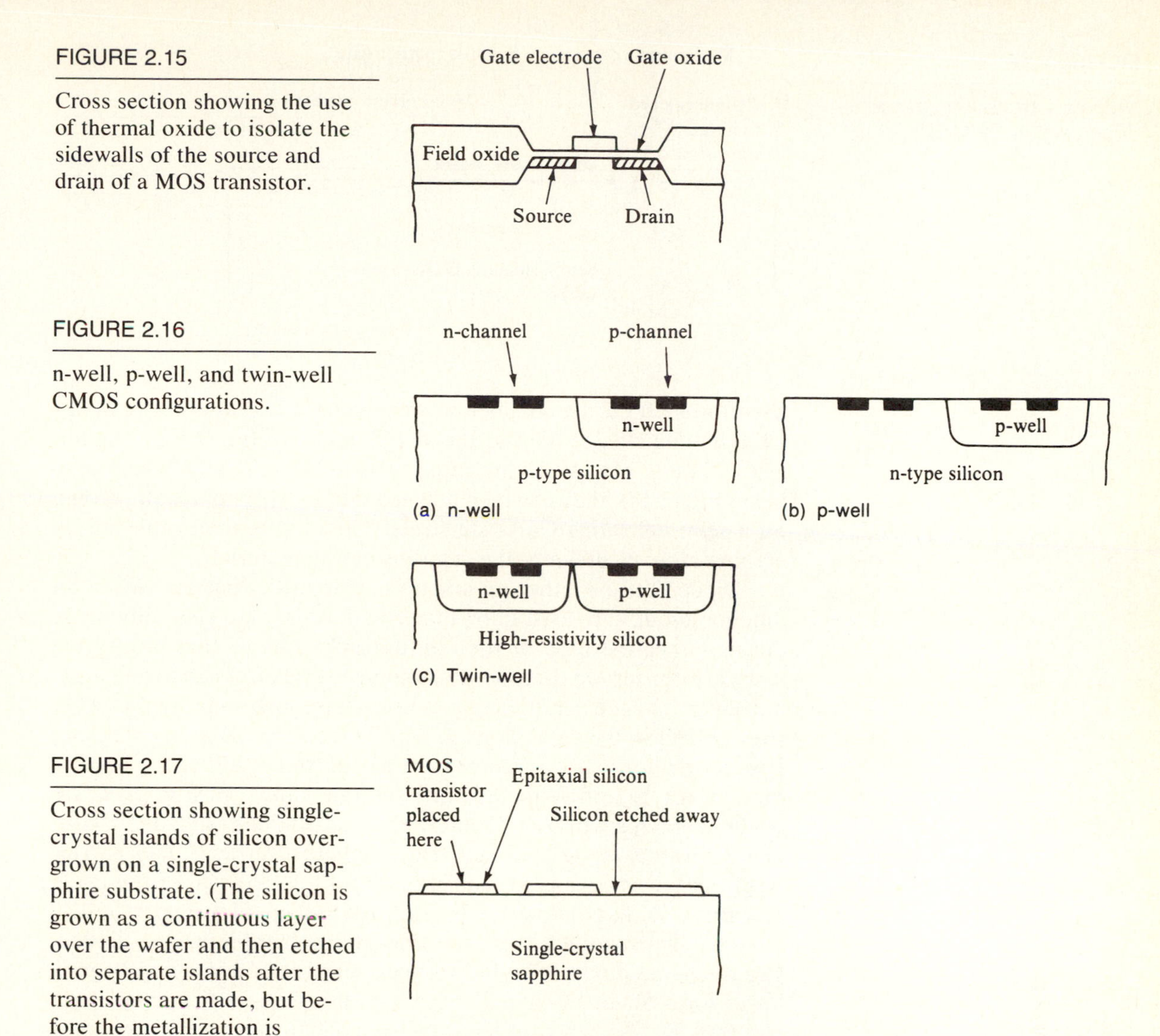

FIGURE 2.15

Cross section showing the use of thermal oxide to isolate the sidewalls of the source and drain of a MOS transistor.

FIGURE 2.16

n-well, p-well, and twin-well CMOS configurations.

FIGURE 2.17

Cross section showing single-crystal islands of silicon overgrown on a single-crystal sapphire substrate. (The silicon is grown as a continuous layer over the wafer and then etched into separate islands after the transistors are made, but before the metallization is added.)

or twin-well CMOS structures shown in Fig. 2.16. Note that MOS transistors are self-isolating, but in order to reduce capacitance from source–drain to substrate and thus improve high-frequency performance, MOS transistors on insulating substrates are sometimes used. They are usually built in silicon epitaxially overgrown on single-crystal sapphire (SOS), as shown in Fig. 2.17.

An alternative to the MOS transistor is the MESFET, shown in Fig. 2.18. The source and drain are made as in a conventional MOS transistor. However, instead of an electrode over a thin gate oxide, a Schottky diode placed where the gate oxide would normally be is

FIGURE 2.18

MESFET transistor such as is often used in GaAs ICs.

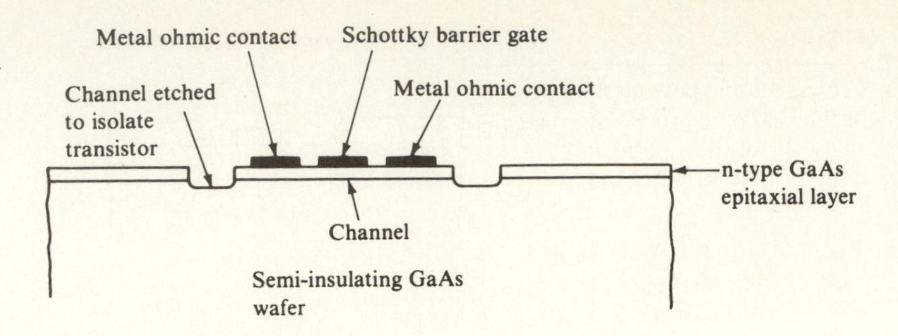

used to pinch off the channel between source and drain. The MESFET is thus closely akin to the JFET, and like the JFET, it is not self-isolating. MESFETs are often used in GaAs ICs, in which case (as in Fig. 2.18), they can be made in a thin n-layer epitaxially grown on a semi-insulating GaAs substrate. Isolation is then achieved by etching grooves down to the semi-insulating material.

Diodes can be either pn junction or Schottky barrier. When pn junction diodes are used in bipolar circuits, they are generally made during either the base or the emitter diffusions so that processing steps are conserved. In order to minimize series resistance and maximize capacitance per unit area, emitter–base junctions are preferred over collector–base junctions in bipolar circuits. To minimize junction interactions, the collector is often shorted to either the base or the emitter. Schottky diodes, like the one shown in Fig. 2.19, are made by having a barrier material, such as platinum silicide, in contact with a bare, very clean semiconductor surface. The barrier height is lower on p-type than it is on n-type material of the same resistivity so that the reverse breakdown voltage of p-type Schottky diodes is substantially less than that of comparable n-type diodes. For the same diode area, the forward current at a given voltage is lower for a Schottky than it is for a pn junction. A Schottky diode can thus be used in parallel with the collector–base junction of a bipolar switching transistor to keep the transistor from saturating during that part of the switching operation when the collector–base junction is forward-biased.

FIGURE 2.19

Schottky diode. (Rectification occurs at the metal–semiconductor interface.)

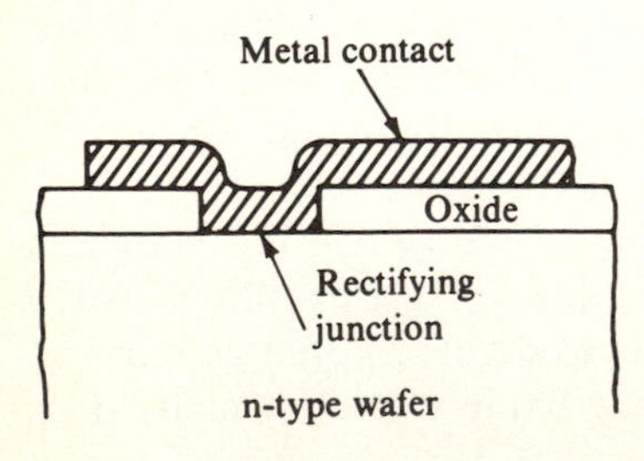

IC resistors are usually made from a thin, high sheet resistance Si or GaAs diffused layer, but sometimes a high-resistance metal film such as nichrome is used. In the early days of the IC, the base diffusion was also the resistor diffusion. Now, in order to get higher-value resistors into smaller areas, separate higher sheet resistance diffusions are usually used.

Capacitors may be either pn junction, MOS, or some other metal–insulator–metal combination. In order to conserve space, they may be between two levels of leads or in the walls of grooves etched in the wafer, as is the case with trench capacitors (Fig. 2.20).

In order to interconnect the many components of an IC, some leads must cross others. Fig. 2.21 shows the three ways in which crossing is done. The oldest method is the diffused crossunder, in which current of one lead goes under the other by traveling through

FIGURE 2.20

Integrated circuit MOS capacitors: (a) conventional, (b) trench, and (c) stacked.

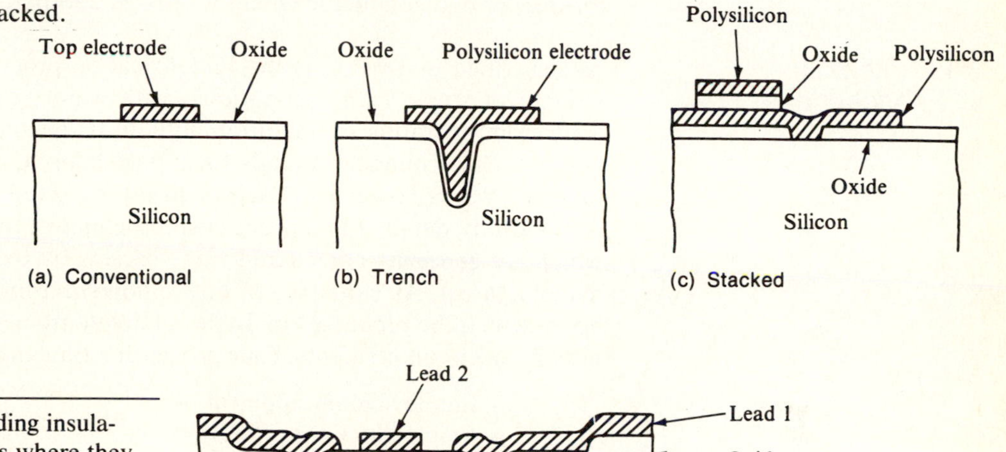

FIGURE 2.21

Methods of providing insulation between leads where they cross one another.

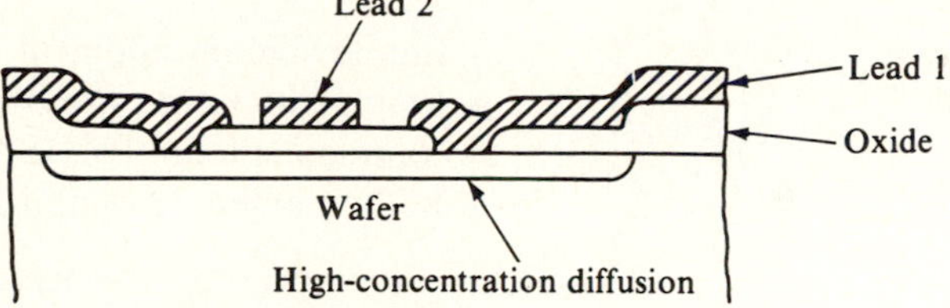

(a) Diffused crossunder method

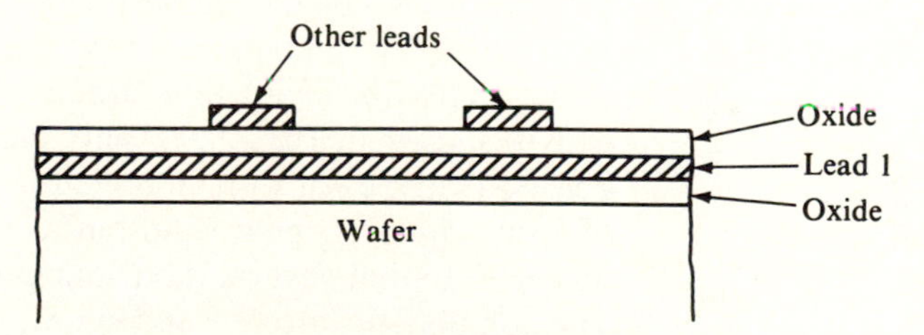

(b) Conventional double-level method

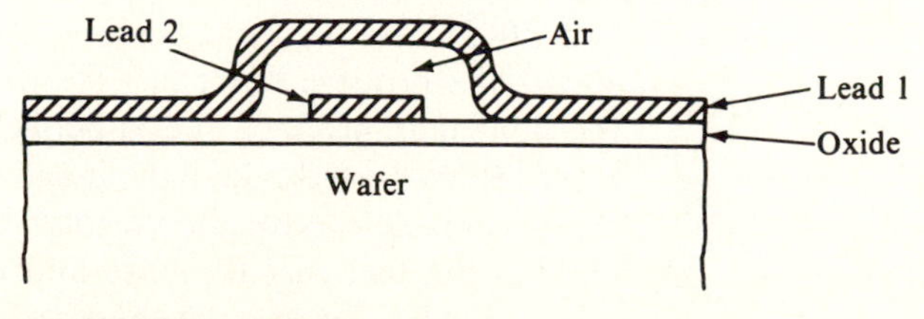

(c) Air-bridge method

a high-conductivity path diffused into silicon. When a large number of leads exists, unless some leads are stacked on top of others, leads on the surface, and not active components in the semiconductor, limit IC size. Multiple layers of leads (double-level metal, triple-level metal) use either a CVD insulating layer such as SiO_2 or an organic layer such as a polyimide for insulation. When the capacitance of overlapping leads or a heavily diffused tunnel is too great, air crossovers are sometimes used. They make use of an unsupported loop (bridge) of one conductor where it crosses another.

2.4.2 Steps in IC Fabrication

As described in Table 2.1, the IC fabrication process consists of a number of steps, many repetitious, with wafers moving back and forth from operation to operation until an IC is completed. A tabulation of the sequence of steps for a particular IC may be referred to as a process flowchart. Each of the steps listed in the flowchart is more fully detailed in a process specification (the exact details of which are generally considered trade secrets by the semiconductor manufacturer). As examples of how a flowchart maps out the manner in which the processes of Table 2.1 are combined to produce the necessary circuit elements, four processing flows will be discussed:

A simple silicon bipolar flow
A simple aluminum PMOS flow
A complex silicon CMOS flow
A gallium arsenide digital IC flow

These flows include only the salient features; it should be understood that the processes followed in commercial manufacture may include many more steps, often proprietary, designed to improve yield and enhance device performance.

For the first example, suppose that a simple silicon TTL (transistor–transistor logic) gate circuit is to be made. The electrical circuit is shown in Fig. 2.22. This circuit has few components, and consequently, even with large feature sizes such as contact openings of $5\mu m \times 5\mu m$, it can be laid out as a square approximately 0.7 mm on a side so that several thousand can be placed on a 100 mm wafer. The components are few and simple; only transistors, resistors, and lead crossunders are used. The transistors are pn junction isolated and have a subcollector preepitaxial diffusion. The starting material is a 10–20 Ω-cm p-type slice with a (111) orientation. Table 2.5 shows the process flowchart for this simple TTL IC. Steps 1–7 of the flowchart produce the subcollector diffusion and the epitaxial layer. Steps 8–10 produce the localized p-diffusions that will provide electrical isolation for the various devices. Not shown on this flowchart is the fact that the base and the resistors each have a design value of 120 Ω/sq. so that they can both be made by the base diffu-

FIGURE 2.22

TTL gate. (There are no diodes in this circuit.)

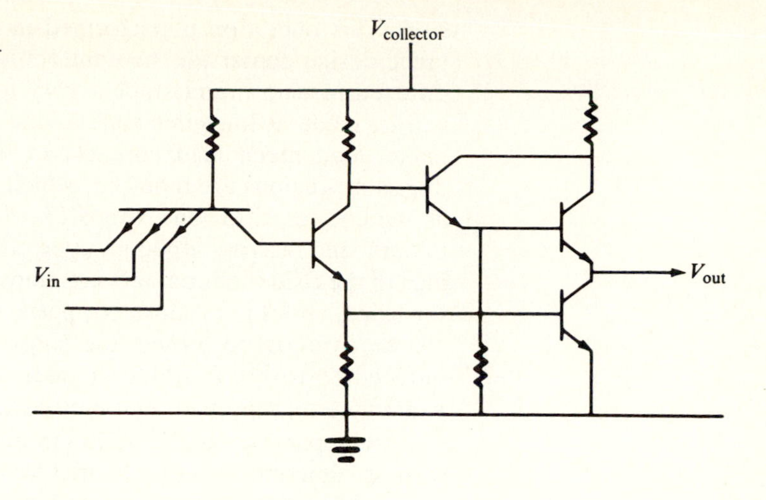

TABLE 2.5

Simple Silicon TTL IC Process Flowchart

Step	Operation
1	Clean slice.
2	First oxidation: Grow approximately 10,000 Å of thermal oxide.
3	Apply photoresist; expose, using first mask of set; develop to remove resist where subcollector diffusion will be.
4	Etch oxide; remove resist; clean wafer.
5	Diffuse antimony into openings (windows).
6	Remove oxide; clean wafer.
7	Grow a 5–6 μm thick, 0.2–0.5 Ω-cm, n-type epitaxial layer.
8	Second oxidation: Grow approximately 10,000 Å of thermal oxide.
9	Repeat steps 3 and 4 using second mask (isolation mask) of set.
10	Diffuse boron for isolation (also to regrow oxide).
11	Repeat steps 3 and 4 using third mask (base mask).
12	Diffuse boron for bases and resistors (also to regrow oxide).
13	Repeat steps 3 and 4 using fourth mask (emitter mask).
14	Diffuse phosphorus for emitters, crossunders, and collector contacts (also to regrow oxide).
15	Repeat steps 3 and 4 using fifth mask (contacts mask).
16	Evaporate aluminum over whole slice.
17	Repeat step 3 using sixth mask (leads mask) to define aluminum pattern.
18	Etch to remove unwanted aluminum; remove resist.
19	Sinter metal to ensure good electrical contact; clean wafer.
20	Vapor-deposit protective overcoat (PO) layer of SiO_2.
21	Repeat step 3 using seventh mask (PO mask) to define window openings over bonding pads.
22	Etch to remove oxide over bonding pads; remove resist.
23	Multiprobe wafer.

sion. This operation is performed in steps 11–12. The sheet resistance design center for the emitters is 5 Ω/sq. The crossunders are diffused resistors, and since a very low resistance is needed, they can be made at the same time as the emitter diffusion. In order to make good metal lead contacts to the n-type collectors, heavily doped n^+-regions are required, which can be made at the same time as the emitter diffusions. Steps 13–14 thus provide emitters, crossunders, and heavily doped n^+-type collector contact regions. Openings in the oxide for contacts are made in step 15. Aluminum is used for leads, and it is evaporated, patterned, and sintered in steps 16–19. An overcoat to protect the chips from scratches and moisture-induced corrosion is added in steps 20–22. Finally, in step 23, the wafers are multiprobed, and faulty chips are marked with ink.

A comparably simple MOS circuit (now outdated) can be made with an aluminum gate p-channel MOS (PMOS) process, the flowchart of which is shown in Table 2.6. Because of the sensitivity of a MOS device to the cleanup before gate oxidation and the growth of the thin oxide itself, the pregate cleanup and the gate oxide growth steps are substantially more involved than the cleanups and oxide growth steps of the previously discussed TTL process flow. Step 8, to ion implant the threshold adjust, is a step not found in bipolar fabrication, although ion implantation for other purposes may be used.

TABLE 2.6

Simple Aluminum Gate PMOS Process Flowchart

Step	*Operation*
1	Clean slice.
2	Grow field oxide.
3	Perform source–drain OR.*
4	Perform boron source–drain diffusion.
5	Perform gate OR.
6	Clean pregate.
7	Grow gate oxide.
8	Ion implant threshold adjust.
9	Perform contact OR.
10	Metallize.
11	Pattern and etch metal.
12	Sinter.
13	Deposit protective overcoat (PO).
14	Pattern and etch PO.
15	Multiprobe wafer.

*OR is a commonly used abbreviation for oxide removal. An OR generally involves the steps of (1) applying the resist, (2) exposing and developing the resist, (3) etching the oxide, and (4) removing the resist and cleaning the wafer.

The construction of a twin-well, self-aligned, silicon gate CMOS, as outlined in the process flowchart of Table 2.7, is quite complex compared to the processes covered by the first two example flows. The process flow of Table 2.7 does, however, still use the same basic processes described in Table 2.1, and no individual devices are used in the chip that were not described in section 2.4.1. The twin-well construction was shown in Fig. 2.16. Typically, the starting slice resistivity is about 50 Ω-cm. The orientation, as for essentially all MOS circuits, is (100). Steps 1–5 produce the n-well, and steps 6–10 the p-well. Oxide isolation at the sidewalls of the sources and drains, as was shown in Fig. 2.15, is used in this IC. The oxide pattern is defined and the oxide is grown in steps 11–16. The region where this oxide is grown is sometimes referred to as a "moat," although the oxide grown is the "field" oxide. Since the thermal oxide thickness is about twice that of the silicon consumed, the surface of a localized oxidation will rise above the rest of the wafer surface. If a planar surface after oxidation is desired, the surface before oxidation must thus be recessed. In this flow, step 14 provides for such a recess. To prevent inversion of p-silicon under the field oxide, a high-concentration boron implant (step 15) is made before field oxidation (channel stop). The circuit described by this flow uses trench capacitors as shown in Fig. 2.20b. They are built in steps 17–25. The gate oxide is grown and the polysilicon gates are formed in steps 26–31. After the gate regions are cleared of thick oxide, a sacrificial oxide (dummy oxide) is grown and stripped before gate oxidation to clean the silicon (step 27). In the self-aligned gate process described here, the polysilicon gate electrode material is used to help define the location of sources and drains, which are made in steps 32–38. This ensures that the gate does not substantially overlap the sources and drains and cause poor performance. After the sources and drains, an interlevel CVD oxide to cover the polysilicon gates and leads is added, and holes are cut in it where the top level of metal is to contact the polysilicon leads (steps 39–40). The top metal is added, patterned, and sintered in steps 41–43. The protective overcoat is deposited and opened up over the bonding pads in steps 41–42. The wafer is thinned, after which its backside is coated with gold to improve electrical conductivity and multiprobe is performed (steps 46–48).

For the final example, consider a gallium arsenide digital IC. High-performance bipolar and MOS transistors are difficult to fabricate in gallium arsenide. The most commonly used GaAs transistor is the MESFET (Fig. 2.18), and it is described in this flow. Both enhancement- and depletion-mode MESFETs can be constructed, and digital circuits are designed using only these transistors and diodes for circuit elements. Analog circuits, such as microwave am-

TABLE 2.7

Twin-Well, Self-Aligned, Silicon Gate CMOS Process Flowchart

Step	Operation
1	Clean slice.
2	Grow thin oxide (stress buffer between silicon and nitride).
3	Deposit silicon nitride (oxidation mask).
4	Perform n-well nitride removal.*
5	Ion implant n-well (as an alternative to diffusion deposition).
6	Grow oxide over n-well.
7	Remove nitride (and expose remainder of wafer, which will be p-well).
8	Ion implant p-well.
9	Remove oxide; clean wafer.
10	Drive-in and anneal ion implants.
11	Grow thin oxide (buffer).
12	Deposit nitride (mask for moat oxidation).
13	Remove moat nitride.
14	Etch silicon to depth of half that of oxide to be grown.
15	Implant channel stop.
16	Grow field oxide.
17	Deposit oxide for trench mask (for trench capacitor).
18	Perform trench OR (do not remove resist).
19	Etch trench.
20	Remove resist.
21	Diffuse to dope trench walls (n^+).
22	Remove oxide grown during trench diffusion.
23	Grow trench capacitor oxide.
24	Deposit polysilicon to fill in trench; form capacitor plate.
25	Pattern polysilicon.
26	Perform gate OR.
27	Grow dummy oxide.
28	Remove (strip) dummy oxide.
29	Grow gate oxide.
30	Deposit polysilicon for gate electrode and first level of interconnection.
31	Pattern polysilicon.
32	Pattern n-tub sources and drains.
33	Implant n-tub sources and drains.
34	Strip resist.
35	Pattern p-tub sources and drains.
36	Implant p-tub sources and drains.
37	Strip resist.
38	Anneal sources and drains.
39	Deposit CVD interlevel oxide.
40	Via OR (cuts holes for contacts between two levels of interconnection).
41	Deposit second level of metallization.
42	Perform second-level metal patterning and removal.
43	Sinter contacts.

TABLE 2.7 (continued)

Step	*Operation*
44	Deposit protective overcoat.
45	Remove protective overcoat at bonding pads.
46	Thin wafers (backgrind).
47	Deposit gold on backside of wafer (to improve electrical contact of chip to lead frame).
48	Multiprobe wafer.

*A nitride removal includes essentially the same steps as an oxide removal.

TABLE 2.8

Gallium Arsenide Digital MESFET Process Flowchart

Step	*Operation*
1	Clean slice.
2	Add CVD silicon oxide layer.
3	Apply photoresist; expose and develop to remove resist where transistors will be located.
4	Ion implant silicon through CVD oxide, using resist as mask (channel implant).
5	Repeat steps 3 and 4 for source and drain implants.
6	Strip resist; clean wafer.
7	Cap whole wafer with more CVD oxide.
8	Anneal to activate implanted ions.
9	Strip CVD oxide; clean wafer.
10	Add CVD silicon nitride layer.
11	Apply photoresist; expose and develop to remove resist where contact windows will be located.
12	Etch nitride from contact windows.
13	Deposit Ge:Au alloy; deposit Ni layer.
14	Remove resist along with metal on top of it; clean wafer.
15	Alloy to form contacts.
16	Apply photoresist; expose and develop to remove resist where transistor gates will be located.
17	Etch GaAs to thin gate region.
18	Deposit Ti:Pd:Au layer.
19	Remove resist along with metal on top of it.
20	Apply photoresist; expose and develop to remove resist where first-level leads will be located.
21	Deposit first-level metal; remove resist along with unwanted resist.
22	Apply photoresist; expose and develop to remove resist where second-level metal (air-bridge metal) will contact first level.
23	Deposit thin metal layer.
24	Apply photoresist; expose and develop to define air-bridge metal.
25	Electroplate exposed pattern.
26	Remove top layer of resist (down to metal of step 23).
27	Etch away exposed thin metal.
28	Remove underlying resist (resist added in step 22).
29	Electrically test circuits.

plifiers, may use diffused and thin-film resistors, metal–insulator–metal capacitors, and inductors formed by pancake-spiraled leads. GaAs ICs are used only when very-high-frequency operation is required, and to minimize lead crossover capacitance, air bridges as in Fig. 2.21c are common. The starting material is a semi-insulating slice with a (100) orientation. In some processes, as, for example, that used in Fig. 2.18, an n-epitaxial layer is added, and individual devices are isolated by etching away the n-layer between them. In other processes, such as the one described by the flowchart of Table 2.8, localized ion implantation into the semi-insulating GaAs is used to provide the necessary thin n-regions (steps 1–4). Steps 5–6 produce the sources and drains. Excessive arsenic is lost when any processing temperature is over about 600°C unless the wafer is covered with an impervious layer such as CVD SiO_2 or silicon nitride. Thus, "capping" layers (step 7) are added from time to time during GaAs wafer fabrication. After the capping, the ion implants are annealed, and the cap is removed (steps 8–9). A nitride layer, used for surface passivation, is added in step 10. Resist is applied, contact openings are cut in the nitride, and metal is deposited. The resist is then removed, and with it, all of the metal except that in the contact openings. This "lift-off" process (described in Chapter 5) is done in steps 11–14. The contacts are sintered in step 15. To improve the MESFET's electrical performance, the implanted channel (step 4) must be reduced in thickness, which is done by etching a slight depression in the GaAs where the gate electrodes are to be applied (steps 16–17). The gate metal is then deposited, and the resist defining the gate areas is stripped (another example of lift-off) in steps 18–19. The first-level metal pattern is then deposited and defined in steps 20–21. In this flow, the second level of metal is only used in the air-bridge crossovers. They are made in steps 22–28. First, resist is applied and patterned to open up points of contact to the first-level metal (step 22). A thin layer of metal is deposited over the resist, after which another coat of resist is applied and patterned to define the air-bridge metal (steps 23–24). The continuous, thin metal coating provides electrical contact to all of the bridge sites on the wafer and allows them to be simultaneously electroplated. The electroplating is used to build up the thickness of the bridge metal and provide enough strength to keep it from sagging. After plating, the top layer of resist is removed, and the thin layer of metal between the thick bridge metal sites is etched away. Then, the remaining resist layer (the one of step 22) between the first-level metal and the crossover metal is removed. Finally, in step 29, the wafers are tested.

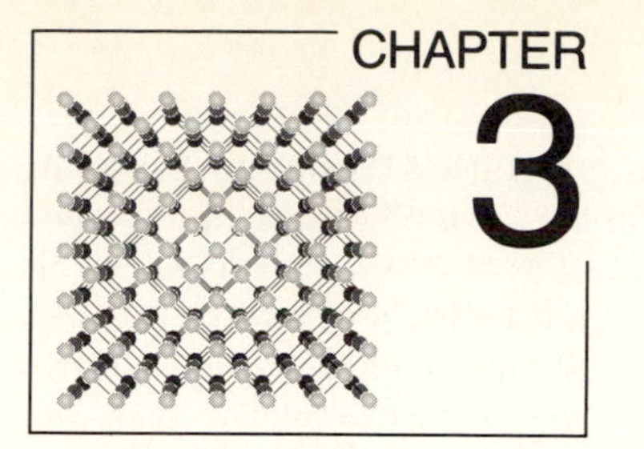

CHAPTER

3

Thermal Oxidation of Silicon

3.1 INTRODUCTION

Silicon dioxide (SiO_2, silica) exists both in crystalline form and in vitreous form. The crystalline form occurs in nature (most commonly as the mineral quartz) and can be produced artificially from a high-temperature, high-pressure, alkaline, aqueous solution of SiO_2. Vitreous silica (fused silica, fused quartz) forms when crystalline silica is melted and allowed to cool, when a silicon-bearing species such as SiH_4 or $SiCl_4$ reacts with oxygen at elevated temperature, or when silicon is directly oxidized in an atmosphere containing oxygen or water vapor. A silicon surface with its oxide removed will reoxidize in air at room temperature, but because the growth is self-limiting, the maximum thickness is only about 40Å.[1] This native oxide, while somewhat porous, forms a protective coating that prevents corrosion or further oxidation in most room-temperature ambients.

When the temperature is raised several hundred degrees C, the oxidation rate is much greater, and the oxide becomes denser and more durable. This higher-temperature thermal oxide is used in IC manufacturing. It is formed by the direct oxidation of the silicon wafer surface at elevated temperature in either an oxygen or a steam ambient through the following reaction:

$$Si + O_2 \rightarrow SiO_2$$

or

$$Si + 2H_2O \rightarrow SiO_2 + 2H_2$$

---

[1]The thickness–time curve is dependent on surface cleanliness, the amount of impurities in the air, and the method of measurement. Typically, the thickness as measured by ellipsometer will be about 5Å in 5 minutes, 20Å in 15 hours, and perhaps 40Å after a year. Using X-ray photoelectron spectroscopy, the thickness is estimated to be about half as much. (See, for example, S.I. Raider et al., *J. Electrochem. Soc. 122*, p. 413 and references therein, 1975.)

While many materials oxidize by the metal's diffusing to the oxide surface and reacting with oxygen, in the case of silicon, the oxygen diffuses through the oxide and reacts at the silicon surface. This reaction has been deduced from the final location of radioactive materials placed on the oxide surface and of oxygen isotopes introduced into the oxidizing atmosphere partway through an oxidation cycle (1, 2, 3).

The oldest semiconductor use of thermal oxide has been as a diffusion mask.[2] The requirements are that the oxide be impervious to the diffusion species for at least as long as the time required for the diffusion and that there be no pinholes in the oxide. Because of the nature of the diffusants, they can penetrate the oxide in two ways. One is by straight diffusion, and the other is by reacting with and gradually dissolving the oxide. Of the standard group IIIA and VA dopants, only gallium has a high enough diffusion coefficient in the oxide to prevent masking. Most of the other dopants react in time with the oxide to form silicates that are liquid at diffusion temperatures. The formation will begin at the outer surface, form a well-defined interface, and gradually move through the oxide. The silicate-glass phase will be rich in dopant, and when the glass has totally consumed the pure oxide, it will be touching the silicon and become a diffusion source. Thus, the thickness of the oxide must be tailored to the required diffusion time and temperature.

While other materials, such as CVD oxide or nitride, can be used as a mask, no commercially proven substitute for a thermal oxide currently exists to provide silicon pn junction passivation or MOS gate oxides.[3] The prime importance of thermal oxide stems from its ability to properly terminate the silicon bonds at the silicon–oxide interface. Thus, the deleterious electrical effects of a bare silicon surface or of one covered only with room-temperature native oxide are minimized. The major requirement of such a passivating layer is that it have few electrical charges either in its bulk or at the silicon–layer interface.

Another major application of thermal oxide as a MOS gate insulator requires the same properties needed for good junction pas-

---

[2]The use of thermal oxide as a mask for diffusion predates the planar process by a few years. When concerns about masking first arose, the primary consideration was to keep diffusing impurities from the back of the slice. (See, for example, C.J. Frosch, Chap. 5, "Silicon Diffusion Technology," in F.J. Biondi, ed., *Transistor Technology*, Vol. III, D. Van Nostrand Co., Princeton, N.J., 1958.)

[3]That CVD insulators may eventually be used for MOS gate insulation does seem possible. (See, for example, T. Hosaka et al., "Ultra-Thin Reliable CVD–SiO_2 Films for VLSIs," *Electrochem. Soc. Ext. Abst. 86–2*, abst. no. 409, October 1986.)

sivation. In addition, because the electric field in a gate insulator can approach 10^7 V/cm, far fewer mobile ions in the insulator can be tolerated, and defects such as pinholes, thin spots, and localized high-conductivity regions will cause device failure.

The physical properties of fused silica and of the CVD oxide discussed in Chapter 4 closely match those of thermal oxide. However, for surface passivation, the thermal oxide is the only material that provides satisfactory electrical interface properties. The key factor that enables the thermal oxide to provide the necessary silicon bond termination is its in situ growth, in which the oxidizing reaction occurs at the silicon surface and minimizes the structural damage associated with the abrupt Si–SiO_2 lattice change. Chemically deposited silicon oxide, as well as other materials such as CVD silicon nitride, can, however, be used for diffusion masking.

3.2 STRUCTURE

As shown in Table 3.1, crystalline SiO_2 exists in ten or more crystalline forms. It is of interest in IC fabrication only in that, occasionally, contamination will cause small regions of thermal oxide to crystallize. Where crystallization occurs, the desirable interface properties of the thermal oxide are lost, and silicon dopants can more easily penetrate the oxide.

The structure of both single-crystal and amorphous silica is such that an assembly of silicon atoms is surrounded tetrahedrally by four oxygen atoms as shown in Fig. 3.1 (4). The various crystallographic forms arise from the relative positions of the tetrahedra when they are joined together in a three-dimensional lattice. Through the Si–O–Si link, substantial rotation of one tetrahedron with respect to an adjacent one can occur without changing either the bond length or the angle between bonds. It is thus possible to lose long-range order, become amorphous, and still have close resemblance to the basic crystalline structure. Also, this ability to twist and distort probably allows the silica lattice to match up with and properly terminate the silicon surface bonds. A comparison of angles and distances is given in Table 3.1.

The interface structure has been widely studied (5) and appears to make quite an abrupt transition from single-crystal silicon to amorphous SiO_2. Fig. 3.2 is a transmission electron micrograph (TEM) of an interface. Over short distances, the surface appears atomically flat but does show steps. This step roughness increases with higher oxidation rates or lower oxidation temperatures. Early spectroscopic studies also led to the conclusion of a very abrupt transition. Knowing that the transition from silicon to amorphous oxide takes place in 5–10 Å still does not answer the question of exactly how the two bond together. Fig. 3.3 shows one speculation.

TABLE 3.1

SiO_2 Crystallographic Data

Crystal Class	*Description*
α-quartz*	Hexagonal; stable at room temperature
β-quartz	Hexagonal; stable from 575°C to 870°C
α-tridymite	Orthorhombic; stable at room temperature
β-tridymite	Hexagonal; stable from 163°C to 1470°C
α-cristobalite	Tetragonal; stable at room temperature
β-cristobalite	Cubic; stable from 218°C to 1710°C (melting point)
Keatite	Tetragonal; stable around 400°C and 5000 psi
Coesite	Stable at 750°C and 35,000 atm
Fibrous	Orthorhombic

Bond Angle and Length	*Thermal Oxide*	*Fused Silica*	*Quartz*
Si–O length (Å)	1.62	1.62	1.62
O–Si–O angle (°)	147 ± 10†	145 ± 10†	

*May be either right- or left-handed.
†There is a ±10° distribution about the mean.

Sources: Data on crystal class from M. Cannon Sneed and Robert C. Brasted, *Comprehensive Inorganic Chemistry,* Vol. 7, D. Van Nostrand Co., New York, 1958. Data on bond angles and lengths from Naoyuki Nagasima, *Jap. J. Appl. Phys. 9,* pp. 879–888, 1970.

FIGURE 3.1

Si–O bonding details. (*Source:* Adapted from W.H. Zachariasen, "The atomic arrangement in glass," *J. Am. Chem. Soc. 54,* pp. 3841–3851, 1932.)

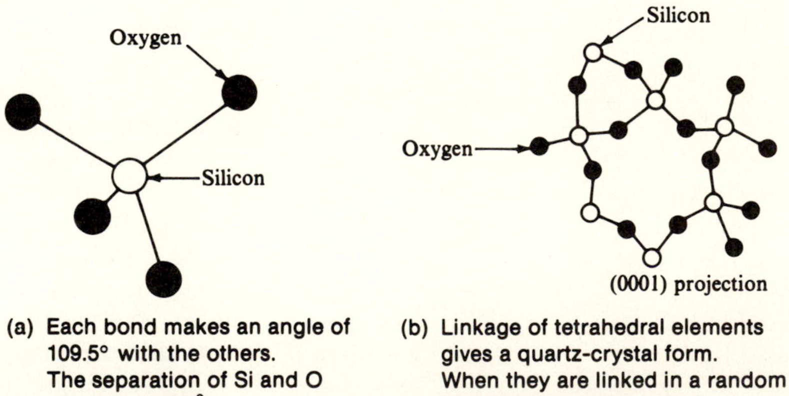

(a) Each bond makes an angle of 109.5° with the others. The separation of Si and O atoms is 1.62/Å.

(b) Linkage of tetrahedral elements gives a quartz-crystal form. When they are linked in a random manner, amorphous silica results.

FIGURE 3.2

High-resolution (lattice-image) transmission electron micrograph (TEM) showing transition from single-crystal Si to amorphous SiO_2. (The section is normal to a (100) surface.) (*Source:* Courtesy of Dr. H.L. Tsai, Central Research Laboratories, Texas Instruments Incorporated.)

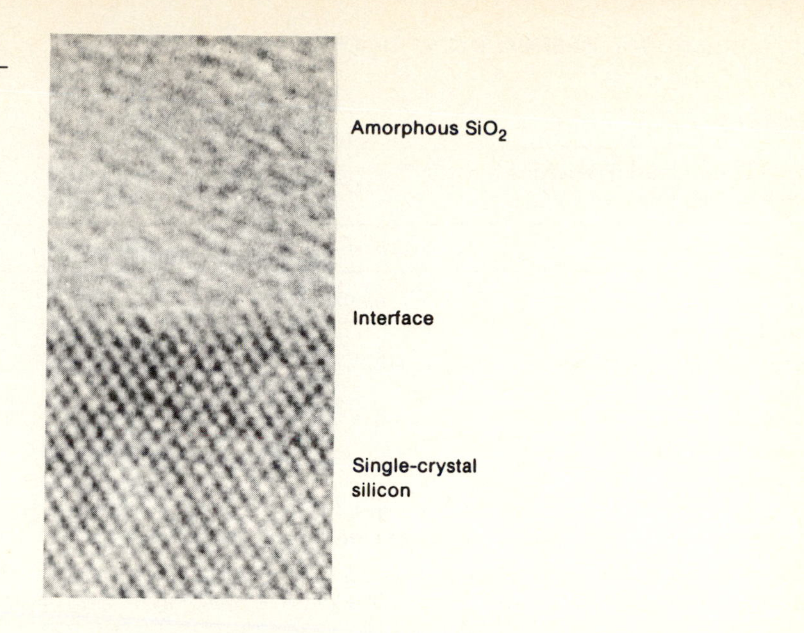

FIGURE 3.3

Suggested model for Si–SiO_2 transition region, viewed looking into silicon (100) face. (*Source:* Adapted from C.R. Helms, *Semiconductor Silicon 81,* p. 455.)

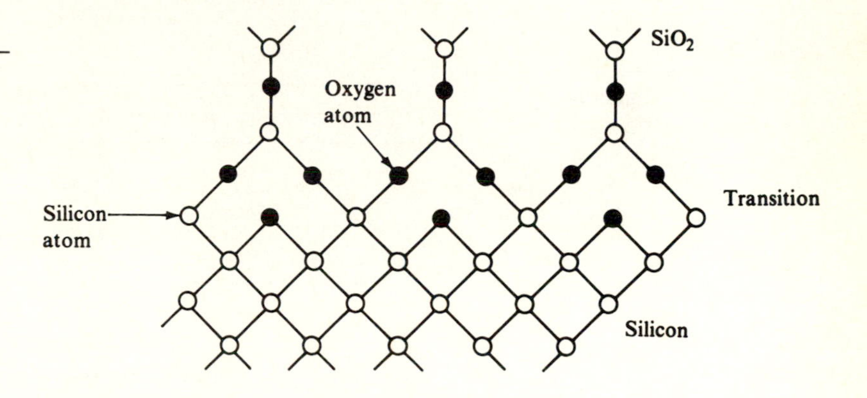

3.3 NONELECTRICAL PROPERTIES

Table 3.2 summarizes many of the properties of thermal oxide and compares them to both silicon and other forms of silica.

3.3.1 Density

The density value reported for thermal oxide is close to that of fused silica and is almost the same as that of silicon (2.22 versus 2.33 grams/cc for silicon). These densities, coupled with their respective molecular weights of 60 and 28, dictate that for every volume V of silicon oxidized, $2.2V$ of oxide will be generated. In terms of thickness, as shown in Fig. 3.4, for an oxide thickness of t, $0.45t$ of silicon

TABLE 3.2

Room-Temperature Properties of Silicon and Silicon Dioxide

Property	*Thermal*	*Fused*	*Quartz*	*Silicon*
Molecular weight	60.08	60.08	60.08	28.06
Specific gravity	2.22 (1)	2.22	2.66†	2.33
Linear expansion coefficient (per °C)	0.5×10^{-6}*	0.5×10^{-6}		2.3×10^{-6}
Young's modulus (dynes/cm^2)	6.6×10^{11} (2)	7.3×10^{11} (3)		16.9×10^{11}‡
Poisson's ratio		0.17		

*Fused silica value.
†Coesite, metastable at room ambient, has a specific gravity of 3.0. Cristobalite, at 2.2, is the same as fused silica.
‡In any direction in a (111) plane. The range in other directions is from $13–18.8 \times 10^{11}$ dynes/cm^2.

Sources: (1) Naoyuki Nagasima, *Jap. J. Appl. Phys. 9*, pp. 879–888, 1970. (2) R.J. Jaccodine and W.A. Schlegel, *J. Appl. Phys. 37*, pp. 2429–2434, 1966. (3) H.J. McSkimin, *J. Appl. Phys. 24*, pp. 988–997, 1953.

FIGURE 3.4

Consummation of silicon during thermal oxidation.

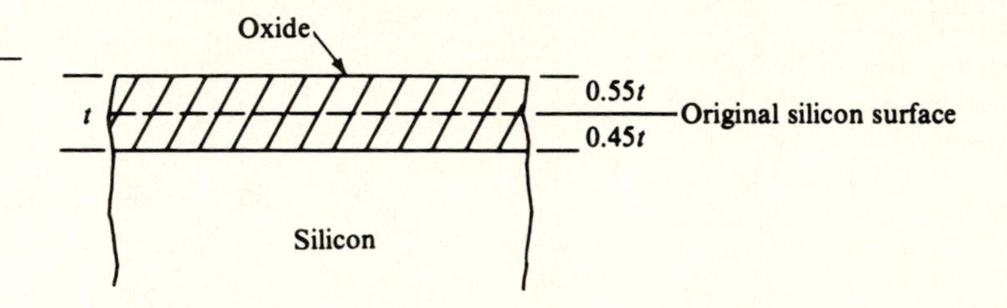

will be consumed, and the top of the oxide will rise $0.55t$ above the original silicon surface.

Because the oxide occupies more space as it grows than does the oxide consumed, appreciable compressive stress in the oxide near the interface is generated. The free top surface allows reduction in a direction normal to the growth surface, but since the oxide is tightly bonded to the silicon, no slipping can occur along the interface in the other two directions. Thus, high intrinsic oxide stresses can arise during growth (6, 7). However, above about 950°C, silicon dioxide has a low enough viscosity for the stress to relax, and little residual stress remains. For oxidations at temperatures below 960°C, the relaxation is much slower, and intrinsic stresses of up to 7×10^9 dynes/cm^2 have been reported. Lower stresses have been found in wafers oxidized in wet oxygen than in those oxidized in dry oxygen (8). This fact is consistent with a lowering of silica viscosity when it contains H_2O (9).

TABLE 3.3

Effect of Oxidation Temperature on Oxide Density as Deduced from Index of Refraction Measurements

Temperature (°C)	*Density*
600	2.286
700	2.265
800	2.253
900	2.236
1000	2.224
1150	2.208

Source: Adapted from E. A. Irene et al., *J. Electrochem. Soc. 129*, p. 2594, 1982.

As the oxidation temperature decreases, there is, based on index of refraction measurements, a slight increase in density (8–10). This change is shown in Table 3.3. By annealing after oxidation in an inert atmosphere at temperatures above about 1000°C, the density and index of refraction return to their high-temperature oxidation values (9). The density decrease after annealing can be independently deduced by measuring the increase in thickness, assuming that care has been taken to ensure that the ambient was indeed nonoxidizing. The thickness change after a 700°C oxidation is about 3%, which is in reasonable agreement with index of refraction measurements (9). The presumed reason for the increasing density at the lower growth temperatures is the reduced amount of viscous flow, which, in turn, prevents the relief of the strain in the oxide that occurs during growth. An increase in the density of oxides grown at high pressures has been observed, but they are also grown at low temperatures, which is apparently the reason for the density increase.

3.3.2 Expansion Coefficient

A very large difference exists between the thermal expansion coefficients of silicon and thermal oxide. Therefore, as an oxidized slice cools from oxidation temperature (usually in the 800°C–1200°C range), the oxide film will be in compression. If residual internal stress occurs, which is also compressive, the two will be additive. If there is an oxide layer on each side of the wafer, the stress on the two sides will be balanced, and there will be no wafer bowing. However, if the oxide is stripped from one side, as is the usual case, substantial bowing may occur. The stress due to thermal mismatch can be up to 3×10^9 dynes/cm^2 in the oxide when it is cooled from 1200°C to room temperature (12). Because of the much greater thickness of the slice, the maximum stress in it will be from 10 to 100 times less.

The stress σ_f in the oxide of a circular slice–oxide sandwich can be determined from the bow in the wafer from Eq. 3.1 (13) as follows:

$$\sigma_f = \frac{\delta E(T^2/t)}{3(1 - \nu)r^2} \qquad 3.1$$

where δ is the deflection of the center of the wafer, E is Young's modulus of the wafer, T is the thickness of the wafer, t is the thickness of the oxide film, ν is Poisson's ratio of the wafer, and r is the radius of the wafer. The stress σ_s in the slice due to bowing caused by the stress σ_f in the oxide film is given by Eq. 3.2 (12) as follows:

$$\sigma_s = \frac{-4\sigma_f t}{T} \qquad 3.2$$

Assuming no viscous flow occurs during cooldown (which would reduce the stress), the stress due to thermal mismatch alone can be calculated from Eq. 3.3 as follows:

$$\sigma_f = \frac{(\alpha_f - \alpha_s)E_f\Delta T}{1 - \nu_f} \qquad 3.3$$

where α_s is the expansion coefficient of the wafer (substrate) and α_f is the expansion coefficient of the oxide film. E_f is Young's modulus of the film, ΔT is the oxidation temperature minus room temperature, and ν_f is Poisson's ratio of the oxide. If the cooldown is slow, or if subsequent annealing occurs at temperatures lower than the oxidation temperature, viscous flow relaxation can reduce the mismatch stress. Thus, for higher-temperature oxidations, Eq. 3.3 would be expected to overestimate stress.

EXAMPLE ☐ Neglecting viscous flow in the oxide or slip in the silicon, use Eq. 3.3 to calculate σ_f at room temperature if the oxidation were done at 1200°C.

From Table 3.2, $E_f = 6.6 \times 10^{11}$ dynes/cm^2 and $\nu_f = 0.17$. Also from the table, $\alpha_f = 5 \times 10^{-7}$/°C and can be used since the value changes little in going from room temperature to 1200°C. However, the room-temperature value given for silicon is too low. Looking up α_s in a handbook shows that 4.5×10^{-6} is a better value over the whole temperature range than is 2.3. $\Delta T = 1175$°C. Substituting these values in Eq. 3.3 gives

$$\sigma_f = 3.7 \times 10^9$$

which compares well with, but is higher than, the reference 12 value of 3×10^9 dynes/cm^2. ☐

Any intrinsic stress remaining in the oxide can be determined by subtracting the calculated stress due to thermal mismatch (Eq. 3.3) from the measured total stress (Eq. 3.1).

3.3.3 Optical Properties

The refractive index varies with oxidation temperature, as is shown in Fig. 3.5 for growth on two different crystallographic orientations. The value saturates at high growth temperature at 1.4620 for a wavelength of 5461 Å, which is slightly higher than the value of 1.4601 for fused silica (10). The infrared absorption peaks associated with P–O, B–O, and As–O stretching frequencies can be used to estimate the amount of phosphorus, boron, or arsenic incorporated into the thermal oxide during the diffusion process. This procedure has, however, now been largely superseded by X-ray fluorescence (14). The O–H line at wavenumber 3650 can be used to estimate the amount of hydroxyl ions present in steam-grown oxides (15). The

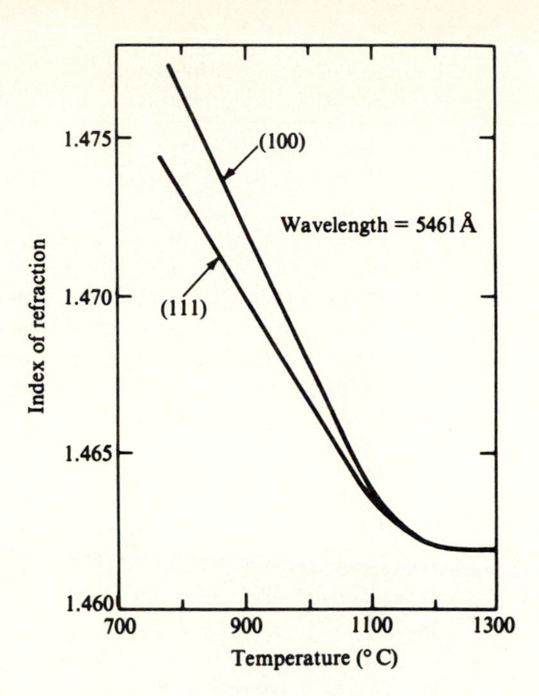

FIGURE 3.5

Index of refraction of thermal oxide as a function of growth temperature and orientation. (*Source:* Adapted from E.A. Taft, *J. Electrochem. Soc. 125*, p. 968, 1978.)

locations of both silica lattice and impurity absorption peaks are given in Table 3.4. The optical transmissivity is very low for wavelengths longer than about 5 μm, but it is very high from 3.5 μm down to less than 0.2 μm.

3.3.4 Chemical Properties

Silica is only slightly soluble in water at room temperature. The percentage concentration C can be calculated from Eq. 3.4[4] as follows:

$$C = 0.382 \times (13.6 + T) \times 10^{-3} \tag{3.4}$$

where T is the temperature in degrees C. The dissolution rate is so slow that atmospheric pressure water-etching is normally inconsequential. However, at higher pressures, etching can become pronounced in a steam atmosphere and must be considered in high-pressure oxidation processes. Silica will dissolve in aqueous HF and in heated solutions of bases such as KOH and can be plasma-etched in fluorine compounds such as CF_4. (Etching details are covered in Chapter 6.)

The reactivity of thermal oxide (and CVD oxide as well) with various metals is of concern in that IC metallization primarily consists of thin metal strips running over thin oxide. Slight interaction

[4]See G.W. Alexander et al., *J. Phys. Chem. 58*, p. 453, 1954.

TABLE 3.4

Oxide Optical Absorption Lines

Wavenumber (per cm)	*Wavelength (μm)*	*Assignment*	*Reference*
458	21.8	Si–O–Si, bending	(1)
720	13.9	B–O–B	(2)
810	12.3	O–Si–O, bending	(2)
930	10.8	As–O, stretching	(3)
1075	9.30	Si–O	(4)
1096	9.12	Si–O, stretching	(1)
1165	8.58	Si–O	(4)
1190	8.40	Si–O, stretching	(1)
1220*	8.20	Si–O	(4)
1330	7.52	P–O, stretching	(3)
1390	7.19	B–O, stretching	(3)
2240	4.46	Si–H, possibly	(3)
2260	4.42	Si–O, overtone	(3)
3650	2.74	O–H, stretching	(3)

*Observed in cristobalite. When a line is seen in silicon, it is taken as indication of small cristobalite regions (5).

References: (1) Naoyuki Nagasima, *Jap. J. Appl. Phys. 9,* pp. 879–888, 1970. (2) D.M. Brown et al., *J. Crystal Growth 17,* pp. 276–287, 1972. (3) W.A. Pliskin, pp. 506–529, in Howard R. Huff and Ronald R. Burgess, eds., *Semiconductor Silicon/73,* The Electrochemical Society, Princeton, N.J., 1973. (4) M.L. Naiman et al., *Electrochem. Soc. Ext. Abst. 83-1,* abst. no. 124, pp. 196–197, 1983. (5) F. Shimura et al., *Appl. Phys. Lett. 37,* p. 483, 1980.

is desirable so that the metal will adhere well, but excessive reactions can consume the metal or the oxide. Molten aluminum (melting point = 660°C) will reduce silicon dioxide, and at temperatures as low as 500°C, aluminum leads will react enough with thermal oxide to noticeably reduce the oxide thickness. Titanium begins to react at 400°C and, by 500°C, will have reduced the oxide thickness (16). There is also evidence that electolytic action can occur at metal–silica interfaces when metallic ions from MOS electrodes are injected into the SiO_2 (17).

Oxides of arsenic, boron, phosphorus, and antimony, as well as oxides of many other metals, will interact with silicon oxide at elevated temperatures to give mixed oxides with softening temperatures well below that of pure SiO_2. These oxides are useful if, for example, rounding the sharp corners of an etched oxide by heating to around 1000°C is desired. However, if keeping a phosphorus-glass diffusion source separated from the silicon surface by a layer of thermal oxide is desired, it is a distinct disadvantage.

Heating thermal oxide in the 1000°C temperature range in an ammonia atmosphere will give a "nitrided" oxide with an index of refraction higher than that of the oxide. The mechanism appears to be the addition of nitrogen atoms to the lattice and/or a nitride growth at the silicon–oxide interface, and not an oxygen atom replacement. Thin "thermal nitride" films can also be grown directly on a bare silicon surface (18, 19). These films contain oxygen and perhaps should be characterized as oxynitrides.

3.4 ELECTRICAL PROPERTIES

3.4.1 Bulk Electrical Conduction

As shown in Table 3.5, the electron mobility in silicon dioxide is reasonably high, but because of the large bandgap, current flow is limited by the ability to inject carriers into the oxide (electrode-limited conduction).[5] The emission near room temperature, with electric fields greater than about 3×10^6 V/cm and for oxide thicknesses above about 50 Å, is best characterized as Fowler–Nordheim (20, 21). For lower fields, Schottky emission has been suggested (22). In many cases, in the low-field region, motion of ions already in the oxide or trap filling may allow a current flow relatively independent of voltage that will be large enough to mask other effects.

When Fowler–Nordheim (F–N) emission occurs,

$$J = AE^2e^{-B/E} \tag{3.5}$$

where J is the current density, E is the electric field, and A and B are constants. Since this conduction depends on the electrode-to-insulator barrier height, A and B depend on the electrode material. For current flow described by Eq. 3.5, a plot of log (J/E^2) versus

TABLE 3.5

Oxide Electrical Properties

Property	*Value*
Dielectric constant	3.9
Dielectric strength	$\sim 10^7$ V/cm*
Bandgap	~9 eV
Resistivity	10^{12}–10^{16} Ω-cm†
Electron mobility	20–40 cm²/V·s
Hole mobility	$\sim 2 \times 10^{-5}$ cm²/V·s

*Function of growth conditions and thickness.
†Function of applied field.

Source: Bandgap, electron mobility, and hole mobility from data listed in D.J. DiMaria, "Defects and Impurities in Thermal SiO_2," in S.T. Pantelides, ed., *The Physics of SiO_2 and Its Interfaces*, Pergamon Press, New York, 1978.

[5]Another example of electrode-limited conduction is Schottky emission from metal contacts to high-resistivity silicon (discussed in Chapter 10). Bulk-limited conduction includes the familiar electronic conduction in semiconductors as well as ionic conduction observed in ionic crystals at elevated temperatures.

$1/E$ will be a straight line. Fig. 3.6 shows typical room-temperature data for several electrode materials. For polysilicon, the most often used MOS gate electrode material, as empirically determined, $A = 2 \times 10^{-6}$ A/V^2 and $B = 2.7 \times 10^8$ V/cm (23). The temperature dependence, as shown in Fig. 3.7, is relatively insensitive to temperatures up to perhaps 200°C. It is this behavior that helps separate F–N tunneling from Schottky emission, which shows an exponential current rise with temperature. For very thin oxides, there are deviations from the F–N prediction of I–V behavior. However, by using the WKB approximation[6] to solve for electron tunneling through various barrier shapes, good agreement between predicted and experimentally observed I–V curves is obtained (24).

Despite the fact that current is electrode limited, an effective resistivity can be calculated for a given field strength and is useful in comparing the thermal oxide to other materials:

$$\rho' = \left(\frac{V}{L}\right)\left(\frac{A}{I}\right) = \frac{E}{J} = \frac{1/E}{J/E^2} \tag{3.6}$$

where ρ' is the effective resistivity, L is the thickness of the oxide, and A is the cross-sectional area of the contact. From Fig. 3.6,[7] for an aluminum contact and a field strength E of 5.5×10^6 V/cm, $\rho' = 1.8 \times 10^{18}$ Ω-cm. If the field were reduced to 10^5 V/cm, the effective

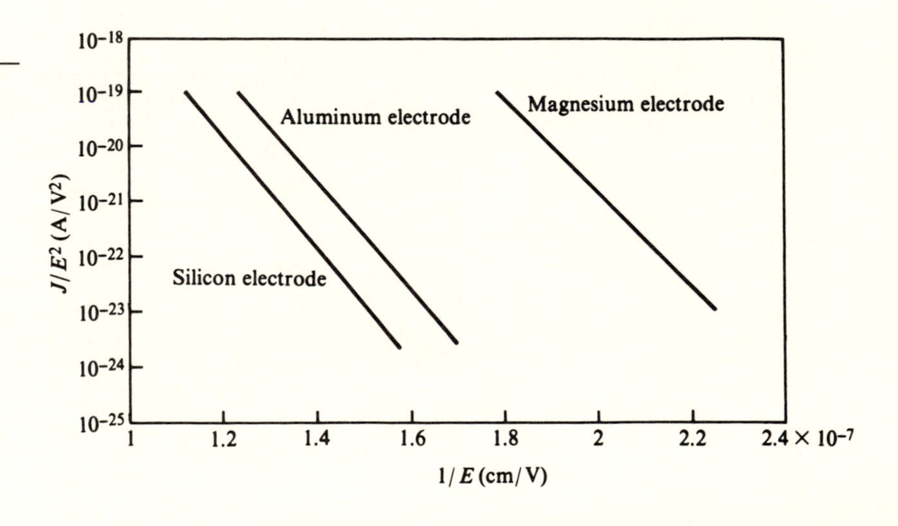

FIGURE 3.6

Fowler–Nordheim plot of thermal oxide *I–V* characteristics. (*Source:* Adapted from M. Lenzlinger and E.H. Snow, *J. Appl. Phys. 40,* p. 278, 1969.)

[6] A method of approximating the solutions of a second-order linear equation that was first applied to the Schrödinger equation by G. Wentzel, H.A. Kramers, and L. Brillouin in 1926.

[7] Values reported by different observers may vary from one another by a factor of at least 10.

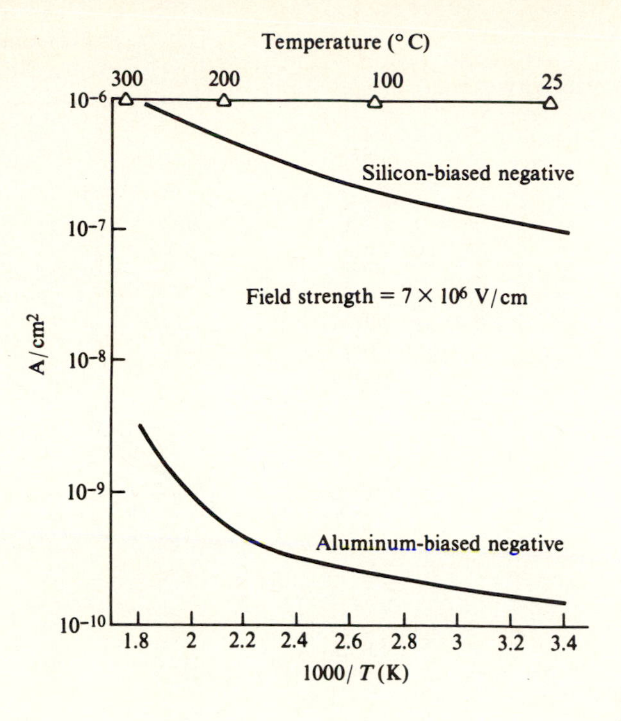

FIGURE 3.7

Variation of current through oxide as a function of temperature. (Upper curve is for electron injection from silicon; lower curve, for injection from aluminum.) (*Source:* Data from C.M. Osburn and E.J. Weitzman, *J. Electrochem. Soc. 49,* p. 603, 1972.)

resistance would increase by several orders of magnitude. Alternatively, a specific contact resistance R_c in Ω-cm² can be calculated from Eq. 3.7 as follows:

$$R_c = \frac{V}{J} = \frac{E \times L}{J} = \frac{(1/E) \times L}{J/E_2} \qquad 3.7$$

If the oxide in the example above were 1000 Å thick, R_c would be 1.8×10^{13} Ω-cm², compared to a value of about 10^6 Ω-cm² for a Schottky barrier on 1 Ω-cm silicon. With such high values, the actual magnitude would ordinarily be of little concern, but EEPROM devices depend on injecting charge onto a floating gate and having that charge not leak away over the course of several years.

Above about 400°C, the conduction mechanism apparently changes, and the current becomes linearly dependent on voltage (25). Data are shown in Fig. 3.8.

3.4.2 Surface Electrical Conductivity

The surface sheet resistance of thermal oxide is a strong function of humidity, as shown in Fig. 3.9. Because of a finite conductivity, charge can move out from a contact (charge spreading) and, if severe enough, create an inversion layer in the silicon beneath. It has been

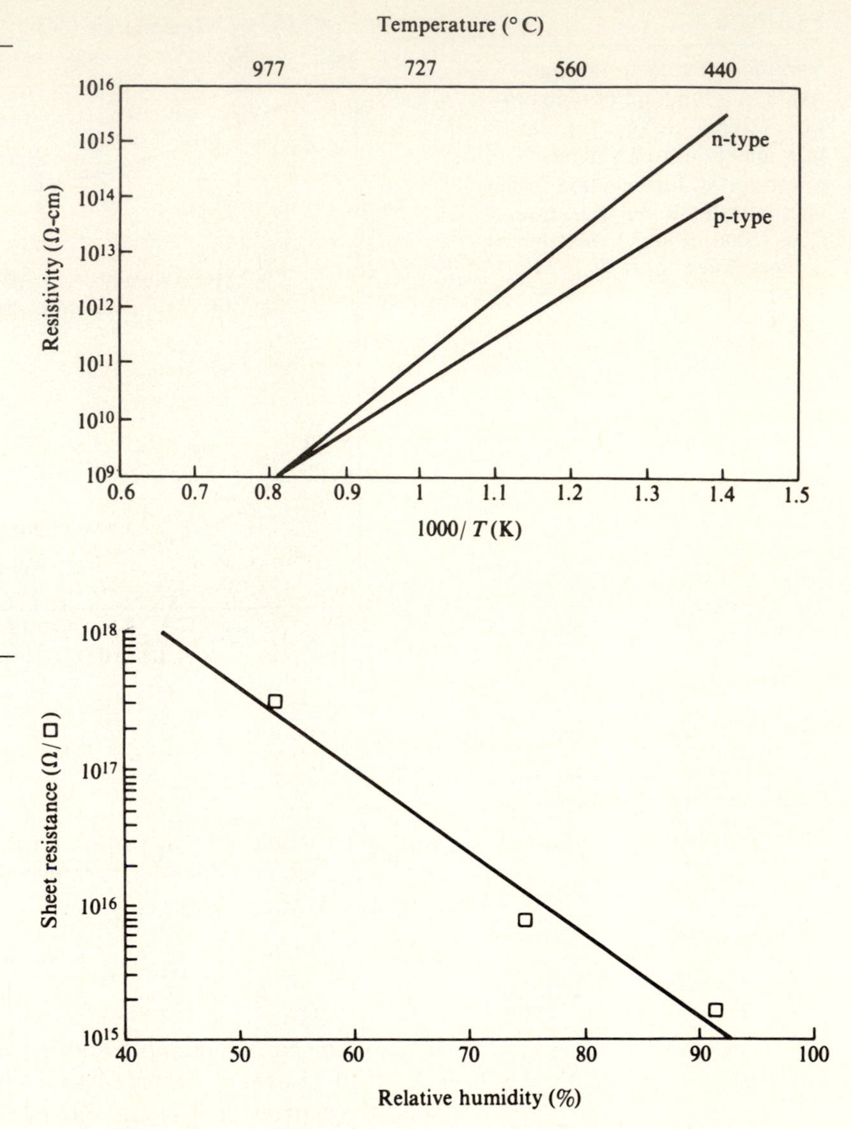

FIGURE 3.8

High-temperature resistivity of thermal oxide. (*Source:* From data in J.K. Srivastava et al., *J. Electrochem. Soc. 132*, p. 955, 1985.)

FIGURE 3.9

Thermal SiO_2 surface resistance versus humidity. (*Source:* E.H. Snow in Tech. Doc. Rpt. RADC AF 30(602)–3776, December 1965.)

shown that the potential V on the outer oxide surface in the vicinity of an electrode is given by Eq. 3.8 (26) as follows:

$$V = V_o \operatorname{erfc}\left(\frac{x}{2\sqrt{t/RC}}\right) \tag{3.8}$$

where V_o is the voltage on the contact, x is the distance out from the contact, t is the time after the voltage is applied to the contact, R is the sheet resistance of the surface, and C is the oxide capacitance.

For any part of the area where V is greater than the threshold voltage V_t, an inversion layer will occur.

3.4.3 Dielectric Breakdown Strength

Dielectric strength is on the order of 10^7 V/cm. Except under conditions of severe electrical overstress, bipolar ICs will not have fields close to this value. MOS devices, which depend on thin gate oxides for good electrical performance, usually operate very close to breakdown. Thus, any process fluctuations that cause a reduction in dielectric strength will generally cause yield losses and may cause long-term reliability problems as well. The distribution of breakdown fields for MOS capacitors in Fig. 3.10 shows the three peaks often observed. The cluster of very low breakdowns is ascribed to faulty fabrication involving, for example, pinholes in the oxide (23) and gross contamination on the surface before oxidation. The move to better cleanrooms has led to the continuing decrease of the number of units in this category. The next cluster, sometimes merging with and thus giving a long tail to the maximum breakdown distribution, is apparently caused by a reduction in barrier height brought about by such things as silicon surface asperities,[8] small regions of metallic precipitates (often decorated crystal defects) such as Ni, Cu, and Sn in the underlying substrate (27), or locally high-sodium

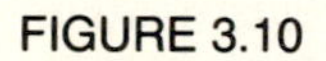

FIGURE 3.10

Breakdown field distribution for MOS capacitors.

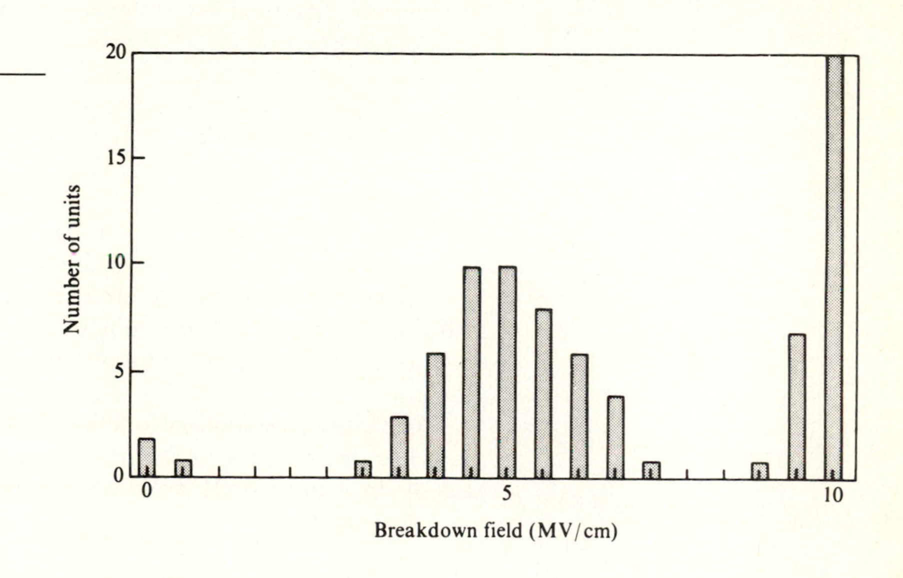

[8]This effect gives orders of magnitude-increased current for the same field, and it has been suggested that deliberate texturing could be used to reduce the voltage needed to program EEPROMs.

concentrations in the oxide itself (28). Unlike the behavior of the low-breakdown peaks, the peak of the maximum-breakdown distribution increases as the oxide thickness decreases. Otherwise, it appears to be relatively insensitive to processing parameters (29, 30).

The effect of asperities is generally observed only when the oxide is grown on polysilicon (31). Crystal defects are minimized by oxidizing in a very clean environment and using wafer gettering whenever possible (see Chapter 8 for a discussion of gettering). Good slice surface cleanup before oxidation will minimize local areas of heavy metal precipitates (21). Further improvement can sometimes be obtained by oxidizing, stripping the oxide, and then reoxidizing (32). This step is built into most MOS processes in that either a dummy gate oxide or the thick field oxide is removed and a gate oxidation is then performed. Plasma cleaning of the silicon surface before oxidation may also help (33). Oxidizing in an ambient with a small amount of chlorine-bearing compound present[9] will also minimize low breakdowns due to metallic contamination (34). Wet oxidation shifts the high-field peak distribution up somewhat from those obtained with dry oxidation but gives oxides with higher mobile ion contamination (35).

In addition to the initial breakdown failures just described, when an otherwise good oxide is subjected to high fields for long periods of time, failures will occur. This phenomenon is referred to as time-dependent dielectric breakdown and is one limit to MOS device lifetime. It is sometimes called a latent defect since it shows up only after prolonged operation. However, it is probably better characterized as a wear-out mechanism. The cause(s) is not clearly defined, but apparently it is a barrier height reduction caused by the accumulation of charges near an electrode (36–39). The charge can come from holes, electrons, or ions injected into the oxide from the contacts. Since the failure rate is more severe the thinner the oxide, and as devices have shrunk in size and the gate oxide has been scaled down in thickness, more and more attention has been given to this problem.

The failure rate appears to follow a log normal distribution[10] (40, 23, 41) so that cumulative failures give a straight line when they are plotted on a probability scale axis against time on a logarithmic

[9]This process is almost always used in MOS manufacture in order to minimize sodium- and potassium-induced threshold drift.

[10]A log normal distribution is one that shows the typical bell-shaped curve when the independent variable is plotted on a log scale. The log standard deviation is given by $s_{\log}$ = log (time for 50% failures/time for 15.9% failures).

axis.[11] The failure rate increases exponentially both with operating temperature and gate voltage. A model of form

$$TF(x\%) = Ke^{U/kT}e^{f(E)} \qquad 3.9$$

fits thin-oxide data reasonably well (41). $TF(x\%)$ is the time for $x\%$ of the devices to fail. K is a constant, U is an activation energy that is dependent on the applied electric field E, k is Boltzman's constant, and $f(E)$ is a temperature-dependent function of the applied field. Table 3.6 shows estimates for 100 Å oxides and illustrates the point that, with the current state of processing, thin oxides must have reduced applied voltages in order to exhibit the long lifetimes expected of semiconductor devices. An alternative to time-to-breakdown as a performance criterion is charge-to-breakdown. In this case, the total charge required for failure is used as a measure of performance.

3.4.4 Mobile Ion Transport

It has long been observed that, under the influence of an electric field, some metal ions can move rather easily through glass[12] and can similarly move through high-purity silica. Na (sodium) was identified in the 1960s (42) as a contaminant that was usually found distributed throughout semiconductor oxides and that would easily migrate under the influence of an electric field. Fields such as those shown in Fig. 3.11, which are required for device operation, are large enough, and in many cases of the required polarity, to cause these mobile ions to pile up at the oxide–silicon interface, where the extra charge drastically affects device operation.

TABLE 3.6

Time for 10% of 100 Å Oxide Capacitors to Fail

Field (MV/cm)	*Room Temperature*	*85°C*
	Time (hours)	*Time (hours)*
8	1.5	0.66
7	250	25
6	20,000	550
5	3,250,000	2500 (3.5 months)

Source: Data extrapolated from information in reference 41 in end-of-chapter listing.

[11] This kind of distribution is often observed for semiconductor device failure rates.

[12] A demonstration of 50 years ago was to immerse a burning light bulb in a container of a molten (300°C) mixture of $NaNO_3$ and $NaNO_2$ and then, by biasing the light filament negatively, deposit metallic sodium on the inside of the light bulb.

FIGURE 3.11

Origin of fields that may cause ion drift.

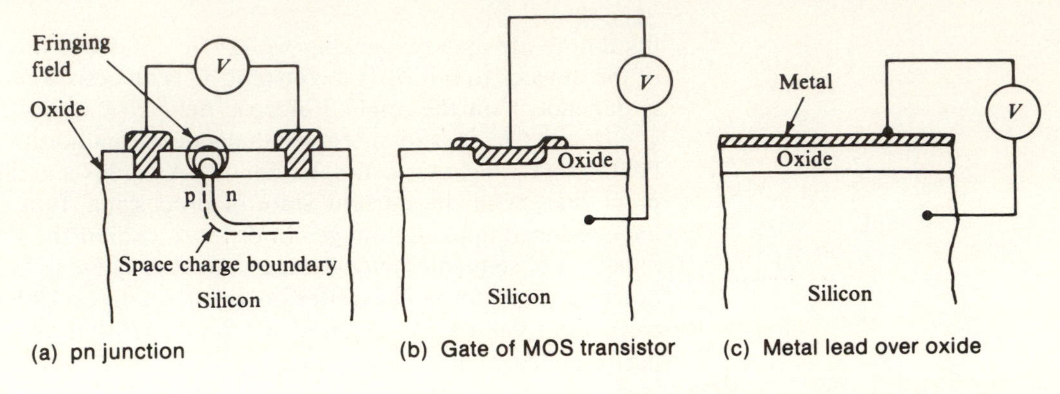

In addition, the other alkali metals K^+ (potassium), and Li^+ (lithium) are quite mobile. Cs (cesium), apparently because of its large ionic radius, is virtually immobile. Lithium moves about as sodium but is not as widely distributed as sodium and potassium. Sodium is to be found nearly everywhere, and that fact combined with the fact that it is more mobile than potassium makes it the most troublesome of the three. About a hundred degree higher temperature is required for potassium to drift comparably to sodium (43, 44).

One-dimensional ionic motion by diffusion and field-induced drift is described by Eq. 3.10 as follows:

$$\frac{D\partial^2 N}{\partial x^2} - \frac{v\partial N}{\partial x} = \frac{\partial N}{\partial t} \tag{3.10}$$

where D is the diffusion coefficient, N is the ionic concentration, v is the velocity given to the ion by the field, and t is the time. The velocity $v = \mu E$ where E is the electric field and μ is the mobility. The mobility is related to the diffusion coefficient D through the expression $\mu = qD/kT$; thus, for a singly charged ion, $v = qDE/kT$ where k is Boltzmann's constant and T is the temperature. Experimentally, it has been demonstrated that field-aided diffusion is not what limits the transport of sodium and potassium ions in thermal silicon oxide. Rather, the ions are apparently initially trapped at the metal–oxide interface, and the rate of transfer to the silicon–oxide interface depends primarily on the rate of emission of ions into the oxide (42–46). Thus, diffusion coefficients determined without considering ion trapping at the interfaces will be substantially understated.

The number of ions transported from one interface to another N_i in ions/cm^2 is given by Eq. 3.11 as follows:

$$N_i = N_O K t^{1/2} \qquad 3.11$$

where K is a constant that depends on field, temperature, and the ion species. It should be remembered that in actual practice a fixed number of ions will usually be moved so that N_O of Eq. 3.11 will be decreasing with time. A MOS transistor threshold voltage will shift as the charge is added to the oxide–silicon interface and can be used to calculate the ion density.[13] The shift ΔV is given by Eq. 3.12 as follows:

$$\Delta V = \frac{-N_i q}{C_{ox}} \qquad 3.12$$

where C_{ox} is the capacitance per unit area of the oxide and q is the electronic charge. Note that electronic charge is measured in coulombs (C).

The first method developed to combat sodium migration (introduced before the actual ion causing the trouble was identified) was the use of heavy phosphorus doping in the top region of the oxide (49). The heavy phosphorus layer both getters existing Na to it and prevents subsequent contamination by blocking Na movement through it (50). In the making of bipolar circuits, getting occurs normally during the emitter cycle of most npn transistors. The only requirement is that the heavy phosphorus layer not be subsequently etched away. In order to maintain good long-term stability, a several-percentage concentration of phosphorus should be present. However, if the concentration gets too high, reactions with moisture and aluminum lead systems can occur and cause a different kind of failure. The use of a thin layer of 4% P_2O_5 will reduce the migration time by several orders of magnitude. The mechanism for retarding positive ion motion is thought to be ion trapping by a negatively charged P–O complex residing at an Si site in the SiO_2 lattice (46).

It was next discovered that growing oxides in the presence of chlorine[14] would produce an oxide free of sodium and with immunity from rapid sodium migration (51, 52). The chlorine sources first used were Cl_2 and HCl, but very shortly afterward trichloroethylene was

---

[13]The change in the shape of a MOS capacitor C–V plot can also be used to determine N_i.

[14]The development of chlorine treatments seems to be an example of independent studies in two different laboratories. Further, in both cases, these studies appear to have been initiated as the result of closely examining the results of some other investigation. Bell Northern (51a) introduced sodium contamination from NaCl and from Na_2CO_3 and observed that there was not nearly as much contamination from the NaCl source. RCA (51b) grew oxides from a water source with HCl in it (looking for gettering of heavy metals) and observed a lower sodium contamination level.

suggested since it is not corrosive as is HCl (53). Considerably later, trichloroethane was also introduced as being equally as effective and less toxic (54). In general, the more chlorine incorporated into the oxide, the greater the protection, but too high a concentration will produce a chlorine-rich phase separation at the oxide–silicon interface and, in some cases, a blistering of the oxide (55). The region to be avoided is roughly the one bounded by oxidation temperatures above 1100°C and HCl concentrations greater than 8%. Even at low concentrations, the chlorine is still primarily segregated near the Si–SiO_2 interface (56). The chlorine incorporation does not prevent the sodium from moving through the oxide but rather apparently traps and neutralizes it in the high-chlorine concentration region near the interface.

Simultaneously with these activities, great effort has been made to provide a sodium-free processing environment. This effort includes the use of purer materials, HCl cleaning of furnace tubes, and improved cleanroom procedures to minimize contamination from manufacturing personnel. Another method for providing sodium protection is to use an overcoat material that is impervious to sodium migration. Silicon nitride is such a material and is often used. Low-temperature plasma-assisted CVD nitride that can be applied after metallization can be used as a combination scratch protection for leads and moderately effective sodium barrier. High-temperature CVD nitride is much more effective but must be used before metallization.

3.4.5 Interface Charges

In addition to the mobile charges just discussed, a variety of charges also reside at or near the Si–SiO_2 interface. The most common of these charges are the fixed positive oxide charge Q_f (or Q_{ss})[15] and a variable interface charge Q_{it} located in traps variously referred to as surface traps, interface traps, surface states, or fast states (43). The number of charges N_{it} (or N_{ss}) associated with each can range from 10^{10}–10^{12} charges/cm^2, and anything above 10^{10} charges/cm^2 will often adversely affect performance for both bipolar and MOS devices. In the case of MOS, Q_f causes the turn-on voltage V_t to shift by an amount $-Q_f/C_{ox}$, where C_{ox} is the capacitance per unit area of the oxide.

EXAMPLE ☐ Determine ΔV_t for $N_{it} = 10^{11}$ charges/cm^2 if the gate oxide thickness = 1000 Å.

[15]Old terminology. New symbols were adopted as an outgrowth of a joint Electrochemical Society and IEEE committee. (See, for example, Bruce E. Deal, *J. Electrochem. Soc. 127*, p. 979, 1980.)

Capacitance is given by $A\varepsilon\varepsilon_0/X$ where ε is the oxide dielectric constant, ε_0 is the permittivity of free space, X is the oxide thickness, and A is the area. A consistent set of units uses permittivity $= 8.85 \times 10^{-14}$ farads/cm (abbreviated F/cm), X in cm, and A in cm^2. Thus,

$$C_{ox}/\text{cm}^2 = \frac{3.9(8.85 \times 10^{-14}\ \text{F/cm})}{10^{-5}\ \text{cm}}$$

$$= 3.45 \times 10^{-8}\ \text{F/cm}^2$$

$$Q_{it} = qN_{it} = 1.6 \times 10^{-19}\ \text{C} \times 10^{11}/\text{cm}^2$$

$$= 1.6 \times 10^{-8}\ \text{C/cm}^2$$

$$\Delta V_t = -\frac{1.6 \times 10^{-8}\ \text{C/cm}^2}{3.45 \times 10^{-8}\ \text{F/cm}^2}$$

$$= -0.46\ \text{C/F} = -0.46\ \text{V}$$

□

This value is comparable to other terms in the expression for V_t and thus would not result in a catastrophic shift in V_t. However, if Q_f/q were 10^{12}, then its contribution to V_t would be 4.6 V, and very poor transistors would result. If Q_f/q were in the 10^{10} to 5×10^{10} range, then its effect would be negligible.

The kind and amount of charge at any given time due to the interface states will depend on the energy distribution of the states, whether they act as donors or acceptors, and the position of the Fermi level (57). These kinds of states give rise to surface recombination and thus can be particularly troublesome for bipolar devices requiring high values of H_{fe} at low collector currents. With no surface space charge region, the surface recombination velocity S is given by Eq. 3.13 (58) as follows:

$$S = \sigma_c v_{th} N_{it} \qquad 3.13$$

where σ_c is the capture cross section of the state for a carrier (10^{-16} to 10^{-15} cm^2), v_{th} is the thermal velocity of carriers ($\cong 10^7$ cm/s at room temperature), and N_{it} is the density of states/cm^2. It has been shown that, in general, when the fast states are uniformly distributed in energy over the central portion of the bandgap, only those within a few kT of mid-gap will contribute. Under these circumstances, πkTD_{it} where D_{it} is the density per unit area per unit energy is substituted for N_{it}.

EXAMPLE □ If $D_{it} = 10^{12}/\text{cm}^2 - \text{eV}$ for a particular thermal oxide, calculate the surface recombination velocity when $\sigma_c = 10^{-15}\text{cm}^2$.

$k = 8.62 \times 10^{-5}$ eV/K (since D_{it} is in terms of eV, this value must be used rather than 1.38×10^{-16} erg/K), and $S = 811$ cm/s, which is extraordinarily high. However, as will be seen shortly, various anneals are available that will reduce N_{it} by nearly two orders of magnitude. *Also,* the high end of the estimated range of capture cross section was used in this calculation. The actual measured values at planar surfaces range from 250 cm/s down to 1–10 cm/s after the proper anneals. These small values found after annealing have little effect on device performance. □

Both the fixed charge and the interface charge depend on the orientation of the surface on which the oxide is grown (59). Their values decrease in order from the (111) to (110) to (100) orientation, and since lower Q_f values allow lower-threshold MOS transistors to be built, most MOS ICs use (100) material. Typical Q_f values for a 1200°C dry oxygen oxidation (courtesy of George Brown, Texas Instruments Incorporated) are as follows:

Orientation[16]	Q_f/cm^2
(100)	1.1×10^{10}
(115)	1.3×10^{10}
(110)	4.2×10^{10}
(111)	1.5×10^{11}

However, depending on the oxidation temperature, oxidation ambient, cooldown ambient, and subsequent heat treatment, the values may be higher by at least one order of magnitude. The values may also decrease as processing improves. For example, since 1967 (60), the decrease has been in the 25%–50% range.

Line AC of Fig. 3.12 shows the effect of oxidation temperature using dry oxygen on Q_f for (111) orientation. Line BC shows the effect of a subsequent anneal in an inert atmosphere. The effect is reversible in that if, after an anneal, the wafer is heated in dry oxygen, Q_f will increase to the value given by line AC for the new heat-treatment temperature. The inset in Fig. 3.12 shows how long it takes at temperature to move from a value set by curve AC to the BC value. Also, at higher temperatures, continued heating in nitrogen will cause Q_f to again increase in value (43). Fig. 3.13 shows the effect of oxidizing and cooldown atmospheres on Q_f for (100) orientation wafers. The trend is the same for (111), although the differences are less (57).

N_{it} usually shows the same dependence on oxide growth conditions as does Q_f. However, it can be reduced to a very low level by a subsequent hydrogen anneal in the 300°C–450°C range. Such anneals can be done directly in hydrogen or forming gas or can take place during the aluminum sintering operation commonly used to

[16]See Appendix A for a discussion of crystal plane notation.

FIGURE 3.12

Effect of heat treatment on fixed charge density for (111) surfaces. (*Source:* Bruce E. Deal, *J. Electrochem. Soc. 121*, p. 198C, 1974. Reprinted by permission of the publisher, The Electrochemical Society, Inc.)

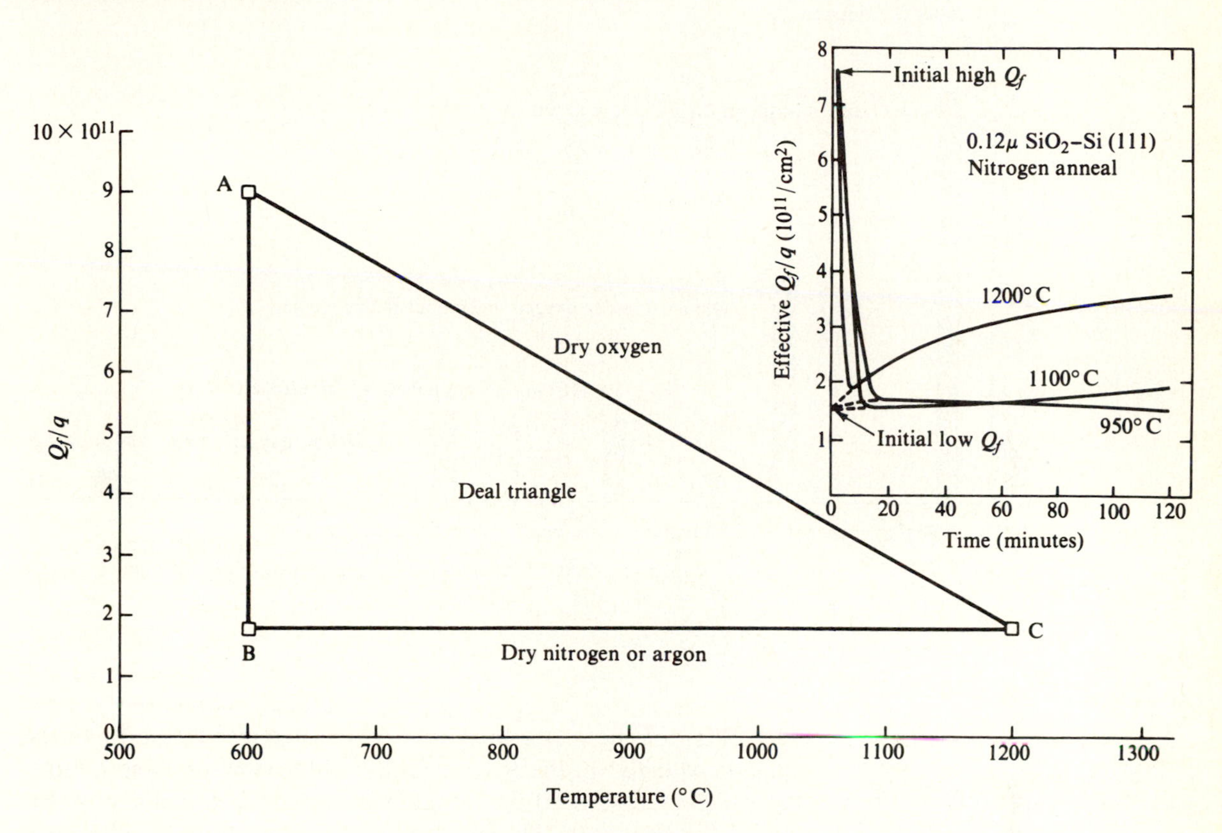

produce good ohmic contacts. In the latter case, it is presumed that hydrogen is produced by the reduction of moisture trapped on the oxide surface before the aluminum was deposited (61). Since silicon nitride, often used as a sodium barrier, is also relatively impervious to hydrogen, anneals to reduce N_{it} after its application are quite difficult (62). For temperatures in the 700°C and above range, noticeable amounts of hydrogen will diffuse through nitride, but, of course, such high temperatures cannot be combined with aluminum metallization.

FIGURE 3.13

Effect of processing on Q_f for (100) wafers. (*Source:* Adapted from Reda R. Razouk and B.E. Deal, *J. Electrochem. Soc. 126,* p. 1573, 1979.)

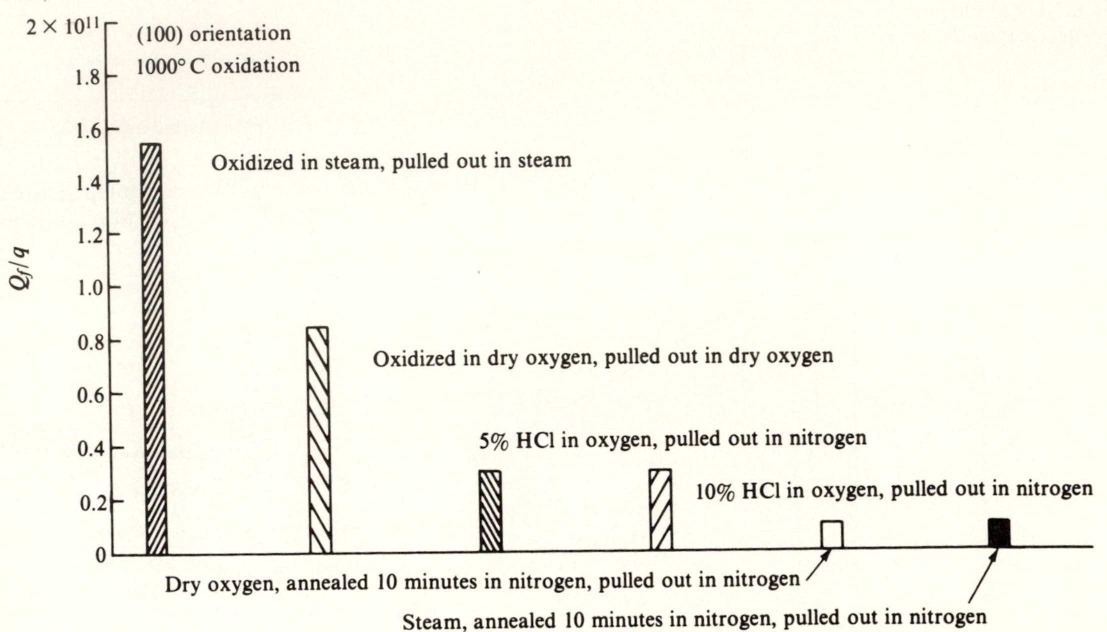

3.5 KINETICS OF OXIDATION

Over much of the oxide thickness range of interest, the thickness depends roughly on the square root of the oxidation time t. Early workers therefore used the simple parabolic relation $AX = \sqrt{t}$ to predict thickness. In this equation, A is a constant, and X is the thickness of oxide grown in a time t. While not strictly accurate, this expression sufficed for the needs of most bipolar processing. However, as early as 1959, it was realized that there was an initial region of about 60 Å where the growth appeared parabolic, followed by a linear region, and then by another parabolic region (1). The very thin region was of little interest at that time and was generally disregarded. However, with the advent of VLSI, the requirement for smaller devices has now led to reduction of MOS gate oxide thicknesses to the 200 Å range and the desire to further reduce them to 100–150 Å. Early efforts to include the nonparabolic behavior over part of the growth range of interest led to a power law expression of the form $AX^n = t$ where n ranges from 1 to 2. The expression was supplanted by an expression of the form $AX + BX^2 = t$ where A and B are constants, which for small X gives a linear behavior and for large X, a parabolic behavior. However, until the Deal–Grove model, which is discussed in the next section, there was no good explanation of why these two regions should exist.

The oxide growth rate depends on the crystallographic orientation of the surface being oxidized; the oxidizing ambient—for example, oxygen, steam, or a mixture thereof—and its partial pressures; the temperature; the amount of group IIIA or VA impurity in the silicon; the amount and kind of impurity in the oxidizer; the kind of preoxidation silicon surface cleanup; and the amount of stress in the oxide and the silicon. A full kinetic model must provide for all of these effects, as well as the initial rapid oxidation region. The original basic Deal–Grove model neglected orientation, doping, stress, cleanups, oxidizing gas impurities, and the thin, initially high oxidation rate region but through the years has been modified to account for most of these effects.

3.5.1 Deal–Grove Model

In 1965, B.E. Deal and A. S. Grove (63) developed the model mentioned in the previous section. It predicted very well the oxide growth rate over a wide range of conditions. The system that was assumed is shown in Fig. 3.14. As mentioned earlier, it has been shown that the oxidizing species diffuses through the oxide to the silicon surface and reacts. The equilibrium concentration of diffusant (oxidant) in the oxide is assumed to be C^*, and the actual concentration at the ambient–oxide surface is C_0. The concentration at the silicon surface is C_i. The flux of material crossing the ambient–oxide interface will be proportional to $C^* - C_0$; that diffusing from one side of the oxide to the other is given by $-D\partial C/\partial x$ where D is

FIGURE 3.14

Geometry used in development of Deal–Grove model.

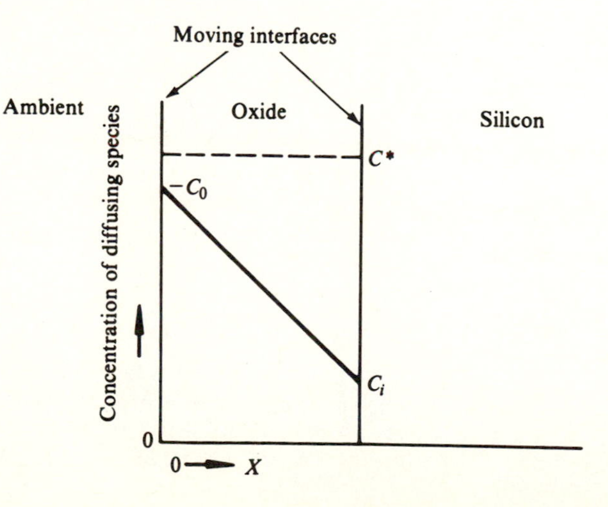

the diffusion coefficient in the oxide. In the steady-state case, assumed in the derivation, $\partial C/\partial x = (C_i - C_0)/x$ where x is in a direction normal to the silicon surface. The amount of oxidant consumed at the silicon interface is assumed to be proportional to C_i—that is, by a first-order reaction. For steady state, all three fluxes must be equal. Thus,

$$h(C^* - C_0) = \frac{D(C_0 - C_i)}{x} = kC_i \qquad 3.14$$

where h and k are rate constants.

Eq. 3.14 can be solved for the flux F as follows:

$$F = \frac{kC^*}{1 + (k/h) + (kx/D)} \qquad 3.15$$

The oxide thickness increase with time (dx/dt) is proportional to the flux, and the thickness at a time t is given by

$$t = \frac{1}{N}\int_0^X F(x)dt \qquad 3.16$$

where N is the number of oxidant molecules per unit area of oxide. Since an initial native oxide exists on the surface when high-temperature growth begins, and since the derivation is based on steady-state conditions, which may require a small amount of oxide growth, let $X = X_i$ at $t = 0^+$. Eq. 3.16 has as a solution

$$t = \frac{X - X_i}{K_l} + \frac{X^2 - X_i^2}{K_p} \qquad 3.17$$

where

$$\frac{1}{K_l} = \frac{N[(1/h) + (1/k)]}{C^*}$$

and

$$\frac{1}{K_p} = \frac{N}{2DC^*}$$

K_l is the linear rate constant, and K_p is the parabolic rate constant. An alternative form of Eq. 3.17

$$X^2 + AX = Bt + X_i^2 + AX_i \qquad 3.18$$

is usually used, and it is the one originally formulated by Deal and Grove. It can be rewritten as follows:

$$X^2 + AX = B(t + \tau) \qquad 3.19$$

where $\tau = (X_i^2 + AX_i)/B$ and is the time corresponding to the initial thickness[17] X_i. A comparison of Eq. 3.19 to Eq. 3.17 shows that B/A is the linear rate constant K_l and that B is the parabolic rate constant K_p.

Steam (or wet O_2) oxidation proceeds at a more rapid rate than does dry oxidation and can be explained on the basis of the relative values of C^* and D for the two processes. Both C^* and C^*D over the entire range of practical oxidation temperatures are larger for H_2O than they are for O_2, even though $D_{\text{oxygen}} > D_{\text{water}}$ (63). The combination of Eq. 3.17 and Eq. 3.19 predicts that under these circumstances, the rate of oxidation in steam will be greater than the rate in dry oxygen.

The nature of the oxygen diffusion mechanism is somewhat uncertain. Some data indicate that it is composed of a molecular oxygen permeation component and a lattice diffusion component and that their relative magnitudes change with temperature (64). Small deviations from Eq. 3.20 in the oxidation rate in the thick-oxide regime have been ascribed to a two-component transport, one of which is independent of oxide thickness (65).

When X is very large, $X^2 >> AX$ and $t >> \tau$ so that Eq. 3.19 reduces to

$$X^2 = Bt \tag{3.20}$$

Growth is in the parabolic region, and the rate is diffusion limited. When X is small, Eq. 3.19 is of the form

$$X = \left(\frac{B}{A}\right)(t + \tau) \tag{3.21}$$

and the thickness increases linearly with time. A characteristic time $A^2/4B$ is often taken as the demarcation between the two regimes (63).

3.5.2 Initial Stage of Oxidation

Unlike oxides grown in steam, which closely follow the model just described over their whole thickness range, oxides grown in dry oxygen initially behave differently. A thickness–time plot of an oxide grown in dry oxygen has the general character shown in Fig. 3.15. Point 0′, which marks the low end of the linear portion of the curve, may be at thicknesses of 50–100 Å. For less thickness, the oxide

[17]The native oxide thickness is probably no more than 20 Å, but when fitted to the experimental data, X_i for dry oxidations is sometimes in the 100–200 Å range. The growth behavior of thinner oxides is not properly portrayed by Eq. 3.17.

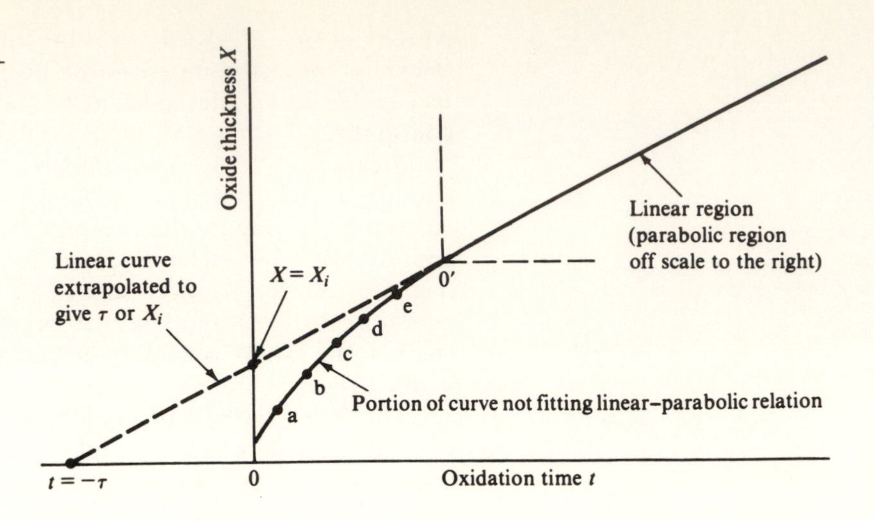

FIGURE 3.15

Thermal oxide thickness–time plot showing a nonlinear region between the origin and the linear portion.

grows in a nonlinear fashion and faster than predicted by Eq. 3.19 when values of A and B determined from thicker oxides are used. The 50–100 Å is an appreciable portion of the total thickness of the thin gate oxides used in high-density MOS circuits, so models that provide a better prediction of thickness versus time in this region are useful. Several alternatives have been proposed. Some are just fitted parameters, and the list includes linear, parabolic, cubic, logarithmic, and inverse logarithmic forms (66). Others, based on various assumptions about the oxidation mechanism, have been proposed (67–72). The region of high-stressed oxide near the interface that was discussed earlier has been suggested as also being a region in which oxygen transport is restricted. Thus, as growth progresses and the layer gets thicker, oxygen reaching the interface will decrease more rapidly than predicted by Eq. 3.14 (67–70). For layers thicker than that of the high-stress region, growth will then proceed as described by Eq. 3.19.

Approaches involving alternate oxidation transport paths or reactions have also been proposed (65, 71–73). It has been pointed out that since the alternate reactions for thicker oxides reduce to Eq. 3.19, the corrections tend to be either terms added to or multiplied by Eq. 3.19 (74). In such form, all of the suggested equations look much alike and have enough constants to fit the experimental data reasonably well. Thus, using a "best fit" criterion to choose the mechanism actually responsible for the enhanced oxidation is difficult. An empirical expression for the oxidation rate that gives a good fit to the data and is used in the computer program SUPREM III (75, 76) is as follows:

$$\frac{dX}{dt} = Ce^{-X/L} + \frac{B}{2X + A} \qquad 3.22$$

where L is a characteristic length of about 70 Å. Based on Eq. 3.19, the rate would be

$$\frac{dX}{dt} = \frac{B}{2X + A} \qquad 3.23$$

3.5.3 Reaction Path

At the beginning of this chapter, oxidation was said to occur by the reaction $Si + O_2 \rightarrow SiO_2$. However, various paths ultimately result in this reaction. The three parallel paths along which oxidation is currently thought to proceed (77) are as follows:

(1) $Si + O_2 = SiO_2$ + dislocations + lattice strain

(2) $Si + Si_V = (Si - Si_V)_{complex}$
$(Si - Si_V) + O_2 = O + (Si - O)_{complex}$
$(Si - O) + Si_V = (Si - O - Si_V)_{complex}$
$(Si - O - Si_V) + O = SiO_2$

(3) $2Si + O_2 = O + Si_I + (Si - O)_{complex}$
$Si + (Si - O) = (Si - O - Si)_{complex}$
$(Si - O - Si) + O = SiO_2 + Si_I$

where Si_V is a silicon vacancy and Si_I is an interstitial silicon atom.

If some portion of the silicon is oxidized through path 2, the rate will depend on the availability of vacancies, and anything that will change their concentration should affect the oxidation rate. Since some vacancies carry a charge, their population will depend on the Fermi level, which can be changed, even at oxidation temperature, by heavy doping of the silicon. As will be discussed in a later section, heavily doped silicon does have a different oxidation rate than does lightly doped silicon. This rate difference can be explained on the basis of vacancies and is considered to be good evidence of path 2's existence. Simultaneous oxidation and diffusion can produce an enhanced diffusion rate that has been explained on the basis of silicon interstitials that are generated during oxidation. Path 3 is a method by which the interstitials could be produced.

3.5.4 Determination of Rate Constants

To determine A and B from experimental curves of oxide thickness versus time using Eq. 3.17, 3.18, or 3.19 is not straightforward. However, as an example, Eq. 3.19 can be rewritten as follows:

$$X = \frac{B(t + \tau)}{X} - A \qquad 3.24$$

Then, in a plot of X versus $(t + \tau)/X$, $-A$ will be the intercept with the X axis, and B will be the slope of the line. Since the line should be straight, a standard least-squares linear curve fitting routine can be used. To neglect τ may be tempting, but unfortunately, to do so leads to substantial error in A. τ can, however, be estimated by extrapolating from the linear region back to the time axis as shown in Fig. 3.15.

An alternative method is based on the fact that the only reason for concern in the extrapolation of X_i or τ is that, for thin oxides grown in dry oxygen, a deviation from the basic linear–parabolic equation occurs. If it is assumed that the departure occurs for the portion of the curve below 0′ in Fig. 3.15, then if the origin of the axes were shifted to O′, the points for the remainder of the curve should have a best fit to Eq. 3.24 with $\tau = 0$. Thus, if the data points a, b, c, and so on, are sequentially used as the beginning of the curve, and if the data are fitted sequentially to equations of the form (78)

$$X - X_a = \frac{B(t - t_a)}{(X - X_a)} - A \qquad 3.25a$$

$$X - X_b = \frac{B(t - t_b)}{(X - X_b)} - A \qquad 3.25b$$

the correlation coefficient should become progressively better until the starting point moves out of the box, at which point it should become constant. This method allows A, B, and X_i or τ to be determined.

Yet another approach (79) is to differentiate Eq. 3.18 or 3.19 to get the expressions

$$\frac{dX}{dt} = \frac{B}{A + 2X} \qquad 3.26a$$

$$\frac{dt}{dX} = \frac{A}{B} + \frac{2X}{B} \qquad 3.26b$$

Now, independently of X_i or τ, a plot of dt/dX versus X gives a line with a dt/dX intercept of A/B and a slope of $2/B$. This method is appropriate, however, only if the data are good enough to allow dependable values of dX/dt to be determined.

3.5.5 Calculation of Oxide Thickness

Solving for X from Eq. 3.18 gives

$$X = \sqrt{Bt + \frac{A^2}{4} + X_i^2 + AX_i} - \frac{A}{2} \quad 3.27a$$

$$X = \sqrt{B(t + \tau) + \frac{A^2}{4}} - \frac{A}{2} \quad 3.27b$$

Thus, if the rate constants just discussed are to be used, either Eq. 3.27a or 3.27b can be used to calculate the thickness X versus time.

For straight modeling work, equations such as the ones just described, which relate to the physical processes that occur, need not be used. One could, for example, use something as simple as $t = a + bX + cX^2$ after having first evaluated the constants from experimental data by using straightforward polynomial curve fitting routines. Table 3.7 lists representative rate constants for several conditions.

3.5.6 Temperature Behavior of Rate Constants*

The parabolic constant is given by $2DC^*/N$ and hence would be expected to have the same temperature dependence as the combination of solubility limit and diffusion coefficient of the oxidizer in silica. In the case of dry oxygen, the evidence is that the diffusant

TABLE 3.7

Oxidation Rate Constants

Temperature (°C)	B_{100} (Å²/min.)	B/A_{100} (Å/min.)	X_i (Å)	B_{111} (Å²/min.)	B/A_{111} (Å/min.)	X_i (Å)
Dry Oxidation						
800	660	0.43	70	1,350	0.9	68
900	5,590	2.08	85	6,500	4.6	88
1000	28,600	8.65	113	26,500	16.4	118
Wet Oxidation (640 mm H_2O)						
900	0.24×10^6	25	0	0.25×10^6	42	0
1000	0.52×10^6	111	0	0.52×10^6	194	0
1100	0.87×10^6	497	0	0.86×10^6	821	0

Sources: Data from Hisham Z. Massoud et al., *J. Electrochem. Soc. 132*, p. 1745, 1985, and Bruce E. Deal, *J. Electrochem. Soc. 125*, p. 576, 1978.

*See reference 63.

TABLE 3.8

Activation Energy

Item	Activation Energy (eV)*	Reference
O_2 diffusion in silica	1.17	(1)
H_2O diffusion in silica	0.79	(2)
B (dry O_2)	1.23	(3)
B (steam)	0.78	(4)
B/A (steam)	1.94	(3, 4)
B/A (dry O_2)	2.0	(3)

*The activation energy in eV can be converted to kcal/mol by multiplying by 23.1.

References: (1) A.J. Moulson and J.P. Roberts, *Trans. Faraday Soc. 57,* p. 1208, 1961. (2) F.J. Norton, *Nature 171,* p. 701, 1961. (3) B.E. Deal and A.S. Grove, *J. Appl. Phys. 36,* p. 3770, 1965. (4) Don L. Kendall et al., *J. Electrochem. Soc. 125,* p. 1514, 1978.

is O_2; in wet oxygen or steam, it is H_2O. The solubility of water in fused silica (C^*) was found to be essentially constant over the 900°C–1200°C temperature range. No comparable data exists for oxygen, but since the activation energy of B for oxidation with oxygen and water closely matches the activation energies for oxygen and water diffusion coefficients (Table 3.8), it is concluded that C^* is relatively temperature-insensitive in both cases. The linear constant B/A, given by $(C^*/N)(hk)/(h + k)$, appears not to depend on transfer from the gas stream to the oxide. Therefore, $h >> k$, and the temperature behavior will follow that of k, the rate constant at the silicon–oxide interface. The observed value of approximately 2 eV matches reasonably well with that of the energy required to break an Si–Si bond (1.8 eV).

More detailed observations have suggested that dry oxides grown above 1000°C behave differently than do those at lower temperatures and that there are different activation energies for the two ranges (80). Since 1000°C is roughly the temperature at which silicon dioxide begins to show appreciable plastic flow, it has been suggested that the increased low-temperature interface and oxide strains that occur during growth are the probable cause.

3.5.7 Effect of Crystal Orientation on Rate

Crystal orientation will affect only the linear rate constant since, in the parabolic region, growth is limited by nutrient diffusion through the growing oxide. However, orientation effects, which are substantial, can be observed throughout the total growth cycle. Fig. 3.16 shows, for a dry oxidation, how the thickness varies over a range of orientations. The plot is based on thickness versus the angle that the oxidized surface was tilted away from the (100) plane toward the

FIGURE 3.16

Oxide thickness versus crystal orientation for a one hour dry oxidation at 1000°C.

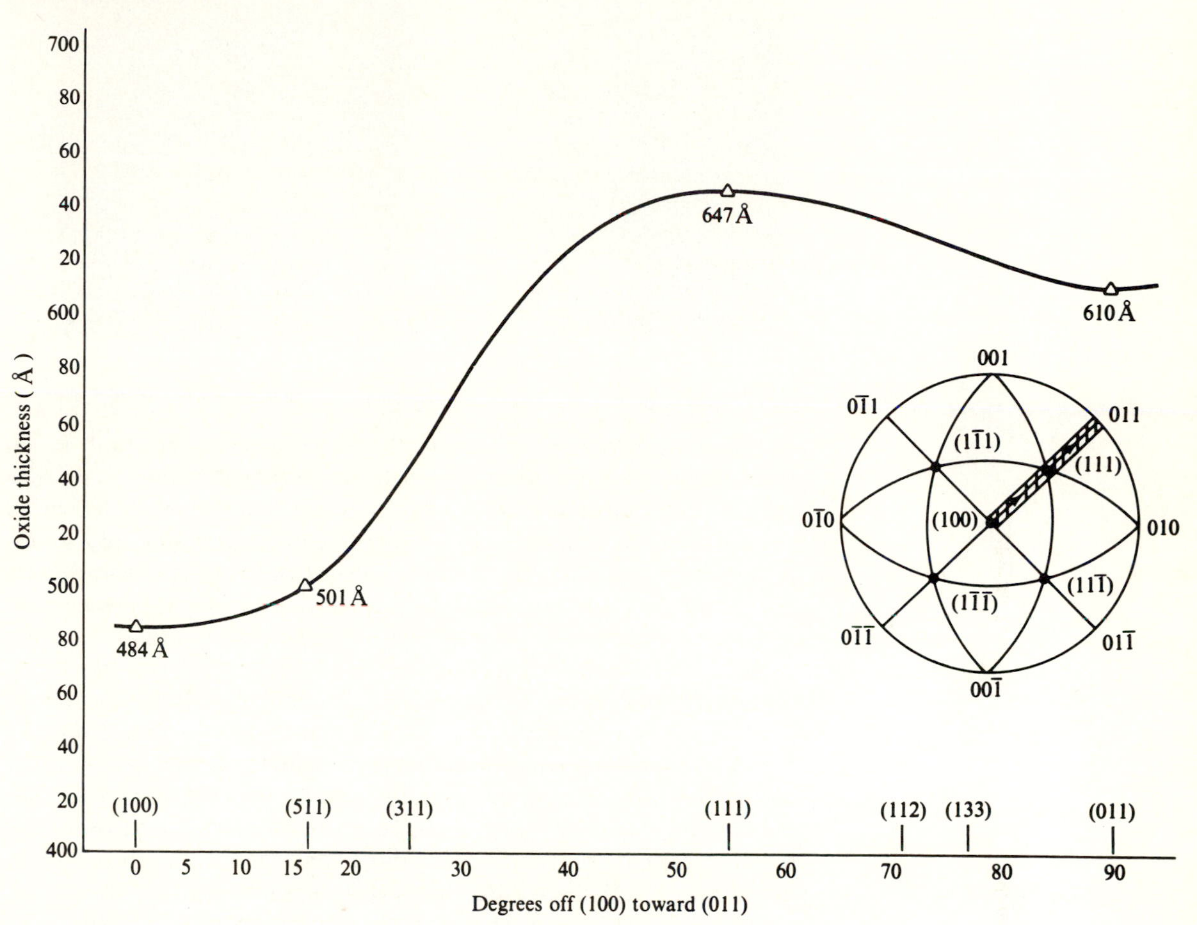

(011) plane. The location of this series of planes can be visualized from the standard projection shown in the inset. (For more details on interpreting standard projections, see Appendix A.)

The effects of crystal orientation on oxidation rate become important in the development of high-density memory designs, wherein use is made of the third dimension—that is, of going into the bulk silicon—for capacitors and transistors. Fig. 3.17 shows both an engineer's drawing (left) and a high-magnification cross-sec-

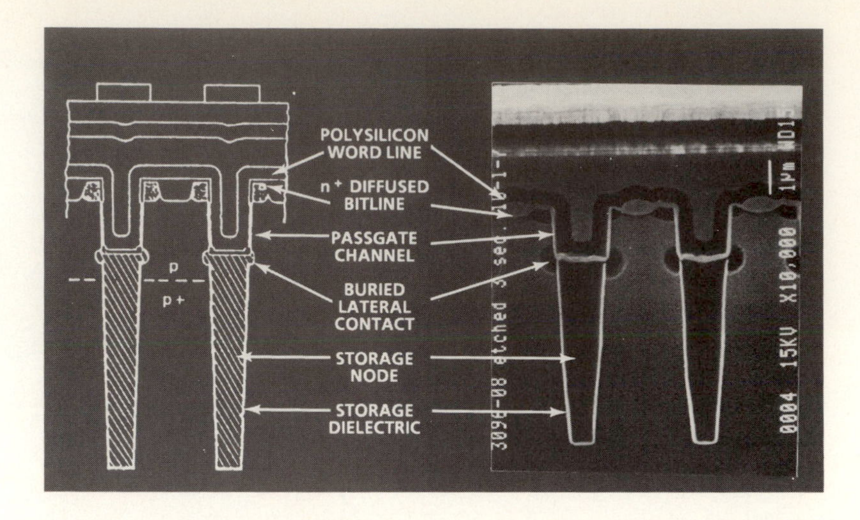

FIGURE 3.17

Engineer's drawing (left) and high-magnification cross-sectional SEM micrograph of a trench transistor crosspoint DRAM cell. (*Source:* Courtesy of Texas Instruments Incorporated.)

tional SEM[18] micrograph of a trench transistor crosspoint DRAM cell. In this structure, a capacitor and a transistor are both fabricated in an ~1 μm diameter well (trench) that is etched vertically ~7 μm deep into the silicon. The storage (capacitor) dielectric is silicon oxide that is thermally grown on the inner walls of the approximately 1 μm diameter trench. During the growth of this dielectric, all of the crystallographic planes shown along the periphery of a (001) stereographic projection (see Appendix A), and also shown in the stereographic projection in Fig. 3.16, are being oxidized at different growth rates. Fig. 3.18 (top) is a horizontal cross-sectional view of the storage node area—that is, the view looking from the top (100) surface down, but at the storage node level. The polycrystalline silicon storage node is readily visible as is the silicon dioxide dielectric around the trench wall. It is also evident that this oxide is much thicker in the four <110> directions than in the four <100> directions. Thickness differences corresponding to oxidation on (100), (510), (310), and (110) planes can be seen. Fig. 3.18 (bottom) shows another horizontal cross-sectional view of a defective (leaky) storage capacitor. The defects occur as would be expected—at the thin-oxide areas, near the <100> directions.

The oxidation rate versus orientation function was examined before 1961 in the hope that it would shed light on the oxidation reaction. The oxidations were done with pressures of from 40 atm to 150 atm and at temperatures between 500°C and 800°C. Over this range,

[18]SEM work by Dr. H.L. Tsai, Central Research Laboratories, Texas Instruments Incorporated.

FIGURE 3.18

Top: Horizontal cross-sectional view of trench capacitor showing oxidation of inside wall of trench etched into (100) silicon. *Bottom:* Same view of trench capacitor showing defects in the oxide, located azimuthally near the thin oxide (100) direction.

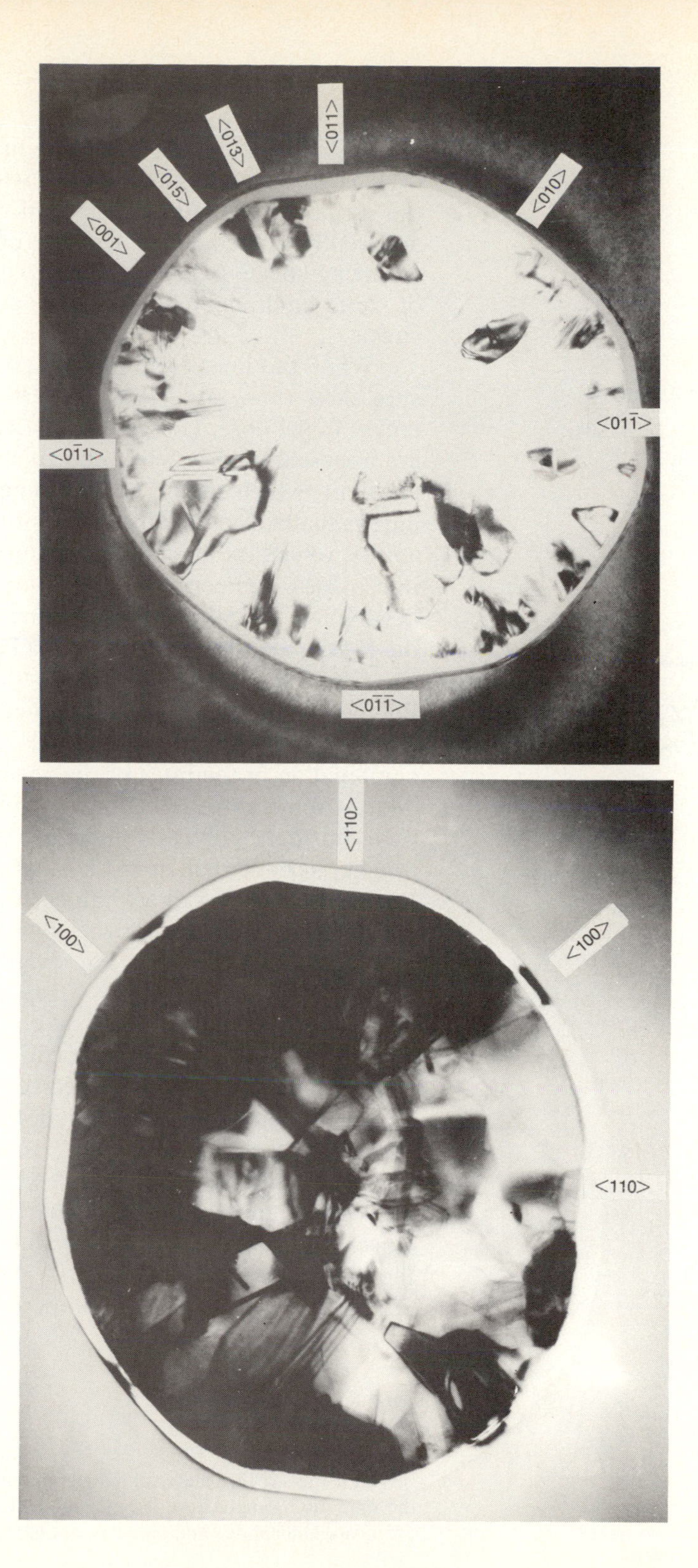

the oxidation rates observed were in the order (110) > (311) > (111) (81). These rates do not correlate with the number of bonds leaving the surface or the number of atoms on the surface. Rather, they are apparently related to the number of sites actually available to the oxidizing specie (which will be less than the total number because of steric hindrance) and to the relative ease of bonding to various sites.

While the linear rate constants measured at atmospheric pressure show $(B/A_{100}) < (B/A_{111}) < (B/A_{110})$ (82), most actual rates increase in the order (100) < (110) < (111), as shown in Fig. 3.16, and are apparently so ordered due to the impact of the parabolic rate constant, which should be independent of orientation. Under conditions that accentuate the effect of the linear rate constant—for example, oxidation at low temperature or the growth of only a very thin oxide—higher growth rates for (110) orientation are observed (83–84). At partial oxygen pressures below 0.07 atm, a reversal of the (111) and (100) oxidation rates has been reported (85).

3.5.8 Polycrystalline Silicon Oxidation

Polycrystalline or amorphous[19] silicon oxidation rates are somewhat different from that of single crystals. The overall rate is due to the combination of the rates of the variously oriented small crystallites and, in some cases, to doping segregation along the grain boundaries. Further, polycrystalline or amorphous silicon has a much rougher surface than does the single-crystal surface generally used. During oxidation, rather than getting smoother, the oxide–polysilicon interface gets yet rougher (86–88). An example of the amount of roughness is shown in Fig. 3.19. The exact amount depends on the starting roughness and the doping level. It is this roughness that gives rise to increased electrical conduction of thermal oxide grown on polycrystalline silicon by locally increasing the electric field.

While the various published rate behaviors may disagree somewhat in detail, in general, for both wet and dry oxidation over a substantial temperature range, the thickness–time curve for undoped polysilicon will lie between the undoped (111) and the (100) curves as shown in Fig 3.20. For thin oxides, the thickness will be closest to the (111) orientation and, for longer times, will approach the (100) orientation thickness (89–91). The thickness can be approximated by Eq. 3.28 as follows:

$$X = at^n \tag{3.28}$$

[19]Even if the starting material is amorphous, it will have crystallized by the time the oxidation temperature is reached. LPCVD silicon layers grown below about 575°C are amorphous and above about 625°C are polycrystalline.

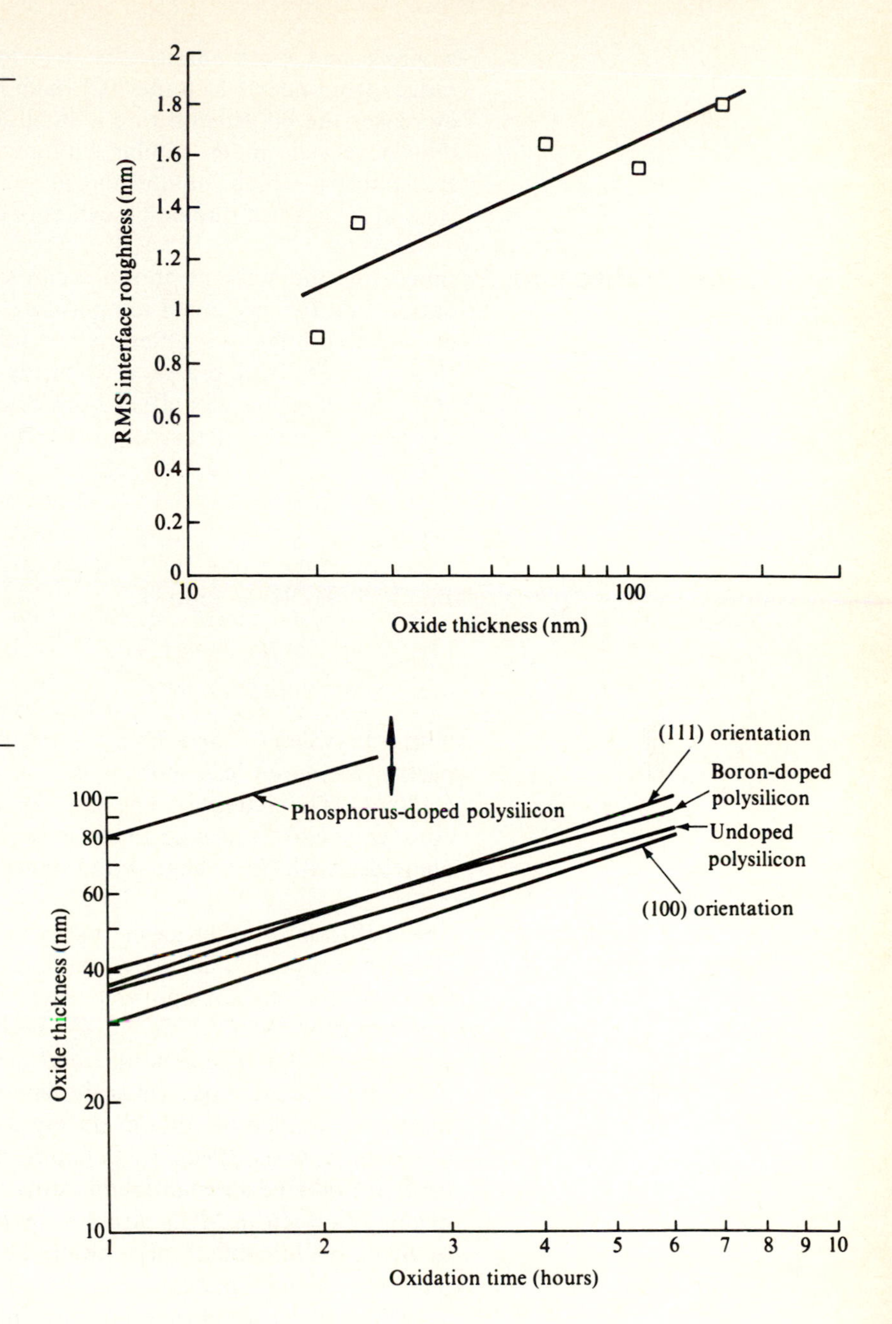

FIGURE 3.19

Trend of increasing roughness for thermal oxide grown on phosphorus-doped polysilicon at 800°C pyrogenic oxidation. (*Source:* Data from L. Faraone and G. Harbeke, *J. Electrochem. Soc. 133*, p. 1410, 1986.)

FIGURE 3.20

General trends for oxidation behavior of polysilicon. (Values are reasonable for a 900°C dry oxidation.)

where a and n are empirically determined constants. Note that n ranges from near 1 to under 0.5 have been reported. Boron doping increases the polysilicon rate a small amount and is comparable to that observed under similar circumstances for single crystals.[20] Phosphorus, which, unlike boron, segregates at the grain boundaries, may produce rate enhancements of ten times or more.

3.5.9 Effect of Pressure on Oxidation

Increasing the ambient pressure can substantially increase the oxidation rate for any given temperature. Thus, low-temperature oxidations that might not otherwise be feasible because of excessively slow rates are quite practical when the pressures are raised to 10–20 atm. The pressure will affect both rate constants through C^*. Over the normal range of temperatures, and up to at least 10 atm, the constants for steam are directly proportional to the partial pressure or, in the case of elevated pressures, to the total pressure P (92). In dry oxygen, however, only the parabolic constant B is linear with pressure above 1 atm. The linear constant can be approximated by Eq. 3.29 (93, 94) as follows:

$$\frac{B}{A} = \left(\frac{B}{A}\right)_{1\ \text{atm}} P^n \qquad 3.29$$

with n between 0.7 and 0.8 for P between 1 atm and 1000 atm. For partial pressures less than 1 atm, n is also <1 for temperatures $<1100°C$. The activation energies for both the linear and parabolic constants show a change in slope in the 900°C–950°C range, which coincides with the change in the oxide viscous behavior.

3.5.10 Effect of Ambient Impurities on Oxidation Rate

The inclusion of other gases may substantially affect the oxidation rate. For example, because of the large difference in oxidation rates between dry oxygen and wet oxygen or steam, even very small amounts of moisture in the "dry" oxygen can substantially increase the rate (82, 95). The general trend of the effect of moisture on the oxidation rate using oxygen is shown in Fig. 3.21. Some data suggest that the presence of sodium during oxidation will change the rate (96), which, in the 1960s, was of some concern since many oxidation tubes introduced substantial amounts of sodium. Since sodium cannot be tolerated in MOS circuits, techniques have been developed to virtually eliminate it, and the impact on the oxidation rate is now only of academic interest.

Chlorine or a chlorine-bearing species such as HCl, trichloroethane, or trichloroethylene is often used during oxidation. All af-

---

[20] Impurity-enhanced oxidation is discussed in more detail in a later section.

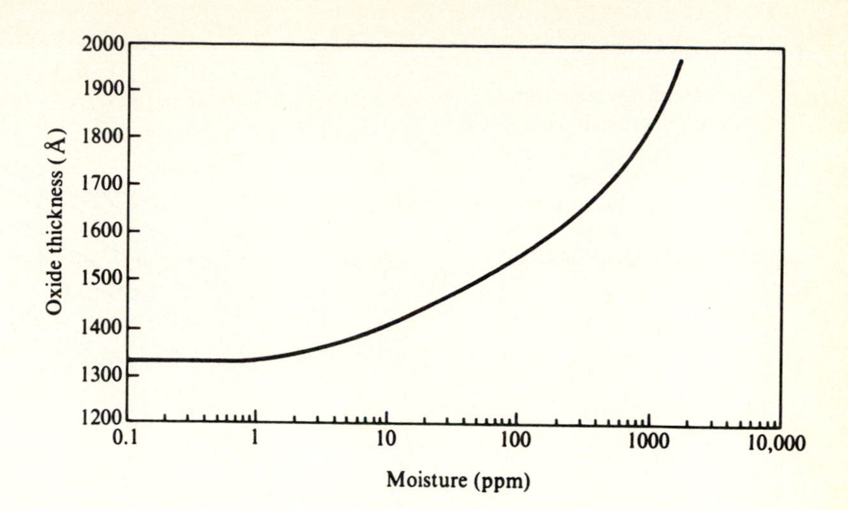

FIGURE 3.21

General trend of effect of moisture on oxidation rate using oxygen for a 1000 minute oxidation at 900°C. (*Source:* Data from E.A. Irene and R. Ghez, *J. Electrochem. Soc. 124,* p. 1757, 1977.)

fect the rate. Representative behaviors are shown in Figs. 3.22–3.24. They have similar characteristics in that a small amount enhances the rate, but the effect soon saturates. Small quantities of fluorine are reported to affect the rate much more than does HCl. When NF_3 is used, quantities of 100–500 ppm, depending on temperature, will produce a peak in the oxidation rate that can be as much as five or ten times as great as without the NF_3. Once the peak is passed, a competing fluorine etching reaction apparently begins (97).

3.5.11 Effect of Silicon Doping on Oxidation Rate

Heavily n- or p-doped silicon has been observed to increase the oxidation rate (98–102). The effect of n-doping is much more pronounced than is the effect of p-doping. In either case, the effect is primarily on the linear rate constant B/A. For n-dopants, the ratio of $(B/A)_{doped}$ to $(B/A)_{undoped}$ may be as much as 20. For p-dopants, the ratio is in the range of 2. The n-doped parabolic constant B increases, but by only about a factor of 2 near 800°C and by less at higher temperatures (101). For p-doping, B increases a few percent as the temperature goes from 800°C to 1150°C (100). Within experimental error, that all of the enhancements are independent of orientation is apparent.

A reasonably satisfactory model for n-behavior is based on the supposition that the linear rate constant can be expressed as the sum of two terms, one of which represents the effect of silicon vacancies (102). Thus,

$$\frac{B}{A} = R + KC_v \tag{3.30}$$

FIGURE 3.22

Oxide thickness versus time for oxidation in various O_2/HCl mixtures. (*Source:* D.W. Hess and B.E. Deal, *J. Electrochem. Soc. 124*, p. 735, 1977. Reprinted by permission of the publisher, The Electrochemical Society, Inc.)

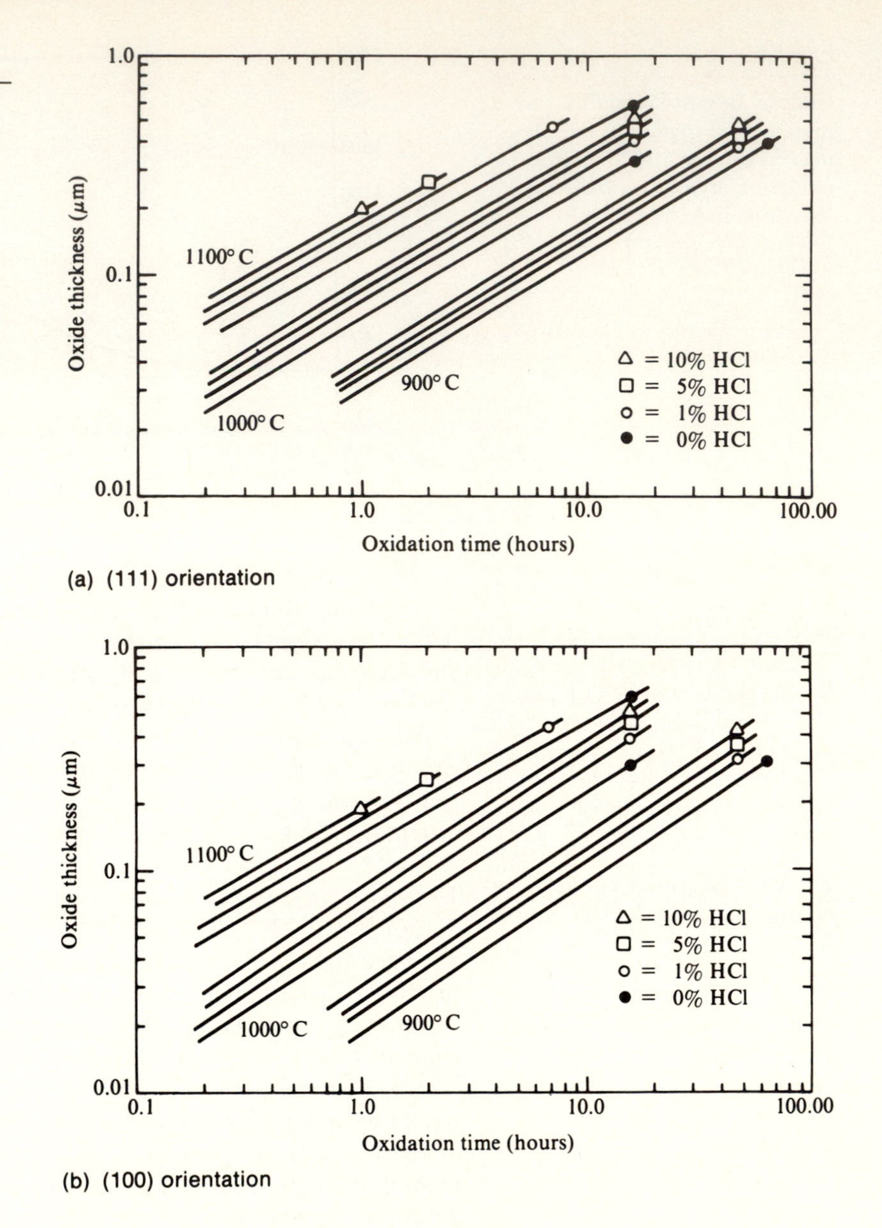

where R is the term that does not depend on vacancy concentration, C_v is the total concentration of vacancies, and K is a constant of proportionality. Vacancies exist as neutrals, single-charged deep acceptors, double-charged deep acceptors, and single-charged donors. The concentration of neutrals is a function only of temperature. The charged vacancy concentration will also depend on the net concentration of electrically active impurities when the doping level is in-

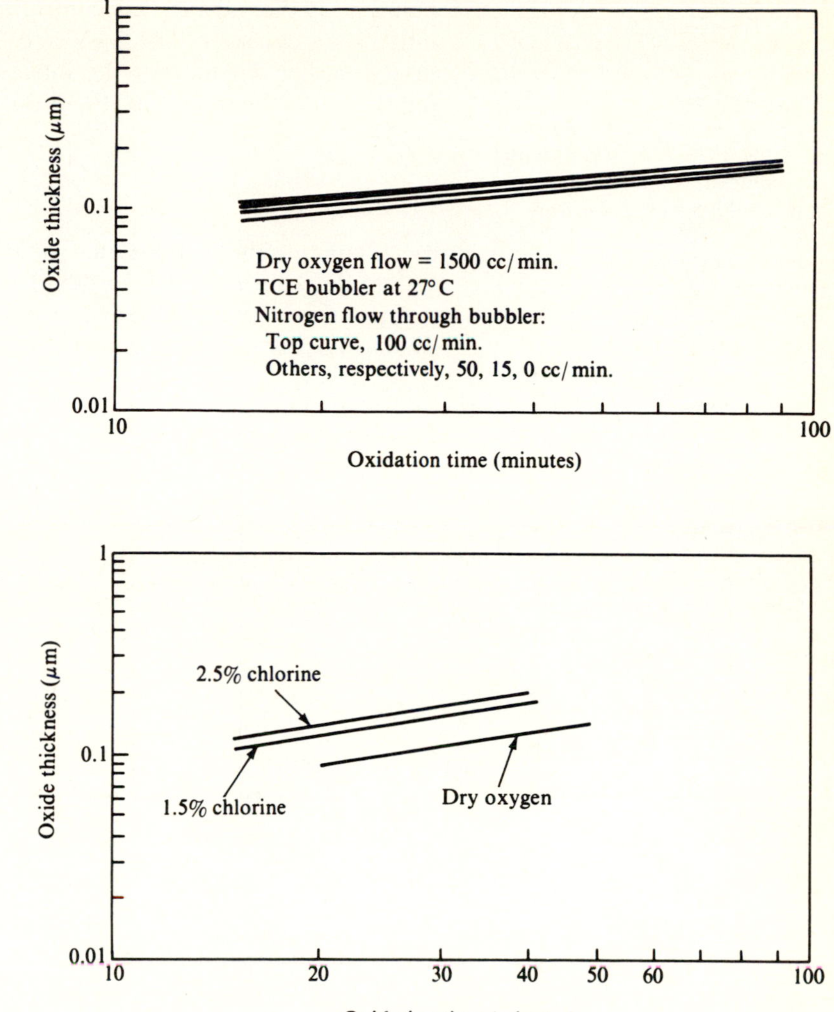

FIGURE 3.23

Oxide thickness versus time for oxidation at 1200°C, (100) orientation, with TCE addition. (*Source:* Adapted from Takeshi Hattori, *Jap. J. Appl. Phys. 17,* p. 69, 1978.)

FIGURE 3.24

Oxide thickness versus time for oxidation at 1200°C, with (100) orientation, with chlorine addition. (*Source:* Adapted from R.J. Kriegler et al., *J. Electrochem. Soc. 119,* p. 388, 1972.)

creased to the point where it is greater than n_i at oxidation temperature.[21] For the intrinsic case (no heavy doping), Eq. 3.30 can be written as

$$\left(\frac{B}{A}\right)^i = R + KC_v^i = C_1 e^{-2/kT} \qquad 3.31$$

[21] n_i is the intrinsic carrier concentration. Thus, vacancy enhancement occurs when doping is heavy enough for the silicon to be extrinsic at the oxidation temperature.

where C_v^i is the vacancy concentration while the silicon is intrinsic and C_1 is a constant. The 2 eV activation energy for B/A was discussed earlier. Combining the intrinsic case expression (Eq. 3.31) and the general expression (Eq. 3.30) gives

$$\frac{B}{A} = \left(\frac{B}{A}\right)^i \left[1 + \left(\frac{K}{C_1}\right)\left\{\frac{C_v}{C_v^i} - 1\right\} C_v^i e^{2/kT}\right] \qquad 3.32$$

By calculating the increase in C_v and using one set of experimental data to determine the KC_v^i product, curves such as the ones shown in Fig. 3.25a may be plotted. For p-doping, the agreement is not as good as it is for n-doping. Experimentally, a larger effect is observed

FIGURE 3.25

(a) Change of B/A with silicon doping level. (*Source:* C.P. Ho and J.D. Plummer, *J. Electrochem. Soc. 126*, p. 1523, 1979. Reprinted by permission of the publisher, The Electrochemical Society, Inc.) (b) Change of oxide thickness with doping level. (*Source:* Adapted from C.P. Ho et al., *J. Electrochem. Soc. 125*, p. 665, 1978.)

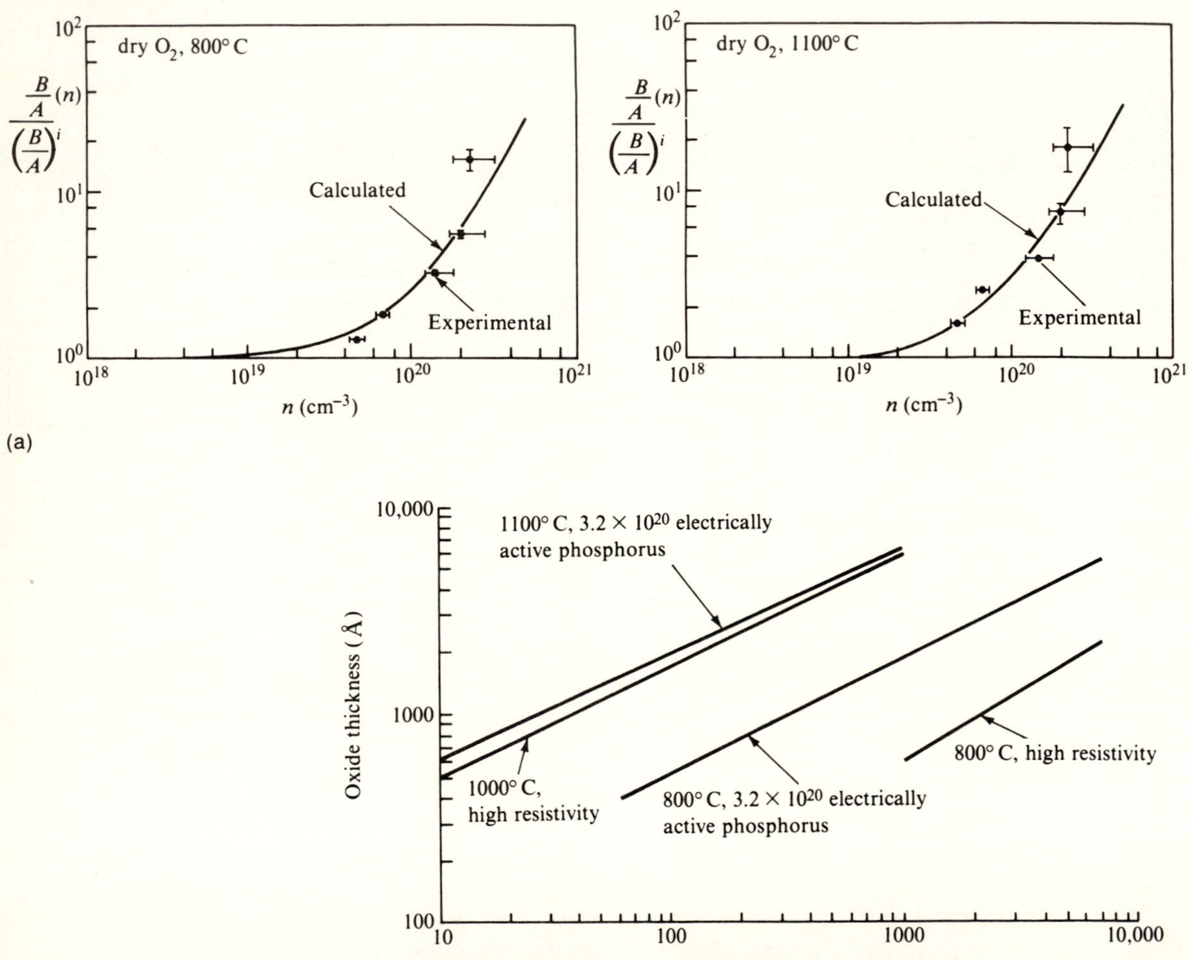

than that predicted by Eq. 3.32 (102). It is possible that in this case the dopant pileup in the oxide is the driving force rather than vacancies (remember that for p-material, heavier doping will only increase the density of the singly charged donor vacancy). (Dopant pileup is discussed in the next section.)

Even though the change of B/A appears to be independent of wet or dry oxidation and of orientation, the actual oxide thickness will still depend on these things because the B/A to B ratio differs from case to case. Fig. 3.25b shows the effect of doping on oxide thickness for two temperatures and illustrates the difference in rates observed even though the B/A increase with doping is virtually the same (Fig. 3.25a) for the two cases.

The polycrystalline silicon oxidation rate increases with doping. All data regarding p-doping seem to agree that polycrystalline behavior and single-crystal behavior are similar. In the case of very heavy n-doping, some data indicate comparable behavior (90, 103), while other data show an increased enhancement over that predicted by Fig. 3.25 (104). The oxidation rate at high pressure also shows an n-doping dependence. On the basis of very limited data, it does not appear to be as large as the atmospheric pressure effect (105).

Table 3.9 shows some trends of heavy doping effects on oxide thickness compared to a lightly doped wafer under the same conditions. Note that if, for example, the effect of heavy doping is to be minimized, high temperatures, (111) orientation, and a dry oxygen ambient should be chosen (101). Conversely, if the differences in oxide thickness over lightly and heavily doped regions are to be accentuated, then steam oxidations at low temperatures on (100) material are more appropriate. By taking advantage of a maximized effect, it is possible, for example, to grow substantially thicker oxide over an n-channel source and drain than over the gate or to simultaneously grow a thin gate oxide and a much thicker interlevel polysilicon oxide (101, 104).

TABLE 3.9

Impact of Heavy Doping on Oxide Thickness

n	*p*	*(111)*	*(100)*	*High Temp.*	*Low Temp.*	*Dry*	*Wet*	*Thin*	*Thick*	*Effect*
X		X			X	X		X		Medium increase
X		X		X		X			X	None
X			X		X		X	X		Large increase
	X	X			X		X	X		None
	X	X			X	X		X		Small increase

3.5.12 Impurity Profile at the Oxide Interface

While the oxide is growing, silicon is being consumed and the impurities that were in the silicon must be redistributed. The manner in which this distribution occurs depends on the segregation coefficient m of the impurity and the relative diffusion coefficients of the impurity in silicon and in SiO_2 (106, 107). m is defined as the ratio of the equilibrium concentration of impurity on the silicon side to that on the oxide side of the interface. In principle, as oxidation proceeds, three cases are possible. The impurities could be rejected by the oxide ($m > 1$), and a pileup of impurity in the silicon would then occur (Fig. 3.26a). If $m < 1$, dopant will be depleted from the silicon and built up in the oxide (Fig. 3.26b). If $m = 1$, the dopant in the oxide and silicon will be uniform across the interface, but a depletion of impurity from the silicon will still occur (Fig. 3.26c) because the oxide volume is greater than that of the consumed silicon. A high-oxide diffusivity, combined with a rapid transfer across the oxide–ambient boundary, can keep the concentration in the oxide at the interface well below the equilibrium value, which can result in a reduction of the concentration in the silicon near the interface (Fig. 3.26d) even though $m > 1$.

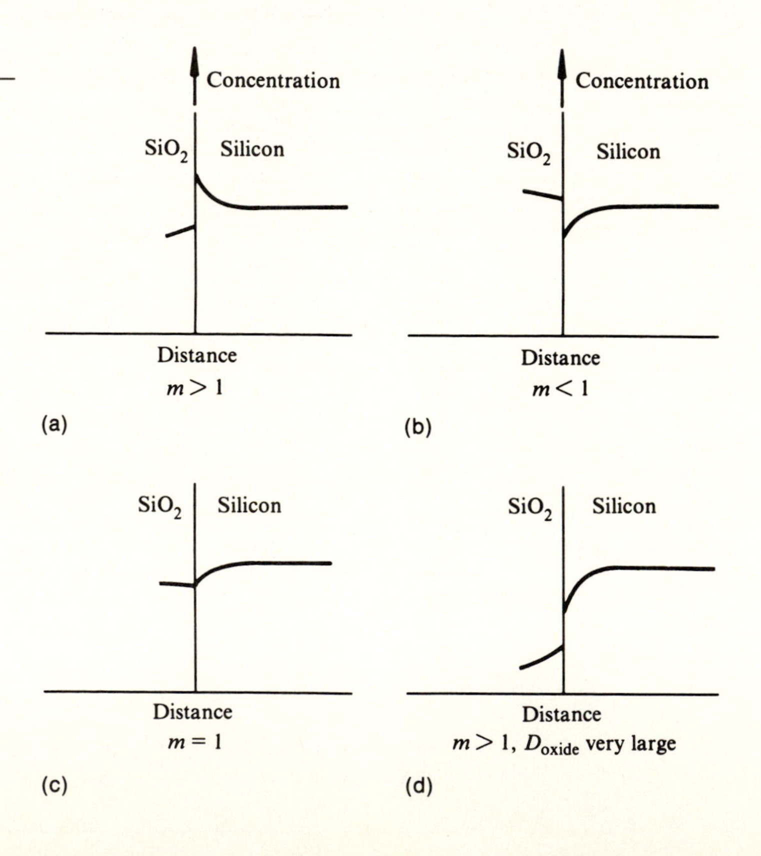

FIGURE 3.26

Effect of impurity segregation coefficient in oxide on impurity distribution after oxidation.

TABLE 3.10

Segregation and Diffusion Coefficients of Impurities in SiO_2

Element	*m*	*Reference*	*Diffusion Coefficient* (cm^2/s)* 1100°C	1200°C
Ga	20	(1)	5.3×10^{-11}	5.8×10^{-10}
B	†	(2)	4.6×10^{-17}	3.2×10^{-16}
In	>1	(1)		
P	10	(1)	1.4×10^{-15}	2.9×10^{-15}
Sb	10	(1)	9.9×10^{-17}	1.5×10^{-14}
As	10	(1)	2.6×10^{-16}	4.4×10^{-15}

†*m values for boron:*

900°C	1200°C	
0.16	0.6	wet oxidation, (100)
0.1	0.5	wet oxidation, (111)
0.5	1	dry oxidation, (100)

Pressure (bar)	1	5	10	15	20	(100), 920°C
m	0.3	0.55	0.7	0.9	0.95	

*Diffusion coefficient values have been chosen arbitrarily from a tabulation of reported values in M. Ghezzo and D.M. Brown, *J. Electrochem. Soc. 120,* p. 146, 1973. See also Yasuo Wada and Dimitri A. Antoniadis, *J. Electrochem. Soc. 128,* p. 1317, 1981; R. Singh et al., *J. Electrochem. Soc. 131,* p. 2645, 1984. Note that the various values listed in the literature may vary orders of magnitude, often because of sensitivity to concentration or diffusing ambient.

References: (1) A.S. Grove et al., *J. Appl. Phys. 35,* p. 2695, 1965. (2) Richard B. Fair and J.C.C. Tsai, *J. Electrochem. Soc. 125,* p. 2050, 1978; P. Deroux-Dauphin and J.P. Gonchond, *J. Electrochem. Soc. 131,* p. 1418, 1984.

Table 3.10 gives *m* values and diffusivities of some common silicon dopants. Gallium is one that has out-diffusion from the silicon, even though the high value of *m* would imply a pileup (108). Because of its importance in device fabrication, boron segregation has been extensively studied. m_{boron} is dependent on the phase that forms at the interface, which, in turn, is dependent on the oxidation conditions. *m* equals 0.6 at 1200°C for wet oxidation, 1 for very dry oxidation, and for a high-boron-content glass at the surface[22] will be over 2. It is also silicon wafer orientation dependent and increases with oxidation temperature and pressure (109, 110). As in most other process-related data, a substantial spread in the *m* values is reported.

[22]This case will not normally occur during an oxidation alone but is likely to occur during a diffusion cycle.

3.6

OXIDE THICKNESS CHARTS

Figs. 3.27–3.31 show representative curves of oxide thickness for oxidations under various conditions. The curves do not cover all circumstances of interest, but they do show the trends. Eq. 3.27 and values of B and B/A will allow behavior outside the range of the curves to be calculated as long as the thickness is greater than perhaps 300 Å. Alternatively, a computer program such as SUPREM, which calculates in a similar fashion, can be used. Fig. 3.27 shows thicknesses for oxidation temperatures in the 700°C–1200°C range in dry oxygen. It also shows the difference in thickness due to slice surface orientation. Fig. 3.28 shows similar curves for steam oxidation at 640 torr. In this case, the steam is generated by burning hydrogen in oxygen directly in the oxidation tube. However, whether the moisture comes from steam or from oxygen bubbled through hot water, for the same pressure, the behavior seems to be the same. Fig. 3.29 shows curves for dry oxygen oxidation at high pressures, and the curves in Fig. 3.30 pertain to high-pressure steam oxidation. Fig. 3.31 shows thicknesses for very short times for the three most common slice surface crystallographic orientations. It also shows the (111)–(100) crossover discussed in the section on orientation effects.

FIGURE 3.27

Oxide thickness versus oxidation time for silicon oxidation in dry oxygen. (*Source:* Courtesy of Bruce E. Deal.)

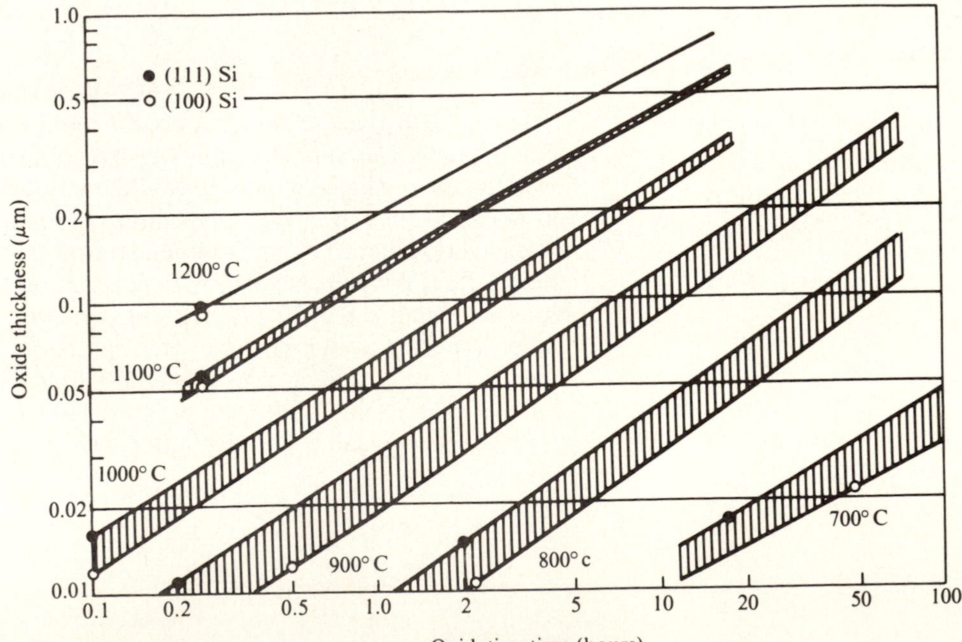

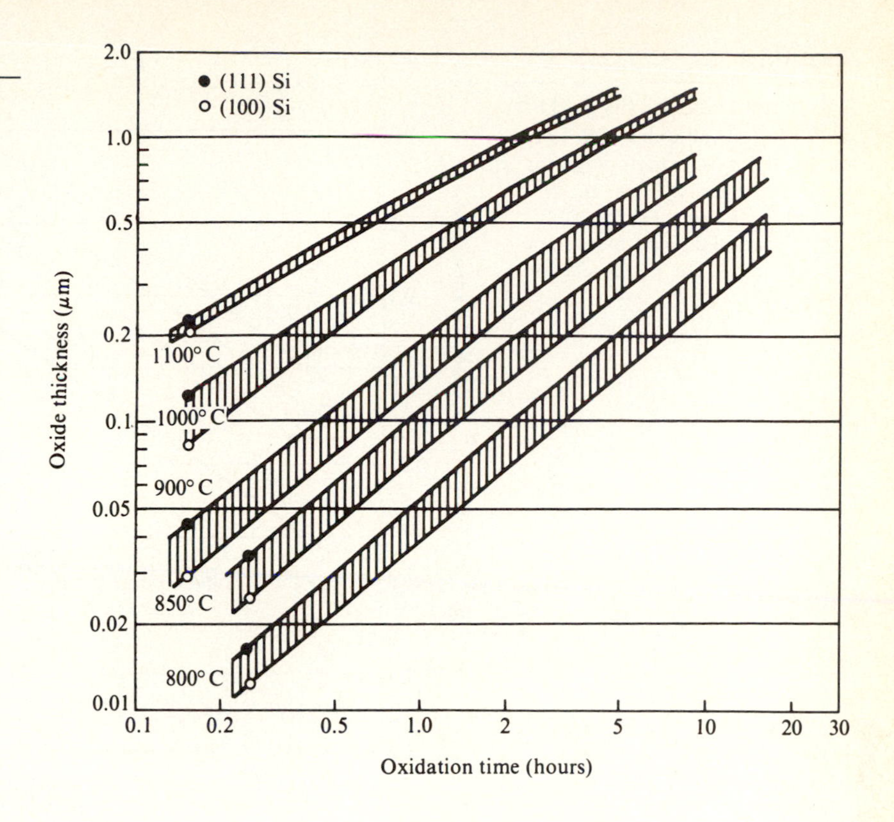

FIGURE 3.28

Oxide thickness versus oxidation time for silicon oxidation in pyrogenic H_2O (640 torr). (*Source:* Bruce E. Deal, *J. Electrochem. Soc. 125*, p. 576, 1978. Reprinted by permission of the publisher, The Electrochemical Society, Inc.)

3.7 PREPARATION OF THERMAL OXIDE

3.7.1 Preoxidation Surface Cleanup*

The kind of cleanup given the surface before oxidation can affect not only the quality of oxide but also the thickness. Some of the many varieties of surface contaminants that may be found on slices before oxidation, along with probable effects on the oxide, are listed in Table 3.11. How many contaminants remain on the oxide depends on the efficiency of the cleanup. A cleaning sequence should proceed in the following order:

1. Remove gross organics and particulates.
2. Remove organic films (to prevent masking of next step).
3. Remove surface oxide.
4. Remove surface-adsorbed ions and plated-metal contaminants.
5. Reform contamination-free native oxide (usually done as part of step 4).

*See references 111–113.

FIGURE 3.29

Oxide thickness versus oxidation time for silicon oxidation in dry oxygen at high pressure. (*Source:* Liang N. Lie et al., *J. Electrochem. Soc. 129,* p. 2828, 1982. Reprinted by permission of the publisher, The Electrochemical Society, Inc.)

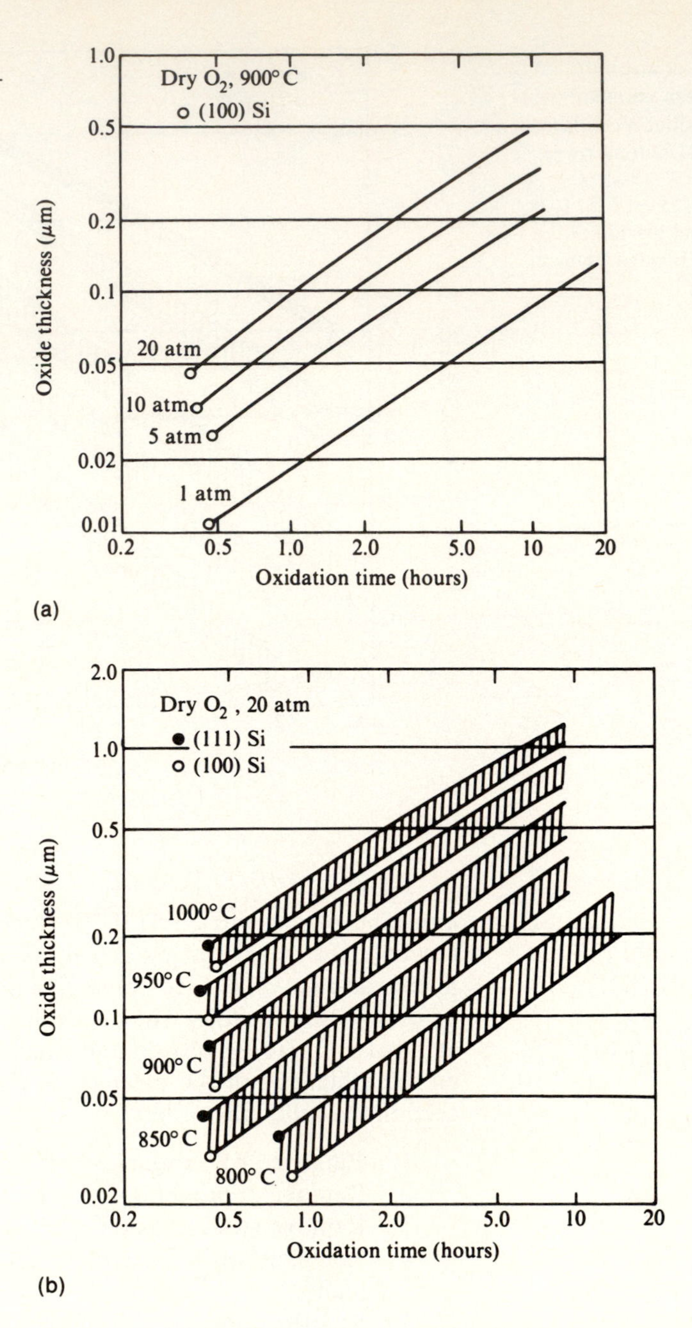

FIGURE 3.30

Oxide thickness versus time for thermal oxidation of silicon in pyrogenic steam at high pressure. (*Source:* Reda R. Razouk et al., *J. Electrochem. Soc. 128,* p. 2214, 1981. Reprinted by permission of the publisher, The Electrochemical Society, Inc.)

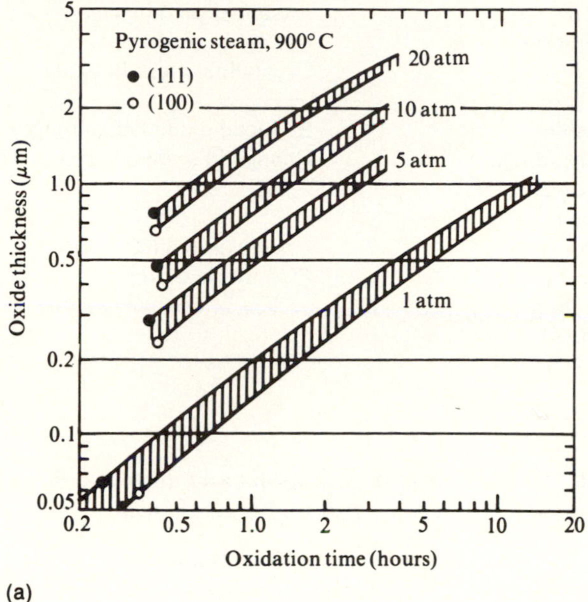

(a)

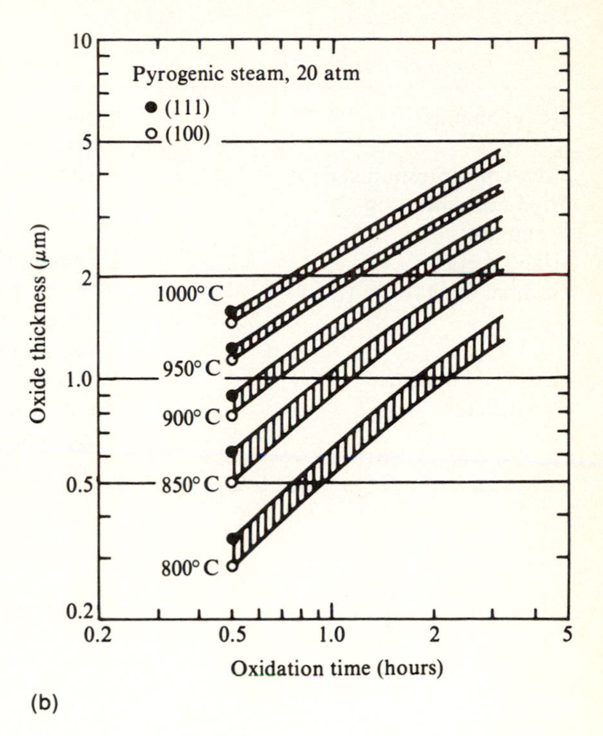

(b)

FIGURE 3.31

Oxide thickness versus oxidation time during initial stages of silicon thermal oxidation. (*Source:* H.Z. Massoud, "Thermal Oxidation of Silicon in Dry Oxygen-Growth Kinetics and Charge Characterization in the Thin Regime," Ph.D. dissertation, Stanford University, 1983.)

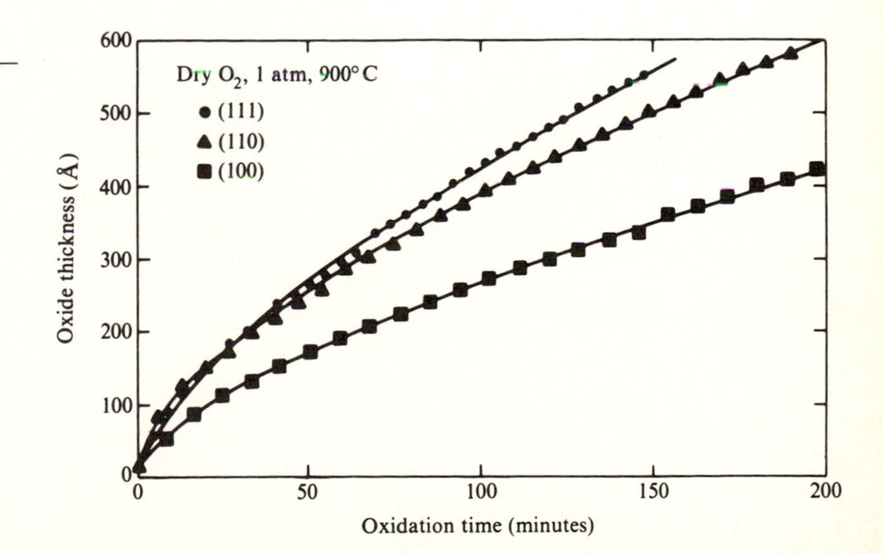

TABLE 3.11

Summary of Slice Surface Contaminants

Contaminant	*Possible Sources*	*Effect*
Heavy metals	Grinding, polishing, any prior cleanup	Reduced lifetime, S-pits (saucer pits) in slice
Low-concentration sodium	Any prior operation	Charge in oxide
High-concentration sodium		Crystalline areas in oxide
Alkali metals	Grinding, polishing	Reduced oxide field strength
Organic films	Plasticizer from storage containers, organic vapors in room	Change in oxidation rate
Adsorbed fluorine	Etch containing HF	Change in oxidation rate
Particulates	Particles in air, in cleaning solutions, or from equipment	Pinholes in oxide

6. Remove all traces of cleaning solutions and all particulates from the surface of the reformed native oxide film.

Heavy metals in particular are difficult to remove since they tend to replate onto the surface from the cleaning solutions. Small particles also are very difficult to remove because of the increasing importance of attractive forces between the particle and the surface relative to the forces that can be exerted on the particle by liquid flowing across the surface. Thus, mechanical scrubbing is almost a necessity for complete small-particle removal. However, mechanical scrubbing has its pitfalls in that the scrub brush can damage the wafer surface and may only slide particulate matter from side to side in oxide windows and not remove it at all. An alternative in many circumstances is ultrasonic agitation of the cleaning bath (114). The intensity must be high enough to produce cavitation but not so high that wafers are broken.

One old, but reasonably good, all-purpose cleaning solution is a sulfuric acid and hydrogen peroxide mixture (affectionately called "piranha"). However, a much better cleaning routine, often referred to as the "RCA cleanup," is as follows (108):

1. Remove gross organics with perchloroethylene (tetrachloroethylene).
2. Remove residual organic films with an H_2O_2–NH_4O (basic) solution.
3. Remove metallics with an H_2O_2–HCl (acidic) solution.

4. Rinse thoroughly in deionized water.
5. Dry.

A single solution that removes organics and some heavy metals very well is choline, water, and a surfactant. However, the acidic RCA solution is more effective for other metals, so that the choline solution followed by HCl–H_2O_2 is potentially better than the RCA sequence. Table 3.12 gives typical compositions of these cleaning solutions. In general, the relative amounts can be changed substantially and not affect performance. Solutions containing hydrogen peroxide lose their peroxide and thus must be periodically replaced.

Hot, concentrated sulfuric acid alone can be used for cleaning but, for maximum performance, must be so hot that a significant amount of acid fumes are released. By mixing it with hydrogen peroxide, the operating temperature can be reduced to a little over 100°C. Usually no external heat is required since the H_2SO_4 heat of solution will provide the initial heating and the heat of decomposition of the H_2O_2 will keep the solution hot over its working lifetime (115).

Depending on the final cleanup, substantial differences in the composition of the native oxide are observed. For example, HF leaves a thin oxide on the surface that quickly adsorbs large quantities of carbon. A hydrogen-peroxide-based cleaner will leave a

TABLE 3.12

Preoxidation Surface Cleaning Solutions

Cleanup	*Composition*				*Conditions*
Choline clean* (parts by volume)	Choline $(CH_3)_3N(CH_2Ch_2OH)OH$	H_2O_2	H_2O	Surfactant (NCW-601)	50°C ± 5°, 10 min. ultrasonic clean, 10 min. DI water rinse, spin dry.
	3	1	95	1	
RCA clean, basic† (parts by volume)	NH_4OH	H_2O_2	H_2O		Room temperature, 10 min. clean time, 10 min. DI water rinse, spin dry.
	1	1	5		
	1	2	7		
RCA clean, acid† (parts by volume)	HCl	H_2O_2	H_2O		Room temperature, 10 min. clean time, 10 min. DI water rinse, spin dry.
	1	1	6		
	1	2	8		
Sulfuric acid clean (percent)	H_2SO_4	H_2O_2			
	60	40			
	70	30			

*See *Electrochem. Soc. Ext. Abst. 81–2*, 1981.
†See *RCA Review 31*, p. 187, 1970.

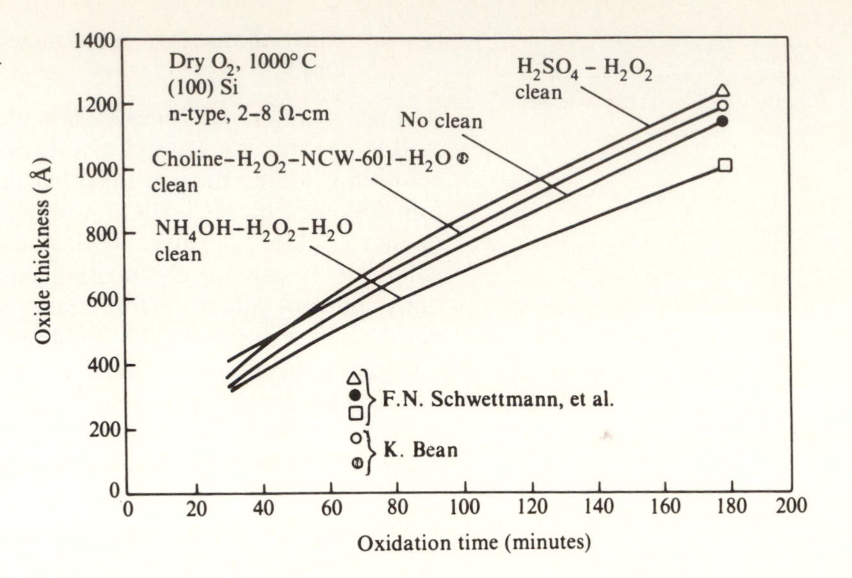

FIGURE 3.32

Oxidation rate versus preoxidation cleanup. (*Source:* Adapted from F.N. Schwettmann et al., *The Electrochemical Soc. Ext. Abst.*, Vol. 78–1, and Ti-Ken Bean, *The Electrochemical Soc. Ext. Abst.*, Vol. 81–2.)

thinner, cleaner, and apparently rather volatile oxide (116). Presumably through the kind of oxide left, the cleanup can affect the linear oxidation rate constant, as shown by the examples in Fig. 3.32.

The equipment used with the cleaning solutions can be broken into two general categories: Spin–spray and tank immersion. Tank immersion may or may not be combined with ultrasonic agitation. Drying is generally by hot nitrogen, which, in the case of spin–spray cleaning, can be introduced into the spin rinser immediately after the last water rinse. Great care must be taken to ensure that no equipment, including that used for water rinsing and drying, generates and distributes particles onto the slice surface. The cleaning solutions, water, and nitrogen should all pass through point-of-use filters.

3.7.2 Atmospheric Pressure Equipment

Atmospheric pressure reactions are primarily carried out in oxidation tubes heated in a tubular diffusion furnace (for details on construction of furnaces, see Chapter 8). Fig. 3.33 is a schematic of an oxidation tube and the normal input oxidizing gases. The combination of large-diameter wafers and the popularity of rapid thermal anneal processing has prompted investigations of lamp-heated oxidations (117–119). Such a heat source seems feasible and appears particularly applicable to thin oxides where the high thermal inertia of ordinary oxidation furnaces makes the control of short oxidation times difficult.

Wet oxidation is done by bubbling oxygen through water held near the boiling point. Steam oxidation can be done either by having

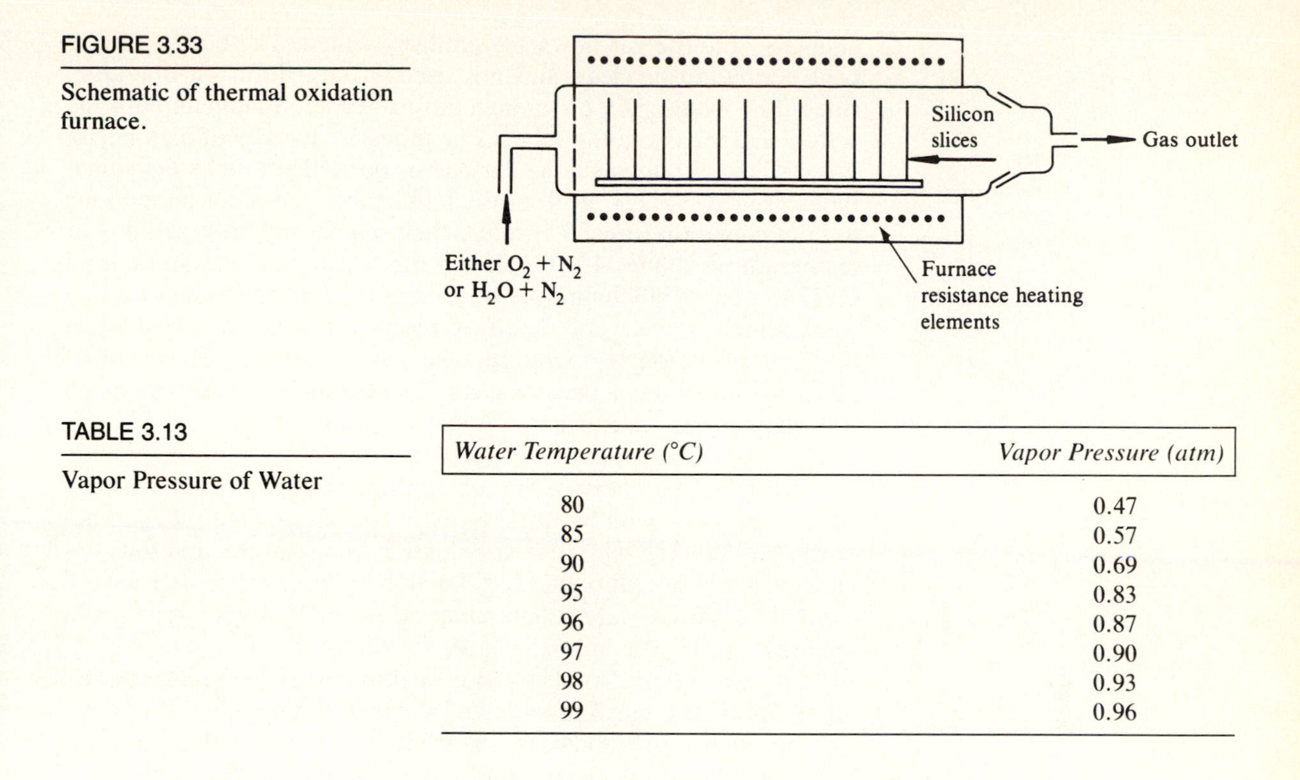

FIGURE 3.33

Schematic of thermal oxidation furnace.

TABLE 3.13

Vapor Pressure of Water

Water Temperature (°C)	*Vapor Pressure (atm)*
80	0.47
85	0.57
90	0.69
95	0.83
96	0.87
97	0.90
98	0.93
99	0.96

an external steam generator or by introducing oxygen and hydrogen into the front of the oxidation tube in stoichiometric proportions (pyrogenic or burnt hydrogen system). In either a wet oxygen or a steam system, the line from the water source to the furnace must be wrapped with heating tape and kept hot so that no condensation occurs. Assuming that the carrier gas (oxygen or an inert gas such as helium or argon) is saturated, the partial pressure of water vapor (P_w) can be determined from the temperature of water in the bubbler (see Table 3.13). If there is a question of saturation, it can be calculated from Eq. 3.33 (120) as follows:

$$P_w = \left(\frac{n_w}{n_w + n_{\text{gas}}}\right) P_{\text{total}} \tag{3.33}$$

where P_{total} will be essentially atmospheric pressure if the tube exhaust is not throttled. The n's are the number of moles of water vapor and carrier gas, respectively, and can be measured.

The slices are held perpendicular to the gas flow in either silicon or fused silica boats. The separation of slices is much less than it is for diffusion and is usually only enough to keep the wafers from physically contacting each other. The tubes are sized for the slice diameter, and the gas flows are similarly scaled. The boats must be kept scrupulously clean and not used in any diffusion operation since they would pick up enough diffusant to contaminate both the wafers and the oxidation tubes. The tubes are usually of high-purity fused silica, although silicon carbide or polysilicon tubes are sometimes used (121–122). Both of the latter tubes are more impervious to contamination transfer through their walls, and both sag less at temperatures above 1100°C. Silicon tubes can be given an internal CVD coating of silicon nitride to protect them from oxidation.

Sodium moves very rapidly through the walls of fused silica tubes at oxidation temperature, and early in the development of MOS devices, when it became clear that sodium in the thermal oxide was deleterious, a concentric tube arrangement was introduced. Thus, as sodium and other impurities diffused through the outside tube to the space between the tubes, they could be swept away by gas flowing between the tubes. Sodium also causes the tubes to devitrify and weaken with time at oxidation temperature, and thus impurities can move through the tube walls with even greater ease.[23] Table 3.14 shows wafer contamination levels that can occur from impurities originally in the furnace.

Contamination can also come in through the gas lines. As already discussed, small amounts of water in the oxygen of a "dry" oxidation system will cause a substantial increase in the oxidation rate. Oxygen, water vapor, and hydrocarbons in the inert gas must also be considered. Studies of flatband voltage versus these contaminants have indicated that little shift occurs when oxygen is below 200 ppm, H_2O is below 120 ppm, and total CO_2 is below 60 ppm (123). These numbers are rather high, and as the shift to thinner oxides and higher operating fields continues, they will probably need to be reduced. Contamination of the oxidation chamber by oxygen, water, and hydrocarbons can occur not only from contaminated incoming gases or plumbing leaks but also from backstreaming from the end of the oxidation tube. Tubes operated with no endcaps or with loose-fitting caps may have enough leakage to exceed the numbers just given. When HCl is used in the system, small leaks can let in enough moisture to allow the HCl to corrode metal plumbing and transport the metallic chlorides into the furnace. Thus, other chlorine-containing gases are often used during oxidation instead of HCl.

TABLE 3.14

Wafer Contamination from Impurity Transfer through Tube Walls

Impurity	*Contamination Level (atoms/cc)*
Cu	1.4×10^{14}
Fe	1.4×10^{14}
Ni	1.3×10^{14}
Cr	3.2×10^{13}
Zr	2.0×10^{14}
Ta	6.0×10^{13}

Source: Data from Paul F. Schmidt, *J. Electrochem. Soc. 130*, pp. 196–199, 1983.

[23]Therefore, tubes should never be handled with bare hands. Enough sodium from the body will transfer where the fingers touch to cause devitrification and indeed will often produce identifiable fingerprint devitrification patterns.

3.7.3 High-Pressure Oxidation Processing

A major processing disadvantage of atmospheric pressure oxidation is the time required at high temperature to grow the thick oxides used, for example, in the oxide isolation described in the next section. From Fig. 3.27, note that a 1 μm thick oxide grown in steam on a (100) surface requires approximately 2 hours at 1100°C, 5 hours at 1000°C, or 15 hours at 900°C. The desire to minimize cycle time leads to higher-temperature oxidations. In turn, more diffusion of impurities that may already have been diffused into the wafer results, the incidence and size of stacking faults increases, and the likelihood of slip occurring during heatup or cooldown also increases.

To provide reduced time and/or temperature during oxidation and minimize the effects just discussed, high-pressure oxidation can be used. The oxidation time for a fixed thickness at a given temperature will be inversely proportional to the pressure. Thus, the 1 μm layer previously alluded to would take only 3 hours to grow at 5 atm pressure or 1.5 hours at 10 atm. Alternatively, for a given oxidation time, the temperature can be reduced about 30°C for each atmosphere of pressure increase. Again, referring to the earlier example, if 5 hours is an acceptable time, by using 10 atm, the growth temperature could be dropped from 1000°C to about 700°C.

3.7.4 High-Pressure Equipment

Early high-pressure oxidation studies, beginning in 1959, were made using a bomb-type apparatus in which the slices and a measured amount of water were sealed in a heavy-walled metal container and heated to the desired oxidation temperature (1, 2, 81). The amount of water included would then determine the operating pressure. The attainable pressure can be quite high, and studies up to 500 atm have been made. More recent versions of the bomb approach have used externally connected high-pressure gases to supply the internal pressure (124). By using this approach, dry as well as wet oxidations can be performed. Such equipment is cumbersome, however, and not very amenable to production usage. To achieve the rates and temperatures currently deemed appropriate for production, pressures of only 10–20 atm are required. By using a fairly conventionally designed oxidation furnace inside a cold-walled high-pressure container, much more satisfactory production equipment for this pressure range is obtained (105, 125–128). An alternative method, suitable for a few atmospheres, is to use a thick-walled fused silica tube in an ordinary furnace (129). Extra safety precautions should be taken since mishandling of the tube can introduce surface damage that will reduce the breaking strength. For dry oxidation, high-pressure oxygen can be introduced directly into the high-pressure chamber. For steam oxidation, an external high-pressure steam generator or pyrogenic steam internally formed can be used. Pyrogenic steam is currently the most common source.

3.7.5 Selective Oxidation

When a pn junction is used for component isolation, as shown in Fig. 3.34a, the heavy doping near the surface allows only a very narrow depletion layer to form and hence forms a high-capacitance junction. Such high capacitance at bipolar collector–base junctions, MOS source and drain junctions, and CMOS well isolation junctions can seriously degrade high-frequency performance. On the bottom of the diffusion, where the concentration is lighter and the space charge region much wider, the total capacitance is usually less than it is on the periphery, even though the area may be greater. If a sidewall oxide isolation, as shown in Fig. 3.34b, can be achieved, device performance will improve. The process used in this case is usually referred to as LOCOS (local oxidation of silicon) (130) or dielectric isolation.[24] It depends on the fact that silicon nitride is impervious to oxygen diffusion and can be used as a mask against oxidation. Thus, if a bare silicon wafer is covered with a nitride layer that then has openings etched in it, a subsequent oxidation step will grow oxide only where the nitride is removed.

Silicon nitride is assumed to oxidize by one or both of the following reactions:

$$Si_3N_4 + 6H_2O \rightarrow 3SiO_2 + 6H_2 + 2N_2$$
$$Si_3N_4 + 6H_2O \rightarrow 3SiO_2 + 4NH_3$$

Examples are shown in Fig. 3.35. The rate is slow enough for nitride to be a very practical masking material. In order to speed up the silicon oxidation cycle at low temperatures where little dopant diffusion occurs, high pressure is often used. At pressures of about 150 atm, the nitride oxidizes too rapidly to be practical, but at the lower

FIGURE 3.34

Two methods of isolating components.

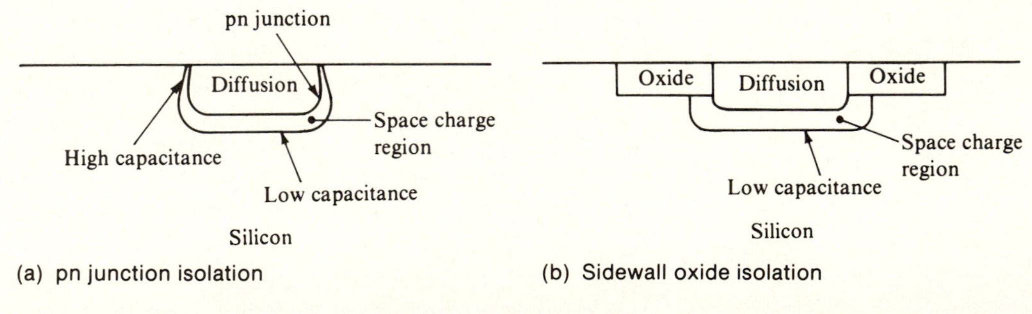

[24]Dielectric isolation is somewhat of a misnomer since the term usually refers to a completely different process that provides for full oxide isolation.

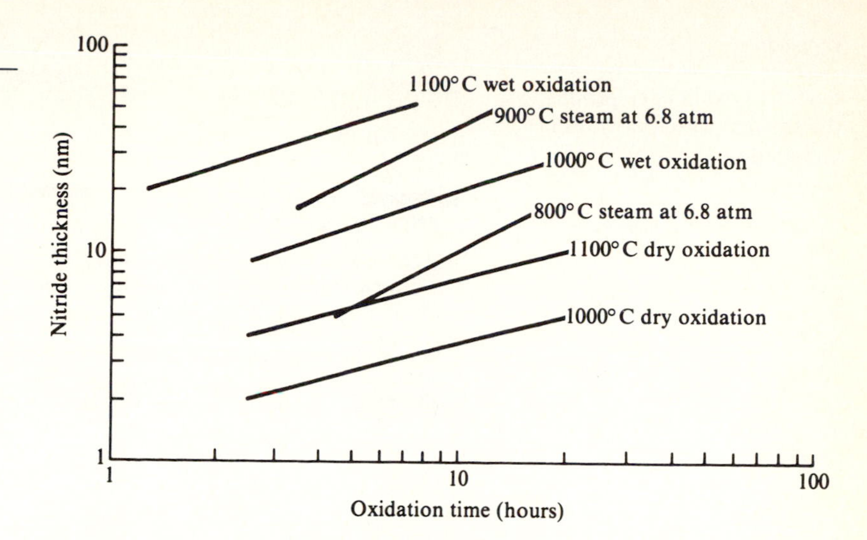

FIGURE 3.35

Amount of silicon nitride oxidized under various conditions. (Nitride used in high-pressure study was deposited at 800°C; the other was annealed at 1200°C after deposition.) (*Source:* Atmospheric data from I. Franz and W. Langheinrich, *Solid State Electronics 14,* p. 499, 1971; high-pressure data from H. Miyoshi et al., *J. Electrochem. Soc. 125,* p. 1824, 1978.)

pressures typically used, the ratio of silicon to nitride oxidation rates is close to that at atmospheric pressure (131–132).

Because of the high stress at the nitride–silicon interface, dislocations will often be generated in the silicon at the corners of windows during oxidation (133–136). To prevent this problem, a thin layer of thermal oxide (pad or buffer oxide) is generally used between the nitride and the silicon. Unfortunately, oxygen can diffuse through this oxide along the interface and cause additional oxide growth under the nitride, giving rise to the enhanced "birdbeak" structures shown in Fig. 3.36. Reducing the pad oxide thickness reduces the length of the birdbeak (136). Since the oxide is thicker than the silicon it consumes, a recess is often first etched in the silicon so that after oxidation the surface is approximately flat. The recess changes the character of the oxide profile as is shown in Fig. 3.36c. Minimizing the effect hinges on an optimum choice of recess depth, nitride thickness, and pad oxide thickness. In general, thin pads help but may lead to more crystal defects. Typically, thicknesses of 300–500 Å are used. Thick nitride causes more defects, but a minimum thickness is dictated by the amount of nitride oxidized during the silicon oxidation cycle. Typical thicknesses of 1000–2000 Å are used. By using an additional nitride sidewall protection layer, the enhanced birdbeak can be largely eliminated (135, 137–138), as shown in Fig. 3.36d. However, additional defects may be introduced into the silicon with such a process (139).

In addition to crystal defects and an exaggerated birdbeak profile, a "white ribbon" is also sometimes seen just inside the thick oxide frame after a MOS gate oxidation. It is caused by a very narrow band of nitride that prevents oxidation. The origin of the band

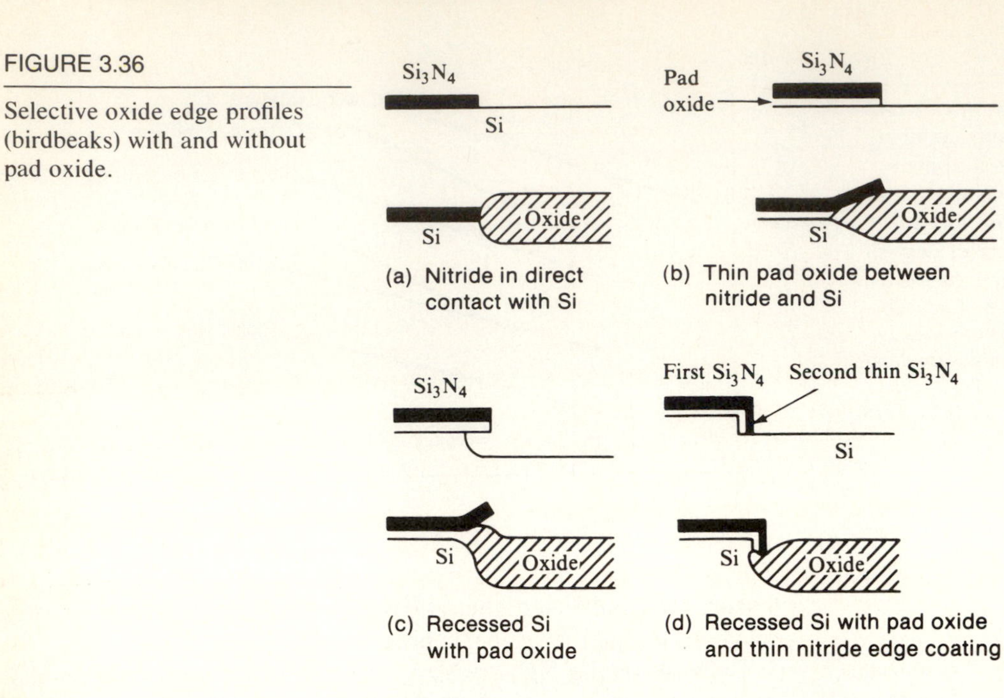

FIGURE 3.36

Selective oxide edge profiles (birdbeaks) with and without pad oxide.

has been ascribed to a transport of nitrogen from the masking nitride to the silicon–oxide interface (140). Thus, if the nitride is stripped and the pad oxide is removed in dilute HF, the nitride at the oxide–silicon interface remains unless substantial overetching occurs during the oxide removal step. Alternatively, the pad oxide removal could be followed by a subsequent nitride etch. However, such a second nitride removal step is seldom used. The growth and subsequent removal of a sacrificial oxide to remove the nitride is a more common step.

3.7.6 Two- and Three-Dimensional Oxidation

When a silicon wafer is uniformly oxidized, the oxidation process can be described by means of one-dimensional equations such as Eq. 3.15. When holes are cut in a uniformly thick oxide and then reoxidized; when local areas are initially oxidized, as was described in the preceding section; and when trenches (grooves) are being oxidized for either trench capacitors or trench isolation, the geometries are such that at least two- and sometimes three-dimensional oxidation solutions are required to fully describe the behavior. During local reoxidation, which occurs, for example, when a gate oxide is grown, an observable thickness variation near the thick/thin step is at least partially described by a two-dimensional analysis (141). During birdbeak development, oxygen must diffuse laterally along the pad oxide and react with silicon beneath the oxide (131, 142). When protuberances or the walls of small holes or grooves are oxidizing, two- or three-dimensional solutions are required if the oxide thickness is more than perhaps 25% of the initial radius (143). Much of the difference between one- and two- or three-dimensional solutions arises because of the difference in area between growing surface and the surface presented to the ambient. In addition, growth in confined space usually increases stress and cannot be handled in one dimension.

3.7.7 Oxide Physical Defects

In addition to the various ionic and interface electrical charges, which are defects in their own right, a variety of other defects exist as well. Pinholes in the oxide are generally introduced during lithographic steps, but they can also be produced during oxide growth wherever nonvolatile particulates are attached to the wafer. Even if a particulate does not produce a discernible hole, it may well produce a region in the oxide that has a lower breakdown strength than the main body of the oxide (144) (as was discussed in an earlier section). Pinholes in a previous oxidation can allow small heavily doped regions to occur in unwanted positions during a subsequent diffusion. During the next oxidation, these areas will oxidize more rapidly and produce oxide "spots." These defects are not a result of any misprocessing during thermal oxidation.

Heavy sodium contamination, followed by long anneals, will produce devitrification (145). Oxide defects that look like small flowers can be formed during subsequent deposition steps if local high concentration of the dopant oxides form separate phases. Also, long anneals after a high-concentration phosphorus diffusion have been shown to promote local SiO_2 crystallization (146). Oxides grown on heavily boron-doped wafers have small separate phase regions identified as amorphous B_2O_3, and high-temperature oxidations have some occluded silicon regions of perhaps 0.05 μm diameter (100). During oxidations in the 1100°C–1200°C range in atmospheres containing 10% HCl, small regions containing a separate, chlorine-rich phase have been reported.[25] After several-hour oxidations, bubbles in the oxide sometimes occur and can be severe enough to cause oxide–silicon separation (147–148). The separation is apparently caused by a chlorine gas buildup at the interface.

Heavy metals, as discussed in the next section, lead to crystallographic defects in the silicon wafer. In addition, however, metal precipitates that form near the silicon surface may extend into the oxide and cause a premature breakdown of MOS gate oxides, apparently because of an electric field enhancement from the small,

[25]To minimize this problem, the normal concentration range is from 2% to 4%.

rod-like precipitates. Both iron and copper have been shown to cause this problem (149).

3.7.8 Defects Generated in Silicon during Oxidation

Two kinds of crystallographic defects commonly occur in silicon wafers after oxidation. First, saucer pits (S-pits) (150) that have densities in the 10^4–10^6/cm^2 range may occur. Second, oxidation-induced stacking faults (OSF or OISF) may occur, and although they occur at a much lower density, they are more deleterious to device yield.

S-pits, also sometimes referred to as haze or fog, can be seen after the oxide is stripped and the wafer is given a light etch in a delineant such as Wright etch.[26] Their presence is a sign of furnace or wafer contamination with a precipitate-forming metal. For a given amount of contamination, the haze density will vary with the furnace removal rate. Slow rates will allow more and larger precipitates to grow. Thus, there is a conflict in processing requirements. Slip minimization requires a slow cooldown cycle, while S-pit suppression requires rapid cooldown. Cu, Ni, S, Ca, Fe, K, and Zn all have been reported as possible S-pit initiators (151). A small portion of S-pit defects have, in turn, been shown to initiate the more damaging oxidation-induced stacking faults (152). Methods to minimize or eliminate S-pits and stacking faults are to keep the wafer and furnace as free as possible of contaminants and to use heavy phosphorus doping, mechanical damage, misfit dislocation, or oxygen precipitate intrinsic gettering to trap the metals (32, 154).

Stacking faults are more likely to occur in (100) material than in (111), and n-type (100) material has more stacking faults than p-type. It has been suggested that the higher density of bulk defects in the p-type (100) material provides for more internal gettering (32). High-temperature oxidations favor a high fault density. By the use of high-pressure oxidation, the temperature can be lowered, and as a consequence, the OSF density may be lowered as well (155–156). The growth of the faults is fed by the supply of silicon interstitials generated during oxidation. The experimentally observed fault length L_{sf} is described by Eq. 3.34 as follows:

$$L_{sf} = Kt^{2/3}\, e^{-2.37\text{eV}/kT} \qquad 3.34$$

where K is a constant and t is the oxidation time (151). Thus, lowering the temperature or shortening the oxidation time minimizes the size of the faults.

[26] For the formulation of this and other suitable etchants, see Chapter 6.

CHAPTER 3 KEY IDEAS

- □ Despite the fact that the bulk chemical, physical, and electrical properties of vapor-deposited SiO_2 are close to those of thermally grown SiO_2, the electrical properties of the oxide–silicon interface are substantially different, and, thus far, only the thermal oxide is adequate for pn junction passivation and MOS transistor gates.
- □ Under the influence of an electric field, sodium, potassium, and a few other ions are particularly mobile in SiO_2.
- □ If mobile ions such as sodium are present in the oxide, the electrical fields present in devices during normal operation are often enough to cause the ions to migrate to the silicon–oxide interface and adversely affect device performance.
- □ For oxide growth using water vapor as the oxidant or when O_2 is the oxidant and oxides more than a few hundred Å thick are grown, the Deal–Grove model is quite satisfactory.
- □ For dry oxides thinner than a few hundred Å, marked deviations from the Deal–Grove model occur, and several alternatives have been proposed.
- □ The oxidation rate depends on the wafer surface orientation, and (111) surfaces oxidize more rapidly than (100) surfaces.
- □ Heavily doped silicon oxidizes more rapidly than does high-resistivity silicon.
- □ The most common defects introduced into silicon during oxidation arise from metallic contamination either on the surface or in the silicon bulk.

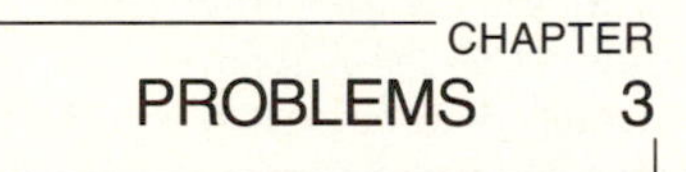

CHAPTER 3 PROBLEMS

1. If a (111) silicon wafer is oxidized, how much silicon is consumed while 3000 Å of thermal oxide is grown? How much is consumed if the orientation is (100)?
2. If a sidewall oxide isolated transistor (Fig. 3.34b) is to be made with an isolating oxide thickness of 1.2 μm, how deep should the depression in the silicon be if the oxide surface is to be level with the main silicon surface? Draw a sketch to show the steps.
3. A (100) 150 mm diameter 600 μm thick silicon wafer has 1 μm of oxide thermally grown on one side. Calculate the center deflection of the wafer if the stress in the oxide is 2×10^9 dynes/cm^2. Note that neither Young's modulus E nor Poisson's ratio ν for silicon is constant for all directions lying in a (100) plane. For purposes of this problem, use values of $E = 1.5 \times 10^{12}$ dynes/cm^2 and $\nu = 0.2$. (For more details on the anisotropy of these properties, see J.J. Wortman and R.A. Evans, *J. Appl. Phys. 36*, p. 153, 1965.)
4. Calculate the capacitance of a MOS gate that has a cross section of 2 μm by 2μm if the gate oxide is 400 Å thick and there are no spurious interface or oxide charges.
5. If a MOS transistor is contaminated with potassium, how much, in atoms/cm^2, is present at the oxide–silicon interface if the threshold voltage has changed by -0.2 V? Assume that the gate oxide is 400 Å thick.
6. Using the data from Fig. 3.28, determine A and B for a 1100°C wet oxidation of a (100) wafer. Are the numbers consistent with the data of Table 3.7?
7. Using the data from Fig. 3.27, determine A, B, and τ for a 1000°C dry oxidation of a (100) wafer.

8. If the only uncontrolled variable in an oxidation is the moisture content of the oxygen used in a "dry" oxidation, estimate the maximum amount of water vapor that would be tolerable if a nominal 1200 Å thick oxide is to be reproducible to within 10%.
9. Calculate the time required to grow a 1 μm thick layer of dry oxide at 1100°C on a wafer doped with 2×10^{20} atoms/cc of phosphorus.
10. If 10 μm by 10 μm holes are cut in a 1 μm thick oxide grown on a (100) high-resistivity silicon wafer, how thick will the initially 1 μm thick oxide be by the time 1500 Å has been regrown in the windows? The second oxide is grown at 1000°C in dry O_2.

CHAPTER 3 REFERENCES

1. M.M. Atalla, "Semiconductor Surfaces and Films; the Silicon–Silicon Dioxide System," in Harry C. Gatos, ed., *Properties of Elemental and Compound Semiconductors*, Interscience Publishers, New York, 1959.
2. J.R. Ligenza and W.G. Spitzer, "The Mechanism for Silicon Oxidation in Steam and Oxygen," *J. Phys. Chem. Solids 14*, p. 131, 1960.
3. W.A. Pliskin and R.P. Gnall, "Evidence for Oxidation Growth at the Oxide–Silicon Interface from Controlled Etch Studies," *J. Electrochem. Soc. 111*, pp. 872–873, 1964.
4. W.H. Zachariasen, "The Atomic Arrangement in Glass," *J. Am. Chem. Soc. 54*, pp. 3841–3851, 1932.
5. S.A. Schwarz and M.J. Schulz, Chap. 2, "Characterization of the Si–SiO_2 Interface," in Norman G. Einspruch and Robert S. Bauer, eds., *VLSI Microelectronics Microstructure Science*, Vol. 10, Academic Press, New York, 1985.
6. J.R. Patel and Norio Kato, "X-Ray Diffraction Topographs of Silicon Crystals with Superposed Oxide Films II. Pendellosung Fringes: Comparison of Experiment with Theory," *J. Appl. Phys. 44*, pp. 971–977, 1977.
7. E.P. EerNisse, "Stress in Thermal SiO_2 during Growth," *Appl. Phys. Lett. 35*, pp. 8–10, 1979.
8. M.V. Whelan et al., "Residual Stresses at an Oxide–Silicon Interface," *Appl. Phys. Lett. 10*, pp. 262–264, 1967.
9. E.A. Irene et al., "A Viscous Flow Model To Explain the Appearance of a High Density Thermal Oxide at Low Oxidation Temperatures," *J. Electrochem. Soc. 129*, pp. 2594–2597, 1982.
10. E.A. Taft, "The Optical Constants of Silicon and Dry Oxygen Oxides of Silicon at 5461 Å," *J. Electrochem. Soc. 125*, pp. 968–971, 1978.
11. E.A. Irene et al., "Residual Stress, Chemical Etch Rate, Refractive Index, and Density Measurements on SiO_2 Films Prepared Using High Pressure Oxygen," *J. Electrochem. Soc. 127*, pp. 396–399, 1980.
12. R.J. Jaccodine and W.A. Schlegel, "Measurement of Strains at the Si–SiO_2 Interface," *J. Appl. Phys. 37*, pp. 2429–2434, 1966.
13. R.Glang et al., "Determination of Stress in Films on Single Crystal Silicon Substrates," *Rev. Sci. Inst. 36*, pp. 7–10, 1965.
14. Carlo Grilletto, "An X-Ray Fluorescence Technique for the Rapid Determination of Phosphorus in PSG Film," *Solid State Technology*, pp. 27–30, February 1977.
15. W.A. Pliskin, "Use of Infrared Spectroscopy for Characterization of Dielectric Films on Silicon," in Howard R. Huff and Ronald R. Burgess, eds., *Semiconductor Silicon/73*, Electrochemical Society, Princeton, N.J., 1973.
16. C.Y. Ting et al., "Interaction between Ti and SiO_2," in Kenneth E. Bean and George A. Rozgonyi, eds., *VLSI Science and Technology/84*, Electrochemical Society, Princeton, N.J., 1984.
17. D.J. Silversmith, "Dielectric Loss, Electrode Deterioration, and the Sodium Problem in MOS Structures," *J. Electrochem. Soc. 119*, pp. 121–122 and earlier references therein, 1972.
18. Judith A. Nemetz and Richard E. Tressler, "Thermal Nitridation of Silicon and Silicon

Dioxide for Thin Gate Insulators," *Solid State Technology*, pp. 209–216 and references therein, September 1983.

19. R.P. Vasquez and A. Madhukar, "A Kinetic Model for the Thermal Nitridation of SiO_2/Si," *J. Appl. Phys. 60*, pp. 234–242, 1986.
20. M. Lenzlinger and E.H. Snow, "Fowler–Nordheim Tunneling into Thermally Grown SiO_2," *J. Appl. Phys. 40*, pp. 278–283, 1969.
21. C.M. Osburn and E.J. Weitzman, "Electrical Conduction and Dielectric Breakdown in Silicon Dioxide Films on Silicon," *J. Electrochem. Soc. 119*, pp. 603–609, 1972.
22. N.J. Chou, "D.C. Conduction in SiO_2 Films at Elevated Temperatures," *J. Elect. Mat. 1*, pp. 344–347, 1972.
23. Yu-Pin Han et al., "Ultra-Thin Gate–Oxide Characteristics and MOS/VLSI Scaling Implications," *IEDM*, pp. 98–102, 1982.
24. Chi Chang et al., "Charge Tunneling and Impact on Thin Oxide Device Reliability," *Electrochem. Soc. Ext. Abst. 83–1*, abst. no. 394, 1983.
25. J.K. Srivastava et al., "Electrical Conductivity of Silicon Dioxide Thermally Grown on Silicon," *J. Electrochem. Soc. 132*, pp. 955–963, 1985.
26. W. Shockley et al., "Mobile Electric Charge on Insulating Oxides with Application to Oxide Covered P–N Junctions," *Surface Science 2*, pp. 277–287, 1964.
27. P.S.D. Lin et al., "Adverse Effects of Substrate Stacking Faults on Leakage and Breakdown Strength of Thin Oxide Capacitors," *Electrochem. Soc. Ext. Abst. 82–2*, abst. no. 215, 1982.
28. C.M. Osburn and D.W. Ormond, "Sodium Induced Barrier Height Lowering and Dielectric Breakdown on SiO_2 Films on Silicon," *J. Electrochem. Soc. 121*, pp. 1195–1198, 1974.
29. C.M. Osburn and D.W. Ormond, "Dielectric Breakdown in Silicon Dioxide Films on Silicon I: Measurement and Interpretation," *J. Electrochem. Soc. 119*, pp. 591–597, 1972.
30. C.M. Osburn and D.W. Ormond, "Dielectric Breakdown in Silicon Dioxide Films on Silicon II: Influence of Processing and Materials," *J. Electrochem. Soc. 119*, pp. 597–603, 1972.
31. M. Sternheim et al., "Properties of Thermal Oxides Grown on Phosphorus In-Situ Doped Polysilicon," *J. Electrochem. Soc. 130*, pp. 1735–1739, 1983.
32. P.W. Koob et al., "Reduction of Surface Stacking Faults on N-Type (100) Silicon Wafers," *J. Electrochem. Soc. 133*, pp. 806–810, 1986.
33. S. Iwamatsu, "Effects of Plasma Cleaning on the Dielectric Breakdown in SiO_2 Film on Si," *J. Electrochem. Soc. 129*, pp. 224–225, 1982.
34. C.M. Osburn, "Dielectric Breakdown Properties of SiO_2 Films Grown in Halogen and Hydrogen-Containing Environments," *J. Electrochem. Soc. 121*, pp. 809–815, 1974.
35. Richard G. Cosway, "Effect of Oxidation Ambient on the Electrical Properties of Thin Silicon Dioxide," *J. Electrochem. Soc. 132*, pp. 3052–3054, 1985.
36. C.M. Osburn and N.J. Chou, "Accelerated Dielectric Breakdown of Silicon Dioxide Films," *J. Electrochem. Soc. 120*, pp. 1377–1384, 1973.
37. S. Holland et al., "Time-Dependent Breakdown of Thin Oxides," *Electrochem. Soc. Ext. Abst. 86–2*, abst. no. 396, 1986.
38. J.J. Tzou et al., "Charge Generation and Breakdown in SiO_2," *Electrochem. Soc. Ext. Abst. 86–2*, abst. no. 397, 1986.
39. T.S. Taylor et al., "Wearout of Thin Silicon Dioxide Layers due to Trap Generation Associated with Tunneling Currents," *Electrochem. Soc. Ext. Abst. 86–2*, abst. no. 398, 1986.
40. D.L. Crook, "Method of Determining Reliability Screens for Time-Dependent Dielectric Breakdown," *Proc. Reliability and Physics of Failure Symposium*, pp. 1–7, 1979.
41. J.W. McPherson and D.A. Baglee, "Acceleration Factors for Thin Oxide Breakdown," *J. Electrochem. Soc. 132*, pp. 1903–1908, 1985.
42. E.H. Snow et al., "Ion Transport Phenomena in Insulating Films," *J. Appl. Phys. 36*, pp. 1664–1673, 1965.
43. Bruce E. Deal, "The Current Understanding of Charges in the Thermally Oxidized Silicon Structure," *J. Electrochem. Soc. 121*, pp. 198C–205C, 1974.
44. G.F. Derbenwick, "Mobile Ions in SiO_2: Potassium," *J. Appl. Phys. 48*, pp. 1127–1130, 1977.
45. S.R. Hofstein, "An Investigation of Instability and Charge Motion in Metal–Silicon Oxide–Silicon Structures," *IEEE Trans. on Electron Dev. ED–13*, pp. 222–227, 1966.
46. M. Kuhn and D.J. Silversmith, "Ionic Contami-

nation and Transport of Mobile Ions in MOS Structures," *J. Electrochem. Soc. 118*, pp. 966–970, 1971.
47. J.M. Eldridge and D.R. Kerr, "Sodium Ion Drift through Phosphosilicate Glass–SiO_2 Films," *J. Electrochem. Soc. 118*, pp. 986–991, 1971.
48. T.W. Hickmott, "Thermally Stimulated Ionic Conductivity of Sodium in Thermal SiO_2," *J. Appl. Phys. 46*, pp. 2583–2595, 1975.
49. D.R. Kerr et al., "Stabilization of SiO_2 Passivation Layers with P_2O_5," *IBM J. of Res. Dev. 8*, pp. 376–384, 1964.
50. P. Balk and J.M. Eldridge, "Phosphosilicate Glass Stabilization of FET Devices," *Proc. IEEE 57*, pp. 1558–1563, 1969.
51. (a) R.J. Kriegler et al., "The Effect of HCl and Cl_2 on the Thermal Oxidation of Silicon," *J. Electrochem. Soc. 119*, pp. 388–392, 1972. (b) P.H. Robinson and F.P. Heiman, "Use of HCl Gettering in Silicon Device Processing," *J. Electrochem. Soc. 118*, pp. 141–143, 1971.
52. References in J. Monkowski, "Role of Chlorine in Silicon Oxidation: Part I," *Solid State Technology*, July 1979; "Part II," August 1979.
53. Mao-Chieh Chen and John W. Hile, "Oxide Charge Reduction by Chemical Gettering with Trichloroethylene during Thermal Oxidation," *J. Electrochem. Soc. 119*, pp. 223–225, 1972.
54. Edmond J. Janssens and Gilbert J. Declerck, "The Use of 1,1,1,-Trichloroethane as an Optimized Additive To Improve the Silicon Thermal Oxidation Technology," *J. Electrochem. Soc. 125*, pp. 1696–1703, 1978.
55. D.W. Hess and R.C. McDonald, "Investigation of Silicon Etching and Silicon Dioxide Bubble Formation during Silicon Oxidation in HCl–Oxygen Atmospheres," *Thin Solid Films 42*, pp. 127–131, 1977.
56. Ronald L. Meek, "Residual Chlorine in O_2:HCl Grown SiO_2," *J. Electrochem. Soc. 120*, pp. 308–310, 1973.
57. Reda R. Razouk and Bruce E. Deal, "Dependence of Interface State Density on Silicon Oxidation Process Variables," *J. Electrochem. Soc. 126*, pp. 1573–1581, 1979.
58. A.S. Grove, *Physics and Technology of Semiconductor Devices*, John Wiley & Sons, New York, 1967.
59. P. Balk et al., "Orientation Dependence of Built-In Surface Charge on Thermally Oxidized Silicon," *Trans. IEEE 53*, pp. 2133–2134 (L), 1965.
60. B.E. Deal et al., "Characteristics of the Surface-State Charge (Q_{ss}) of Thermally Oxidized Silicon," *J. Electrochem. Soc. 114*, pp. 266–274, 1967.
61. (a) P. Balk, "Effect of Hydrogen Annealing on Silicon Surfaces," *Electrochem. Soc. Ext. Abst. 14* (1), abst. no. 109, 1965. (b) P. Balk, "Low-Temperature Annealing in the Al–SiO_2–Si System," *Electrochem. Soc. Ext. Abst. 14* (2), abst. no. 111, 1965.
62. B.E. Deal, E.L. MacKenna, and P.L. Castro, "Characteristics of Fast Surface States Associated with SiO_2–Si and Si_3N_4–Si Structures," *J. Electrochem. Soc. 116*, pp. 997–1005, 1969.
63. B.E. Deal and A.S. Grove, "General Relationship for the Thermal Oxidation of Silicon," *J. Appl. Phys. 36*, pp. 3770–3778, 1965.
64. J.A. Costello and R.E. Tressler, "Isotope Labeling Studies of the Oxidation of Silicon at 1000°C and 1300°C," *J. Electrochem. Soc. 131*, pp. 1944–1947, 1984.
65. A.G. Revesz et al., "Structure of SiO_2 Films as Revealed by Oxygen Transport," *J. Electrochem. Soc. 133*, pp. 586–592, 1986.
66. A.C. Adams et al., "The Growth and Characterization of Very Thin Silicon Dioxide Films," *J. Electrochem. Soc. 127*, pp. 1787–1793, 1980.
67. William A. Tiller, "On the Kinetics of the Thermal Oxidation of Silicon IV: The Two-Layer Film Approximation," *J. Electrochem. Soc. 130*, pp. 501–506, 1983.
68. A. Fargeix and G. Ghibaudo, "Dry Oxidation of Silicon: A New Model of Growth Including Relaxation of Stress by Viscous Flow," *J. Appl. Phys. 54*, pp. 7153–7158, 1983.
69. J.K. Srivastava and E.A. Irene, "Measurement of the Effect of Intrinsic Film Stress on the Overall Rate of Thermal Oxidation of Silicon," *J. Electrochem. Soc. 132*, pp. 2815–2816, 1985.
70. Eugene A. Irene et al., "Silicon Oxidation Studies: Orientation Effects on Thermal Oxidation," *J. Electrochem. Soc. 133*, pp. 1253–1256, 1986.
71. J. Blanc, "A Revised Model for the Oxidation of Si by Oxygen," *Appl. Phys. Lett. 33*, pp. 424–426, 1978.
72. S.M. Hu, "Thermal Oxidation of Silicon: Chem-

isorption and Linear Rate Constants," *J. Appl. Phys. 55*, pp. 4095–4105, 1984.

73. C.J. Han and C.R. Helms, "A Physical Model for Si Oxidation Kinetics in the Thickness Range from 30 Å to 1 μm," *Electrochem. Soc. Ext. Abst. 86–1*, abst. no. 157, 1986.
74. Joseph Blanc, "On Modeling the Oxidation of Silicon by Dry Oxygen," *J. Electrochem. Soc. 133*, pp. 1981–1982, 1986.
75. Hisham Z. Massoud et al., "Thermal Oxidation of Silicon in Dry Oxygen Growth-Rate Enhancement in the Thin Regime: Parts I and II," *J. Electrochem. Soc. 132*, pp. 2685–2700, 1985.
76. H.Z. Massoud, C.P. Ho, and J.D. Plummer, Stanford University Tech. Rpt., in J.D. Plummer, ed., *Computer Aided Design of Integrated Circuit Fabrication Processes for VLSI Devices*, 1982.
77. J.D. Plummer, "The Role of the Si/SiO_2 Interface in Silicon Oxidation Kinetics," in Howard R. Huff and Rudolph J. Kriegler, eds., *Semiconductor Silicon/81*, Electrochemical Society, Pennington, N.J., 1981.
78. E.A. Irene and Y.J. van der Meulen, "Silicon Oxidation Studies: Analysis of SiO_2 Film Growth Data," *J. Electrochem. Soc. 23*, pp. 1380–1384, 1976.
79. M.A. Hopper et al., "Thermal Oxidation of Silicon," *J. Electrochem. Soc. 122*, pp. 1216–1222, 1975.
80. Hisham Z. Massoud, James D. Plummer, and Eugene A. Irene, "Thermal Oxidation of Silicon in Dry Oxygen—Accurate Determination of the Kinetic Rate Constants," *J. Electrochem. Soc. 132*, pp. 1745–1753, 1985.
81. Joseph R. Ligenza, "Effect of Crystal Orientation on Oxidation Rates of Silicon in High Pressure Steam," *J. Phys. Chem. 65*, pp. 2011–2014, 1961.
82. E.A. Irene, "The Effects of Trace Amounts of Water on the Thermal Oxidation of Silicon in Oxygen," *J. Electrochem. Soc. 121*, pp. 1613–1616, 1974.
83. Hideo Sunami, "Thermal Oxidation of Phosphorus-Doped Polycrystalline Silicon in Wet Oxygen," *J. Electrochem. Soc. 125*, pp. 892–897, 1978.
84. Eugene A. Irene et al., "Silicon Oxidation Studies: Silicon Orientation Effects on Thermal Oxidation," *J. Electrochem. Soc. 133*, pp. 1253–1256, 1986.
85. S.I. Raider and L.E. Forget, "Reversal of Relative Oxidation Rates of <111> and <100> Oriented Silicon Substrates at Low Oxygen Partial Pressure," *J. Electrochem. Soc. 127*, pp. 1783–1787, 1980.
86. M. Sternheim et al., "Properties of Thermal Oxides Grown on Phosphorus In-Situ Doped Polysilicon," *J. Electrochem. Soc. 130*, pp. 1735–1740, 1983.
87. E.A. Irene, E. Tierney, and D.W. Dong, "Silicon Oxidation Studies: Morphological Aspects of the Oxidation of Polycrystalline Silicon," *J. Electrochem. Soc. 127*, pp. 705–713, 1980.
88. L. Faraone and G. Harbeke, "Surface Roughness and Electrical Conduction of Oxide/Polysilicon Interfaces," *J. Electrochem. Soc. 133*, pp. 1410–1413, 1986.
89. Hideo Sunami, "Thermal Oxidation of Phosphorus-Doped Polycrystalline Silicon in Wet Oxygen," *J. Electrochem. Soc. 125*, pp. 892–897, 1978.
90. C.Y. Lu and N.S. Tsai, "Thermal Oxidation of Undoped LPCVD Polycrystalline Silicon Films," *J. Electrochem. Soc. 133*, pp. 446–447, 1986.
91. C.Y. Lu and N.S. Tsai, "Thermal Oxidation of Heavily Boron Implanted Polycrystalline Silicon Films," *J. Appl. Phys. 59*, pp. 3574–3576, 1986.
92. Reda R. Razouk et al., "Kinetics of High Pressure Oxidation of Silicon in Pyrogenic Steam," *J. Electrochem. Soc. 128*, pp. 2214–2220, 1981.
93. Liang N. Lie et al., "High Pressure Oxidation of Silicon in Dry Oxygen," *J. Electrochem. Soc. 129*, pp. 2828–2834, 1982.
94. C. Camelin and G. Demazeau, "High Pressure Dry Oxidation Kinetics of Silicon—Evidence of a Highly Stressed Structure," *Appl. Phys. Lett. 48*, pp. 1211–1213, 1986.
95. E.A. Irene and R. Ghez, "Silicon Oxidation Studies: The Role of H_2O," *J. Electrochem. Soc. 124*, pp. 1757–1761, 1977.
96. A.G. Revesz and R.J. Evans, "Kinetics and Mechanism of Thermal Oxidation of Silicon with Special Emphasis on Impurity Effects," *J. Phys. Chem. Solids 30*, pp. 551–564, 1969.
97. M. Morita et al., "Fluorine Enhanced Thermal

Oxidation of Silicon in the Presence of NF_3," *Appl. Phys. Lett. 45*, pp. 1312–1314, 1984.

98. B.E. Deal and M. Sclar, "Thermal Oxidation of Heavily Doped Silicon," *J. Electrochem. Soc. 112*, pp. 430–435, 1965.
99. C.P. Ho et al., "Thermal Oxidation of Heavily Phosphorus-Doped Silicon," *J. Electrochem. Soc. 125*, pp. 665–671, 1978.
100. E.A. Irene and D.W. Dong, "Silicon Oxidation Studies: The Oxidation of Heavily B- and P-Doped Single Crystal Silicon," *J. Electrochem. Soc. 125*, pp. 1146–1151, 1978.
101. Charles P. Ho and James D. Plummer, "Improved MOS Device Performance through the Enhanced Oxidation of Heavily Doped n^+ Silicon," *IEEE Trans. on Electron Dev. ED-26*, pp. 623–630, 1979.
102. C.P. Ho and J.D. Plummer, "Si/SiO_2 Interface Oxidation Kinetics: A Physical Model for the Influence of High Substrate Doping Levels. I. Theory, II. Comparison with Experiment and Discussion," *J. Electrochem. Soc. 126*, pp. 1516–1522 and pp. 1523–1530, 1979.
103. Krishna C. Saraswat and Harinder Singh, "Thermal Oxidation of Heavily Phosphorus-Doped Thin Films of Polycrystalline Silicon," *J. Electrochem. Soc. 129*, pp. 2321–2326, 1982.
104. John. J. Barnes et al., "Low Temperature Differential Oxidation for Double Polysilicon VLSI Devices," *J. Electrochem. Soc. 126*, pp. 1779–1785, 1979.
105. Makoto Hirayama et al., "High-Pressure Oxidation for Thin Gate Oxide Insulator Process," *IEEE Trans. on Electron Dev. ED-29*, pp. 503–507, 1982.
106. A.S. Grove et al., "Redistribution of Acceptor and Donor Impurities during Thermal Oxidation of Si," *J. Appl. Phys. 35*, pp. 2695–2707, 1964.
107. B.E. Deal et al., "Observation of Impurity Redistribution during Thermal Oxidation of Silicon Using the MOS Structure," *J. Electrochem. Soc. 112*, pp. 308–314, 1965.
108. A.S. Grove et al., "Diffusion of Gallium through a Silicon Dioxide Layer," *J. Phys. Chem. Solids 25*, pp. 985–992, 1964.
109. Richard B. Fair and J.C.C. Tsai, "Theory and Direct Measurement of Boron Segregation in SiO_2 during Dry, Near Dry, and Wet O_2 Oxidation," *J. Electrochem. Soc. 125*, pp. 2050–2058 and references therein, 1978.
110. P. Deroux-Dauphin and J.P. Gonchond, "The Influence of High Pressure Oxidation on Boron Redistribution in LOCOS Structures," *J. Electrochem. Soc. 131*, pp. 1418–1423, 1984.
111. J.A. Amick, "Cleanliness and the Cleaning of Silicon Wafers," *Solid State Technology*, pp. 47–52, November 1976.
112. F.N. Schwettmann et al., "Variation of Silicon Dioxide Growth Rate with Preoxidation Clean," *Electrochem. Soc. Ext. Abst. 78–1*, abst. no. 276, 1978.
113. Werner Kern and David. A. Puotinen, "Cleaning Solutions Based on Hydrogen Peroxide for Use in Silicon Semiconductor Technology," *RCA Review 31*, pp. 187–206, 1970.
114. A. Mayer and S. Shwartzman, "Megasonic Cleaning: A New Cleaning and Drying System for Use in Semiconductor Processing," *J. Elect. Mat. 8*, p. 885, 1979.
115. F. Pintchovski et al., "Thermal Characteristics of the H_2SO_4–H_2O_2 Silicon Wafer Cleaning Solution," *J. Electrochem. Soc. 126*, pp. 1428–1430, 1979.
116. R.C. Henderson, "Silicon Cleaning with Hydrogen Peroxide Solutions: A High Energy Electron Diffraction and Auger Electron Spectroscopy Study," *J. Electrochem. Soc. 119*, pp. 772–775, 1972.
117. Gary Grant et al., recent news paper presented at Electrochemical Society Meeting, New Orleans, October 1984.
118. J. Nulman, J.P. Krusius, and A. Gat, "Rapid Thermal Processing of Thin Gate Dielectrics—Oxidation of Silicon," *Electron Dev. Lett. EDL-6*, pp. 205–207, 1985.
119. Yoshiyuki Sato and Kazuhide Kiuchi, "Oxidation of Silicon Using Lamp Light Radiation," *J. Electrochem. Soc. 133*, pp. 652–654, 1986.
120. Y. Ota and S.R. Butler, "Reexamination of Some Aspects of Thermal Oxidation of Silicon," *J. Electrochem. Soc. 121*, pp. 1107–1111, 1974.
121. W. Dietze et al., "The Preparation and Properties of CVDSilicon Tubes and Boats for Semiconductor Device Technology," *J. Electrochem. Soc. 121*, pp. 1112–1115, 1974.
122. Santos Mayo and William H. Evens, "Thermodynamic Considerations in the Use of Polysilicon Oxidation Tubes for Clean SiO_2 Preparation," *J. Electrochem. Soc. 125*, pp. 106–110, 1978.

123. Thomas G. Wolfe, Ray Roberge, and George A. Brown, "Effect of Gaseous Growth Ambient Impurities on the Electrical Properties of Gate Oxides," *Electrochem. Soc. Ext. Abst. 85–2*, abst. no. 200, 1985.
124. L.P. Trombetta et al., "Electrical Properties of Silicon Dioxide Films Fabricated at 700°C III: High Pressure Thermal Oxidation," *J. Electrochem. Soc. 132*, pp. 2706–2713, 1985.
125. P.T. Panousis and M. Schneider, "High Pressure Steam Apparatus for the Accelerated Oxidation of Silicon," *Electrochem. Soc. Ext. Abst. 73–1*, abst. no. 53, 1973.
126. Robert Champagne and Monte Toole, "High-Pressure Pyrogenic Oxidation in the Production Environment," *Solid State Technology*, pp. 61–63, December 1977.
127. R.J. Zeto et al., "Pressure Oxidation of Silicon: An Emerging Technology," *Solid State Technology*, pp. 62–69, July 1979.
128. E. Bussmann, "High-Pressure Oxidation in n-Channel MOS Technology," *Semiconductor International*, pp. 162–165, April 1983.
129. Kenneth E. Bean et al., "Semiconductor Processing Facility for Providing Enhanced Oxidation Rate," U.S. Patent 4,599,247, July 8, 1986.
130. E. Kooi and J.A. Appels, "Selective Oxidation of Silicon and Its Device Applications," in Howard R. Huff and Ronald R. Burgess, eds., *Semiconductor Silicon/73*, Electrochemical Society, Princeton, N.J., 1973. (See also the earlier references contained therein.)
131. R.J. Powell et al., "Selective Oxidation of Silicon in Low-Temperature High-Pressure Steam," *IEEE Trans. on Electron Dev. Ed-21*, pp. 636–640, 1974.
132. H. Miyoshi et al., "Selective Oxidation of Silicon in High Pressure Steam," *J. Electrochem. Soc. 125*, pp. 1824–1829, 1978.
133. E. Bassous et al., "Topology of Silicon Structures with Recessed SiO_2," *J. Electrochem. Soc. 123*, pp. 1729–1737, 1976.
134. Kenji Shibata and Kenji Taniguchi, "Generation Mechanism of Dislocations in Local Oxidation of Silicon," *J. Electrochem. Soc. 127*, pp. 1383–1387, 1980.
135. I. Magdo and A. Bohg, "Framed Recessed Oxide Scheme for Dislocation-Free Planar Si Structures," *J. Electrochem. Soc. 125*, pp. 932–936, 1978.
136. Y. Tamaki et al., "Evaluation of Dislocation Generation on Silicon Substrates by Selective Oxidation," *J. Electrochem. Soc. 130*, pp. 2266–2270, 1983.
137. D. Kahng et al., "A Method for Area Saving Planar Isolation Oxides Using Oxidation Protected Sidewalls," *J. Electrochem. Soc. 127*, pp. 2468–2471, 1980.
138. M. Inuishi et al., "Defect-Free Process of a Bird's Beak Reduced LOCOS," *Electrochem. Soc. Ext. Abst. 86–2*, abst. no. 273, 1986.
139. R.C.Y. Fang et al., "Defect Characteristics and Generation Mechanism in a Bird Beak Free Structure by Sidewall Masked Technique," *J. Electrochem. Soc. 130*, pp. 190–196, 1983.
140. E. Kooi et al., "Formation of Silicon Nitride at a Si–SiO_2 Interface During Local Oxidation of Silicon and during Heat Treatment of Oxidized Silicon in NH_3 Gas," *J. Electrochem. Soc. 123*, pp. 1117–1120, 1976.
141. Lynn O. Wilson, "Numerical Simulation of Gate Oxide Thinning in MOS Devices," *J. Electrochem. Soc. 129*, pp. 831–837, 1982.
142. J.C.H. Hui and W.G. Oldham, "Two-Dimensional Oxidation Mechanisms in Local Oxidation Process," *Electrochem. Soc. Ext. Abst. 83–1*, abst. no. 384, 1983.
143. Lynn O. Wilson and R.B. Marcus, "Oxidation of Silicon: Cylinders and Spheres," *Electrochem. Soc. Ext. Abst. 86–2*, abst. no. 547, 1986.
144. K. Yambe et al., "Thickness Dependence of Dielectric Breakdown of Thermal SiO_2 Films," *Electrochem. Soc. Ext. Abst. 83–1*, abst. no. 309, 1983.
145. R.L. Meek and R.H. Braun, "Devitrification of Steam Grown Silicon Dioxide Films," *J. Electrochem. Soc. 119*, pp. 1538–1544, 1972.
146. E.I. Alessandrini and D.R. Campbell, "Catalyzed Crystallization in SiO_2 Thin Films," *J. Electrochem. Soc. 121*, pp. 1115–1118, 1974.
147. J. Monkowski et al., "The Structure and Composition of Silicon Oxides Grown in HCl/O_2 Ambients," *J. Electrochem. Soc. 125*, pp. 1867–1873, 1978.
148. D.W. Hess and R.C. McDonald, "Investigation of Silicon Etching and Silicon Dioxide Bubble Formation during Silicon Oxidation in HCl-Oxygen Atmospheres," *Thin Solid Films 42*, pp. 127–131, 1977.
149. Kouichirou Hondo et al., "Breakdown in Silicon

Oxides (II)—Correlation with Fe Precipitates," *Appl. Phys. Lett. 46*, pp. 582–584 and references therein, 1985.

150. D. Pomerantz, "A Cause and Cure of Stacking Faults in Silicon Epitaxial Layers," *J. Appl. Phys. 38*, pp. 5020–5026, 1967.
151. C.W. Pearce and V.C. Kannan, "Saucer Pit Defects in Silicon," *Electrochem. Soc. Ext. Abst. 83–1*, abst. no. 307, 1983.
152. C.W. Pearce and R.G. McMahon, "Role of Metallic Contamination in the Formation of 'Saucer' Pit Defects in Epitaxial Silicon," *J. Vac. Sci. Technol. 14*, pp. 40–43, 1977.
153. W.T. Stacy et al., "The Role of Metallic Impurities in the Formation of Haze Defects," in Howard R. Huff and Rudolph J. Kriegler, eds., *Semiconductor Silicon/81*, Electrochemical Society, Pennington, N.J., 1981.
154. S.P. Murarka et al., "A Study of Stacking Faults during CMOS Processing: Origin, Elimination, and Contribution to Leakage," *J. Electrochem. Soc. 127*, pp. 716–724, 1980.
155. Makoto Hirayama et al., "High-Pressure Oxidation for Thin Gate Insulator Process," *IEEE Trans. Electron Dev. ED–29*, pp. 503–507, 1982.
156. L.E. Katz and L.C. Kimerling, "Defect Formation during High Pressure, Low Temperature Steam Oxidation of Silicon," *J. Electrochem. Soc. 125*, pp. 1680–1683, 1978.
157. S.P. Murarka and G. Quintana, "Oxidation Induced Stacking Faults in N- and P-Type (100) Silicon," *J. Appl. Phys. 48*, pp. 46–51, 1977.

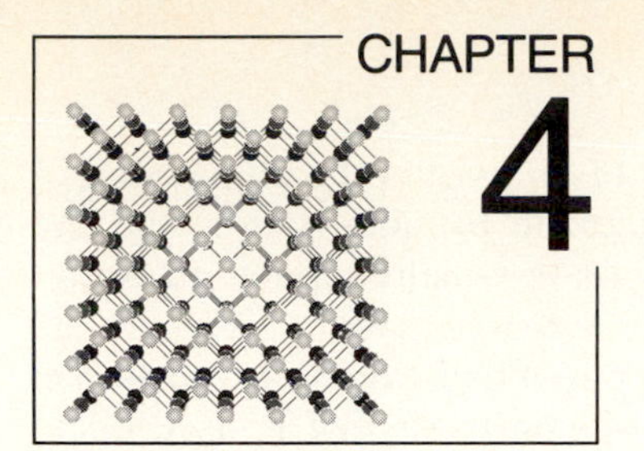

CHAPTER

4 Deposition of Inorganic Thin Films

4.1 INTRODUCTION

As was briefly discussed in Chapter 2, much of an integrated circuit is composed of thin layers of either conducting or insulating films deposited over the semiconductor wafer surface. In general, the conducting films are metal, but they may also be metal silicides or polycrystalline silicon (poly, or polysilicon). Low-temperature polysilicon deposition for interconnects and MOS gates will be discussed in this chapter, but the metals and silicides will be considered in Chapter 10. Amorphous silicon, used in some solar cells, requires somewhat different deposition conditions and will not be discussed. Thermally grown silicon dioxide is universally used on silicon ICs, but other thin nonmetallic films are used for diffusion masks, diffusion sources, surface passivation (GaAs), capping[1] (GaAs), insulation between metal leads, capacitance dielectrics, and protective overcoats. In the case of gallium arsenide, since thermal oxides do not grow well and do not possess the required characteristics, all insulating films used in GaAs ICs must be deposited rather than grown in situ from the wafer surface itself as in silicon processing. Examples of peripheral applications involving other materials are the use of silicon carbide for X-ray projection printing masks, silicon carbide for wear-resistance coatings to thermal printheads, and ferroelectrics for polarizable MOS gates.

The common ways of depositing thin layers of insulating materials are vacuum evaporation and sputtering (often referred to as physical vapor deposition), coating the wafer with a liquid film that will form an inorganic insulating material when heated,[2] and chem-

[1]Capping is the adding of an impervious layer to the GaAs wafer to prevent evaporation of arsenic from the surface during subsequent high-temperature operations.

[2]Sometimes called spin-on glass, so named for the way it is applied. (See Chapter 5 for a discussion of methods of applying spin-on materials to wafers.)

ical vapor deposition (CVD).[3] Less often used is the conversion of thin-deposited metal films to their oxide by anodization and the coating of wafers with a frit slurry of a low-melting-point glass that will fuse and form a continuous layer when heated.

The primary materials used for protective overcoats are silicon nitride and doped silicon dioxide. Silicon monoxide is used occasionally in specialized applications such as overcoating thin-film resistors. Thin-film diffusion sources can be either doped CVD SiO_2 or a spin-on material that will thermally decompose into volatiles and a doped SiO_2 film. When thermal silicon dioxide cannot be used as a diffusion mask, either CVD SiO_2 or silicon nitride is usually satisfactory. Insulation between metal leads[4] is usually provided by doped CVD SiO_2. GaAs surface passivation is also usually done with CVD SiO_2. In some capacitor applications, a dielectric with a dielectric constant higher than that of SiO_2 is desirable. Currently, no preferred alternative exists, but TiO_2 is an example of a material with a substantially higher dielectric constant.

While a multitude of film properties must be considered, many properties depend on the choice of film material, and some depend on deposition conditions. Of these properties, the most important ones are listed in Table 4.1. The etching characteristics of most of the films are discussed in Chapter 6. The electrical properties, such as resistivity, interface charge density, and dielectric strength, for various thin-film dielectrics have been extensively studied with the goal of providing a usable substitute for silicon thermal oxide passivation. Thus far, they have not proven satisfactory for most such applications, but they are widely used for protective coatings against both mobile ions[5] and mechanical damage and as isolation between levels of interconnects.

For ordinary IC applications, the thermal expansivity is of interest only as it affects the mechanical stress in the film. The breaking strength of most films is in the 10^{10} dynes/cm^2 range, and for films to remain crack free, the film stress should be less than about 10^9

---

[3]CVD in this discussion is considered to include all chemical forms of deposition from the gas phase. Later in the chapter, a distinction will be made between various CVD techniques.

[4]The terms *leads* and *interconnects* are used interchangeably in this and the remaining chapters to mean thin-film electrical interconnections between the various circuit components comprising an integrated circuit.

[5]The effect of mobile ions is discussed in Chapters 3 and 11.

TABLE 4.1

Important Properties of Films

Property	*Comment*
Etch rate	Rate must be high enough to etch film in a reasonable time, but not so high as to be uncontrollable.
Ease of etching	A readily available and relatively nontoxic solvent is desirable.
Electrical resistivity	For most nonconductive applications, any insulating material is satisfactory. (The criteria for conductive layers are discussed in Chapter 10.)
Interface charge density	This property is of importance when material is deposited directly on a pn junction or is to be used as a MOS gate dielectric.
Dielectric strength	This property determines minimum thickness for a given breakdown voltage.
Thermal expansivity	Difference in expansivity of wafer and film contributes to internal stress.
Thermal conductivity	This property is of concern only in special applications.
Hardness	This property determines scratch resistance.
Mechanical stress	Excessive internal stress can crack film or wafer.
Effectiveness as a mobile ion barrier	This property is very important for protective coatings.
High-temperature flow characteristics	Some films are reflowed after window etching to make step coverage easier. Thus, low-temperature flow is necessary.

dynes/cm^2 over the whole range of temperatures to which the films are subjected. The total stress[6] σ_f in the film is given by

$$\sigma_f = \frac{(\alpha_f - \alpha_s)E_f\Delta T}{1 - \nu_f} + \sigma_{if} \qquad 4.1$$

where α_s is the wafer (substrate) expansivity, α_f is the film expansivity, ΔT is the temperature excursion, E_f is Young's modulus of the film, ν_f is Poisson's ratio for the film, and σ_{if} is the internal stress of

[6] A positive stress is tensile; a negative stress is compressive.

the film. σ_{if} is process dependent and can be highly variable. (A method of experimentally determining σ_f was discussed earlier in section 3.3.2.)

4.2 SPUTTER DEPOSITION

Sputtering as a method of depositing films has been known at least since 1852[7] and thus ranks as comparable in age to the evaporation method. However, probably because it does not require as high a vacuum as evaporation, sputtering seems to have been used much more extensively than evaporation prior to 1900. But, since the high pressures (typically 10–100 mtorr) allowed substantial contamination to occur, not until recently, when the procedure of an initial hard-vacuum pumpdown followed by introduction of a very-high-purity sputter gas was introduced, did sputtering become a widely accepted tool for semiconductor fabrication. Its main advantage is that deposition is independent of the temperature of the substrate, and films can therefore be deposited at room temperature. Unfortunately, however, lack of adhesion and/or film cracking may prevent the use of such films unless the temperature of the substrate is raised to about 200°C.

Fig. 4.1 is a schematic of a sputter deposition chamber. As

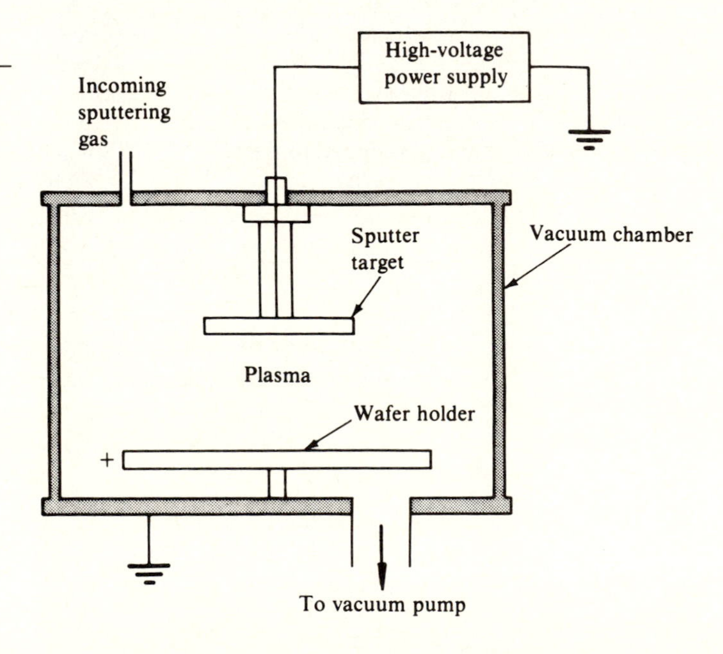

FIGURE 4.1

Schematic of a sputter deposition chamber. (The chamber may be very large so that many wafers can be placed on the wafer holder, or it may be just large enough for one slice to be sputtered at a time.)

[7]W.R. Grove, *Phil. Trans.*, p.1, 1852.

shown, a plasma is formed between the target (material to be deposited) and a metallic platen holding the wafers. Positive ions from the plasma are accelerated to the target, where they dislodge (sputter) molecules from it. Some of these molecules then strike and condense on the wafer surface. The ions used for sputtering are positive, and the target is negatively biased. This negative bias can be done directly by a DC power supply or accomplished by a charge buildup on the target when it is fed by a capacitance-coupled RF voltage. (For a further discussion of this voltage, see Chapter 6.) The use of an RF system also makes possible the sputtering of insulating materials such as silica.

The ions used for sputtering come from the residual gas in the chamber and, in principle, could come from any gas. Usually, an inert gas is preferred so that it does not react with the target material. Further, some ions are much more efficient than others at sputtering. In most cases, the efficiency increases in the order of He–N–Ne–N_2–Ar–Kr–Xe, and the difference between He and Ar is as much as 10 times (1). This sequence is in the same order as the increases in the atomic or molecular weights, which is 4–14–20–28–40–84–131.

In sputtering two-component materials such as SiO_2 or Si_3N_4, the target can be the same as the material to be sputtered, which is sometimes a problem to construct, or the target may be made of one component while the sputter gas contains the other. Thus, sputtering a silicon target in a nitrogen ambient can give silicon nitride. This process is called reactive sputtering. The compound formation can occur at the target surface before sputtering so that it is the compound that is sputtered, or it can occur by a reaction at the substrate surface. Ordinarily, reactions would not be expected in the plasma because of the low density of reactants. When some materials are subjected to a plasma containing oxygen, oxidation occurs, and it appears that, in the case of the reactive sputtering of titanium to form TiO_2, the oxide forms on the titanium surface and then is sputtered away (2). The electrical and physical properties of reactively sputtered SiO_2 and Si_3N_4 have been extensively studied (3, 4), but the mechanism of their formation does not seem to have been considered. Even when conventional sputtering is used, the addition of an appropriate gas such as oxygen during oxide sputtering or nitrogen during nitride sputtering appears to help ensure stoichiometry and will often improve the electrical properties of the film (5, 6).

The kind of sputtering equipment configuration depends on the electrode arrangement. The simplest type is diode sputtering, either DC or RF, as was shown in Fig. 4.1. Diode sputtering has the disadvantage of having a relatively slow deposition (sputtering) rate because of the limited number of ions produced in the plasma. By installing a filament to supply additional electrons to the plasma (triode sputtering), the number of ions and thus the sputtering rate can be substantially increased (7). Another way of increasing the ion density is by using a magnetic field to spiral the electrons toward the cathode.[8] The electron path length is then increased, and thus the ion generation efficiency. This method is referred to as magnetron sputtering and is commonly used (8, 9).

4.3 EVAPORATION*

Evaporating material in a hard vacuum (typically $<10^{-5}$ torr) and allowing it to hit a substrate in the chamber, as shown schematically in Fig. 4.2, is a coating method that has been known for about as long as sputtering. However, it did not become widely used until much later, probably because of the lack of good vacuum pumps and good ways of heating the evaporate. The wrapping of wire of the desired evaporate around a twisted tungsten filament (introduced after 1900) provided a simple evaporation source for most metals and greatly increased the ease of evaporation deposition. Thus, by the time aluminum metallization was applied to semiconductors, evap-

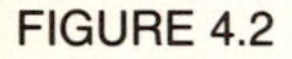

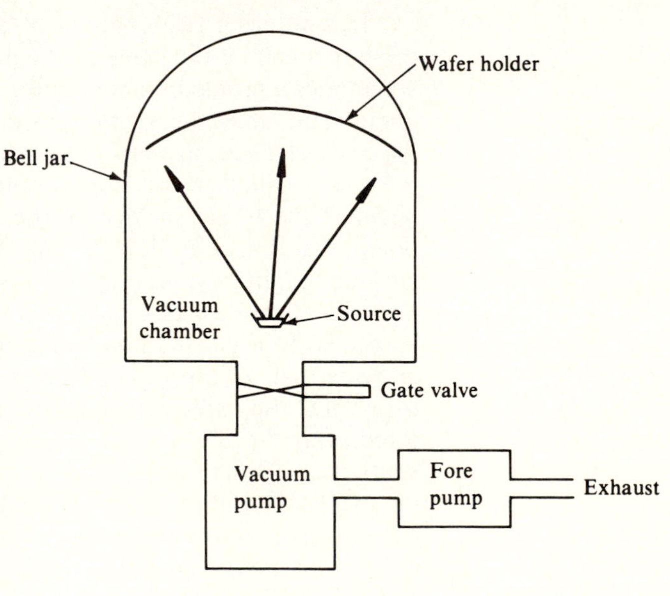

FIGURE 4.2

Schematic of a simple evaporation system.

[8]In electrical engineering parlance, the cathode is the most negative electrode. However, some plasma process engineers and some literature define the electrode coupled to an RF power supply as the cathode regardless of whether it is positive or negative with respect to the other electrode. (See the plasma section of Chapter 6 for more discussion of electrode voltages.)

*See reference 10.

oration was a well-established technique and, indeed, was much favored over sputtering.

The high vacuum is necessary so that the evaporant collides with a negligible number of gas molecules on its way to the substrate (wafers in this case). The rate of evaporation is dependent on the vapor pressure of the source (evaporant) material, so for a reasonable rate, the source temperature must be at least a few hundred degrees C. Thus, in most cases, a high vacuum is also required to minimize chemical reactions between the material being evaporated and the residual gases. For a minimal number of collisions, the mean free path L of the molecules in the chamber must be considerably greater than the distance from source to substrate. Several expressions have been developed for the mean free path, but the most common one is a modified Clausius derivation that gives

$$L = \frac{T/273}{\xi^2 \psi \pi \sqrt{2}} \tag{4.2}$$

where T is the temperature in K, ψ is the number of gas molecules/cc, and ξ is the diameter of the gas molecules in cm. ξ can be estimated from Arnold's rule (see Chapter 2 in reference 10): ξ in Å $= (V_m)^{1/3}$, where V_m is the molar volume in cm^3 at the boiling point but is generally in the 2–5 Å range. At 0°C, $\psi = 2.3 \leftarrow 2.7 \times 10^{19}P$, where P is in atmospheres.

The deposition rate r will be given by the difference between the rate of arrival from the source and the rate of evaporation from the wafer surface. Unless the wafer is very hot, evaporation from its surface will be negligible, and the deposition rate will be proportional to the source evaporation rate. The source evaporation rate r_s is proportional to the equilibrium partial pressure of the source material at the evaporation temperature and is usually expressed in terms of grams/s. The substrate deposition rate is related to r_s through the deposition efficiency η, which is the ratio of the solid angle subtended by the substrate to the emission solid angle of the source. Thus, for evaporation from a small (point) source in which material is emitted in all directions (a 4π solid angle),

$$\eta = \frac{A}{4\pi Z^2} \tag{4.3}$$

where Z is the distance from source to substrate and A is the substrate area projected onto a plane perpendicular to Z. If the source is a small crucible, then material is emitted only into a hemisphere, and Eq. 4.3 becomes

$$\eta = \frac{A}{2\pi Z^2} \tag{4.4}$$

Since the different components of multicomponent materials usually have different vapor pressures, it is difficult to deposit such materials by evaporation. Nevertheless, through the use of molten sources of an appropriate composition or by flash evaporation of a continuous stream of feed material, the composition can be reasonably well controlled. However, when aluminum interconnect leads doped with a small percentage of materials such as copper or silicon became necessary, sputtering proved to be a more practical method of deposition. (Metal deposition is covered in more detail in Chapter 10.) It is also difficult to vaporize high-melting-point materials such as fused silica, but by using an electron beam impinging on the surface of the material to be evaporated, satisfactory rates can be obtained. When evaporated quartz (fused silica) was first used as a protective overcoat for ICs, it was generally electron beam (E-beam) evaporated. However, RF sputtering is now the most common method.

4.4 ION BEAM DEPOSITION

Ion beam deposition uses a plasma to generate ions that can then be accelerated as desired toward the sputtering target as shown in Fig. 4.3 (11). By separating the ion generation, which must be done in low vacuum, from the main chamber, the wafers may be kept in a hard vacuum, thereby minimizing film contamination by the internal gas ambient. Further, a separate ion generator allows high power to be applied to it without the possibility of the wafers' overheating or suffering lattice damage from incident electrons. It is also possible to use the ion beam itself as the deposition material. By deflecting a focused beam, patterned depositions can be made directly on a wafer. (See also Sec. 5.14.)

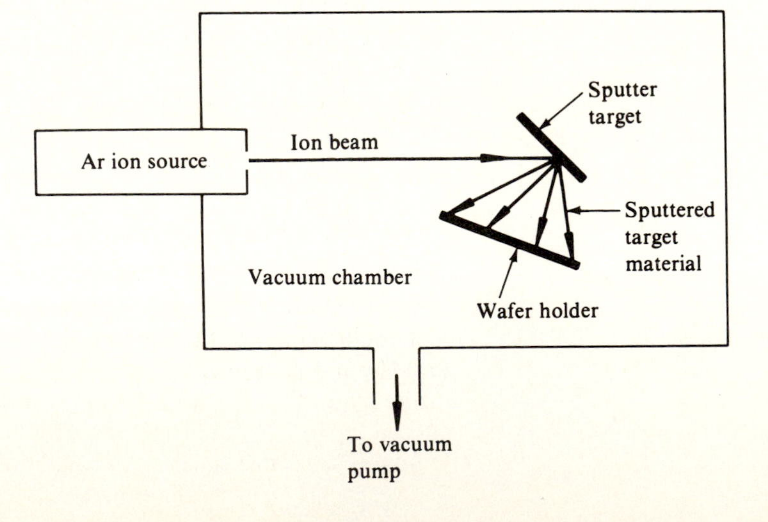

FIGURE 4.3

Schematic of an ion beam deposition system. (The pressure in the vacuum chamber may be quite low—for example, 10^{-5} torr.)

4.5 CHEMICAL VAPOR DEPOSITION

Chemical vapor deposition (CVD) is done by choosing either a compound that will decompose (pyrolyze) under the influence of heat to give the desired film or a set of compounds that will react to give the film. In both cases, the reaction must take place only at the wafer surface. If the reaction takes place partially or totally in the gas stream above the wafer, "fluff" and other forms of particulates will form. Such conditions generally give films with low density and many pinholes. In almost every case, CVD reactions that are useful in IC manufacturing can be characterized as surface catalyzed. That is, even though the reactants may be present in the gas ambient above the wafers and at the same temperature as the wafers, the reaction proceeds on the hot wafer surface and on any other hot surfaces, but not in the gas stream. However, some processes used in other industries do deliberately form particles in a high-velocity gas stream and then direct the stream to the surface to be coated. There, a portion of the particles stick, coalesce, and form a useful coating.

Specific reactions will be discussed in the following sections. However, an example of decomposition is the heating of silane (SiH_4) to temperatures above about 400°C, at which point the silane decomposes to give silicon and hydrogen (12). An example of two components that react is the use of silane plus oxygen to give SiO_2. In this particular case, a stream of silane in air will ignite spontaneously, and under some circumstances, silane–air mixtures will violently explode. However, when properly diluted, the mixture is well behaved and is widely used to produce silica films at temperatures as low as 400°C.

CVD techniques can be used to deposit amorphous, polycrystalline, or single-crystal layers. Single-crystal deposition is referred to as epitaxy and is discussed in detail in Chapter 7. This chapter is devoted solely to amorphous and polycrystalline depositions.

The CVD process has several distinct subdivisions. One, done at atmospheric pressure, is referred to as CVD or APCVD. When lower pressures are used, where diffusion effects are minimized, the operation is LPCVD (low-pressure CVD). If a plasma is used to generate ions or radicals that recombine to give the desired film, the process is PECVD (plasma-enhanced CVD). In some cases, photon energy is used instead of or as an adjunct to the thermal energy usually required for the reaction. These processes are referred to as photon-enhanced CVD. Silica and silicon nitride are often deposited by CVD. Silicon nitride is also widely deposited by PECVD. Polycrystalline silicon layers are generally deposited by LPCVD. Photon-enhanced CVD is currently still largely experimental. The choice of which process to use depends on factors such as the maximum temperature that can be used and the amount of stress and the

FIGURE 4.4

Typical operating temperatures for depositions of various kinds of silicon dioxide and nitride.

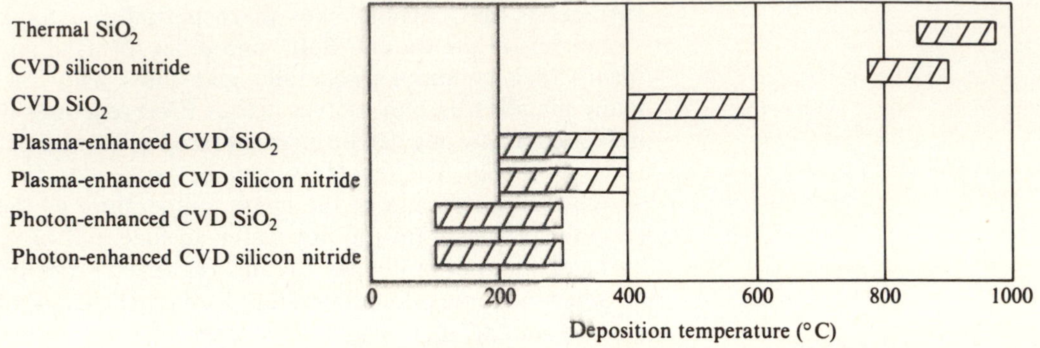

number of pinholes that can be tolerated. The maximum temperature, in turn, depends on where in the process the layer is to be deposited. Fig. 4.4 shows typical temperature ranges for the deposition of various oxides and nitrides. If the goal is only to minimize diffusion, any method other than thermal oxidation will probably be satisfactory. If the layer is to be deposited over refractory contacts, the upper end of the CVD oxide temperature range will probably suffice. But, if deposition is to be over aluminum metallization, either the low end of the CVD oxide range or else PECVD will be required. If a coating is to be put over a unit with bonding wires attached, a very low temperature (perhaps photon-enhanced deposition) may be required.

4.5.1 Atmospheric Pressure and Low-Pressure Depositions

The deposition rate will depend on the rate of reactants arriving at the surface and the rate at which the surface reaction proceeds. Depending on which of these two factors is limiting, the observed rate is said to be, respectively, mass-transport controlled or kinetically controlled. Mass-transport limitations can occur either because of a limit on the concentration at the gas inlet or because the reactants can get to the surface only by diffusing through some sort of relatively stagnant gaseous layer (diffusion-limited). The kinetic region's behavior is largely dominated by effects at the growing interface and generally shows an Arrhenius behavior: rate $= r = ae^{-E/RT}$, where a is a constant, E is the activation energy in kcal/mol, R is the universal gas constant, and T is absolute temperature.[9]

EXAMPLE ☐ If a process has an activation energy of 1 eV,[10] what is the activation energy in kcal/mol?

From Appendix B, 1 eV = 3.832×10^{-20} cal and 1 molecule = 1 mol/6.023×10^{23}. Thus, 1 eV/molecule = 3.832×10^{-20} cal/ 1 mol/6.023×10^{23} = 23.08×10^{3} cal/mol = 23.08 kcal/mol. ☐

For APCVD, the temperature dependence of the deposition rate for the various processes is generally characterized by a kinetic-controlled region at lower temperatures and a mass-transport-limited region at higher temperatures. Thus, a rate versus $1/T$ plot has the characteristics shown in Fig. 4.5a. In the kinetic-controlled region, if the temperature can be adequately controlled, the deposition rate across the wafer(s) can be quite uniform, as shown in Fig. 4.5b. However, in the mass-transport-limited region, where the rate depends on the flux of reactants, a substantial change of deposition rate can occur along the flow direction, as is also shown in Fig. 4.5b. The actual shape of the curve can vary widely. Some deposition conditions give a linear decrease with distance, and some give an

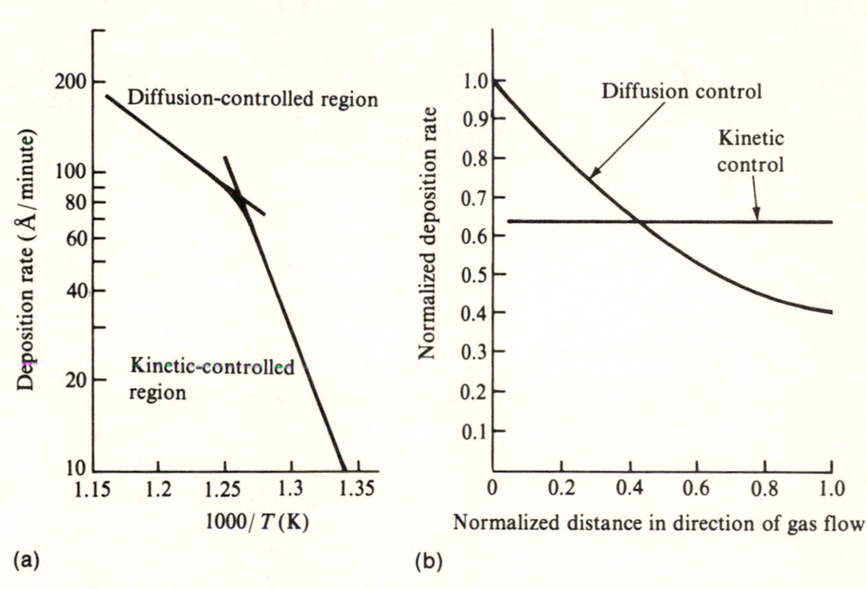

FIGURE 4.5

Effect of deposition temperature on deposition rate and uniformity.

[9]The more common expression found in solid-state physics literature for these kinds of relations is $r = ae^{-E/kT}$, where E is in eV per molecule (or per atom or electron, depending on the situation) and k is Boltzmann's constant in eV/K. In discussions of chemical processes, however, R instead of k is normally used, and the activation energy is expressed in kcal/mol instead of eV.

[10]Ordinarily, when eV is used in this context, the phrase "per molecule," while being implicit, is not written.

exponential decrease. By reactor gas flow or hardware changes, the profiles can be substantially flattened and, in some cases, may even be caused to tilt up at the trailing edge.

The gas flow in APCVD and LPCVD reactors can be broken into two general categories of flow: parallel to the wafer surface, as shown in Fig. 4.6a, and perpendicular to the surface of a series of wafers, as shown in Fig. 4.6b. The parallel flow, as actually embodied in CVD reactors, takes many forms. One of the oldest places wafers along a horizontal tube as shown in Fig. 4.7a. However, pancake and barrel configurations as shown in Figs. 4.7b and 4.7c, respectively, have been widely used in epitaxial CVD. Also used are single-wafer reactors as shown in Fig. 4.7d, which are most practical for wafers with diameters above about 100 mm. Parallel flow is used almost exclusively for APCVD, and perpendicular flow is normally found only in low-pressure systems. Modeling of deposition rates, which requires information on both the flow characteristics and the chemical reaction path, has been done for both kinds of systems. In the case of parallel flow, most of the work has been done for the specific case of silicon epitaxial growth. Modeling of flow as in Fig. 4.6b has been done for polysilicon and silicon nitride depositions.

In order to gain insight into the specific details of flow, several studies using either smoke patterns or holography have been made. In the case of parallel flow down a long, rectangular or circular tube, if the susceptor is the hottest part of the system, spiraling of the gas as it moves along (13), as well as one or more circulating gas cells,

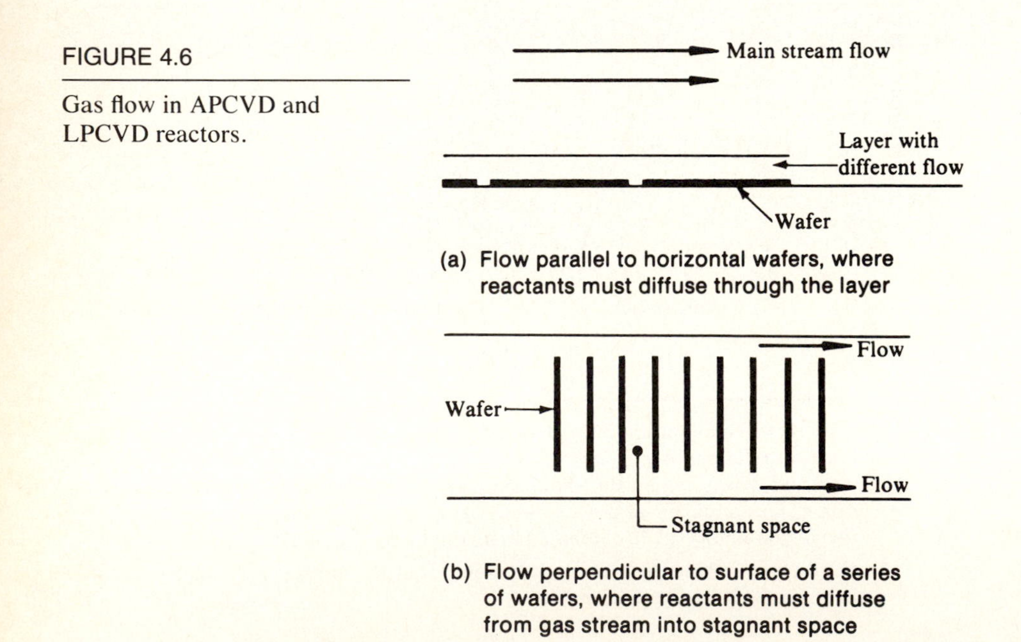
Main stream flow
Layer with
different flow
Wafer
(a) Flow parallel to horizontal wafers, where reactants must diffuse through the layer
Flow
Wafer
Flow
Stagnant space
(b) Flow perpendicular to surface of a series of wafers, where reactants must diffuse from gas stream into stagnant space

FIGURE 4.6

Gas flow in APCVD and LPCVD reactors.

FIGURE 4.7

Examples of reactor configurations in which gas flows across wafer surface.

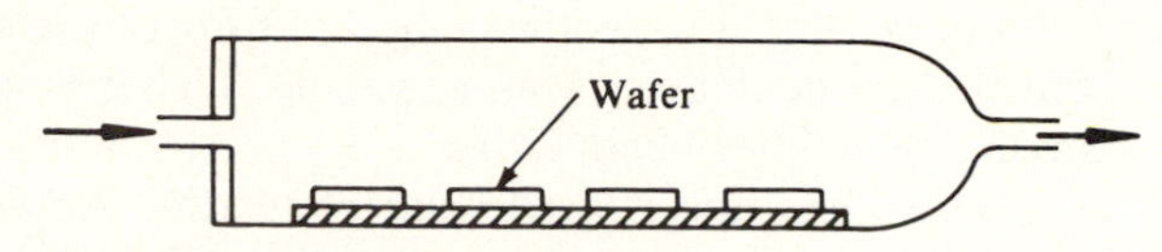

(a) Horizontal tube configuration

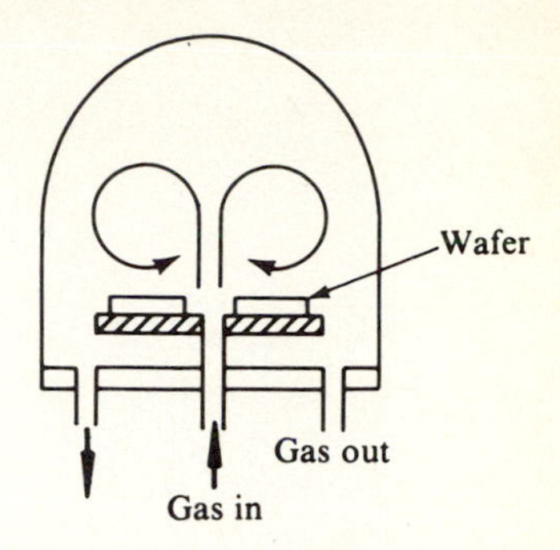

(b) Pancake configuration

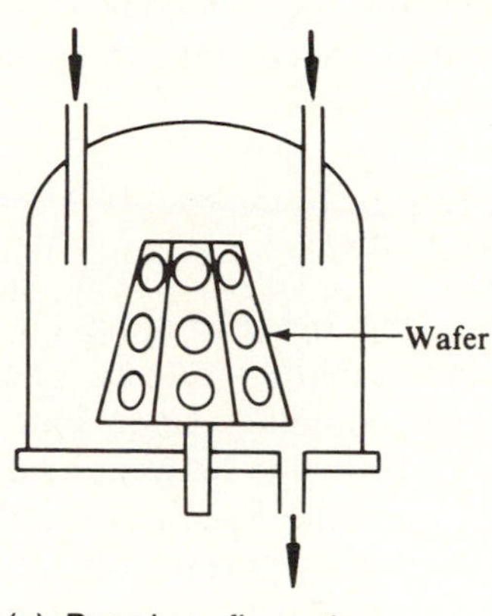

(c) Barrel configuration

Wafer

(d) Single-wafer configuration

which in cross section appear as shown in Fig. 4.8 (14), can occur. The circulating cells, sometimes referred to as conduction rolls, are predicted mathematically and can lead to deposition nonuniformities (15). Often, a layer just above the susceptor/wafers will occur that appears to be laminar, and turbulent flow will fill the rest of the chamber (16, 17). The value of the Reynolds number[11] Re for a particular set of circumstances can be used to judge whether or not laminar flow will occur. In general, for a value greater than about 2000, turbulence will occur. However, under the influence of steep temperature gradients, as would be found in the gas above a heated substrate, turbulence is more likely, and it has been suggested that

FIGURE 4.8

Cross section of flow pattern in a tubular horizontal reactor. (*Source:* Adapted from R. Takahashi et al., *J. Electrochem. Soc. 119*, p. 1406, 1972.)

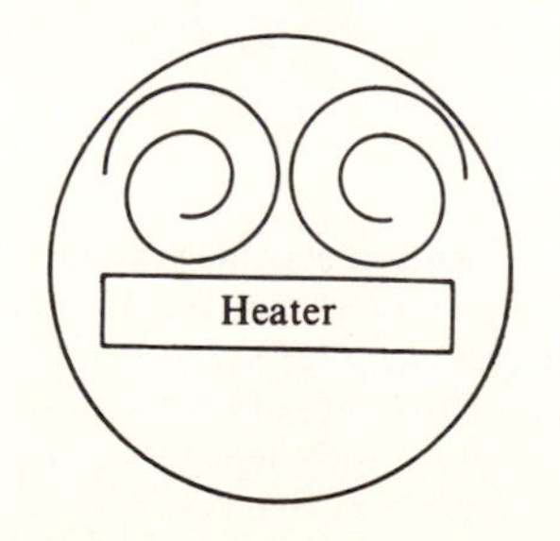

[11]A series of dimensionless ratios are regularly used in fluid flow analysis, one of which is the Reynolds number Re. It is defined as $Re = \rho V \ell / \mu$, where ρ is the gas density, V is its velocity, μ is its absolute viscosity, and ℓ for this case is the free distance above the substrate/wafers.

the ratio of the Grashof number[12] Gr to Re^2 is then a good criterion. For a given gas velocity, the Reynolds number for argon and nitrogen is nearly two orders of magnitude greater than it is for helium and hydrogen, and it has been observed in some studies that there is much more turbulence when argon and nitrogen are used (16). Observations of flow patterns in the geometry of Fig. 4.6b show that with increasing flow, eddies begin to form at the edges of the wafers but that by decreasing wafer spacing, higher velocities can be used before the eddies begin (18).

When gas flows over a flat plate (the substrate and wafers of parallel flow reactors), it is normally assumed that the gas velocity V is zero at the surface of the plate. The velocity increases normal to the plate until at some distance δ it equals the velocity in the main stream. δ increases with distance along the plate in the direction of flow and is approximately given by

$$\delta = \left(\frac{x\mu}{\rho V}\right)^{1/2} = \left(\frac{x\ell}{Re}\right)^{1/2} \qquad 4.5$$

where x is the distance from the inlet along the gas stream, ρ is the gas density, and μ is the absolute viscosity. The region between the plate and δ is called the boundary layer and, when thin, is often assumed to be stagnant. If flow is between two plates or is, for example, in a tube, and when the length of the flow path is enough, the upper and lower boundary layers will meet, making the flow "fully developed." The boundary layer then cannot be considered stagnant; rather it has a velocity that ranges from a maximum in the center to zero at the surfaces.

Note that the values of gas properties such as ρ and μ to be used in Eq. 4.5 and similar calculations are the ones appropriate for the reactor chamber operating temperature. Hence, ordinary room-temperature handbook values will not suffice. Some high-temperature values are readily calculable, but more often they are available in the form of polynomial expressions that have been fitted to experimental points. A useful example is the viscosity of hydrogen, given in micropoise by (19)

$$\ln\left(\frac{\mu}{S}\right) = A \ln T + \frac{B}{T} + \frac{C}{T^2} + D \qquad 4.6$$

---

[12]The Grashof number is defined as $Gr = \rho^2 g \ell^3 \beta \Delta T/\mu^2$, where g is the gravitational constant and β is the gas expansion coefficient. For an ideal gas, the expansion coefficient β is just $1/T$, so in the ratio of Gr to Re^2, given by $Gr/Re^2 = g\ell\Delta T/TV^2$, no gas property is involved and the kind of gas used should not affect the onset of thermal-gradient-induced turbulence. Experimentally, turbulence for ratios greater than 0.5 (when expressed in cgs units) have been reported (13).

where $S = 88.0$, $A = 0.68720$, $B = -0.61732$, $C = -111.49$, $D = -3.9001$, and T is the temperature in K.

EXAMPLE □ If the inlet flow of hydrogen into a 100 mm diameter tube furnace operating at atmospheric pressure and 1100°C is 100 liters/minute, what are the density and the velocity of the gas in the tube if the boundary layer thickness is considered to be negligible?

Since the volume of a gas at constant pressure varies as $V_1/T_1 = V_2/T_2$, the density will vary inversely with temperature; that is, $\rho = (T_1/T_2)(0.090)$ gram/liter, or $\rho = (273/1373)(0.090) = 0.018$ gram/liter. The change in flow rate due to the increased temperature in the tube is the same as the change in gas volume. Thus, the new flow rate is $100(T_2/T_1) = 500$ liters/minute. The velocity equals flow rate divided by cross-sectional area, or 5×10^5 cm^3/minute/78.5 cm^2 = 6369 cm/minute = 106 cm/s. □

Even though reactants are carried along by the flow, they must diffuse through the stagnant layer to reach the wafer surface. When the temperature is high enough for no kinetic limitations, the choice of assumptions made about the nature of the layer gives rise to various expressions for deposition rate. If no stagnant layer is considered, a linear drop in rate along the gas path is predicted because of reactant gas depletion (20). If a uniformly thick stagnant layer is considered, the falloff in rate is exponential (21). There are also solutions assuming the boundary layer thickness increases as in Eq. 4.2 (22), where it increases but is not assumed to be stagnant (23) and where it is assumed that the flow is fully developed before the first wafer (24). Models for the barrel-type reactor of Fig. 4.7c are also available but are considerably more complex than those for the simple tubular geometry (19, 25, 26) of Fig. 4.7a.

In any of these cases, since it is assumed that there is no reaction rate limitation, the flux J of reacting species moving from the gas stream to the surface will determine the deposition rate. Travel through the boundary layer will be by diffusion, and the flux in a simple one-dimensional case will be given by

$$J = \frac{-D\partial C}{\partial X} \tag{4.7}$$

where D is the diffusion constant (coefficient) and C is the concentration of reactant. (Diffusion behavior is discussed in detail in Chapter 8.)

The gaseous diffusion constants are very large compared to those encountered in solid-state diffusion and more often are calculated instead of being experimentally measured. As an example of the values to be expected, D for $SiCl_4$ in H_2 at 1500 K is 5.7 cm^2/s. Several equations have been developed, although most of them give comparable results (25). Simple theory predicts that D should vary as T^b/P where $b = 1.5$ and P is the gas pressure. Experiment and more elaborate theory substantiate the $1/P$ dependency but yield a b value of between 1.75 and 2.

The perpendicular wafer geometry of Fig. 4.6b, desirable because of the relatively inexpensive equipment and the large number of slices that can be processed per run, behaves differently from the other geometries in that no flow occurs between wafers. Not only must reactants move from the main gas stream to the edges of the wafers, but also they must then diffuse down between the wafers and produce a uniform deposit. Even if the system were operating in the kinetically controlled region, a problem of radial reactant depletion over large-diameter wafers would still exist unless the spacing between wafers were inordinately large. However, by operating at reduced pressure, diffusion can be substantially increased, and the geometry becomes practical (27). For an LPCVD system operating at 0.5 torr (atmospheric pressure = 760 torr), the diffusion constant, and consequently the flux of reactant, is approximately 1500 times that at atmospheric pressure. Note, however, that if the surface reaction generates more moles of gaseous products than are consumed, their outward flow will impede the inward diffusion flow of fresh reactant. For example, in the decomposition of tetraethylorthosilicate (TEOS) to produce SiO_2, 1 mol TEOS diffusing to the wafer surface produces 1 mol solid SiO_2 on the surface and liberates 4 mol ethylene and 2 mol water. This effect is sometimes called convection via chemical reaction and can cause severe radial deposition nonuniformity unless a large percentage of the reactor gas is the inert carrier (28).

Like the barrel configuration, the modeling of the tubular LPCVD system is considerably more involved than is the modeling for the simple flat-plate horizontal reactor. The problem is further complicated because, in many cases, tubular LPCVD is used to deposit compound layers such as silicon nitride, and the equations to be solved must then include many more reactions and reactants. Also often included is consideration of the fact that while the final deposition reaction is surface catalyzed, there may be a partial decomposition of reactants in the gas stream (29–31).

Several equipment configurations have been used to minimize the effects of reactant depletion along the direction of flow. One, used in tube deposition chambers where the wafers lie flat,[13] is to have wafers on a stand tilted with respect to the gas flow so that

---

[13]This geometry is largely obsolete.

those downstream project higher into the stream. This geometry increases the velocity of gas since the cross-sectional area is decreased. The increased velocity then decreases the thickness of the boundary layer so that more reactant can reach the wafer surface (20). The temperature of the wafers can be increased along the path of gas flow so that the increased deposition rate due to the higher temperature can at least partially compensate for a decrease due to depletion. Tubular LPCVD systems usually use this approach (32). Another method, applicable to pancake geometries, and used in some plasma processing equipment, provides an inward radial flow so that the depleting reactants are concentrated as they approach an exhaust port in the center of the wafer holder (33).

In the kinetically controlled region, the deposition rate is limited, not by material transport, but by the rate of some chemical reaction. During deposition, a whole series of reactions must occur, including adsorption and desorption of one or more species on the deposition surface. If the limiting reaction is one that occurs in the gas phase (in other words, the production of an intermediary through $A + B \rightarrow C$) the deposition rate r of species C will be of the form

$$r = K[A][B] \tag{4.8}$$

where $[A]$ and $[B]$ denote the concentration of species A and B and K is the experimentally determined rate constant. When a surface adsorption step is limiting, as appears to be the case in most CVD processes, the rate equation is much more complex and will vary depending on the specific adsorption details. In this case, some variation of the Langmuir–Hinshelwood model can often be used for rate predictions.[14]

4.5.2 Plasma-Enhanced CVD

When a gas is excited by a high enough electric field—for example, in the reaction chamber of a plasma deposition reactor such as the one shown in Fig. 4.9—a glow discharge (plasma) is formed. In the plasma, high-energy electrons exist that can impart enough energy to reaction gases for reactions that normally take place only at high temperature to proceed near room temperature. The reactor looks superficially like the sputtering equipment shown in Fig. 4.1, but there are some substantial differences. In plasma-enhanced CVD (PECVD), the inlet gas contains the reactants for deposition, and the anode instead of being sputtered away remains unaffected. The voltage applied to sputtering electrodes may be either RF or DC,

[14]For more details on reaction rates predicted for the more complex reactions, refer to a standard chemistry text or to Don W. Shaw, Chapter 1, in C.H.L. Goodman, ed., *Crystal Growth–Theory and Techniques*, Plenum Press, London, 1971.

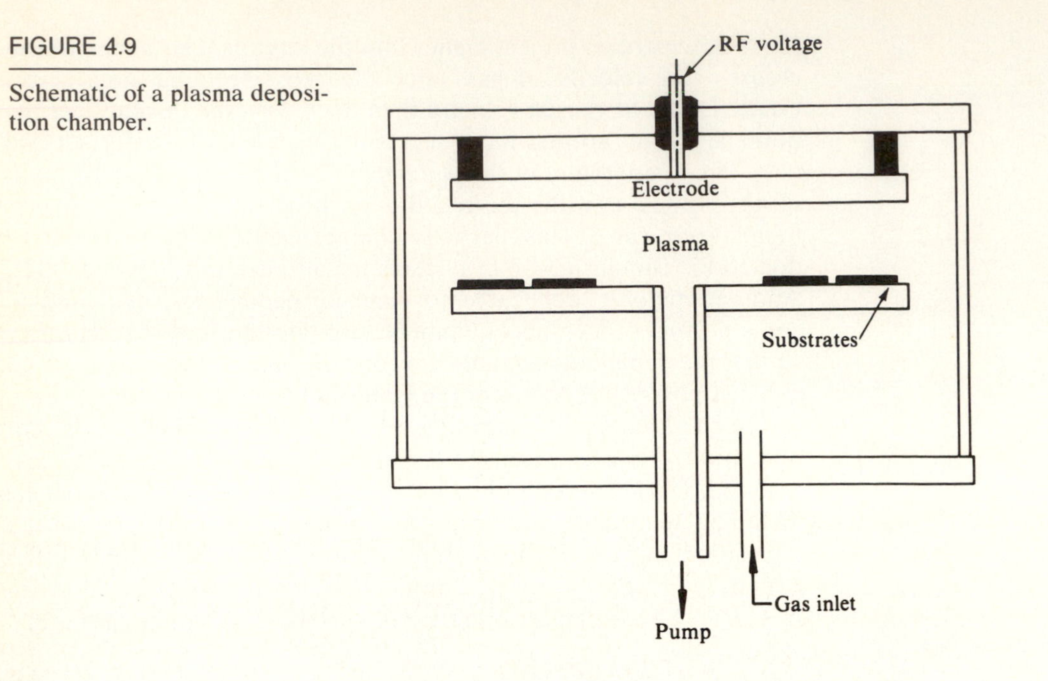

FIGURE 4.9

Schematic of a plasma deposition chamber.

depending on the specific mode of operation, but plasma deposition requires RF voltage.

The glowing (plasma) region will contain, in addition to the free electrons, normal neutral gas molecules, gas molecules that have become ionized, ionized fragments of broken-up gas molecules, and free radicals.[15] Deposition occurs when the molecules of incoming gases are broken up in the plasma and then the appropriate ions are recombined at the surface to give the desired film. Thus, molecule *AB* could be broken into ions *A* and *B*, and molecule *CD* into ions *C* and *D*. *A* and *D* could then react at the wafer surface to give film *AD*. The actual reactions that occur are thought to be considerably more complicated than those encountered in ordinary CVD, and modeling of the deposition process is not very extensive (35).

The fraction of the gas molecules ionized will generally be from 0.1% to 1%; the range of electron energies, from 1 eV to 20 eV; and the density of ions, in the 10^9–$10^{12}/\text{cm}^3$ range (34). The gas plasma is separated from the electrodes and the confining chamber by a dark region (ion sheath). All electrons, ions, neutral molecules, neutral

[15]Free radicals, usually designated with an asterisk after their chemical symbol (e.g., F^*), are neutral species whose electrons have absorbed extra energy and hence are quite active chemically. Since the electrons rather quickly lose their energy, the life of a radical is short.

fragments, and free radicals may leave the plasma, either by diffusion or because of additional velocity imparted to them either by impact with other particles or by the applied RF field. Since the gas is excited by an RF field, the electrons with their light mass will reach a much higher velocity during a cycle than will the ions. Electrons will thus have very high effective temperatures, while the ions have effective temperatures not far above room temperature. The plasma can be used as an adjunct or alternative to high temperatures for providing the energy necessary for a reaction to proceed. Therefore, PECVD ordinarily will have lower operating temperatures than will APCVD or LPCVD and will be used when lower temperatures are needed.

The low temperatures of the gas and the substrate reduce the mobility of the reactants impinging on the wafer deposition surface. Thus, plasma-deposited layers are usually amorphous and often contain higher percentages of extraneous reaction components than would be expected of high-temperature CVD depositions. A negative voltage of a few volts exists between the plasma and the wafer, and if any positive ions are present in the plasma, they will have gained substantial energy by the time they strike the wafer. This energy may help in producing a more defect-free film, but it can also cause damage to the substrate. In addition, even though the wafer is negative with respect to the plasma, some of the high-energy electrons in the plasma escape and strike the wafer, causing radiation damage. Usually, some of the subsequent high-temperature processing steps will anneal out the damage. However, by using an alternative reactor geometry in which the plasma is located away from the wafers, plasma-produced neutral molecule fragments and free radicals can still reach the wafers, while the electrons and ions are prevented from reaching them (36). Such reactors are often referred to as "afterglow" and usually use a microwave cavity to produce the plasma. Because of the limited size of the cavity, the plasma products must be dispersed over a wide area and large volume in order to cover the area of typical production wafers (37).

4.5.3 Photon-Enhanced CVD

Energy for the deposition reaction can sometimes be obtained by absorption of photons rather than from high temperatures. The two ways in which this absorption can occur are by the direct absorption of energy by the reactants and by a "photosensitizer" that absorbs the light energy and then transfers energy to the reactants by collision. In the case of SiH_4 and NH_3 reacting to form Si_3N_4, both silane and ammonia will directly absorb photon energy when the wavelength is less than about 2200 Å. If mercury vapor is also present, it will very strongly absorb at 2537 Å and transfer enough energy to the reactants for the reaction to proceed. The rate varies linearly with UV intensity, and it appears that the mercury is adsorbed on the surface where it absorbs energy from the photons and then transfers it to adsorbed reactants. In other cases, it is likely that mercury vapor absorbs energy and then transfers it to some gaseous species to produce a precursor that then further reacts on the surface. Processes for SiO_2, Si_3N_4, and amorphous silicon have been reported (38–41).

4.6 DEPOSITION OF SPECIFIC MATERIALS

4.6.1 CVD Silicon Dioxide

The impetus for much of the early work on CVD silicon dioxide seems to have been the desire to produce an oxide that could be used as a capacitor dielectric. However, even before 1946, the burning of methyl silicate was used to produce an in situ SiO_2 optical coating on glass (42). In 1961, SiO_2 produced by the pyrolytic decomposition of tetraethoxysilane [$Si(OC_2H_5)_4$], ethyltriethoxysilane [$C_2H_5Si(OC_2H_5)_3$), and other alkoxysilanes was proposed as a diffusion mask for germanium[16] (43). Since then, several other reactions have been studied, including the most commonly used reactions such as the pyrolytic decomposition of tetraethoxysilane (TEOS, tetraethylorthosilicate, ethyl silicate) and the SiH_4–oxygen reaction.

For TEOS deposition, a carrier gas such as nitrogen is bubbled through the TEOS to provide a vapor, which is then fed into a reactor containing the wafers. The wafers as well as the gas must be kept at an elevated temperature. If the vapor is heated to the cracking temperature and the products are allowed to condense on a cool substrate, hard polymers are formed that have many of the characteristics of silica. However, only when the wafers are also heated is pure SiO_2 deposited (44). The reactor may be tubular with the slices placed normal to the tube axis (45) as shown in Fig. 4.6b, or it may have a pancake configuration with the wafers lying flat on the heater. Temperatures for deposition can range from about 400°C to at least 1100°C. However, at low temperatures, the deposition rate is low, and the deposit is soft and porous. At the higher temperature ranges, a dark film is sometimes obtained, probably because of elemental carbon. Temperatures near 800°C provide a reasonable deposition rate and good-quality films. In this, and indeed in any of the other deposition processes as well, if the wafer surface is not scrupulously clean, excessive pinholes will occur in the film. Thus, having a good cleaning procedure is very important. Further, since deposition on all of the hot surfaces will occur, if the reactor is not periodically

[16] Germanium, like gallium arsenide, lacks a native oxide suitable for surface passivation, diffusion masking, and so on.

cleaned, particulates will flake off and get on the wafers while they are in the reactor. To minimize this effect, wafers can be placed vertically or even held upside down in the reactor.[17]

The reaction of compounds such as silicon tetrachloride ($SiCl_4$) or silicon tetrabromide ($SiBr_4$) with oxygen will also, in principle, produce silica. However, in order to maintain better control of the film properties, most processes using the halides do not use pure oxygen but rather use some compound such as H_2O, CO_2, or NO_2 to supply the oxygen (46, 47). By the use of an oxygen source such as CO_2, which is in itself nonoxidizing, a silica film can be grown directly after an epitaxial deposition without an extensive purging of the epitaxial reaction by-products. With an oxidizer and no purging, various products other than pure SiO_2 are formed. The overall CO_2 reaction can be written as

$$SiCl_4 + 2CO_2 + 2H_2 \rightarrow SiO_2 + 2CO + 4HCl \quad 4.9$$

However, the actual reaction must proceed with some intermediate steps since without excess hydrogen, no silica is formed (46). The use of $SiBr_4$ instead of $SiCl_4$ allows the use of a lower deposition temperature and was of interest for germanium before the introduction of silane-produced silica as described next.

The use of silane, which allows SiO_2 to be deposited at a lower temperature, was first reported in 1967 (48). With O_2 used, the reaction often assumed is

$$SiH_4 + 2O_2 \rightarrow SiO_2 + 2H_2O \quad 4.10$$

However, when the temperature is below about 500°C and the silane concentration is low, no water has been detected (49), thus leading to the reaction of Eq. 4.11 as more probable in that temperature range:

$$SiH_4 + O_2 \rightarrow SiO_2 + 2H_2 \quad 4.11$$

Above about 600°C, water vapor has been observed (50).

As in the $SiCl_4$ reaction, oxygen can be supplied by other oxygen sources, such as CO_2 (51). The actual reaction that occurs is not clear since, while the production of water in an intermediate step is anticipated, the experimental data indicate otherwise (52). An advantage of using CO_2 instead of oxygen is that, as before, if a deposited oxide is to be added immediately after epitaxy, there is no

---

[17]An "upside down" reactor heated by infrared heat lamps was used for silane oxide deposition as early as 1967.

oxygen to react with residual gases from the epitaxial deposition and cause particulates (51). By adding HCl—that is,

$$SiH_4 + CO_2 + HCl \rightarrow SiO_2 + \cdots \quad 4.12$$

a deposited oxide on silicon is produced that has interface properties close to those of thermal oxide (53).

4.6.2 Plasma Silicon Dioxide

Silicon dioxide films were plasma deposited as early as 1965 (54), but it was not until 1987 that machines specifically designed for depositing plasma oxide were marketed. Plasma films generally have a lower intrinsic stress than CVD films do and thus are less likely to crack when they are deposited in thick layers. The stress of plasma silica can be changed from compressive to tensile by changing the operating conditions, as is shown in Fig. 4.10 for one machine configuration (55). Step coverage will generally be better than is obtained with atmospheric pressure CVD (56). As is common with PECVD materials, stoichiometry can deviate somewhat from SiO_2, and a substantial amount of hydrogen may be incorporated in it (57). The usual deposition temperature range is from 200°C to 400°C, and for material deposited over this range, the etch rate decreases as the temperature increases.

Reactions that can be used include the following (58):

$$SiH_4 + O_2 \rightarrow SiO_2 + 2H_2 \quad 4.13$$

$$SiH_4 + 2N_2O \rightarrow SiO_2 + 2H_2 + 2N_2 \quad 4.14$$

$$SiH_4 + CO_2 \rightarrow SiO_2 + CH_4 \quad 4.15$$

$$Si(OC_2H_5)_4 \rightarrow SiO_2 + \cdots \quad 4.16$$

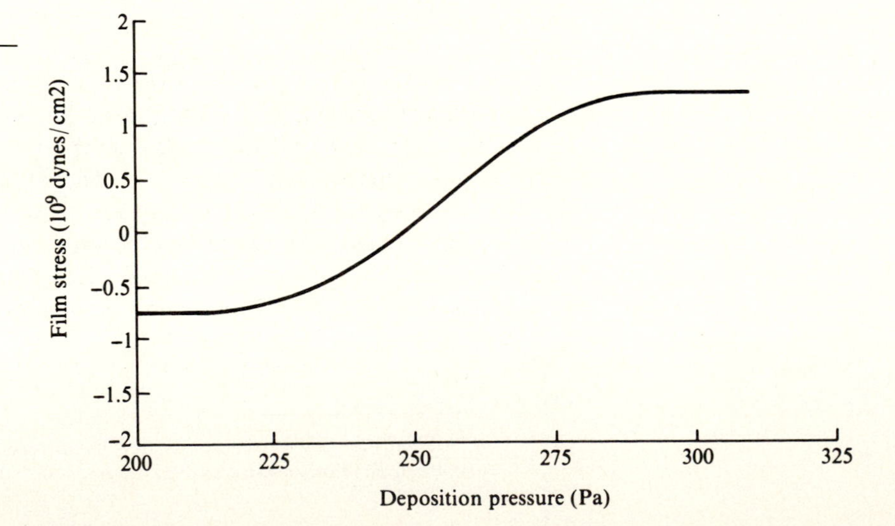

FIGURE 4.10

Film stress versus deposition pressure for PECVD silicon oxide deposition. (*Source:* Adapted from E.P. van de Ven et al., Proc. 4th IEEE V-MIC Conference, June 1987.)

When pure oxygen is used (reaction in Eq. 4.13), excessive particulates and pinholes as well as general nonreproducibility are observed (54, 59). The reaction in Eq. 4.9 is the one generally used and, because of the nitrogen, may leave a small percentage of nitrogen in the film (57).

4.6.3 Doped Silicon Oxide

Doped oxides are used for diffusion sources (see Chapter 8), ion migration barriers (see Chapter 3), isolation between conductor layers (see Chapter 10), and as a planarizer to fill in, for example, the trenches of trench capacitors. The doping for diffusion sources will generally be either antimony, arsenic, phosphorus, or boron. For ion migration protection, a small percentage of phosphorus is required. Isolation between conductive layers (interlevel oxide) usually will have phosphorus to act as an ion migration barrier, but the prime reason for doping is to decrease the reflow temperature.[18] As will be discussed in Chapter 10, a major difficulty in interlevel connections is the running of leads over abrupt steps such as those that occur in contact holes in the interlevel oxide. If the softening point of the interlevel oxide can be reduced enough, then it can be reflowed after the windows are cut in order to round off the abrupt edges. Similarly, after an oxide is deposited over a series of leads, the abrupt shoulders can be smoothed by reflowing.

Doped oxide films can be made by introducing dopant compounds such as phosphine (PH_3) or diborane (B_2H_6) along with SiH_4 and an oxygen source. The dopants then react to give an oxide (P_2O_5 or B_2O_3 in this case) that will co-deposit with the SiO_2 to form the appropriate mixed silicate. An extensive list of dopants that have been used in conjunction with TEOS depositions is summarized in Table 4.2 (60). PH_3 and B_2H_6 are commonly used when silane is the source of silicon. The reactions of dopant and the silica source are not independent in that the introduction of a dopant will often substantially change the oxide deposition rate. The deposition rate of silica from undoped TEOS is negligible at 600°C, but the addition of phosphine (PH_3) leads to a usable rate (60). Further, the introduction of a second dopant may materially affect both of their deposition rates. For example, when PH_3 or B_2H_6 is used singly in conjunction with SiH_4 in a plasma reactor, the concentration of boron or phosphorus in the silica increases with an increase in the B_2H_6/SiH_4 or

---

[18]The term *reflow temperature* is somewhat subjective but is the temperature at which the viscosity of the glass is low enough for surface tension to appreciably smooth sharp corners and edges in the original glass surface. This expression does not correspond to any of the terms *annealing point*, *softening point*, or *deformation point*, which are widely used in the glass industry and which are defined in terms of specific viscosities.

TABLE 4.2

Materials Used with TEOS to Produce Doped CVD Oxide

Dopant	*Dopant Source**	*Deposition Pressure*	*Deposition Temperature (°C)*	*Film Application*
B	$B(OCH_3)_3$	AP	650–730	Diffusion source
	$B(OCH_3)_3$ $(+O_2)$	LP	680, 750	Trench filling, diffusion source
	$B(OC_3H_7)_3$	LP	750	Diffusion source
	$B(OC_3H_7)_3$ $(+O_2)$	AP	475, 800	Diffusion source
P	$PH_3 + O_2$	LP	650	Diffusion source
	$POCl_3 + O_2$	LP	725	Flow glass
	$PO(OCH_3)_3$	AP	740, 800	Diffusion source
	$PO(OCH_3)_3$	LP	650–800	Flow glass
	$P(OCH_3)_3$	LP	750	Diffusion source
	$P(OC_2H_5)_3$	LP	750	Diffusion source
B + P	$B(OCH_3)_3 + PH_3/O_2$	LP	620–700	Trench filling, flow glass
	$B(OCH_3)_3 + PO(OCH_3)_3$	LP	680	Flow glass
	$B(OCH_3)_3 + P(OCH_3)_3$	LP	650–725	Flow glass
As	$AsCl_3 + O_2$	AP	500–700	Diffusion source
	$As(OC_2H_5)_3(+O_2)$	LP	700–730	Trench doping
	$AsO(OC_2H_5)_3 + O_2$	LP	700–730	Trench doping
Sb	$SbCl_5 + O_2$	AP	500	Diffusion source
	$Sb(C_2H_5)_3 + O_2$	AP	250–500	Dielectric
Sn	$Sn(C_2H_5)_4$	AP	700	Diffusion source for GaAs
	$Sn(CH_3)_4$	AP	500	Diffusion source for GaAs
Zn	$Zn(C_2H_5)_2$	AP	700	Diffusion source for GaAs

*See the source of this table for references to the individual processes.

Source: F.S. Becker and S. Röhl, *J. Electrochem. Soc. 134,* p. 2923, 1987. Reprinted with the permission of the publisher, The Electrochemical Society, Inc.

PH_3/SiH_4 ratio. However, when they are used together, an interaction occurs, and an increase in the diborane ratio causes a reduction in the amount of phosphorus deposited (61). Also, the rate–temperature dependence will generally be different for the various dopants and the oxide so that as temperature is changed, for fixed inputs, the oxide composition will change.

Typically, with an SiO_2 deposition having a small percentage of boron and a small percentage of phosphorus (BPSG for *b*oro*p*hoso*s*ilicate *g*lass), satisfactory reflow can be achieved at temperatures of between 800°C and 900°C. Fig. 4.11 shows a cross section of reflowed BPSG glass. In this case, a 5500 Å film of BPSG

(4% B, 4% P) was deposited over 5000 Å thick polysilicon leads, reflowed at 800°C, and densified at 900°C in wet nitrogen. There is a limit to how much the reflow temperature can be reduced since phosphorus contents above 7%–8% cause corrosion problems with aluminum leads and boron contents above about 4% give a glass unstable in high humidity (62). In some cases, the as-deposited film may be unstable and require additional high-temperature heat treatment to stabilize the composition. For example, in using PH_3 + B_2H_6 in conjunction with SiH_4 to form a glass with over about 3% boron and phosphorus, the boron may precipitate as boric oxide (63). For some compositions, annealing in a particular temperature range may cause a phase separation. Alternative compositions using arsenic or germanium instead of boron have been examined (62), but the ease of preparing BPSG makes it the current frontrunner for interlevel applications.

4.6.4 CVD Silicon Nitride

Amorphous CVD silicon nitride films are deposited in the 800°C–900°C temperature range and are generally referred to as "high-temperature nitride." Such nitride can be quite close in composition to Si_3N_4, although silicon-rich films can be produced. The other "nitride" is made by plasma deposition at temperatures in the 200°C–300°C range and is referred to as "plasma nitride." Its composition is not stoichiometric, but its properties are close enough to those of Si_3N_4 to be very useful.

It was established very early on that, unlike SiO_2, little sodium transport occurred in silicon nitride films. Thus, if they could be used instead of or in conjunction with SiO_2, considerable silicon device protection against sodium migration could be achieved. However, because of the Si–Si_3N_4 interface properties, it was found that where Q_f or N_{it} were important, the nitride could seldom be used in direct contact with silicon. Silicon nitride is relatively impermeable to oxygen; however, it will oxidize, although considerably slower than will silicon. A comparison is shown in Fig. 4.12, from which it can be seen that the differential is enough that a nitride covering of modest thickness can be used to mask areas of silicon that are to remain oxide free. Unlike SiO_2, since silicon nitride provides an efficient barrier against gallium diffusion, it can be used as a diffusion mask when gallium diffusions are made. Of more significance is the fact that it can be used to cap GaAs and prevent gallium from escaping from the GaAs arsenide surface at high temperatures.

The common reaction used for preparing high-temperature nitride is (64, 65)

$$3SiH_4 + 4NH_3 \rightarrow Si_3N_4 + 12H_2 \qquad 4.17$$

FIGURE 4.11

Example of how reflowing smooths out sharp corners left after a CVD layer is applied over leads. (*Source:* Photograph courtesy of Novellus Systems, Inc., San Jose, California.)

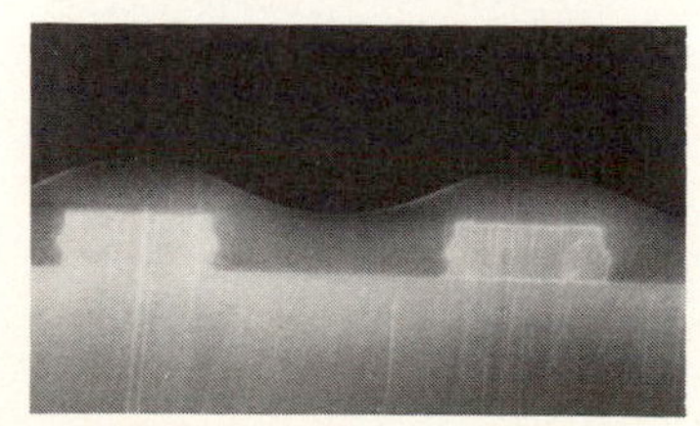

At 700°C, the deposition rate is only a few Å per minute, but by 900°C, the rate can, with appropriate flow rates, be a few thousand Å per minute (66). Temperatures above 1000°C–1100°C are not used because the higher temperatures favor localized crystallite formation in the amorphous film. In order to ensure that good stoichiometry is achieved, it is customary to use a substantial excess of ammonia—for example, a silane-to-ammonia ratio of 1:15 to 1:20. Hydrogen is commonly used as a carrier gas, but nitrogen or an inert gas such as helium has also been used (67). Various properties of silicon nitride are listed in Table 4.3. Also included for comparison are properties of nitride prepared by the PECVD process discussed in the next section.

Other silicon-bearing species such as $SiCl_4$ (68, 69) and SiH_2Cl_2 (27) can also be combined with ammonia to give silicon nitride. The overall reaction for $SiCl_4$ is

$$3SiCl_4 + 4NH_3 \rightarrow Si_3N_4 + 12HCl \quad 4.18$$

As is true in most of the surface reactions involved in CVD processing, the intermediate steps are not totally clear, but $SiCl_4$ and NH_3 begin reacting at room temperature or below, perhaps first by

$$SiCl_4 + 8NH_3 \rightarrow Si(NH_2)_4 + 4NH_4Cl \quad 4.19a$$

or by

$$SiCl_4 + 6NH_3 \rightarrow SiN_2H_2 + 4NH_4Cl \quad 4.19b$$

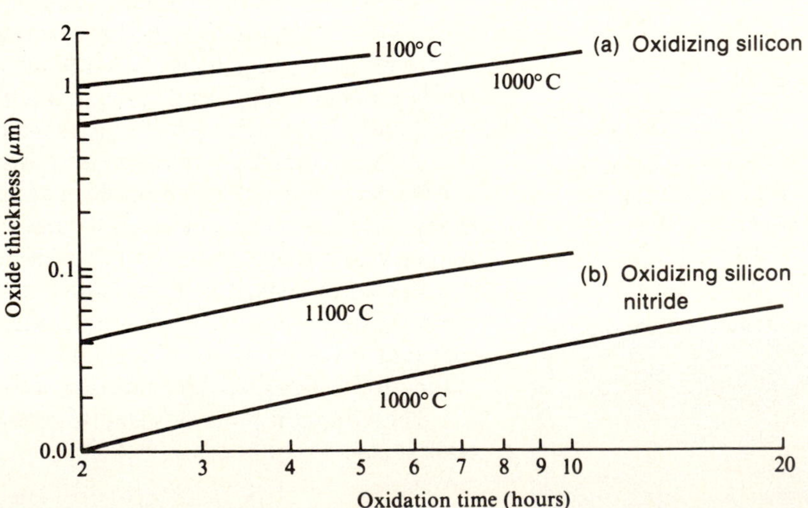

FIGURE 4.12

Oxidation rate of silicon and high-temperature silicon nitride in a wet oxygen atmosphere. (*Source* (a): Data from Bruce Deal, *J. Electrochem. Soc. 125,* p. 576, 1978. *Source* (b): Data from I. Fränz and W. Langheinrich, *Solid-State Electronics 14,* p. 499, 1971.)

TABLE 4.3

Properties of Silicon Nitride

Property	*High-Temperature CVD*	*PECVD**	*Reference*
Density (grams/cm^3)	2.8–3.1	2.4–2.8	(1, 2)
Hardness (Knoop)	1000–3500		(3)
(Moh)†	9$^+$		(3)
Stress (10^9 dynes/cm^2)‡	12–18	−2 to +5	(1, 2)
Expansion coefficient (10^{-6}/°C)	4	4–7	(1)
Refractive index	2.0§	2.0–2.2	(3, 2)

*PECVD properties can change substantially with deposition conditions. The values given here are typical ranges.
†Decreases to near 7 as deposition temperature drops from 900°C to 800°C.
‡+ = tension; − = compression.
§As film becomes silicon rich, the index will increase, reaching the value for silicon in limit of no nitrogen.

References: (1) J.R. Hollahan and R.S. Rosler, p. 350, in John L. Vossen and Werner Kern, eds., *Thin Film Processes*, Academic Press, New York, 1978. (2) A.C. Adams, *Solid State Technology*, p. 135, April 1983. (3) Kenneth E. Bean et al., *J. Electrochem. Soc. 114*, p. 733, 1967.

Then, as the temperature increases, these reaction products break down and rearrange to eventually give Si_3N_4 (68). Film deposition over the 550°C–1200°C temperature range has been reported, but temperatures of around 900°C are typical. The substantial amount of HCl produced by the reaction in Eq. 4.18 has been reported to help clean the deposition equipment and the wafer surface and to minimize ion contamination (70).

While all of the CVD nitride processes just described were originally done at atmospheric pressure, commercial low-pressure deposition equipment (LPCVD) using SiH_2Cl_2–NH_3 and SiH_4–NH_3 feed gases can also be used (27). By using ammonia and either $SiCl_4$ or SiF_4 in an LPCVD system operating at temperatures of from 1400°C to 1500°C, deposits of substantially pure α–Si_3N_4 can be obtained (71). Such deposits are, however, not applicable to Si or GaAs IC manufacturing because of the high temperatures involved (and perhaps because of the properties of crystalline silicon nitride, which have not been extensively studied).

4.6.5 Plasma Silicon Nitride

Silicon nitride deposition from a plasma was first reported in 1965 and used silane and ammonia in an RF-induced plasma (54). In addition, silane–nitrogen (72), N_2–$SiBr_4$ in a DC discharge plasma (54), and N_2–SiI_4 in a microwave afterglow (73) have also been reported. The common reactants now used are silane and ammonia, with argon or helium as a carrier/diluent. The first deposition equipment (reactor) was either for single-slice processing or of tubular design that held only a few slices and thus was not very amenable to volume production. However, a parallel-plate RF plasma deposition machine described in 1974 allowed production quantities to be economically processed and opened the way for the widespread application of plasma deposition processing[19] in the semiconductor industry (33).

Silicon nitride film production using the parallel-plate concept was apparently the earliest serious application of PECVD to the semiconductor industry (33, 74, 75). The main advantage is that deposition is at a much lower temperature than is straight CVD so that it can be used to cover, for example, metallized wafers. As in the case of the plasma oxide, however, many of the properties are considerably different from those of high-temperature CVD. Some of them were listed in Table 4.3, where they can be compared with the high-temperature values. Also, as in PECVD oxide, substantial hydrogen will be incorporated in the film when silane and/or ammonia are used.

4.6.6 APCVD and LPCVD Polysilicon

Applications of polysilicon to IC construction include its use for leads and MOS gate electrodes (see Chapter 10 for more details), as a fuse material in making ROMs (read-only memories), as load resistors in some MOS circuits, and as a diffusion source for emitter or sources/drains. In addition, active devices will occasionally be made in thin-film polysilicon (76, 77). All of these applications require layers that are generally less than a micron thick so that relatively slow deposition rates (generally associated with lower temperatures) are quite acceptable.

Polycrystalline silicon (polysilicon) occurs on occasion when the growing of single-crystal epitaxial silicon is attempted, and it will generally have rather large grain size. These occurrences are sporadic and undesirable. When polysilicon films are to be used in their own right, a uniform and generally small grain size is required, which usually implies a considerably lower deposition temperature than that used for epitaxy. Further, the applications themselves normally require relatively low-temperature depositions in order to minimize diffusion and/or wafer damage. In principle, any of the silicon epitaxy reactors discussed in Chapter 7 can be used, along with any of the reactions listed. Typically, however, for such thin applica-

---

[19]Tubular plasma etch equipment for photoresist removal was already widely used.

tions, low-pressure tube depositions with silane as the silicon source are now used. An LPCVD reactor provides for a very economical deposition (78), and silane decomposition proceeds at temperatures as low as 400°C. When much thicker polycrystalline layers are required—for example, in dielectrically isolated wafers (79)—speed of deposition, as well as an inexpensive feedstock is desirable, and the process used is much like that of high-temperature epitaxy. Typically, the hydrogen reduction of $SiCl_4$ or $SiHCl_3$ at temperatures as high as 1200°C in pancake-type reactors is used.

The deposition rate for LPCVD polysilicon depends on temperature, silane partial pressure, and whether or not a doping gas such as phosphine, arsine, or diborane is present. Fig. 4.13 shows growth rates r versus deposition temperature that have been reported. These rates can be characterized by

$$r = ae^{-E/RT} \tag{4.20}$$

where E is typically about 35 kcal/mol and which, as discussed earlier, is indicative of operation in the kinetically controlled region. Below about 580°C, the deposits are amorphous; above this temperature, the texture, grain size, and orientation depend on deposition conditions, with the grains generally increasing in size with deposition temperature (80–82). Fig. 4.14 shows the range over which var-

FIGURE 4.13

LPCVD polysilicon growth rate versus $1/T$. (*Source* (a): Data from Richard S. Rosler, *Solid State Technology 63,* pp. 63–70, April 1977. *Source* (b): Data from G. Harbeke et al., *J. Electrochem. Soc. 131,* p. 675, 1984. *Source* (c): Data from P. Joubert et al., ibid., *134,* p. 2541, 1987.)

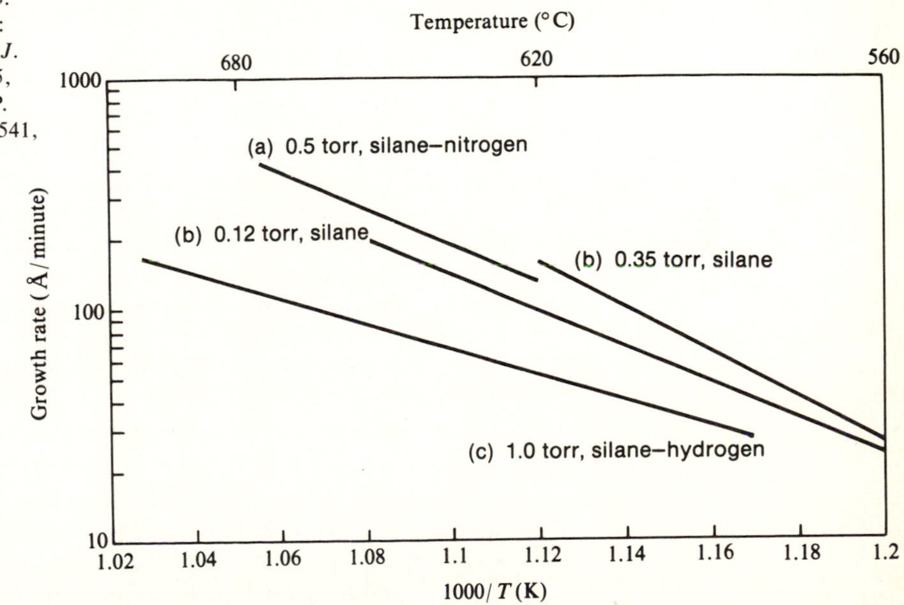

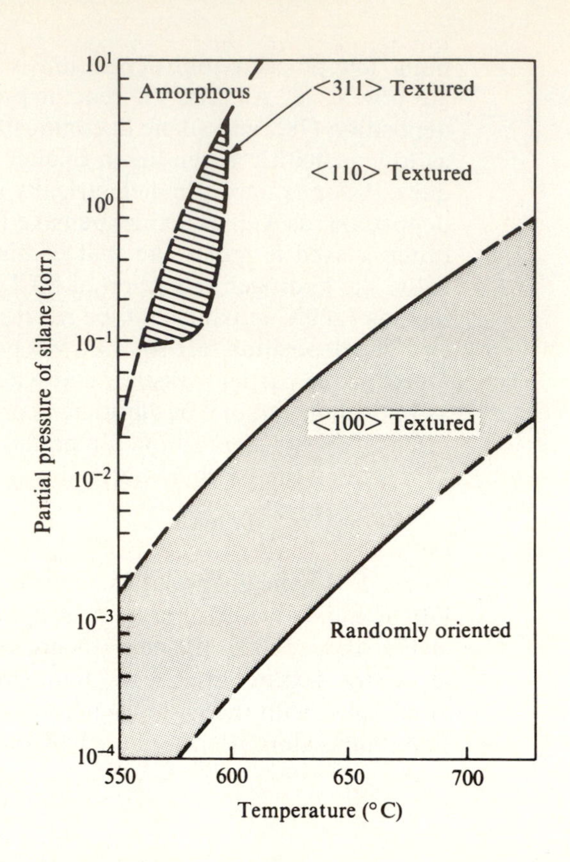

FIGURE 4.14

Morphology versus temperature of silicon deposited on SiO_2 from silane. (*Source:* P. Joubert et al., *J. Electrochem. Soc. 134,* p. 2541, 1987. Reprinted by permission of the publisher, The Electrochemical Society, Inc.)

ious crystallographic features have been reported to exist versus deposition temperature and silane partial pressure. By using molecular beam epitaxial techniques (evaporation of silicon in a very hard vacuum), polycrystalline silicon depositions down to 400°C have been observed (83). The CVD amorphous phase is unstable at higher temperatures and, if deposited, is generally converted to polysilicon during subsequent processing. One hour in nitrogen at 800°C significantly reduces the amount of amorphous material (80), while 30 minutes at 900°C is reported to give very stable, fine-grained material (81).

A problem with depositing polycrystalline silicon on amorphous substrates such as SiO_2 that was recognized early on is that of obtaining uniform nucleation and coverage (84). The details of the beginning of nucleation have since been studied for silicon on both SiO_2 and silicon nitride (85). This problem is quite pronounced with high-temperature $SiCl_4$ depositions but is hardly noticeable with low-temperature silane depositions. In fact, when high-temperature $SiCl_4$ depositions are desired for their higher rates, beginning with a short silane pyrolysis is sometimes helpful (86).

The addition of the doping gases phosphine or arsine to silane reduces the deposition rate, usually to impractical levels, as well as produces greater variation in layer thickness. Thus, n-type doping is done after deposition, either by diffusion or by ion implantation (see Chapters 8 and 9 for details of these processes) (87–89). The rate decrease appears to be caused by the dopant atoms that compete for sites where the silane reduction would normally take place. Diborane causes an increase in deposition rate (87) but not enough to cause any process problems, so p-doping can be done directly in LPCVD reactors.

Besides the problem of introducing the doping impurity into the polysilicon, the problem of the behavior of the dopant and carriers at the grain boundaries must also be addressed. The resistivity of a poly layer is never as low as that of a single-crystal layer of the same thickness even when the total impurity concentration is the same (90–92). One possible reason is that doping impurities that segregate at grain boundaries are electrically inactive. Another reason is that grain boundaries trap carriers, which cause potential barriers at the boundaries and, in turn, reduces the carrier mobility in polycrystalline material. The contribution of each of these effects is still in question, but, in general, the larger the grains, which implies fewer boundaries, the lower the resistivity.

4.6.7 Plasma Polysilicon

Polysilicon has been plasma deposited using both dichlorosilane (SiH_2Cl_2) (93) and silane (94) as feed material. The amorphous-poly temperature regions appear comparable to those of LPCVD poly. The stress in PECVD films, both poly and low-temperature amorphous, is usually compressive (93, 95). In principle, the rate should be less sensitive to temperature, and doping should be easier. However, despite the wide use of plasmas to produce hydrogenated amorphous silicon films, LPCVD continues to be the prime technique for IC poly applications.

4.7 OTHER CVD FILM MATERIALS

The materials already described include those most commonly used in IC fabrication, but by no means include all materials that can be deposited chemically from the vapor. Table 4.4 lists other less commonly deposited materials that have been reported. Some materials, such as tantalum pentoxide, have been examined for their dielectric properties. Others, such as Ge_3N_4, have been of interest as possible passivating layers for germanium or gallium arsenide pn junctions.

TABLE 4.4

Less Commonly Deposited Materials

Material	*Reactants*	*Method*	*Reference*
Aluminum oxide	$Al_2(CH_3)_6$* + O_2	CVD	(1)
	$Al(OC_3H_7)_3$† pyrolysis	CVD	(2)
	See reference 2 for other choices.		
Tantalum pentoxide	TDDAA‡ + O_2	CVD	(3)
	$TaCl_5 + H_2 + CO_2$	CVD	(4)
Aluminum nitride	$AlCl_3 \cdot NH_3$ pyrolysis	CVD	(5)
Boron nitride	$B_2H_6 + NH_3$	CVD	(6)
	$B_2H_6 + NH_3$	Plasma CVD	(7)
Germanium nitride	$GeCl_4 + NH_3$	CVD	(8, 9)
	$GeH_4 + NH_3$	Plasma CVD	(10)
Silicon carbide	$SiCl_4 + C_3H_8$	CVD	(11, 12)

*Trimethyl aluminum.
†Aluminum isopropoxide.
‡Tantalum dichloro-diethyoxy-acetylacetonate, $TaCl_2\ (OC_2H_5)_2C_5H_7O_2$.

References: (1) M.T. Duffy and Werner Kern, *RCA Review 31*, p. 754, 1970. (2) Werner Kern and Vladimir S. Ban, Chap. III–2, in John L. Vossen and Werner Kern, eds., *Thin Film Processes*, Academic Press, New York, 1978. (3) E. Kaplan et al., *J. Electrochem. Soc. 123*, p. 1570, 1976. (4) W.H. Knausenberger and R.N. Tauber, *J. Electrochem. Soc. 120*, p. 927, 1973. (5) T.L. Chu and R.W. Kelm, Jr., *J. Electrochem. Soc. 122*, p. 995, 1975. (6) M.J. Rand and J.F. Roberts, *J. Electrochem. Soc. 115*, p. 423, 1968. (7) S.B. Hyder and T.O. Yep, *J. Electrochem. Soc. 123*, p. 1721, 1976. (8) H. Nagai and T. Niimi, *J. Electrochem. Soc. 115*, p. 671, 1968. (9) Takehisa Yashiro, *J. Electrochem. Soc. 119*, p. 780, 1972. (10) D.B. Alford and L.G. Meiners, *J. Electrochem. Soc. 134*, p. 979, 1987. (11) K.E. Bean and P.S. Gleim, *J. Electrochem. Soc. 114*, p. 1158, 1967. (12) K. Furukawa et al., p. 231, in Ext. Abst. of Conference on Solid State Devices and Materials, Tokyo, 1987.

4.8 SPIN-ON GLASSES

Spin-on glasses, often called organic silicates, can be an alkoxysilane, such as tetraethoxysilane, in a solvent, an acyloxysilane in a solvent, or a silicon polymer such as a polysiloxane in a solvent. In each case, the liquid may be applied as a thin film to the wafer in the same manner as photoresist (see Chapter 5) and then heated to convert it to a silica film. When tetraethoxysilane (TEOS) based solutions are used, a catalytic agent (initiator) along with heat is ordinarily used to convert the TEOS to silica. When a doped film is made, the doping agent itself—for example, $POCl_3$—may suffice as the initiator (96). By reacting TEOS with acetic anhydride (acetyl oxide) in a suitable solvent, an acyloxysilane results. These materials can be decomposed into silica in the 200°C temperature range without the use of an initiator (97). The use of polysiloxanes increases the shelf life of a spin-on over that provided by acyloxysilanes and provides for thicker films (97). However, bake temperatures of from 600°C to 900°C are required (98).

These materials were originally used as diffusion dopant sources (see Chapter 8) and thus had to also have a doping impurity added to them. They have also been used to mask diffusions and for leads protection. An acyloxysilane rather than a silicon polymer is more appropriate as a leads protector since heating to about 200°C will drive off the solvent and most water. (Complete densification does not occur, however, until it is heated to about 800°C.) More recently, spin-ons have been examined for use as an interlevel insulator. In any of these applications, they must remain crack free over the operating temperature range and not produce enough stress in the wafer to introduce defects. The primary advantage of spin-on glasses is the simplicity of applying a layer to a wafer, but when they are used as an interlevel insulator, they have the additional advantage of topography planarization.

While most of the spin-on glasses are silicon based, formulations exist that use other materials. One, for example, decomposes into TiO_2 instead of SiO_2 upon heating (99) and is useful when a dielectric constant higher than that of SiO_2 is desired. For boron doping applications, a boron–nitrogen based polymer that reacts with moisture and oxygen to give B_2O_3 has been developed (100).

4.9 GLASS FRITTING

By using a glass frit applied by squeegee or electrophoretic deposition and then fired, a wider glass composition range than can be obtained from any of the other methods described is available. Frit also allows thick layers to be applied easily and holes and grooves in the wafer to be filled. However, the firing temperature must be high enough for the glass to soften and flow and hence severely restricts its use. The most common application, and one not related to ICs, is the covering of exposed junctions of high-voltage discrete devices.[20] Thus, the composition is chosen not only on the basis of expansion coefficient and softening point but also on passivating characteristics. Typically, a zinc–borosilicate glass is used. The general procedure is to formulate the glass, crush it, ball-mill it to a 5–10 μm powder, and then either mix it with a binder and squeegee it over the surface or disperse it along with suitable ions in a dielectric liquid such as isopropyl alcohol and electrophoretically deposit it (101–103).

[20]Because of low breakdown voltages associated with the curvature of planar pn junctions, high-voltage junction devices are usually mesa structures and would have exposed junctions were it not for a coating of silicone or deposited glass.

SAFETY

CHAPTER 4

Many of the reactants used in chemical vapor deposition, as well as some of the reaction products, are very poisonous and/or pyrophoric and/or explosive. Thus, care must be taken to ensure that leaks do not exist in the piping or deposition chamber and that the CVD reaction products are safely burned or vented. When new processes are being developed, all reactants should be carefully studied with respect to health hazards. In particular, metallorganics should be viewed with great caution. CVD equipment should, in general, be housed in rooms with high-capacity emergency exhaust facilities (scram rooms), and gas cabinet enclosures should be continuously exhausted.

Silane and hydrogen are both very explosive. If a small leak occurs in a silane line, the issuing silane will generally spontaneously ignite and burn, leaving a white silica powder telltale mark. However, if there is not enough air for the silane to burn as, for example, in a small enclosure, probably nothing will happen until the enclosure door is opened, at which time a violent explosion will occur. At least one death has occurred due to an explosion following the raising of the door of a fume hood with the blower turned off in which silane had been flowing all night. To minimize the likelihood of such occurrences, silane is usually supplied diluted with an inert gas such as argon to the point where an explosion is unlikely.

Hydrogen is sometimes used as a carrier gas and is explosive when it is mixed with air in concentrations by volume of from 4% to 75% hydrogen (104). It will not spontaneously ignite at room temperature but has an auto-ignition temperature of about 585°C. Being lighter than air, hydrogen tends to collect in stagnant volumes near room ceilings. Vents and good air circulation will minimize the risk of explosion in the event that a leak does occur. Hydrogen monitors should be installed wherever hydrogen is likely to escape into confined rooms. While wafer-fab facilities will generally have piped-in hydrogen, laboratories often use individual cylinders. (For precautions in the handling of hydrogen cylinders, the reader should consult reference 104.)

Caution should also be exercised in the design of the piping system so that no possibility of the feeding of a gas under high pressure back into a cylinder of low-pressure gas exists. To be absolutely safe, there should be no direct pipe connections. As an example of the seriousness of such episodes, a cylinder of silane became contaminated with nitrous oxide, presumably from cross-connected piping, and three people were killed as they tried to bleed off the mixture from the silane cylinder (105). It has sometimes been the practice to depend on pressure regulators to reduce the pressure to the point where no cross contamination should occur. However, if the regulators fail, as they do on occasion, then contamination can still occur. Even feeding the various gases separately into an atmospheric pressure reaction chamber is no absolute assurance of safety. If the exhaust vent clogs, pressure can build up in the chamber and, unless the chamber explodes first, feed gases back into low-pressure lines.

High-pressure gas cylinders and cylinders containing explosive or poisonous gases should be handled so that the likelihood of the cylinder falling and breaking off the valve does not exist. If cylinder venting should be necessary and there is any possibility of an explosion, then a remote system to open the valve, coupled with sandbags and personnel flack clothing, should be used.

While not normally considered a problem, fine silicon dust such as might be produced by gas phase reactions is explosive when it is mixed with air. The likelihood of such an explosion is minimal for particle sizes above perhaps 10 μm (106).

Halide compounds such as $SiCl_4$, $SiHCl_3$, and SiH_2Cl_2 are more noxious than poisonous in small quantities, although in the presence of moisture from the air or body membranes, HCl (or HBr or HF) will be formed and will cause respiratory tract or lung damage. $SiHCl_3$ is flammable, and SiH_2Cl_2 may self-ignite if a bottle containing it is broken. Once it is ignited, there is no good way of fighting it (107). The sometimes recommended use of a coarse water sprinkle does not necessarily contain it and produces large amounts of HCl. Phosphine, diborane, arsine, and stibine, all occasionally used as an oxide dopant, are very toxic and, because of the toxicity, are generally supplied diluted with an inert carrier gas. Their reported safe limits in air being breathed are given in Table 4.5. Also shown is the limit for ammonia, which is about three orders of magnitude higher.

TABLE 4.5

Safe Limits for Some Gases Used in CVD

Gas	*PEL* (ppm)*	*IDLH† (ppm)*
Arsine	0.05	6
Diborane	0.1	40
Phosphine	0.3	200
Stibine	0.1	40
Ammonia	100	

*PEL = Permissible exposure limit (maximum breathing air concentration for an 8 hour day).

†IDLH = Immediately dangerous to life or health (maximum limit to which one can be exposed for 30 minutes and survive with no impairment).

Source: NIOSH/OSHA Pocket Guide to Chemical Hazards.

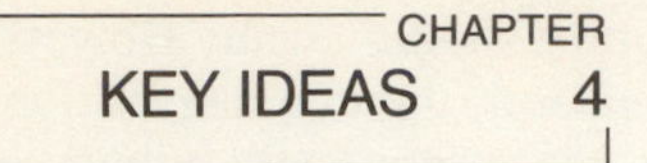

CHAPTER 4 KEY IDEAS

- ☐ Sputtering or evaporation are best suited for metal depositions.
- ☐ Chemical vapor deposition is most commonly used for dielectric depositions.
- ☐ In a few specialized applications, either glass frit or spin-on organic silicates are used instead of CVD.
- ☐ Low-pressure CVD is used to minimize gaseous diffusion effects and allow closer spacing of wafers inside a deposition chamber.
- ☐ Plasma-deposited materials are generally not stoichiometric and may contain a high percentage of reactant products.
- ☐ Sputtering and evaporation can deposit material on cold wafers. CVD and plasma CVD generally can deposit only on heated wafers.

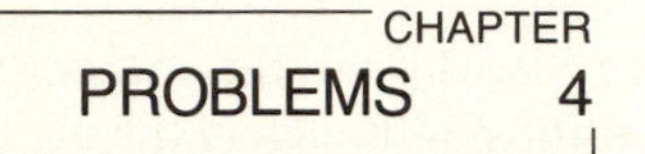

CHAPTER 4 PROBLEMS

1. Based on Eq. 4.1, describe an experiment that will allow the separation of intrinsic stress and stress introduced by differential thermal expansion. Comment: Beware of changing experimental conditions such as deposition temperature or substrate material that may, in turn, change σ_{if} in an unpredictable fashion.
2. Calculate the mean free path in an evaporation chamber at a chamber pressure of 10^{-4} torr. Consider the gas molecule diameter to be 4 Å.
3. Which of the following materials would you suggest be deposited by sputtering and which by evaporation?

   Aluminum
   Platinum
   Aluminum oxide
   Silicon
   Silicon nitride
   Explain your choices.
4. For a process that is kinetically controlled, plot rate versus temperature if the rate is 1 μm/minute at 1100°C and the activation energy is 35 kcal/mol.
5. Plot the following CVD temperature and rate data and mark the region that is diffusion controlled:

T (°C)	r (μm/min.)
1250	2.2
1200	2.0
1150	2.1
1100	1.8
1050	1.3
1000	0.8
950	0.6

6. Since in a low-pressure deposition system the gaseous diffusion constant is much larger than it is in an atmospheric system, would deposition in the diffusion-controlled region be expected to also be much higher? Explain your answer.
7. Some CVD reactor designs in which the chamber is a long tube with the exhaust at one end have several gas inlets positioned along the length of the tube. What would be a possible advantage of such an inlet system?
8. Sketch the gas piping for a high-temperature silicon nitride deposition system. (Be sure to read the safety section.)

CHAPTER 4 REFERENCES

1. J.L. Vossen and J.J. Cuomo, Chap. II–1, "Glow Discharge Sputter Deposition," in John L. Vossen and Werner Kern, eds., *Thin Film Processes*, Academic Press, New York, 1978.
2. K.G. Geraghty and L.F. Donaghey, "Kinetics of the Reactive Sputter Deposition of Titanium Oxides," *J. Electrochem. Soc. 123*, pp. 1201–1207, 1976.
3. J. Kortlandt and L. Oosting, "Deposition and Properties of RF Reactively Sputtered SiO_2 Layers," *Solid State Technology*, pp. 153–159, October 1982.
4. C.J. Mogab et al., "Effect of Reactant Nitrogen Pressure on the Microstructure and Properties of Reactively Sputtered Silicon Nitride Films," *J. Electrochem. Soc. 122*, pp. 815–822, 1975.
5. S. Suyama et al., "The Effect of Oxygen–Argon Mixing on Properties of Sputtered Silicon Dioxide Films," *J. Electrochem. Soc. 134*, pp. 2260–2264, 1987.
6. G.J. Kominiak, "Silicon Nitride by Direct RF Sputter Deposition," *J. Electrochem. Soc. 122*, pp. 1271–1273, 1975.
7. T.C. Tisone, "Low Voltage Triode Sputtering with a Confined Plasma," *Solid State Technology*, pp. 34–40 and references therein, December 1975.
8. Vance Hoffman, "High Rate Magnetron Sputtering for Metallizing Semiconductor Devices," *Solid State Technology*, pp. 57–61 and references therein, December 1976.
9. Ted Van Vorous, "Planar Magnetron Sputtering: A New Industrial Coating Technique," *Solid State Technology*, pp. 62–66, December 1976.
10. Various chapters, in Carroll F. Powell, Joseph H. Oxley, and John M. Blocher, Jr., eds., *Vapor Deposition*, John Wiley & Sons, New York, 1966.
11. George R. Thompson, Jr., "Ion Beam Coating: A New Deposition Method," *Solid State Technology*, pp. 73–77, December 1978.
12. E.A.V. Ebsworth, *Volatile Silicon Compounds*, Macmillan, New York, 1963.
13. Vladimir S. Ban, "Transport Phenomena Measurements in Epitaxial Reactors," *J. Electrochem. Soc. 125*, pp. 317–320, 1978.
14. R. Takahashi et al., "Gas Flow Pattern and Mass Transfer Analysis in a Horizontal Flow Reactor for Chemical Vapor Deposition," *J. Electrochem. Soc. 119*, pp. 1406–1412, 1972.
15. S. Rhee et al., "On Three-Dimensional Transport Phenomena in CVD Processes," *J. Electrochem. Soc. 134*, pp. 2552–2559, 1987.
16. L.J. Giling, "Gas Flow Patterns in Horizontal Epitaxial Reactor Cells Observed by Interference Holography," *J. Electrochem. Soc. 129*, pp. 634–644, 1982.
17. Vladimir S. Ban and Stephen L. Gilbert, "The Chemistry and Transport Phenomena of Chemical Vapor Deposition of Silicon from $SiCl_4$," *J. Crystal Growth 31*, pp. 284–289, 1975.
18. J. Monkowski and J. Stach, "System Characterization of Planar Source Diffusion," *Solid State Technology*, pp. 38–43, November 1976.
19. C.W. Manke and L.F. Donaghey, "Analysis of Transport Processes in Vertical Cylinder Epitaxy Reactors," *J. Electrochem. Soc. 124*, pp. 561–569, 1977.
20. P.C. Rundle, "The Epitaxial Growth of Silicon in Horizontal Reactors," *Int. J. Electronics 24*, pp. 405–413, 1968.
21. F.C. Eversteyn et al., "A Stagnant Layer Model for the Epitaxial Growth of Silicon from Silane in a Horizontal Reactor," *J. Electrochem. Soc. 117*, pp. 925–931, 1970.
22. S.E. Bradshaw, "The Kinetics of Epitaxial Silicon Deposition by the Hydrogen Reduction of Chlorosilanes," *Int. J. Electronics 21*, pp. 205–227, 1966.
23. J.C. Gillis et al., "Fluid Mechanical Model for CVD in a Horizontal RF Reactor," pp. 21–29, in *CVD '84* (Vol. 84–6), Electrochemical Society, Pennington, N.J., 1984.
24. W.H. Shepherd, "Vapor Phase Deposition and Etching of Silicon," *J. Electrochem. Soc. 112*, pp. 988–994, 1965.
25. E. Fujii et al., "A Quantitative Calculation of the Growth Rate of Epitaxial Silicon from $SiCl_4$ in a

Barrel Reactor," *J. Electrochem. Soc. 119*, pp. 1106–1113, 1972.

26. H.A. Lord, "Convective Transport in Silicon Epitaxial Deposition in a Barrel Reactor," *J. Electrochem. Soc. 134*, pp. 1227–1235, 1987.
27. Richard S. Rosler, "Low Pressure CVD Production Processes for Poly, Nitride, and Oxide," *Solid State Technology*, pp. 63–70, April 1970.
28. William Velander and Daniel White, Jr., "Induced Convective Effects on Intrawafer Uniformity in LPCVD," *J. Electrochem. Soc. 134*, pp. 951–956, 1987.
29. K.F. Jensen and D.B. Graves, "Modeling and Analysis of Low Pressure CVD Reactors," *J. Electrochem. Soc. 130*, pp. 1950–1957, 1983.
30. Karl F. Roenigk and Klavs F. Jensen, "Low Pressure CVD of Silicon Nitride," *J. Electrochem. Soc. 134*, pp. 1777–1785, 1987.
31. Makarand G. Joshi, "Modeling of LPCVD Reactors," *J. Electrochem. Soc. 134*, pp. 3118–3122, 1987.
32. M.L. Hammond, "Introduction to Chemical Vapor Deposition," *Solid State Technology*, pp. 61–65, December 1965.
33. A. Reinberg, "RF Plasma Deposition of Inorganic Films for Semiconductor Applications," *J. Electrochem. Soc. 121*, p. 85C, abst. no. 6, 1974, or *Electrochem. Soc. Ext. Abst. 74–1*, abst. no. 6, 1974.
34. Russ A. Morgan, *Plasma Etching in Semiconductor Fabrication*, Elsevier, New York, 1985.
35. S. Rhee and J. Szekely, "The Analysis of Plasma-Enhanced Chemical Vapor Deposition of Silicon Films," *J. Electrochem. Soc. 133*, pp. 2194–2201, 1986.
36. Stephen Dzioba et al., "Downstream Plasma Induced Deposition of SiN_x on Si, InP, and InGaAs," *J. Electrochem. Soc. 134*, pp. 2599–2603, 1987.
37. John E. Spencer et al., "New Directions in Dry Processing Using the Flowing Afterglow of a Microwave Discharge," *Electrochem. Soc. Ext. Abst. 86–2*, abst. no. 285, 1986.
38. P.D. Garner et al., "Characteristics of the Low-Temperature-Deposited SiO_2 $Ga_{0.47}In_{0.53}$ as Metal/Insulator/Semiconductor Interface," *Thin Solid Films 117*, pp. 173–190, 1984.
39. C.H.J. v.d. Brekel and P.J. Severin, "Control of the Deposition of Silicon Nitride by 2573 Å Radiation," *J. Electrochem. Soc. 119*, pp. 372–376, 1972.
40. J.W. Peters et al., "Low Temperature Photo-CVD Silicon Nitride: Properties and Applications," *Solid State Technology*, pp. 121–126, September 1980.
41. "LPCVD System Accommodates Different Sizes and Types of Substrates," *Semiconductor International*, p. 198, April 1988.
42. H.A. Tanner and L.B. Lockart, Jr., "Germane Reflection Reducing Coatings for Glass," *J. Opt. Soc. Am. 36*, pp. 701–706, 1946.
43. E.L. Jordon, "A Diffusion Mask for Germanium," *J. Electrochem. Soc. 108*, pp. 478–481, 1961.
44. Julius Klerer, "On the Mechanism of the Deposition of Silica by Pyrolytic Decomposition of Silanes," *J. Electrochem. Soc. 112*, pp. 503–506, 1965.
45. V.D. Wohlheiter and R.A. Whitner, "High Production System for the Deposition of Silicon Dioxide," *Electrochem. Soc. Ext. Abst. 75–1*, abst. no. 180, pp. 424–425, 1975.
46. W. Steinmaier and J. Bloem, "Successive Growth of Si and SiO_2 in Epitaxial Apparatus," *J. Electrochem. Soc. 111*, pp. 206–209, 1964.
47. Myron J. Rand and John L. Ashworth, "Deposition of Silica Films on Germanium by the Carbon Dioxide Process," *J. Electrochem. Soc. 113*, pp. 48–50, 1966.
48. N. Goldsmith and W. Kern, "The Deposition of Vitreous Silicon Dioxide Films from Silane," *RCA Review 28*, pp. 153–165, 1967.
49. K. Strater and A. Mayer, "The Oxidation of Silane, Phosphine, and Diborane during Deposition of Doped-Oxide Diffusion Sources," pp. 469–480, in Rolf R. Haberecht and Edward L. Kern, eds., *Semiconductor Silicon*, Electrochemical Society, New York, 1969.
50. T.L. Chu et al., "Silica Films by the Oxidation of Silane," *Trans. Metall. Soc. AIME 242*, pp. 532–538, 1968.
51. William J. Kroll, Jr. et al., "Formation of Silica Films on Silicon Using Silane and Carbon Dioxide," *J. Electrochem. Soc. 122*, pp. 573–578, 1975.
52. A.K. Gaind et al., "Preparation and Properties of SiO_2 Films from SiH_4–CO_2–H_2," *J. Electrochem. Soc. 123*, pp. 111–117, 1976.

53. A.K. Gaind et al., "Preparation and Properties of CVD Oxides with Low Charge Levels from SiH_4–CO_2–HCl–H_2 Systems," *J. Electrochem. Soc. 123*, pp. 238–246, 1976.
54. H.F. Sterling and R.G.G. Swann, "Chemical Vapor Deposition Promoted by R.F. Discharge," *Solid-State Electronics 8*, pp. 653–654, 1965.
55. E.P. van de Ven et al., "High Rate PECVD to Reduce Hillock Growth in Aluminum Interconnects," in Proc. 4th IEEE V-MIC Conference, 1987.
56. Akira Takamatsu et al., "Plasma-Activated Deposition and Properties of Phosphosilicate Glass Film," *J. Electrochem. Soc. 131*, pp. 1865–1870, 1984.
57. A.C. Adams et al., "Characterization of Plasma-Deposited Silicon Dioxide," *J. Electrochem. Soc. 128*, pp. 1545–1551, 1981.
58. John R. Hollahan, "Deposition of Plasma Silicon Oxide Thin Films in a Production Planar Reactor," *J. Electrochem. Soc. 126*, pp. 930–934, 1979.
59. L.G. Meiners, "Electrical Properties of SiO_2 and Si_3N_4 Dielectric Layers on InP," *J. Vac. Sci. Technol. 19*, pp. 373–379, 1981.
60. F.S. Becker and S. Röhl, "Low Pressure Deposition of Doped SiO_2 by Pyrolysis of Tetraethylorthosilicate (TEOS)," *J. Electrochem. Soc. 134*, pp. 2923–2931, 1987, and extensive references therein.
61. Jerry E. Tong et al., "Process and Film Characterization of PECVD Borophosphosilicate Films for VLSI Applications," *Solid State Technology*, pp. 161–170, January 1984.
62. K. Nassau et al., "Modified Phosphosilicate Glasses for VLSI Applications," *J. Electrochem. Soc. 132*, pp. 409–415, 1985.
63. W. Kern et al., "Optimized Chemical Vapor Deposition of Borophosphosilicate Glass Films," *RCA Review 46*, pp. 117–152, 1985.
64. V.Y. Doo et al., "Preparation and Properties of Pyrolytic Silicon Nitride," *J. Electrochem. Soc. 113*, pp. 1279–1281, 1966.
65. Tetsuya Arizumi et al., "Thermodynamical Analyses and Experiments for the Preparation of Silicon Nitride," *Jap. J. Appl. Phys. 7*, pp. 1021–1027, 1968.
66. Kenneth E. Bean et al., "Some Properties of Vapor Deposited Silicon Nitride Films Using the SiH_4–NH_3–H_2 System," *J. Electrochem. Soc. 114*, pp. 733–737, 1968.
67. V.Y. Doo et al., "Property Changes in Pyrolytic Silicon Nitride with Reactant Composition Changes," *J. Electrochem. Soc. 115*, pp. 61–64, 1968.
68. M.J. Grieco et al., "Silicon Nitride Films from $SiCl_4$ Plus NH_3: Preparation and Properties," *J. Electrochem. Soc. 115*, pp. 525–531, 1968.
69. V.D. Wohlheiter and R.A. Whitner, "A High Production System for the Deposition of Silicon Nitride," *J. Electrochem. Soc. 119*, pp. 945–948, 1972.
70. E.A. MacKenna and P. Kodama, "The Suppression of Ionic Contamination during Silicon Nitride Deposition," *J. Electrochem. Soc. 119*, pp. 1094–1099, 1972.
71. J.J. Gebhardt et al., "Chemical Vapor Deposition of Silicon Nitride," *J. Electrochem. Soc. 123*, pp. 1578–1582, 1975.
72. R. Gereth and W. Scherber, "Properties of Ammonia-Free Nitrogen–Si_3N_4 Films," *J. Electrochem. Soc. 119*, pp. 1248–1254, 1972.
73. M. Shiloh et al., "Preparation of Nitrides by Active Nitrogen," *J. Electrochem. Soc. 124*, pp. 295–300, 1977.
74. Richard S. Rosler et al., "A Production Reactor for Low Temperature Plasma-Enhanced Silicon Nitride Deposition," *Solid State Technology*, pp. 45–50, June 1976.
75. A.K. Sinha et al., "Reactive Plasma Deposited Si–N Films for MOS–LSI Passivation," *J. Electrochem. Soc. 125*, pp. 601–608, 1975.
76. M. Dutoit and F. Sollberger, "Lateral Polysilicon p–n Diodes," *J. Electrochem. Soc. 125*, pp. 1648–1651 and references therein, 1978.
77. R. Pennell et al., "Fabrication of High Voltage Polysilicon TFTs on an Insulator," *J. Electrochem. Soc. 133*, pp. 2358–2361, 1986.
78. Richard S. Rosler, "Low Pressure CVD Production Processes for Poly, Nitride, and Oxide," *Solid State Technology*, pp. 63–70, April 1970.
79. K.E. Bean and W.R. Runyan, "Dielectric Isolation: Comprehensive, Current and Future," *J. Electrochem. Soc. 124*, pp. 5C–12C, 1977.
80. T.I. Kamins, "Structure and Stability of Low Pressure Chemically Vapor-Deposited Silicon

Films," *J. Electrochem. Soc. 125*, pp. 927–932, 1978.

81. G. Harbeke et al., "Growth and Physical Properties of LPCVD Polycrystalline Silicon Films," *J. Electrochem. Soc. 131*, pp. 675–682, 1984.
82. P. Joubert et al., "The Effect of Low Pressure on the Structure of LPCVD Polycrystalline Silicon Films," *J. Electrochem. Soc. 134*, pp. 2541–2545, 1987.
83. Makoto Matsui et al., "Low Temperature Formation of Polycrystalline Silicon Films by Molecular Beam Deposition," *J. Appl. Phys. 52*, pp. 995–998, 1982.
84. Earl G. Alexander and W.R. Runyan, "A Study of Factors Affecting Silicon Growth on Amorphous SiO_2 Surfaces," *Trans. Metall. Soc. AIME 236*, pp. 284–290, 1966.
85. W.A.P. Claassen and J. Bloem, "The Nucleation of CVD Silicon on SiO_2 and Si_3N_4 Substrates Part III," *J. Electrochem. Soc. 128*, pp. 1353–1359, 1981, and Parts I and II referenced therein.
86. V.J. Silvestri, "Growth Kinematics of a Polycrystalline Trench Refill Process," *J. Electrochem. Soc. 133*, pp. 2374–2376, 1986.
87. F.C. Eversteyn and P.H. Put, "Influence of AsH_3, PH_3, and B_2H_6 on the Growth Rate and Resistivity of Polycrystalline Silicon Films Deposited from a SiH_4–H_2 Mixture," *J. Electrochem. Soc. 120*, pp. 106–110, 1973.
88. B.S. Meyerson and W. Olbricht, "Phosphorus-Doped Polycrystalline Silicon Via LPCVD," *J. Electrochem. Soc. 132*, pp. 2361–2365, 1984. (See also the following article beginning on p. 2366.)
89. Andrew Yeckel and Stanley Middleman, "A Model of Growth Rate Nonuniformity in the Simultaneous Deposition and Doping of a Polycrystalline Silicon Film by LPCVD," *J. Electrochem. Soc. 134*, pp. 1275–1281, 1987.
90. M.M. Mandurah et al., "Phosphorus Doping of Low Pressure Chemically Vapor-Deposited Silicon Films," *J. Electrochem. Soc. 126*, pp. 1019–1023, 1979.
91. J.R. Monkowski et al., "Comparison of Dopant Incorporation into Polycrystalline and Monocrystalline Silicon," *Appl. Phys. Lett. 35*, pp. 410–412, 1979.
92. Junichi Murota and Takashi Sawai, "Electrical Characteristics of Heavily Arsenic and Phosphorus Doped Polycrystalline Silicon," *J. Appl. Phys. 53*, pp. 3702–3708, 1982.
93. T.I. Kamins and K.L. Chiang, "Properties of Plasma-Enhanced CVD Silicon Films," *J. Electrochem. Soc. 129*, pp. 2326–2331, 1982.
94. W.R. Burger et al., in J. Bloem et al., eds., Proc. 4th European Conference on Chemical Vapor Deposition, The Netherlands, 1984.
95. J.P. Harbison et al., "Effect of Silane Dilution on Intrinsic Stress in Glow Discharge Hydrogenated Amorphous Silicon Films," *J. Appl. Phys. 55*, pp. 946–951, 1984.
96. K.M. Mar, "Diffusion in Silicon from a Spin-On p-Doped Silicon Oxide Film," *J. Electrochem. Soc. 126*, pp. 1252–1257, 1979.
97. K.D. Beyer, "A New Paint-On Diffusion Source," *J. Electrochem. Soc. 123*, pp. 1556–1560, 1976.
98. Pei-Lin Pai et al., "Material Characteristics of Spin-On Glasses for Interlevel Dielectric Applications," *J. Electrochem. Soc. 134*, pp. 2829–2834, 1987.
99. Bulent E. Yoldas, "Diffusion of Dopants from Optical Coatings and Single Step Formation of Antireflective Coatings and p–n Junctions in Photovoltaic Cells," *J. Electrochem. Soc. 127*, pp. 2478–2481, 1980.
100. B.H. Justice et al., "A Novel Spin-On Dopant," *Solid State Technology*, pp. 1553–1559, October 1984.
101. A.H. Berman, "Glass Passivation Improves High Voltage Transistors," *Solid State Technology*, pp. 29–32, March 1976.
102. Yutaka Misawa, "Properties of ZnO–B_2O_3–SiO_2 Glasses for Surface Passivation," *J. Electrochem. Soc. 131*, pp. 1862–1865, 1984.
103. Masaru Shimbo et al., "Electrophoretic Deposition of Glass Powder for Passivation of High Voltage Transistors," *J. Electrochem. Soc. 132*, pp. 393–398, 1985.
104. *Matheson Gas Data Book*, The Matheson Company, 1980.
105. *Electronic News*, p. 30, September 5, 1988.
106. R.K. Eckhoff et al., "Ignitability and Explosibility of Silicon Dust Clouds," *J. Electrochem. Soc. 133*, pp. 2631–2637, 1986.
107. David Ranier, "Dichlorosilane—Leak Spill and Fire Control," *Electrochem. Soc. Ext. Abst. 85–2*, abst. no. 299, 1985.

CHAPTER 5

Lithography

5.1 INTRODUCTION

The steps required to go from a circuit function existing only in a designer's mind to a pattern etched in oxide or metal are shown in Fig. 5.1. First, the circuit must be designed, and then the size of the various circuit elements, from transistor bases to lead widths, must be determined and their exact positions on the chip established. All of this information is then converted into computer data files that can be used by the pattern generator (optical or E-beam) to determine which positions on the reticle to expose. A reticle is generally printed at 5× or 10× the size of the finished chip. Its image is then reduced to the correct size during printing onto the mask to be used for wafer printing. When an E-beam wafer writer is used to write the image on the wafer directly from the digitized information, as Fig. 5.1 shows, pattern generation and stepping operations are bypassed. When steppers are used for printing, a separate reduction step is ordinarily not needed.

Masks are not produced in the wafer-fab area but rather in a separate mask shop. Accordingly, only a general discussion of their fabrication will be given. However, since the mask quality drastically affects the wafer-fab yields, it is very important that methods of evaluating quality be considered in some detail.

5.2 PHOTOMASKS

5.2.1 Masks for Optical Printing

Making masks for optical printing starts with square glass plates somewhat larger than the slice to be printed. The plates are first coated with a material opaque in the wavelength region used to expose resist. Chromium is most commonly used. The coating process is very critical since if any pinholes exist in the metal, they will remain as part of the finished mask. The metal is then coated with resist,[1] and the resist is exposed to the appropriate image, which

[1]Called resist because it "resists" the chemicals used to etch into the unexposed areas.

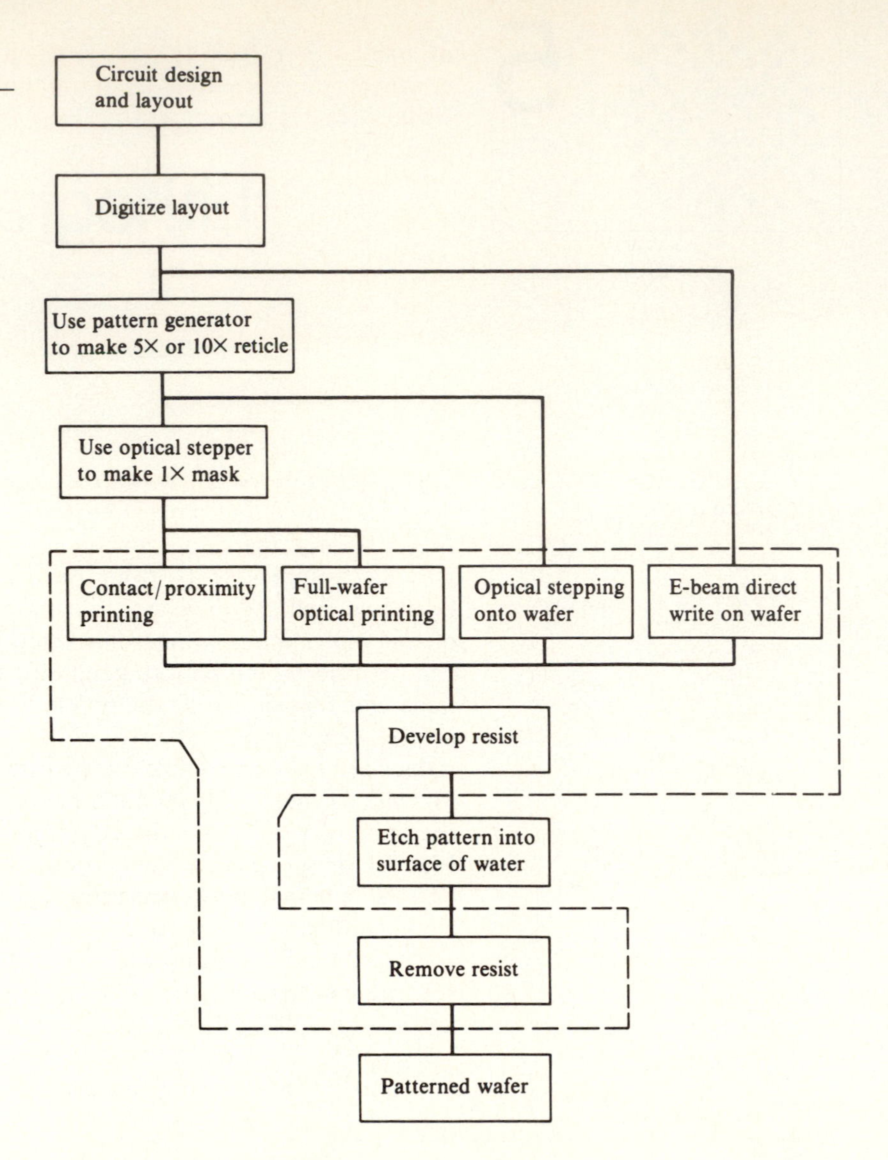

FIGURE 5.1

Path from circuit design to a patterned wafer. (This chapter considers only the operations within the dashed box.)

may be by either an optical or an electron beam (E-beam) pattern generator. The optical generator builds up the image by exposing a short line segment, moving a small distance (stepping), exposing another short segment, and continuing until the whole mask pattern has been exposed. The light used is very intense so that the individual exposures can be quite short. E-beam printing requires a resist that is sensitive to an electron beam rather than to light. Since masks are several inches square, and since electron beams can be deflected over only a very small angle without distortion, some combination of small-area E-beam scan combined with stage travel is used to cover the complete mask. The combination of low beam energy and relatively insensitive resist makes E-beam writing very slow. However, because diffraction effects are much less than they are in optical pattern generators, E-beam writing is indispensable for high-resolution masks.

The first plate made by either generator is usually referred to as a reticle and will include only one chip pattern if a mask is to be made. If the reticle is to be used directly in a stepper, it will include as many whole chip patterns as will fit into the stepper field of view. This aspect is discussed in some detail in section 5.11. The reticle geometry will generally be drawn at either 5× or 10× actual size, with 5× as the preferred size because of better manufacturing cost effectiveness. A mask made by printing enough 1× patterns from the reticle to cover a full wafer is referred to as a master. Multiple copies of the master can be made by contact printing and are much less expensive than masters but have many more defects. Reticles are used directly in step and repeat-projection printers to print chip by chip on the wafer. The full-size mask is used with contact printers, with proximity printers, or by 1:1 scanning projection printers. Projection printers will generally use masters, and contact printers will use copies.

Soda lime glass blanks, which have a rather large thermal expansion coefficient, were used as mask substrates until registration demands required a lower expansivity. Borosilicate glass was the next step and was used extensively for several years. Fused silica is now widely used for new designs. If an old product is to be switched to fused silica, a full set of new masks must be made; otherwise, the set will not thermally track. Thus, replacement masks for products that have not been converted will still be of borosilicate glass. Bright chromium is the normal metallization, but to reduce reflections from the mask, chromium with an antireflecting front surface (AR chrome) is sometimes used. When visual level-to-level alignment is used and masks for the different levels are not accurately sized to one another (or if the alignment markers are not properly placed), slight adjustments may be made by the operator in an effort to get the "best alignment." In such cases, it is helpful to have masks that are partially transparent (see-through) in the visible region, while remaining completely dark in the near UV regions where the resist is sensitive. This feature is most useful in the case of positive resist masks, which usually are mostly opaque. CVD iron oxide is a material that works satisfactorily for this application (1). Fortunately, however, sizing and placement control has now improved, and such operator judgment is seldom needed. Indeed, much alignment is now done completely automatically. This better control has been achieved not only through improved mask making techniques but by better wafer-fab practices as well.

5.2.2 X-Ray Masks

The making of X-ray masks is complicated by the fact that no thick, rigid, mask blank material that is transparent to the X rays used in X-ray lithography is available. Very thin membranes must be used, and they have problems of both durability and dimensional stability. Membrane materials that have been used include polyimide, parylene, titanium, silicon, silicon carbide, silicon nitride, and boron nitride (2, 3). The inorganic films are generally preferred because of their greater strength and dimensional stability. A gold layer about 1 μm thick is ordinarily used as the X-ray opaque material. The steepness of the gold edges will affect printing sharpness, and selective electroplating of thick gold onto a thin, patterned gold layer is sometimes used instead of initially depositing a thick gold layer and then etching the pattern into it.

The steps in making an X-ray mask using CVD membrane material are shown in Fig. 5.2. Transparent, stress-free silicon carbide layers on silicon can be obtained by depositing stoichiometric SiC from a mixture of 0.89% $SiCl_4$, 0.37% C_3H_8, and 98.74% hydrogen

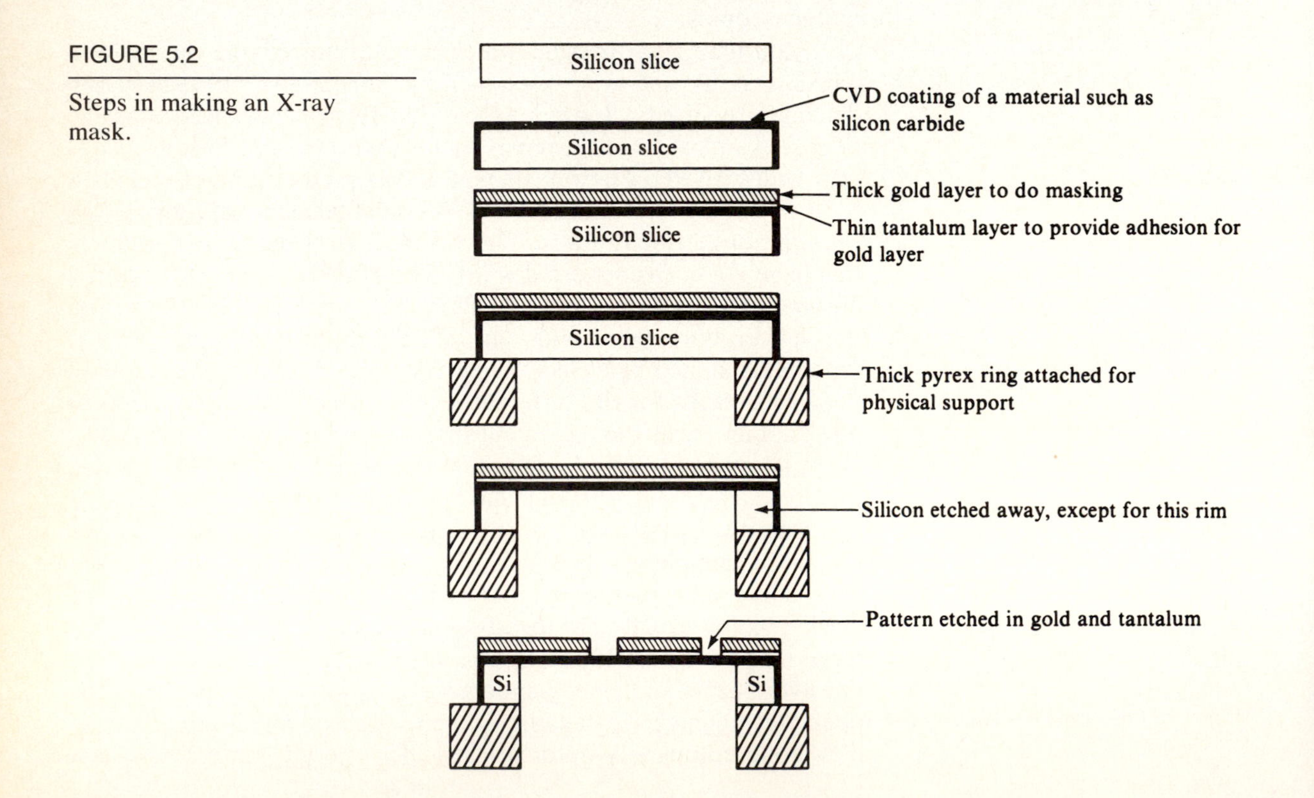

FIGURE 5.2

Steps in making an X-ray mask.

at 1200°C. The slice and layer are then attached to a pyrex ring for support, and the silicon is etched away, leaving only a peripheral ring for support of the film. The gold, or other efficient absorber such as tungsten, can be deposited before or after the silicon is removed, but to prevent subsequent distortion, patterning should be done after silicon removal.

5.2.3 Mask Protection

Masks used in contact printing often adhere to the resist so that when the wafer and mask are separated, small areas of resist remain on the mask, thus leaving holes (defects) in the wafer resist. If the resist is not removed, it may produce a defect on the next wafer exposed. To minimize such occurrences, release coatings applied over the resist or to the mask surface have been used with reported defect reductions of from 10% to 40%. One example of a release agent applied to the resist is polyvinyl alcohol with a surfactant such as Triton X-100 and a lubricant such as calcium stearate spun onto the wafer just after resist soft-bake (4). Release agents applied to the mask have included parylene (5) and fluoropolymers (6).

Masks used for projection printing do not touch other surfaces but nevertheless will show a gradual increase in flaws. In this case, the flaws are dust particles settling on the mask surface. The particles can originate from the operators handling the masks, from the storage containers, or from the interior of the printing machinery itself. Ordinarily, particles on the back of the mask cause little trouble since they will not be in focus. Furthermore, steppers usually use 5× or 10× size masks so that a particle that does alight will be reduced in size and may thus be smaller than the limit of resolution of the lens and not print. Where such particles are a problem, a thin-membrane pellicle can be placed above the front surface of the mask so that particles will be caught by it and held out of focus. An approximate expression for the pellicle–mask separation d_p required for a 10% or less intensity variation is (7)

$$d_p = \frac{nfP}{280}$$

where n is the index of refraction of the material between the mask and pellicle (1 for air and approximately 1.5 for glass), f is the condenser f-number, and P is the diameter of the particle on the pellicle. Some representative light source f-numbers are shown in Table 5.1. Depending on whether or not deliberate overexposure is being used, even a 10% reduction in intensity could cause a thinning of the resist. While the thin spot might not be enough to cause the mask to fail during an etch operation, it could fail to give protection during an ion implant step. The effectiveness of the pellicle or the image

TABLE 5.1

Printer Light Source *f*-Numbers

Printer	*f-Number*
Perkin–Elmer 200 (1X)	*f*/4*
Ultratech 900 (1X)	*f*/4
GCA 4800 (10X)	*f*/25†

*Aperture 4.
†Effective *f*-number after lens reduction.

reduction in minimizing the printability of particles depends inversely on how well optimized the system is for high resolution. Thus, unless the maximum lens resolution is required for the geometries being printed, best overall performance may occur when the total system is adjusted for less than maximum resolution.

Since pellicles represent more reflecting surfaces in the optical path and more opportunities for light scattering and attenuation, the pellicle composition and thickness must be carefully chosen. Nitrocellulose is a common pellicle material. The thickness will ordinarily be only a few microns and may be adjusted to give minimum reflection at the printing wavelength. Alternatively, antireflection coatings, either organic or inorganic, can be used to ensure minimum reflection loss.

5.3 MASK DEFECTS

Mask defects can be broadly described as follows:

Pinholes
Notches
Bridges[2]
Missing geometry
Geometry of wrong size
Runout/runin
Particulates
Scratches

Some of these defects are shown schematically in Fig. 5.3. Pinholes usually stem from holes in the chromium film when it was originally deposited on the mask blank. Mask chipping and scratching can occur during transfer and other handling and produce defects ranging from small pinholes to long tears cutting across whole chip patterns.

FIGURE 5.3

Mask defects.

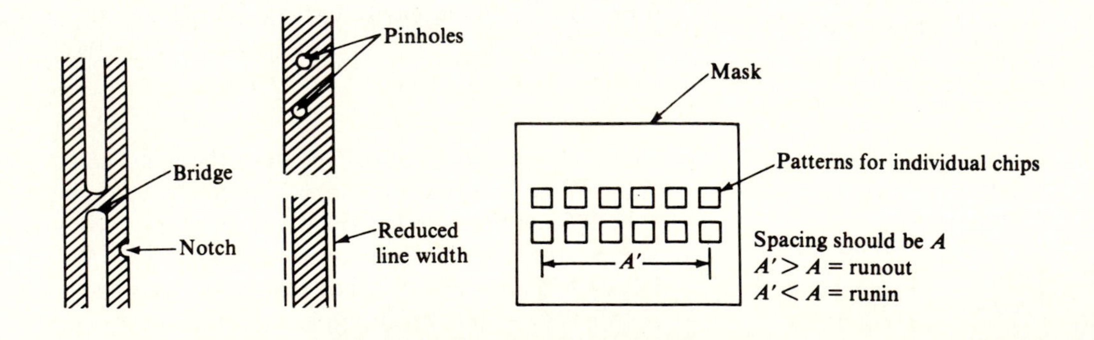

[2]Holes and notches are often called clear defects. Bridges are opaque defects.

The harder the glass used for the mask blank, the less likely it is to chip. Thus, borosilicate glass is more resistant than soda lime glass. Bridging is a mask printing or developing problem. Missing geometry occurs because of missing data fed to the pattern generator or because of poor chromium–glass adhesion, which allows some of the chromium to be washed away during cleaning. Geometry of the wrong size could occur on a new mask because of errors in describing the size, but the usual cause is a resist exposure problem or a metal etch problem. Runout/runin is discussed in more detail in a later section. Particulates are unique in the list of defects in that they are the only defects that can accumulate and then be easily removed. They arise from mask storage containers and the printing equipment, from chipping of glass off the mask edges, and from the ambient air in which the mask is handled. While every effort must be made to keep particles from getting on the mask, efficient mask cleaning on a routine basis is also required.

5.3.1 Mask Inspection

Inspection for the various defects just discussed must be done not only as masks leave the mask print shop but also regularly in the wafer-fab area. To simplify these inspections, several specialized sets of measuring equipment have evolved. Even the standard optical microscope, if used for examining masks with attached pellicles, must be equipped with long working length objectives. Otherwise, the pellicle spacing will prevent all except the lowest power objectives from being brought close enough to the mask surface to focus.

The simplest line-width measuring equipment consists of a microscope and either a filar or image-shearing eyepiece. The width of the image is measured, and accuracy depends on being able to recognize the actual edge of the object from its image. Experiments have shown that a human observer using a filar eyepiece will undersize a clear line and oversize an opaque line. When the observer uses an image-shearing eyepiece, the reverse occurs (8). Better results are obtained if a photoelectronic scan of the image is used. In this case, edge-point detection can be preset to occur at a given fraction of the intensity change observed when going from dark to light. For zero transmission of the "dark" region, the fraction is calculated to be 25% (8). If there is some transmission, the number is higher. The most accurate measurements use either a laser or an electron beam to directly detect the line edges as the mask is moved along and a laser interferometer to measure the stage motion between the two detected edges (9, 10).

Pinholes, bridges, scratches, particulates, and missing geometry can all, in principle, be detected by a thorough microscopic examination. However, to do a thorough job means that each piece of geometry on the whole mask must be inspected.[3] The time consumed by such an inspection is overwhelming, and the possibility of missing a defect is quite large. One early inspection machine used complementary colors of light for simultaneously viewing chip patterns in adjacent rows. Portions of the pattern that were alike appeared as either black or white. Differences between the two patterns were in one of the illuminating colors and easy for an operator to see (11). Chips having differences were then further visually inspected to determine which one was flawed. Another approach, which looked at individual chip patterns, made use of the fact that masks are generally composed of rectangular segments. Spatial optical filtering was used to remove rectangular components of the Fourier spectrum and leave only that from curved objects (most defects would appear to be curved). This filtering enhanced the image of the defects relative to that of the proper geometries and made the defects easy to see (12). Such machines are difficult to automate, however, and sometimes give spurious differences. The most successful approach has been to optically scan each of two adjacent chip patterns and compare the results (13). Unlike the color machine, precise optical alignment of the images is not necessary since the two patterns can be digitized and computer overlayed for best fit. A search for differences can then be made without alignment interference. None of the methods just described will detect missing geometries caused by faulty inputs to the pattern generator since all of the chip patterns will have the same defect. It is possible, however, with some machines, to compare the digitized scan data from a single chip pattern with the original digitized layout pattern data and detect missing patterns.

A completely different approach can be used to look for particulates on pellicles, on the back of masks, on mask blanks, and on unpatterned wafers. With this approach, an intense light is directed to the surface under inspection, and the light scattered from any surface irregularities is detected and mapped. Two variations of the approach have proven quite successful. One covers the whole surface with light and detects light scattered from the surface with a

---

[3]When visual inspection is performed with an optical microscope, care must be exercised to ensure that no contaminants are introduced during inspection. Contaminants can come from such things as sneezes and flecks of dandruff, skin, or cosmetics, and even from the breath of a person who has recently smoked. Inspection stations using robots or other forms of mechanical wafer transport can produce particles from the wearing of linkages and should themselves be inspected regularly. (See, for example, Brian Hardegen and Andrew Lane, "Testing Particle Generation in a Wafer Handling Robot," *Solid State Technology*, pp. 189–195, March 1985.)

very sensitive television camera. The other scans the surface with a laser light beam and detects scattered light with an integrating sphere and light detector (14).

5.3.2 Mask–Wafer Overlay Accuracy

Overlay inaccuracies can arise from a random misplacement of a feature within a die, from a random misplacement of a die within the whole mask array, or from a systematic change in die-to-die spacing as the mask is traversed (runout/runin). Even if a set of masks is perfectly matched as manufactured, the various levels printed on a given wafer may not match when contact or full-wafer scanning projection printing is used. This mismatch can arise for several reasons. One is due to thermal expansion of the mask and/or wafer. Printing of all levels may not be at the same temperature because of drifts in cleanroom or equipment temperature. Thus, since the expansion coefficients of mask and wafer are generally different, level-to-level matching will suffer. Also, a temperature difference may exist between wafer and mask, which may increase from wafer to wafer during a run due to the high-intensity light source used during exposure. Differences in chromium coverage (which affects reflectivity) from one mask to another can affect the temperature rise during exposure. Hence, data taken for one mask geometry may not be appropriate for another (15). Data indicate that if precautions are not taken, up to a 4°C rise in mask temperature can occur during a printing sequence. Thus, the amount of this runout will vary from wafer to wafer within a run. Temperature gradients may also develop across the mask during exposure due to an uneven distribution of cooling air.

If the mask and wafer features are overlayed at the center of the wafer, the runout R for a change in temperature of the mask and/or wafer between two levels of printing is given by

$$R = r(\Delta T_m \alpha_m - \Delta T_w \alpha_{Si}) \qquad 5.1a$$

where r is the wafer radius, ΔT_m is the temperature change of the mask between exposures, ΔT_w is the temperature change of the wafer between exposures, α_m is the expansion coefficient of the mask, and α_{Si} is the expansion coefficient of silicon. If the wafer temperature remains constant, but the temperature of different levels of masks varies by Δt, Eq. 5.1a reduces to

$$R = r\Delta t \alpha_m \qquad 5.1b$$

The expansion coefficients of various mask glasses and the amount of runout across a 150 mm wafer that might be expected are shown in Table 5.2. For a 1°C temperature differential across the mask, the runout for a 150 mm slice is given by line 2 of Table 5.2.

TABLE 5.2

Thermal Expansion Effects

Effect	*Material*			
	Silicon	*Soda Lime*	*Borosilicate*	*Fused Silica*
Expansion Coefficient (10^{-6}/°C)	2.3	9.4	4.5	0.5
Runout μm (1°C across mask)		0.7	0.34	0.037
Runout μm (4°C change; see text)		2.1	0.7	−0.55

The runout for a 4°C excursion of both mask and wafer is given in line 3 of Table 5.2. (A ±2°C temperature control of a cleanroom is reasonable.) Note that the use of fused silica with its low expansion coefficient does not eliminate runout in the case just discussed. However, its use does make temperature control during mask making, as well as temperature control of the mask during printing, of less concern.

Bending of the mask can also cause a small amount of runout. Most mask blanks will not be perfectly flat when freely supported, but when the image is being printed onto them, whether by E-beam, optical pattern generator, or direct contact, the plate will be pulled flat against a carefully ground chuck. However, when the mask is being used (except for contact printing), it will again be supported only at its edges. To estimate the runout that can occur, assume that the mask has spherical curvature and that the maximum deflection is much less than the thickness of the mask. Also suppose that over the area of a wafer of diameter D, the maximum mask deflection (at the center of mask and wafer) is δ. It can then be shown that the runout, which is half the difference ΔD between the distance D along the curved mask and the projected distance D' on a flat surface, is given by

$$R \cong \frac{\delta^2}{D} \tag{5.2}$$

Bowing of the wafer can cause misregistration problems similar to those caused by the mask. Furthermore, a dirt particle or burr on the chuck surface can cause severe local distortion and consequent local misregistration. Thus, care should be given to maintaining very flat and clean chuck surfaces. By knowing the vacuum pressure used to pull the wafer against the chuck and the maximum wafer deflection, which will be the diameter of the trapped particle, an estimate of the diameter of the distorted area can be made. Fig. 5.4 is a plot of the calculated distorted area diameter and surface elongation versus particle size for a single wafer size.

FIGURE 5.4

Wafer distortion due to a particle between the chuck and wafer. (*Source:* Jere D. Buckley, Director of Advanced Studies, The Perkin–Elmer Corporation. From material presented at a Wafer Flatness Workshop at TI on October 4, 1984.)

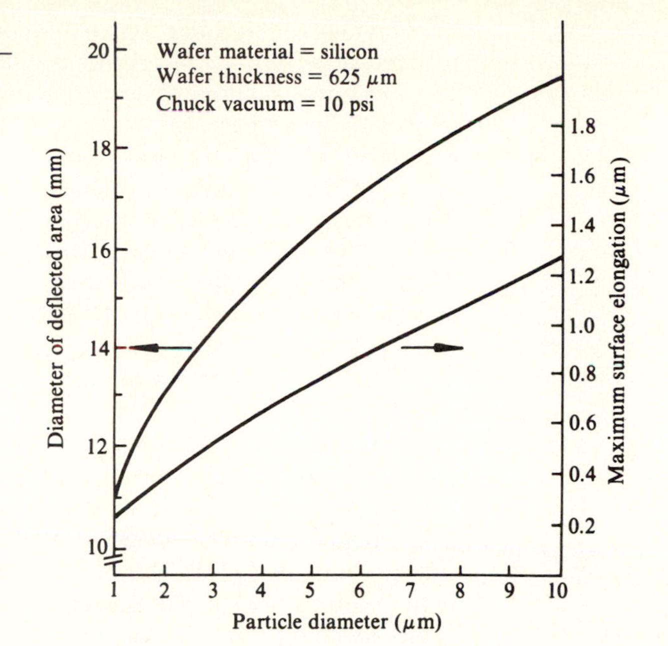

TABLE 5.3

Required Registration

Device	*Typical Feature Size (μm)*	*Minimum Feature Size (μm)*	*Registration (μm)*
64K DRAM	2.5	1.8	±0.5
256K DRAM	1.5	1.1	0.3
1M DRAM	1.3	1.0	0.2
4M DRAM	1.0	0.8	0.2

Table 5.3 gives the approximate level-to-level registration required in order for no yield loss to occur because of registration. Thus, for no yield loss on a wafer, the total runout must be less than the numbers shown in the table.

5.3.3 Overlay Accuracy Measurements

Alignment accuracy is generally considered to include only the accuracy with which a marker from one level matches that of another. For contact, proximity, or full-wafer projection printing, there are only two such markers per level per wafer, and it is easy to measure deviations. Even if the markers do match, geometries in the various chips in the wafer may not because of local distortions in the mask or wafer and, if printing is by optical projection, because of the various distortions in the optical system. In addition, as mentioned in the previous section, the alignment markers may be misplaced with respect to the actual chip patterns. Overlay (registration) accuracy considers how accurately the geometric patterns of one level match their corresponding patterns on other levels. To achieve the necessary measurement precision, special test patterns are used, and to speed up the measurement process, many of the test patterns are designed so that misregistration can be determined by automatic testing. In some cases, these patterns are incorporated into the chip or put in the scribe lines. In other cases, special wafers covered with test patterns are run from time to time to evaluate a printing machine.

Fig. 5.5 shows four patterns that have been used. The first, a set of verniers, shown in Fig. 5.5a, can be put on each chip, with the two halves of the verniers going on two consecutive levels. They are quite sensitive and are directly and individually read visually (16). The next pattern set, shown in Fig. 5.5b, is run on test wafers and requires individual measurement with optical line-width measuring equipment (17). The last two use separate test wafers and require a diffusion or ion implant step, but the degree of misalignment can be read on automatic test equipment in terms of electrical resistance. With these two patterns, x, y, and Θ can be determined for each pattern, and misregistration over the whole wafer can be quickly mapped out (18–19). When the patterns of Fig. 5.5c are used, the corner-contacted square resistor, along with all of the contact arms 1 through 8, are defined by the first printing operation. The second one positions the voltage-sensing contacts that connect arms 1, 3, 5, and 7 to the square resistor. How far they are from being perfectly centered can be determined by electrical measurements and is a measure of misregistration. The resistor and all of the contacts except V_1 and V_2 of the pattern of Fig. 5.5d are defined during one printing step. The V_1 and V_2 contacts are added during a later printing step, and by determining R and R_1–R_4, misregistration can be calculated.

5.4 PHOTORESISTS

The first resist used in semiconductor fabrication produced a negative image and was thus called negative resist. As shown in Fig. 5.6a, areas where the light strikes become polymerized and more difficult to remove so that when the resist is "developed" (subjected to a solvent), the polymerized region remains. Later, positive resist, based on a different chemistry was sometimes used instead. As shown in Fig. 5.6b, exposure to light changes the material so that in that region the resist is soluble in a different solvent and can be more easily removed. Since the exposed regions are removed, the resist is referred to as positive.

FIGURE 5.5

Mask patterns for use in determining level-to-level pattern printing misregistration.

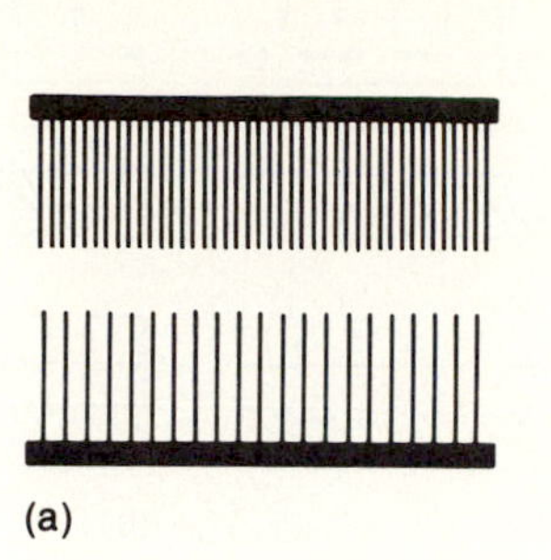

(a)

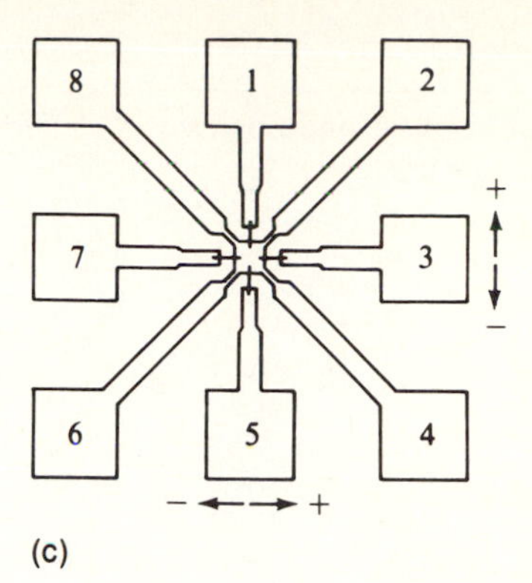

(c)

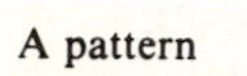

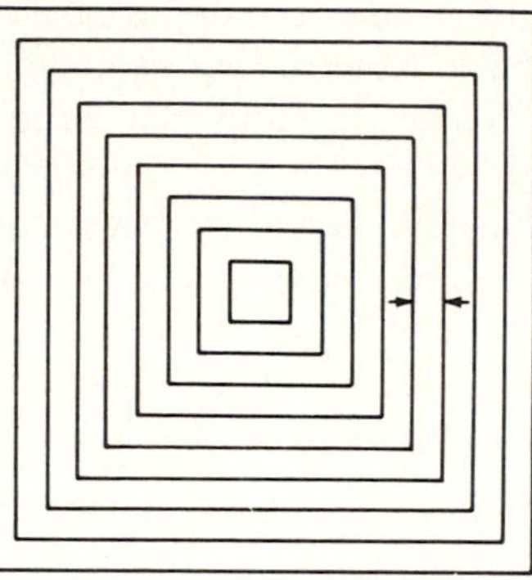

B pattern

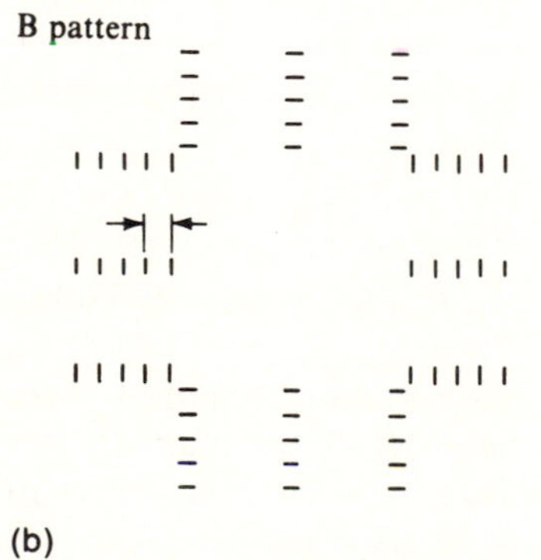

(b)

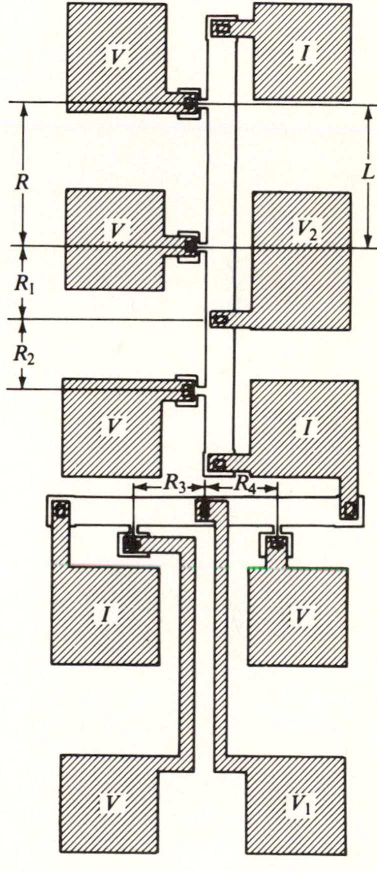

(d)

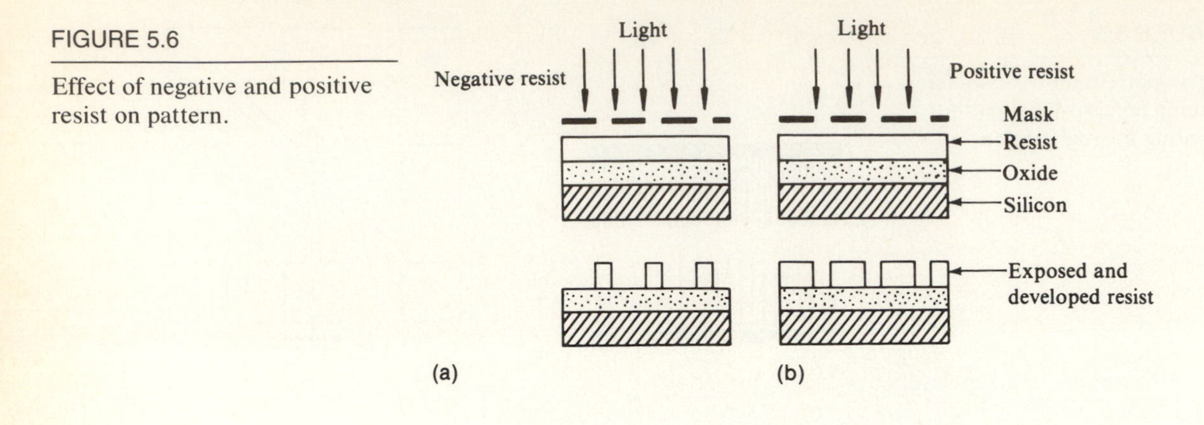

FIGURE 5.6

Effect of negative and positive resist on pattern.

In general, resists have several components. First is the resin, which has the property of withstanding the etch solution. The resin may or may not be photosensitive but is, in any event, not very sensitive. Thus, a sensitizer is added, which is itself very light-sensitive and, in addition, in the case of negative resist, transfers energy to the basic resin during exposure. Since the exposure process requires changes in molecular bonding, substantial photon energy is required, which, in turn, generally dictates the use of ultraviolet light. Since neither of the basic components may adhere well to the surface being coated, an adhesion promoter may also be needed. The result can be a very viscous material that is difficult to spread uniformly. The viscosity must then be reduced with a thinner. The major resist properties that a user must consider are the following.

1. *Sensitivity:* How much light energy is required for a good exposure.
2. *Spectral response:* Sensitivity versus wavelength of the exposing light.
3. *Contrast:* The difference in amount of light to produce a good image and the threshold amount of light where an image just begins to form.
4. *Resolution:* Size feature that can be reproduced in the resist.
5. *Adhesion:* How tenaciously the resist sticks to the substrate.
6. *Etch resistance:* How well the resist will protect the substrate from the etchants being used.
7. *Ease of processing:* How difficult it is to apply, develop, and strip the resist.
8. *Pinhole density:* How many holes in the resist occur because of its inherent properties.
9. *Toxicity of resist and related processing chemicals:* The allowable level of fumes in air breathed by operators.

These properties will be discussed in more detail throughout the remainder of this chapter. The myriad of resists that have been and continue to be developed are in response to these requirements, and the fact that work continues on new resists attests to the fact that resists meeting all of the requirements have yet to be found.

5.4.1 UV Negative Resist Formulation*

Polyvinyl cinnamate was used as the basic resin in most of the resists of the 1960s. However, most negative resists now are based on cyclized polyisoprene polymers. A wide range of sensitizers may be used, including quinones, azido compounds, and nitro compounds for polyvinyl cinnamate and azides for cyclized polyisoprene. A small quantity of novolac resin can be added to improve adhesion and reduce swelling during developing. Solvents/developers include nitrobenzene, acetic acid, and furfural for polyvinyl cinnamate and xylene and benzene for polyisoprene (cyclized rubber).

Regardless of the resin or additives, negative resists suffer from swelling during development. Even though the developer does not dissolve the exposed resist, it is absorbed and causes swelling. During subsequent rinse operations, the developer is removed, and the resist then shrinks. If the resist lines are close together (as required for high resolution), the swelling can cause them to touch, and during the subsequent shrinkage, they may still remain stuck together. Long, narrow lines well-anchored at the ends can become wavy during swelling and, if surface adherence is good, will retain their waviness after shrinkage. Alternatively, the lines may be pulled loose from the substrate.

5.4.2 UV Positive Resist Formulation†

Positive resists function very differently from negative resists. The sensitizer and the main body resin do not interact, so the change in solubility is all due to the sensitizer. The sensitizer breaks down under the influence of light and increases the solubility rate in alkaline solutions by a factor of about a thousand. Since the presence of the sensitizer inhibits dissolution, it is often referred to as an inhibitor. While negative resists use only 2% or 3% sensitizer, in positive resists, 20% may be added. The most common resin is phenol formaldehyde novolac, and the most common sensitizer is napthoquinone diazide. Ethylene glycol monomethyl is a possible solvent; diluents can be butyl acetate, xylene, cellosolve acetate, or some combination thereof.

Positive resists have a broader optical sensitivity than negative resists and can utilize more of a conventional UV lamp's output.

---

*See references 20–22.

†See references 20, 23.

They do not show the oxygen effect[4] of negative resists, but if they are exposed in a vacuum, the breakdown of the inhibitor forms structures that are not soluble in alkaline solutions. Moisture normally present in the atmosphere or in the film is required for an alternate reaction to take place that does produce the normal alkaline solubility. By exposing the resist to the desired pattern in a vacuum and then flood-exposing the entire surface with UV, the previously unexposed resist becomes soluble and that exposed in vacuum remains insoluble. Thus, a positive resist can behave as a negative resist without swelling problems, although the main reason for image reversal processing is to obtain higher resolution than is otherwise obtainable with most positive resists. In practice, positive resist image reversal processing generally involves the addition of an extra photosensitive material to the resist before flood-exposing (24).

5.4.3 E-Beam, X-Ray, and Ion Beam Resists*

Because the usual UV resists are too insensitive for practical X-ray and E-beam use, new formulations are required. The task of combining sensitivity, etch resistance, and thermal stability into one resist has proven to be much more difficult, and completely satisfactory resists are not yet available. The earliest E-beam resist (the same resists are generally used for E-beam, X-ray, and ion beam) was poly–(methyl methacrylate), usually referred to as PMMA. It is a vinyl polymer positive resist that degrades when subjected to radiation. Its resolution is good, but its sensitivity and thermal stability are poor. Another group of positive resists is represented by poly–(butene-1 sulfone), or PBS. It is much more sensitive but has poor plasma etch resistance. Suitable developers for either of these resists are methyl ethyl ketone or methyl isobutyl ketone. Various negative resists are available, and several are based on poly–(glycidyl methacrylate), or PGMA—for example, poly–(glycidyl methacrylate-co-ethyl acrylate), or COP. Sensitivity and etch resistance for these resists are generally good, but like other negative resists, since they suffer from swelling during development, line spacings must be greater than for positive resists. When these resists are used for ion beam lithography, sensitivity is two to three orders of magnitude higher than it is for E-beams. However, the ion beam density is much less, so the time for exposure is comparable. However, it does not appear feasible to increase ion beam resist sensitivity. So few ions per unit area are now required for exposure that increased sensitivity could lead to large statistical variations in exposure.

---

[4]See section 5.4.8 on reciprocity.

*See reference 25.

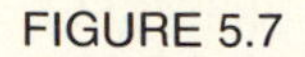

FIGURE 5.7

Inorganic resist composition and steps in its exposure and development.

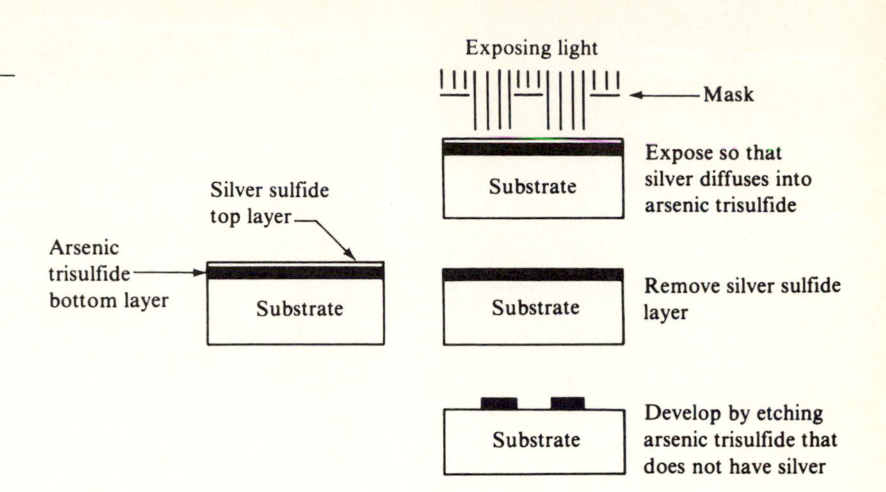

5.4.4 Inorganic Resists

The resists discussed thus far have all been organic-based. It is possible, however, to develop patterns in some inorganic materials (26). In many cases, less normalized energy differential is required between "light" and "dark" for exposure than is required for their organic counterparts. Thus, for a given quality optical imaging system, smaller features can be printed when inorganic resist is used. A considerable increase in processing complexity occurs, however. As currently implemented (see Fig. 5.7), an inorganic resist requires two separate layers. The bottom layer is a chalcogenide glass such as arsenic trisulfide or germanium selenide deposited to a thickness of about 200 nm. On top of the glass is a very thin (100–200 Å) layer of silver or a silver-bearing compound such as silver sulfide or silver selenide. Exposure to light at room temperature causes the silver to diffuse vertically downward into the glass (photodoping). The addition of silver to the top layer of the glass makes it resistant to alkaline etches so that it will remain while the unexposed regions are etched away. Undercutting is not a problem, apparently because the columnar structure of the glass makes it etch much more slowly in a horizontal direction than in a vertical direction. Since these chalcogenides have high optical absorptivity, no standing waves can form when there are reflective surfaces. They are useful over the wavelength range from UV to X ray.

5.4.5 Plasma-Developable Resists

In the continuing effort to provide an all-dry lithographic process, both dry application and dry developing have been considered. Dry application could, in principle, be CVD, plasma-assisted CVD, sputtering, or evaporation. Dry developing implies a resist with a differential plasma or reactive ion etch rate between exposed and unexposed regions. Thus far, no success has been achieved with dry resist application, and while there has been limited success with development, no production-worthy plasma-developable resists currently exist.

The first reports of a plasma develop process used a resist with a volatile monomer in a polymer base. Exposure to light caused some linking of the monomer and polymer. After exposure, heating drove off much of the monomer in the unexposed region and thus left a thinner layer. Subsequent plasma etching completely removed the thinner unexposed region and left a thin layer in the exposed region (27, 28). Another approach, directed more toward X-ray applications, used a resist made by mixing an organometallic monomer with a host resist polymer. Exposure caused the metal ions to form bonds that later, when exposed to an oxygen plasma, allowed the metal to oxidize and form a protective surface (29). A related approach has been to expose optical and E-beam resist in the normal fashion and then treat it with an inorganic compound that reacts with either the exposed or the unexposed resist and renders that part more difficult to plasma etch (30). In the case of ion beam printing, the metal ions needed for such reactions are implanted in the resist during exposure.

5.4.6 Resist Contrast

Resist contrast is a measure of the sharpness of the transition from exposure to nonexposure. Fig. 5.8 is an idealized plot of the relative thickness of resist remaining after development as a function of the exposure energy. E_i is the amount of energy required for gelling to

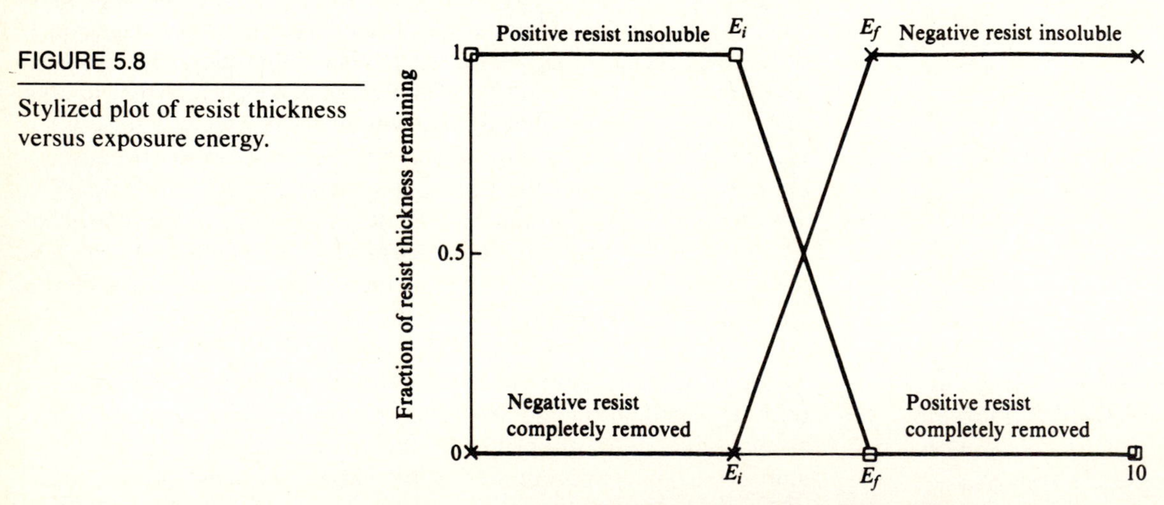

FIGURE 5.8

Stylized plot of resist thickness versus exposure energy.

FIGURE 5.9

Normalized AZ1350J thickness remaining after development versus exposure energy (with postexposure-bake).
(*Source:* Adapted from W.G. Oldham and A.R. Neureuther, *Solid State Technology*, p. 106, May 1981.)

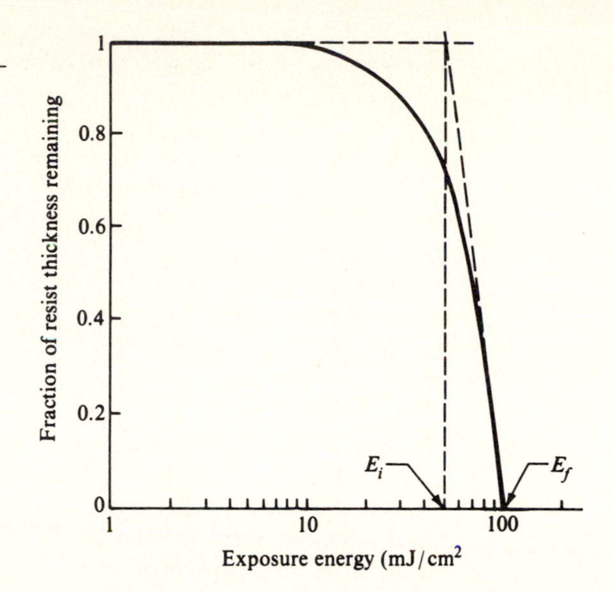

just begin for negative resist or for positive resist to just begin to break down. E_f is the minimum energy at which full thickness of negative resist remains after development or where positive resist is completely removed. Fig. 5.9 is an example of the behavior of an actual resist.[5] In this case, well-defined points E_i and E_f do not exist, but a tangent to the curve can be drawn and E_i and E_f determined as shown. The swing between intensities E_i and E_f as determined in this way does not move from zero to full-thickness resist but will usually be greater than half the full thickness, which is considered adequate for most purposes. The resist contrast γ is a measure of the steepness of the curve in transition from E_i to E_f. A commonly used definition is as follows:

$$\gamma = \frac{1}{\log_{10} E_f - \log_{10} E_i} = \frac{1}{\log_{10} (E_f/E_i)} \qquad 5.3$$

A plot of this equation is given in Fig. 5.10. For positive resists, γ's from 2 to 4 are typical.

The intensity I of the exposing light is related to E through $E = It$, where t is the time. Thus, the light intensity must swing from I_f in the light portions of the image down to at least I_i in the dark portions for satisfactory resist images to be formed. Fig. 5.11a shows

[5]The curve in Fig. 5.9 is still an approximation in that even with no exposure, some resist will be dissolved by the developer solution.

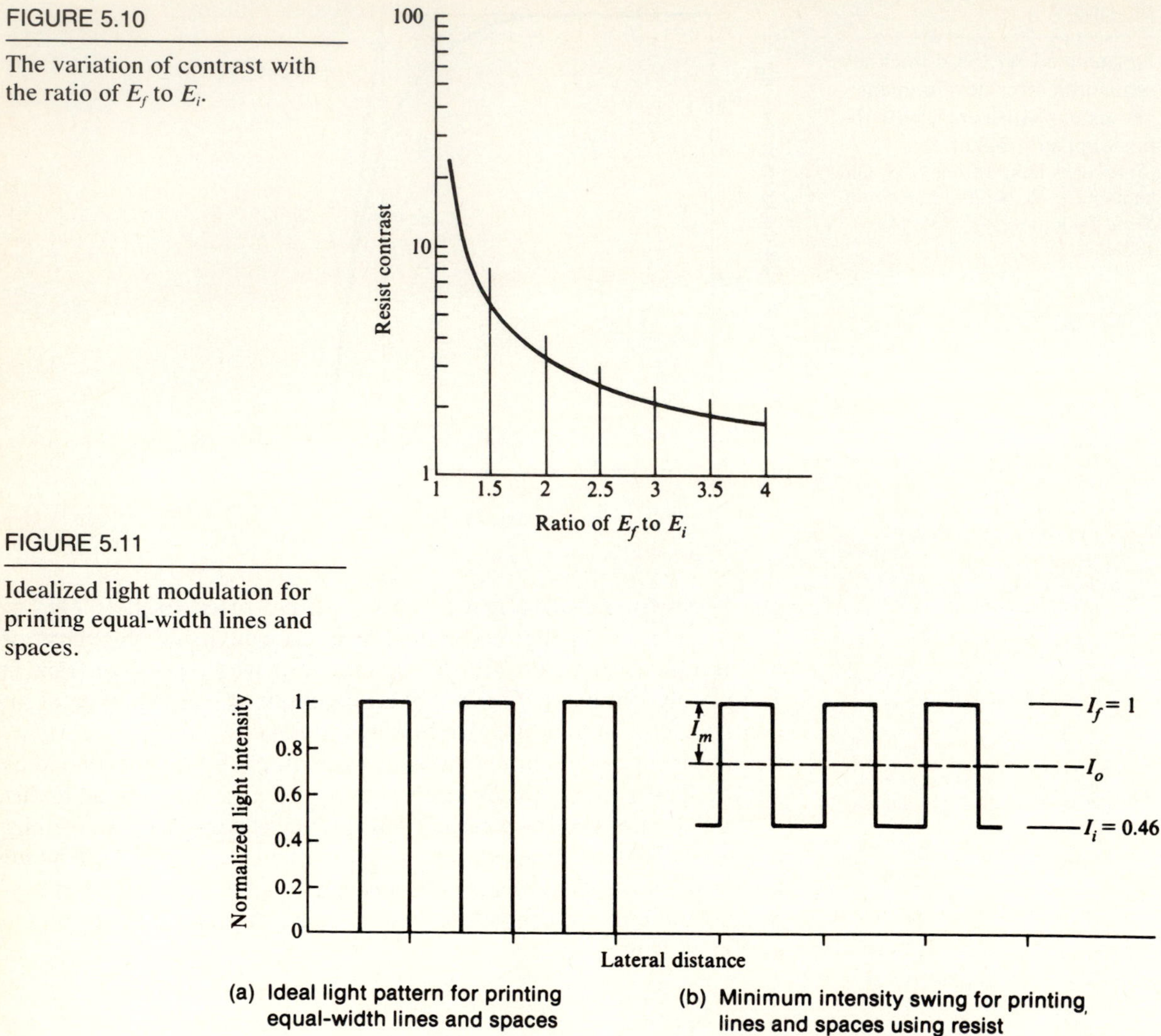

FIGURE 5.10

The variation of contrast with the ratio of E_f to E_i.

FIGURE 5.11

Idealized light modulation for printing equal-width lines and spaces.

the case of the intensity varying from $I_f = E_f/t$ to zero. The case of the light intensity varying from I_f on down to an amplitude that will just barely prevent the resist from responding ($I_i = E_i/t$) is shown in Fig. 5.11b for the case of resist with a contrast ratio of 3. This square wave of period L can be expressed as

$$I = I_o + I_m \Sigma f \sin \frac{2\pi(2n - 1)x}{L} \qquad 5.4$$

where I_o is the average intensity, I_m is the amplitude of the modulating light, and f is given by $4/\pi(2n - 1)$, where n goes from 1 to infinity.[6] The wave modulation M_i of the image is defined as

$$M_i = \frac{I_m}{I_o} \qquad 5.5$$

Fig. 5.11b shows that when the lines will just print, $I_i = I_o - I_m$ and $I_f = I_o + I_m$. By using these values and Eq. 5.5, the expression for M_{ic}, the critical modulation below which printing will not be acceptable, becomes

$$M_{ic} = \frac{I_f - I_i}{I_f + I_i} \qquad 5.6$$

By using Eq. 5.3, M_{ic} can also be expressed in terms of the resist contrast γ as

$$M_{ic} = \frac{10^{1/\gamma} - 1}{10^{1/\gamma} + 1} \qquad 5.7$$

This equation would predict that for a resist contrast of 3, an M_i of 0.3 would be acceptable. However, experimentally, it is found that for positive resist, an M_i of 0.6 is desirable, although 0.4 can sometimes be tolerated. Fig. 5.12 is a plot of Eq. 5.7 over the contrast range from 1 to 5.

With good chrome masks, the source modulation M_s is 1—that is, the "black" areas are black, and the "clear" areas are clear.[7] Thus, the modulation transfer function (MTF) of the optics of the printing system necessary to print the desired image is equal to M_{ic}. MTF will be discussed in more detail in a later section, where it will be shown that as the feature size gets smaller than a few microns, MTF begins to decrease and will limit the ability to print small features. If the contrast of the resist were improved, then printing could be done with a lower M_i (and hence lower printer MTF). It does continue to improve, but in addition, the effective contrast of a given resist can sometimes be improved by adding a layer of photobleachable material on top of the resist (31–33).

A photobleachable material is one whose optical transmission decreases as the amount of incident light increases, as, for example,

---

[6]When the higher-frequency components are removed, the initially square wave becomes rounded, until finally, for $n = 1$, only a sine wave remains.

[7]This was not true of the older emulsion masks. They had a noticeable transmissivity in the dark regions. Too high an intensity light source would cause "burn-through" and resist exposure in the dark regions.

FIGURE 5.12

Image modulation required for pattern formation versus resist contrast.

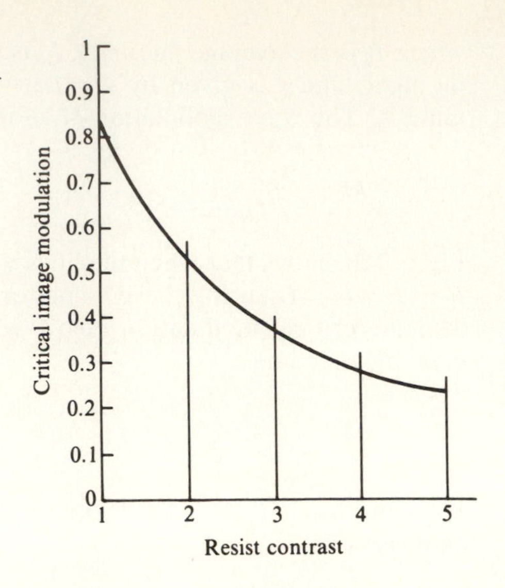

the various diarylnitrones that are converted to oxaziridine by absorbed light. A layer of such material on top of the resist will first bleach in the regions of highest light intensity and begin exposing the resist before the darker areas begin to transmit. Thus, the darks are made darker and the lights lighter.

5.4.7 Sensitivity

Sensitivity is the measure of how much energy is required to allow the resist to develop properly. However, how to define sensitivity properly is not always straightforward. It is commonly defined as an energy value equal to, or closely related to, the E_f discussed in the previous section. In the case of negative resist, it might be the point where 50% or 70% of the resist remains after development. A practical definition is that it is the energy required to give a "good" pattern. As the exposure energy increases, the development (resist dissolution) rate will first be very low and will then increase very rapidly. Sensitivity also can be defined as the energy required to raise the ratio of the exposed dissolution rate to the initial rate to some arbitrary value. In either case, higher sensitivity means a lower energy to expose, and it is generally desirable since for a given light source, throughput improves with higher sensitivity resist. Also in either case, the sensitivity is dependent on the develop conditions and on the wavelength range of the source included in the energy measurement. In the case of E-beams and ion beams, sensitivity is usually expressed either as the number of coulombs (C) of charge required or as the number of particles required. In either case, the sensitivity depends on the particle energy. Table 5.4 lists representative sensitivities of various resists. As mentioned earlier, the exact

TABLE 5.4

Resist Sensitivity

Resist	*Sensitivity* UV (mJ/cm^2)	Deep UV (mJ/cm^2)	X-Ray (mJ/cm^2)	E-Beam (C/cm^2)	Ion Beam (C/cm^2)	Reference
PMMA* (positive)		500–1500	500–1000	2×10^{-4}	4×10^{-6}	(1, 2)
PBS† (positive)			200	1.6×10^{-6}		(1)
COP‡ (negative)			175	4×10^{-7}	6×10^{-8}	(1)
PGMA (negative)				10^{-6}		(1)
Ge_xSe_{1-x} (negative)		50 (220 nm)				(2)
Hunt HNR (negative)	22§					(3)
Kodak 732 (negative)	20§					(4)
Hunt HPR (positive)	50§					(3)
Kodak 809 (positive)	150§					(4)

*Al X-ray target, 20 kV electrons, 100 kV protons.
†Mo X-ray target, 20 kV electrons.
‡Pd X-ray target, 20 kV electrons, 100 kV protons.
§Based on the multiline spectrum of an Hg light source.

References: (1) M.J. Bowden, p. 39, in ACS Symposium Series 266, *Materials for Microlithography,* American Chemical Society, Washington, D.C., 1984. (2) E. Ong et al., p. 71, in D.A. Doane and A. Heller, eds., *Inorganic Resist Systems,* Electrochemical Society, Pennington, N.J., 1982. (3) Philip A. Hunt Chemical Corporation product literature. (4) Kodak product literature.

value will depend on specific conditions. In general, negative resist is much more sensitive than positive resist, and X-ray resists are much slower (less sensitive) than UV resists.

5.4.8 Reciprocity

Over the exposure region where the product of exposure time and light intensity for proper exposure remains constant, a photosensitive material is defined as exhibiting reciprocity (the Bunsen and Roscoe reciprocity law). With the range of exposure times normally used, there is no evidence of reciprocity failure in positive resist. However, most negative resists, during exposure, are sensitive to oxygen in the resist. If oxygen is available from the ambient to diffuse into the resist during exposure, the photo-initiated products of the sensitizer react with the oxygen rather than forming cross-links in the polymer (34). The result is that as exposure time is increased, there is a reduction in the number of cross-links for a given amount of incident energy and thus a reduction in sensitivity. In contact printing, the tight contact between mask and wafer protects the resist from ambient oxygen, and there is no "oxygen effect." However, if negative resist is used with a projection printer with the wafer in air, a significant reduction in sensitivity will occur. A nitrogen blanket can be used to exclude the air and allow normal exposure. Also, the higher the exposure intensity, the shorter the exposure time and the less the effect.

When very high intensities are used—for example, when a series of short, high-intensity pulses are used instead of a continuous lower-intensity exposure—reciprocity failure can occur for other reasons in both negative and positive resist. Examples of pulse exposure are eximer-laser deep UV sources and laser-initiated X-ray sources.

5.4.9 Etch Resistance

As more vigorous etching environments develop, as in plasma etching, it becomes increasingly difficult for the resist to provide the necessary protection during etching. In general, the more resistant the resist, the more difficult to develop. Consequently, to be able to harden the resist after development is useful. Higher-temperature postbakes help, as does a flooding with high-intensity UV, which increases the amount of cross-linking.

5.5 RESIST PROCESSING FLOW

The processing flow is essentially the same for all standard organic resists, positive, negative, optical, E-beam, or X-ray. The steps of this basic flow are listed in Table 5.5. The general aspects of these steps are discussed in the following sections, but substantial differences in the behavior of different resists may exist. Therefore, the manufacturer's recommended procedures for the specific resist being used should always be consulted.

In order to prevent unwanted resist exposure, all operations from application through developing must be done in light with the ultraviolet and blue removed. Thus, yellow light is left and is obtained from specially coated fluorescent bulbs—hence the name "yellow room" for that portion of the wafer-fab facility used for lithography operations. The total cycle time from resist application to stripping can be less than an hour but, because of queuing, is often much longer. Ordinarily, the longer the time from application to etch, the poorer the adhesion. Long-term storage of resist-coated wafers in ordinary yellow light may produce a film over the whole surface that will not develop away. For negative resist, nitrogen storage may aggravate the problem since no oxygen is present to reduce sensitivity.

5.5.1 Adhesion Promoters and Adhesion

Perfect adhesion, which minimizes undercutting, is generally the goal, but a gentle lifting (separation of the resist from the surface) during etching gives tapered sidewalls, which are sometimes desirable. However, such processes are difficult to control and are sel-

TABLE 5.5

Resist Processing Flow

Step	*Comments*
Adhesion promoter application	Sometimes used.
Resist application	Done by dispensing resist onto static or slowly rotating wafer and then rapidly spinning.
Resist dry	To remove solvents.
Resist prebake (soft-bake)	To slightly harden resist and improve adhesion.
Expose	Defines image in resist.
Postexpose-bake	Sometimes used to suppress effect of standing waves.
Resist developing	Removes unwanted resist.
Rinse	Displaces developer in resist.
Resist postbake (hard-bake)	To improve adhesion and etch resistance done at a higher temperature than prebake.
Etch oxide*	Resist must not react with etch or peel from surface.
Resist removal	Done with oxygen plasma or hot acid.

*Sometimes, another insulator or metal is used. Sometimes, photoresist is used as an ion implant mask rather than as an etch mask.

dom used. Clean, dry surfaces of undoped oxide, silicon nitride, polysilicon, and most metals will give satisfactory adhesion. Adherence to heavily phosphorus-doped oxide and to gold is generally rather poor, and various promoters (primers) are sometimes used.

A wafer prebake (drying) cycle before resist application is recommended for good adhesion,[8] and if a promoter is also desired, a 1% solution of hexamethyldisilizane (HMDS) in xylene spun onto the wafer and allowed to dry a few seconds before resist application can be used for negative resist. For positive resist, a solvent such as EGMEA must be used.[9] As an alternative method of application, the wafer can be exposed to HMDS vapor just before resist spinning, and the possibility of solvent-induced problems can thus be avoided. Phosphorus, carbonyl, or sulfur-containing additives to the basic promoter are suggested to improve adhesion to gold metalli-

[8]For example, 2 hours at 200°C for negative resist and 300°C for positive resist is suggested by Philip A. Hunt Chemical Corporation.

[9]Xylene and some other negative resist solvents will cause gelatinous clusters to form in positive resist—thus, the use of ethyl glycol monoethyl ether acetate (EGMEA).

zation (35). Higher soft-bake temperatures for positive resist and higher hard-bake temperatures for negative resist will help adhesion. Higher soft-bake temperatures reduce optical sensitivity. Higher hard-bake temperatures can cause unwanted flow and pattern distortions and also make it more difficult to remove the resist during the stripping operation.

5.5.2 Resist Application*

The main requirements for resist application are that the resist be pinhole-free and of uniform and reproducible thickness. Integrated circuit applications in general require tighter thickness tolerance and a thinner layer than do printed circuit board applications. Hence, roller application, dipping, spraying, or the use of sheets of dry resist is not appropriate. The use of thin layers also requires that much more attention be given to the removal of particles and gelatinous masses from the resist before it is dispensed. Resist is carefully filtered by the manufacturer, but because subsequent contamination may occur and long storage times may produce more gel, filtration is also often done just before application.

The standard application method now in use puts a metered amount of resist onto the center of the wafer and then spins the wafer to fling off the excess resist and produce a uniform covering. This operation is referred to as spinning, although the literature of 1970 and before often called it "whirling." It is a seemingly simple process but is very difficult to accurately model. Table 5.6 lists many of the variables that affect the thickness.

Theoretical modeling predicts the thickness t to vary as Eq. 5.8, which gives

$$t = KS\left(\frac{\nu}{\omega^2 R^2}\right)^{1/3} \tag{5.8}$$

where K is a constant, S is the fraction of solids in the resist, ν is the kinematic viscosity of the resist, ω is the angular velocity of the spinner, and R is the wafer radius (36). Experimentally, S and ν are intertwined since the viscosity is ordinarily adjusted by adding solvent to the resist, which, in turn, lowers the percentage of solids in the resist. Fig. 5.13 gives experimental data for dynamic viscosity versus solids content for a particular resist. Curves such as this one can be used to determine the amount of solvent necessary to lower resist viscosity to some desired lower value. One equation for a particular resist combines S and ν (40) so that

---

*See references 36–39.

TABLE 5.6

Resist Application Variables

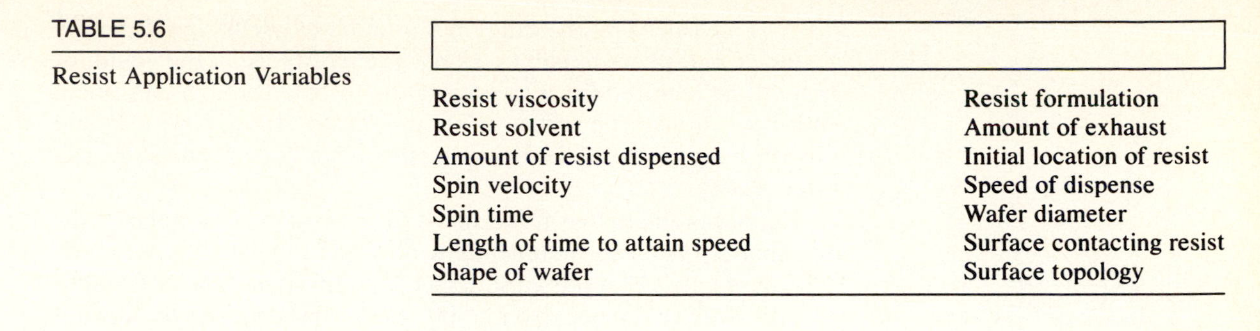

Resist viscosity	Resist formulation
Resist solvent	Amount of exhaust
Amount of resist dispensed	Initial location of resist
Spin velocity	Speed of dispense
Spin time	Wafer diameter
Length of time to attain speed	Surface contacting resist
Shape of wafer	Surface topology

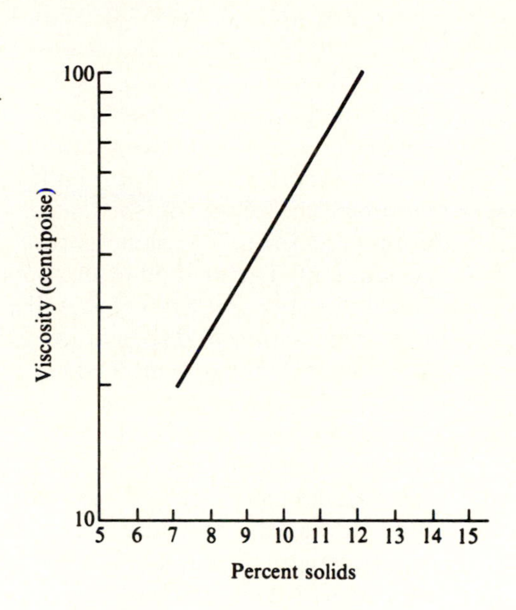

FIGURE 5.13

Dynamic viscosity versus solids content for a negative resist.

$$t = \frac{K(\ln \nu + 0.2258)^2}{\omega^{1/2}} \qquad 5.9$$

The dynamic viscosity is defined as the shearing stress divided by the rate of shearing strain and has the units of poise. Kinematic viscosity is dynamic viscosity divided by the density of the fluid and is expressed in stokes. Note that

$$1 \text{ poise} = 1 \text{ dyne}\cdot\text{s/cm}^2 = 1 \text{ gram/cm}\cdot\text{s}$$
$$1 \text{ centipoise} = 0.1 \text{ poise}$$
$$1 \text{ stoke} = 1 \text{ cm}^2\text{/s}$$

Since the density is relatively insensitive to solids content, the ratio of kinematic to dynamic viscosity remains reasonably constant. Viscosity is measured by the drag on a rotating or oscillating drum, by the time for a spherical ball to drop a given distance in the liquid, or by the time required for a given volume to flow through an orifice. Both flow through an orifice (Canon–Fenske viscometer) (41) and drag on a drum (Brookfield viscometer) have been used to characterize resist.

Eq. 5.8 predicts that if the resist behaves in a Newtonian fashion—that is, viscosity independent of shear rate—then $t \cong \omega^{-2/3}$. Experimentally observed variations of the thickness t with the spin velocity ω have ranged from t inversely proportional to the 0.4 power of ω up to 0.7 (36–38). Data for some resists show that the exponent is a function of the viscosity (36), with a value near -0.66 for low viscosities and near -0.5 for high viscosities. Modeling is complicated by the fact that the resist properties change during spinning because of solvent evaporation.

The diameter of the wafer appears to have less effect on thickness than the range of thickness variations normally observed during processing. The same is true of the ramp rate (spin spindle acceleration). The method of dispensing the resist, the amount dispensed, the speed with which it is dispensed, and the speed of wafer rotation at the time of dispense are all important factors in providing a reproducible process, but the interactions are so complex that only some general statements can be made. The resist should be dispensed while the wafer is either stationary or just barely rotating. A substantial amount of excess resist should be dispensed. Despite the safety requirements for rapid exhaust of solvents, removing the solvents too rapidly can cause the resist to prematurely dry and give poor coverage. If the gas flow above the wafer produces eddies, resist drops leaving the wafer edge may be transported back to the middle of the slice. If the dispense nozzle is too far from the wafer, it can cause splattering. If the wafer is irregularly shaped, as is some GaAs, the resist will usually have a nonuniform thickness over the wafer. The resist formulation contributes to resist striations,[10] with some varieties showing almost none and others being heavily striated. Edge beading is a buildup of resist along the edge of the wafer and may be up to 50 μm wide and 2 μm high. The height is reduced as the spin speed and acceleration are increased. Some resist dispense equipment has resist solvent nozzles located so that solvent can be applied to the edge of the wafer and can dissolve the bead.

---

[10]Resist striations are small thickness variations that arise during resist application. (See, for example, P. Frasch and K.H. Saremski, *IBM J. Res. Develop. 26*, p. 561, 1982.)

5.5.3 Soft-, Postexpose-, and Hard-Bake

The soft and hard resist bakes are done respectively just before exposure and just after development. Soft-bake (prebake) temperatures are in the 70°C–90°C range, and depending on whether a conventional convection furnace or infrared or microwave heating is used, the time may range from 30 minutes down to 4 or 5 minutes. Prebakes are necessary to drive solvent out of the resist so that it can be properly exposed. The subsequent behavior of the resist is dependent on the prebake cycle in that some chemical changes take place at the elevated temperatures. As mentioned in a previous section, prolonged baking will reduce the resist sensitivity (speed) but will also improve adhesion. If negative resist baking is done in an atmosphere high in NO_2 or ozone—that is, with a high smog content—an insoluble scum may form, which will interfere with developing and the subsequent etching step (42).

When interference effects (discussed in a later section) cause the resist to be exposed in layers, a postexposure-bake will cause the exposed resist reaction products to diffuse from layer to layer and thus allow for more uniform development of the entire resist film.

The hard-bake (postbake) is done after development in order to increase both adherence and etch resistance. Hard-bake temperatures will vary with the resist, but they are generally higher than the prebake temperatures. Positive resists must be postbaked at lower temperatures than negative resists—for example, 90°C–120°C versus 130°C–140°C. If the temperature is too high, the resist will be difficult to strip and also may flow and deform the pattern.

5.5.4 Developing

Negative resist can be developed by a solvent such as xylene, but proprietary mixtures formulated for the various resists are often used. Immediately after development, there is usually a rinse in n-butyl acetate to displace the developer in the remaining resist and reduce swelling and to wash away any remaining reaction products. Positive resist requires an alkaline developer, which can be NaOH or KOH. Concern over metallic ion contamination has prompted the use of organic bases, mostly of proprietary nature. Buffering is often used, which makes the developer less sensitive to CO_2 in the air reacting with it. Rinsing is also necessary with positive resist. In this case, since the reaction products are water soluble, deionized water is used for the rinse.

5.5.5 Resist Removal

When only silicon and silicon oxide or nitride are present, very vigorous stripping reactions can be used, but when resist is being removed from metal, much more care must be exercised. As in the case of developers, several proprietary stripping solutions are available. Plasma etching in an oxygen atmosphere is often used and provides a dry process. Burning of the resist at elevated temperatures can be used if the resist is free of ash-forming material. Hot sulfuric acid is very effective. A variety of organic solvents can also be used. Some materials that will remove positive resist are methyl ethyl ketone, acetone, and cellosolve. Phenol- or cresol-based solutions are used for both positive and negative resists. Stripping procedures may vary somewhat from level to level since some processing steps make the resist difficult to remove. Examples are resist that has been used as an ion implant mask and resist used during some plasma metal etching processes.

5.6 LIGHT SOURCES

5.6.1 Arc Sources

High-pressure mercury arc lights are the normal source of radiation for optical printing systems (43, 44).The pressure during operation is 35–40 atm, and since the envelope must be fused silica in order to transmit the UV, the size must of necessity be rather small so that the envelope does not explode. The arc size is approximately 2 mm in diameter by 5 mm long and radiates light in a toroidal pattern as shown in Fig. 5.14a. This light must be collected and spread uniformly over the area of the mask, which may be as large as 200 mm by 200 mm. Two approaches, both shown in Fig. 5.14b, have been used (45). One approach surrounds the source with a parabolic mirror and then uses either additional mirrors and/or lenses to transform the beam to the proper size and divergence. The other approach surrounds the bulb with a series of lenses, each of which collects a portion of the output. Early steppers used only one collection lens, but that number was soon increased in order to collect more of the total light. Methods of combining the outputs of the several collection lenses include the use of fiber optics (46) and the use of mirrors and prisms.

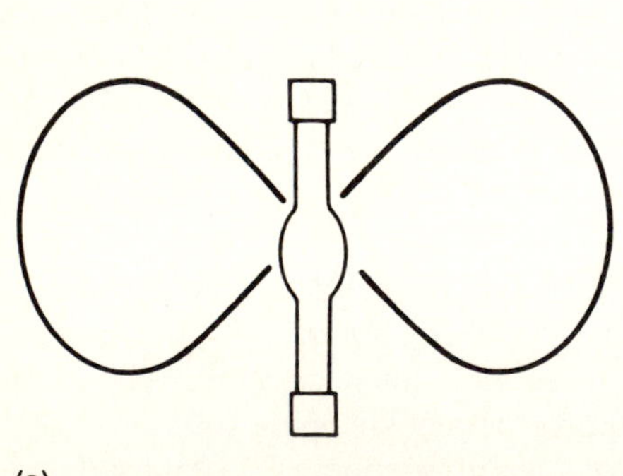

(a)

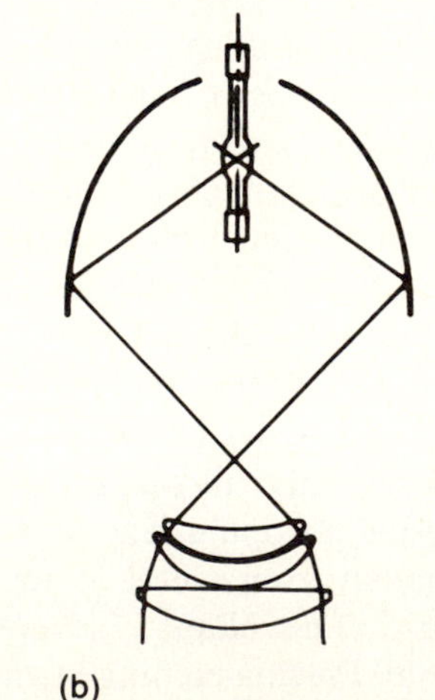

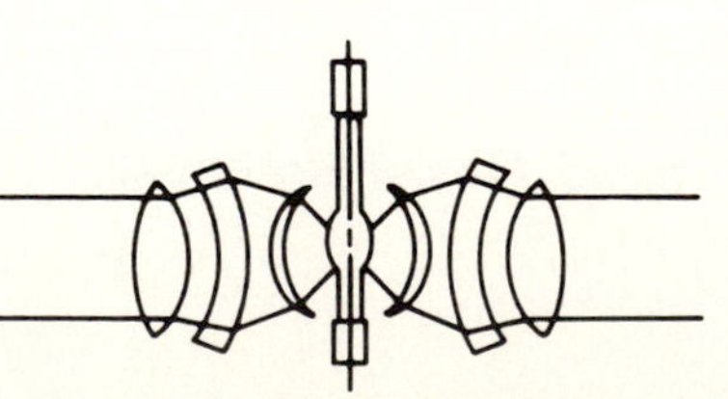

(b)

FIGURE 5.14

Light source optics.

The output spectrum varies somewhat with construction details and length of use, but a typical spectrum is shown in Fig. 5.15. With age, the intensity decreases, primarily because the interior of the bulb becomes coated with a thin layer of electrode material. Further, since the metallic coating absorbs more at the shorter wavelengths, a shift in the spectrum also occurs. Because of the high photon energy required to expose resist, only the 436 nm and shorter wavelengths are of use in lithography. For contact print systems, lens systems with low resolution, or reflective optics systems, the energy from all lines can be used at once. For high resolution, the lenses are corrected for just one of the wavelengths, and filters are used to remove the rest. The single frequency, while making lens design easier, also increases exposure time and makes standing waves between various reflecting surfaces in the optical path more pronounced. To combat interference, at the expense of optical resolution, lenses are sometimes corrected to both 436 nm and 405 nm. In order to minimize heating of either the mask or the wafer during exposure, a filter is used to remove the longer wavelength part of the spectrum. To minimize the energy absorbed by the filter, it is usually designed as a beam-splitting mirror (cold mirror) so that the desired wavelengths are reflected and the long-wavelength IR is transmitted.

The extension of optical printing to shorter and shorter wavelengths in order to provide high pattern resolution has made it increasingly difficult to find satisfactory light sources. Reference to Fig. 5.15 shows that with a conventional high-pressure mercury arc, 0.3 μm (300 nm) is about the lower limit. By adding xenon, a usable output is obtained down to 250 nm. Deuterium lamps produce a continuum in the 200–350 nm region, but their power output is even lower. Pulsed sources can be used to momentarily increase the power output, and even in the 400 nm range, the bulbs are often idled at a lower power than that used when exposures are being made.

FIGURE 5.15

(a) High-pressure mercury emission spectrum in the visible and UV regions. (b) Additional lines from the introduction of xenon into the mercury arc.

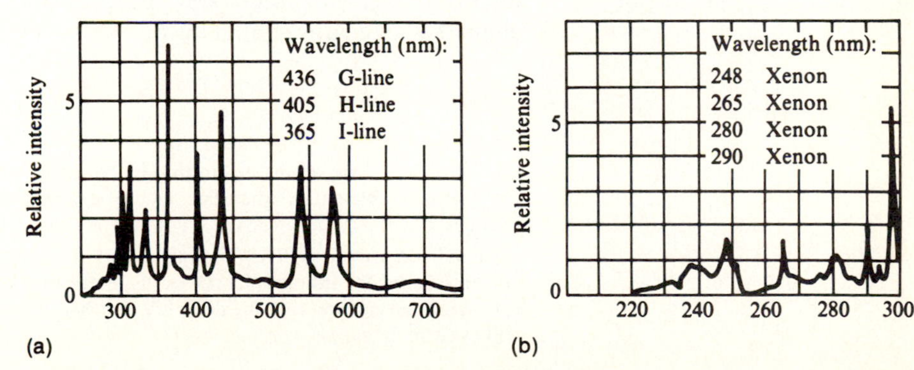

TABLE 5.7

Excimer Laser Materials

Material	*Wavelength (nm)*
Xenon fluoride (XeF)	351
Xenon chloride (XeCl)	308
Krypton fluoride (KrF)	249
Argon fluoride (ArF)	193
Fluorine (F_2)	157

5.6.2 Excimer Lasers

The sources currently capable of producing the highest power output and the shortest wavelengths are excimer lasers[11] (47). Table 5.7 lists some excimer lasing materials and their wavelengths (48). Such sources have been used for contact (49–51), proximity (52), and projection printing (53). These sources are capable of 10–20 W and thus, even with relatively insensitive resists, could be used for high-throughput systems. Because of the high power associated with each laser pulse, reciprocity effects will probably dictate more total energy to expose the resist than would be predicted from standard resist sensitivity calculations.

The high degree of spatial coherence present in most lasers generates image noise and has prevented their widespread use as light sources for equipment requiring good image definition. The excimer lasers are multimode, have poor spatial coherence, and hence can produce low noise images.

The shorter wavelengths of deep UV require fused silica masks and either fused silica or reflecting optics. Below about 200 nm, not even fused silica is appropriate. At 157[12] nm, LiF has been used for a lens material (light source collimator) and CaF_2 for the mask (50). Because of the high attenuation of the 157 nm light at atmospheric pressure, the light path must be in vacuum. (Such short-wavelength UV is often referred to as "vacuum ultraviolet.")

5.7 CONTACT PRINTING

Contact printing is a lensless system in which the mask is held in direct contact with the wafer. Before exposing, the mask must be aligned with features already on the surface, such as oxide holes.

[11]An excimer is two atoms or molecules that are unstable in the ground state but that form a stable entity when excited. An example of the reaction to produce an excimer is $Xe^* + Cl_2 \rightarrow XeCl + Cl$, where Xe^* is an excited atom.

[12]The 1500 Å wavelength UV is only about 100 times as long as the soft X rays currently used for lithography. It seems likely that there will be a continual narrowing of this gap and that the deep UV and X-ray proximity printing technologies will eventually merge.

For this operation, the mask and wafer are slightly separated so that they can be moved relative to each other without damaging either. A split-field microscope, with its two objectives positioned to look diametrically opposite each other near the edges of the wafer, is used to simultaneously view the mask and the wafer below it at these two positions. Using x, y, and Θ movements, an operator can shift the relative positions until alignment markers[13] on the two overlap. After alignment, the mask and wafer are clamped together, the microscope is moved away, and the resist is exposed with UV light. The equipment is relatively inexpensive, but the process suffers from a serious difficulty. Since the mask comes in direct contact with the resist on the wafer, sometimes the resist will stick to the mask and be pulled away from the wafer, thus causing a defect on the wafer and a flaw on the mask that may not be removed before the next use. Additionally, abrasive particles or epi spikes on the wafer will permanently damage the mask so that it degrades during each use. Economics do not justify throwing away a mask after only one use, and as the size of the chip increases, the same number of defects per unit area causes increasingly lower yields. (The impact of defects introduced during processing on yield is covered in detail in Chapter 11.) Thus, the move to progressively larger chips has caused the gradual disappearance of contact printing, even though it does have the potential for fine-line definition. The good definition occurs because of lack of diffraction effects if the mask is in direct contact with a thin resist. If the wafer is so wavy that good contact does not occur or if the resist is thick, some diffraction effects can still occur, and, of course, the properties of the resist itself affect the resolution actually observed. Using deep UV and conventional resist, a resolution of 0.5 μm has been reported (54).

In order to minimize wafer and mask damage, instead of a "hard clamp" of the mask and wafer, reduced pressure is sometimes used, in which case, the machine operation mode is referred to as "soft contact." If the mask is kept slightly separated from the slice, then no contact damage occurs, although resolution suffers. This type of printing is referred to as proximity printing and requires that some additional features be built into the printing machine.

5.8 OPTICAL PROXIMITY PRINTING

The additional mechanical requirement of a proximity printer is that provisions be made to hold the mask parallel to and a specified distance away from the wafer. To minimize mask damage, a wide sep-

[13]Alignment markers are used in all forms of printing and take many shapes. In most cases, alignment is based on overlaying and centering one geometric pattern, such as a square or a cross, within a slightly larger one.

aration is desirable, but to minimize diffraction and divergence effects, the smaller the gap the better. Since the separation will be only a few microns, the Fraunhofer far-field approximations cannot be used to calculate diffraction effects. Rather, the more complex Fresnel equations must be solved for a complete picture of the theoretical diffraction effects.

As a rough estimate, the minimum line width W_m that can be satisfactorily printed is

$$W_m = (d\lambda)^{1/2} \tag{5.10}$$

where λ is the wavelength of the light being used and d is the mask–wafer separation (55). Eq. 5.10 is graphed in Fig. 5.16 for wavelengths of 436 nm, 249 nm, and 10 Å (typical of the length of X rays used in X-ray lithography). Actual line-width values will also have a dependence on the kind of resist used and the way in which it is developed. Experimental values for 249 nm UV exposure onto AZ1350J resist are also shown in the figure.

To minimize penumbra and its associated line broadening, a well-collimated (coherent) light source is desirable, but since such light also enhances diffraction effects, light a few degrees off collimation may be used in practice to strike a balance between the two effects (56). The penumbra contribution of a total beam divergence of 2θ to an image of width W is given by

$$\Delta W = 2d \tan \theta \tag{5.11}$$

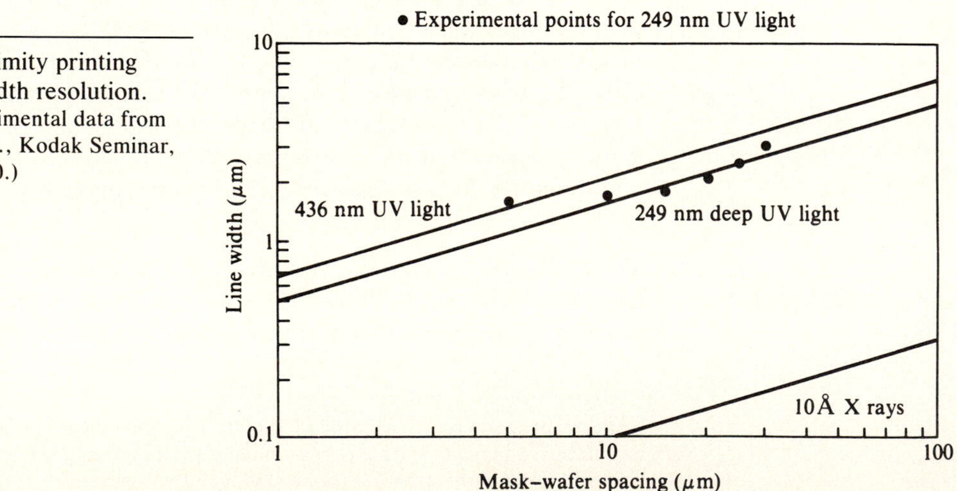

FIGURE 5.16

Effect of proximity printing gap on line-width resolution. (*Source:* Experimental data from T. Kaneko et al., Kodak Seminar, San Diego, 1980.)

Thus, if, for example, a nominal 2 μm line is to be widened no more than 10% because of divergent light when printed with a mask–wafer spacing of 10 μm, the angle of divergence must be less than 0.5°.

5.9 FUNDAMENTAL OPTICAL LIMITATIONS OF PROJECTION PRINTING

The two kinds of printing discussed thus far use optical elements only in the light source, and their performance is little affected by optical lens limitations. However, the projection printing discussed in the next two sections depends on optical elements to form the printed image. Performance limitations caused by the inability to construct perfect systems are discussed in these sections. In addition, theoretically perfect optical projection systems still have limitations, some of which are discussed here.

5.9.1 Resolution

Resolution is a measure of the ability of a system to form separate images of closely spaced objects. What constitutes a separate image depends somewhat on the use of the image. For example, if all that is required is that the eye be able to surmise that probably two objects exist, then an image intensity like the one shown in Fig. 5.17a would suffice. However, if an image is to print two separate objects on photoresist, then the image should be more like the one in Fig. 5.17b. The two objects are clearly discernible as separate images, but whereas the object had equal lines and space, the resist image has wide lines and a narrow space. If the object is to print a faithful reproduction, the image must be yet better defined, as in Fig. 5.17c. Note that because of the clipping action of the resist's exposure behavior, a square wave can be printed even though the exposing light intensity does not have square shoulders. It is this property that allows optical printing to be used for much finer geometries than was originally expected. A criterion[14] for resolution that is often used to

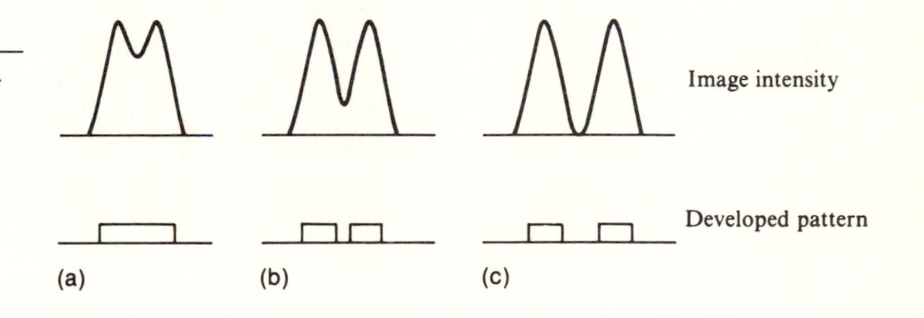

FIGURE 5.17

Effect of increasing object separation when spacing is near diffraction limit.

[14]Remember that this is the diffraction limit—that is, the limit for a perfect lens. Design or manufacturing flaws reduce this number. Lens manufacturers often rate their resolution limits at about 1.5 times that predicted by Eq. 5.12.

describe lens capability and that corresponds roughly to Fig. 5.17a is the Rayleigh limit

$$S_r = \frac{0.6\lambda}{NA} \tag{5.12}$$

where S_r is the separation of two barely resolvable objects, λ is the wavelength of the light being used, and NA is the numerical aperature of the lens. NA is a lens design parameter and is given by

$$NA = n \sin \phi \tag{5.13}$$

where n is the index of refraction of the medium outside the lens and ϕ is the half-angle of the two most divergent rays of light that can pass through the lens. The f-number F is defined as

$$F = \frac{f}{D} \tag{5.14}$$

where f is the focal length of the lens and D is the diameter of the entrance pupil. When the image is formed near the focal point, as in the case of printing equipment and microscopes, $F \cong 1/2NA$.

The modulation transfer function (MTF) of an optical system is the ratio of the modulation of an optical image (see Eq. 5.5) to that of the object. For a diffraction-limited optical system, it is a function of the spatial frequency of the object,[15] the light wavelength, the numerical aperture of the lens being used, and the degree of spatial coherence of the light source. Spatial frequency is often expressed in terms of lines/mm so that, for example, a series of 5 μm lines and spaces would have a period of 10 μm and hence would be 100 lines/mm. The light source is spatially coherent if it is a point source and incoherent if it is infinite in extent. Practically, if the source is just large enough to fill the entrance pupil of the lens, then the light is assumed incoherent. A common description of the degree of incoherence is the ratio of the diameter of the light source to entrance pupil. Thus, a point source is 0% incoherent, and a source as large or larger than the entrance pupil is 100% incoherent.

Fig. 5.18 shows how MTF varies with spatial frequency for various degrees of light coherence (57). The two limiting cases are 0% and 100% spatial light coherence. The cutoff frequency ζ for incoherence is given by

$$\zeta = \frac{2NA}{\lambda} = \frac{1}{F\lambda} \tag{5.15}$$

---

[15]The remaining discussion assumes that the intensity of the object is sinusoidal rather than square wave.

FIGURE 5.18

Modulation transfer function (MTF) versus normalized spatial frequency.

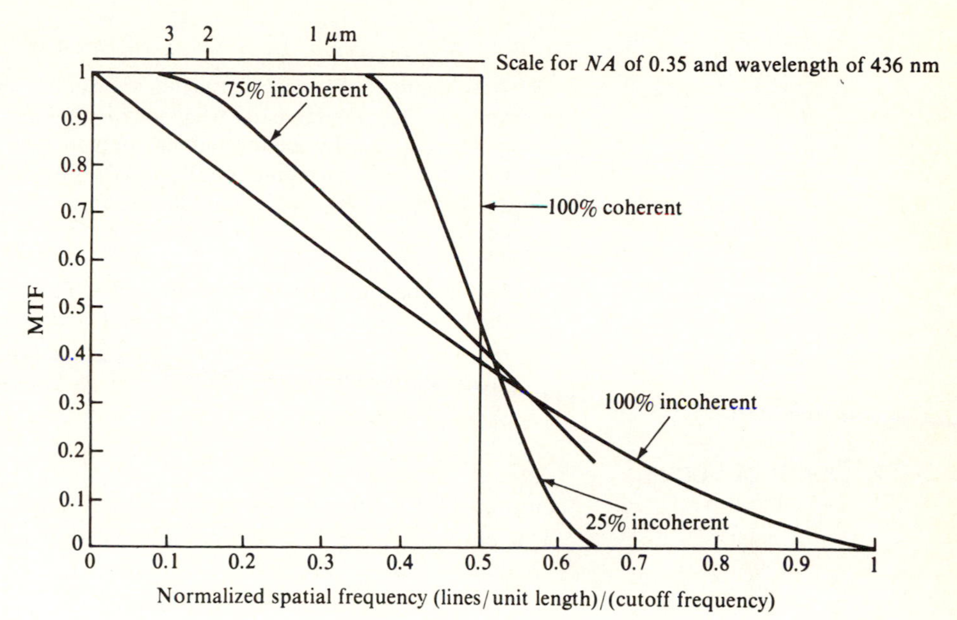

That is, by the time the wavelength $L = 1/\zeta$ is reduced to $\lambda/2NA$, there is no contrast between the two lines. Note that this value is comparable to the Rayleigh spacing S_r of Eq. 5.12, or $0.6\lambda/NA$. The MTF derivation assumes a sinusoidal-varying intensity, whereas, in printing, the object and the desired image are square waves. However, since a square wave can be described by a Fourier series, it is possible to calculate how a square wave object intensity $I(x)$ will fare by multiplying each term of the Fourier series that describes it by the appropriate $\mathrm{MTF}_{k\nu}$ and then summing:

$$I(x) = I_o + \Sigma\, \mathrm{MTF}_{k\nu} I_k \sin(2\pi k\nu x) \qquad 5.16$$

where ν is the fundamental spatial frequency of $I(x)$.

EXAMPLE ☐ Using this concept, calculate what the image intensity profile will be for a mask with 1 μm lines and spaces projected through a lens system with an NA of 0.35, a 100% incoherence factor, and a wavelength of 436 nm.

The first four terms of the square-wave expansion are as follows:

$$I_0 + I_0\left(\frac{4}{\pi}\right)\sin\left(\frac{2\pi x}{L}\right) + I_0\left(\frac{4}{3\pi}\right)\sin\left(\frac{6\pi x}{L}\right) + I_0\left(\frac{4}{5\pi}\right)\sin\left(\frac{10\pi x}{L}\right)$$

The cutoff frequency $\zeta = (2 \times 0.35)/0.436 = 1.6/\mu m$. The wavelength of the mask pattern is a 1 μm line plus a 1 μm space, or 2 μm. The fundamental frequency is thus 0.5/μm, giving a normalized spatial frequency (actual frequency/cutoff frequency) of 0.31. The third harmonic is 0.93, and the fifth is 1.53. The MTF for each of these frequencies can be read from Fig. 5.18. For the lowest frequency (0.31), MTF ≅ 0.6. For the others, it is negligible. Thus, even though a square wave of light and dark is on the mask, only a sinusoidal light pattern will be projected onto the wafer in this case. Since the MTF for the first harmonic is 0.6, the sine-wave image modulation will also be 0.6, and the normalized intensity will swing between 0.25 and 1.0 (see Problem 7 at the end of the chapter). □

5.9.2 Depth of Focus

An optical image degrades as the system is defocused, and the amount of defocusing that can be tolerated is called depth of focus. The amount to be tolerated of course depends on the use of the image and thus is somewhat subjective. The Rayleigh depth of field criterion is one that is widely used for comparing lenses, and it is defined as

$$\delta = \frac{\pm\lambda}{(2NA)^2} \tag{5.17}$$

where δ is the depth of field. In lithography, the limiting item is how much line-width change can be tolerated. In some cases, the nonlinear resist behavior allows much more defocus tolerance than would be predicted from the Rayleigh expression, as shown in Fig. 5.19. Comparing the depth of field expression with the resolution expression shows that resolution is obtained at the expense of field depth. Thus, the higher the *NA* of the lens being used, the more difficult it is to keep the whole wafer properly focused. For a given resolution, a shorter wavelength and smaller *NA* will improve depth of field.

5.9.3 Variation in Position of Best Focus

After the depth of field for a particular lens has been established, there remains the task of providing an environment that allows the whole wafer to stay within this depth of field during exposure. The first step is to determine each item that causes the image focus to be other than exactly on the wafer surface. Then, a budget can be assigned to each of these items and improvements made as necessary to ensure that the printing is satisfactory. A list of such items along with some typical values is given in Table 5.8. The total focus de-

FIGURE 5.19

Effect of changing NA on depth of focus.

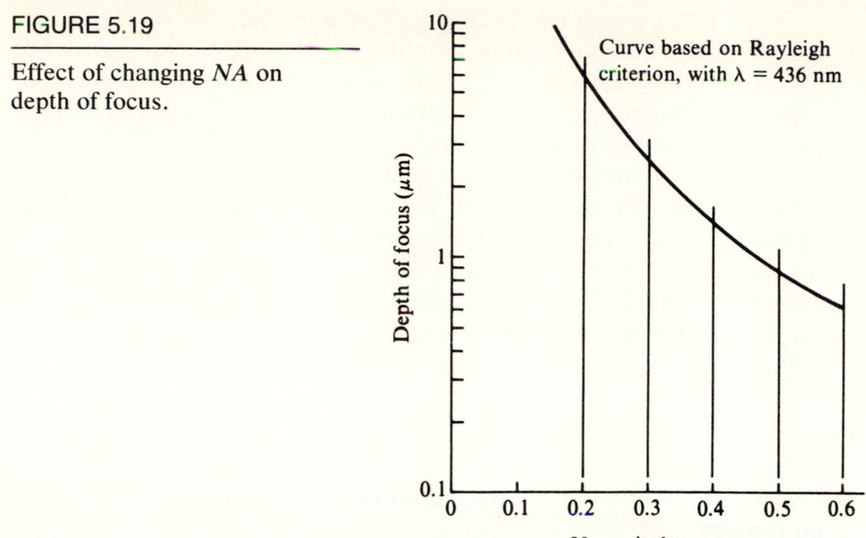

TABLE 5.8

Factors Affecting Image Focus on Slice

Item	*Typical Value*	*Comments*
Process-Related		
Wafer topography height variation	±1 μm	Due to metallization, vias, and so on; can be reduced with multilayer resist.
*Incoming-Slice-Related**		
Slice flatness	±3 μm	With steppers, the area per step is reduced, and deviation over a step can be less than 0.5 μm.
Slice taper (150 mm)	20 μm	Applicable only to printers with backside reference.
*Mask-Related**		
Mask bow (5 × 5 inch plate)	±5 μm	Printers using reduction lenses reduce the effective value by square of reduction.
*Machine-Related**		
Autofocus accuracy	±0.25 μm	Stepping printers.
Best focus setting	±0.05 μm	Applicable only to printers with autofocus.
Chuck flatness (150 mm)	±0.3 μm	
Stage travel	±0.05 μm	Per step on stepping printers.

*These items are expected to improve with time.

viation can be estimated (58) by assuming that the values are all independent and thus can be combined by taking the square root of the sum of the squares. If there are constant, systematic errors, as, for example, the "best focus" setting, they should be linearly added to the root of the sum of the squares of the others.

The curvature of the slice-holding chuck is listed in Table 5.8, but another element of chuck flatness exists that can cause local printing defects as well as damage to the silicon itself. If the chuck is damaged, resulting in a raised burr, or if a particle is trapped between the chuck and the slice, the wafer surface may be enough out of focus to cause printing flaws. To minimize this problem, many chuck designs allow very little actual contact area between the chuck and the wafer.

5.10 FULL-WAFER PROJECTION PRINTING

As an alternative to proximity printing, which is difficult to control if the gap is narrow and which is troubled with excessive diffraction if the gap is wide, projection printing has evolved. The early machines (late 1960s) used full-wafer projection through a multiple-element lens onto 2 inch diameter wafers. It was difficult to get quality lenses that could cover even this small area, and as the slice diameter increased, alternative approaches were required. Excluding diffraction effects, which for a given wavelength and numerical aperture are independent of design, lens performance problems can be lumped into two general categories. One is excessive background light from internal reflections and scattering. This excessive light reduces the contrast of the image. The other is a series of aberrations that either blur or geometrically distort the image. These aberrations can be broadly categorized as follows:

Spherical aberration
Coma
Astigmatism
Field curvature
Chromatic aberration
Distortion

Chromatic aberration is eliminated by the use of reflective optics. The others can be made very small for an image lying in a narrow ring concentric with the optical axis of the reflecting mirrors. Despite the restriction of using only the narrow ring, a full-size mask image can be projected onto a wafer by synchronously moving the mask and wafer while the mask is illuminated by a segment of a circular band of light as shown in Fig. 5.20. In 1973, Perkin–Elmer introduced a 3 inch diameter full-wafer scanning projection printer based on this concept (59). Since that time, all full-slice printers

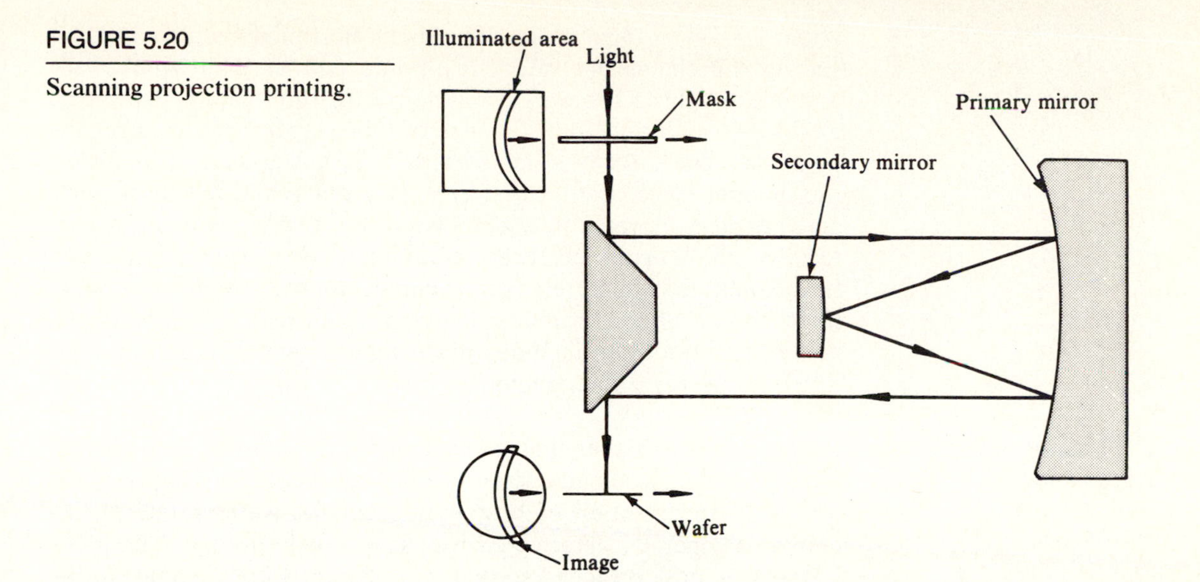

FIGURE 5.20

Scanning projection printing.

TABLE 5.9

Estimated Percentage of Wafers Printed by Various Technologies

Year	*Contact/ Proximity*	*Full-Slice Projection*	*Steppers*	*E-Beam*
1975	90%	10%	0	0
1980	70%	29.8%	0.2%	0
1985	25%	65%	9.7%	0.3%
1990	15%	60%	24.6%	0.4%

have used that basic design, although newer large-slice machines have two primary mirrors symmetrically placed in order to further reduce distortion. The wafer size capability has been increased to 150 mm, and the resolution has been improved from 3 μm to 1 μm. As Table 5.9 shows, full-wafer projection printing continues to be the major method used, with contact and proximity printing decreasing and stepper printing increasing.

5.10.1 Operation

Because the image is scanned instead of being projected at one time, exposure time is generally described in terms of scan speed rather than total time. Greater slit widths (width of the ring of light used) allow more light to the resist. Thus, for a given resist, a faster scan speed can be used, giving a higher machine throughput. The use of wider slits allows more aberration so that resolution suffers. For example, in going from zero slit width (diffraction limit) to 2.7 mm on a Perkin–Elmer 200 series printer, the resolution changes from about 1.5 μm to about 2 μm (60). Also, often a light source aperture setting will change the effective *f*-number of the light source, the amount of light to the wafer, and hence the throughput. It will also affect the resolution since the size of the aperture stop changes the light coherence. As was mentioned in the discussion on pellicles, the amount of separation between the pellicle and the mask for a given degree of protection depends on the source *f*-number. To determine the impact of these parameters on image quality and throughput, manufacturers' specifications for the specific equipment must be consulted. In addition, as an aid in using their products, resist literature often includes performance curves based on some of these specialized parameters.

5.10.2 Focusing Accuracy

Regardless of whether manual or automatic focusing is used, two factors determine the accuracy of the final focus. One factor is the choice of the position of best focus. On full-wafer printers, this choice is generally determined by using a focus wedge. The focus wedge is a mask designed so that it is tilted at a known angle to the optical axis and contains a series of identical resolution patterns. After an exposure is made at what is assumed to be the proper focus, the pattern with the best image is located. The combination of its position on the mask with the known angle of inclination of the mask to the optical axis allows the position of best focus to be determined relative to the initial setting. After this position is determined as accurately as possible, the problem of reproducing it from wafer to wafer still remains. If automatic focusing is used, an error will be associated with it. With manual focusing, the problem of reproducing the best focus setting exists. Some full-slice printers position the front of the wafer against three reference pads. If buildup of resist on the pads occurs, focus will be affected. Other designs eliminate this problem by using air gage sensors instead of pads so that the wafers never contact the reference surfaces. Since wafers are almost never completely flat, if the distribution of bow changes, focusing may also need to change.

5.10.3 Registration

Table 5.10 lists potential causes of misregistration. Misalignment is the amount of misregistration between alignment markers of differ-

TABLE 5.10

Causes of Misregistration During Scan Projection Printing

Misalignment	Magnification change
Internal mask errors	Wafer expansion
Optical distortion	Mask expansion

ent levels. It can occur because of inaccuracies in automatic aligning and may be exacerbated by processing steps that cause the edge slope of alignment markers to degrade. Manual alignment is subject to operator error and will, in general, vary from operator to operator. The misplacement of alignment markers with respect to the rest of the mask patterns will cause misregistration error when alignment is perfect. Unlike other mask errors, this error can be corrected by an operator's aligning by pattern geometry rather than by markers. Fig. 5.21a shows how feature overlay would appear with x-y or θ misalignment.

Mask errors were discussed earlier in the section on mask making. Optical distortion can be pincushion, barrel, or random and will cause square arrays to appear as in Fig. 5.21b. If the same printer can be used on all levels, no overlay problem will occur. Unfortunately, this luxury is usually available only during the R&D phase. Two different types of printers—for example, steppers and full-slice printers—may have different kinds of dominant distortion and hence may overlay more poorly than two printers of similar nature. Distortion can be caused not only by errors in mirror grinding but also by temperature changes or other mechanical stresses that cause flexing of the mirror.

Magnification changes, as illustrated by Fig. 5.21c, are generally negligible compared to the others since all 1:1 projection printers are designed to be telecentric,[16] at least on the wafer side. Wafer or mask thermal expansion appears as an apparent magnification change. Expansion effects can be minimized by judicious mask cooling and a temperature-controlled printroom. The cooling air must be carefully filtered; otherwise, particles will collect on the mask and/or wafer and cause printing defects that can easily overshadow the expansion problems.

FIGURE 5.21

Types of misregistration.

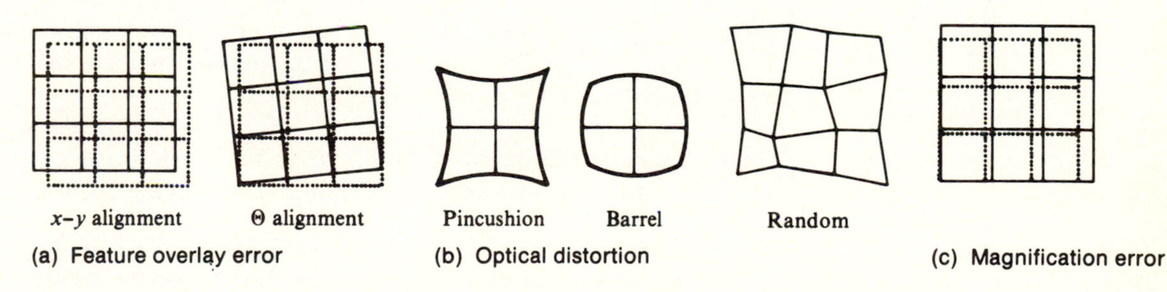

[16]A telecentric design is one that produces unity magnification even if the mask and/or wafer is substantially out of focus. Such a design does not, however, prevent image distortion when defocused.

5.11 STEPPER PRINTING

As discussed in an earlier section, the capability to design and construct optical projection lenses to provide both high resolution and full coverage for large-diameter wafers does not currently exist. One method of circumventing this problem is to design lenses with a relatively small field and then print the full-wafer surface by a series of adjacent exposures.[17] In the late 1970s, GCA introduced a system based on their mask-making equipment that used a large, multielement lens to project an area of only 1 cm^2 onto a wafer at one time. A carefully constructed stage then moved the wafer to successive positions until the entire wafer was exposed.

The stepper provides improved resolution but, because of the multiple steps involved, is slower than full-slice printing. Table 5.11 summarizes characteristics of many of the lenses currently available. The table shows that these lenses have predominately a 5× or 10× reduction ratio and cover fields of from 1×1 cm to 2×2 cm. The resolutions of several of the lenses go below 1 μm. The reduction lens used in most stepping printers is comprised of a large number of refracting elements in order to compensate for various optical aberrations. Fig. 5.22a shows the complexity to be found in such dioptric lens designs. By using a catadioptric design,[18] which is more appropriate to a 1:1 stepper, the lens system can be substantially simplified and still provide comparable image quality. An example, often referred to as a Wynne–Dyson design (61), is shown in Fig. 5.22b.

The reduction lens used in most steppers has an advantage over the lens used in 1:1 printing systems in that there is much less sensitivity to dust on the mask. Many particles that would be troublesome in a 1:1 printer are reduced to the point that they are below the resolution of the lens and hence will not print. The thickness of the mask itself is enough to keep particles on the back surface out of focus enough to offer reasonable protection. For the front surface, pellicles (see the section on mask protection) can be used to keep particles from being in focus. Some steppers have built-in particle detectors for periodically scanning the mask or pellicle surface for particles large enough to cause a yield loss.

5.11.1 Focus

Steppers can be refocused at each site so that they are relatively insensitive to wafer surface undulations even though their high-

---

[17]The actual lens field is, of course, circular, but the field that includes the patterns is either square or rectangular. Hence, the effective field is the largest square or rectangle that can be inscribed in the circle.

[18]An optical system using both refractive and reflecting elements.

TABLE 5.11

Optical Stepper Lenses

Wavelength (nm)	*NA*	*Advertised Resolution*	*Field Size (mm)*	*Lens Reduction*	*Manufacturer*
365	0.32	0.9	14 × 14	5X	Zeiss
365	0.35	0.75	11 × 11	5X	Tropel
365	0.35	0.75	15 × 15	5X	Tropel
365	0.35	0.8	10 × 10	10X	Nikon
365	0.42	0.7	10 × 10	10X	Zeiss
405	0.28	1.25	20 × 20	5X	Zeiss
405	0.30	1.0	10 × 10	5X	Cerco
405	0.35	1.0	17 × 17	5X	Wild
405–436	0.30	1.25	10 × 10	5X	Cerco
436	0.16	2.0	30 × 30	2.5X	Nikon
436	0.20	1.75	20 × 20	5X	Zeiss
436	0.24	1.25	20 × 20	5X	Tropel
436	0.28	1.25	20 × 20	5X	Zeiss
436	0.28	1.25	10 × 10	10X	Zeiss
436	0.30	1.2	20 × 20	5X	Nikon
436	0.30	1.1	14 × 14	5X	Zeiss
436	0.32	1.0	15.5 × 15.5	5X	Tropel
436	0.32	1.0	10 × 10	10X	Tropel
436	0.35	1.0	14 × 14	5X	Canon
436	0.35	1.0	15 × 15	5X	Nikon
436	0.35	1.0	14 × 14	5X	Tropel
436	0.35	1.0	10 × 10	10X	Zeiss
436	0.38	0.9	14 × 14	5X	Zeiss
436	0.38	0.9	10 × 10	10X	Zeiss
436	0.43	0.8	15 × 15	5X	Canon
436	0.48	0.7	15 × 15	5X	Canon
436	0.45	0.75	15 × 15	5X	Nikon
436	0.60	0.6	5 × 5	10X	Nikon
390–450	0.28	1.25	27.5 × 10.6	1X	Ultratech
390–450	0.315	1.0	27.5 × 10.6	1X	Ultratech
390–450	0.40	0.7	31.1 × 11.1	1X	Ultratech

Source: Data from Pieter Burggraaf, "Wafer Steppers and Lens Options," *Semiconductor International,* pp. 56–63, March 1986, and Pieter Burggraaf, "Stepper Lens Options for VLSI," *Semiconductor International,* pp. 44–49, February 1988.

FIGURE 5.22

Cross section showing relative simplicity of a catadioptric design.

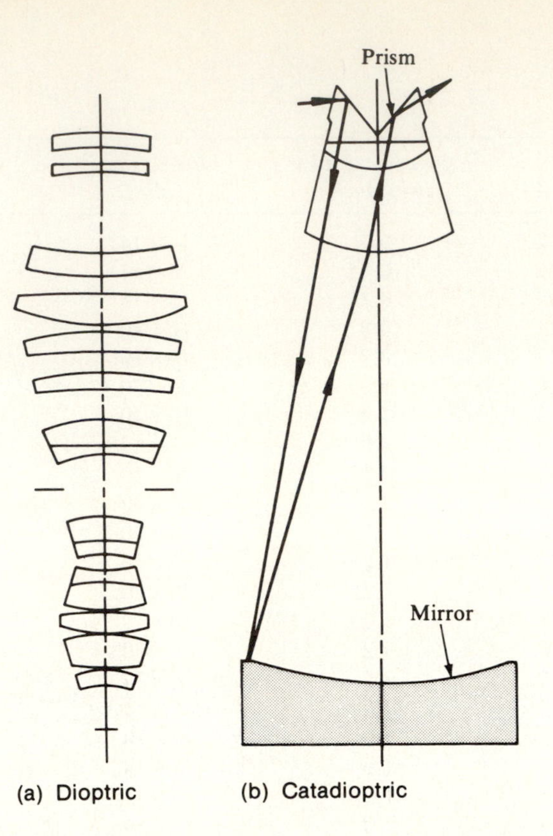

resolution lenses have a relatively small depth of field. However, without wafer leveling, automatic focus cannot compensate for the amount of taper typically found in wafers. (This item is discussed further in the section on wafer flatness.)

5.11.2 Alignment

Three stages of alignment can be involved. The first is a rough prealignment step, which positions the wafer so that sensors can detect its alignment markers. The next is global alignment, which uses only two marks per wafer, just as do the full-wafer printers just discussed. Each site is then indexed from these markers by precision wafer-stage motion. When steppers were first introduced for making masks from 10× reticles, screw rotation was used to determine the position of the stage, and relatively large cyclic errors occurred. Precision verniers etched in fused silica glass were next used to reduce the error, but all later machines have had laser interferometry and are much more accurate. Since global alignment is usually done with auxiliary alignment optics, if the mask position is not accurately known relative to the optical axis of the alignment optics, errors will occur. By placing alignment markers on each chip pattern, each site can be aligned individually. Aligning each site is more time consuming than is single global aligning and, if the added precision is not required, is often not used. However, some machines are designed to always do site-by-site alignment first, and only if an insufficient signal is available from a given site, do they resort to positioning this site from the global markers.

5.11.3 Registration

In order to fully utilize the high-resolution lenses now available, a method of maintaining level-to-level registration to accuracies approaching 20% of the minimum feature size is necessary. In many cases, a lack of registration and not lack of resolution limits feature size. Various elements causing lack of registration are given in Table 5.12. While no value is given for magnification change when a dedicated machine is used, a change in barometric pressure will cause a change in the index of refraction of air. This change can cause a change in magnification of multiple-element lenses that incorporate a light path through air. The effect is usually small, but most manufacturers provide some form of compensation. A distinction must be made between machines using global alignment and machines using site-by-site alignment. Global alignment has additional sources of error and, consequently, is substantially less accurate than is site-by-site alignment. It is, however, faster to use since only one alignment per wafer rather than one per step is required. It is also a simpler and less expensive system to build into the stepper. As the data of Table 5.12 show, with the current stage of development, the range of misregistration of machines using global alignment is such

TABLE 5.12

Causes of Misregistration during Stepper Operation

	Exhibited by		
Item	*Dedicated Stepper*	*Multiple Steppers*	*Global Alignment**
Wafer expansion	X†	X†	
Mask expansion	X	X	
Internal mask error	X	X	
Misalignment	$X \pm 0.1$	$X \pm 0.2$	
Lens distortion		$X \pm 0.15$	
Magnification change		$X \pm 0.15$	
Reticle rotation			$X \pm 0.2$
Stage orthogonality			$X \pm 0.2$
Stage motion precision			$X \pm 0.2$
Global offset			$X \pm 0.1$
RMS 2σ limits (μm)	± 0.1	± 0.3	± 0.35

*Additional errors due to global alignment.
†X is determined by temperature control. See earlier discussions.

as to begin limiting yields when these machines are used for geometries of less than perhaps 2 μm.

5.11.4 Throughput

The time T to expose one wafer can be calculated from

$$T = O + NE + Nt \qquad 5.18$$

where O is the time for overhead functions such as loading and unloading the slice, leveling (if done), prealigning, and initial focusing; N is the number of exposures; E is the time for each exposure; and t is the time for the stage to travel from one site to the next. Typical values are $O = 10$ s, $E = 0.5$ s,[19] $t = 0.4$ s. N will, of course, depend on the wafer size and the size of step that is used. Substituting these numbers into Eq. 5.18 gives

$$T(\text{s}) = 10 + 0.9N \qquad 5.19$$

Thus, to maximize throughput, the steps must be minimized, and the machine is usually programmed to step only those sites that will print whole chips on the wafer. Just interchanging one lens for one with a larger field—that is, going from 1×1 cm to 2×2 cm—will help throughput in that N of Eq. 5.18 will be reduced. But if the light source remains the same, then E will be commensurately increased. However, t will remain the same so that an overall gain is realized.

Even though the field size may be a nominal 1 or 4 cm^2, the actual field used may be considerably smaller because of the fill factor. When chips are smaller than the maximum field size, either the field must be stopped down or else multiple chip patterns must be used to fill the field. However, since each field must have an integral number of chip patterns, it may not be possible to have a complete fill. A plot of field size versus chip size for the particular case of square chips and a 1×1 cm maximum field size is shown in Fig. 5.23a. Rectangular chips will show the same trends. As the chip dimensions decrease from 1×1 cm, less and less of the full field is used until the dimensions are reduced enough for additional whole chips to be added. At this point, the fill factor abruptly rises and then again begins to decrease, producing the sawtooth pattern seen in the figure.

As the step size decreases, more steps are required to cover a wafer, and since the time to expose a wafer is almost linearly dependent on the number of steps, wafer throughput versus chip size, as shown in Fig. 5.23b, has the same sawtooth character. Circuit designers should keep this curve in mind since near critical areas,

---

[19]This time will decrease as light sources and resist sensitivity are improved.

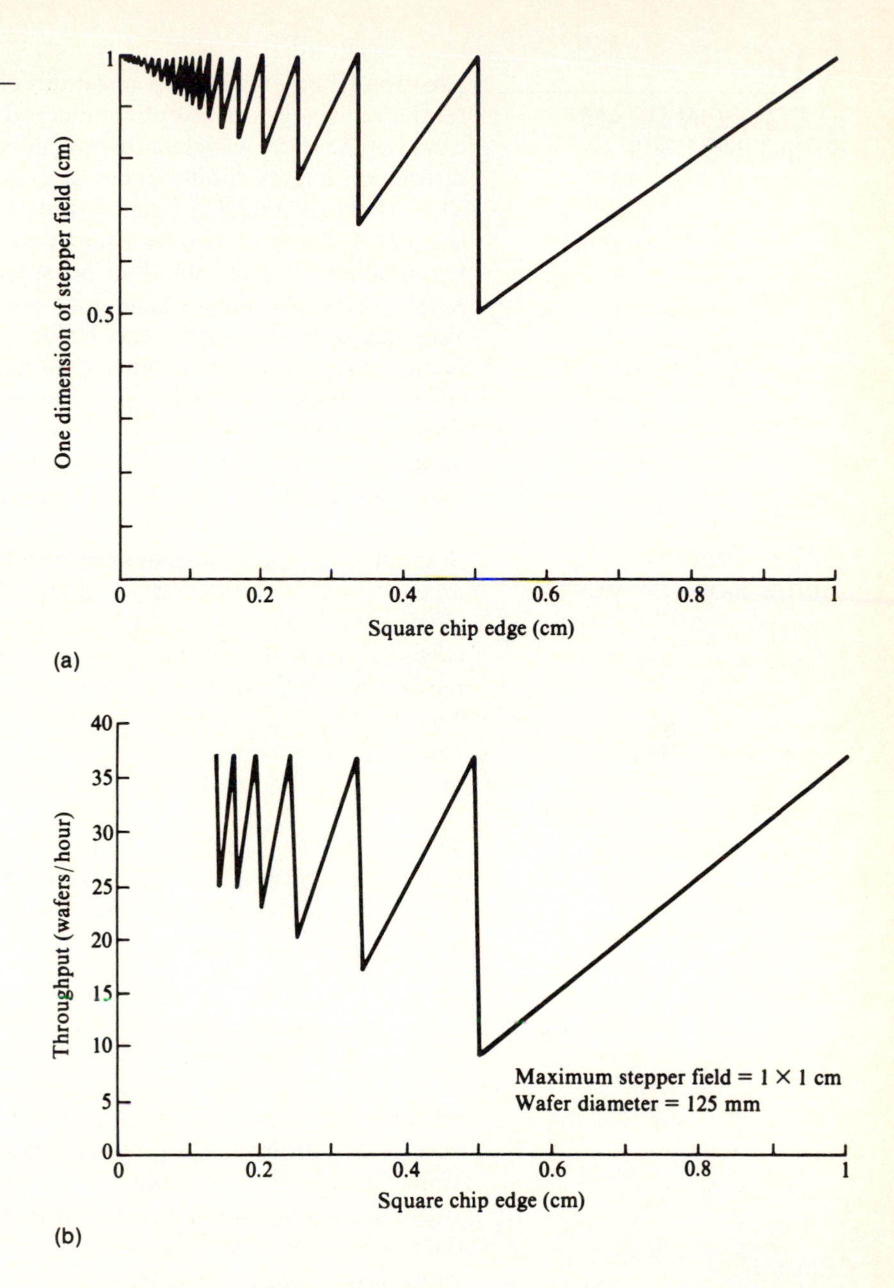

FIGURE 5.23

(a) Size of optimum stepper field versus chip size. (b) Stepper throughput in wafers/hour versus chip size, assuming stepper field is filled as much as possible.

slight changes in chip dimensions can radically affect throughput. The stepping pattern used in covering a wafer may also have an effect on the number of steps required and should be carefully considered. For example, it may be advantageous to linearly shift each row of patterns with respect to the previous one in order minimize the steps required in each row.

5.12 ELECTRON BEAM PRINTING*

Electron beam exposure of resist directly on a wafer offers two potential advantages over optical methods. One advantage is that because of the short wavelength associated with an electron beam, few diffraction effects should occur at geometries currently visualized. The other advantage is that by directly scanning the E-beam over the wafer, the mask can be eliminated. The cycle time from design to completed wafer can thus be substantially reduced. Unfortunately, these advantages have not been realized in a timely fashion. Whereas optical systems can be designed that are very close to being diffraction limited, other limitations in E-beam printing have severely limited its high-resolution potential. The relatively low throughput and high expense of a quality E-beam system and attendant facilities have prevented its widespread use even in applications where a premium is placed on quick turnaround.

5.12.1 Pattern Generation

Since an electron beam cannot be accurately deflected over the total area of a wafer, a combination of stage-travel, electromagnetic, and electrostatic deflection is used to cover the complete wafer. In some systems, a small spot beam is swept back and forth as the stage moves in serpentine fashion over the slice. Some systems form a variable-length beam as the stage travels, and others, as in optical steppers, stop the stage and expose an area by various scanning methods and then move to the next field.

5.12.2 Resolution

The smallest feature that can be exposed is the diameter of the electron beam. Diffraction, electron–electron interactions, and E-beam lens design all contribute to the beam diameter. The wavelength λ associated with an electron beam is very short and is approximately given by

$$\lambda = \frac{12.3}{\sqrt{V}} \qquad 5.20$$

where λ is in Å and V is the accelerating voltage in V. Thus, a beam accelerated through 30 kV would have a wavelength of 0.07 Å and would have negligible diffraction when printing a 0.5 μm feature. Beam expansion due to electrons interacting with other electrons in the beam is dependent on the electron density and is manageable by controlling beam current. Focusing and deflection optics are a major limitation, but beam diameters of from 0.125 μm to 0.2 μm are cur-

*See references 62–63.

rently available. Thus, none of these limitations would prohibit the printing of 0.5 μm features.

What is a limitation is the scattering of electrons by the resist and the underlying substrate. Not only does scattering tend to make individual exposures cover more area, but also it gives rise to proximity effects in which the scattering from one feature affects the exposure of other parts of the same feature as well as that of adjacent features (64–66). This problem is minimized by programmed reduction of beam intensity in adjacent regions and the use of as low a beam energy as possible.

5.12.3 Alignment

Since a scanning beam is already available, it, along with suitable detectors, is used to locate alignment markers. In general, as in the other systems, a flat finder is first used to provide rough azimuthal orientation. Then, some coarse markers near the edge of the wafer are used to determine the position well enough for the small markers on individual chip fields to be located. Quoted overlay tolerances, which include alignment and distortion, vary from 0.15 μm to 0.5 μm.

5.13 X-RAY PRINTING

X-ray wavelengths are very much shorter than that of the UV light used in conventional lithography and thus, when they are used for exposing photoresist, have minimal diffraction effects. The wavelengths that are ordinarily used are in the 5–10 Å range. Because X rays cannot be readily focused, printing must be either contact or proximity. For the same reason, the only way to get a reasonably collimated beam is to have a small source and have the wafer far from the source. However, small-size sources limit the intensity of the X-ray beam, as do large distances. Typical values are a source diameter of 2–4 mm, a source-to-wafer distance of 30–40 cm, and a mask–wafer spacing of 10–40 μm (67–68). Fig. 5.24 is a schematic diagram of an X-ray printing system. Since the soft (low-energy, long-wavelength) X rays suffer appreciable attenuation when they pass through a 40 cm air path, either a helium-filled or vacuum chamber is provided. Because many of the masks are opaque to visible light (see the earlier section on mask making), conventional alignment systems have, in many cases, been replaced by indirect systems. Furthermore, the main purpose of X-ray lithography is to allow finer geometries to be printed than are possible with UV (probably line widths less than 0.5 μm). The finer geometries require more precise alignment and hence more elegant systems than are often found on UV systems.

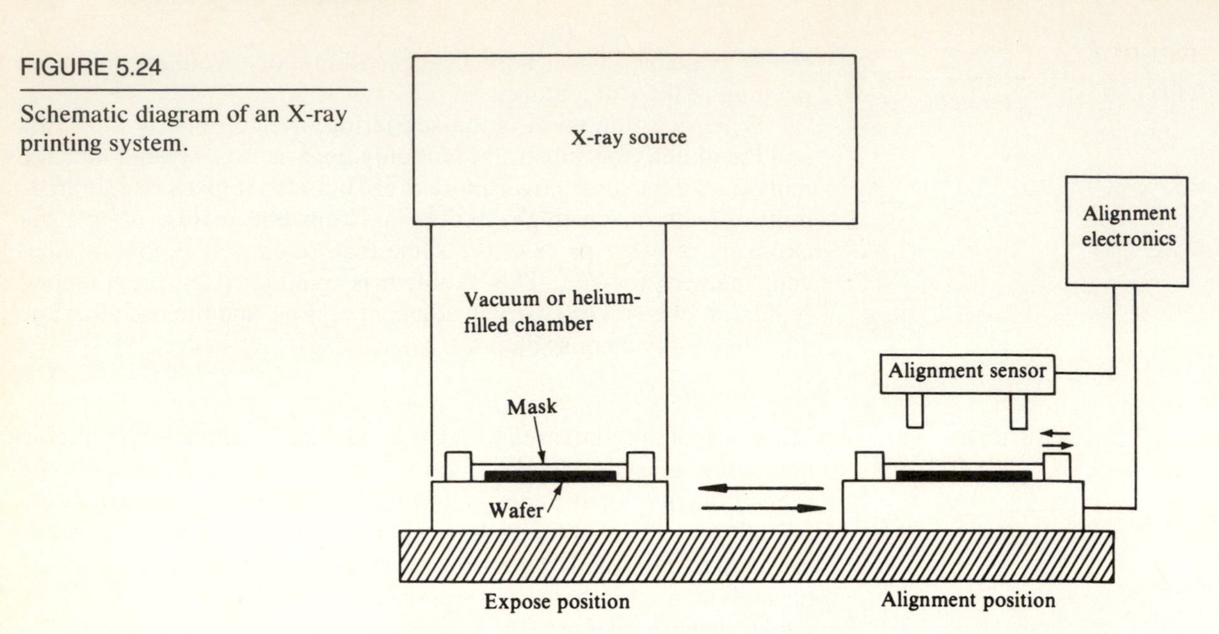

FIGURE 5.24

Schematic diagram of an X-ray printing system.

5.13.1 Image Distortion

The lack of a parallel beam causes the image to be slightly larger than the mask, as shown in Fig. 5.25. The magnification varies with the mask–wafer separation, and if the separation is not uniform, image distortion will occur. A variable separation can arise from a nonflat wafer, a bowing mask, or a mask–wafer tilt. The magnification m is given by

$$m = \frac{B}{B - d} \tag{5.21}$$

where B is the source-to-wafer distance and d is the mask–wafer separation. For a typical mask–wafer separation of 40 μm and a 40 cm wafer–source separation, $m = 1.0001$. For an image covering a full 100 mm wafer, this amounts to a 10 μm increase in image size over mask size.

5.13.2 Resolution

Factors that affect X-ray printed line widths (see Fig. 5.25) are diffraction, shadowing from the finite-size X-ray source, leakage of X rays through the edges of the mask, and electrons generated when the X rays are absorbed in the resist. The width of the diffraction blur δ_d is approximately (69)

$$\delta_d = 0.4\left(\frac{\lambda d}{2}\right)^{1/2} \tag{5.22}$$

FIGURE 5.25

Factors affecting resolution of X-ray printing.

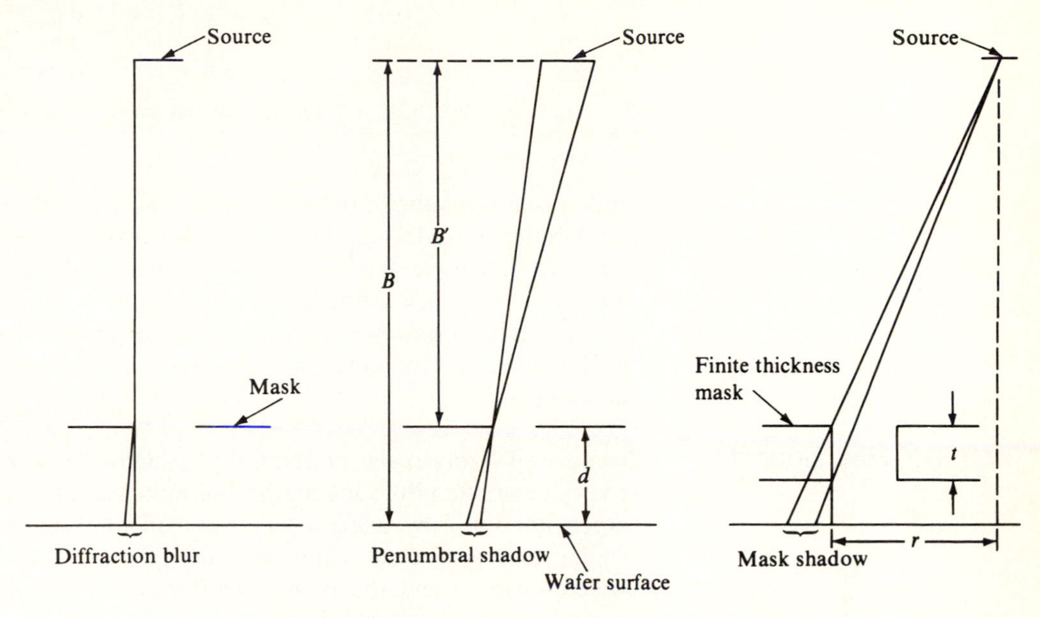

where d is the wafer–mask separation. The width δ_s of the penumbral shadow is

$$\delta_s = \frac{S_d d}{B} \tag{5.23}$$

where S_d is the source diameter and B is the distance from source to wafer. Since $d << B$, the distance from the source to mask B' is assumed equal to B, and the mask edge shadow δ_m can then be approximated as (2)

$$\delta_m = \frac{rt}{B} \tag{5.24}$$

where r is the distance from the centerline of the source–wafer to a particular mask edge and t is the thickness of the masking material. The range δ_e of the electrons generated in the resist has been estimated as

$$\delta_e = 10^{-23}\lambda^{-7/4} \tag{5.25}$$

where λ is in meters. Table 5.13 compares these factors for some typical system parameters. From the values shown, it can be con-

TABLE 5.13

Values for Components of Resolution Limitations

Wavelength (Å)	δ_d (μm)	δ_s (μm)	δ_m (μm)	δ_e (μm)
5	0.04	0.3	0.05	0.19
10	0.057	0.3	0.05	0.056
20	0.08	0.3	0.05	0.017

B = 40 cm, d = 40 μm, S_d = 3 mm, r = 20 mm, and t = 1 μm.

cluded that for submicron geometries, the penumbral shadow can be a limiting item. Its significance can, however, be reduced by going to a much narrower gap, such as 10 μm. To minimize scattering and diffraction effects, it appears that a window in the 10–20 Å wavelength region is optimum. For shorter wavelengths, resist scattering becomes a problem; for longer ones, diffraction becomes a limitation.

5.13.3 X-Ray Sources

Because of X-ray resist insensitivity, a high-intensity X-ray source is very desirable (70). One method of increasing power input to the target (anode) is by using a water-cooled rotating anode. The impinging electron beam is thus constantly presented with a freshly cooled surface, and the power input can be substantially greater than it is with a stationary water-cooled anode. However, some stationary targets have a thin-walled conical depression cooled by high-velocity water that can handle very-high power inputs (71) and are not plagued with the mechanical problems of rotating anodes. Pulsed laser-powered X-ray sources also have potential for X-ray lithography (69). With such systems, in which the individual pulses are from 1–100 ns long, the peak power may be hundreds of times greater than the power normally incident on the X-ray mask and resist. Thus, if the resist shows a problem with reciprocity, exposure times may be considerably longer than anticipated. The mask absorbing layer can be subjected to substantial stress and heating by such pulses, and it has been suggested that the peak energy per pulse should be limited to about 10 mJ/cm^2. Synchrotron radiation is also a possible X-ray lithography source (72, 73). Conventional synchrotrons are of course very large, but the small, storage-ring models may prove practical.

5.13.4 X-Ray Steppers

Because of the difficulty in producing the large-area, very-thin membranes used in X-ray masks, steppers offer some advantages. With their use, the movement to ever larger slices will pose no additional mask-making problem. However, as long as X-ray sources remain relatively weak and the resist relatively insensitive, the requirement for many exposures per wafer will dictate a low throughput.

5.14

ION BEAM PRINTING

5.14.1 Focused Ion Beams

Focused ion beam printing (74, 75) uses a focused beam of ions, accelerated to perhaps 150 kV, that is scanned over a small area just as in E-beam lithography. Currently, the available ion beam density is much lower than the electron beam density. However, because the ions are so much heavier than the electrons, resist sensitivity is much greater for ions. The net result is that exposure times are comparable. Ions that have been used include protons and doubly charged silicon. A double-charged ion has the advantage of providing greater resist penetration for the same accelerating voltage.

As an alternative to E-beam exposure, focused ion beams appear to offer several advantages. Diffraction effects, as in E-beam optics, are negligible, because of the higher mass of ions, scattering of the beam by the resist is minimal, and no problem occurs with the proximity effect. Beams of 0.1 μm diameter can be formed and can, in principle, be printed. Printing speeds may be substantially slower than those of E-beam systems, however, because of the slower scan speeds associated with longer ion beam transit times. To prevent charge buildup from causing a widening of the beam and thus setting a lower limit on line width, conductive resists may also be necessary.

5.14.2 Masked Ion Beam Printing

Masked ion beam printing, analogous to optical proximity printing, E-beam flood-printing, or X-ray printing, uses a beam of 1–2 cm^2 to sweep over a mask and expose the resist beneath (76, 77). It avoids difficulties in precision ion beam focusing and location but introduces the problem of masking. Some way of patterning the mask (presumably E-beam) is required. Either a stencil mask (holes in an ion beam impervious material) or an impervious material supported on a thin ion beam transparent material may be used. The stencil mask requires either that no ring-and-dot type geometries be used or that a complementary mask pair be used. The thin membrane approach has all of the problems associated with X-ray masks plus scattering of the beam by thin amorphous materials. One solution to scattering is to use a thin silicon membrane oriented so that channeling can occur. Alignment, as with X-ray printing, is more difficult than it is with either optical or E-beam stepping and must be done indirectly.

5.15

PRINTER SELECTION

Many factors must be considered in choosing the most suitable printer for a particular operation. The final choice should be one that satisfies the technical requirements, achieves a satisfactory chip cost, and is compatible with the long-term plans for the wafer fab-

TABLE 5.14

Factors Affecting Choice of Printer

Initial Investment
Printer cost
Printer facility cost
Mask retooling cost
Printer Operating Cost
Repair and maintenance cost
Mask cost
Operator labor
Depreciation
Facility upkeep
Printer Throughput
Inherent speed
Uptime
Setup time
Process Applicability
Wafer diameter
Chip size
Defects generated by printer relative to defects in incoming mask
Process yield of each printer process
Compatibility of printer processes
Resolution of printer
Time until obsolescence

rication facility. Table 5.14 lists factors that must be considered. The first three groups can be classified as purely economic; the last group contains technical issues. Of prime importance in printer selection is technical capability: Immediately rule out machines that cannot print wafers of the desired diameter with the required resolution. Fig. 5.26 shows the current (1988) resolution limits for each type of printing technology.

Most of the remaining items contained in the four groups are interrelated so that numerous trade-offs must be evaluated. Which ones are most important will vary with the situation, but as examples, consider the following. Contact printing is capable of good resolution, high throughput, and low downtime but has high mask usage, high defect density, and marginal overlay capability. Initial cost and operating cost are low; the yield will, however, be low because of the high defect density introduced; and the mask cost could be high. E-beam machines and their associated facilities are very expensive, and the throughput is low. However, they allow quick turnaround from design to wafer printing, and the printing defects are low. Stepper throughput is low relative to full-wafer scanners, but scanner mask costs for large-diameter wafers may be excessive.

In some cases, the use of two or more different kinds of printers in the same wafer-fab area, each assigned to its own particular levels, may be advantageous. For example, steppers might be used on levels requiring the smallest feature size and full-wafer scanners on the others. If there is some level, such as the final lead pattern of a gate array, that is customized and needs quick turnaround for good customer responsiveness, then it might be printed with E-beam machines. This approach, referred to as "mix and match" makes use of the strengths of each printer type. Its successful implementation requires that differences in distortion between the machines are acceptable and that the mask alignment markers are compatible. In some cases, multiple sets of markers may be required since, for example, stepper site-by-site alignment requires a marker on each die and full-wafer printers need only two markers per wafer. Also, printers by different manufacturers may use completely different alignment sensing. For example, one might use Fresnel zones and another the scanning of crosses.

5.16 LINE-WIDTH CONTROL

As mentioned earlier, providing line-width control is one of two major goals of a good lithography process (the other is maintaining adherent pinhole-free resist layers). Control is lost when the feature size becomes resolution limited, and this aspect has already been discussed. However, even if the feature size is well above the resolution limit, several things limit size reproducibility. Ordinarily,

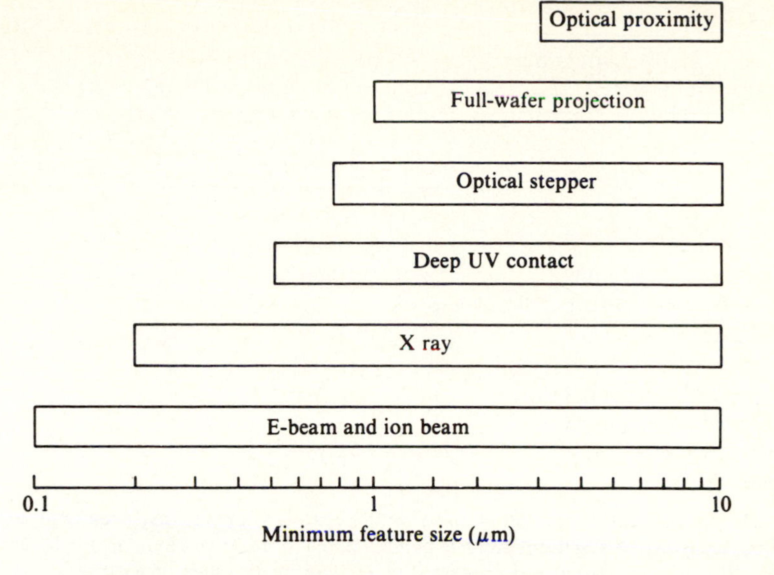

FIGURE 5.26

Printing technology versus current (1988) minimum feature size.

±10% of a line width on an IC is considered satisfactory, and processes are designed with this percentage as a minimum goal. Not all of this tolerance can be used for the pattern width variation since pattern undercutting and initial mask tolerance all contribute to the final feature size. Undercutting can be both an etching and a lithographic problem since undercutting is accentuated if the resist peels during etching. Mask control 2σ limits are typically 0.2 μm plate-to-plate and 0.1 μm within a plate (78). Thus, for features of 2 μm and less, all of the desired tolerance is used in the mask alone when 1:1 printing is done. However, when pattern reduction and stepping printing are used, the numerical (not the percentage) variation of the image line width is reduced by the same factor as the image reduction being used.

5.16.1 Process-Initiated Control Problems

Table 5.15 lists processing factors that affect line width. In addition, the table shows, for positive resist, the effect of a change in these factors on feature size. Not shown in the table is the effect of resist and developer compositional variability. A distribution of line widths will occur just because of this normal variability. Moreover, on occasion, errors in formulation that occur must be considered. When they do, line width may change, and such changes must be separated from the variables listed in Table 5.15. In order to study how the various items shown in the table affect line width, curves can be constructed that plot line width versus each of the individual variables, with all of the others held constant, and the sensitivity determined. These kinds of curves are referenced in Table 5.15 and

TABLE 5.15

Process Variables Affecting Feature Width

Item	Direction of Change	Effect on Oxide Feature Width* for Positive Resist	Reference
Exposure time	+	+	Fig. 5.27
Develop time	+	+	Fig. 5.29
Developer concentration	+	+	Fig. 5.30
Time between exposure and develop	+	−	Fig. 5.31
Resist thickness	+	−	Fig. 5.32
Printer defocus	+	†	Fig. 5.33‡
Soft-bake time or temperature	−	+	
Hard-bake temperature	+	§	
Resist adhesion	−	+	
Printer *NA*‖	+	+	

*Note that if the feature is a cut in an oxide, a negative value for resist removal, as shown in Fig. 5.27, gives a larger feature. However, if a metal pattern is being defined, a negative value means a narrower lead line width.
†Depending on the amount of exposure, the line width can widen or narrow.
‡The intensity changes described in this graph must be translated into line-width changes.
§If the resist flows, the line will be too narrow; if the high temperature causes gas to be trapped and poor adhesion, the line will be wider.
‖This item is not the maximum system *NA* but rather that set by the aperture actually used during exposure.

show, for example, line width versus resist thickness. Alternatively, a "best" value for some particular variable can be plotted against a range of some other variable. An example would be the best exposure time as the resist thickness is varied. The first approach is most useful in developing a wide-latitude process, while the latter is more useful in providing operational information for a factory. Unfortunately, neither approach gives a clear picture of the interactions between multiple variables. For example, depending on the soft-bake time or temperature, the best exposure time versus resist thickness curve may vary substantially. One approach to clarifying the situation, and one that minimizes the number of experiments, is to use a set of statistically designed experiments to examine the interactions among a whole series of variables (79). The series of curves shown in the next few sections (and referred to in Table 5.15) are of the "change one variable at a time" variety in order to give more insight into the individual process variables.

The amount of exposure will affect line width, and a plot of line width versus exposure is often used as the criterion for proper exposure. Alternatively, a resolution test pattern can be exposed for various times and the exposure chosen on the basis of best resolution. The energy for proper exposure is in the 10 mJ/cm^2 range for negative resist and in the 40 to 60 mJ/cm^2 range for positive resist. Table 5.16 gives conversion factors to other units.

For most negative resists, underexposure will cause poor resist adherence and the resist lines to be narrower than normal, while overexposure will cause them to be wider. For positive resists, the opposite is true; overexposure produces narrower lines.[20] Typical data for positive resists are given in Fig. 5.27. In working with a specific resist, curves applicable to this particular resist should be used. Because of the nonlinearity of the exposure/line-width curve, the sensitivity of line width to exposure is less for line widths less than nominal. Thus, better control can be achieved by using an un-

TABLE 5.16

Conversion of Energy and Intensity Units

Unit		*Conversion Factor*
1 millijoule	=	1 milliwatt-second
1 millijoule	=	10^4 ergs
1 calorie	=	4.186 joules
1 coulomb		No conversion*

*Used in E-beam exposure.

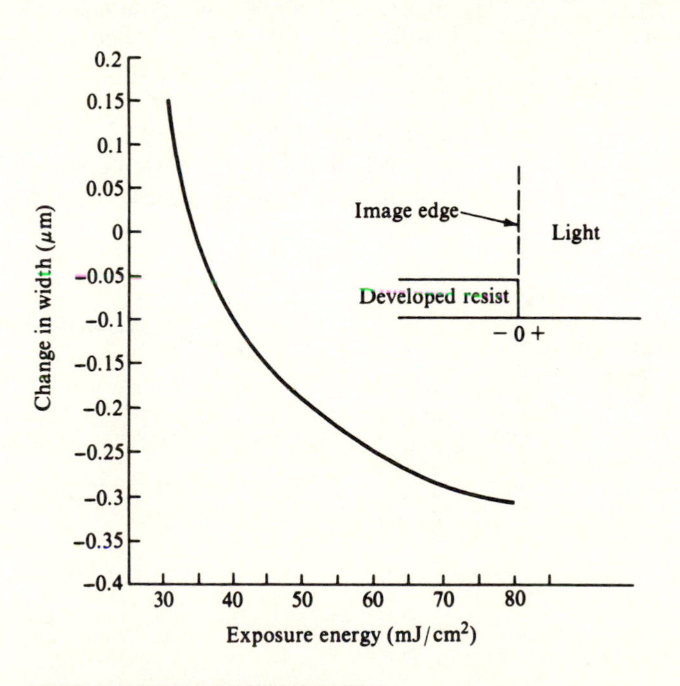

FIGURE 5.27

Location of resist edge versus amount of exposure for positive resist, 1.8 μm thick. (*Source:* Data from Terry V. Nordstrom, *Semiconductor International*, p. 158, September 1985.)

[20]Note that these are narrower *resist* lines. In the literature, a *line width* will usually refer to the etched feature and not to the resist.

dersized mask (80). Because of interference effects changing the amount of energy absorbed by the resist, the correct exposure and, in turn, the line width can be very sensitive to rather small changes in resist and oxide thicknesses. The kinds of intensity changes to be expected are discussed in a later section. When lines of different widths are being exposed simultaneously by projection printing, the narrower ones may receive substantially less energy per unit area and hence be underexposed. The magnitude to be expected is shown in Fig. 5.28. The smaller the *NA* of the lens, the sooner the effect becomes noticeable. The decreasing intensity arises because more and more of the light incident on a slit is diffracted outside its boundary as it becomes narrower. Behavior for other lens and line-width combinations can be calculated from Eq. 5.16.

The development time affects the line width as shown in Fig. 5.29 for positive resist. Note that at the nominal size opening (where the resist size matches the mask size), the change in size with time is very rapid. Better line-width control can often be maintained by making the mask opening somewhat smaller and by developing longer (81). The strength of the developer (the dilution) can affect line width. Fig. 5.30 shows data for positive resist. Even the length of time between exposure and develop has an effect on line width, as Fig. 5.31 shows, which again underscores the fact that cycle time through the lithography process should be minimized. For a given exposure, changing resist thickness will change line width, as shown in Fig. 5.32. For this reason, resist thickness must be kept under good control.

When the image is defocused, it becomes smeared, and line-width control deteriorates. Therefore, much attention is paid to

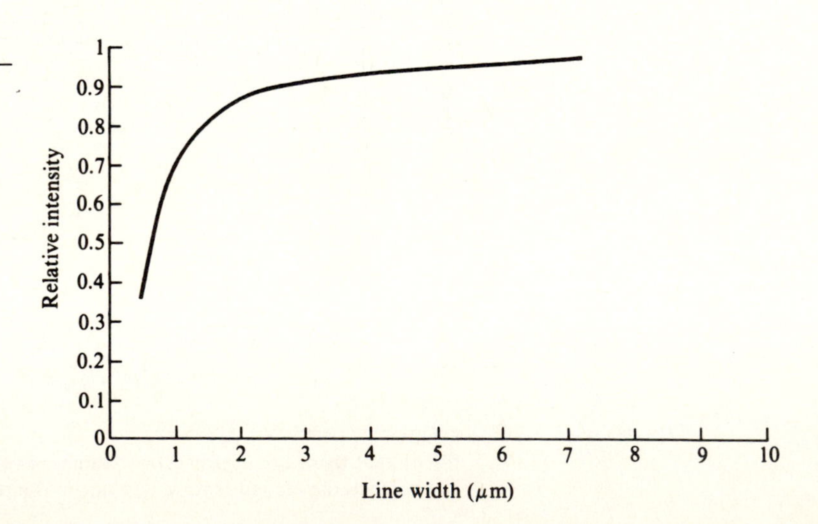

FIGURE 5.28

Calculated light intensity at middle of a line image versus line width for 0.28 *NA* system, 530 nm light. (*Source:* Data from J.H. Altman, *Photographic Science and Engineering 10,* p. 140, 1966.)

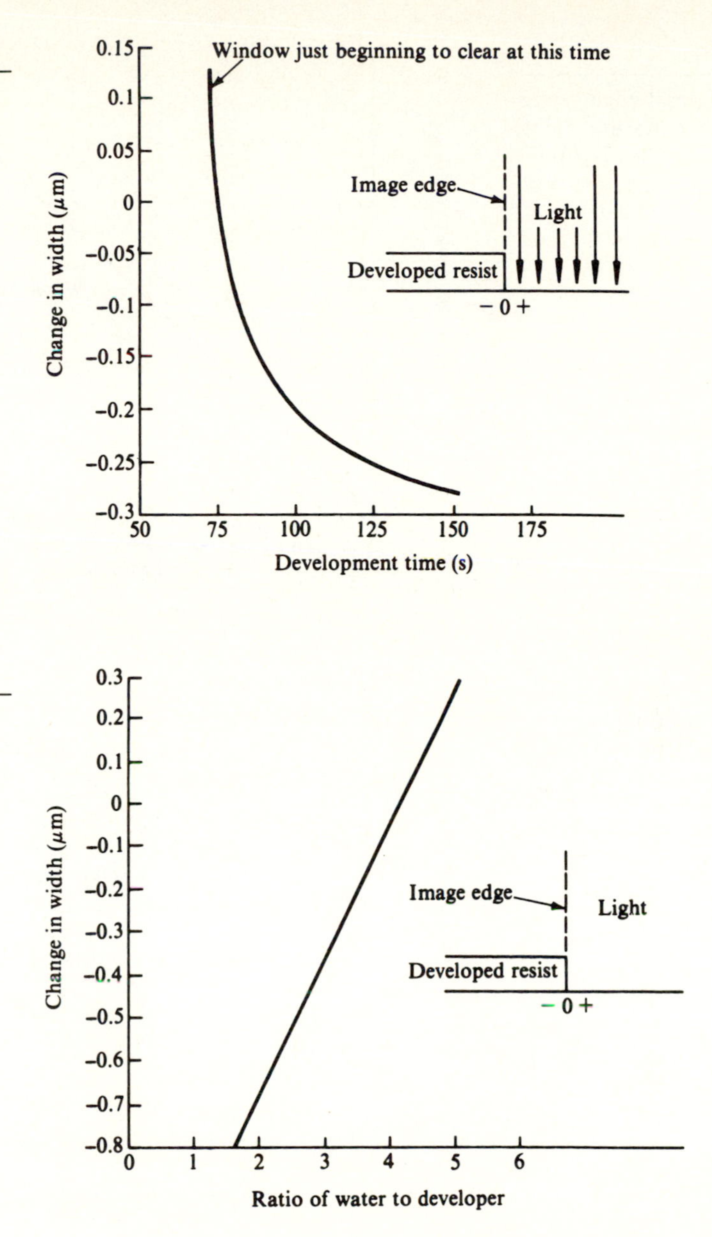

FIGURE 5.29

Location of resist edge versus development time for AZ1350J positive resist, 580 nm thick, developed with 1:1 AZ developer: H_2O. (*Source:* Data from F.H. Dill et al., *IEEE Trans. on Electron Dev. ED-22*, p. 456, 1975.)

FIGURE 5.30

Effect of developer dilution on location of resist edge for AZ1350J positive resist, 2 μm thick, 50 s exposure, 60 s develop. (*Source:* Data from David J. Elliot, *Solid State Technology*, p. 66, September 1977.)

depth of focus of the lenses being used and to the flatness of the wafers being printed. Fortunately, because of the resist characteristics, the useful depth is often substantially greater than the Rayleigh criterion of Eq. 5.17. Fig. 5.33 shows how defocusing changes the light intensity near the mask edge. The idealized curve shows the abrupt light–dark transition that is usually visualized. As shown

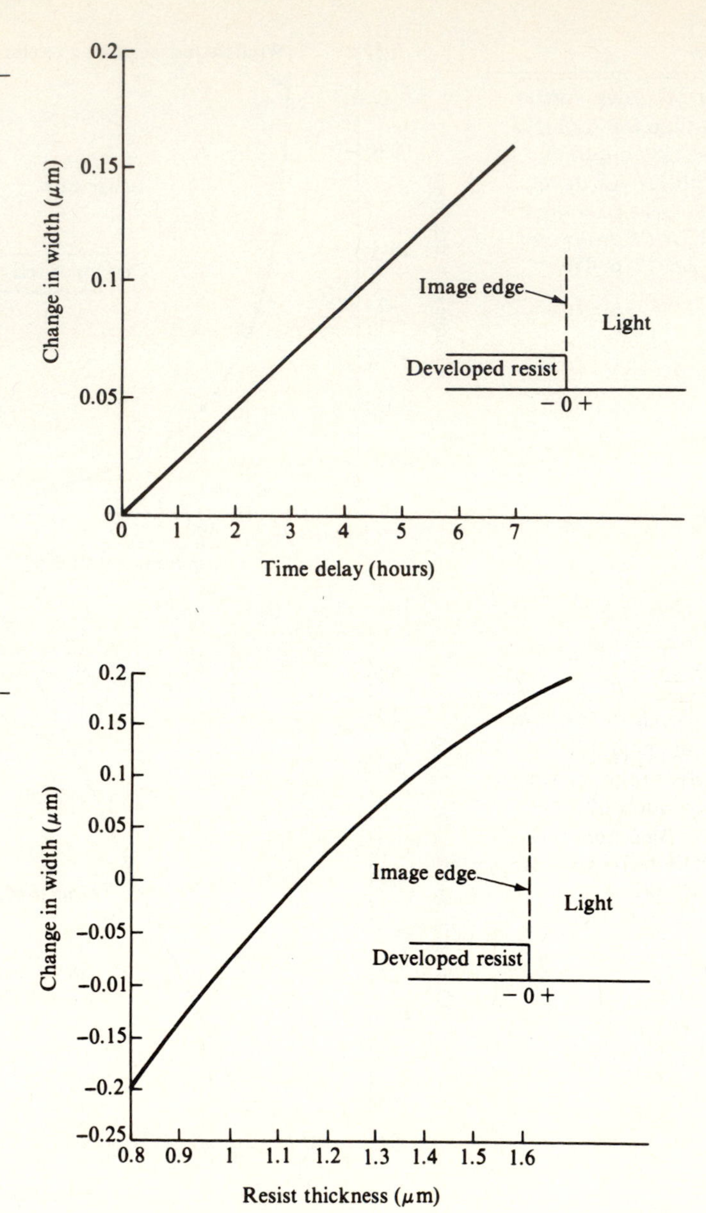

FIGURE 5.31

Location of resist edge versus time between exposure and develop for Kodak 809 positive resist. (*Source:* Data from D.W. Frey et al., *Proc. Kodak Microelectronics Seminar,* p. 40, Dallas, October 1981.)

FIGURE 5.32

Location of resist edge versus resist thickness for positive resist, 35 mJ/cm² exposure. (*Source:* Data from Philip A. Hunt Chemical Corporation product literature.)

by the effect of diffraction curve, the transition is always somewhat sloped. Defocusing makes the curve even more sloping. However, a pivot point (A of Fig. 5.33) somewhere between $I = 0.25$ and $I = 0.5$ (depending on the light coherence) is independent of the amount of defocus. Thus, if the exposing intensity and resist can be matched so that the resist cutoff energy (E_f of Fig. 5.8) equals the intensity I_f

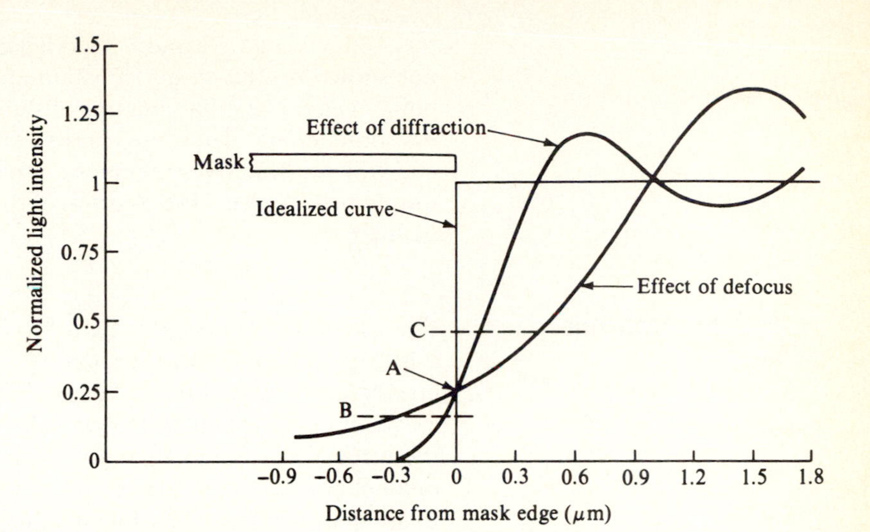

FIGURE 5.33

Effect of defocus on an edge image for coherent illumination, 436 nm wavelength, 0.36 *NA*. (*Source:* Data from Stephen H. Rowe, *J. Optical Soc.*, p. 711, 1969.)

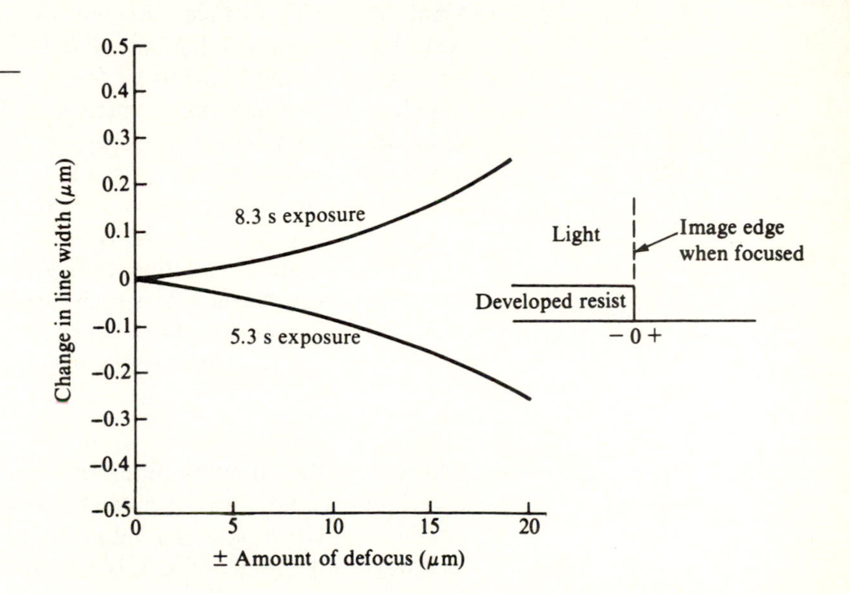

FIGURE 5.34

Location of resist edge versus amount of defocus for *f*/3 system, 0.9 μm of negative resist. (*Source:* Data from J.W. Bassung, Proc. SPIE 100, *Semiconductor Microlithography II*, 1977.)

at point A times the exposure time, then over a broad range of defocus, no change in line width should occur. Increasing exposure energy causes I_f to be reached lower on the intensity axis so that it will move from C to A to B. If, for the exposure time chosen, I_f is at B, a change from the idealized to the defocused curve causes this intensity to be reached well under the edge of the mask, and a negative resist line will widen with defocus. If E_f is at C, then the line will narrow. Experimental data are given in Fig. 5.34. The two exposure times of 5.3 s and 8.3 s bracket the pivot point A. Other data not shown on this graph substantiate the thesis that, with I_f set at point A, a substantial amount of defocus can occur with negligible line broadening. There are limits, of course. Fig. 5.33 is idealized and will not hold over an infinite range of defocus. Furthermore, as the slope decreases, the position of the resist boundary becomes less defined.

5.16.2 Interference Effects

The use of monochromatic light[21] to expose the resist, coupled with a highly reflecting surface such as silicon or metal beneath it, can give standing waves in the resist (82–85). When standing waves do exist, the light intensity through the resist in a direction normal to the wafer surface will vary in a cyclic fashion, causing resist development normal to the surface to vary in a similar fashion and leading to undulating instead of straight walls as shown schematically in Fig. 5.35. Since there is a minimum in intensity of a standing wave at a highly reflective surface, adjacent to this surface, the resist will always be underexposed. If an oxide is between the reflecting surface and the resist, the intensity of the standing wave at the oxide–resist interface will depend on the thickness of the oxide. It can range from minimum to maximum intensity, and since it does depend on the oxide thickness, it is often possible to choose the oxide thickness so that maximum intensity occurs at the interface. In this way, the resist next to the oxide is more likely to be fully exposed. While printers using broadband illumination are, in principle, not subject to standing waves, the use of a resist with a narrow spectral transmissivity can still cause them.

FIGURE 5.35

Effect of standing waves on the cross section of developed resist.

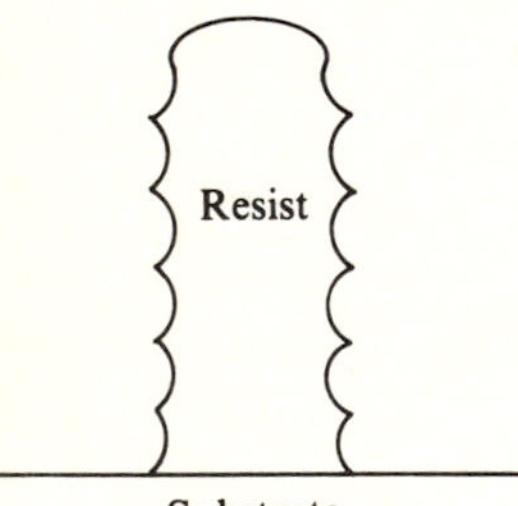

The E-field E_I of the incident light can be expressed as

$$E_I = E_O e^{i\omega t} e^{-2\pi i N z/\lambda} \qquad 5.26$$

where N is the complex index of refraction $n - ik$, λ is the wavelength in vacuum, and k is given by $\alpha\lambda/4\pi$, where α is the absorption coefficient. The n, k, and α values for various materials encountered in lithography are given in Table 5.17. The wave is considered to be traveling in the $-z$ direction, which is normal to, and toward, the wafer surface. The primary reflecting surface (the silicon or metal) is at $z = 0$.

[21]High-performance lenses using all refractive elements can only be designed for use with monochromatic light. Lenses using partially reflecting optics, such as full-wafer scanners and 1:1 steppers, can use a broader light spectrum and are much less susceptible to interference effects.

TABLE 5.17

Index of Refraction and Absorption Coefficient

Material	*n*	*α (/cm)*	*k*	*Wavelength*	*Reference*
Si	4.6	2×10^4	0.07	450 nm	(1)
Au	1.4	5×10^5	1.88	450 nm	(2)
Al	0.49	1.2×10^6	4.32	450 nm	(2)
Cr	0.6		1.55	(not known)	(3)
Air	1.0				
SiO_2	1.46			450 nm	
AZ1350J resist	1.68		0.02	450 nm	(4)
KTFR (negative resist)	1.54			Na light	(5)
KMER (negative resist)	1.55			Na light	(5)

References: (1) H.R. Philipp and E.A. Taft, *Phys. Rev. 120,* p. 37, 1960. (2) O.S. Heavens, *Optical Properties of Thin Films,* Dover, New York, 1965. (3) Reference in J.H. Altman and H.C. Schmitt, *Proc. Kodak Photoresist Seminar,* p. 12, Los Angeles, 1968. (4) Frederick H. Dill et al., *IEEE Trans. Electron Dev. ED-22,* p. 456, 1975. (5) Kenneth G. Clark, *Solid State Technology,* pp. 52–56, June 1971.

Instead of exponential notation, the following expressions can also be used. First,

$$E_I = E_O \cos(\omega t - \beta z) \tag{5.27}$$

And, the E-field E_R of the wave reflected back at the silicon or metal interface is given by

$$E_R = RE_O \cos(\omega t + \beta z) \tag{5.28}$$

where $\beta = 2\pi(n - ik)/\lambda$ and R is the reflection coefficient such that

$$R = \left(\frac{N_1 - N_2}{N_1 + N_2}\right)^2 \tag{5.29}$$

N_2 is the complex index for the medium doing the reflecting, and N_1 is for the medium in which the wave is traveling. In the case of either oxide or resist, k is small and, for purposes of Eq. 5.28, can be disregarded. Thus, β is real and given by $2\pi n/\lambda$.

When E_I and E_R are added together, they have a standing wave component

$$E_S = E_{SO} \sin \omega t \sin\left(\frac{2\pi nz}{\lambda} + \theta\right) \tag{5.30}$$

It has the same wavelength λ/n as the incoming wave, but its phase θ may not match that of either E_I or E_R, as is shown in Fig. 5.36.

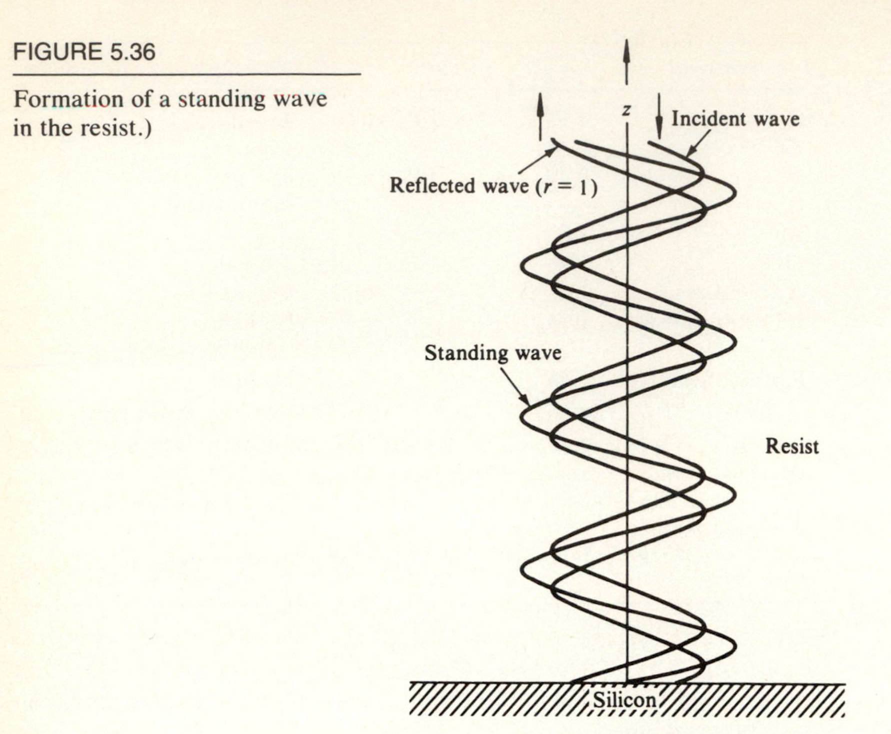

FIGURE 5.36

Formation of a standing wave in the resist.)

The minima and maxima are each separated by $\lambda/2n$. The minima will occur whenever $\sin(2\pi nz/\lambda + \theta) = 0$ —that is, when

$$\frac{2\pi nz}{\lambda} + \theta = m\pi \tag{5.31}$$

where m is zero or a positive integer. If the phase shift is 180° (π), as is the case of Si or GaAs or when the reflector is metallic and perfect, the first minimum will be at the interface, and the others will be located at

$$z = \frac{m\lambda}{2n} \tag{5.32}$$

Similarly, the maxima (antinodes) will be located at

$$z = \frac{(2m + 1)\lambda}{4n} \tag{5.33}$$

Otherwise, the position of the first minimum will be given by

$$z = \frac{\lambda\theta}{2\pi n} \tag{5.34}$$

where θ is given by

$$\theta = \frac{\sin(-2\pi nz/\lambda) + r\sin(2\pi nz/\lambda + \phi)}{\cos(-2\pi nz/\lambda) + \cos(2\pi nz/\lambda + \phi)}$$

The intensity I of the light is proportional to the square of the E-field so that $I(z)$ is given by

$$I(z) = I_o \sin^2 \beta z \qquad \text{or} \qquad I_o \sin^2 (\beta z + \phi) \tag{5.35}$$

depending on whether the reflector is silicon or metallization.

Resist that is exposed by light with an intensity varying in the manner of Eq. 5.33 will reflect this variation in the manner that it develops. The resist near the wafer interface will be poorly exposed. If it is positive resist, complete removal may not be achieved, and bridging will occur between closely spaced resist lines. If the resist is negative, the bottom part, being underexposed, may adhere poorly as well as being undercut during development. If the resist line is narrow, undercutting may be severe enough to allow the whole line to lift off. Depending on the resist thickness, there may be anything from maximum to minimum intensity at the surface. Thus, the top layer may range from fully developed to underdeveloped.

When the waves reflected from the wafer surface that are re-reflected from the resist are considered, it is found that, while the location of the nodes of the standing wave does not change, the amplitude may. It will depend on the distance between the two reflecting surfaces.

If the removal rate of resist is assumed to be proportional to the product of $I(z)$ and exposure time, the removal rate at any depth below the original surface can be calculated. Numerical computer analysis techniques based on this approach have been developed (86) that give good simulations of the experimentally observed profiles of resist exposed by standing waves.

5.16.3 Minimizing Effect of Standing Waves

When ordinary resist lies directly on silicon or metal, the position of the first node cannot be adjusted, and without additional precautions, a layer of underdeveloped resist near the interface must be accepted. If the resist is on top of a layer of oxide, the optical properties of the two are so close that the standing wave can be considered to be continuous across the oxide–resist interface. If the thickness of the oxide is z_1, there will be an antinode at its surface if $z_1 = (2m + 1)\lambda/4n$ and a node if $z_1 = m\lambda/2n$. Thus, when resist is on top of an oxide, the oxide thickness can be varied to give an antinode at the interface and ensure proper exposure at the interface.

By using a very thin highly absorbing layer between the substrate and the resist, standing waves can be eliminated. This technique is not widely used, however. Dye can be incorporated into the resist to absorb additional light energy and minimize standing waves (87) and light scattering from rough surfaces. Some commercially available resists have such a dye in them. The incorporation of dye may cause the resist sidewalls to be excessively tapered during development. However, by using an additional additive that forms a thin patternable skin on the top surface of the resist during softbake, nearly straight sidewalls can be maintained (88).

A bake after exposure, but before developing, can cause the sensitizer to diffuse enough to smooth out the undulations so that the profile is not so ragged and so that the bottom of the resist layer will develop properly (89). A light plasma etch can be used after development to remove any scum left in the bottom of windows due to poor exposure and development.

5.16.4 Resist over Steps

Line-width variations will usually occur where resist lines cross steps. These variations have been ascribed to light scattering at the step exposing resist in unexpected regions and to changes in the interference effects causing variable development (90). In either case, reducing reflectivity helps, as does making the resist thicker. Reflectivity of metal or polysilicon can sometimes be decreased by changing the grain structure and surface roughness. Care must be taken to ensure that such an approach does not adversely affect the electrical current carrying capability of the conductor. A complete, albeit more complex, solution is to use multilayer resist as described next.

5.16.5 Multilayer Resist

Topographical changes in going from the top level of metallization through a contact opening down to the silicon surface can easily be a micron on wafers with multilevel metallization. Such large distances use up an appreciable part of the focus budget when printing is done with high *NA* lenses and must be avoided whenever possible. Exposure difficulties arise in trying to pattern metal going over high steps, and thick layers of resist are needed to prevent resist failure at steps. Thick resist, in general, gives poorer optical resolution than does thin resist, and in the case of E-beam printing, thick resist presents more resist to scatter the electrons and thus reduce resolution. To simultaneously correct these problems and minimize problems with reflective substrates, multilayered resist can be used (91–95).

In this process, shown in Fig. 5.37, a thick layer of one kind of resist is applied to the wafer to smooth out surface variations. A second layer of a thin resist of different composition and spectral

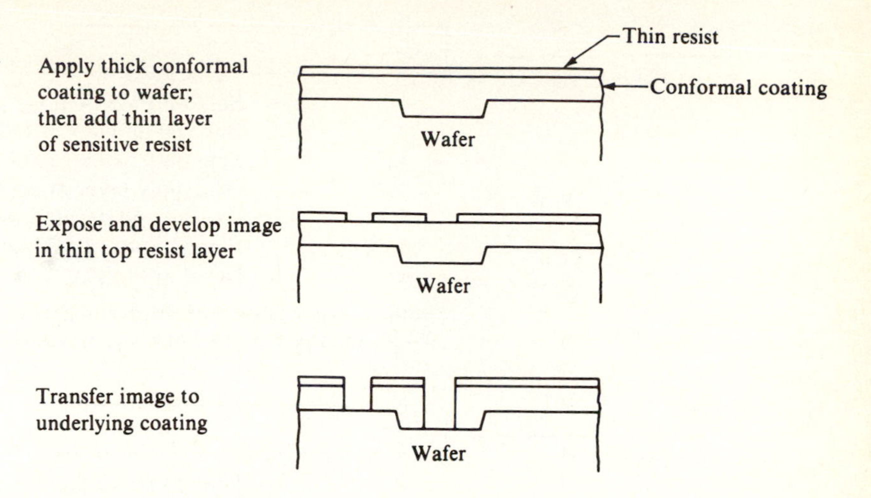
Thin resist
Apply thick conformal coating to wafer; then add thin layer of sensitive resist
Conformal coating
Wafer
Expose and develop image in thin top resist layer
Wafer
Transfer image to underlying coating
Wafer

FIGURE 5.37

Use of two levels of resist to improve image resolution over rough topography.

response is then added. Because of the different spectral responses, the pattern in the top layer can be exposed independently of the bottom one. After development, the pattern is then transferred to the bottom layer. A major problem with this approach is interfacial mixing of the resists during processing. To minimize this difficulty, the basic "bilayer" approach is sometimes modified by adding a third layer ("trilayer" approach) of a relatively inert material such as plasma-deposited polysilicon between the two organic resists. After patterning the top layer, conventional etching can be used to remove the middle layer. In either process, the bottom layer can be patterned by plasma etching or by using the top layer as a contact mask, flood-exposing with the proper wavelength light, and developing.

Tables 5.18 and 5.19 show two bilayer[22] and one trilayer[22] flow and illustrate different approaches for minimizing interactions between the resist layers and for transferring the pattern from the top to the bottom layer.

As lateral feature size becomes smaller and approaches the vertical step size, more attention must be paid to the manner in which the thick bottom layer covers the surface contours. Viscosity is important, as well as thermal flow during baking operations. The geometric layout of the features can affect the way the resist flows during its application, and indeed it may be helpful to add dummy geometries in some cases in order to promote more uniform coverage (96).

---

[22]Sometimes, these flows are referred to as "bilevel" and "trilevel," and sometimes the general approach is referred to as "portable conformal masking."

TABLE 5.18a

Bilayer Resist Flow 1 (91)

Step	Operation
1	Spin on PMMA 1.5–2 μm thick.
2	Bake 60 minutes at 150°C.
3a*	Spin on 0.5–0.9 μm of Hunt HPR 204 resist; bake.
3b*	Or: Apply layer of germanium selenide inorganic resist.
4	Expose and develop top resist.
5	Flood-expose PMMA with deep UV light.
6	Develop PMMA.

*Both of these resists are sensitive to light in the 400 nm range and opaque in the 200–250 nm range where PMMA is most sensitive.

TABLE 5.18b

Bilayer Resist Flow 2 (93)

Step	Operation
1	Spin on polyimide for a postbake thickness of 1.3 μm.
2	Bake 60 minutes at 350°C.
3	Spin on silicon containing positive resist; bake.
4	Expose and then develop with sodium hydroxide solution.
5	Transfer resist pattern to polyimide using oxygen reactive ion etching.

TABLE 5.19

Trilayer Resist Flow (92)

Step	Operation
1	Apply HMDS.
2	Spin on PMMA 1 μm thick.
3	Bake 15 minutes at 85°C.
4	Spin on PMMA 1 μm thick.
5	Bake 60 minutes at 160°C.
6	Apply Si layer by plasma CVD or by E-beam evaporation.*
7	Bake 60 minutes in air at 100°C.
8	Apply HMDS.
9	Spin on 0.45 μm of thinned AZ1350J resist.
10	Bake 15 minutes at 85°C.
11	Expose and develop AZ1350J.
12	Bake 30 minutes at 100°C.
13	Reactive ion etch silicon in CF_4.
14	Reactive ion etch PMMA with O_2 or flood-expose and develop.

*See reference 94 for a flow that uses a spin coat of TiO_x as the intermediate layer.

5.17 WAFER FLATNESS

Because wafer flatness can be one of the larger contributors to focus problems, some attention must be given to it. Wafer flatness may be described in several different ways, and some are better suited than others for use with a particular printing machine. Flatness measurements are sometimes made at various stages of processing to see whether additional warpage is being introduced, but most measurements are made on the slice just after it is polished.

Total indicated runout (TIR), also sometimes called total indicated reading, is made while the back of the slice is clamped to a flat surface such as a vacuum chuck. As shown in Fig. 5.38a, it is the difference between the highest point above a selected reference plane lying in the front surface of the slice and the lowest point below the reference plane. This measurement is most useful for wafers to be used for contact or proximity printing.

A similar measurement is the nonlinear thickness variation

FIGURE 5.38

Methods of defining wafer flatness.

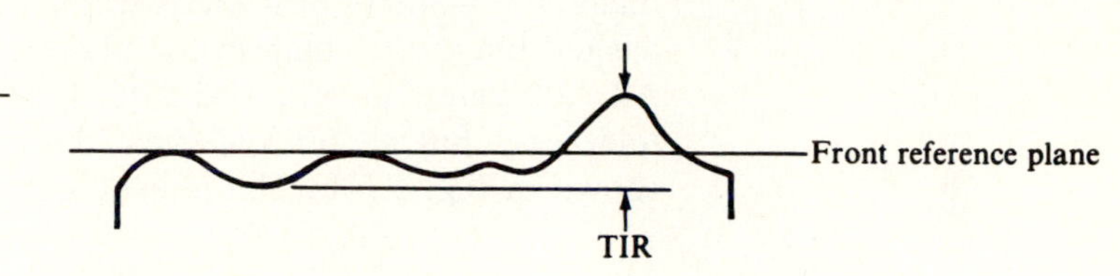

(a) TIR (total indicated runout) or PV (peak to valley)

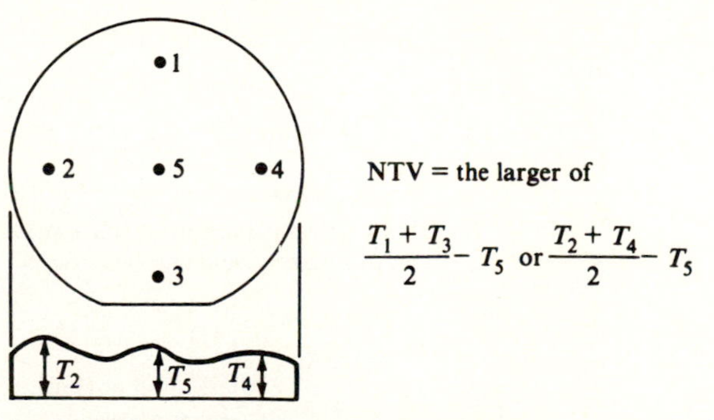

NTV = the larger of

$$\frac{T_1 + T_3}{2} - T_5 \text{ or } \frac{T_2 + T_4}{2} - T_5$$

(b) NTV (nonlinear thickness variation)

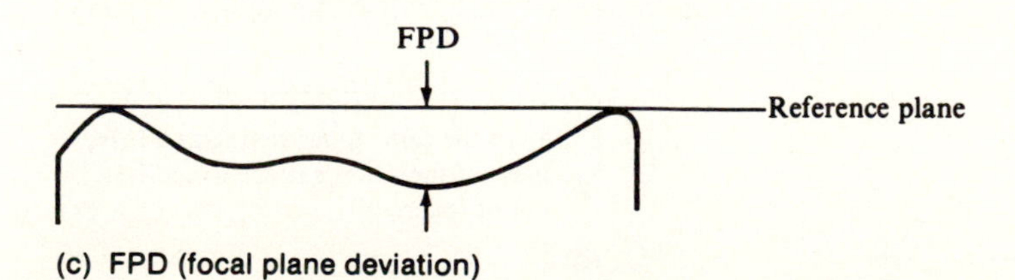

(c) FPD (focal plane deviation)

(NTV). However, instead of scans over the whole slice, measurements are made at five points, and NTV is calculated as shown in Fig. 5.38b.

Focal plane deviation (FPD) has been tailored to the evaluation of slices to be used with full-wafer scanning projection printers. Instead of using an arbitrary reference plane, it uses the plane established by a replica of the pins the front surface of the wafer is held against during printing. FPD is defined as the maximum deviation (+ or −) between the slice surface and the reference plane (Fig. 5.38c). It can range in value from (TIR)/2 to TIR, depending on the shape of the surface. Its utility is that, in principle, if the allowable depth of field is less than FPD, then the whole wafer will print. The surface contour of a wafer will have either the general appearance shown in Fig. 5.39a or the one shown in Fig. 5.39b. Then, if the printer is adjusted so that the plane of best focus lies in the plane of the reference pins, both wafers should print satisfactorily if neither exceeds the FPD limit. If, however, an "a" wafer were used to set focus, the plane of best focus would be as shown by the horizontal dashed line rather than in the plane defined by the reference pins. All "a" wafers meeting the FPD spec will be well within the focus tolerance, but a "b" wafer can be substantially out of focus, as shown in Fig. 5.39b. The polishing process determines the relative

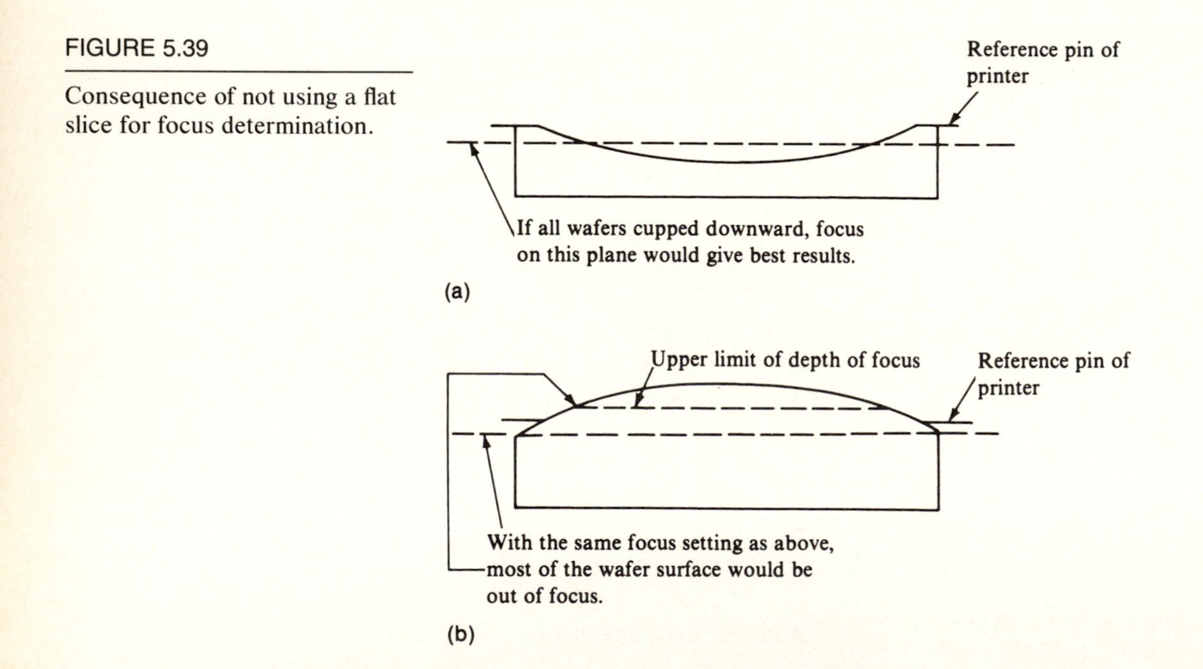

FIGURE 5.39

Consequence of not using a flat slice for focus determination.

numbers of "a" and "b" wafers, but over a long period of time, the numbers appear to be about equal.

If a printer is being used that references from the back of the wafer, a measurement that takes both wafer taper and front surface contour into account is required. Total thickness variation (TTV), which is the difference between the high and low points on the front surface of the wafer as measured while the back is held clamped to a reference plane, has been used. As can be seen from Fig. 5.40a, the front surface variation is generally overshadowed by wafer taper. For application to steppers, which are the only machines that use backside reference (and not all of them do), two other measurements have been used. They are local slope (LS) and local FPD (LFPD). The requirement for such information can be seen from Fig. 5.41. Even though a stepper may expose only 1 cm^2 at a time, if there were 10 μm of taper across a 100 mm slice and the back of the slice was the reference surface, the variation in height over each site would be 1 μm. Local slope at a particular location is determined by taking four closely spaced slice elevation readings, as shown in Fig. 5.40b. The total change in elevation over a stepper site is then approximated by the product of LS and the step size. Because of the sensitivity of LS to the separation distance, local FPD is a better measurement. In local FPD, shown in Fig. 5.40c, a

FIGURE 5.40

Flatness measurements that include taper.

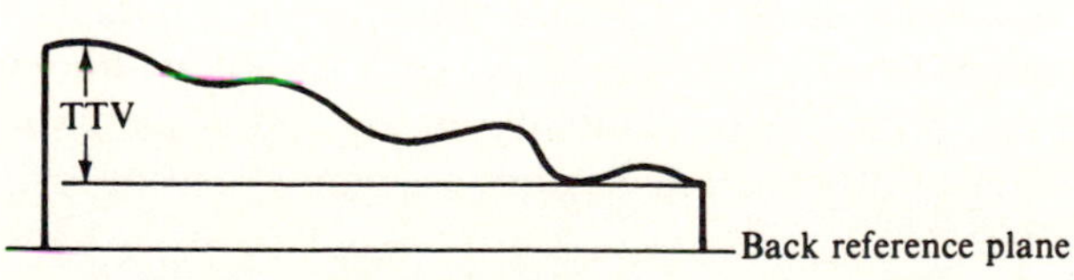

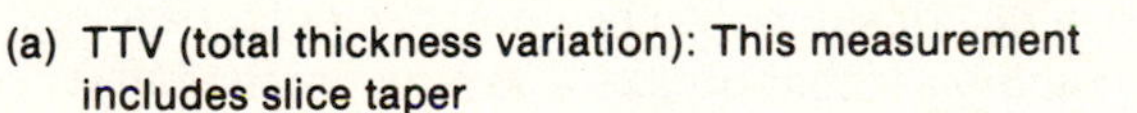
(a) TTV (total thickness variation): This measurement includes slice taper

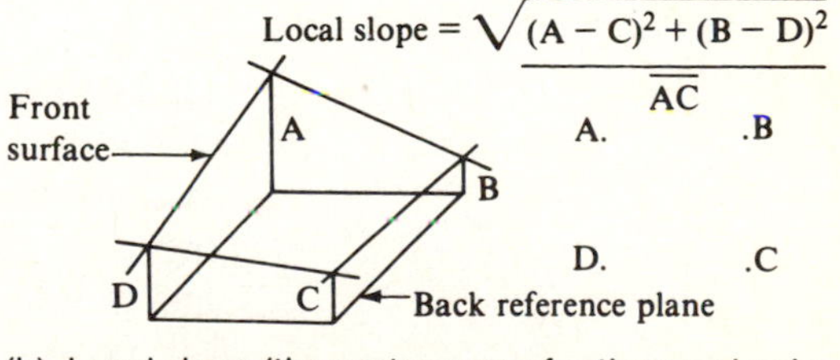

(b) Local slope (the vector sum of orthogonal pairs of elevations from a reference plane divided by the pair separation)

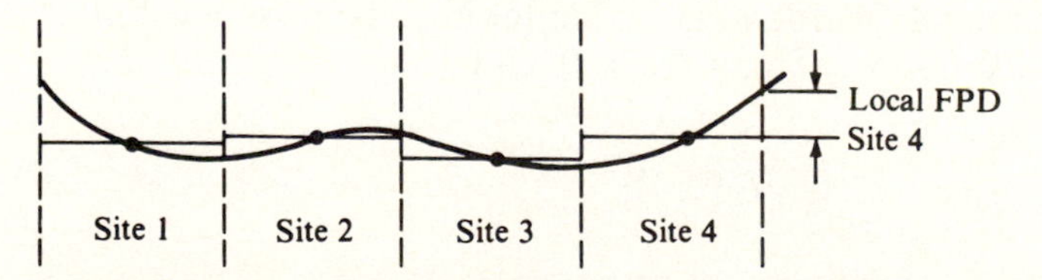

(c) Local FPD (the FPD of each stepper site referenced against a plane through the center of each site)

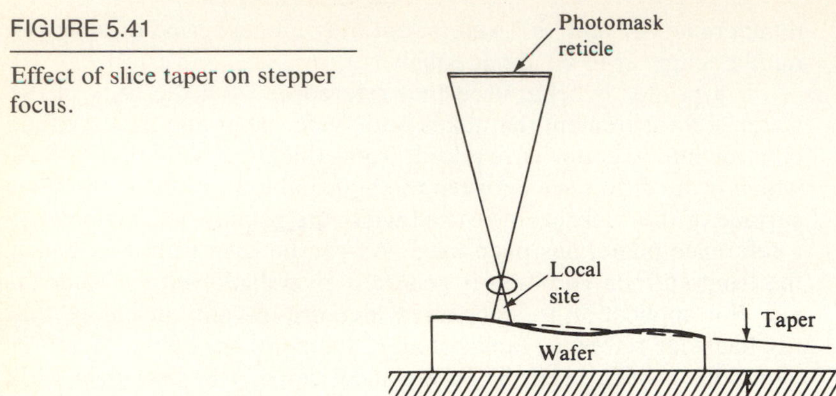

FIGURE 5.41

Effect of slice taper on stepper focus.

reference plane through the center of each stepper site is used. The center is chosen because stepper autofocusing is done on the center of each site. Flatness measuring machines are available that make LFPD measurements on the same sites to be used in actual printing so that wafers can be precisely evaluated.

5.18 LITHOGRAPHY PROCESS MODELING

While using desk-top computers to model various steps of the lithography process is very convenient, modeling of the whole process is much more complex and requires more computing power. A widely used integrated model is SAMPLE (simulation and modeling of profiles for lithography and etching), available from the University of California at Berkeley (33, 85, 91). SAMPLE combines a model of resist behavior during development with optical models of the printing equipment to predict properties such as line width and edge slope. Because of the unpredictability of negative resist swelling, the model has generally been restricted to organic positive resist and to inorganic resists.

Positive resist can be considered to be composed of a base resin and a photosensitive inhibitor that breaks down under the influence of light and that prevents resist dissolution when present. The optical behavior can be characterized by three constants (92). One, C, relates the rate of destruction of inhibitor at some point z within the resist to the intensity of light I at z. That is,

$$\frac{-\partial M(z,t)}{\partial t} = I(z,t)M(z,t)C \qquad 5.36$$

where M is the relative concentration of inhibitor at time t. The other two constants A and B relate the optical absorption coefficient to the fractional amount of inhibitor remaining:

TABLE 5.20

SAMPLE Inputs

Printer lens *NA*
Illuminator coherence
Amount of defocus
Optical properties of substrate (could, for example, be silicon, polysilicon, or metal)
Optical properties of thin-films between wafer and resist (for example, SiO_2 or nitride)
Thickness of thin films
Resist optical properties
Resist thickness
Exposure dose (and wavelengths)
Resist development properties (constants discussed earlier)
Development time

$$\alpha = AM(z,t) + B \tag{5.37}$$

The optical printer is described in terms of the lens MTF, which can be calculated from the wavelength and numerical aperture of the lens. Alternatively, a measured MTF can be used by the program. Using the MTF, the optical properties of the resist and any other layers present, and the underlying reflecting surface, the intensity of the light at any point z can be calculated. From $I(z,t)$ and the total exposure time, the amount of inhibitor remaining in the resist is calculated. An etching model (92) is then used to calculate the profile of the resist remaining after development (which is really a dissolution or etching process).

Inputs required for the simulator are given in Table 5.20. From these inputs, the actual line widths can be calculated, as well as the vertical profile of the resist (as in Fig. 5.35). Thus, the effect of any of the variables of Table 5.20 can be quickly modeled. One can, for example, study the effect of defocus on line width or the effect of oxide thickness on edge scalloping.

CHAPTER 5

SAFETY

Since the conventional short-arc UV bulb operates at 35–40 atm, the chance of a bulb's exploding always exists. Under normal circumstances, it will be shielded by the light collection optics so that the only damage from an explosion is to the optics. When cool, mercury bulbs have less than an atmosphere of pressure and hence pose little safety hazard during handling and installation. Xenon/mercury bulbs may have a few atmospheres of pressure when cold and thus must be handled with care.

High-intensity UV generates ozone in the air, but with normal ventilation, the ozone should not cause a health problem. The odor threshold for ozone is about 0.01 ppm in air, whereas the OSHA-proposed maximum allowable for an 8 hour period is 0.10 ppm.

Since the high-intensity UV itself can be an eye hazard, caution should be exercised to ensure that shielding is always in place. Particular care should be taken when the high-power excimer-laser sources are used. It should also be remembered that the eye does not see in the UV range so that if the visible portions are filtered out or if lasers are being used, there will be no warning prior to a burned cornea.

Most of the solvents used in and with resists are combustible and should be kept away from open flames. Where they are used in quantity, special ventilation is necessary. With prolonged contact, some of the solvents may cause skin irritation. The toxicity of the newer deep UV resists should be carefully reviewed before using.

The possibility of pressure buildup in containers should not be overlooked. In particular, the diazoquinone in positive resists slowly decomposes to give off nitrogen, particularly when it is stored above 70°F. Storage at 50°F is recommended by some manufacturers.

CHAPTER 5 KEY IDEAS

- Large-feature geometry photomasks are made with optical pattern generators because of their speed. E-beam writing, while slow, is predominately used for the small-feature geometry masks, although laser optical generators show promise as an alternative.
- Masks for X-ray lithography are very difficult to make and currently are a major limitation to X-ray usage.
- Contact printing, while inexpensive and capable of printing very fine geometries, is practically obsolete because of its inability to print defect-free patterns.
- E-beam direct wafer writing is capable of the finest feature size definition currently visualized, but it is so slow as to be economically undesirable in most cases.
- Optical projection printing is capable of printing features at least as small as 0.75 μm and is a relatively economical process.
- Because of mask defects and lack of wafer flatness, the full resolution of printer lenses is seldom used.
- Negative resists are generally at least five times more sensitive than positive resists.
- Positive resists do not exhibit swelling during development and thus are preferred over negative resists when the spacings between long resist runs is very narrow.
- Multilayer resists afford a good way of minimizing the effect of a very narrow depth of focus in printing rough wafer topography while using a high-resolution lens.
- Optical interference producing standing waves in the resist can cause regions of it to be improperly exposed. The use of a dye in the resist to increase optical absorption minimizes the effect.
- Quite good computer models of the lithography process are available.

CHAPTER 5 PROBLEMS

1. Show that if a flat wafer of diameter D is uniformly bowed so that the center deflection is δ, its radius of curvature is approximately $D^2/8\delta$.
2. Draw a mask that will allow the letter *I* to be etched into the surface of a silicon wafer if negative resist is to be used. Draw a mask that will allow a relief letter *L* to remain after etching if positive resist is used.
3. If AZ1350J resist is to be properly exposed in 1 s, what should the light intensity in $mJ/cm^2 \cdot s$ be? If a resist has a contrast ratio of 4 and the sensitivity of AZ1350J, what is the minimum swing in exposure energy that will give a satisfactory image?
4. If, using a standard light source, the exposure time for a particular resist is 2 s, what would you expect the exposure time to be if a newly developed laser light source 1000 times as intense were used? Explain your answer.
5. What functions do wafer prebake, resist softbake, and resist hard-bake serve?
6. Why is polyisoprene-based negative resist not as suitable for fine-line geometry as novalac-based positive resist?
7. Show that for an image modulation of 0.6, the normalized intensity swings between 0.25 and 1.0.
8. Prove that for a given resolution, the use of a lens with the shortest wavelength and smallest numerical aperture gives the greatest depth of field.
9. Using the Rayleigh criterion for resolution limits, first plot resolution versus NA from $NA = 0.2$ to $NA = 0.6$ for a wavelength of 436 nm. Next, plot on the same graph the reported resolution of the lenses listed in Table 5.11 for all lenses with an NA of 0.4 or greater. Do they all seem reasonable?
10. If standing waves cause a resist development problem such as shown in Fig. 5.35, what will be the spacing of the undulations if the exposing wavelength is 0.436 μm?
11. Sketch the surface profile of a wafer that has both an FPD and an LFPD of 0.5 μm.

CHAPTER 5 REFEFENCES

1. Mike V. Sullivan, "Iron Oxide See-Through Masks," *J. Electrochem. Soc. 120*, pp. 545–550, 1973.
2. R.K. Watts, K.E. Bean, and T.L. Brewer, "X-Ray Lithography with Aluminum Radiation and SiC Mask," in Robert Bakish, ed., *Electron and Ion Beam Science and Technology 78–5,* Electrochemical Society, 1978.
3. Phil E. Mauger and Alex R. Shimkunas, "Additive X-Ray Mask Patterning," *Semiconductor International*, pp. 70–74, March 1986.
4. Dervin L. Flowers and Henry G. Hughes, "On the Use of Photomask Coatings in Photolithography," *J. Electrochem. Soc. 124*, pp. 1599–1602, 1977.
5. D.L. Flowers, "Lubrication in Photolithography-Resist Protect Coats," *J. Electrochem. Soc. 124*, pp. 1608–1612, 1977.
6. Eugene R. Blome, "Photolithographic Mask Release Coating," *Semiconductor International*, pp. 75–78, November 1979.
7. Ron Hershel, "Pellicle Protection of IC Masks," *Semiconductor International*, pp. 97–106, August 1981.
8. John M. Jerke et al., "Accurate Linewidth Measurements at the National Bureau of Standards," *Proc. Kodak Microelectronics Seminar*, pp. 51–59, Monterey, Calif., 1978.
9. Kiwao Nakazawa et al., "Measuring System for Fine-Line Lithography," *Proc. Kodak Microelectronics Seminar*, San Diego, 1980.
10. Peter H. Singer, "Linewidth Measurement: Ap-

proaching the Submicron Dimension," *Semiconductor International*, pp. 48–54, March 1983.

11. I. Tanabe, "Production of High Quality MOS LSIs," *Proc. Kodak Microelectronics Seminar*, pp. 91–99, 1975
12. Masana Minami et al., "A New Mask Defect Inspecting Device," *Proc. Kodak Microelectronics Seminar*, pp. 67–80, Monterey, Calif., 1976.
13. K. Levy and P. Sandland, "Automatic Mask Inspection: What Can It Find, What Will It Miss, and What Can It Do?" *Proc. Kodak Microelectronics Seminar*, pp. 84–94, 1977.
14. L.K. Gailbraith, "Advances in Automated Detection of Wafer Surface Defects," *Proc. ASTM Symposium*, January 1982.
15. Mikio Makito et al., "Analysis of Registration Errors in 1:1 Projection Mask Aligners," *Proc. Kodak Microelectronics Seminar*, San Diego, 1980.
16. G.L. Resor and A.C. Tobey, "The Role of Direct Step-on-the-Wafer in Microlithography Strategy for the 80's," *Solid State Technology*, p. 101, August 1979.
17. H.R. Rottmann, "Overlay in Lithography," *Semiconductor International*, pp. 83–94, December 1980.
18. David S. Perloff, "A Four-Point Electrical Measurement Technique for Characterizing Mask Superposition Errors on Semiconductor Wafers," *IEEE J. Solid State Cir. SC-13*, pp. 436–444, 1978.
19. T.J. Russell, T.F. Leedy, and R.L. Mattis, "A Comparison of Electrical and Visual Alignment Test Structures for Evaluating Photomask Alignment in Integrated Circuit Manufacturing," *IEDM Technical Digest*, 1977.
20. William S. DeForest, *Photoresist Materials and Processes*, McGraw-Hill Book Co., New York, 1975.
21. Donald J. Sykes, "Recent Advances in Negative and Positive Photoresist Technology," *Solid State Technology*, pp. 53–57, August 1973.
22. Alvin Stein, "The Chemistry and Technology of Negative Photoresists," Philip A. Hunt Chemical Corporation brochure no. 1014/0535R1.
23. Alvin Stein, "The Chemistry and Technology of Positive Photoresists," Philip A. Hunt Chemical Corporation brochure no. 1015/0535R1.
24. John Hock et al., "Resist Image Reversal for Next-Generation VLSI Circuit Fabrication," *Semiconductor International*, pp. 164–168 and references therein, September 1987.
25. *Materials for Microlithography*, ACS Symposium Series 266, American Chemical Society, Washington, D.C., 1984
26. Daryl Ann Doane and Adam Heller, eds., *Proc. Inorganic Resist Systems Symposium 82–9*, Electrochemical Society, Pennington, N.J., 1982.
27. T.C. Penn, "Forecast of VLSI Processing—A Historical Review of the First Dry-Processed IC," *IEEE Trans. on Electron Dev. ED-26*, pp. 640–643, 1979.
28. J.N. Smith et al., "A Production-Viable Plasma Developable Photoresist," *Semiconductor International*, pp. 41–47, 1979.
29. Gary N. Taylor and Thomas M. Wolf, Chapter 4, "The Status of Dry-Developed Resists for Each Lithographic Technology," in Norman G. Einspruch and Graydon B. Larrabee, eds., *VLSI Electronics*, Vol. 6, Academic Press, New York, 1983.
30. G.N. Taylor, L.E. Stillwagon, and T. Venkatesan, "Gas-Phase-Functionalized Plasma-Developed Resists: Initial Concepts and Results for Electron-Beam Exposure," *J. Electrochem. Soc. 131*, pp. 1658–1664, 1984.
31. B.F. Griffing and P.R. West, "Contrast Enhanced Lithography," *Solid State Technology*, pp. 152–157 and references therein, May 1985.
32. Don R. Strom, "Optical Lithography and Contrast Enhancement," *Semiconductor International*, pp. 162–167, May 1986.
33. A.R. Neureuther and W.G. Oldham, "Resist Modeling and Profile Simulation," *Solid State Technology*, pp. 139–144, May 1985.
34. Shigeki Shimizu and George R. Bird, "Chemical Mechanisms in Photoresist Systems," *J. Electrochem. Soc. 126*, pp. 273–277, 1979.
35. J.N. Helbert, "Photoresist Adhesion Promoter for Gold Metallization Processing," *J. Electrochem. Soc. 131*, pp. 451–452, 1984.
36. B.D. Washo, "Rheology and Modeling of the Spin Coating Process," *IBM J. Res. Develop. 21*, pp. 190–198, 1977.
37. Patrick O'Hagen and William J. Daughton, "An Analysis of the Thickness Variance of Spun-On Photoresist," *Proc. Kodak Microminiaturization Seminar*, 1977.

38. G.F. Damon, "The Effects of Whirler Acceleration on the Properties of the Photoresist Film," Kodak pub. P195, pp. 34–41, a collection of papers from the 1965 and 1966 Kodak seminars.
39. Hans Denk, "Optimization of Photoresist Processing with the IBM 7840 Film Thickness Analyzer," *Proc. Kodak Microelectronics Seminar*, pp. 28–34, Monterey, Calif., 1976
40. Kodak pub. G-96 (positive resist 809).
41. Mary Long, "Quantitative Evaluation of a Photoresist," *Proc. Kodak Microelectronics Seminar*, pp. 60–69, San Diego, 1970.
42. Burke Leon, "The Nature of Negative Photoresist Scumming: Part II," *Solid State Technology*, pp. 58–61, May 1977.
43. Ted C. Bettes, "UV Exposure, Systems, and Control," *Semiconductor International*, pp. 83–95, April 1982.
44. Jerry Bachur, "Deep UV Exposure Technology," *Solid State Technology*, pp. 124–127, February 1982.
45. Hans Schellenberg, "Optical Mask Alignment and Exposure Systems," *Proc. Kodak Photoresist Seminar*, pp. 54–63, Philadelphia, 1969.
46. Jeanne Roussel, "Step-and-Repeat Wafer Imaging," *Solid State Technology*, pp. 67–71, May 1978.
47. James J. Ewing, "Rare-Gas Halide Lasers," *Physics Today*, pp. 32–39, May 1978.
48. R.A. Lawes, "Use of Excimer Lasers in Photolithography," *Semiconductor International*, pp. 76–77, July 1986.
49. K. Jain, C.G. Wilson, and B.J. Lin, "Ultrafast Deep UV Lithography with Excimer Lasers," *IEEE Elec. Dev. Lett. EDL3*, pp. 53–55, 1982.
50. H. Craighead et al., "Contact Lithography at 157 nm with an F_2 Excimer Laser," *J. Vac. Sci. Tech. B1*, pp. 186–189, 1983.
51. Elmar Cullmann, "Excimer Laser Applications in Contact Printing," *Semiconductor International*, pp. 332–334, May 1985.
52. Karl Suss product literature.
53. K. Jain and R.T. Kerth, "Excimer Laser Projection Lithography," *Appl. Opt. 23*, pp. 648–630, 1984.
54. T. Kaneko, T. Umegaki, and Y. Kawakami, "A Practical Approach to Sub-Micron Lithography," *Proc. Kodak Microelectronics Seminar*, San Diego, October 1980.
55. R.K. Watts and J.H. Bruning, "A Review of Fine-Line Lithographic Techniques: Present and Future," *Solid State Technology*, pp. 99–105, May 1981.
56. Richard C. Heim, "Practical Aspects of Contact/Proximity Photomask/Wafer Exposure," Proc. SPIE 100, in *Semiconductor Microlithography II*, pp. 104–114, 1977.
57. Michael C. King, "Future Developments for 1:1 Projection Photolithography," Semicon/Europa, Zurich, 1978, and "The Characteristics of Optical Lithography," *Proc. Kodak Microelectronics Seminar*, San Diego, October 1980.
58. John W. Bossung, "Projection Printing Characteristics," Proc. SPIE 100, in *Semiconductor Microlithography II*, 1977.
59. David A. Markle, "A New Projection Printer," *Solid State Technology*, pp. 50–53, June 1974.
60. Perkin-Elmer product literature.
61. David A. Markle, "Submicron 1:1 Optical Lithography," *Semiconductor International*, pp. 137–142 and references therein, May 1986.
62. Y. Lida and K. Mori, "A Modified S.E.M. Type EB Direct Writing System and Its Application on MOS LSI Fabrication," *J. Electrochem. Soc. 128*, pp. 2429–2434, 1981.
63. H.C. Pfeiffer, "Direct Write Electron Beam Lithography—A Production Line Reality," *Solid State Technology*, pp. 223–227, September 1984.
64. J.S. Greeneich, "Impact of Electron Scattering on Linewidth Control in Electron-Beam Lithography," *J. Vac. Sci. Technol. 16*, pp. 1749–1953, 1979.
65. D.F. Kyser and C.H. Ting, "Voltage Dependence of Proximity Effects in Electron Beam Lithography," *J. Vac. Sci. Technol. 16*, pp. 1759–1763, 1979.
66. Nao-aki Aizaki, "Proximity Effect Dependence on Substrate Material," *J. Vac. Sci. Technol. 16*, pp. 1726–1933, 1979.
67. Shoichiro Yoshida et al., "X-Ray Lithographic System," *Proc. Kodak Microelectronics Seminar*, San Diego, October 1980.
68. W.D. Buckley et al., "The Design and Performance of an Experimental X-Ray Lithography System and the Fabrication, Patterning, and Dimensional Stability of Titanium Membrane X-Ray Lithography Masks," *Proc. Kodak Microelectronics Seminar*, October 1980.

69. H.A. Hyman et al., "Intense-Pulsed Plasma X-Ray Sources for Lithography: Mask Damage Effects," *J. Vac. Sci. Technol. 21*, pp. 1021–1026, 1982.
70. Armand P. Neukermans, "Current Status of X-Ray Lithography," *Solid State Technology*, pp. 185–188 and references therein, September 1984.
71. J.R. Maldonada et al., "X-Ray Lithography Source Using a Stationary Solid Pd Source," *J. Vac. Sci. Technol. 16*, pp. 1942–1945, 1979.
72. E. Spiller et al., "Application of Synchrotron Radiation to X-Ray Lithography," *J. Appl. Phys. 47*, pp. 5450–5459, 1976.
73. H. Aritomi et al., "X-Ray Lithography by Synchrotron Radiation of the SOR-Ring Storage Ring," *J. Vac. Sci. Technol. 16*, pp. 1939–1941, 1979.
74. John A. Doherty et al., "Focus Ion Beams in Microelectronic Fabrication," *IEEE Trans. on Components, Hybrids, and Manufac. Tech.*, CHMT6, pp. 329–333, 1983.
75. Kenji Gamo and Susumu Namba, "Ion Beam Lithography," *Ultramicroscopy 15*, pp. 261–270, 1984.
76. J.N. Randall, "Prospects for Printing Very Large Scale Integrated Circuits with Masked Ion Beam Lithography," *J.Vac. Sci. Technol. A*, pp. 777–783, 1985.
77. John L. Bartelt, "Masked Ion Beam Lithography: An Emerging Technology," *Solid State Technology*, pp. 215–220, May 1986.
78. Ronald C. Bracken and Syed A. Risvi, Chapter V, "Microlithography in Semiconductor Device Processing," in Norman G. Einspruch and Graydon B. Larrabee, eds., *VLSI Electronics: Microstructure Science*, Academic Press, New York, 1983.
79. Michael Johnson and Karen Lee, "Optimization of a Photoresist Process Using Statistical Design of Experiments," *Solid State Technology*, pp. 281–284, September 1984.
80. Terry V. Nordstrom, "Reticle Sizing for Optimized CD Control," *Semiconductor International*, pp. 158–161, September 1985.
81. Frederick H. Dill et al., "Modeling Projection Printing of Positive Resist," *IEEE Trans. on Electron Dev. ED-22*, pp. 456–464, 1975.
82. J.H. Altman and H.C. Schmitt, "On the Optics of Thin Films of Resist over Chrome," *Proc. Kodak Photoresist Seminar 2*, pp. 12–19, Los Angeles, 1968.
83. S. Middelhoek, "Projection Masking, Thin Photoresist Layers, and Interference Effects," *IBM J. Res. Develop. 14*, pp. 117–124, 1970.
84. Dietrich W. Widmann, "Quantitative Evaluation of Photoresist Patterns in the 1-μm Range," *Appl. Opt. 14*, pp. 931–934, 1975.
85. J.D. Cuthbert, "Optical Projection Printing," *Solid State Technology*, pp. 59–69, August 1977.
86. William G. Oldham et al., "A General Simulator for VLSI Lithography and Etching Processes: Part 1—Application to Projection Lithography," *IEEE Trans. on Electron Dev. ED-26*, pp. 717–722, 1979.
87. T.R. Pampalone and F.A. Kuyan, "Improving Linewidth Control over Reflective Surfaces Using Heavily Dyed Resist," *J. Electrochem. Soc. 133*, pp. 192–196, 1986.
88. T.R. Pampalone and F.A. Kuyan, "Contrast Enhancing Additives for Positive Photoresist," *J. Electrochem. Soc. 135*, pp. 471–476, 1988.
89. Edward John Walker, "Reduction of Photoresist Standing-Wave Effects by Post-Exposure Bake," *IEEE Trans. on Electron Dev. ED-22*, pp. 464–466, 1975.
90. Dietrich W. Widmann and Hans Binder, "Linewidth Variations in Photoresist Patterns on Profiled Surfaces," *IEEE Trans. on Electron Dev. ED-22*, pp. 467–471, 1975.
91. E. Ong et al., "Bilevel Resist Processing Techniques for Fine-Line Lithography," *Proc. Kodak Microelectronics Seminar*, pp. 91–97 and references therein, Dallas, 1981.
92. E. Bassous et al., "A Three-Layer Resist System for Deep U.V. and RIE Microlithography on Nonplanar Surfaces," *J. Electrochem. Soc. 130*, pp. 478–484 and references therein, 1983.
93. Y. Saotoma et al., "A Silicon Containing Positive Resist (SIPR) for a Bilayer Resist System," *J. Electrochem. Soc. 132*, pp. 909–913, 1985.
94. H. Umezaki et al., "Deep-UV Contact Lithography Using a Trilevel Resist System for Magnetic Bubble Devices with Submicron Minimum Features," *J. Electrochem. Soc. 132*, pp. 2440–2444, 1985.

95. C. Lyons and W. Moreau, "Aqueous Processable Portable Conformable Mask (PCM) for Submicron Lithography," *J. Electrochem. Soc. 135*, pp. 193–197 and references therein, 1988.
96. L.K. White, "A Modeling Study of Superficial Topography for Improved Lithography," *J. Electrochem. Soc. 132*, pp. 3037–3041, 1985.
97. W.G. Oldham and A.R. Neureuther, "Projection Lithography with High Numerical Aperature Optics," *Solid State Technology*, pp. 106–111, May 1981.
98. F.H. Dill et al., "Characterization of Positive Photoresist," *IEEE Trans. on Electron Dev. ED-22*, pp. 445–452, 1975.
99. R.E. Jewett et al., "Line-Profile Resist Development Simulation Techniques," *Polymer Eng. Sci. 17*, pp. 381–384, 1977.

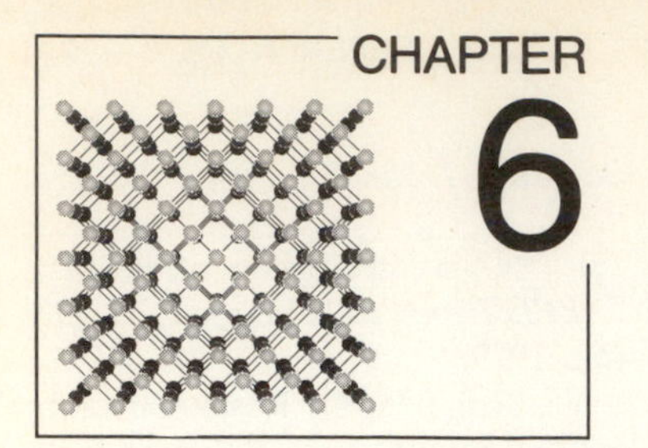

CHAPTER

6

Etching

6.1 INTRODUCTION

In many IC manufacturing steps, whole wafers are completely coated with a layer or layers of various materials such as silicon dioxide, silicon nitride, or a metal. The unwanted material is then selectively removed by etching through a mask to leave, for example, holes in a thermal oxide where diffusions are to be made or long stripes of aluminum for electrical interconnects between individual circuit elements. In addition, various patterns must sometimes be etched directly into the semiconductor surface. Examples are circular holes or short grooves where trench capacitors are to be made in silicon; mesas that are required in the silicon dielectric isolation process (discussed later); and small, flat depressions in GaAs where the gate metal is to be deposited.

Possible kinds of etching are wet chemical, electrochemical, pure plasma etching, reactive ion etching (RIE), ion beam milling, sputtering, and high-temperature vapor etching. Wet etching, in which the wafers are immersed in aqueous etching solutions, is the oldest and, when applicable, the least expensive process. Electrochemical etching is done in aqueous-based solutions and has very limited applicability. Plasma etching and combination plasma/RIE are relatively new, are performed in a low-pressure gaseous plasma, and are most commonly used in fine-geometry applications. Plasma etching generally involves fewer safety hazards and spent chemical disposal problems, but the additional cost of plasma equipment is a deterrent to its use when fine-line definition (less than approximately 3 μm) is not necessary. Ion beam milling, done in vacuum, is seldom used. Sputtering, done in a relatively low vacuum, is quite slow, produces surface damage, and is used only for a surface cleanup. Vapor etching generally requires temperatures in the order of 1000°C and is mostly used as an in situ cleanup before epitaxial depositions.

A few etching procedures in IC manufacturing do not involve any local masking. These procedures include etching whole semiconductor slices to remove damage and/or to polish the surface and etching slices or chips to delineate crystallographic defects. In addition, before the advent of planar technology, a variety of germanium and silicon etching steps were used for removing damage from junctions. Because of this application, a body of literature exists that relates to junction cleanup etches (see, for example, reference 1); however, since these etches do not relate to modern IC technology, they will not be discussed in this context. Most IC etching does require material removal in selected regions only and hence requires a series of related processing steps, which are

1. Coat the wafer with an adherent and etch-resistant photoresist.
2. Selectively remove the resist to leave the desired pattern.
3. Etch to transfer the mask pattern to the underlying material.
4. Remove (strip) the photoresist and clean the wafer.

Of these steps, all but step 3 were discussed in Chapter 5 on lithography.

Etching completely through thin layers presents a somewhat different set of problems from the problems encountered in etching partway through thick layers. In the latter case, the thickness removed must generally be determined by time and etch rate. In the case of thin layers, the etch used is ordinarily selective and will slow or stop its etching action when the next layer is reached. It may, however, continue to etch laterally and give undercutting. Wet etching is particularly prone to this problem. Etches that behave thusly are isotropic; that is, they etch at the same rate in all directions. For etching to be truly isotropic, the material itself must be isotropic, as is, for example, thermal SiO_2, and the etch rate must be controlled by surface kinetics. When the material is anisotropic, as is Si or GaAs, diffusion-controlled etching will remove the effects of any anisotropic surface kinetics and may still give isotropic etching. More likely, however, diffusion effects will cause enhanced diffusion near mask window edges (2). Fig. 6.1a shows an isotropic profile when etching is done through a window into a very thick material. In this geometry, the center of curvature for the undercut is assumed

FIGURE 6.1

(a) Isotropic etching showing undercutting of an amount equal to depth of etch. (b) Isotropic etching of a thin layer that has just barely been etched through. (c) Etching at a later time showing increased undercutting and steeper sidewall slope.

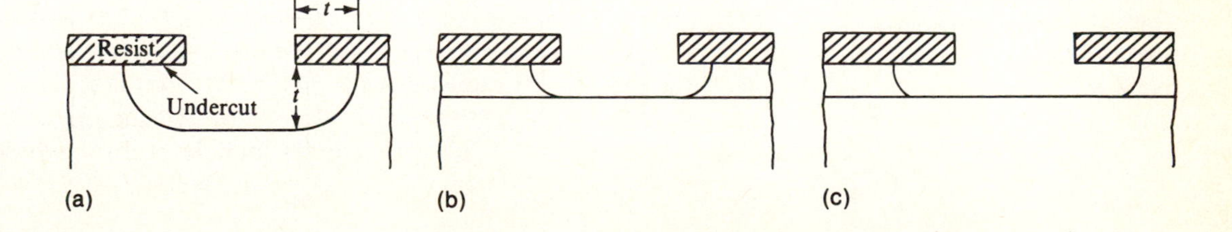

to be at the mask edge. Fig. 6.1b shows the profile when etching is done through a thin layer. As undercutting progresses, the radius becomes larger, and the etching edge steeper (3), as Fig. 6.1c shows.

Anisotropic etching implies that a substantial difference exists in etch rates in different directions. There are, however, two usages of the term within this context. In one, which is commonly encountered in thin-film etching discussions, anisotropic etching means appreciable etching in a direction normal to the wafer surface and essentially no etching laterally. In the other, which may arise during the etching of single crystals, anisotropy is a measure of the differing etch rates in different crystallographic directions. Etches that show this latter behavior are anisotropic because their rate is limited, not by diffusion of etching reactants to and from the etching surface, but by differing reaction rates on different crystallographic planes. Such etches are not applicable to thin-film applications since most thin films are either amorphous or fine-grained polycrystalline material. An anisotropic etch mechanism sometimes operative in combination plasma/RIE restricts lateral etching and is independent of crystallography. As depicted in Fig. 6.2a, it depends on the sidewalls becoming coated during etching with a polymer that resists further etching (4). Etching will also be highly anisotropic, as shown in Fig. 6.2b, if the etching molecules or ions arrive at the etching surface in a parallel beam and have negligible surface mobility.

Many pattern-etching applications involve sequentially etching through multiple layers. Sometimes, the same etch formulation is used for all of the layers, but seldom will the etch rates be the same. When they are not, substantial undercutting of one or more of the layers may occur. The advantage of a single etch is that it requires only one etch step and is much less time consuming; however, when a single etch is used with wet etching, it is only applicable to large geometries. To minimize undercutting, the layers can be etched in sequence, using an etch specific to each layer, with each etched layer acting as the mask for the next one. Such an approach presumes that a series of etches is available that will attack the desired

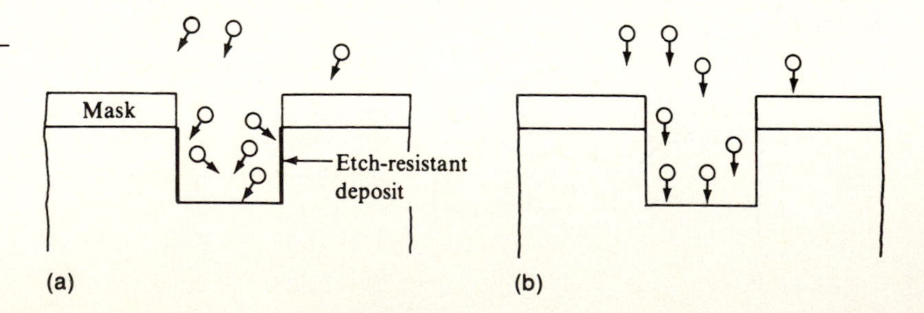

FIGURE 6.2

(a) Anisotropic etching by sidewall passivation. (b) Anisotropic etching by directed reactive ions.

layer at a useful rate and all of the rest either very slowly or not at all. Thus, the difference in etch rates (selectivity) between different materials is a very important factor. The individual rate values are also of importance since if the rate is too slow, the process may not be economically feasible and if it is too fast, the process may not be controllable.

In addition to rate, selectivity, and anisotropy, other etchant properties of interest are the effect of the etch on the mask material (whether it erodes the masking material or causes it to peel up from the surface) and the contaminants left behind. Etchant purity and cleanliness are particularly important because many liquid etchants can have heavy metal contaminants that will plate onto semiconductor surfaces and particulates that are difficult to remove during the rinse cycles. Plasma etching may leave solid reaction products in the windows, and sputter etching and ion milling sometimes transfer material from some part of the etching machinery hardware to the etching surface.

6.2 WET ETCHING

Wet etching in its simplest form consists of a beaker of etchant adjacent to several beakers of rinse water. The wafers are loaded into some sort of plastic holder and immersed in the etchant beaker for a specified time. They are then quickly transferred to the first rinse water beaker in order to abruptly stop etching (quench) and then moved in sequence to the others. To ensure that the final rinse water remains essentially free of etchant, the first beaker used is dumped after each use, the remaining beakers are moved forward, and a fresh one is added at the end. For reliable finished parts, it is imperative that no etch reactants remain on the surface. Otherwise, either the electric fields present during operation will cause surface ion migration and subsequent electrical failure or moisture diffusing to the chip surface through plastic packages will react with the ions and produce lead corrosion.

Several versions of wet etchers (often referred to as wet processors) are now available (5), but they all incorporate the same etch–rinse steps used in the multiple-beaker approach. Usually, the etch bath is temperature controlled, and often the etching solution will be continually recirculated and filtered. The subsequent rinse operation uses hot deionized water and may have nitrogen bubbling through it for agitation. The resistivity of the water can be monitored to determine the time required for rinsing, based on the time for the water resistivity to recover to its value before wafer insertion. This time will be a function of water flow, the specific design of the rinse tank, and the etchant used. Of the common etchants, dilute HF is the easiest to remove as measured by a resistivity monitor; nitric acid, the next easiest; and sulfuric acid, the most difficult. The spread of times is from 30% to 50%. For low flow rates, the time decreases as flow rate increases but becomes progressively less dependent on flow as the flow increases. Typically, times long enough to allow 4–6 changes of water in the tank are required. The resistivity of the rinse water is, however, misleading in that it does not measure the amount of ions that remain adsorbed on the semiconductor surface. Since fluorine ions are particularly difficult to remove from silicon, silicon wafers may require several additional minutes of water rinsing over that indicated by a resistivity monitor.

One early approach to automating the wet etching process used a single wafer chamber in which first the etchant and then the rinses were introduced (6). A more common approach is to use multiple tanks (much like the beakers) with a robot arm to transfer the wafer carrier from tank to tank. A sketch of a fume hood with an included etch tank is shown in Fig. 6.3. Not shown are the rinse tanks inside the hood. To minimize deionized water usage, the water rinse tanks are often cascaded so that water flows from tank to tank, with the pure water input going into the last tank used. Yet another method opens the bottom of the tank containing the wafer holder and etch solution when etching is complete so that the etchant is very rapidly

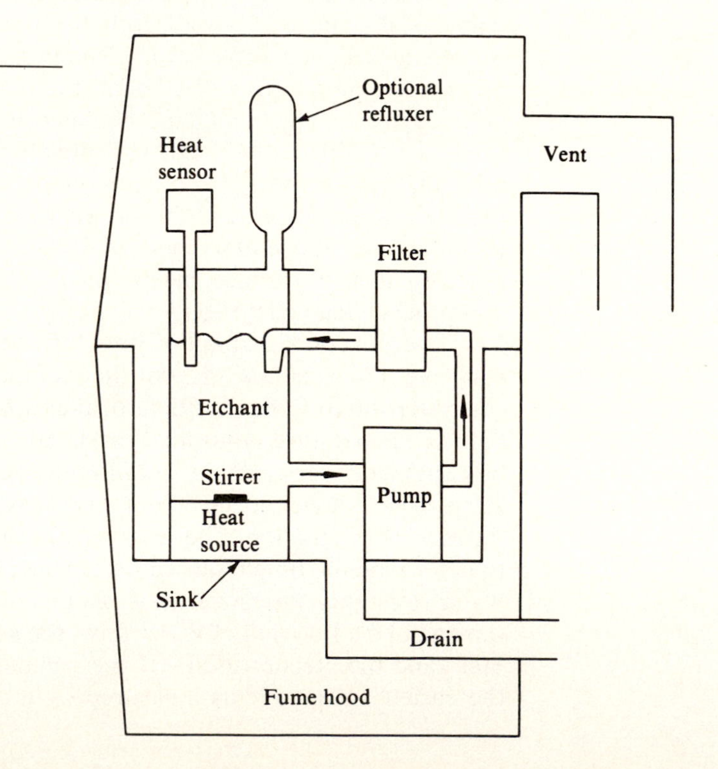

FIGURE 6.3

Fume hood used for wet etching.

removed (dumped). The tank is then closed and flooded with water, and the dump step is repeated. The rinses can be repeated as many times as are necessary to ensure complete etchant removal. In some machines, the etch and then the rinse water are sequentially sprayed over the wafers, after which the holder containing the wafers is rapidly spun to remove the water (spin–rinse etching). During etching of deep, narrow geometries under stagnant conditions, diffusion of reactants into the small openings becomes a major limitation. Spray jets directed normal to the surface offer some improvement, but the thin, stagnant layer on the surface still must be penetrated by diffusion. It has been suggested that further improvement should be possible by making use of the fact that the etch solution and the reaction by-products have different densities and using a centrifugal field to promote circulation (7). However, since extensive undercutting generally occurs in wet etching, small geometries are almost always plasma etched.

6.2.1 Etch Formulation

Etch composition is often described in terms of parts by volume—for example, nitric, hydrofluoric, and acetic acids in the ratio of 1:1:2. Solids added are then specified in terms of grams per volume of solution. When composition is given in this way, one question that remains is the composition of the starting materials. Not only must the normal manufacturing spread in composition be considered, but also the fact that many of the acids of commerce are sold in different concentrations. For example, HF is commonly available as 49% by weight HF, remainder water, and also as 53% HF, remainder water. When water is part of the formulation, it must be made clear whether the water in the acid is included in the water specified. For example, does 1 part HF, 1 part water mean 1 part of 49% HF plus its water and an additional 1 part water, or does it mean a solution of 50% HF and 50% water?

Solution normality is another way to express concentration. A 1 *normal* solution is one that contains 1 gram equivalent weight (GEW) of solute per liter (l) of solution. A GEW is defined as the gram molecular weight divided by the number of available H^+ or OH^- ions per molecule. Molarity is also used, with a 1 *molar* solution containing 1 gram molecular weight of solute per liter of solution. A 1 *molal* solution is one with 1 gram molecular weight of solute per 1000 grams of solvent. Under most circumstances, when water is the solvent and the solute is a solid that singly ionizes, 1 molar and 1 molal solutions are nearly equivalent. *Mole* (abbreviated as mol) is a measure of the number of molecules, atoms, or ions present, with 1 mol defined as the number of molecules in 1 gram molecular weight, or 6.02×10^{23}. The weight of 1 mol of an ion is given by the sum of the atomic weights of the ion constituents. So, for example, 1 mol of NH_4 ions $= 14 + (4 \times 1) = 18$ grams.

EXAMPLE ☐ How many grams of glacial acetic acid (CH_3COOH) are required for 1 gram equivalent weight? How much of the acid must be added to 1 liter of water to give a 1 normal solution? How much for a 1 molar solution?

First, the number of available hydrogen ions per molecule must be determined. The formula as written above (CH_3COOH) gives no insight. However, when it is written in the alternate form of $HC_2H_3O_2$, convention dictates that the beginning number of hydrogens is the available number. Hence, there is one hydrogen available per molecule, and a gram equivalent weight is the gram molecular weight, or 60 grams. Since glacial acetic acid is $99^+\%$ pure acid, assume that it is 100%. Since a 1 normal solution is 1 GEW per liter of solution, 60 grams (57 ml) must be added to 943 ml of water (assuming that the volume of the mixture equals the sum of the component volumes). Since, in this case, a gram equivalent weight and a gram molecular weight are the same, a 1 normal solution is the same as a 1 molar solution. ☐

In the particular case of the hydronium ion, the concentration is usually expressed in terms of pH, which is $-\log_{10}$ of the concentration in moles per liter of solution. Thus, 1 gram molecular weight of strong acid such as HCl in a liter of solution will completely ionize, produce an H_3O^+ concentration of 1 mol per liter and have a pH of 0. Experimentally, it is observed that by the time the ionizing material being added to the water is reduced to the point that only water is present, the water itself will very slightly dissociate into equal quantities of hydronium and hydroxide ions, each with a concentration of about 10^{-7} mol.[1] Since the number of + and − ions are equal, the water solution is neutral. By the definition just given, it also has a pH of 7. Thus, acid solutions have a pH range of 0–7. When the hydrogen concentration is less than 10^{-7}, the OH^- ions are in the majority, and the pH of basic solutions ranges from 7 to 14.

Many reactions, including most etchants, will depend on the pH of the solution. As etching proceeds, the reactants are depleted so that the pH and the etch rate change. In order to provide a means of initially adjusting the pH, as well as to stabilize it during etching, buffered solutions are often used. Buffering entails adding a salt of the etching acid to the acid, and it is applicable to water solutions of weak acids.[2] Thus, for example, the salt NH_4F could be added to

[1]The dissociation constant of water at room temperature is 10^{-14} so that if the OH^- ions have a concentration of 1 mol per liter of water solution, the concentration of H_3O^- ions will be 10^{-14}, and the solution will have a pH of 14.

[2]For more details on buffering, pH, and the different ways of expressing concentration, see an elementary chemistry textbook.

the weak acid HF. A weak acid is one that does not completely dissociate in water at room temperature and includes all but six of the common acids. These six "strong" acids (perchloric, hydriodic, hydrobromic, hydrochloric, nitric, and sulfuric) completely dissociate and thus do not lend themselves to buffering. For a weak acid of concentration $[C_a]$ above a small fraction of a mole, the hydronium ion concentration $[H_3O^+]$ without buffering is given by[3]

$$[H_3O^+] \cong \{K_a[C_a]\}^{0.5} \qquad 6.1$$

where K_a is the dissociation constant of the acid. For a solution with a salt of concentration $[C_s]$ added, the hydronium ion concentration is

$$[H_3O^+] \cong \frac{K_a[C_a]}{[C_s]} \qquad 6.2$$

It can be shown that for the $[C_a]/[C_s]$ ratio near 1, the pH is quite insensitive to change when either a strong acid or a base is added to the solution (hence, the term "buffering").

6.2.2 Nonselective Silicon Etching

Silicon etches that are nonselective with regard to crystallographic orientation and to crystallographic damage and that thus tend to produce smooth surfaces are primarily based on $HF–HNO_3–H_2O$ mixtures. The etching is a two-step process in which the silicon is first oxidized by the nitric acid and the oxide is then dissolved by HF. The overall reaction is assumed to be (8)

$$18HF + 4HNO_3 + 3Si \rightarrow 3H_2SiF_6 + 4NO + 8H_2O \qquad 6.3$$

Acetic acid is sometimes used either with or instead of additional water. Its role appears to be primarily as a diluent (9), although there is also the suggestion that the surface tension of the mixture is reduced so that better wetting and a smoother etched surface are achieved (10). This behavior is different from that observed during the etching of germanium, in which case the acetic acid is described as a moderator that prevents violent reaction (11). HF in water alone will also very slowly etch silicon (12). In this case, the OH^- ion from water dissociation oxidizes the silicon. Since addition of HF will depress the OH^- ion, higher HF concentrations reduce the reaction rate.

For the $HF–HNO_3–CH_3COOH$ system, Fig. 6.4a shows how the rate varies with composition.[4] At each end of the $HF–HNO_3$ con-

---

[3]Brackets [] in chemical formulae are used to denote concentration.

[4]For instructions on reading triangular graphs, see Appendix B.

centration range, the rate is determined by the concentration of the low-concentration species. In the low HNO_3 region, the rate is limited by oxidation and is sensitive to crystal orientation, crystal defects, and the presence of lower oxides of nitrogen, such as NO. In this region, the reaction may begin either slowly or not at all unless an initially damaged surface is present or a small amount of sodium nitrite (9) is added to the solution. In the low HF region, dissolution of the oxide is the limiting step, and the rate becomes diffusion controlled. These two regions are also characterized as giving rough and polished surfaces, respectively. The general location of the smooth–rough boundary is shown in Fig. 6.4b.

The compositions of three nonselective silicon etches are as follows:

CP4: 3 HF, 5 HNO_3, 3 acetic, 0.06 bromine (32 μm/minute)
CP4A: 3 HF, 5 HNO_3, 3 acetic (25 μm/minute)
Planar: 2 HF, 15 HNO_3, 5 acetic (3.5–5.5 μm/minute)

All parts are by volume, 49% HF, 70% HNO_3, glacial acetic (100%). CP4 (chemical polish no. 4) was originally used for germanium and is a very rapidly acting etch. By leaving out the Br (CP4A), the rate is reduced, and a more tractable silicon etch results. As with many other data presented, published etch rates vary substantially. For example, the rates from Fig. 6.4a and the ones just tabulated differ by about a factor of 2. Such discrepancies are not necessarily the

FIGURE 6.4

(a) Effect of composition of HF–HNO_3–acetic acid etching mixture on room-temperature etch rate of silicon. (b) Etchant composition region where silicon polishing occurs. (*Source:* B. Schwartz and H. Robbins, *J. Electrochem. Soc.* *123*, p. 1903, 1976. Reprinted by permission of the publisher, The Electrochemical Society, Inc.)

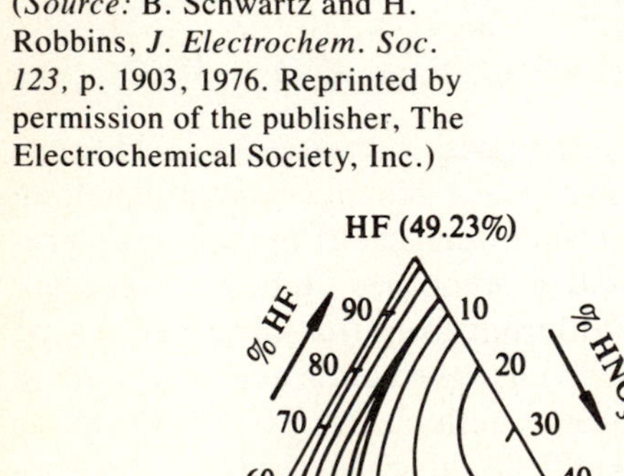

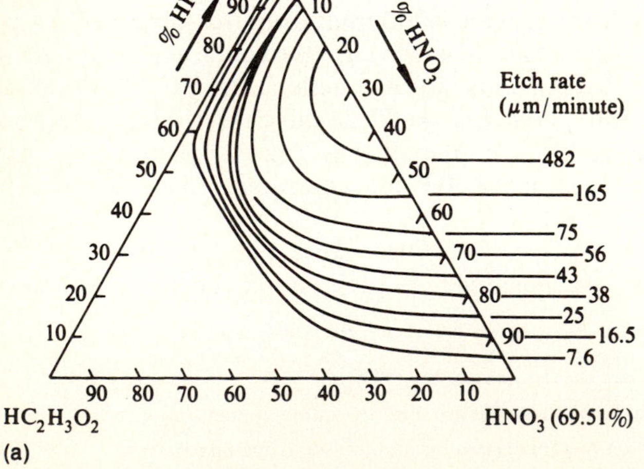

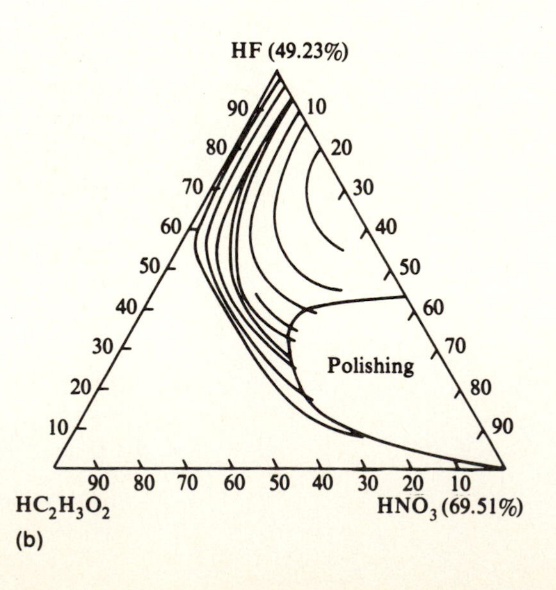

FIGURE 6.5

Temperature dependence of silicon etch rate for a 20% HF, 45% HNO_3, 35% acetic acid solution. (*Source:* Replotted from B. Schwartz and H. Robbins, *J. Electrochem. Soc. 108,* p. 365, 1961.)

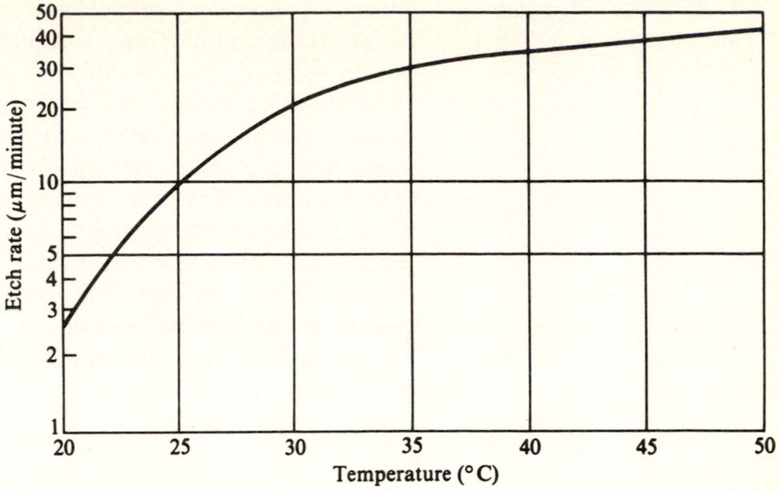

results of poor data but can stem from the fact that these etch rates are dependent on a number of difficult-to-control variables, such as a buildup of NO or the temperature at the etching surface. The effect of temperature for a composition near that of CP4A is shown in Fig. 6.5. Based on this curve, etching done at 30°C could easily be twice that done at 25°C. In the diffusion-controlled region, rate is sensitive to variables such as stirring and wafer spacing. (Dependency on these variables is often considered proof of a diffusion-controlled process.)

6.2.3 Polycrystalline Silicon Etching

Polycrystalline silicon, particularly the fine-grained, low-temperature poly used for MOS gates and leads, etches very rapidly in most etches designed for single-crystal silicon. In pattern etching polysilicon, the polysilicon will usually have been deposited over SiO_2 so that HF-bearing etches may offer little polysilicon/oxide selectivity. In addition, undercutting makes control of the gate and lead geometries difficult. Thus, plasma etching is generally used in these applications (see section 6.3). However, a dilute (15%) KOH–water solution has been used, as has ethylenediamine–pyrocatechol–water (13). Also, by working far into the high nitric acid portion of the polish region—for example, 6 HF, 100 HNO_3, 40 H_2O—slow (<1 μm/minute), smooth edge definition and minimal oxide attack are possible (14). Choline, used as a sodium-free alkaline developer for positive resist and for wafer cleaning, can also be used for etching polycrystalline silicon. A 5% solution etches phosphorus-doped polysilicon at a rate of about 250 Å/minute at room temperature, while it etches LPCVD nitride and thermal oxide at rates of <1 Å/minute.

6.2.4 Anisotropic Silicon Etching

The etches described in section 6.2.2 are "hot" etches that are good for polishing and etching isotropically. There are also etches composed, for example, of chromic oxide, water, and HF that etch single-crystal silicon very slowly but regions with crystallographic damage much more rapidly. In a sense, they are anisotropic, but some silicon etchants are available that are truly anisotropic and etch extremely slowly in <111> directions and at a useful rate in other directions. This property can be used to form a variety of patterns in a silicon wafer that would be very difficult to achieve in any other manner. The slow etch rate in <111> directions is a consequence of the diamond lattice. A "(111) plane" is really a double layer bound together by more atomic bonds than are found between other planes, as is illustrated in Fig. 6.6. The figure shows a view of

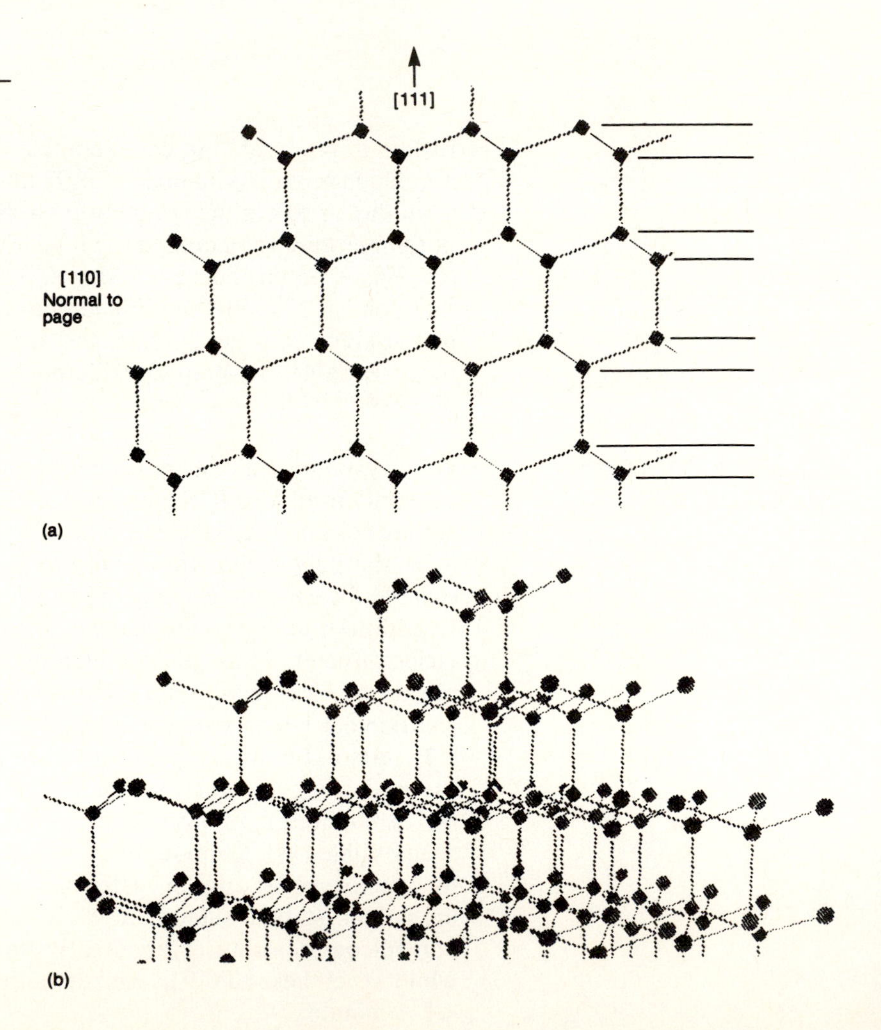

FIGURE 6.6

Views of diamond or zinc blende lattice showing double layers of atoms encountered in moving through lattice in a [111] direction: (a) Looking in a [110] direction. (b) Looking in a slightly different direction, with perspective.
(*Source:* Program for drawing lattice courtesy of Dr. Anthony Stephens, Texas Instruments Incorporated.)

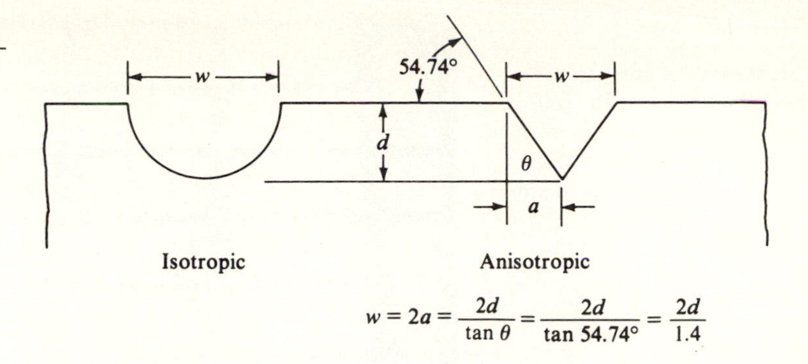

FIGURE 6.7

Relative width of grooves of depth d etched into (100) silicon surfaces when isotropic and anisotropic etches are used.

the diamond lattice when looking in a [110] direction and a perspective view in which the line of sight is slightly off a [110] direction so that the bonds tying the double layer together can be more easily seen.

One of the oldest applications of anisotropic etching, also referred to as orientation-dependent etching (ODE), is the etching of mesas on silicon wafers as one step in the production of dielectrically isolated ICs. Mesas can be etched using an isotropic etch, but because of undercutting of the mask, they cannot be as closely spaced as when an anisotropic etch is used, as is shown in Fig. 6.7 for the particular case of (100) oriented wafers (15). Such circuits are used for radiation hardening and for high-voltage ICs (16, 17). More recently, silicon has been used for a variety of micromechanical applications, many of which require anisotropic etching (18).

Any structures etched using anisotropic etches will have the etched features bounded predominately by (111) planes, although in some cases facets of planes with intermediate etch rates will appear. To consider the shape of the features that can be etched, both the traces of the (111) planes intersecting the wafer surface and the angles of the (111) planes with the wafer surface must be considered. Anisotropic etching on a (111) wafer is not possible since the first plane encounter is the slow-etching (111) plane. Etching on (110) wafers is possible, but such wafers are not currently used for IC production. The widely used (100) wafers are, however, quite amenable to anisotropic etching. Traces of {111} planes intersecting a (100) surface make right angles with each other. If a mask with a slot opening in it is oriented azimuthally on the wafer[5] so that the slot edge is parallel to a (111) trace, a trench bounded by $(\bar{1}1\bar{1})$ and $(\bar{1}\bar{1}1)$ planes on the sides can be etched. If etching is allowed to

[5]Wafers that conform to SEMI standards will have an edge flat ground parallel to a (110) plane that can be used for orienting the mask.

FIGURE 6.8

(a) SEM view of anisotropically etched grooves in (100) silicon. (b) Cross section of a polycrystalline silicon wafer with dielectrically isolated single-crystal islands.

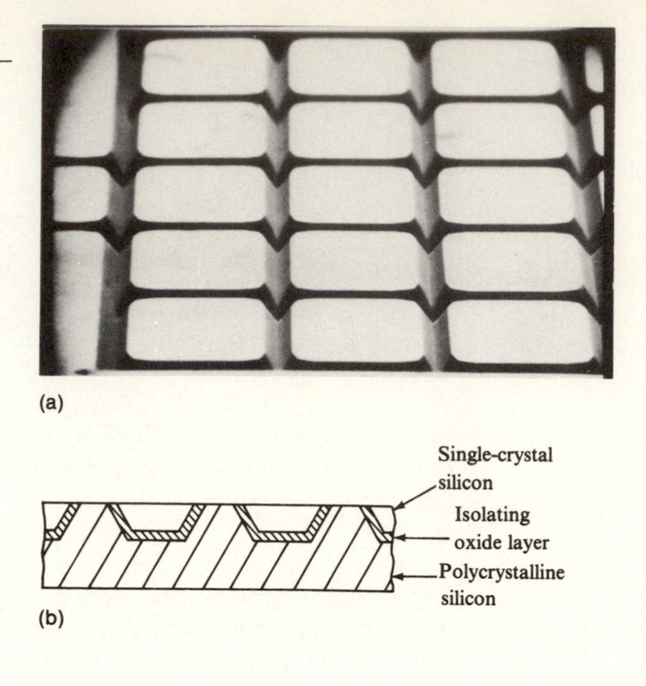

continue to completion, the trench will be V-shaped, with both the width and depth determined by the mask opening. Etching can be stopped earlier, however, leaving a flat-bottomed trench. In this case, the depth is a function of etching time. If, instead of a single slot opening, the mask is patterned with many slots at right angles to each other, a series of mesas can be etched. An example of the kind of etching required for dielectrically isolated (DI) ICs is shown in Fig. 6.8a. After the DI process is completed, the wafer has a series of regions isolated from the main body of the wafer by oxide layers, as shown in cross section in Fig. 6.8b.[6]

At outside corners (corners of mesas where two (111) planes intersect), etching may not stop abruptly. Instead, it will continue until an intermediate etching rate set of planes such as the {331} are exposed and, in cases in which the mesas are small, may produce pronounced corner faceting as shown in Fig. 6.9. This effect can be minimized by using a corner-compensation mask design that has

[6]After mesa etching, a thermal oxide about 2 μm thick is grown on the etched surface. Next, a polysilicon layer about as thick as the original wafer is deposited over the oxide. The wafer is then turned polysilicon side down, and the opposite side (the single-crystal side) is ground off and polished until only single-crystal islands isolated by the oxide layer and immersed in a polysilicon matrix are left. This leaves the structure shown in Fig. 6.8b.

FIGURE 6.9

Exposure of higher-index planes at corners during orientation-dependent etching of (100) silicon. (The etch used was KOH–propanol–water, and the exposed corner planes are of the {331} family. Other etchants may expose other planes.)

"ears" sticking out from each corner to slow down etching at the corners (19–21).

When [110] wafers are used, since several of the {111} family of planes are perpendicular to the [110] surface, slots with vertical walls can be etched out by aligning the edge of the mask with the traces of the perpendicular (111) planes (22). However, since these {111} planes do not intersect one another at right angles, rather than being able to have square or rectangular mesas, they must be rhombic-shaped. It is also not possible to etch deep, short slots because other low-angle planes of the {111} family come in from the ends and limit the depth. However, when the slots go to the wafer edge, the depths can be up to hundreds of microns deep since the difference in etch rates in the (110) and (111) directions is at least 600 to 1 (23). Fig. 6.10 shows an example of a series of deep vertical grooves.

Table 6.1 lists several silicon anisotropic etches that have been reported. None of them involve the HF–HNO_3 system previously discussed. The KOH reaction, which is also often used for damage removal even though it gives a faceted surface, probably proceeds as

$$Si + H_2O + 2KOH \rightarrow K_2SiO_3 + 2H_2 \qquad 6.4$$

The advantages of a KOH etch for damage removal (which does not require the anisotropic feature) are that both the fresh and spent etches pose less of a safety hazard than do the HF–HNO_3 mixtures and that it is often more economical.

EXAMPLE ☐ Prove the assertion that square or rectangular mesas can be etched into (100) silicon and that the walls of the mesas will make angles of 54.74° with the (100) surface.

Since the (111) planes are the slowest in etching, the mesas

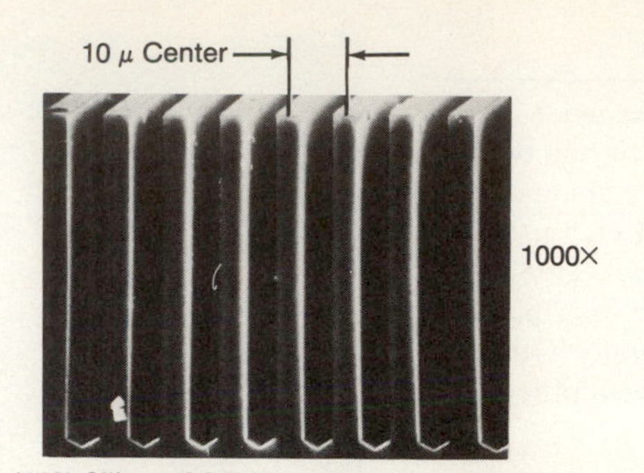

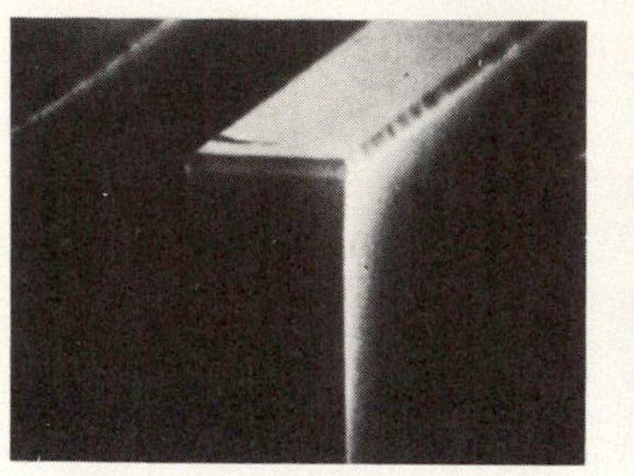

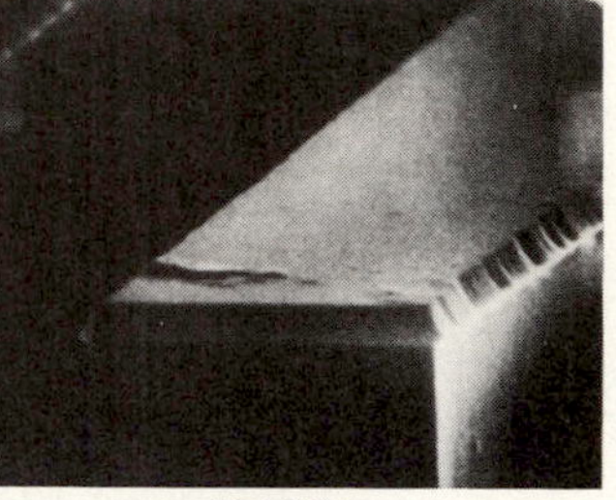

FIGURE 6.10

Cross-sectional views of etching into a (110) silicon plane using 80°C KOH and water. (In the higher magnification view (*bottom*), the silicon dioxide layer used as mask during the etching can be seen. Photoresist cannot be used since it is rapidly attacked by the alkaline etch at this temperature.)

will be bounded by them. There is no (100) projection in Appendix A, but there is a (001) projection, and it is equivalent in that the wafer surface could just as easily have been labeled (001) as (100). Looking at the (001) projection shows that four (111) planes have their traces on the (001) plane perpendicular to each other: $(\bar{1}\bar{1}1)$, $(\bar{1}11)$, (111), and $(1\bar{1}1)$. To find the angle that each makes with the (001) plane, either calculate it from the equation given in Appendix A or read it from the table of angles between planes. Either method will verify that 54.74° is the correct angle. □

6.2.5 Resistivity-Sensitive Silicon Etching

This form of etching is often referred to as concentration-dependent or selective etching. Some etchants have large differences in etch rates depending on whether the resistivity is high or low and on whether the silicon is n- or p-type. This kind of behavior has long been used to delineate high–low resistivity and pn junction regions for characterization (24). However, the low etch rate region may also be used as an etch stop and produce structures such as very thin membranes (not applicable to mainstream IC production, but sometimes useful in sensor construction or for processing silicon-on-insulator structures). The general approach is to use the slow

TABLE 6.1

Silicon Anisotropic Etches

Solution Components	*Reference*
KOH, water, normal propanol	(1)
KOH, water	(2)
Hydrazine, pyrocatechol	(3)
Ethylenediamine, pyrocatechol, water	(4)
Ethylenediamine, water	(4)
Hydrazine, iso–2–propyl alcohol	(5)
Hydrazine, water	(5)

Details for Specific Etches

Etch	*Composition*	*Comments*
(100) ODE: Etches [100] ~100 × [111] direction	KOH–normal propanol–H_2O KOH, 250 grams Normal propanol, 200 grams H_2O, 800 grams	~1 μm/minute at 80°C, [100]; "stops" at p^{++}-interface; etches Si_3N_4 at 14 Å/hour; etches SiO_2 at 20 Å/minute.
(110) ODE: Etches [110] 600 × [111] direction	KOH–H_2O 50–50 volume	~0.8 μm/minute at 80°C in (110) silicon.
Ethylenediamine: Orientation-dependent and concentration-dependent	Ethylenediamine–pyrocatechol–H_2O EDA, 255 cc H_2O, 120 cc PC, 45 grams	~1.1 μm/minute at 100°C in [100]; "stops" at p^{++}-interface; very slow etching of SiO_2 (~3 Å/minute); negligible etch of Al, Au, Ag, Cu, Ni, or Ta.

Note: Ethylenediamine is $NH_2(CH_2)_2NH_2$; hydrazine, N_2H_4; and pyrocatechol, $C_6H_4(OH)_2$.

References: (1) H.A. Waggener et al., International Electron Devices Meeting, October 1967; J.B. Price, p. 339, in Howard R. Huff and Ronald R. Burgess, eds., *Semiconductor Silicon/73,* Electrochemical Society, Princeton, N.J., 1973. (2) A.I. Stoller, *RCA Review 31,* p. 271, 1970; Kenneth E. Bean, *IEEE Trans. on Electron Dev. ED-25,* p. 1185, 1978; Don L. Kendall, *Ann. Rev. Mater. Sci. 9,* p. 373, 1979. (3) J.M. Crishal and A.L. Harrington, Electrochemical Society Spring Meeting Abstracts, abst. no. 89, p. 202, 1962. (4) R.M. Finne and D.L. Klein, *J. Electrochem. Soc. 114,* p. 965, 1967. (5) D.B. Lee, *J. Appl. Phys. 40,* p. 4569, 1969; Michel J. Declercq et al., *J. Electrochem. Soc. 122,* p. 545, 1975.

etching variety as an etch stop. Thus, for example, if lightly doped n-type silicon is unattacked in a particular etch, a thin n-layer could be epitaxially deposited on a heavily doped substrate and the substrate then etched away. One difficulty with these techniques is that the etched surface is often not smooth, due in part to the etch sensitivity to resistivity nonuniformities.

By changing the concentration of the HF–HNO_3–CH_3COOH mixture used for polish etching to the ratio of 1:3:8, the etch rate of silicon, n- or p-type, increases by about a factor of 15 (from ~0.2 to

~3 μm/minute) when the doping concentration increases above ~10^{19} atoms/cc (25). A KOH–H_2O–isopropyl alcohol solution at 80°C etches silicon with a boron concentration of 2×10^{18} atoms/cc about 25 times faster than when the silicon is doped to a concentration of 2×10^{20} atoms/cc (26). Ethylenediamine–pyrocatechol–water etches high-resistivity silicon but not heavily doped p-type (27, 28). By heavily doping with boron and compensating with germanium, a very effective etch stop is produced. Etch ratios of 1:1500 between the lightly doped and heavily doped regions have been observed.

6.2.6 Electrolytic Silicon Etching

Low-resistivity n-type silicon and p-type silicon can be electrolytically etched when the silicon is made the anode (+ electrode) in a suitable electrolyte, such as 5% by weight HF in water (29, 30). When silicon is electrolytically etched, it is removed in either divalent or tetravalent states, depending on the current density and the electrolyte concentration (31–32). With a high concentration of HF solution and a current density J below some critical value J_c, reactions such as the following are thought to occur (32):

$$Si + 2HF + 2e^+ \rightarrow SiF_2 + 2H^+ \qquad 6.5$$

The SiF_2 then reacts to give an amorphous silicon layer (stain) that is often seen after etching:

$$2SiF_2 \rightarrow Si + SiF_4 \qquad 6.6$$

The reaction in Eq. 6.6 is then followed by

$$SiF_4 + 2HF \rightarrow H_2SiF_6 \qquad 6.7$$

For low HF concentrations, when silicon is dissolved in the divalent state, the reactions can proceed without the formation of amorphous silicon, perhaps as follows:

$$Si + 2H_2O + 2e^+ \rightarrow Si(OH)_2 + 2H^+ \qquad 6.8a$$

$$Si(OH)_2 + 2H_2O \rightarrow Si(OH)_4 + H_2 \qquad 6.8b$$

$$Si(OH)_4 \rightarrow SiO_2 + 2H_2O \qquad 6.8c$$

$$SiO_2 + 6HF \rightarrow H_2SiF_6 + 2H_2O \qquad 6.8d$$

At higher current densities (and higher anode potentials), silicon dissolves in the tetravalent state, possibly as follows:

$$Si + 4H_2O + 4e^+ \rightarrow Si(OH)_4 + 4H^+$$

It continues according to Eqs. 6.8c and 6.8d and is usually smooth and specular. However, in etching p^+ from n-layers, it is reported that the remaining surface will be rough unless H_2SO_4 is added to the HF solution (33). HF/organic mixtures, such as alcohols, glycols, and glycerine, help maintain a smooth surface during etching of p-type silicon (31). Shining light on n-type silicon will increase the supply of holes and thus the etch rate, and, in fact, the lighted area can be used to produce selective etching (31). Under some conditions, etching of n-type silicon will also produce tunnels in the silicon (8, 34).

Early applications of electrolytic etching of silicon were for polishing and for attempting to slice silicon using very thin wire electrodes. The polishing application was only partially successful, and the slicing application was a complete failure. More recently, electrolytic etching has been used for the thinning of wafers in order to leave thin membranes (29, 30) and for the production of porous silicon, which can then be rapidly oxidized and used for electrical isolation between components (35, 36). It has also been reported that when hot KOH-based anisotropic etches are used, the use of an applied potential allows more precise control of the silicon etch rate (37).

6.2.7 Silicon Staining

When a silicon surface is exposed to HNO_3–HF mixtures either during intentional etching or as a result of being contacted by the solution at the completion of contact window etching, stains will often form. In addition, during boron diffusions, relatively insoluble layers at the silicon–oxide interface may form. They will remain after normal contact etching and appear to be a stain.

Films can be produced on silicon either anodically in HF (p-type silicon) or by straight immersion in a concentrated HF solution containing a small amount of nitric acid or sodium nitrite (38, 39). The exact composition and appearance depend on the resistivity and type, but apparently they are polymerized silicon hydrides (39). The manner in which nitric–HF etchants are quenched will affect the amount of staining. A much thicker film is reported when the nitric acid concentration is low and quenching is by large quantities of water than when the nitric acid concentration is first greatly increased and then quenched with water. Also, apparently enough current flow can occur during oxide contact window etching to cause staining of p-type silicon after the window is completely open unless the back of the wafer is covered with oxide or resist (40).

6.2.8 Gallium Arsenide Etching (General)

The etching of gallium arsenide proceeds in a manner analogous to that of silicon: The gallium and arsenic are first oxidized, and then the oxides are dissolved. The GaAs case, however, has a wider choice of oxidants and oxides that are much easier to dissolve. The net result is a wide selection of GaAs etches. They may be broadly categorized as acidic or basic, depending on whether an acid or a base is used to dissolve the oxides. Some selections reported are HCl, HNO_3, H_2SO_4, H_3PO_4, citric acid, NaOH, and NH_4OH. Methanol, an organic, can also be used for oxide dissolution. Common oxidants are hydrogen peroxide and bromine, but $K_2Cr_2O_7$, $KMnO_4$, $K_3Fe(CN)_6$, and $Ce(SO_4)_2$ have also been used. The many choices for each function in the etch have given rise to a large number of etchant formulations. Some are less harsh on masks than others, some give better surface polishing, some give more desirable etch profiles, and some are easier and safer to formulate and use. Table 6.2 gives the constituents of a few of the etches that have been reported.

The two different mechanisms of dissolution are straight chemical and those involving charge transfer. The latter may either require an external source of current or be electroless and derive carriers from reduction of the oxidant (redox reactions) (41). In either case, there is generally a kinetic-limited etching region that is high in oxidant and a diffusion-limited region that is high in oxide etchant. As composition is changed from high oxidant to high etchant, a peak

TABLE 6.2

Gallium Arsenide Etches

Composition	*Comments*	*Reference*
1 NaOCl, 20 H_2O	Used with mechanical wafer rotation for polishing.	(1)
NH_4OH, H_2O_2	Also used with mechanical wafer rotation for polishing.	(2)
3 H_2SO_4, 1 H_2O_2, 1 H_2O	For giving polished surface; not good for patterning; erodes photoresist.	(3)
0.3N–NH_4OH, 0.1N–H_2O_2	Minimal undercutting of SiO_2 masking; gives flat surface.	(4)
1 HCl, 1 acetic, 1 $K_2Cr_2O_7$*	Gentle with photoresist.	(5)
10 citric acid, 1 H_2O_2†	Gentle with photoresist; no enhanced etching at mask edge.	(6)
3 H_3PO_4, 1 H_2O_2, 50 H_2O‡	No enhanced etching at mask edge; can be used with photoresist.	(7)

*12N–HCl, 17N–acetic, 1N–$K_2Cr_2O_7$.
†Citric acid = 50% by weight $C_3H_4(OH)(COOH)_3 \cdot H_2O$ + 50% by weight H_2O.
‡85% by weight H_3PO_4, 30% by weight H_2O_2.

References: (1) A. Reisman and R. Rohr, *J. Electrochem. Soc. 111,* p. 1425, 1964. (2) J.C. Dayment and G.A. Rozgonyi, *J. Electrochem. Soc. 118,* p. 1346, 1971. (3) S. Jida and K. Ito, *J. Electrochem. Soc. 118,* p. 768, 1971. (4) J.J. Gannon and C.J. Nuese, *J. Electrochem. Soc. 121,* p. 1215, 1974. (5) Sadao Adchi and Kunishige Oe, *J. Electrochem. Soc. 131,* p. 126, 1984. (6) Mutsuyuki Otsubo et al., *J. Electrochem. Soc. 123,* p. 676, 1976. (7) Yoshifumi Mori and Naozo Watanabe, *J. Electrochem. Soc. 125,* p. 1510, 1978.

in the etch rate occurs, and the etched profiles change from anisotropic to isotropic.

When the requirement is to etch a shallow, flat depression, such as is required for a MOS gate, neither of these regions is exactly appropriate. Anisotropic etching may leave a re-entrant corner that is difficult to run leads over, and isotropic diffusion-controlled etching will generally leave trenches next to mask edges. As can be seen from Table 6.2, several of the H_2O_2 etches are reported as not exhibiting trenching. This is apparently because a layer forms on the surface of the GaAs during etching and limits the etch rate through the length of time for reactants and etch products to diffuse through it (42).

6.2.9 Gallium Arsenide Etching (Anisotropic)

Gallium arsenide (111) planes, like those of silicon, generally etch slower than other planes. Unlike silicon, however, parallel pairs of planes such as (111) and $(\bar{1}\bar{1}\bar{1})$ are not completely equivalent because in the case of gallium arsenide (zinc blende structure), one layer of a (111) double layer is totally comprised of gallium atoms, while the other layer is totally of arsenic. Thus, depending on whether a (111) plane is approached from the [111] or the $[\bar{1}\bar{1}\bar{1}]$ direction, either a sheet of gallium atoms or a sheet of arsenic atoms will be encountered. By convention, a (111) gallium face is referred to as a (111) A face, and the arsenic face is called the (111) B face. Also by convention, a (111) plane has an A gallium surface, while the $(\bar{1}\bar{1}\bar{1})$ plane has a B arsenic surface. The other gallium face planes are $(\bar{1}\bar{1}1)$, $(\bar{1}1\bar{1})$, and $(1\bar{1}\bar{1})$. As might be expected, the etch rates of the gallium and arsenic faces are different (the gallium face is slower). Because of this difference in etch rates, instead of the square-based pyramidal etch pits found during etching of an Si (100) plane, rectangular pits sometimes occur in (100) GaAs because one pair of opposing (111) planes etches faster than the other pair. During etching of mesas in (100) faces, the edges of the mask openings should, as in

FIGURE 6.11

(a) Octahedron bounded by {111} family of planes. (A (100) plane passes through the corners marked 1–2–3–4.) (b) Cross section parallel to either a [011] or a $[0\bar{1}1]$ direction through a mesa etched into a (100) wafer when mask edges were oriented parallel to traces of (111) planes on the (100) surface and etch rate is same for A and B faces. (c) Cross section when B faces etch more rapidly than A faces.

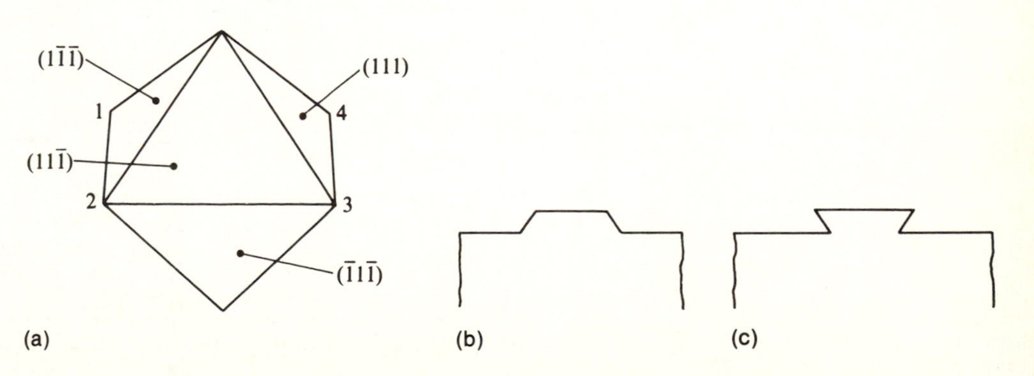

silicon, be aligned along the traces of {111} planes. However, because of the lack of symmetry just discussed, different slopes on the two pairs of mesa sides may exist, as shown in Fig. 6.11. The four planes of the upright pyramid are the planes that would be exposed if a mesa were etched onto (100) GaAs. A vertical cross section through the mesa parallel to either the [011] or the [0$\bar{1}$1] direction will have identical profiles, as shown in Fig. 6.11b. For etches in which the GaAs B faces are rapidly etching and the A faces are very slowly etching, the slow-etching ($\bar{1}1\bar{1}$) and ($\bar{1}\bar{1}1$) planes of the lower half of the octahedron of Fig. 6.11a will be exposed instead of the fast-etching ($11\bar{1}$) and ($1\bar{1}1$) planes in the upper half. Thus, for GaAs, a vertical section parallel to a [01$\bar{1}$] direction will look like the one shown in Fig. 6.11b, while one in the [011] direction will appear as shown in Fig. 6.11c.

Under some etching conditions, the A planes may not be the slowest etching, in which case the section profile may not look like those of Figs. 6.11b and 6.11c. In particular, for some etch compositions, (110), (221), (331), or even (111) B planes are exposed, while in others, etching is nearly isotropic. In the latter case, the profile will resemble that of Fig. 6.1. The edge profiles have been extensively studied for a variety of etchants (43, 44). Fig. 6.12 shows the

FIGURE 6.12

Mesa profiles that are observed when a sulfuric acid, hydrogen peroxide, water mixture is used for etching GaAs and in which the slow-etching planes in (011) section are ($33\bar{1}$) and ($11\bar{1}$). (*Source:* Don W. Shaw, *J. Electrochem. Soc. 128,* p. 874, 1981. Reprinted by permission of the publisher, The Electrochemical Society, Inc.)

Acid	Volume ratios*	Etch rate (100) (μm/minute^{-1})	Cross-sectional profiles	
			(011) section	(0$\bar{1}$1) section
H_2SO_4	1:8:1	14.6	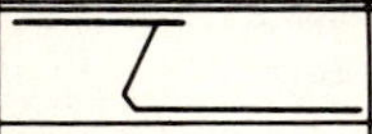	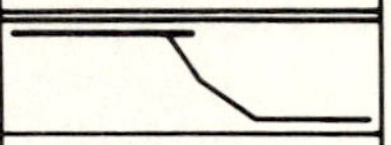
H_2SO_4	1:8:40	1.2	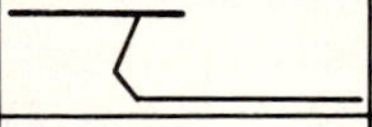	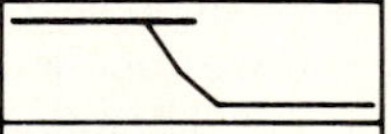
H_2SO_4	1:8:80	0.54	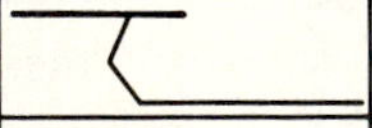	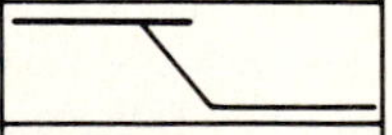
H_2SO_4	1:8:160	0.26	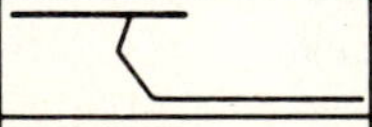	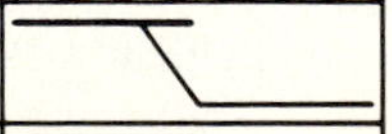
H_2SO_4	1:8:1000	0.038	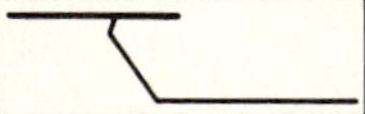	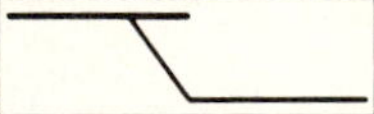
H_2SO_4	1:1:8	1.3		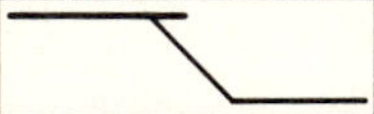
H_2SO_4	4:1:5	5.0	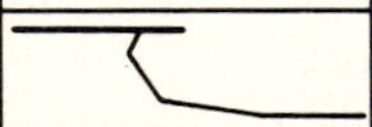	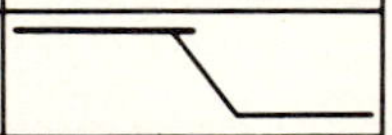
H_2SO_4	8:1:1	1.2	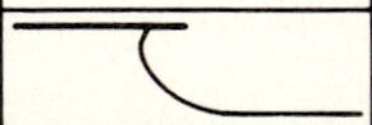	
H_2SO_4	3:1:1	5.9	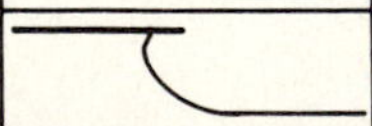	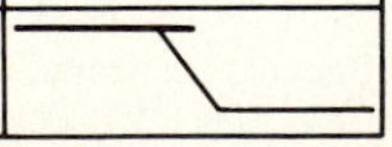

*Acid (concentrated):H_2O_2(30%): H_2O

range of profiles observed for the H_2SO_4–H_2O_2–H_2O system as the relative proportions of the components change.

6.2.10 Preferential Etches

Preferential etches are used to delineate various crystallographic defects such as dislocations, stacking faults, and twin boundaries; polished-over mechanical damage; and swirl patterns in silicon. Many of these defects are associated with slice manufacture and will have been evaluated before the slice reaches a wafer-fab facility. However, since various high-temperature wafer-fab operations have the capability of introducing large quantities of defects, evaluation during processing is very important. (The various defects to be expected are discussed in the chapters on diffusion, oxidation, and epitaxy.)

The six most common silicon etches are given in Table 6.3, along with comments on their general behavior. Preferential gallium arsenide etches seem not to have been as thoroughly studied as those of silicon, but Table 6.4 lists some that have been reported. The performance of the various etches depends in part on the exact

TABLE 6.3

Preferential Silicon Etches for Defect Delineation

Name	*Composition*	*Comments*	*Reference*
Dash	1 HF, 3 HNO_3, 10 CH_3COOH	Delineates defects in (111) silicon; requires long etch times; is concentration-dependent.	(1)
Sirtl	1 HF, 1 (5M–CrO_3)	Delineates defects in (111); needs agitation; does not reveal etch pits in (100) very well.	(2)
Secco	2 HF, 1 (0.15M–$K_2Cr_2O_7$)	Delineates OSF in (100) silicon very well; agitation reduces etch times.	(3)
Wright–Jenkins	60 ml HF, 30 ml HNO_3, 30 ml (5M–CrO_3), 2 grams Cu $(NO_3)_2$, 60 ml CH_3COOH, 60 ml H_2O	Delineates defects in (100) and (111) silicon; requires agitation.	(4)
Schimmel	2 HF, 1 (1M–CrO_3)	Delineates defects in (100) silicon without agitation; works well on resistivities 0.6–15.0 Ω-cm n- and p-types.	(5)
Modified Schimmel	2 HF, 1 (M–CrO_3), 1.5 H_2O	Works well on heavily doped (100) silicon.	(5)
Yang	1 HF, 1 (1.5M–CrO_3)	Delineates defects on (111), (100), and (110) silicon without agitation.	(6)

Note: Agitation is ultrasonic.

References: (1) W.C. Dash, *J. Appl. Phys. 27,* p. 1193, 1956. (2) E. Sirtl and A. Adler, *Z. Metallk. 52,* p. 529, 1961. (3) F. Secco d'Aragona, *J. Electrochem. Soc. 119,* p. 948, 1972. (4) Margaret Wright Jenkins, *J. Electrochem. Soc. 124,* p. 757, 1977. (5) D.G. Schimmel, *J. Electrochem. Soc. 126,* p. 479, 1979. (6) K.H. Yang, *J. Electrochem. Soc. 131,* p. 1140, 1984.

TABLE 6.4

Preferential Gallium Arsenide Etches

Name	*Composition*	*Comments*	*Reference*
Schell	1 HNO_3, 2 H_2O	Shows pits on Ga [111] faces.	(1)
R–C	1 HF, 5 HNO_3, 10 $AgNO_3$ solution*	Shows pits on both {111} faces.	(2)
W–R	2 HCl, 1 HNO_3, 2 H_2O	Shows pits on Ga {111} faces.	(3)
A–B	1 HF, 2 H_2O, Cr + Ag†	Shows pits on As {111} faces.	(4)
KOH	Molten KOH at 400°C	Shows pits on {100} faces.	(5)
Sirtl	1 HF, 1 (5M–CrO_3) in H_2O	Leaves mounds instead of pits.	(6)

*1% by weight silver nitrate in water.
†Add 33% by weight of CrO_3 and 0.3% by weight of $AgNO_3$ to the HF–H_2O solution. By premixing CrO_3 with half of water, and $AgNO_3$ with HF and remainder of water, the two solutions can be stored indefinitely and mixed together just before using.

References: (1) H.A. Schell, *Z. Metallk. 48,* p. 158, 1957. (2) J.L. Richards and A.J. Crocker, *J. Appl. Phys. 31,* p. 611, 1960. (3) J.G. White and W.C. Roth, *J. Appl. Phys. 30,* p. 946, 1959. (4) M.S. Abrahams and C.J. Buiocchi, *J. Appl. Phys. 36,* p. 2855, 1965. (5) Morgan Semiconductor product literature. (6) See Table 6.3.

experimental procedures used, especially including the surface cleanup given before the defect delineation etch. Too short a time will give small and poorly developed features. Excessive times may broaden and smear the features, leave only a replica of the defect etched into the underlying material, or completely remove the layer of interest. Thus, it would be well to study details in the original papers (references in Tables 6.3 and 6.4) before seriously using the etchants. Reference 24 discusses experimental techniques and interpretation of results.

6.2.11 Silicon Dioxide Etching

HF in various dilutions in water and often buffered with ammonium fluoride is the standard silicon dioxide wet etchant. Si and GaAs etch in HF at a minuscule rate and thus provide an etch stop after an overlying oxide layer is patterned. Hot alkaline bases such as sodium or potassium hydroxide will also etch at useful rates. They are, however, very harsh on organic masking materials and also etch Si and GaAs.

In dilute water solutions, HF will partially dissociate and give

$$HF \rightarrow F^- + HF_2^- \qquad 6.9$$

In more concentrated solutions, additional complex ions, perhaps $H_3F_4^-$ or $H_4F_5^-$, are also present. The overall reaction with SiO_2 is (45)

$$SiO_2 + 6HF \rightarrow H_2SiF_6 + 2H_2O \qquad 6.10$$

However, the actual mechanism is quite involved, and the SiO_2 etch rate depends on the HF and HF_2^- concentrations but not on F^- (46). This type of behavior is not unique to the HF–SiO_2 system, however. It is, for example, reported that zinc is attacked by molecular HCl and by molecular acetic acid (47). Since F^- ions alone will not etch oxide, a salt of fluorine alone dissolved in water cannot be used for etching. The pH must be adjusted to provide HF_2^- ions or HF. Stirring or ultrasonic agitation affects the etch rate only in the region where diffusion dominates, which is reported to be for buffer ratios[7] greater than 10:1 (48).

Hydroxides etch oxide through reactions giving water-soluble silicates; for example,

$$2KOH + SiO_2 \rightarrow K_2SiO_3 + H_2O \qquad 6.11$$

Typical room-temperature etch rate curves for thermal oxide in both HF and buffered HF are shown in Fig. 6.13. In both cases, the rate increases with increasing HF concentration, and despite the fact that F^- ions do not by themselves etch SiO_2, the buffered HF has a somewhat higher etch rate, which is apparently due to the increased concentration of HF_2^- ions (46). Increasing temperature increases the etch rate, as shown in Fig. 6.14, with buffered solutions having a slightly higher activation energy.

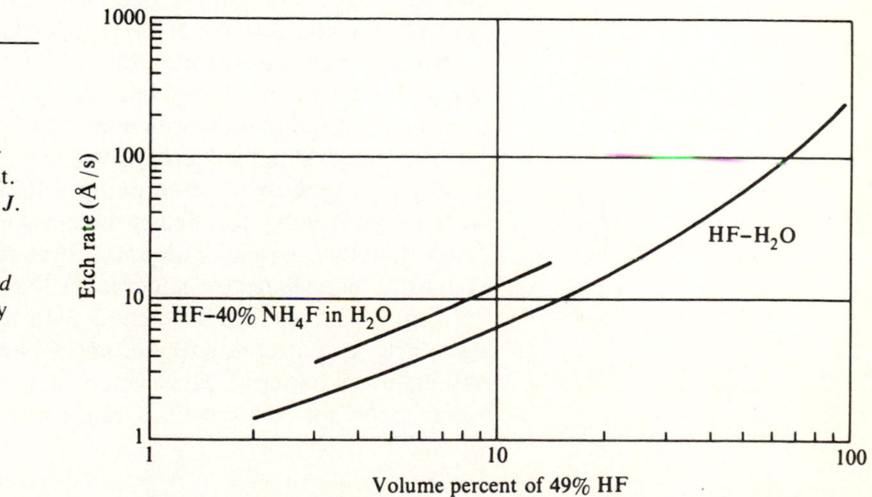

FIGURE 6.13

Etch rate of 25°C of thermal SiO_2 in HF solutions.
(*Source:* From data in Robert Herring and J.B. Price, *Electrochem. Soc. Ext. Abst. 73–2,* abst. no. 160, 1973; G.I. Parisi et al., *J. Electrochem. Soc. 124,* p. 917, 1977; and Chao Chen Mai and James C. Looney, SPC and *Solid State Technology,* p. 19, January 1966.)

[7]Buffer ratio as applied to etching of oxides with HF generally means the volume ratio of a solution of 40% by weight NH_4F in water to 49% HF.

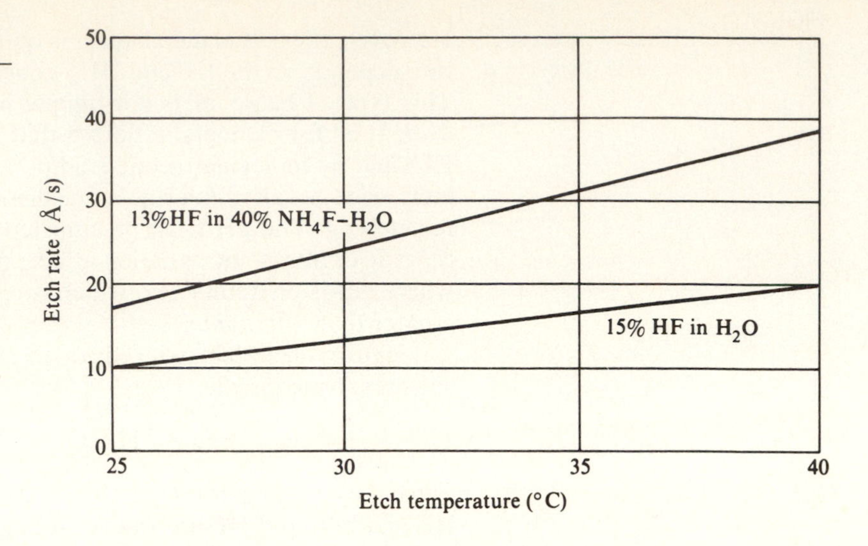

FIGURE 6.14

Etch rate of thermal oxide in HF solutions versus temperature. (*Source:* Data from G.I. Parisi et al., *J. Electrochem. Soc. 124,* p. 917, 1977, and Robert Herring and J.B. Price, *Electrochem. Soc. Ext. Abst. 73–2,* abst. no. 160, 1973.)

A pure CVD oxide as deposited etches faster than a thermal oxide because of structural defects. Subsequent annealing at high temperatures will generally reduce the etch rate differential. Doped oxide etch rates (whether the oxide is doped during CVD growth or by subsequent doping of a thermal oxide) depend on the impurity and the specific etchant. As the silicon content of an oxide increases, the etch rate in HF decreases until in the limit of SiO, nitric acid must be added to the HF. The addition of arsenic or phosphorus to the SiO_2 increases the etch rate in HF or buffered HF. Before the application of X-ray fluorescence analysis, the etch rate was used as a measure of the phosphorus content of phosphorus-doped oxides. Small boron additions decrease the rate, while larger ones increase it (49). By changing the etch composition to one containing HNO_3, such as in P-etch,[8] the lightly boron-doped oxide will etch much faster than the thermal oxide (50). Stress in the films also affects the etch rate; increasing stress increases the rate (51).

Fig. 6.15 gives data for etching with KOH. The caustics are used very little for patterning oxide, and when they are, it is with an inorganic mask material. More often, the oxide etch rate is of interest when the oxide is used as a mask during anisotropic etching of silicon.

[8]Named after W.A. Pliskin (W.A. Pliskin and R.P. Gnall, *J. Electrochem. Soc. 111*, p. 872, 1964), its composition is 15 ml HF, 10 ml HNO_3, 300 ml H_2O.

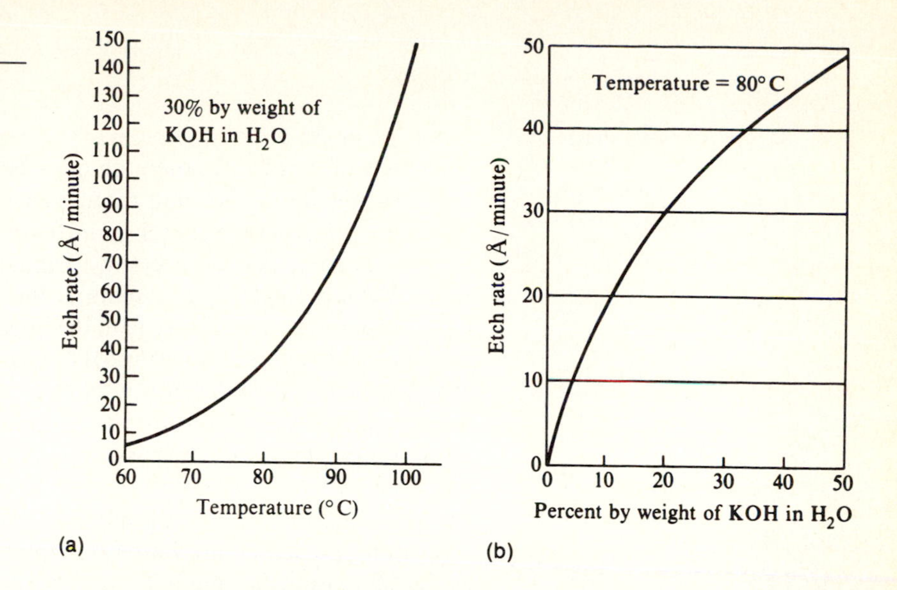

FIGURE 6.15

Etch rate of thermal silicon dioxide by hot KOH. (*Source:* Data courtesy of Richard Yeakley, Texas Instruments, Incorporated.)

6.2.12 Silicon Nitride Etching

Silicon nitride can be wet etched with either HF solutions or with hot phosphoric acid. Phosphoric acid is the "standard" wet nitride etch. In it, the nitride can etch more than 40 times as fast as CVD oxide, which is often used as a mask. The selectivity decreases at high temperatures, but in order to have useful etch rates, high-temperature boiling concentrated H_3PO_4 must be used. For example, 91.5% H_3PO_4 boils at 180°C, etches high-temperature tube nitride[9] at approximately 100 Å/minute, and etches CVD oxide at about 10 Å/minute (52). Under these same conditions, single-crystal silicon etches about 30% as fast as CVD oxide. HF solutions at low temperature etch the nitride much slower than oxide so that undercut control is difficult to maintain if etching nitride over oxide. Typically, the etch ratio is from 1:10 to 1:100, depending on the etch composition and nitride deposition conditions. However, by etching with a low HF concentration HF–H_2O solution at temperatures near the boiling point, comparable etch rates of less than 100 Å/minute have been reported (53, 54). By using an HF–ethylene glycol mixture, comparable rates for tube nitride, thermal oxide, and CVD oxide are observed over a considerably wider temperature and composition range (55). However, wet etch patterning of silicon nitride

[9]"Tube" nitride is deposited at temperatures of ~800°C or above in a quartz tube reactor and has a much lower etch rate than "plasma" nitride deposited at temperatures in the 400°C range in a plasma reactor.

is now seldom done, having been largely superseded by plasma etching.

6.2.13 Metal Etching

Since metal is usually deposited over an insulating layer such as silicon oxide or nitride, the etchant used must not noticeably attack oxide or nitride in the time required to remove the metal. On rare occasions, stripping a metal from bare silicon may be desired, but in these cases, special care and formulations can be used if necessary. When a single metal, such as the commonly used aluminum, is to be patterned, there is usually little difficulty. However, when multiple layers must be removed, considerable undercutting can occur. Much of this undercutting is caused by the enhanced etch rate of a metal overlayed by another appreciably lower in the electromotive series (56). For the case of gold over aluminum and gold over molybdenum, rate enhancements of from 10 to 100 have been observed. Table 6.5 shows the position of some of the metals in the electromotive series and illustrates the point that when gold is used as the top metal, trouble with all other metals is a strong possibility. However, by the proper choice of etchant, this effect can be minimized.

Since metal films are neither single crystal nor amorphous, the differently oriented grains may etch at different rates. In addition, the grain boundaries will generally etch more rapidly. The result is that wet-etched metal edges will often be more ragged than oxide edges. When the film has precipitates of some other phase in it, such as silicon in aluminum doped with silicon, the precipitates may not etch in the primary etch and thus will be left behind scattered on the surface.

Aluminum is the most commonly used IC metallization. Except for nitric acid, in which it is passivated, aluminum can be etched by strong acids or bases; for example,

$$2Al + 6NaOH \rightarrow 2Na_3AlO_3 + 3H_2 \qquad 6.12$$

$$2Al + 6HCl \rightarrow 2AlCl_3 + 3H_2$$

Were it not for a protective layer forming on the surface, it would also react directly with water to give hydrogen and aluminum hydroxide. Since the etchants are all water based, reactions such as those of Eq. 6.12 are not complete in that they do not provide for removal of the protective layer. A typical etch for use on silicon ICs is

20 acetic acid (parts by volume)
3 nitric acid
77 phosphoric acid

TABLE 6.5

The Electromotive Series

Metal	*Volts*
Al	+1.7
Cr	+0.56
Co	+0.28
Ni	+0.23
Ag	−0.8
Pd	−0.8
Pt	−0.86
Au	−1.5

This solution is not applicable to aluminum over GaAs since it will etch GaAs, but a 1 HCl, 2 H_2O mixture can be used instead. (For wet etchants suitable for other metals, see, for example, the extensive tables in reference 54.) As in oxide etching, the trend is away from wet etching and to plasma etching, particularly as more interconnect systems incorporate silicides, which are difficult to wet etch.

6.3 PLASMA ETCHING

Although there are many plasma processing modes, the two embodiments that dominate the semiconductor plasma etch field are the plasma etching mode and reactive ion etching (RIE). Fig. 6.16 shows the major components of a parallel-plate multiple-slice etching machine. Energy is supplied by an RF generator. The wafers lie flat on one of the electrodes, and a plasma is generated between the electrodes. Biasing of the electrodes by the DC voltage from the RF generator power supply is excluded by capacitive coupling the RF to the electrode. The chemical species in the plasma are determined by the source gas(es) used. The operating pressure depends on the plasma mode being used and may range from a few torr to fractions of a millitorr.

In the plasma etching mode, the electrode areas are symmetrical, and the DC voltage between the plasma and either electrode is the same and relatively small (57). The various ions and free radicals that are generated in the plasma diffuse to the electrode and wafer surfaces, where they can react with the material being etched to form volatile products that are pumped away.

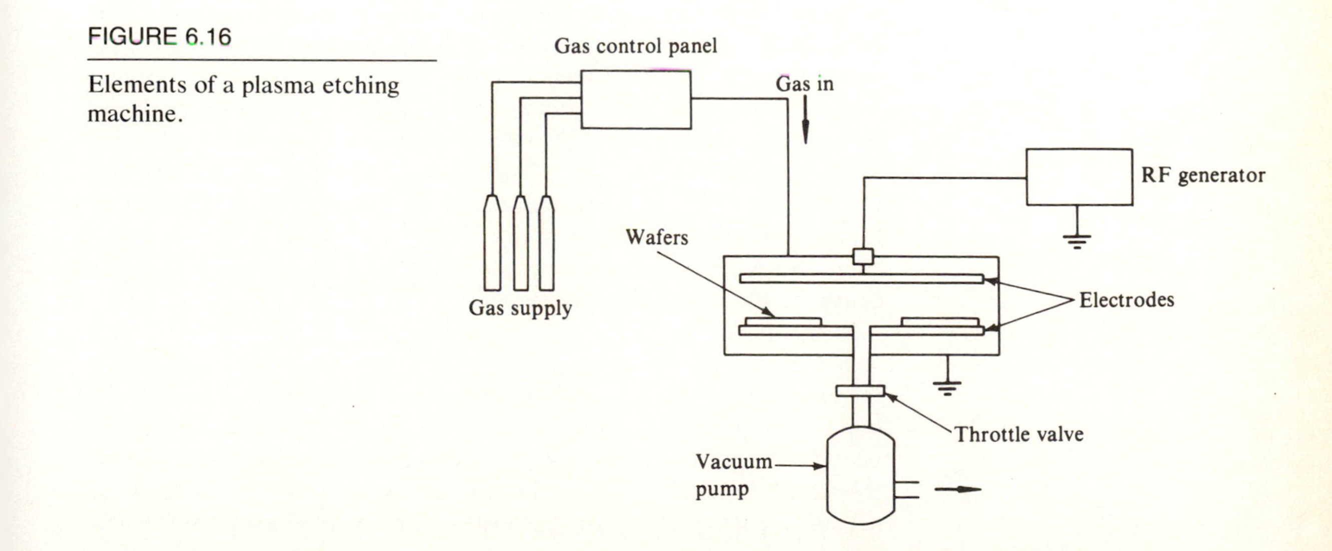

FIGURE 6.16

Elements of a plasma etching machine.

Reactive ion etching uses a negative self-bias DC voltage developed between the plasma and the wafer electrode to accelerate ions from the plasma to the wafers. An asymmetric DC bias between the plasma and the two electrodes is developed when the two electrodes have different areas, with the larger voltage being developed on the smaller electrode (57). Thus, when the electrode on which the wafers rest is deliberately made much smaller than the other electrode, plasma-to-wafer electrode voltages of several hundred volts are readily achievable. In the RIE mode, some etching occurs by free radicals, but more is by ions that are accelerated to the surface. Part of the ion etching is by the ions' chemically reacting at the surface, and part is by physical removal of material when it is struck by the incoming ion (sputtering).

A third kind of etching occasionally used is reactive ion beam etching. In it, ions are produced in a plasma as before. However, they are then extracted from the plasma and accelerated toward the wafer by voltages applied to suitable electrodes. In this manner, the ion density and the accelerating voltage can be set independently, and neutral radicals are essentially eliminated.

Pure plasma mode etching is substantially isotropic without specifically introducing additional reactants to produce sidewall passivation. Without such passivation, this mode is not particularly useful for fine-line etching. Reactive ion and reactive ion beam etching can be quite anisotropic, but their selectivity may not be as good as that of the plasma mode, particularly with regard to photoresist. In addition, because of the high-energy ions, radiation damage to the wafers may occur.

The accelerating voltage used in the RIE process is possible because, while the plasma is conductive and can have a relatively uniform DC potential throughout its volume, the dark space that separates it from the electrodes is quite nonconductive. The voltage difference that develops between the ground electrode and the plasma is greater than the lowest ionization potential of the gas in the chamber (>10–15 V). The voltage between the plasma and the RF-powered electrode can be much greater when the area of the grounded electrode is greater. This electrode is DC isolated from the RF generator by a capacitor and thus cannot transfer charge to the external circuitry. Hence, the negative electron charge that accumulates during half of the RF cycle must be exactly balanced by the positive ion charge during the next half-cycle. Since the electrons are more mobile than the ions, a negative potential will build up until the quantities match. The result is that the electrode becomes negative with respect to both the plasma and the ground electrode. Depending on conditions, the voltage differential may be several hundred volts. The general character of the voltage distribution across the plasma and dark spaces is shown in Fig. 6.17. As the chamber pressure increases, the RF electrode DC voltage decreases, as shown in Fig. 6.18. An ungrounded plate in the plasma will acquire a negative potential (floating potential) with respect to the plasma for the same reason that the hot electrode does. However, the voltage difference between the plasma and the floating

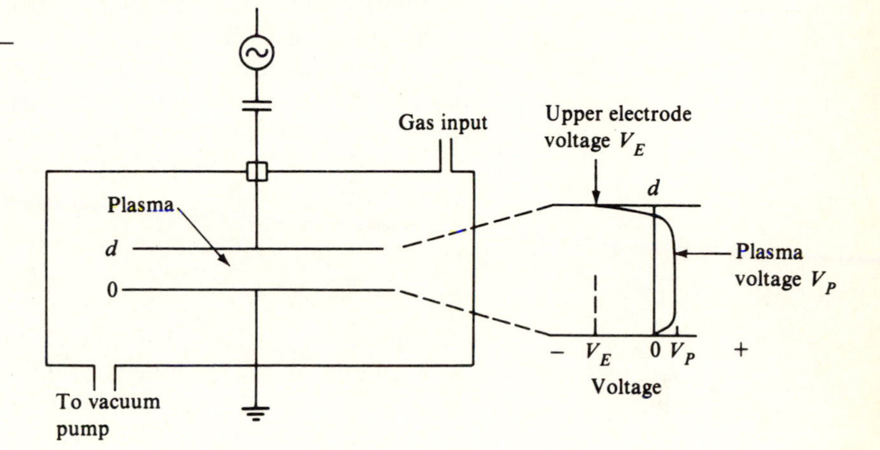

FIGURE 6.17

Voltage distribution across a plasma between parallel plates (electrodes) when plates are fed with an RF source. (The separation of the plates is d. The plate being fed by the RF signal is smaller and much more negative with respect to the plasma voltage than the grounded plate.)

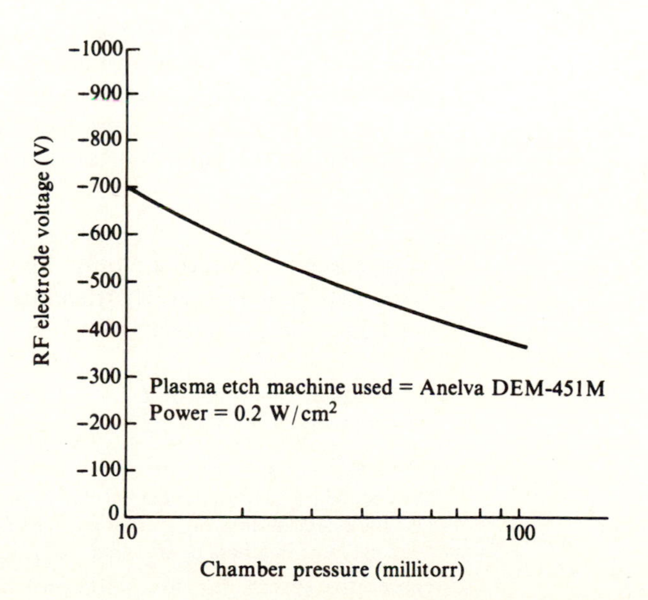

FIGURE 6.18

RF electrode voltage versus chamber pressure.
(*Source:* Data from Shunji Seki et al., *J. Electrochem. Soc. 130,* p. 2505, 1983.)

electrode will generally not be nearly as much as between the plasma and the hot electrode (58).

When wafers are placed on a floating plate, the etching is by free radicals and reactive ions and will be rather isotropic. If the wafers are attached to the RF electrode, any positive ions present will be accelerated from the plasma to the wafers and will more nearly travel in a path normal to the wafer surface, thus approximating the conditions of Fig. 6.2b. The number of ions deviating substantially from normal will depend on the amount of scattering, which, in turn, depends on the mean free path length relative to the width of the sheath (dark space).

Using classical kinetic gas theory and assuming that the molecules and ions in the reactor are in thermal equilibrium at some temperature T_1, the mean free path L of a specific molecule M_1 (or fragment) moving among a large number of M_2 molecules is approximated by[10]

$$L \cong \frac{1}{\xi^2 \psi \pi \sqrt{2}} \qquad 6.13$$

where ξ is the average diameter of the specific molecule M_1 and a molecule M_2 of the main body of molecules and ψ is the number of molecules per cc.[11] The equilibrium number of molecules ψ per cc in a vessel with pressure p and temperature T_1 (K) is given by

$$\psi = \left(\frac{p}{p_{\text{atm}}}\right)\left(\frac{273}{T_1}\right) \times 2.7 \times 10^{19} \text{ molecules/cc} \qquad 6.14$$

However, just because the mean free path is comparable to the dark space width does not mean that all of the ions leaving the plasma reach the wafer without first having a collision and path change. The fraction f of ions that have *not* had a collision and thus a change in direction after traveling a distance X is given by

$$f = e^{-X/L} \qquad 6.15$$

For a sheath of width L, only $1/e$ (37%) of the ions leaving the plasma will reach the wafer traveling in the same direction as when they left.

EXAMPLE ☐ Estimate the value of L for a Cl ion in the sheath for a system operating at 0.4 Pa with chlorine gas as the predominant constitu-

---

[10]See a standard text on the kinetic theory of gases for various corrections, such as for the effect of differing masses of M_1 and M_2 and for the case of only a few of M_1 among huge numbers of M_2 versus comparable quantities of each.

[11]For reactive species, the cross section may be substantially larger than the physical cross section.

ent. Assume that the effect of a DC voltage across the sheath is only to initially direct the ions toward the wafer. Further, assume that the gas temperature in this region is 200°C.

Atmospheric pressure is about 10^5 Pa, and the diameter of a chlorine ion is about 2 Å (59). Assume that the diameter of a chlorine molecule is twice that, or 4 Å. Thus ξ equals 3 Å. From Eq. 6.14, $N = 6.3 \times 10^{13}$ molecules/cc. Substituting these values into Eq. 6.13 gives $L = 4$ cm. □

In addition to the possible anisotropic etching, acceleration of ions also leads to an increased etching rate. This arises both because of normal sputtering and because the increased energy enhances the ion reactivity at the surface. The relative effect of nonreactive argon ions (atomic weight 40) and reactive CF_3 ions (atomic weight 69) is shown in Fig. 6.19.

6.3.1 Etching Reactions

In all etching except that of photoresist (stripping), a halogen species is used. Because of their inherent safety, various Freons[12] or other fluorocarbons, many of which are listed in Table 6.6, are often used as input gases, but $SiCl_4$, BCl_3, Cl_2, HCl, SF_6, and NF_3 are also

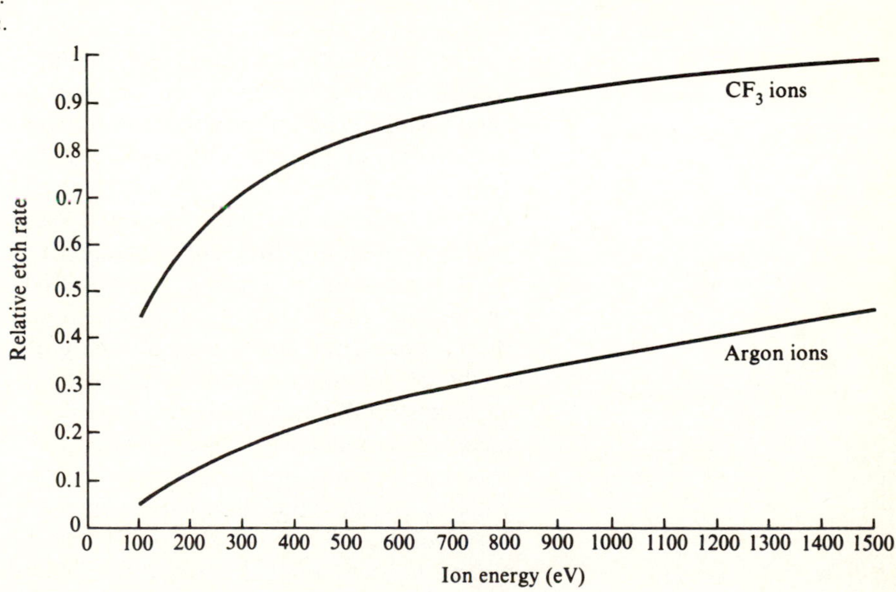

FIGURE 6.19

Relative etch rate of SiO_2 by plasma-generated argon and CF_3 ions versus ion energy. (*Source:* From data of J.M.E. Harper et al., *J. Electrochem. Soc. 128,* p. 1077, 1981.)

[12]Registered trademark of E.I. du Pont de Nemours & Co., Inc.

TABLE 6.6

Fluorocarbon Gases

Formula	*Common Name*	*Chemical Name*
CCl_3F	Freon* 11	Trichlorofluoromethane
CCl_2F_2	Freon 12	Dichlorodifluoromethane
$CClF_3$	Freon 13	Chlorotrifluoromethane
$CBrF_3$	Freon 13B1	Bromotrifluoromethane
CF_4	Freon 14	Tetrafluoromethane
C_2F_6	Freon 116	Hexafluoroethane (perfluoroethane)
C_3F_8	Freon 118	Octafluoropropane (perfluoropropane)
C_4F_8	Freon C318†	Octofluorocyclobutane (perfluorocyclobutane)
$C_2Cl_2F_4$	Freon 114	1,2–Dichlorotetrafluoroethane
C_2ClF_5	Freon 115	Chloropentafluoroethane
$CHCl_2F$	Freon 21	Dichlorofluoromethane
$CHClF_2$	Freon 22	Chlorodifluoromethane
CHF_3	Freon 23	Trifluormethane (Fluoroform)

*Freon is a registered trademark of E.I. du Pont de Nemours & Co., Inc.
†Do not confuse with C_4F_8, octofluoro–2–butene (perfluoro–2–butene), which is considered moderately toxic.

Note: Freons are considered relatively safe, but since chlorinated fluorocarbons containing no hydrogen, when allowed to escape into the air, appear to cause damage to the upper atmosphere, there is a chemical industry drive to replace them with alternative gases.

widely used. H_2, O_2, and inert gases such as Ar are occasionally added to the primary etchant gas. Silicon, silicon dioxide, silicon nitride, many silicides, and some metals can be etched by fluorine compounds, but aluminum must be etched with chlorine in order to have volatile reaction products. Photoresist is stripped by oxidizing it in an oxygen plasma. The specific reactions involved are not necessarily known, and indeed several may take place simultaneously. Further, just because both ions and free radicals may be present does not mean that they both participate in the etching reaction.

For satisfactory etching, several kinds of reactions may be desirable, and to that end there may be a multiplicity of input gases to the plasma. The plasma products may etch the material, but some of the by-products may deposit on the etching surface and either slow or stop the etching. In this case, oxygen, for example, may be included to react with the deposit. Some etchants may react at a reasonable rate with the main material to be etched but hardly at all with its surface oxide film. However, some other plasma gas may be very good at removing the oxide but may otherwise be a slow etcher. By combining the two, a good compromise may be obtained. Occasionally, when reactive ion etching of some component is difficult, physical sputtering may need to be emphasized, in which case, an inert gas that can be ionized may be included in the plasma.

Table 6.7 lists many of the feed gases that have been proposed for plasma etching specific semiconductor materials. Silicon dioxide apparently etches by reactions such as (60)

$$3SiO_2 + 4CF_3^+ \rightarrow 2CO + 2CO_2 + 3SiF_4 \qquad 6.16$$

$$SiO_2 + 2CHF_2^+ \rightarrow 2CO + H_2 + SiF_4$$

Silicon etching in fluorine plasmas appears to be due to the fluorine free-radical interaction (61)

$$Si + 4F^* \rightarrow SiF_4 \qquad 6.17$$

The silicon nitride etching reaction also involves the fluorine radical in the reaction (62):

$$Si_3N_4 + 12F^* \rightarrow 3SiF_4 + 2N_2 \qquad 6.18$$

Many metals such as Ti, W, Ta, and Mo form volatile fluorides and can be etched rapidly with fluorine-containing plasmas. Other metals such as Al do not form volatile fluorides (63). However, in this case, chlorine-based plasmas can be used. In the case of aluminum, $AlCl_3$ is formed and, if the system is kept warm, will volatilize and be pumped out. Otherwise, it collects on the wafers and on equipment walls. If the $AlCl_3$ is exposed to moisture, it will react to form HCl and Al_2O_3. The HCl can cause corrosion problems, and the Al_2O_3 can cause particulate problems. Other chlorine compounds left on the aluminum surface, and even in the overlaying resist, will also cause corrosion if not removed. Thus, a final fluorine-based plasma is often used to replace the chlorine compounds remaining behind with inert fluorine compounds. In addition, the resist can then be stripped in situ in an oxygen plasma to eliminate any chlorine compounds that may not have been removed.

TABLE 6.7

Plasma Etchants

Material/Etchant	*Comments*	*Reference*
Silicon Dioxide		
$CF_4 + O_2$ (10%)	Addition of O_2 increases etch rate.	(1)
CHF_3	Better selectivity over silicon.	(4)
C_2F_6		(5)
C_3F_8	Increased rate over CF_4.	(4)
Silicon Nitride		
$CF_4 + O_2$ (4%)		(6)
CHF_3		(5)
C_2F_6		(5)

(*continues*)

TABLE 6.7 *(continued)*

Material/Etchant	*Comments*	*Reference*
SF_6 + He		(7)
Polysilicon		
$CClF_3$ + Cl_2	Anisotropy and oxide selectivity changes with gas ratio.	(8)
$CHCL_3$ + Cl_2	Use mixture to get uniform initiation; change to Cl_2 for oxide selectivity.	(9)
SF_6	Good oxide selectivity.	(10)
NF_3	Isotropic; high (1 μm/minute) etch rate.	(11)
CCl_4	Somewhat anisotropic.	(11)
CF_4 + H_2	Poor oxide selectivity; hydrogen gives anisotropy.	(12)
C_2ClF_5	Poor oxide selectivity.	(13)
Single-Crystal Si		
Same etchants as polysilicon		
Single-Crystal GaAs		
CCl_2F_2	Gives rough surface.	(22)
CCl_2F_2 + O_2	O_2 minimizes surface film formation.	(23)
Cl_2–$SiCl_4$	Rough surfaces above 10% Cl_2.	(22)
Aluminum		
BCl_3	Low etch rate; rapid break through Al_2O_3.	(27)
CCl_4	Anisotropic; erratic behavior in presence of moisture.	(15)
HBr + Ar HCl + Ar Cl_2 + Ar Br_2 + Ar	Argon sputtering breaks through aluminum oxide.	(16)
BCl_3 + Cl_2	High etch rate of Cl_2 plus rapid etch initiation.	(14)
BBr_3 + Cl_2	As above plus better Al-to-resist selectivity.	(14)
$SiCl_4$		(17)
BCl_3 + Cl_2 + $CHCl_3$ + N_2	Anisotropic; good etch rate.	(21)
Molybdenum		
CF_4		(2, 24)
Tungsten		
Cl_2, Cl_2 + BCl_3	Addition of small amounts of BCl_3 increases etch rate.	(18)
$CBrF_3$ + O_2 + He		(19)
Titanium		
$C_2Cl_2F_4$		(2)
CF_4, $CClF_3$		(25)
$CBrF_3$ + He + O_2	Good selectivity over Si nitride.	(25)
Ti:W		
Cl_2 + O_2		(24)
Ti Silicide		
CF_4 + O_2		
Ta Silicide		
SF_6 + C_2ClF_5	50/50 flow rate gives straight walls and good oxide selectivity.	(3)

TABLE 6.7 *(continued)*

Material/Etchant	*Comments*	*Reference*
$SF_6 + CCl_2F_2$		(26)
W Silicide		
$SF_6 + C_2ClF_5$	50/50 flow rate gives straight walls and good oxide selectivity.	(3)
Cl_2, $Cl_2 + BCl_3$	Addition of small amounts of BCl_3 increases etch rate.	(18)
CCl_2F_2		(20)
Gold		
$C_2Cl_2F_4$	Low oxide etch rate.	(2)
$C_2Cl_2F_4 + O_2$	Addition of oxygen increases gold etch rate.	(2)

Note: This table should not be considered as complete, but only as a starting point in considering possible etchants. The literature on plasma etches is quite voluminous. The references given in each of these references should also be checked.

References: (1) Alan R. Reinberg, *Etching for Pattern Definition,* p. 91, Electrochemical Society, Princeton, N.J., 1976.

(2) W.H. Legat and H. Schilling, *Electrochem. Soc. Ext. Abst. 75–2,* abst. no. 130, p. 336, 1975.

(3) Stephen E. Clark et al., *Solid State Technology,* p. 235, April 1984.

(4) Mitch Hamamoto, *Electrochem. Soc. Ext. Abst. 75–2,* abst. no. 129, p. 335, 1975.

(5) Aaron D. Weiss, *Semiconductor International,* p. 56, February 1983.

(6) D.L. Flamm et al., *Electrochem. Soc. Ext. Abst. 81–2,* abst. no. 270, 1981.

(7) Richard L. Bersin, *Solid State Technology,* p. 117, April 1978; International Plasma Corporation Specification PTE–4A, 1975.

(8) E. Laes et al., *Electrochem. Soc. Ext. Abst. 83–1,* p. 235, 1983.

(9) F. Faili and F. Wong, in G.S. Mathad et al., eds., *Plasma Processing,* pp. 458–469, Proc. Electrochem. Soc. 87–6, 1987.

(10) K.M. Eisele, *J. Electrochem. Soc. 128,* p. 123, 1981.

(11) Douglas H. Bower, *J. Electrochem. Soc. 129,* p. 795, 1982.

(12) J.W. Coburn and J.F. Winters, *J. Vac. Sci. Technol. 16,* p. 391, 1979.

(13) J. Hayes and T. Pandhumsoporn, *Solid State Technology,* p. 71, November 1980.

(14) H.B. Bell and R.W. Light in G.S. Mathad et al., eds., *Plasma Processing,* Proc. Electrochem. Soc. 87–6, 1987.

(15) K. Tokunaga and D.W. Hess, *J. Electrochem. Soc. 127,* p. 928, 1980.

(16) P.M. Schaible et al., *J. Vac. Sci. Technol. 15,* p. 334, 1978.

(17) E.O. Degenkolb, *J. Electrochem. Soc. 129,* p. 1150, 1982.

(18) D.S. Fischl and D.W. Hess, *J. Electrochem. Soc. 134,* p. 2265, 1987.

(19) M.E. Burba et al., *J. Electrochem. Soc. 133,* p. 2113, 1986.

(20) L. Ephrath and R. Bennett, in *VLSI Science and Technology,* Electrochemical Society, Pennington, N.J., 1982.

(21) Ching-Hwa Chen, Lam Research Corporation literature.

(22) C.B. Cooper et al., in G.S. Mathad et al., eds., *Plasma Processing,* Proc. Electrochem. Soc. 87–6, 1987.

(23) A.P. Webb, *Semiconductor International,* p. 154, May 1985.

(24) William Y. Hatta, in G.S. Mathad et al., eds., *Plasma Processing,* Proc. Electrochem. Soc. 87–6, 1987.

(25) C.J. Mogab and T.A. Shankoff, *J. Electrochem. Soc. 124,* p. 1766, 1977.

(26) Jeff Herrmann et al., in G.S. Mathad et al., eds., *Plasma Processing,* pp. 494–506, Proc. Electrochem. Soc. 87–6, 1987.

(27) S.M. Cabral et al., "Characterization of a BCl_3 Parallel Plate System for Aluminum Etching," *Proc. Kodak Microelectronics Seminar,* pp. 57–60, Dallas, 1981.

6.3.2 Plasma Etch Rates

The machine variables available when a typical plasma/RIE etcher is used are operating pressure, wafer temperature, electrode voltage, inlet gas composition, gas flow rate, and loading (area of etchable surface divided by area of electrode). Additionally, the electrode spacing, diameters, and composition can generally be changed before a run. By design, the operating frequency may range from a few hundred kHz to many MHz. The etch rate is so intertwined with these many variables that it is very difficult to make a priori rate predictions. These circumstances are to be contrasted with those of wet etching, where the variables are simply etch composition, temperature, and amount of agitation. Despite the complexity of plasma etching, some general observations can be made.

The etch rate will generally depend on the flux of active species (free radicals and chemically reactive ions), although in some cases a high flux will generate enough by-product to completely stop the reaction. Since the active species are primarily produced by collision of electrons with the neutral gas molecules, any method to increase the number of collisions should increase the flux and thus the etch rate. The generation rate r_i for a species i is given by (64)

$$r_i = K_i n_e N_o \qquad 6.19$$

where K_i is the rate constant for the ith species, n_e is the number of electrons, and N_o is the number of neutral gas molecules (proportional to the operating pressure P). Thus, increasing the pressure or the number of electrons will, in principle, increase r_i and the etch rate. The number of electrons can be increased by a power increase or by adding extra ones from, for example, an auxiliary hot wire filament (an equipment redesign). By applying a properly directed magnetic field to the plasma, the electrons can be made to follow a spiral path and thus travel further and produce more ions before reaching a wall (again an equipment redesign). As pressure initially increases, there is also an increase in the number of ionizing collisions of each electron before it finally leaves the plasma since the mean free path is decreased. As the pressure continues to increase, at some point the mean free path will become so short that the electrons will be unable to acquire enough energy between collisions to ionize the gas, and the etch rate will drop. For a given electrode spacing and power setting, as the pressure is increased from a very low value, there will be no plasma and no etching until the pressure reaches some critical value where the plasma is initiated. Etching will then begin and increase in rate rather rapidly. When substantial etching is by reactive ions, the rate will then decrease apparently because the ion energy is reduced due to a decreasing sheath potential (wafer electrode to plasma voltage) as the pressure increases (65).

The etch rate generally increases with increasing power, as is expected from Eq. 6.19. The rate of etching in the RIE mode will also increase because of increased voltage across the sheath, as was illustrated in Fig. 6.19. The sheath voltage drop increases as the gap decreases, with increasing RF voltage, and as the ratio of the area of the ground electrode to hot electrode increases.

Initially, the etch rate will increase as the gas flow rate is increased since the reactant will not be as readily depleted. The rate will then remain essentially independent of flow until it is high enough for the residency time to be less than the lifetime of the various active species, at which point it will again decrease. The residency time t_r can be calculated from the gas volume divided by the gas flow in the volume. Gas flow f_A is generally measured outside the plasma chamber under standard conditions of temperature and pressure. Inside a chamber with a reduced pressure p_C, the flow f_C will be given by $f_C = 760f_A/p_C$, where p_C is in torr. Thus,

$$t_r = \frac{60Vp_C}{760f_A} \qquad 6.20$$

where V is the volume, f_A is in cm^3/minute, and t_r is in seconds.

Adding a gas diluent to the main processing gas may do no more than slow down the etch rate by reducing the concentration of reactant. However, if etching is by the RIE mode, the additional sputtering by the diluent may increase the removal rate either directly or by the removal of inert reaction by-products. An example of this is the removal of the polymers that sometimes form when halocarbons containing hydrogen are used as the plasma etching gas. The additional gas may also react with the species formed in the plasma, change their relative concentrations, and thus change the etch rate (66). For example, in a CF_4 plasma, both F and CF_2 are formed, but the CF_2 then recombines with F to form CF_3. Added oxygen will also combine with CF_2 to give $CO_2 + F_2$. Thus, when oxygen is present, some of the CF_2 will combine with it instead of with F, thereby raising the F concentration and increasing the rate of any etching depending on the F concentration. Hydrogen added will combine with F to form HF and with CF_3 to form CF_2 + HF so that F is suppressed and the concentration of CF_2 increased.

Even though plasma processes are not in thermal equilibrium and the energy of the etching species is far higher than the thermal energy of the wafer, increasing the temperature of the wafers often slightly increases the etch rate $r(T)$, with the dependency still generally being of the form

$$r(T) = r_o e^{-E/RT} \qquad 6.21$$

where r_o is a constant, E is the activation energy in eV, k is the Boltzmann constant (given in Appendix C), and T is the absolute temperature.

6.3.3 Selectivity

The selectivity ratio is used in this context as the ratio of the etch rate of the layer being deliberately removed by etching to that of a layer exposed to the etch but not to be intentionally removed. Unlike wet etching, where selectivity is generally very good, plasma etch processes must generally be content with ratios of 25 or less. This ratio is important in determining how far into the next layer etching proceeds and how much the windows in masking material are enlarged during pattern etching.

Based on the expected film etch rate and film thickness uniformity, an estimate can be made of the selectivity required in order not to exceed the allowable penetration depth D_s into the next layer. Several approaches and approximations can be made, but a simple procedure is to assume that over a wafer (single-slice etching) or over all wafers of a batch process, the etch rate can be approximated by $E_f = E_o \pm e$ where E_o is the nominal etch rate, and the film thickness can be approximated by $D_f = D_o \pm d$ where D_o is the nominal thickness. The nominal etch time t_o is then D_o/E_o. However, any thin region in the area of high etch rate will be removed in a shorter time

$$t_1 = \frac{D_o - d}{E_o + e} \qquad 6.22$$

Thick-film regions in the slow-etching areas will not be removed until a later time

$$t_2 = \frac{D_o + d}{E_o - e} \qquad 6.23$$

Thus, some portions of the underlying film will have been exposed to etch for a time $\Delta t' = t_2 - t_1$ by the time all of the layer is removed. In some circumstances, the etch time may be extended an amount $\Delta t''$ past $\Delta t'$ in order to provide additional processing margin. The total over-etch time Δt is then given by $\Delta t' + \Delta t''$. To meet the requirement that no more than D_s of the second film be removed, its etch rate E_s must be less than $D_s/\Delta t$, and the necessary selectivity can be determined from these expressions.

Mask material removal during pattern etching can cause the window openings to increase in size. However, if the geometries are large and the enlargement is reproducible, it is usually possible to compensate by down-sizing the original mask openings. With geometries having feature sizes not much different from the resist thick-

FIGURE 6.20

Effect of mask removal during etching on window widening.

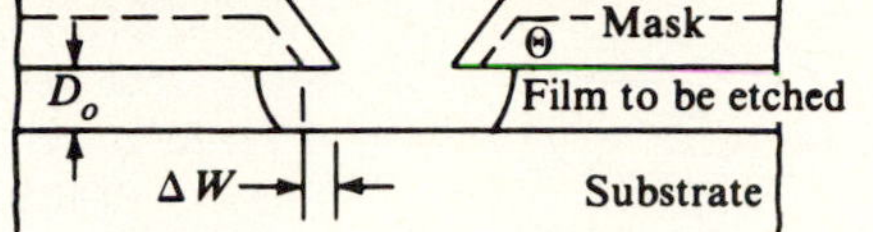
Removal rate normal to top surface
and to inclined surfaces is equal
Mask
Θ
D_o
Film to be etched
Δ W
Substrate
Vertical removal rate is the same
on both top and inclined surfaces
Θ
Δ W

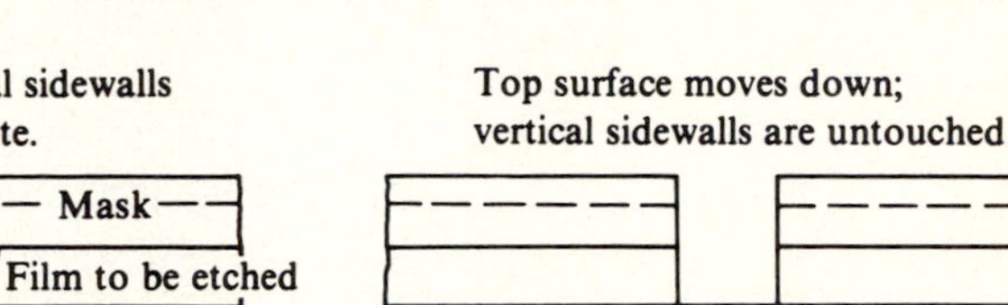
Top surface and vertical sidewalls
are removed at same rate.
Mask
Film to be etched
Δ W
Substrate
Top surface moves down;
vertical sidewalls are untouched
Δ W = 0

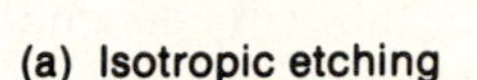
(a) Isotropic etching

(b) Anisotropic etching

ness, this approach is not feasible, and anisotropic etching and/or very high resist etching selectivity is required. Even with anisotropic etching, substantial window enlargement can still occur if the resist walls are sloped and selectivity is low. These effects are shown in Fig. 6.20. For isotropic removal of resist that has a sidewall angle θ with the horizontal, the opening will increase in width by $2\Delta W = 2\Delta R/\sin\theta$. ΔR is the thickness of resist removed from the top and is given by $E_r t$ where E_r is the removal rate of resist. In the limit of $\theta \rightarrow 90°$, $2\Delta W = 2\Delta R$. If removal is anisotropic, $2\Delta W = 2\Delta R \cot\theta$. When the walls are perpendicular, $\cot\theta = 0$, and there is no undercutting. For the maximum ΔW allowable in the process, ΔR can be calculated for the particular etching regime used, and then the minimum resist/film etching selectivity can be determined from $(D_o/\Delta R)(1 + d/D_o)(1 + e/E_o)$.

The straightforward method of improving selectivity is by the choice of gases that do not readily react with one of the layers. For example, the selectivity of SiO_2 relative to aluminum is better when a fluorine-based etchant is used instead of one that is chlorine based since aluminum fluoride is quite nonvolatile and tends to form a protective layer over the aluminum. The mixing of gases can sometimes provide selective mechanisms. For example, the addition of either molecular hydrogen or a hydrogen-bearing compound to fluorine-based etch gases is often used to improve SiO_2 to silicon selectivity. This improved selectivity occurs because hydrogen suppresses the

TABLE 6.8

Plasma Etching Selectivities

Material	*Etch*	*Selectivity with Respect to:*
Thermal SiO_2	C_2F_6 + CHF_3	Si—15:1, resist—5:1
Doped CVD SiO_2	C_2F_6 + CHF_3	Si—30:1, resist—10:1
Polysilicon	Cl_2	SiO_2—25:1, resist—5:1
Aluminum	BCl_3 + Cl_2	SiO_2—25:1, resist—5:1, polysilicon—3:1
Si_3N_4	CF_4 + O_2	CVD SiO_2—1:1, resist—3:1, polysilicon—1:8
Resist	O_2	SiO_2—>1000:1, Si—>1000:1

fluorine atoms, and since Si is etched primarily by fluorine and SiO_2 by the CF_3^+ ion, a reduction in the fluorine atom improves selectivity (67). Excessive hydrogen may, however, cause problems such as unwanted polymer buildup or structural damage. Table 6.8 lists selectivities reported for various etchant/material combinations. The actual values can vary substantially depending on the etch mode being used and on the power input and exact gas composition, but the table can serve as a guide.

6.3.4 End-Point Detection

Because of the poor selectivity associated with plasma etching, end-point detection is usually used to minimize over-etch times. In principle, several approaches are possible. The thickness of the film being etched can be measured optically by interferometer, ellipsometer (68, 69), or light beam displacement (70). If the film is opaque, the difference in reflectivity between the film being etched and the one beneath it can be used to detect the disappearance of the film being etched (69). The presence of a reaction product or the absence of a reacting species can also be monitored and used to determine when the film etching is complete. Detection in these cases can be either by optical (71–75) or by mass (76) spectroscopy.

Thickness measurements require an appropriately sized area to be located in a fixed position in the etch reactor, and mass spectroscopy calls for a rapid sampling of the exit gas stream. Optical spectroscopy can be done independently of the position of the wafer(s) and does not disturb the gas flow. For these reasons, optical spectroscopy is currently used almost exclusively. Emission (rather than absorption) spectroscopy is used to follow either the increased reaction product concentration or the depression of a reactant concentration. For example, since silicon dioxide etching and photoresist removal both produce substantial amounts of carbon monoxide, one of the CO emission bands is used. During silicon etching with a fluorine containing gas, a depression in the concentration of fluorine occurs, and its level is commonly used for end-point detection. There are many emission lines in each spectrum, and the choice of the one to use for control depends on their relative strength, whether or not there are interfering lines from other plasma constituents, and the ease of detecting a particular wavelength. As an example of interference, etching aluminum produces AlCl, which is widely used for control. When the etching gas is one containing no molecular chlorine, such as CCl_4 or $SiCl_4$, an intense band at 261 nm can be used for control. If Cl_2 is added to the input gases (as it sometimes is), a chlorine line at 257 nm produces substantial interference, but another AlCl line such as the one at 522 nm can then be used (74).

6.3.5 Effect of Loading on Etch Rate

It is often observed that as more wafers are put into a plasma reactor or as masking is changed to allow more area on a wafer to be etched, the etch rate declines. Thus, etch rates generally must be experimentally determined for a specific number of wafers and/or photoresist coverage. The effect can also be very insidious in causing the rate to increase when an etching surface is clearing and reducing etching area. The loading effect occurs when there is a substantial depletion of the active species from the plasma during etching. By considering the active species' generation rate, the loss rate due to etching, recombination losses at positions other than on the etching surface, and losses due to the species being swept away by gas flow, an etch rate r versus a normalized etching area α can be calculated. α is given by the ratio of the area of film exposed for etching to the total area available on the reactor platten. For only one active etching species and with various simplifying assumptions, an expression of the form

$$r = \frac{K_1}{1 + K_2\alpha} \tag{6.24}$$

results (77). K_1 and K_2 are constants for the particular system configuration, operating conditions, and etchant used. For multiple-slice reactors, Eq. 6.24 can be written as

$$r = \frac{K_1}{1 + K_2'N} \tag{6.25}$$

where N is the number of slices being etched and K_2' includes a factor to normalize N to the number of slices at full load. Eq. 6.25 can be rewritten as

$$\frac{1}{r} = K_o + KN \tag{6.26}$$

from which it can be seen that $1/r$ versus N will give a straight line if Eq. 6.25 is obeyed. If there are two or more active species, they will almost certainly be depleted at different rates, and a linear relation between $1/r$ and N will no longer exist (78). Further, under these circumstances, the etching character as well as the rate could change with increasing numbers of slices. For r to be relatively insensitive to N or pattern geometry, $K_2\alpha$ or $K_2'N$ must be much less than 1. Since K_2 and K_2' are functions of the etchant gas and reactor geometry, for a given gas and reactor design, little can be done to minimize a loading effect other than by reducing α. In single-slice reactors, this option is not available, and in multiple-slice reactors, severely reducing the number of slices generally makes the operation uneconomical.

6.3.6 Effect of Electrode Material

The electrodes as well as the wafers are subject to sputtering, and when this happens, electrode material is introduced into the plasma. If Si or SiO_2 is used as an electrode (or electrode covering) in fluorine plasmas, the products are volatile and like those from pattern etching polysilicon and SiO_2. However, if an aluminum electrode is used, it may be sputtered and redeposited. It is also possible for the plasma products to react with the electrode material and give deleterious compounds. It is, for example, thought that stainless steel electrodes promote polymerization of CCl_4 and reduce its effectiveness in etching aluminum (79).

6.3.7 Plasma-Induced Damage*

Three kinds of damage have been observed: structural, foreign ion implantation, and rough surfaces. In addition, undesirable residues are sometimes left on the surface. The latter can, in principle, be eliminated by the proper choice of gases, but these same residues may contribute to an anisotropy that is desirable or to a selectivity that is necessary. Rough surfaces are primarily a problem when etching only partially through a material—for example, when etching trenches in silicon or gate depressions in GaAs. However, rough walls can also be a problem after etching, for example, vias completely through a wafer. Roughness can generally be eliminated or reduced to a tolerable level by the choice of etchant or, in some cases, by a change of electrode materials (66).

When RIE is used, the ion impact energy seems to be enough to always produce structural damage. In the case of silicon, either a furnace anneal, a rapid thermal anneal, or a removable thin sacrificial oxide removes or prevents most damage. Much of the concern has arisen over structural damage induced in the contact area and

---

*See references 80–86.

the thin residue of fluorocarbon left over it during the etching of contact windows in thermal oxide over silicon. Structural damage from hydrogen either deliberately introduced with CF_4 to promote selectivity or from hydrogen in the etchant gas, such as in CHF_3, is particularly severe. Amorphous layers are sometimes observed, and deep-level electron traps are found. A shift in the barrier height of Schottky diodes made on such etched surfaces occurs, and there is a reduction in MOS lifetime. A common way to remove both the surface film and the underlying damage is to first subject the etched area to an oxygen plasma to oxidize the film and a thin layer of silicon. The oxidized residue is then stripped in dilute HF.

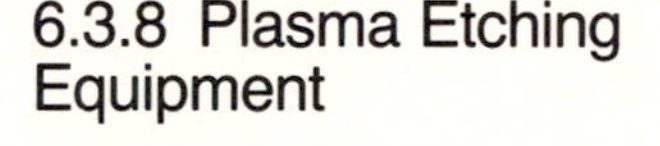

6.3.8 Plasma Etching Equipment

The normal operating frequency for plasma etching equipment is 13.56 MHz (a frequency assigned for industrial use), although lower (100–300 kHz) and higher (microwave, such as 2.45 GHz) frequencies are used in some equipment. The gases used are primarily those listed in Table 6.7. Valves and flow meters required are the standard kind used in the semiconductor industry. The vacuum required varies with the mode of operation and is generally in one of the ranges shown in Fig. 6.21.

The plasma generation chamber and the associated wafer chamber designs vary widely from application to application and between manufacturers. Schematics of some of these designs are shown in Fig. 6.22. The barrel (tubular) reactor shown in Fig. 6.22a was the earliest geometry and was primarily used for stripping resist. To minimize wafer damage, the plasma was confined to the perimeter by a perforated metal shield, and the reactive species diffused to the wafers. The etch rate was much higher near the edge than it was in the center, and the machine was thus not suitable for pattern etching. To minimize the nonuniformity problem, the parallel-plate reactor (Fig. 6.22b) was introduced. Many new designs use single-wafer chambers (Fig. 6.22c). In this case, one wafer will cover the whole bottom electrode, and the gases must escape around the outer edge of one of the electrodes rather than through the center of the wafer platen. The geometry in Fig. 6.22d, shown with the chamber (ground electrode) raised, is an alternative to the parallel plate. This arrangement allows the ground electrode to have a much greater

FIGURE 6.21

Typical pressure ranges for ion milling, reactive ion etching, reactive ion beam etching, and free-radical plasma etching.

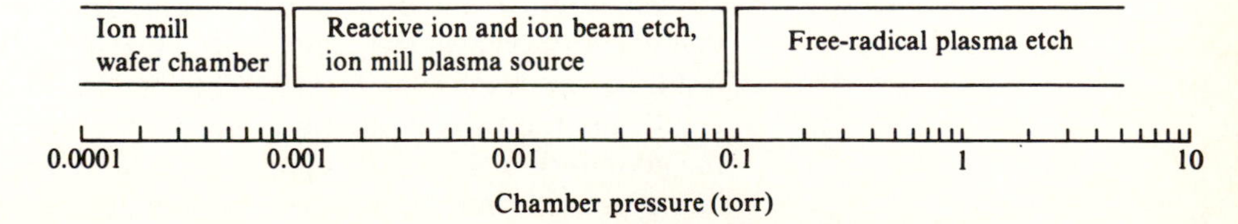

FIGURE 6.22

Various configurations of a reaction chamber.

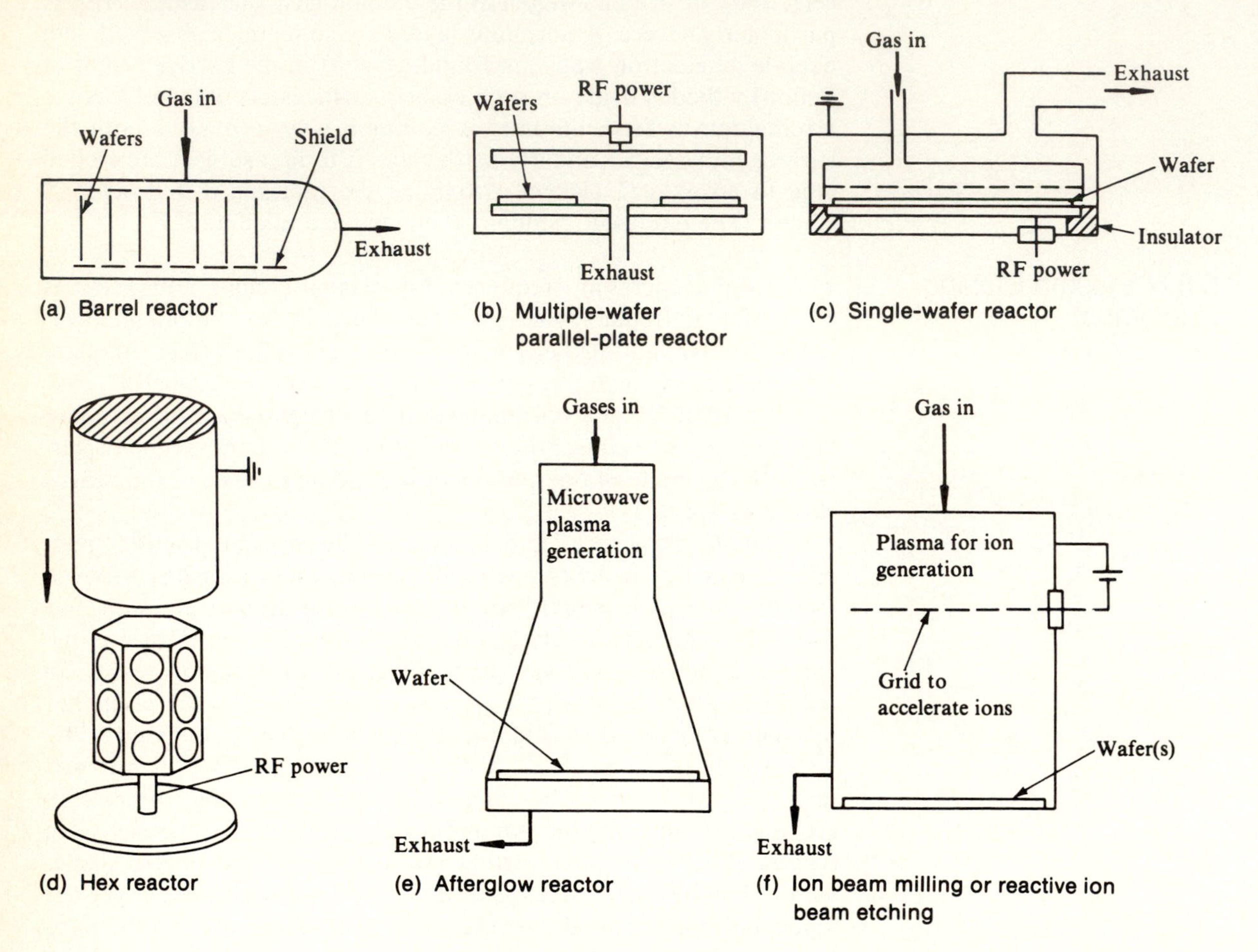

area than the power electrode and thus increase the sheath voltage available for ion acceleration.

For cases where it is desirable to ensure that there are no high-energy ions, the plasma can be generated in a microwave cavity, generally smaller in cross section than a wafer, and then the active species is allowed to diffuse to the slice. This principle is not unlike that of the original barrel reactors, but by using a geometry as shown in Fig. 6.22e, a more uniform etching results.

If, instead, it is desirable to accent the high-energy ion aspect, the geometry of Fig. 6.22f can be used (87). The ions are produced in a plasma as before, but there is now an accelerating grid so that unlike RIE, the energy of the ions can be adjusted independently of the power used to produce them. If the input gas is chosen to give reactive ions, the process is reactive ion beam etching (RIBE). However, if an inert gas such as argon is used, then any substrate (wafer) etching will be strictly by physical erosion, and the process is ion milling.

6.4 VAPOR PHASE ETCHING

At elevated temperatures, some gases will react directly with various semiconductor materials and, for some specialized applications, offer an alternative to either wet or plasma etching. The most common application of vapor phase etching is in the in situ cleanup of silicon surfaces before epitaxy. (This aspect is discussed in Chapter 7.) Some gases that will etch silicon directly at high temperature are Cl_2, Br_2, HCl, $SiCl_4$, and SF_6. Anhydrous HF + water vapor will react with SiO_2 even at room temperature, allowing its removal at rates of tens of Å/s (88).

CHAPTER 6 SAFETY

Wet etching involves the use of numerous acids or bases that can produce severe burns when they come into direct contact with the skin. In addition, fumes can produce skin irritation and/or eye damage. The etching process should be done in a well-vented hood. When acids or other corrosive liquids are handled, an apron, rubber gloves, and safety glasses should be worn. The gloves should not, however, be considered as protection that allows the hands to be immersed in the liquid since small difficult-to-detect pinholes can allow enough leakage to produce severe burns. The work area should be well lighted and free of obstructions that might cause etchant spills. Care must be taken in the mixing, handling, and disposal of etching solutions. In diluting acids or bases, always add the chemical to the water in order to prevent possible excessive heat buildup and explosion. One of the more insidious acids is HF, which is a rather weak acid but which will nonetheless penetrate the skin and produce deep and very painful burns that may not be apparent for several hours. Other acids produce an immediate burning sensation. In addition, concentrated HNO_3 produces a brown discoloration of the skin, while concentrated H_2SO_4 gives a blackening. If an accident involving acid occurs, immediately flush the affected area with copious amounts of water. When the burns appear severe or if HF is involved, medical attention is advisable.

Acids and organic solvents should be stored in separate cabinets and disposed of in separate containers; otherwise, they may react to cause a fire or explosion. Cyanide plating solutions or etchants are reasonably safe when they are kept alkaline, but if mixed with an acid, the highly poisonous hydrogen cyanide gas may be liberated.

The plastic hoods and exhaust ducts commonly used in wet etching operations are quite flammable. Should they catch fire from an immersion heater, for example, the smoke and fumes can be a health hazard (89).

Plasma reactors use high-voltage RF and have high DC voltages associated with the RF generator. Hence, servicing should be done only by qualified personnel, and no electrical safety interlocks should be deactivated. Some of the plasma etchants are Freon based and thus quite nontoxic. However, metal etchants often use chlorine or some other toxic gas, albeit in small quantities. Even when using nontoxic gases, some of the reaction products are toxic and are often either condensed on the chamber walls or trapped and concentrated in the plasma reactor vacuum pump system. Thus, care should be taken in cleaning the reactor, servicing pumps, and disposing of spent oil (90). Some gases, such as NF_3, become reactive at elevated temperatures and should be distributed through lines that have been carefully cleaned of organics and are free of gages and valves with combustible packing glands. The quest for new etching gases continues, but there is no assurance that they will be benign or will have been studied enough for their toxicity to have been properly evaluated when they are introduced.

CHAPTER 6 KEY IDEAS

- ☐ Except when special etches are used for single-crystal material (such as Si or GaAs), wet etching is isotropic.
- ☐ The fact that etching is isotropic causes mask undercutting during etching of individual layers. Since it is difficult to match etch rates for different materials, extra undercutting of some layers will also occur when multilayer structures are being etched.
- ☐ By the proper choice of plasma etching mode (such as RIE) and etchant, plasma etching can be done with minimal undercutting.
- ☐ Wet etching is less expensive, but plasma etching is more appropriate for fine geometries.
- ☐ Because of the wide choice of wet etches, they can usually be more material-selective than plasma etches.
- ☐ Plasma anisotropic etching occurs because of an etch-resistant polymer buildup on vertical surfaces.
- ☐ Vapor phase etching is a high-temperature process and is seldom used except just prior to epitaxial deposition.

CHAPTER 6 PROBLEMS

1. How much 49% HF should be mixed with 500 ml of water to give a 1 normal solution? (The specific gravity of the 49% HF is approximately 1.16.)
2. After silicon slices are etched with CP4A, the slices usually have a pincushion cross section. Based on this evidence, would you judge that the etch region is diffusion or kinetically controlled? Why?
3. If etching is to go to completion, how wide should the mask opening be if a groove 10 μm deep is to be anisotropically etched in a (100) silicon wafer? How should the mask opening be oriented with respect to the wafer flat? (See Chapter 2 for flat orientation according to SEMI standards.)
4. How could the Ga face of a (111) oriented GaAs slice be determined?
5. Etch solutions are often filtered during the etch operation. When SiO_2 is etched with buffered HF, what materials might be removed by the filter?
6. How many molecules of O_2 are in a room-temperature 10 liter chamber at a pressure of 1 millitorr? How many molecules of CCl_2F_2 would there be under the same conditions?
7. Calculate the approximate residency of a gaseous species in a 1 liter chamber at a pressure of 4 millitorr if the flow of gas into the chamber is 100 cm^3/minute when measured at atmoshperic pressure.
8. If 1 μm of CVD SiO_2 over polysilicon is to be etched and no more than 100 Å of silicon are to be removed, what etch selectively will be required? Assume that the oxide film thickness varies by ±10% and that the etching uniformity is also ±10%. In addition, consider that an additional over-etch of 10% is required because of the variability in end-point detection.
9. For a given plasma etch reactor configuration, sketch how the etch rate versus exposed area to be etched will probably vary. Indicate the region over which the concentration of an etching species could be used for end-point detection. Explain your answer.
10. If etch damage is of particular concern, which configuration(s) shown in Fig. 6.22 would be most appropriate? Why?

CHAPTER 6 REFERENCES

1. "Chemical/ Metallurgical Properties of Silicon," in *Integrated Silicon Device Technology*, Vol. X, Research Triangle Institute Tech. Rpt. ASD–TDR–63–316, November 1965.
2. H.K. Kuiken et al., "Etching Profiles at Resist Edges," *J. Electrochem. Soc. 133*, pp. 1217–1226, 1986. (See also following article in journal.)
3. R.G. Brandes and R.H. Dudley, "Wall Profiles Produced during Photoresist Masked Isotropic Etching," *J. Electrochem. Soc. 120*, pp. 140–142, 1973.
4. R. Sellamuthu et al., "A Study of Anisotropic Trench Etching of Si with NF_3–Halocarbon," *J. Vac. Sci. Technol. B5*, pp. 342–346 and references therein, 1987.
5. Chuck Murray, "Wet Etching Update," *Semiconductor International*, pp. 80–85, May 1986.
6. D.R. Oswald, "Automatic Chemical Processing for Silicon Wafers," *J. Electrochem. Soc. 123*, pp. 531–534, 1976.
7. H.K. Kuiken and R.P. Tijburg, "Centrifugal Etching: A Promising New Tool to Achieve Deep Etching Results," *J. Electrochem. Soc. 130*, pp. 1722–1729, 1983.
8. D.R. Turner, "On the Mechanism of Chemically

Etching Germanium and Silicon," *J. Electrochem. Soc. 107*, pp. 810–816, 1960.

9. H. Robbins and B. Schwartz, "Chemical Etching of Silicon II. The System HF, HNO_3, H_2, and $HC_2H_3O_2$," *J. Electrochem. Soc. 107*, pp. 108–111, 1960. (For more information on the HF–HNO_3 system, see the other three papers in this series: Harry Robbins and Bertram Schwartz, "Chemical Etching of Silicon I. The System HF, HNO_3, and H_2O," *J. Electrochem Soc. 106*, pp. 505–508, 1959; B. Schwartz and H. Robbins, "Chemical Etching of Silicon III. A Temperature Study in the Acid System," *J. Electrochem. Soc. 108*, pp. 365–372, 1961; B. Schwartz and H. Robbins, "Chemical Etching of Silicon IV. Etching Technology," *J. Electrochem. Soc. 123*, pp. 1903–1909, 1976.)
10. Margaret Wright Jenkins, "A New Preferential Etch for Defects in Silicon Crystals," *J. Electrochem. Soc. 124*, pp. 757–762, 1977.
11. John N. Shive, Chap. 16, "Intermediate Surface Treatment," in *Transistor Technology*, Bell Telephone Laboratories, Inc., 1952.
12. S.M. Hu and D.R. Kerr, "Observation of Etching of n-Type Silicon in Aqueous HF Solutions," *J. Electrochem. Soc. 114*, p. 414, 1967.
13. A. Reisman et al., "The Controlled Etching of Silicon in Catalyzed Ethylenediamine–Pyrocatechol–Water Solutions," *J. Electrochem. Soc. 126*, pp. 1406–1415, 1979.
14. C. Raetezl et al., "Etching Properties of Different Silicon Films," *Electrochem. Soc. Ext. Abst. 74–2*, abst. no. 142, October 1974.
15. K.E. Bean and W.R. Runyan, "Dielectric Isolation: Comprehensive, Current, and Future," *J. Electrochem. Soc. 124*, pp. 5C–12C, January 1977.
16. G.C. Messenger, *IEEE Trans. on Nucl. Sci. NS–13*, p. 141, 1966.
17. J. Peter Ballantyne et al., "Small Chip Blocks Large Voltages," *Bell Laboratories Record*, pp. 91–94, April 1982.
18. Kurt E. Petersen, "Silicon as a Mechanical Material," *Proc. IEEE 70*, pp. 420–457, 1982.
19. H.A. Waggener et al., "Anisotropic Etching for Forming Isolation Slots in Silicon Beam Leaded Integrated Circuits," Int. Electron Dev. Conference, Washington, D.C., October 18–20, 1967.
20. Kenneth E. Bean, "Anisotropic Etching of Silicon," *IEEE Trans. on Electron Dev. ED-25*, pp. 1185–1193, 1978.
21. M.M. Abu-Zeid, "Corner Undercutting in Anisotropically Etched Isolation Contours," *J. Electrochem. Soc. 131*, pp. 2138–2142, 1984.
22. A.I. Stoller, "The Etching of Deep Vertical-Walled Patterns in Silicon," *RCA Review 31*, pp. 271–275, 1970.
23. Don. L. Kendall, "Vertical Etching of Silicon at Very High Aspect Ratios," *Ann. Rev. Mater. Sci. 9*, pp. 373–403, 1979.
24. W.R. Runyan, *Semiconductor Measurements and Instrumentation*, McGraw-Hill Book Co., New York, 1975.
25. H. Muraoka et al., "Controlled Preferential Etching Technology," pp. 327–338, in Howard R. Huff and Ronald R. Burgess, eds., *Semiconductor Silicon/73*, Electrochemical Society, Princeton, N.J., 1973.
26. C. John Rhee and Jack Saltich, "Integral Heat-Sink IMPATT Diodes Fabricated Using p^+ Etch Stop," *Proc. IEEE 61*, pp. 385–386, 1973.
27. A. Bohg, "Ethylene Diamine–Pyrocatechol–Water Shows Etching Anomaly in Boron-Doped Silicon," *J. Electrochem. Soc. 118*, pp. 401–402, 1971.
28. J.C. Greenwood, Ethylene Diamine–Catechol–Water Mixture Shows Preferential Etching of p–n Junction," *J. Electrochem. Soc. 116*, pp. 1325–1326, 1969.
29. H.J.A. van Dijk and J. de Jonge, "Preparation of Thin Silicon Crystals by Electrochemical Thinning of Epitaxially Grown Structures," *J. Electrochem. Soc. 117*, pp. 553–554, 1970.
30. Ronald L. Meek, "Electrochemically Thinned N/N^+ Epitaxial Silicon—Method and Applications," *J. Electrochem. Soc. 118*, pp. 1240–1246, 1971.
31. A. Uhlir, Jr., "Electrolytic Shaping of Germanium and Silicon," *Bell Syst. Tech. J. 35*, pp. 333–347, 1955.
32. R. Memming and G. Schwandt, "Anodic Dissolution of Silicon in Hydrofluoric Acid Solutions," *Surface Science 4*, pp. 109–124, 1966.
33. C.P. Wen and K.P. Weller, "Preferential Electrochemical Etching of P^+ Silicon in an Aqueous HF–H_2SO_4 Electrolyte," *J. Electrochem. Soc. 119*, pp. 547–548, 1972.
34. M.J.J. Theunissen, "Etch Channel Formation

during Anodic Dissolution of N-Type Silicon in Aqueous Hydrofluoric Acid," *J. Electrochem. Soc. 119*, pp. 351–360, 1972.

35. Y. Wantanabe et al., "Formation and Properties of Porous Silicon and Its Application," *J. Electrochem. Soc. 122*, pp. 1351–1355, 1975.
36. Takasahi Unagami, "Formation Mechanism of Porous Silicon Layer by Anodization in HF Solution," *J. Electrochem. Soc. 127*, pp. 476–483, 1980.
37. H.A. Waggener and J.V. Dalton, "Control of Silicon Etch Rates in Hot Alkaline Solutions by Externally Applied Potentials," *Electochem. Soc. Ext. Abst.*, abst. no. 273, Fall 1972.
38. R.J. Archer, "Stain Films on Silicon," *J. Phys. Chem. Solids 14*, pp. 104–110, 1960.
39. K.H. Beckmann, "Investigation of the Chemical Properties of Stain Films on Silicon by Means of Infrared Spectroscopy," *Surface Science 3*, pp. 314–332, 1965.
40. R.E. Hines, "A Simple Cure for a Form of Window Staining Which Occurs during the Etching of Silicon Integrated Circuits," *Microelectronics and Reliability 11*, p. 537, 1972.
41. H. Kuiken et al., "Etching Profiles at Resist Edges," *J. Electrochem. Soc. 133*, pp. 1217–1226, 1986.
42. E. Kohn, "A Correlation between Etch Characteristics of GaAs Etch Solutions Containing H_2O_2 and Surface Film Characteristics," *J. Electrochem. Soc. 127*, pp. 505–508, 1980.
43. Sadao Adachi and Kunishige Oe, "Chemical Etching Characteristics of (001) GaAs," *J. Electrochem. Soc. 130*, pp. 2427–2435, 1983.
44. Don W. Shaw, "Localized GaAs Etching with Acidic Hydrogen Peroxide Solutions," *J. Electrochem. Soc. 128*, pp. 874–880, 1981.
45. J. Kleinberg, W. Argersinger, and E. Griswold, *Inorganic Chemistry*, Heath and Co., Boston, 1960.
46. John S. Judge, "A Study of the Dissolution of SiO_2 in Acidic Fluoride Solutions," *J. Electrochem. Soc. 118*, pp. 1772–1775, 1971.
47. Robert S. Alwitt and Robert S. Kapner, "A Study of the Dissolution of Zinc in Buffered Acetic Acid Solutions," *J. Electrochem. Soc. 112*, pp. 204–207, 1965.
48. John Lawrence, "Controlled Etching of Silicon Dioxide in Buffered Hydrofluoric Acid," *Electrochem. Soc. Ext. Abst. 72–2*, abst. no. 191, pp. 466–468, 1972.
49. A.S. Tenney and M. Ghezzo, "Etch Rates of Doped Oxides in Solutions of Buffered HF," *J. Electrochem. Soc. 120*, pp. 1091–1095, 1973.
50. Lillian Rankel Plauger, "Etching Studies of Diffusion Source Boron Glass," *J. Electrochem. Soc. 120*, pp. 1428–1430, 1973.
51. Lou Hall, "Etch Rate Characterization of Silane Silicon Dioxide Films," *J. Electrochem. Soc. 118*, pp. 1506–1507, 1971.
52. W. van Gelder and V.E. Hauser, "The Etching of Silicon Nitride in Phosphoric Acid with Silicon Dioxide as a Mask," *J. Electrochem. Soc. 114*, pp. 869–872, 1967.
53. Victor Harrap, "Equal Etch Rates of Si_3N_4 and SiO_2 Utilizing HF Dilution and Temperature Dependence," pp. 354–362, in Howard R. Huff and Ronald R. Burgess, eds., *Semiconductor Silicon/73*, Electrochemical Society, Princeton, N.J., 1973.
54. Werner Kern and Cheryl A. Deckert, Chap. V–I, "Chemical Etching," in John L. Vossen and Werner Kern, eds., *Thin Film Processes*, Academic Press, New York, 1978.
55. Cheryl A. Deckert, "Pattern Etching of CVD Si_3N_4/SiO_2 Composites in HF/Glycerol Mixtures," *J. Electrochem. Soc. 127*, pp. 2433–2438, 1980.
56. J.J. Kelly and C.H. de Minjer, "An Electrochemical Study of Undercutting during Etching of Duplex Metals," *J. Electrochem. Soc. 122*, pp. 931–936, 1975.
57. H.R. Koenig and L.I. Maissel, "Application of RF Discharges to Sputtering," *IBM J. Res. Develop. 14*, pp. 168–171, 1970.
58. J.L. Vossen, "Glow Discharge Phenomena in Plasma Etching and Plasma Deposition," *J. Electrochem. Soc. 126*, pp. 319–324, 1979.
59. Linus Pauling, *The Nature of the Chemical Bond*, Cornell University Press, Ithaca, New York, 1960.
60. Ch. Steinbruchel et al., "Mechanism of Dry Etching of Silicon Dioxide," *J. Electrochem. Soc. 132*, pp. 180–186, 1985.
61. H. Kawata et al., "The Dependence of Silicon Etching on an Applied DC Potential in $CF_4 + O_2$ Plasmas," *J. Electrochem. Soc. 132*, pp. 206–211, 1985.

62. H. Abe et al., "Etching Characteristics of Silicon and Its Compounds," *Jap. J. Appl. Phys. 12*, p. 154, 1973.
63. Alan R. Reinberg, "Plasma Etching in Semiconductor Manufacturing—A Review," pp. 91–110, in Henry G. Hughes and Myron J. Rand, eds., *Etching for Pattern Definition*, Proc. Electrochem. Soc. 87–6, Princeton, N.J., 1976.
64. G.S. Mathad, "Design Considerations for a High Pressure, High Etch Rate Single Slice Reactor," pp. 134–142, in *Proc. 6th Symposium on Plasma Processing, 87–6*, Electrochemical Society, Pennington, N.J., 1987.
65. Russ A. Morgan, *Plasma Etching in Semiconductor Fabrication*, Elsevier Science Publishers B.V., New York, 1985.
66. Daniel L. Flamm et al., "Basic Chemistry and Mechanisms of Plasma Etching," *J. Vac. Sci. Technol. B1*, pp. 23–30, 1983.
67. L.M. Ephrath, "Selective Etching of Silicon Dioxide Using Reactive Ion Etching with CF_4–H_2," *J. Electrochem. Soc. 126*, pp. 1419–1421, 1979.
68. K.L. Konnerth and F.H. Dill, "In-Situ Measurement of Dielectric Thickness during Etching or Developing Process," *IEEE Trans. on Electron Dev. ED-22*, pp. 452–456, 1975.
69. Dave Johnson, "Optical Methods Detect End Point in Plasma Etching," *Industrial Research and Development*, pp. 181–185, October 1980.
70. H.P. Kleinknecht and H. Meier, "Optical Monitoring of the Etching of SiO_2 and Si_3N_4 on Si by the Use of Grating Test Patterns," *J. Electrochem. Soc. 125*, pp. 798–803, 1978.
71. E.O. Degenkolb and J.E. Griffiths, "Simple Optical Devices for Detection of RF Oxygen Plasma Stripping of Photoresists," *Appl. Spectroscopy 31*, pp. 40–42, 1977.
72. Bill B. Stafford and Georges J. Gorin, "Optical Emission End-Point Detecting for Monitoring Oxygen Plasma Photoresist Stripping," *Solid State Technology*, pp. 51–55, September 1977.
73. R.G. Poulsen and G.M. Smith, "Use of Optical Emission Spectra for End-Point Detection in Plasma Etching," pp. 1058–1070, in H.R. Huff and E. Sirtl, eds., *Semiconductor Silicon/77*, Electrochemical Society, Princeton, N.J., 1977.
74. Kwang O. Park and Fredrick C. Rock, "End Point Detection for Reactive Ion Etching of Aluminum," *J. Electrochem. Soc. 131*, pp. 214–215, 1984.
75. S. Mittal et al., "Endpoint Detection for Contacts and Vias in VLSI Technology," pp. 254–266, in *Proc. 6th Symposium on Plasma Processing 87–6*, Electrochemical Society, Princeton, N.J., 1987.
76. George B. Bunyard, "Plasma Process Development and Monitoring via Mass Spectrometry," *Solid State Technology*, pp. 53–57, December 1977.
77. C.J. Mogab, "The Loading Effect in Plasma Etching," *J. Electrochem. Soc. 124*, pp. 1262–1268, 1977.
78. Daniel L. Flamm et al., "Multiple-Etchant Loading Effect and Silicon Etching in ClF_3 and Related Mixtures," *J. Electrochem. Soc. 129*, pp. 2755–2760, 1982.
79. K. Tokunaga et al., "Comparison of Aluminum Etch Rates in Carbon Tetrachloride and Boron Trichloride Plasmas," *J. Electrochem. Soc. 128*, pp. 851–855, 1981.
80. R.G. Frieser et al., "Silicon Damage Caused by Hydrogen Containing Plasmas," *J. Electrochem. Soc. 130*, pp. 2237–2241, 1983.
81. J. Dieleman and F.H.M. Sanders, "Plasma Effluent Etching: Selective and Non-Damaging," *Solid State Technology*, pp. 191–196 and references therein, April 1984.
82. S.J. Fonash, "Damage Effects in Dry Etching," *Solid State Technology*, pp. 201–204 and references therein, April 1985.
83. G.S. Oehrlein et al., "Near-Surface Damage and Contamination after CF_4/H_2 Reactive Ion Etching of Si," *J. Electrochem. Soc. 132*, pp. 1441–1447, 1985.
84. A. Rohatgi et al., "Characterization and Control of Silicon Surface Modification Produced by CCl_4 Reactive Ion Etching," *J. Electrochem. Soc. 133*, pp. 408–416, 1986.
85. G.S. Oehrlein et al., "Investigation of Reactive-Ion-Etching-Related Fluorocarbon Film Deposition onto Silicon and a New Method for Surface Residue Removal," *J. Electrochem. Soc. 133*, pp. 1002–1008, 1986.
86. D. Misra and E.L. Heasell, "A Study of Reactive Ion Etching (CF_4 + O_2 Plasma) Induced Deep Levels in Silicon," *J. Electrochem. Soc. 134*, pp. 956–958, 1987.

87. D.F. Downey et al., "Introduction to Reactive Ion Beam Etching," *Solid State Technology*, pp. 121–127, February 1981.

88. Rinn Cleavelin and Gary T. Duranko, "Silicon Dioxide Removal in Anhydrous HF Gas," *Semiconductor International*, pp. 94–99, November 1987.

89. Peter H. Singer, "Wet Bench Fire Suppression," *Semiconductor International*, pp. 154–157, September 1987.

90. Jean Ohlson, "Dry Etch Chemical Safety," *Solid State Technology*, pp. 69–73, July 1986.

CHAPTER 7

Epitaxy

7.1 INTRODUCTION

The epitaxial growth process involves the deposition of a thin layer of material onto the surface of a single-crystal wafer (substrate) in such a manner that the layer is also single crystal and has a fixed and predetermined crystallographic orientation with respect to the substrate. If the layer material is the same as the substrate, such as silicon on silicon, then it will have the same crystallographic orientation as the original crystal and become a crystallographic extension of the wafer. If deposition is from the vapor phase, this combination is sometimes referred to as *homoepitaxy*, *autoepitaxy,* or *isoepitaxy* but is generally simply called *epitaxy*. If the materials are different, as with silicon on sapphire, the combination is referred to as *heteroepitaxy,* and the orientation of the layer may well be different from that of the substrate.

If the layer is grown from the melt rather than by chemical vapor deposition (CVD), the process is referred to as *liquid phase epitaxy* (LPE). Silicon is seldom grown by LPE; gallium arsenide layers, however, often are. It is also possible to grow epitaxially from the solid phase. For example, the very thin amorphous layer left on top of a single-crystal slice after polishing or the thin amorphous layer formed on a wafer surface because of ion implant damage can be regrown as single crystal by a suitable high-temperature annealing cycle.

The term *epitaxy*, referring to oriented crystallographic overgrowth on a foreign substrate, was apparently coined in the 1920s by the French mineralogist Royer as he observed the overgrowth of water-soluble salts onto the cleaved surfaces of naturally occurring mineral specimens (1). Such studies were of considerable scientific interest and were expanded to include materials evaporated onto mineral surfaces (rock salt was a favorite substrate). However, no industrial applications of these studies took place until silicon and germanium began to be overgrown onto sapphire and spinel substrates in the mid-1960s. During the same time interval (1930–1960), techniques for the vapor phase growth of crystals were becoming well developed, and, indeed, a patent filed in 1951 applied the vapor phase growth of silicon and germanium to semiconductor devices (2). Later, to fulfill a specific need of the fledgling semiconductor industry, when thin layers of single-crystal germanium were vapor grown onto germanium single-crystal slices, the process was labeled as "epitaxy" even though the original definition was considerably different. Thus, when experiments to overgrow silicon and germanium onto single-crystal substrates such as sapphire and spinel began, the new label "heteroepitaxy" was introduced to distinguish that process from the by then established use of epitaxy to describe overgrowth on a substrate of the same material.

The original problem to which an epitaxial layer provided the solution was the reduction of transistor collector resistance. The source of this collector resistance was the layer of high-resistivity semiconductor material between the collector–base junction and the collector contact. The grown junction transistor, shown schematically in Fig. 7.1a, had a very long length of high-resistivity material, and its performance at high current was severely degraded. The mesa transistor, also shown schematically in Fig. 7.1a, had a substantially reduced current path length since the mesa chip thickness was much less than half the length of the grown junction bar. However, for ease of manufacture, the distance from the collector–base junction to the collector contact was still much longer than that required by the device itself.[1] The structure shown in Fig. 7.1b reduces the collector resistance even more by having a large portion of the wafer of low resistivity.

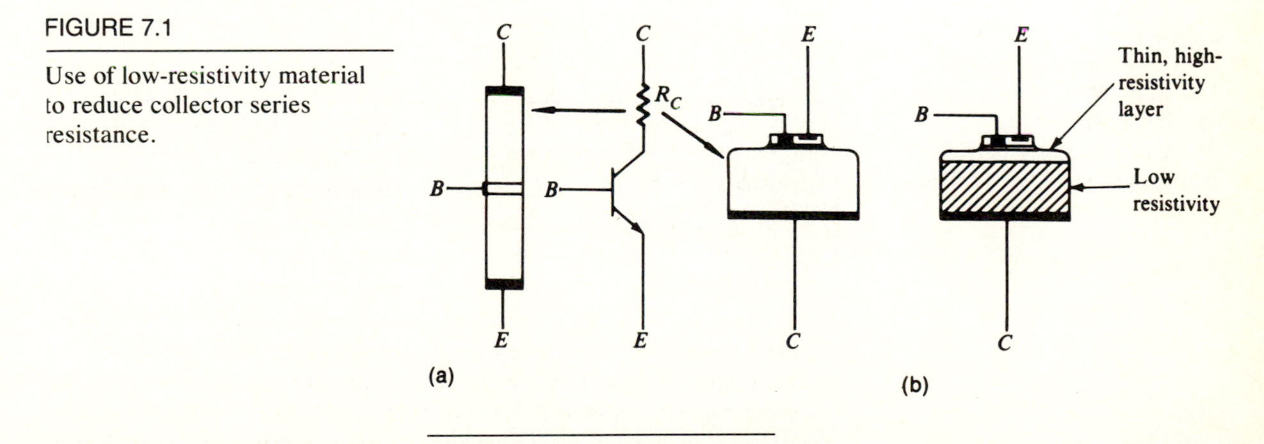

FIGURE 7.1

Use of low-resistivity material to reduce collector series resistance.

[1]The required distance is something just in excess of the width of the collector–base space charge region when the maximum collector–base voltage is applied.

Vapor phase epitaxial growth is the only method currently available for producing layers of high resistivity on low-resistivity substrates such as just described. When doping is done by either solid-state diffusion or ion implantation, the impurities are introduced from the surface so that, of necessity, a higher concentration occurs near the surface than exists deep within the wafer. During growth from the melt, once a dopant is put into the melt, it cannot be easily removed, so the impurity concentration during growth can be abruptly increased but not decreased.[2] However, in vapor phase growth, the dopant, which is deposited from the vapor simultaneously with the semiconductor, can be quickly changed, with respect both to species and to concentration. The time for change is primarily dictated by the time required to change the gas species/concentration in the reactor chamber and can be from a few seconds to a few minutes, depending on reactor design. Thus, in principle, uncompensated sequential layers of widely differing resistivity can be produced, although they are generally not required for IC fabrication.

The mesa transistor structure of Fig. 7.1b can be made by starting with a low-resistivity substrate (slice) and adding a thin higher-resistivity layer of the same conductivity type by epitaxial growth (3). The silicon bipolar IC required a different structure since the goal was to provide physical support and electrical (pn junction) isolation between the many components of the IC. The structure used for some time was a high-resistivity n-layer epitaxially overgrown onto a high-resistivity p-substrate. Component isolation was accomplished by local p-diffusions, which reached from the surface of the n-layer to the p-substrate. This configuration, as shown in Fig. 7.2a,

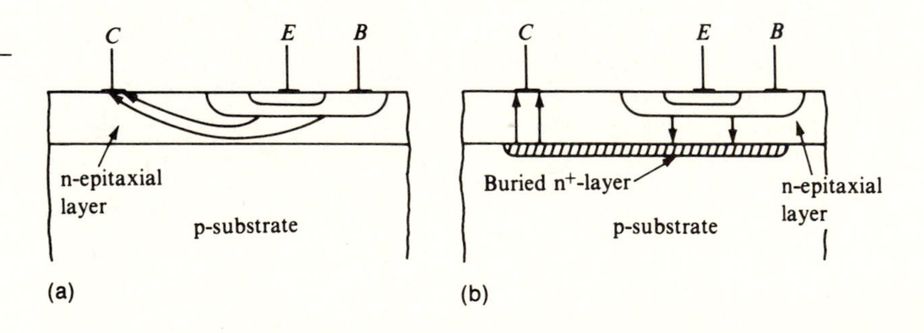

FIGURE 7.2

Use of a local diffusion before epitaxy to reduce the high-resistivity collector current path length and hence collector series resistance in an IC transistor.

[2]The resistivity can be reduced in either diffused structures or melt-grown crystals by adding a compensating impurity. However, since resistivity depends on the difference between the concentration of n- and p-impurities, its value is difficult to control. Further, the addition of compensating impurities in large quantities can substantially reduce carrier mobility.

also had high collector resistance since the path from the collector–base junction to the collector contact on the top of the wafer was long and composed entirely of high-resistivity material. By adding a localized high-concentration n-diffusion before the epitaxy, the structure of Fig. 7.2b can be made, and with it, a substantial portion of the electrical path is now through the low-resistivity diffused layer (4).

The transistors in MOS ICs are self-isolating and therefore can be made in single-resistivity material. However, higher performance and/or smaller chip size can often be achieved by using epitaxy, and beginning in the mid-1980s, high-resistivity layers on low-resistivity substrates of the same type were used in production quantities. Gallium arsenide ICs requiring epitaxy generally use an n-layer on a semi-insulating substrate.

7.2 VAPOR PHASE EPITAXY

As in the chemical vapor deposition (CVD) described in Chapter 4, no solid material should be formed in the gas above the wafer surface; that is, as before, the solid forming reaction must be surface catalyzed. In addition, successful epitaxy requires that each atom permanently added to the surface be in the exact position required to form a defect-free single-crystal layer. A study of vapor phase epitaxy can be roughly divided into the three categories of surface nucleation, surface reaction kinetics, and gas transport of reactants and reaction products to and from the wafer surface.

7.2.1 Nucleation

Classical homogeneous nucleation theory supposes that molecules will collide and at least temporarily bind together and that at any given time a wide distribution of nuclei ranging up in size from one molecule exists. If the nucleus (cluster of molecules) is above some critical size, it will be stable and continue to grow; otherwise, there is a high probability that it will lose molecules and disappear. The change ΔG in the Gibbs free energy of the molecules due to the formation of a spherical cluster (perhaps appropriate for a raindrop forming in air) is given by (5)

$$\Delta G = 4\pi r^2\sigma + \frac{4}{3}\pi r^3\Delta G_v \qquad 7.1$$

where r is the cluster radius, σ is the interfacial energy per unit area between the solid and the vapor, and ΔG_v is the Gibbs free energy difference per unit volume between the vapor and the cluster. ΔG_v is given by

$$\Delta G_v = -\left(\frac{kT}{\Omega}\right)\ln\left(\frac{P}{P_e}\right) \qquad 7.2$$

FIGURE 7.3

Plot of ΔG versus cluster radius r. (It first increases, and then, if the critical radius r^* is reached, rapidly decreases and becomes negative.)

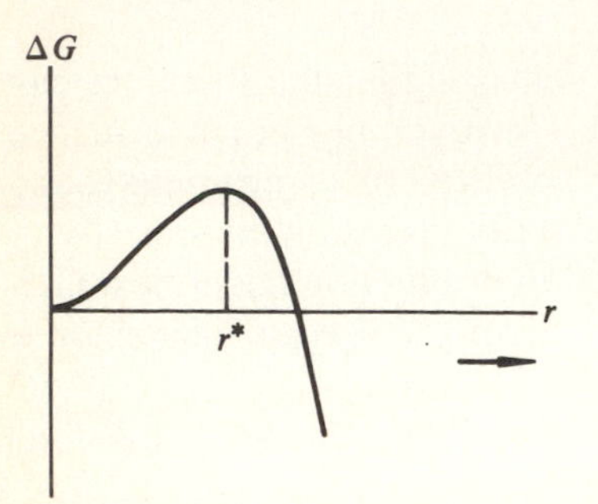

where Ω is the volume per molecule in the crystal, T is the absolute temperature, k is Boltzmann's constant, P is the vapor pressure above the growing cluster, and P_e is the equilibrium vapor pressure. The plot of Eq. 7.1 in Fig. 7.3 shows that initially ΔG increases with r but reaches a maximum and then rapidly decreases. As soon as ΔG begins to decrease, at $r = r^*$, the cluster becomes thermodynamically stable. r^* can be calculated from Eq. 7.1 and is

$$r^* = -\frac{2\sigma}{\Delta G_v} \tag{7.3}$$

Thus, if the absolute value of ΔG_v, as determined by the vapor pressure of the nutrient, is too small, the critical radius will be so large that little or no nucleation will occur.

The spherical homogeneous nucleation case just discussed is not directly applicable to crystal growth, but it does demonstrate the concept of a critical cluster size necessary for growth and that of the necessity of a higher-than-equilibrium vapor pressure. Somewhat closer to crystal growth is a flat-disk cluster of diameter r and height h laying on a flat surface, in which case (5)

$$\Delta G = 2\pi r\varepsilon + \pi r^2(\sigma_1 + \sigma_2 - \sigma_3) + \pi r^2 h \Delta G_v \tag{7.4}$$

where ε is the edge free energy per unit length and where σ_1, σ_2, and σ_3 are respectively the interfacial energy per unit area between the cluster and the substrate, the cluster and the gas ambient, and the substrate and the gas ambient.[3] Eq. 7.4 has the same general character as Eq. 7.1, and thus the earlier conclusions are still applicable. Note that in this case the cluster can collect molecules not only directly from the gas stream but also, through surface diffusion, from the body of molecules that strike the surface around the cluster. If, instead of a circular cluster forming on a broad flat surface, a semicircular cluster forms next to a surface step, it can be shown that the sum of all of the edge and interfacial energies is less than it is for the circular case. Nucleation at a ledge is then favored over nucleation on a flat surface, and a surface with ledges is expected to grow much more rapidly than a smooth one. In many cases, it appears that the observed growth rates of crystals can be explained only by assuming a continuous supply of ledges. This aspect is dis-

[3]The edge free energy of a monolayer high step is usually not equal to hs_2, where h in this case is the thickness of the monolayer. Hence, in Eq. 7.4, for small h, ε instead of hs_2 should be used. The σ_3 term must be subtracted since it represents a free energy change due to covering up a portion of the original surface by the disk.

cussed theoretically in the 1951 classic paper by Burton, Cabrera, and Frank (6).

Screw dislocations in a single-crystal wafer will produce a continuous supply of ledges that will not grow out, regardless of the thickness grown. In addition, from a practical standpoint, with the thin layers normally grown for semiconductor applications, the fact that the surface is almost never exactly on-orientation ensures a supply of atomic steps (ledges) that will not completely disappear during layer growth. This feature can be seen from Fig. 7.4, which shows a series of atomic planes of separation s intersecting the surface at an angle θ. For this case, the distance S along the surface between steps is given by

$$S = \frac{s}{\sin \theta} \tag{7.5}$$

The atomic spacing s is normally considered to be the separation of two adjacent planes of atoms—that is, the spacing of (400) planes when a (100) surface is exposed.[4] However, since evidence exists that in the case of a silicon (100) surface that has been heated above 1000°C, the equilibrium step height is two layers or the spacing be-

FIGURE 7.4

Simultaneous lateral growth from many ledges resulting in a net vertical growth.

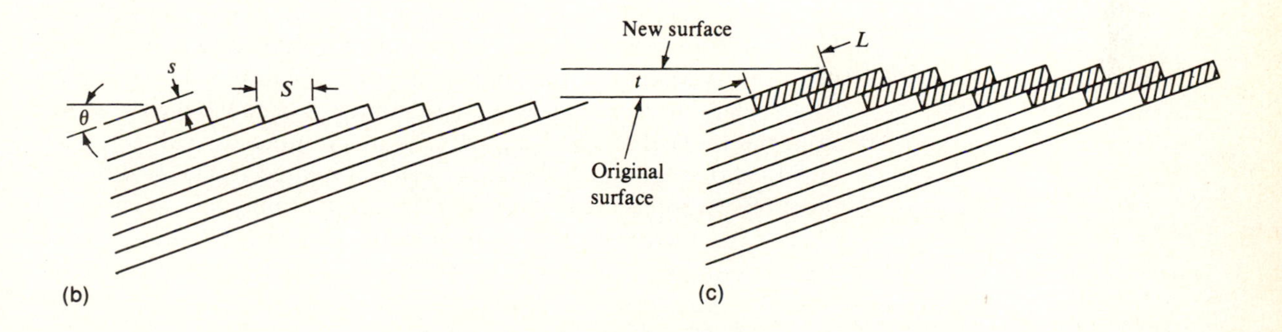

[4]This presumes either a diamond or zinc blende structure. For a discussion of the placement of atoms in these kinds of crystals (typical of silicon and gallium arsenide, respectively), see Appendix A.

tween (200) planes (7), it is possible that multiple-layer step heights also occur on other surfaces.

If growth proceeds by nucleation at the steps, then the steps will each move out laterally as shown in Fig. 7.4a, and the surface of Fig. 7.4b will be transformed into that of Fig. 7.4c at some later time and will have moved upward a distance t. When the thickness t is grown, the length L of the atomically flat surface caused by the steps growing out is given by $L = t/\sin \Theta$. If the surface is misoriented by 1° and a 25 μm layer is grown, a flat of about 1.4 mm would develop on one side of a wafer.

When individual atomic bonding is considered, the picture of crystal growth is somewhat changed from that just described. With the values of vapor pressure normally encountered in vapor phase growth, the number of atoms in a critical cluster may be as low as 1. For such small numbers, a classical thermodynamics approach using properties that appear in Eq. 7.4 is really no longer appropriate, and a statistical mechanics approach is necessary (8). However, the concept of a critical cluster size is retained. In the diamond lattice, enough atomic bonds are available at a ledge on any surface, regardless of its orientation, for a single atom that diffuses to it to be properly oriented and stably bonded. Where no ledges exist and nucleation must take place on a crystal plane, fewer atomic bonds per surface atom can reach out of the plane to attach to an atom on the surface.[5] However, properly oriented single-atom attachment is still sometimes possible. For example, there are two bonds per atom reaching from one (400) plane to the next, and with isoepitaxy, these two bonds are enough to provide orientation so that growth on terraces can, in principle, proceed by a single atom. In the case of (111) orientation, however, only one bond per atom reaches from one (111) double layer to the next. With only one bond, a single atom on a (111) plane surface will neither be tightly held nor uniquely oriented. A cluster of three atoms can provide orientation, but it may be either the correct one or a twinned one (9). Experimentally, it has been observed that the growth rate on a closely oriented (111) plane is considerably less than when orientation is a few degrees away from the (111) (10, 11). Further, it is much more difficult to grow high-quality epitaxy on surfaces closely oriented to (111). Both of these observations are consistent with a greater difficulty in nucleating on (111) terraces (flats between steps) than at (111) steps. However, in the case of growth on (100) planes, only a slight decrease in

---

[5]It should be remembered that the upper bonds in a surface do not dangle (remain unbonded). They will either distort to connect with other lattice atoms without satisfied bonds or else bond to foreign atoms adsorbed on the surface.

rate is observed as the (100) plane is approached, and the quality of epitaxial layers grown on-orientation is found to be quite good. This suggests that nucleation and proper orientation are possible on an atom-by-atom basis on (100) planes.

In summary, the possible actions of a single atom as it is adsorbed on a wafer surface, as shown in Fig. 7.5, are as follows:

1. The atom may strike the surface and then be desorbed before it can diffuse to a ledge or participate in cluster formation.
2. If it has very little energy or if the deposition rate is very high, the atom may be surrounded by other atoms and effectively locked in place without regard to crystallographic location, in which case a polycrystalline or amorphous layer will grow.
3. When the energy is high enough for appreciable motion, but the ledges are widely separated, several atoms may join together on a terrace to form a stable and sometimes properly oriented cluster that can then act as a new ledge for future growth.
4. The atom may diffuse to a ledge, bond properly, and grow epitaxially.
5. On some substrate orientations, the single atom may become epitaxially bonded on a terrace.

Clusters have been observed in silicon-on-silicon molecular beam epitaxy (discussed in a later section), but then such growth is promoted when there are impurities on the surface (12).

When growth is on a single-crystal substrate of the same lattice structure and only a slightly differing lattice constant, there is ap-

FIGURE 7.5

Possible actions of ad-atoms on a crystal surface.

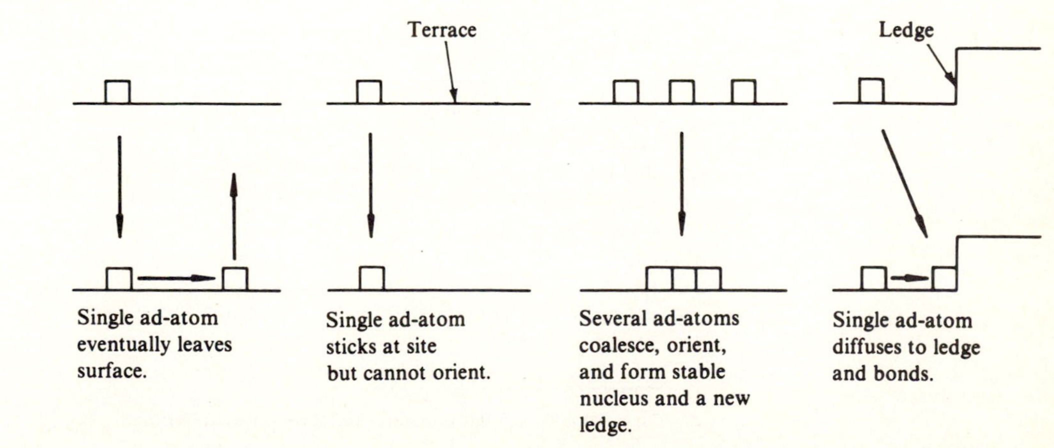
Terrace
Ledge
Single ad-atom eventually leaves surface.
Single ad-atom sticks at site but cannot orient.
Several ad-atoms coalesce, orient, and form stable nucleus and a new ledge.
Single ad-atom diffuses to ledge and bonds.

parently little difference in nucleation behavior from that just discussed. The sites will all be of the proper orientation, although the substrate and layer atomic spacings may not exactly fit. The list of semiconductors having the same crystal structure is actually quite small. Silicon and germanium both have the diamond lattice, while the III–V and some of the II–VI compounds have a zinc blende lattice. Table 7.1 gives lattice spacings of some common semiconductor materials. Note that there are different values for high- and low-resistivity Si and different values depending on whether the heavy doping is n- or p-type. This effect also occurs in the other materials but has not been as extensively studied. There is no set amount of lattice mismatch that can be tolerated. It depends both on the elastic properties of the individual materials and on the density of defects that is acceptable to the user. As an experimental guide, the quality of high-resistivity on low-resistivity Si appears quite satisfactory even though some misfit dislocations occur. Considerably more mismatch occurs between Si and GaAs, and many more misfit dislocations, but GaAs of usable quality can be overgrown on Si. When the mismatch is small, the layer accommodates the misfit by having strain in the first few layers. At some point, the mismatch becomes large enough so that the necessary strain exceeds the yield point of the material, and misfit dislocations occur.[6] When misfit dislocations occur, there are usually not enough of them to reduce the strain to zero. Apparently, there are only enough to reduce the strain to a level less than that of the yield point. For example, in the case of Ge overgrown on GaAs, with no residual strain the dislocation spacing would be expected to be every 0.5 μm, whereas a spacing of 4 μm was experimentally observed (13). When the misfit becomes yet larger, as regions nucleated at different ledges grow together, there can be gross crystallographic defects.

TABLE 7.1

Lattice Spacings for Cubic Semiconductors

Material	*Unit Cell Length (Å)*	*Group*	*Structure*
0.001 Ω-cm B-doped Si	5.4270	IV	Diamond
High-purity Si	5.4309	IV	Diamond
0.001 Ω-cm As-doped Si	5.4315	IV	Diamond
High-purity Ge	5.6576	IV	Diamond
High-purity GaAs	5.6538	III–V	Zinc blende
GaP	5.4504	III–V	Zinc blende
βSiC	4.3596	IV–IV	Zinc blende
βZnS	5.4039	II–VI	Zinc blende

[6]See Appendix A for a discussion of misfit dislocations.

Heteroepitaxial nucleation of a III–V compound semiconductor such as GaAs on Si or Ge is complicated by the fact that in III–V zinc blende crystals, the two interpenetrating sublattices are of different composition. In the case of GaAs, one is of gallium and one is of arsenic. Thus, for growth in a (100) direction (which is an orientation preferred by device designers), there are alternate layers of arsenic and gallium. Under most circumstances, the first layer to nucleate on silicon is arsenic, so over the whole wafer surface, even though stepped, the first layer will be arsenic. As more layers are added and the epitaxy thickness increases, unless each ledge is a multiple of two atomic spacings high, an As layer and a Ga layer will meet over the steps and cause an antiphase boundary defect. One solution to the problem is to use (211) oriented wafers. In this orientation, regardless of the step height, the bonding is such that the arsenic and gallium each have preferred sites, and an antiphase boundary cannot develop (14). Another solution is to use (100) oriented wafers cut a few degrees off-orientation with the tilt toward a (110) plane. Then, if the wafers are annealed a few minutes at 1000°C, regardless of the original step heights, the surface reconstructs itself so that all steps are two atomic spacings high (7). With this sort of surface, the arsenic and gallium layers will match up over the whole wafer and thus produce no antiphase boundaries.

In heteroepitaxy involving materials of different crystal structure, such as silicon on sapphire or other single-crystal oxide, nucleation is much more complex. It is generally assumed that the silicon (or other semiconductor) atoms occupy the position of the metal—for example, aluminum in Al_2O_3 or Al and Mg in spinel—and bond to the oxygen. However, the spacings are almost always very different, and the fit will be very poor. Fig. 7.6 shows that a cluster of five atoms of silicon in a (100) configuration on a $(1\bar{1}02)$ sapphire surface fits moderately well, although from this point on, there is little overlap. It can be theorized that a small stable cluster of the atoms being deposited forms homogeneously on the substrate surface and then has enough surface mobility to become oriented (15, 16). The cluster thus formed will act as a sink for atoms striking the surface nearby and keep the supersaturation so low that no new clusters will form. At some distance away, however, other clusters can form and grow. As shown in Fig. 7.7, at an early stage of growth, a series of single-crystal islands will be oriented mostly in the same direction, which is similar to growth observed on amorphous substrates, except that then the different islands are not oriented, as can be seen in Fig. 7.8. The octahedral equilibrium shape of silicon crystallites is also visible in this figure. The islands of Fig. 7.7 will grow until they coalesce and form a complete layer, at which point nucleation becomes homoepitaxial rather than heteroepitaxial. Un-

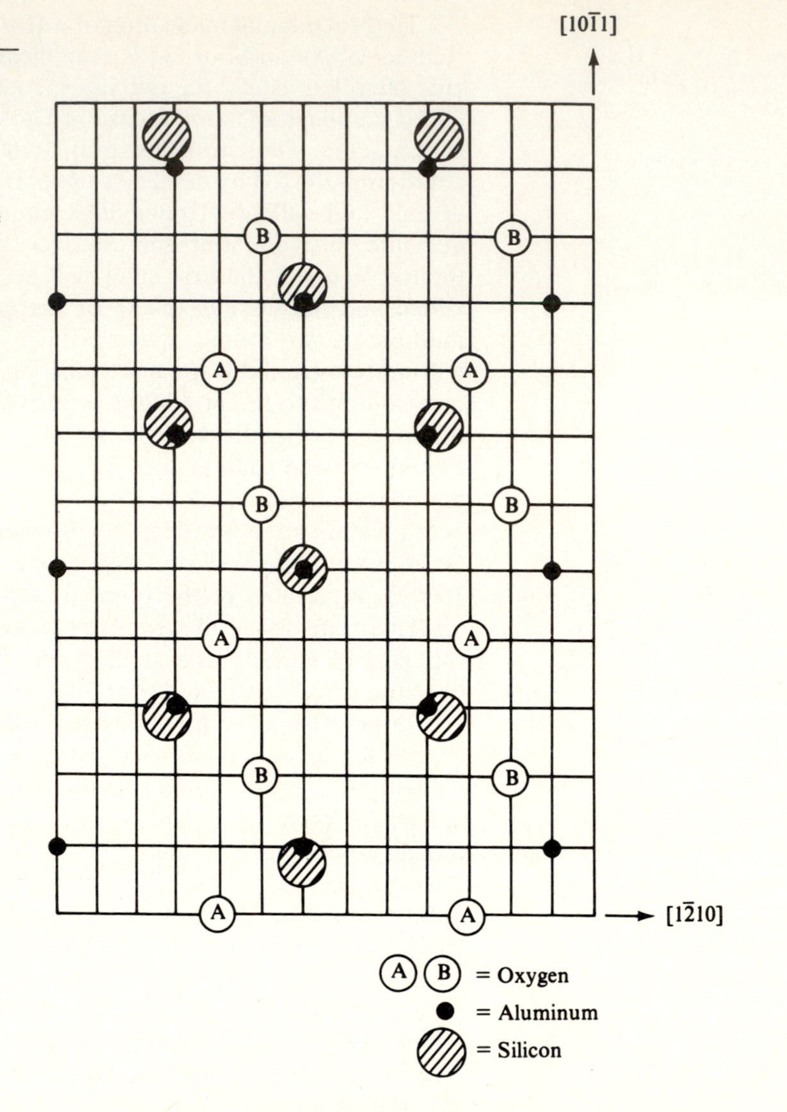

FIGURE 7.6

Overlay of a (100) layer of silicon atoms onto the aluminum atoms of a sapphire (1 $\bar{1}$ 02) surface. (The A and B oxygen positions indicate locations respectively below and above the aluminum atoms. Beyond the cluster of five silicon atoms in the center of the figure, correlation with the aluminum atom positions becomes progressively worse.) (*Source:* Adapted from Arnold Miller and Harold M. Manasevit, *J. Vac. Sci. Technol. 3,* p. 68, 1966.)

fortunately, clusters oriented at different places on the substrate, while all having the same azimuthal orientation, will not be properly spaced one from the other. Thus, various crystallographic defects will occur at the boundaries between the original islands. Because of errors in nucleation, some of the original islands may also not be oriented as the others and cause small misoriented domains.

Surface mobility is a key part of providing for the proper placement of atoms, regardless of whether growth is homoepitaxial or heteroepitaxial. Since the mobility decreases as the temperature decreases, it is reasonable to assume that for a given deposition rate

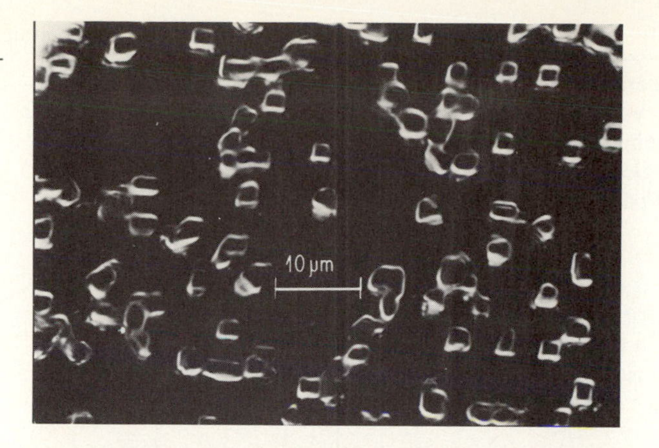

FIGURE 7.7

Silicon single-crystal islands formed on a (100) spinel surface. (Most of the crystallites show (100) symmetry and appear to have the same orientation.) (*Source:* Photograph courtesy of E. Sirtl.)

FIGURE 7.8

Small silicon crystallites that have nucleated and grown on SiO_2. (Note that most of them are octahedral in shape and that the orientation varies from crystallite to crystallite.)

there is a minimum temperature below which epitaxy cannot take place. The surface mobility is proportional to the surface diffusion coefficient D_s, which has the same form of temperature dependence as the bulk coefficient; that is,

$$D_s = D_{so}e^{-E/kT} \tag{7.6}$$

where D_{so} is a constant, E is the activation energy of the diffusion process, k is Boltzmann's constant, and T is the temperature in kelvins. Based on Eq. 7.6, the curve separating the maximum rate for single growth from amorphous growth would be expected to change with temperature as shown in Fig. 7.9. Experimentally, for temperatures above about 1000°C, it has been reported that the curve separating single-crystal silicon from polycrystalline growth has a slope

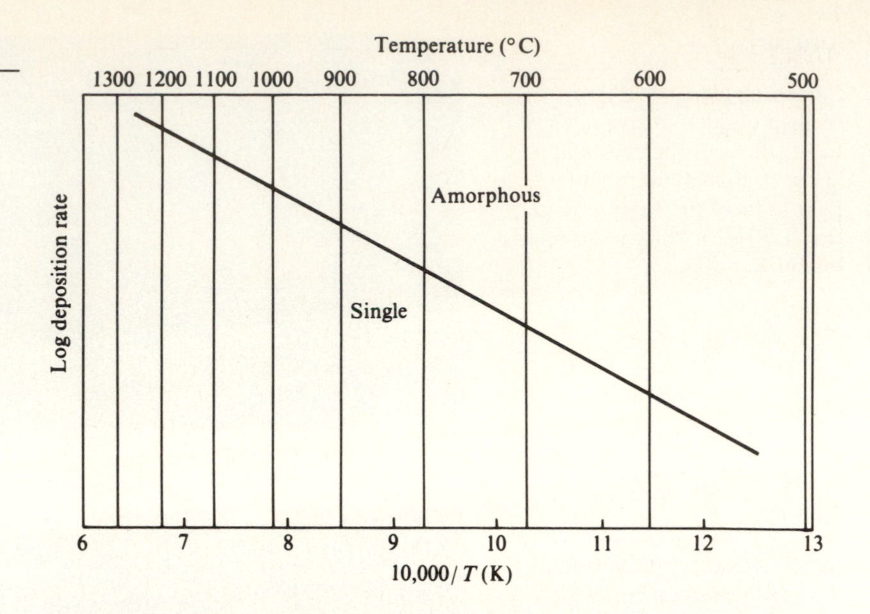

FIGURE 7.9

Trend of line of demarcation between single-crystal and amorphous deposition. (Between the two regions, there will be a band of polycrystalline material.)

of ~5 eV, which is equal to the activation energy of bulk silicon self-diffusion (15). This slope can be explained by assuming that, in the case of homoepitaxy, if bulk diffusion is high enough, the lower temperature limit might be set by bulk diffusion since the atoms could then move to the proper position even after being covered.

7.2.2 Surface Reactions

The previous discussion of nucleation implies that atoms of the species to be grown impinged on the growing surface. However, unless the nutrient is supplied by evaporation, as in MBE, a surface reaction must occur if particle formation above the surface is to be avoided. Such reactions can occur either when the two reactant species are adsorbed at adjacent sites on the surface or when they diffuse to adjacent sites after having been previously adsorbed at remote locations. Under such circumstances, if one component of the gas stream is strongly adsorbed, it may occupy a majority of sites and severely reduce the reaction rate. An example of this is the introduction of phosphine (PH_3) to a silane gas stream in order to dope a silicon layer being grown from silane decomposition. Normally, the silane is adsorbed on the wafer surface and then decomposes. However, phosphine appears to occupy a large percentage of the sites normally occupied by silane and hence reduces the silicon deposition rate (16). Similarly, a major gas stream constituent may be adsorbed on very few sites and hence essentially not participate in the final reaction.

In the case of silicon deposition from $SiCl_4$, because of simultaneous vapor phase reactions, many compounds other than $SiCl_4$

TABLE 7.2

Equilibrium Fractional Coverage of (111) Silicon

Component	*Coverage* θ*	P_e *(dynes/cm²)*†
H	0.63	10
Cl	0.20	0.2
$SiCl_2$	0.16	700
Vacant sites	0.015	
H_2	10^{-4}	10^6
HCl	10^{-7}	2×10^4
$SiHCl_3$	1.7×10^{-10}	4
$SiCl_4$	1.5×10^{-11}	30

*For the Si–Cl–H system at 1230°C with a Cl/H ratio of 0.01.
†P_e = equilibrium pressure; total pressure = 1 atm.

Source: Adapted from A.A. Chernov, *J. Crystal Growth 42*, p. 55, 1977.

are present in the gas stream. Table 7.2 lists a number of these compounds, along with the calculated fractional site coverage Θ for each of them (17). It can be seen that other constituents keep $SiCl_4$ from being adsorbed on the surface, and, hence, it cannot take part in the final reaction to produce elemental silicon. $SiCl_2$, however, is present and has a high adsorption site density. In studying probable reaction paths for cases where some of the reactions occur on surfaces, it is very important to consider surface site coverage. A later section will note that in most silicon halide reactions, $SiCl_2 + H_2 \rightarrow Si + 2HCl$ is listed as the final step, which is in keeping with predictions based on the data of Table 7.2.

If surface reaction chemistry is assumed as the rate-limiting mechanism, the Langmuir–Hinschelwood mechanism can be used to describe the behavior (18). The rate of surface adsorption $d\Theta/dt$ of a given molecular species is proportional to the gas pressure p of the component in question and to the fraction of uncovered sites remaining; that is,

$$\frac{d\Theta}{dt} = k_1 p(1 - \Theta) \tag{7.7}$$

where k_1 is a constant. The rate at which the species leaves is proportional to the number of sites occupied, or

$$\frac{-d\Theta}{dt} = k_2\Theta \tag{7.8}$$

For equilibrium conditions, the number of molecules arriving and leaving must be equal, so Θ can be solved from Eqs. 7.7 and 7.8. By assuming that only two species (A and B) take part in the deposition reaction ($SiCl_2$ and H, for example), the reaction rate (deposition rate) can be calculated from

$$\text{Rate} = K\Theta_A\Theta_B \quad 7.9$$

where Θ_A and Θ_B are the surface fractions covered by A and B and K is a rate constant. By making further assumptions, Eqs. 7.7–7.9 can be solved to give (19)

$$\text{Rate} = \frac{Kb_Ab_B\,p_Ap_B}{(1 + b_A\,p_A + b_Bp_B)^2} \quad 7.10$$

where b_A and b_B are the adsorption coefficients of species A and B. The partial pressures of A and B are p_A and p_B, respectively. Upon extending this approach to include an etching reaction based on HCl, silicon experimental data can be reasonably well fitted.

7.2.3 Material Transport

Growing an epitaxial layer requires a means of transporting the atoms of the growth material to the growth surface. The transport methods can be broadly described as follows:

1. A molecular beam—for example, from the evaporation of silicon in a hard ($\sim 10^{-10}$ torr) vacuum.
2. A vaporized compound(s) transported to the surface via a carrier gas and decomposed at the surface to give the desired material to be grown—for example, silane (SiH_4) decomposed on the surface by heat or plasma to give Si + $2H_2$.
3. A vaporized compound transported to the surface via a carrier gas and reduced at the surface by the carrier—for example, $SiCl_4$ reduced by a hydrogen carrier gas to give Si + HCl + other compounds such as $SiHCl_3$.
4. A closed tube system with a temperature differential containing feed material at the cold end, the substrate to be overgrown at the hot end, and, for example, a silicon-bearing compound that will disproportionate. The disproportionating compound reacts at the cold (seed) end to deposit material and then diffuses to the hot end, where it reacts with the feedstock material. For example, the reaction $2SiI_2 \rightarrow Si + SiI_4$ at the cool end depletes silicon, which is replenished at the hot end via $Si + SiI_4 \rightarrow 2SiI_2$.

All four methods were investigated very early (1960s), and all are still considered useful. The first, molecular beam epitaxy, is primarily a laboratory process. The bulk of the commercial applications use either the second method or the third. The last method, disproportionation, has recently been proposed for low-temperature depositions but since the 1960s has not been used commercially.

In the second and third methods, while the reacting species are brought into the reactor in a straightforward manner, the problem of getting the reactants to the growing surface still remains. If there were complete mixing of the gas over the whole reactor chamber volume, the concentration of a species at the surface would be the same as that anywhere in the chamber and would be directly calculable from the input gas concentrations. Unfortunately, such is not the case. As the gas moves downstream, it can become depleted of reactant and have a buildup of reactant products. Further, because the velocity of the gas goes to zero at the wafer surface, a "stagnant" layer will be adjacent to the wafers. There will be minimal gas mixing in the stagnant region (also referred to as a boundary or a depletion layer), so the reacting species must arrive primarily by diffusion. In some cases, this leads to diffusion-limited growth, as was discussed earlier in Chapter 4, section 4.5.1.

7.3 VAPOR PHASE SILICON EPITAXY

Vapor phase epitaxy at either atmospheric pressure or pressures of 10–100 torr and at temperatures of 950°C–1150°C is used for essentially all silicon epitaxial wafer production. However, interest in lower temperatures continues since they minimize slip[7] and impurity diffusion. Depositions in the 800°C–1000°C range can be done with conventional epitaxial reactors, while depositions from 600°C to 800°C are primarily by molecular beam epitaxy (MBE). Perhaps the most severe problem in any temperature range, but particularly below 900°C, is that of providing a clean, damage-free silicon surface on which to make the epitaxial deposition. Other problems are the incorporation of excessive amounts of substrate dopant in the first part of the layer deposited (autodoping) and the shifting and distortion of shallow depressions in the wafer surface as an epitaxial layer is added (pattern shift and distortion).

7.3.1 Surface Cleaning and Vapor Phase Etching

In order to prevent defects such as stacking faults and spurious polycrystalline growths from originating at the substrate–layer interface, the initial surface must be both clean and damage free. The standard cleanups, such as the RCA procedure discussed in Chapter 3, do a good job of removing most troublesome contaminants. However, the thin layer of oxide remaining after etching and, if deposition is at low temperature, the residual carbon from adsorbed organics

[7]The mechanics of slip generation are discussed in Chapter 8. The way slip is introduced during the epitaxial process will be described later in this chapter.

must all be removed before epi. High-temperature depositions will allow any carbon that does remain to diffuse into the substrate (20). Heavy metal contaminants adsorbed on the surface must be removed, not because they interfere with epitaxial growth, but because they will dissolve in the silicon during the high-temperature epitaxial operation and then precipitate in subsequent oxidation steps to give haze. Aqueous-based cleanups as described in Chapter 3 can be used just before the epitaxial step to remove any metals present. Damage left from the mechanical polishing operation must also be removed, and for this purpose high-temperature HCl vapor phase etching in the epitaxial reactor was introduced (21–23). However, because of advances in substrate polishing techniques, the problem of removing additional material in the reactor is no longer as important as it once was. As processing temperatures have decreased, the major problem has been that of removing the residual oxide.

Vapor phase reactions that have been studied for in situ silicon etching in an epitaxial reactor are listed in Table 7.3. The first four remove SiO_2, with the reactions done under high vacuum being designed primarily for MBE. Even with a hydrogen carrier gas, the

TABLE 7.3

Silicon Vapor Etchants for Preepitaxy Etching

Process	*Reaction*	*Minimum Temperature (°C)*	*Reference**
Thermal etching (high vacuum)	$SiO_2 + Si \rightarrow 2SiO$	800–900	(25, 26–28)
Ga beam etching (high vacuum)	$SiO_2 + 4Ga \rightarrow 2Ga_2O + Si$ $SiO_2 + 2Ga \rightarrow Ga_2O + SiO$	<800	(29)
Si beam etching (high vacuum)	$SiO_2 + Si \rightarrow 2SiO$	700	(27)
Thermal etching (atmospheric pressure)	$SiO_2 + Si \rightarrow 2SiO$	1000	(24, 30)
Hydrogen etching	$SiO_2 + H_2 \rightarrow SiO + H_2O$	1000	(24, 31, 32)
HI + HF in helium	$Si + 4HI \rightarrow SiI_4 + 2H_2$ $SiO_2 + 4HF \rightarrow SiF_4 + 2H_2O$	900	(30)
Hydrogen sulfide	$Si + H_2S \rightarrow SiS_2 + 2H_2$ $SiS_2 + Si \rightarrow 2SiS$	950	(33, 34)
Cl in helium	$Si + 2CL_2 \rightarrow SiCl_4$	1000	(35)
HI in hydrogen	$Si + HI + H_2 \rightarrow mSiI_4 + \cdots$	1000	(36)
Sulfur hexafluoride	$4Si + 2SiF_6 \rightarrow SiS_2 + 3SiF_4$	1050	(37, 38)
HCl in hydrogen	$Si + 2HCL \rightarrow SiCl_2 + H_2$	1100	(22, 23, 39)
Water vapor	$Si + H_2O \rightarrow SiO_2 + 2H_2$ $Si + SiO_2 \rightarrow 2SiO$	1250	(40)
HBr in hydrogen	$Si + H_2 + HBr \rightarrow mSiBr_4 + \cdots$	1250	(41, 42)

*See end-of-chapter references.

third reaction is thermodynamically much more likely to proceed than is the fourth reaction, as can be seen from the equilibrium partial pressure (p_e) of SiO versus temperature shown in Fig. 7.10 for each reaction (24). One difficulty with converting SiO_2 + Si to silicon monoxide at low pressures is that, at the lower temperatures, the reaction progresses so slowly that substantial silicon evaporation and pitting occur in the first regions that become free of oxide (25). If a 1200°C heat cycle is used, then the low-temperature advantage of MBE is partially lost. One approach to speed the reaction at lower temperatures is to deposit a small amount of silicon on top of the residual oxide while the wafer is held at 700°C–900°C (27). Another approach is to deposit a small amount of gallium, which will also react with SiO_2 to give a vaporizable product (29). Sputter etching (not listed in the table) is another possible method for removing oxide, but substantial damage can be introduced that may require extensive high-temperature annealing for removal (43).

The rest of the reactions in Table 7.3 remove silicon and are

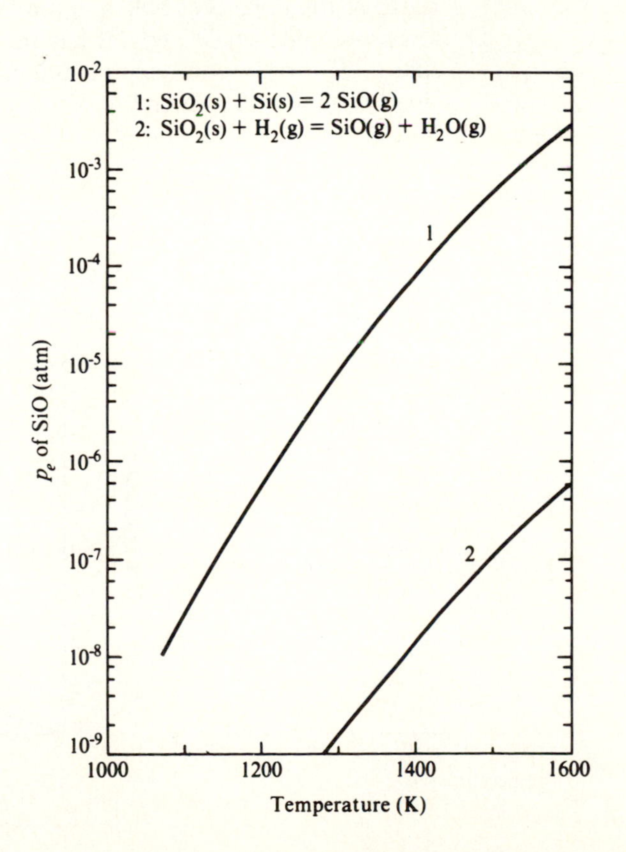

FIGURE 7.10

Equilibrium partial pressure of SiO versus temperature for reactions of SiO_2 with Si and H_2. (*Source:* P. Rai-Choudhury and D.K. Schroder, *J. Electrochem. Soc. 118,* p. 106, 1971. Reprinted by permission of the publisher, The Electrochemical Society, Inc.)

intended for damage removal. However, when used at temperatures above about 1000°C to 1050°C, any oxide present will also be removed, along with the silicon. Thus, even though Table 7.3 lists some reactions as useful down to 900°C, from a practical standpoint, not one of these reactions is satisfactory below 1000°C–1050°C. The minimum temperatures listed in the table were based on keeping a useful etch rate and providing a polishing etch rather than one that produced pitting. The most commonly used silicon vapor etch is HCl, typically at concentrations of between 2% and 4% at a temperature of about 1150°C. Fig. 7.11 shows the general way in which the etch rate varies with temperature and concentration, although the values will depend on the specific reactor design. Depending on the operating conditions, most of the vapor etchants will have polishing regions and regions in which the surface will be rough and pitted. Fig. 7.12 shows the approximate position of the temperature–concentration demarcation for $HCl–H_2$.

An alternative etchant useful at lower temperatures when the surface is already essentially damage free is hydrogen. It was the only etchant available when silicon epitaxy was introduced, but because of the imperfect polishing, it seldom removed enough silicon to ensure good-quality epi. It has now been found that if low-pressure (~25 torr) operation is combined with the use of carefully polished wafers with minimal residual oxide, hydrogen etching can be successfully used down to 950°C (44).

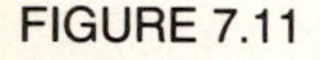

FIGURE 7.11

Effect of temperature and HCl concentration on etch rate of silicon when HCl in hydrogen is used for the etchant. (Temperatures have not been corrected for emissivity.) (*Source:* Adapted from K.E. Bean and Paul Gleim, late newspaper, Electrochemical Society Meeting, Fall 1963.)

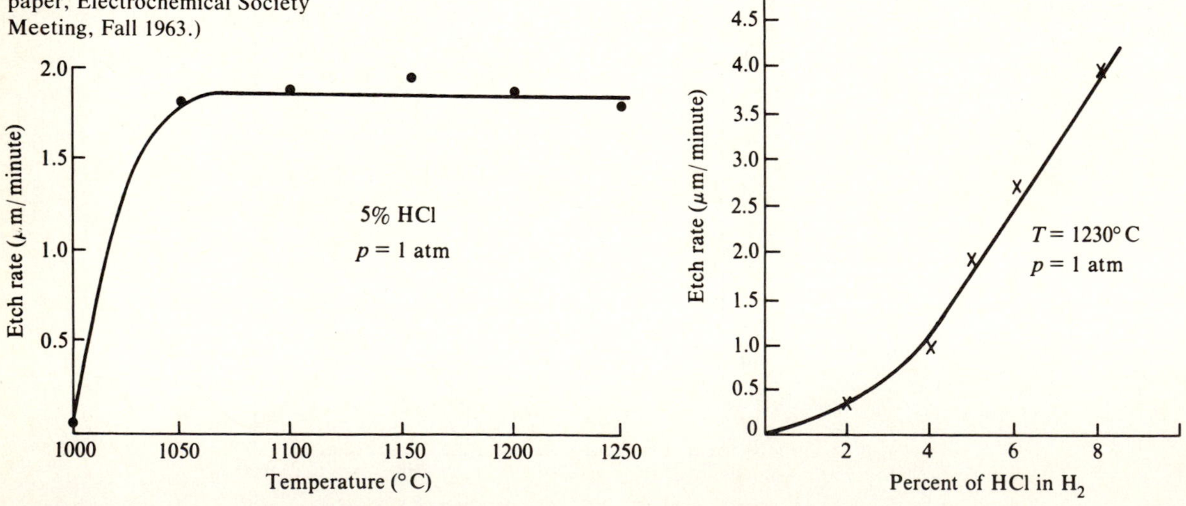

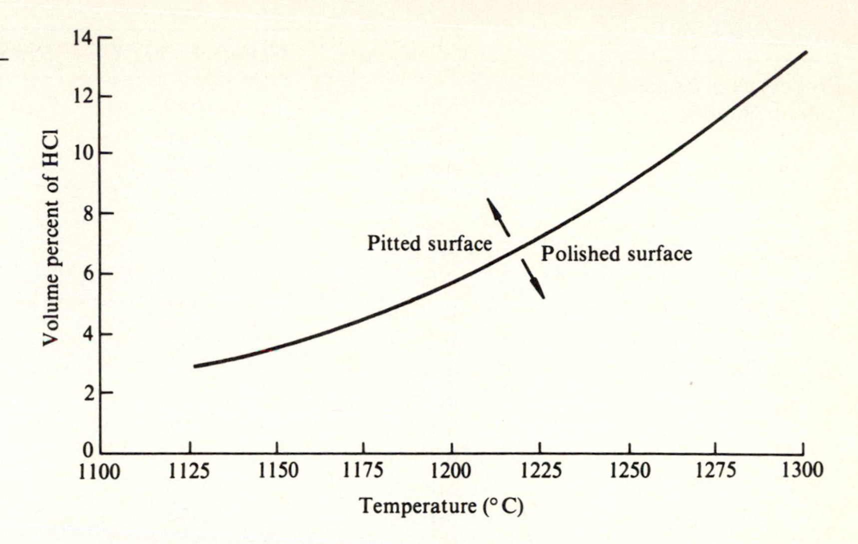

FIGURE 7.12

Effect of temperature and HCl concentration on surface finish of silicon vapor etched in $HCl + H_2$. (*Source:* Data from C.H.J. van den Brekel, *J. Crystal Growth 23*, p. 259, 1974.)

7.3.2 Deposition Reactions

The reactions that have been used to produce silicon at the wafer surface fall into the three general categories of hydrogen reduction, disproportionation, and thermal decomposition. An example of each is as follows:

1. Hydrogen reduction

$$SiCl_4 + 2H_2 \rightarrow Si + 4HCl \qquad 7.11$$

2. Disproportionation

$$SiI_4 + Si \rightarrow 2SiI_2 \quad \text{(higher temperature)} \qquad 7.12$$

$$2SiI_2 \rightarrow Si + SiI_4 \quad \text{(lower temperature)} \qquad 7.13$$

3. Thermal decomposition

$$SiH_4 \rightarrow Si + 2H_2 \qquad 7.14$$

Of these, only hydrogen reduction and thermal decomposition are currently used, with halides and silanes as the most common source compounds. In principle, the fluorine, chlorine, bromine, and iodine series of compounds, listed in Table 7.4, could all be used.[8] However, SiF_4, for example, can be reduced by hydrogen only above 2000°C (45), SiI_4 requires heated lines since it does not boil until

[8]This list is not complete. A variety of more complex H–Cl–Si compounds could conceivably be used.

TABLE 7.4

Properties of Some Silicon Halides

Compound	*Melting Point (°C)*	*Boiling Point (°C)*	*H_f (kcal/mol)**
SiF_4	Sublimes		−370 †
$SiHF_3$	−131	−80	
SiH_2F_2	−122		
SiH_3F		−99	
$SiCl_4$	−70	58	−157 ‡
$SiCl_2$			−38 ‡
$SiHCl_3$	−126	33	−117 ‡
SiH_2Cl_2	−122	8	−75 ‡
SiH_3Cl	−118	−30	−34 ‡
$SiBr_4$	5	153	−95 §
$SiBr_2$			−12 §
$SiHBr_3$	−73	109	−73 §
SiH_2Br_2	−70	66	−46 §
SiH_3Br	−94	2	−19 §
SiI_4	124	290	−32 §
SiI_2			+18 §
$SiHI_3$	8	220	−21 §
SiH_2I_2	−1	150	−11 §
SiH_3I	−57	45	−2 §

*Heat of formation at 298°C.
†E.A.V. Ebsworth, *Volatile Silicon Compounds,* The Macmillan Co., New York, 1963.
‡L.P. Hunt and E. Sirtl, *J. Electrochem. Soc. 119,* p. 1741, 1972.
§L.P. Hunt and E. Sirtl, *J. Electrochem. Soc. 120,* p. 806, 1973.

Source: Adapted from *Comprehensive Inorganic Chemistry,* Vol. 7, Sneed and Brasted, D. Van Nostrand Co., New York, 1958.

290°C, and SiI_2 decomposes at room temperature. Silane (SiH_4), also referred to as monosilane, is widely used, but the next higher silane, Si_2H_6 (disilane) has also been studied (46). Table 7.5 lists the properties of some of the silanes. SiH_4 is quite stable at room temperature, but as the order increases, the stability decreases, with Si_6H_{14} decomposing at room temperature over a period of a few months (47). The thermal decomposition of carbon–silicon compounds such as CH_3SiCl_3 can also provide a source of silicon (48), but, in general, the decomposition of carbon-containing compounds produces carbon along with the silicon and leads to polycrystalline growth.

The final choice of a feed material will depend not only on whether growth can be made to take place but also on the relative complexity of the necessary equipment, on the relative safety of the compound, and on which compounds are readily available at reasonable cost in the required purity range. Table 7.6 lists the most common feed materials, along with the probable reaction path for each. In all cases involving hydrogen reduction, the reactions are

TABLE 7.5

Properties of Silanes

Gas	Melting Point (°C)	Boiling Point (°C)
SiH_4	−185	−112
Si_2H_6	−133	−15
Si_3H_8	−117	53
Si_4H_{10}	−94	80

Source: Compiled from data in E.A.V. Ebsworth, *Volatile Silicon Compounds*, The Macmillan Co., New York, 1963.

TABLE 7.6

Common Source Gases for Epitaxial Silicon

Source	Deposition Temperature* (°C)	Possible Main Reaction Path†
$SiCl_4$	1150–1250	$SiCl_4 + H_2 \rightleftarrows SiHCl_3 + HCl$
		$SiHCl_3 \rightleftarrows SiCl_2 + HCl$
		$SiCl_2 + H_2 \rightleftarrows Si + 2HCl$
$SiHCl_3$	1100–1200	$SiHCl_3 \rightleftarrows SiCl_2 + HCl$
		$SiCl_2 + H_2 \rightleftarrows Si + 2HCl$
SiH_2Cl_2	1000–1100	$SiH_2Cl_2 \rightleftarrows Si + 2HCl$
SiH_4	950–1050	$SiH_4 \rightarrow Si + 2H_2$

*Typical.
†J. Nishizawa and M. Saito, *J. Crystal Growth 52,* p. 213, 1981, and R.F.C. Farrow, *J. Electrochem. Soc. 121,* p. 899, 1974.

much more complex than that indicated by Eq. 7.11, and several intermediates are formed, including small quantities of long-chain, oily, explosive compounds. The equilibrium partial pressures of the various components that exist in an H_2–Cl–Si system as a function of temperature and Cl/H ratio have been calculated (49). Some of these are shown in Fig. 7.13 and can be used as a guide in judging what reactions are probable during deposition. Also of help is the sampling of the gas stream just above the growing silicon surface in order to see what products are present. The possible equations given in Table 7.6 were based on such data (50, 51).

7.3.3 Deposition Rates

Silicon epitaxial deposition rates can range from infinitesimally small up to several microns per minute, depending on the temperature and feed material. However, since layer thicknesses are generally from 1 μm to 10μm, a rate of a few tenths up to a few microns per minute is a range that provides good crystallographic quality and thickness control. Slower rates can make the deposition time so long as to be economically unattractive, and substantially higher rates lead to layer defects.

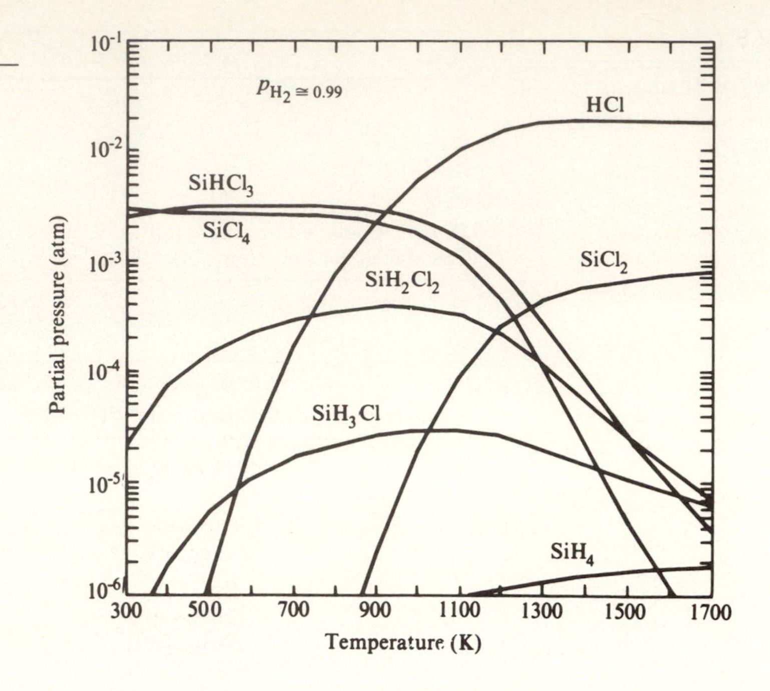

FIGURE 7.13

Equilibrium partial pressure versus temperature for major constituents of the vapor above a free silicon surface when only hydrogen, chlorine, and silicon are present. (*Source:* E. Sirtl, L.P. Hunt, and D.H. Sawyer, *J. Electrochem. Soc. 121*, p. 919, 1974. Reprinted by permission of the publisher, The Electrochemical Society, Inc. *Note:* Recent revisions in values of the thermodynamic data used in calculating these curves will lead to some changes. See L.P. Hunt, *J. Electrochem. Soc. 135*, p. 206, 1988.)

The rate behavior versus temperature and gas flow is shown in Fig. 7.14. At low temperatures, the rate r versus temperature is closely given by an Arrhenius-type equation (region A of Fig. 7.14a):

$$r = Ae^{-E/RT} \tag{7.15}$$

where A is a constant, E is an activation energy, R is the gas constant, and T is the absolute temperature. In this temperature range, reaction rates (kinetics) determine the deposition rate. As the temperature increases, however, a point is reached where the nutrient supply rather than reaction rate is the limiting factor. When the reactants reaching the surfac e are limited by diffusion through the stagnant gas film adjacent to the wafer, the rate becomes relatively temperature insensitive, as shown by the flat portions of the curves of Fig. 7.14a (curves B). When this region is encountered, it is, in principle, possible to increase the rate up to the rate predicted by curve A by increasing the flow rate. The increased flow reduces the thickness of the boundary layer discussed in Chapter 4 and increases the diffusion rate until it is no longer the limiting step. This increase of rate is shown in Fig. 7.14b. Based on simple theory of the reduction in layer thickness with flow, a plot of the rate versus square root of flow should give a straight line until the rate approaches that predicted by Eq. 7.15 (52). In fact, this is not necessarily observed ex-

FIGURE 7.14

Dependency of deposition rate on temperature and flow. (When the temperature becomes high enough for homogeneous nucleation to occur in the gas stream, the rate decreases as indicated by the dashed lines in part a.)

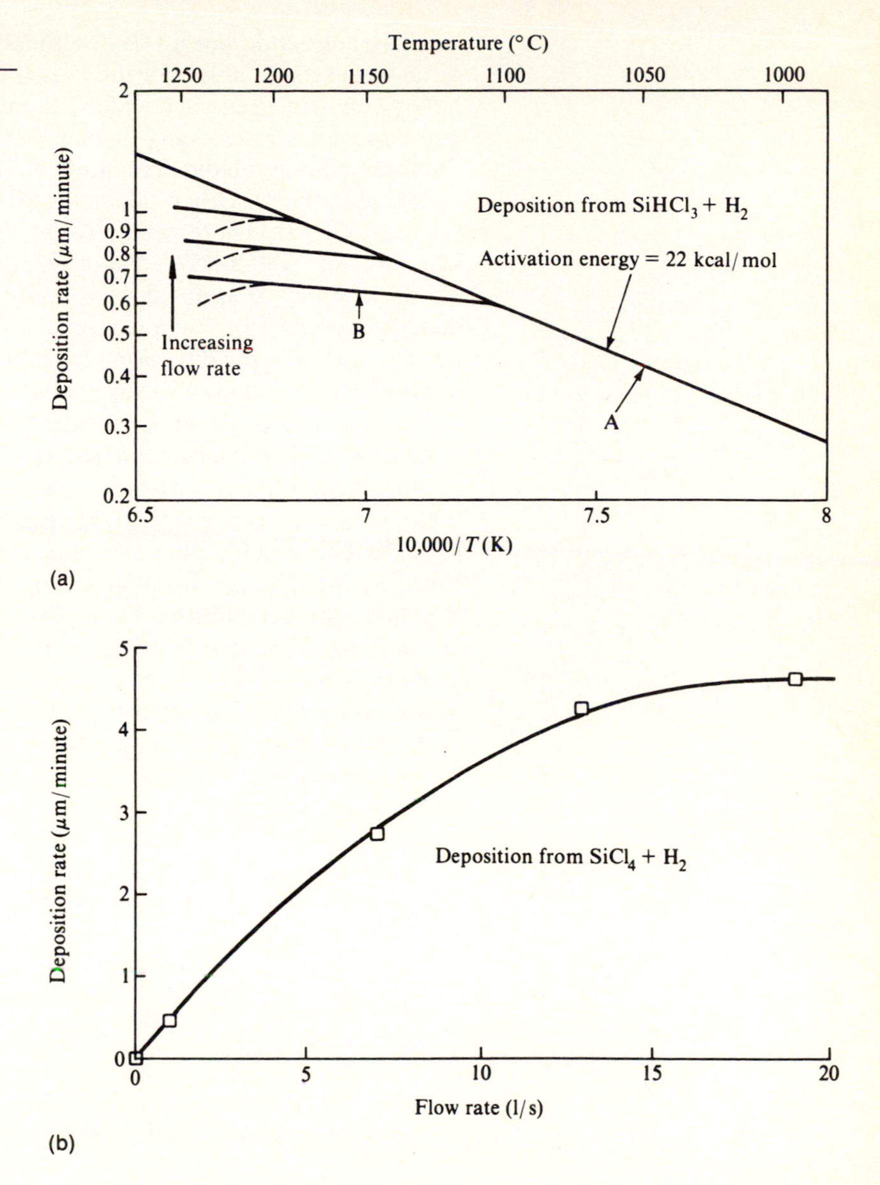

perimentally, but the maximum rate observed does appear to match that predicted by an extrapolation of the Arrhenius curve measured at lower temperatures (53). From a practical standpoint, since it is difficult to maintain uniform temperature over a large multiwafer reactor, diffusion-limited operating conditions are often chosen in order to minimize layer thickness sensitivity to temperature.

The activation energy E is a function of the reaction feed material being used, although the "apparent activation energies" that have been measured vary widely. Some data, based on polycrystalline deposition rates, suggest that there is little difference in E for the three common halide reactants of Table 7.6 and that the value is ~33 kcal/mol (54). Other values for $SiCl_4$ range from 37–50 kcal/mol (31, 53, 55, 56). There seems to be reasonable agreement of ~22 kcal/mol for $SiHCl_3$ (57). Values reported for SiH_2Cl_2 range from ~5–13 kcal/mol at atmospheric pressure (58) to ~24 kcal/mol at reduced pressure (59). The lower values apparently arose from making measurements in the diffusion-controlled region. Reported silane activation energy values also vary widely, ranging from 10–37 kcal/mol (51, 60). It has been demonstrated that a high level of boron can reduce the apparent activation energy (51) and that possibly contamination of the surface with other materials such as oxygen or carbon will increase it. Despite the fact that the deposition rate in the kinetically controlled region has been shown to be orientation dependent, no report has been given of differing activation energies depending on orientation. Thus, the orientation effect is contained in the pre-exponential term (A) of Eq. 7.15. The wide variability in activation energy data leads once again to the statement that many of the properties of interest in semiconductor processing are highly dependent on the experimental conditions. Hence, checking the performance of any new reactor configuration rather than assuming its behavior is advisable.

EXAMPLE ☐ Using the definition of the activation energy E given in Eq. 7.15, verify that the activation energy of the curve shown in Fig. 7.14a really is ~22 kcal/mol.

Take the log of both sides of Eq. 7.15, $\ln r = \ln A - E/RT$. Choosing two values of r, (r_1 and r_2) and determining the corresponding values of $1/T_1$ and $1/T_2$ give, after some manipulation, $E = \ln(r_1/r_2)/[(1/T_2) - (1/T_1)]$. From the curve, for $r = 1$ μm, $T = 1200°C$ and $1/T = 0.68 \times 10^{-3}$ per K. For $r = 0.3$ μm, $1/T = 0.79$. From Appendix C, $R = 2 \times 10^{-3}$ kcal/K·mol. Doing the arithmetic gives $E = (2/0.11) \ln(3.33) = 21.6$ kcal/mol. ☐

As the concentration of $SiCl_4$ or $SiHCl_3$ increases, the rate initially increases but then peaks and eventually decreases and becomes negative so that etching occurs, as shown in Fig. 7.15. Dichlorosilane behavior up to about 12% is also shown. Presumably, it will also etch silicon at higher concentrations. The silane decomposition reaction appears to be first order in that the rate increases linearly with partial pressure. Thus, unlike the halides, where increased concentration soon leads to a reduced deposition rate, rates

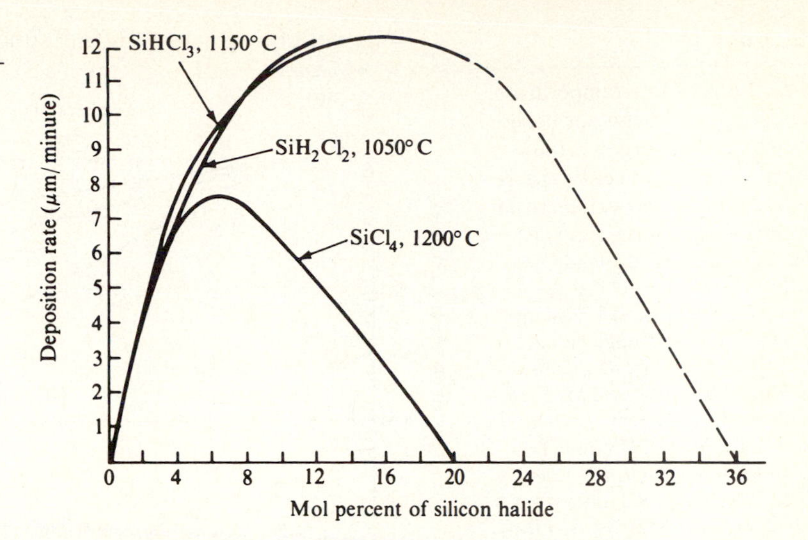

FIGURE 7.15

Effect of increasing halide concentration on deposition rate. (Quantitative comparisons between halides should not be made based on these curves since the data were obtained from different reactors operating under different conditions.)

can be very high and have been reported in excess of 100 μm/minute at 1100°C (15). HCl is sometimes added to silane during deposition, where it reduces the rate as well as gas phase nucleation at higher temperatures (15, 61).

Processing temperatures have gradually decreased since the introduction of the IC because of the desire to minimize high-temperature-induced defects and impurity redistributions due to diffusion. In the case of epitaxy, the specific driving forces have been the desire to reduce substrate up-diffusion because of the trend toward thinner epitaxial layers, the need to minimize wafer slip during epitaxy, and the desire to minimize the effect of deposition on oxygen precipitation behavior. As curve A of Fig. 7.14 predicts, when the temperature is reduced to the 700°C–800°C range, the deposition rates become much less. Fig. 7.16 shows how the low-temperature rates would be expected to drop for activation energies appropriate for the common silicon sources. For these curves, a rate of 1 μm/minute was assumed at 1050°C. Also shown are some experimental points for the atmospheric pressure deposition of $SiCl_4$, SiH_2Cl_2, and silane, as well as two 775°C rates obtained from low-pressure silane depositions. These data and curves indicate that usable rates can be obtained with some sources down at least to the 750°C range. However, as the temperature is lowered, it becomes progressively harder to initially clean the surface of oxide and to keep it clean during deposition. As will be discussed in a later section, nucleation on an oxide surface is much more difficult than nucleation on silicon. Thus, if oxide either is allowed to form during deposition or is not completely removed beforehand, the crystalline

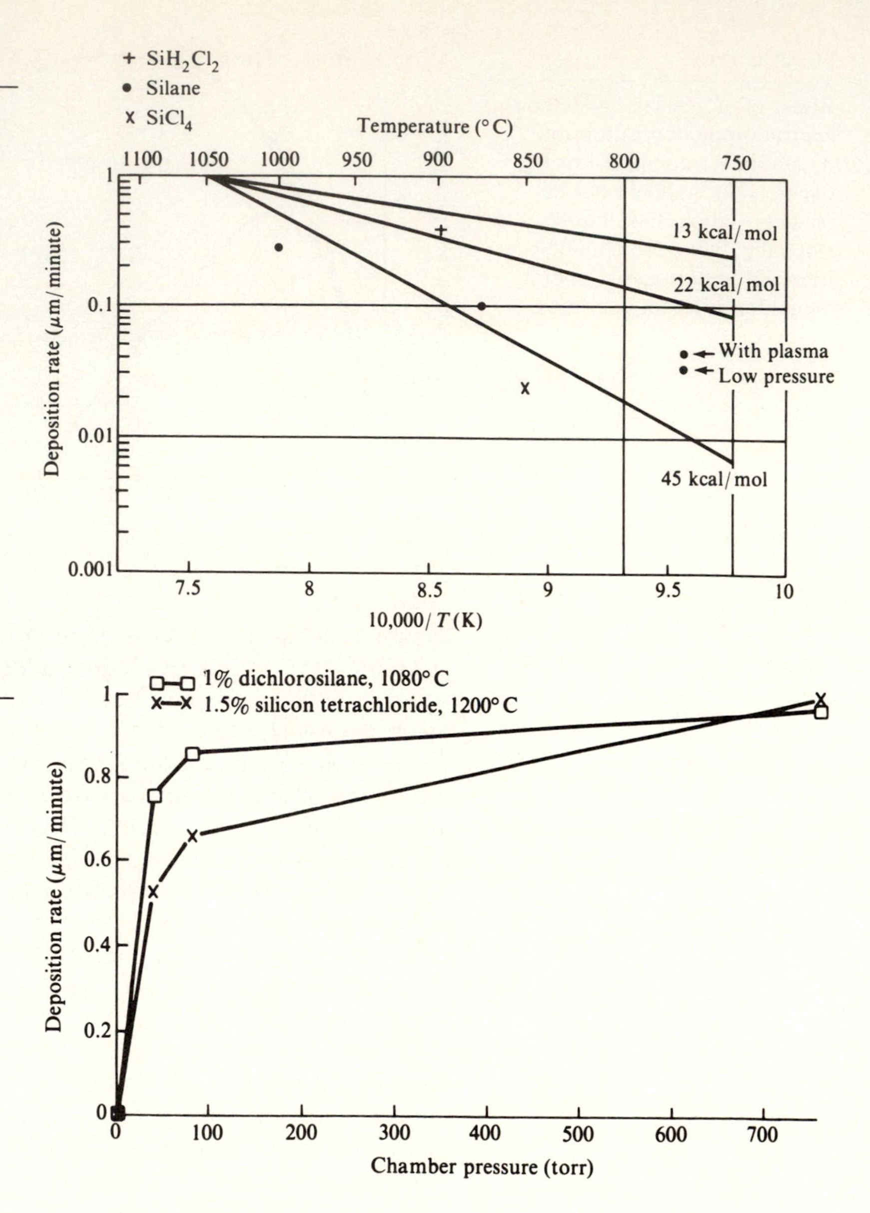

FIGURE 7.16

Projections of low-temperature deposition rate behavior based on activation energies. (Individual points represent experimentally observed values from the following sources: SiH_2Cl_2 data from John D. Borland and Clifford I. Drowley, *Solid State Technology*, p. 141, 1985; Silane data from Hseuh-Rong Chang, *ECS Extended Abst. 85–1*, abst. no. 276, L.D. Dyer, *AIChE J. 18*, p. 728, 1972, J. Bloem, *J. Crystal Growth 18*, p. 70, 1973, and T.J. Donahue et al., *Appl. Phy. Lett. 44*, p. 346, 1984; $SiCl_4$ data from S. Nakanuma, *IEEE Trans. Electron Dev. ED-13*, p. 578, 1966.)

FIGURE 7.17

Deposition rate versus chamber pressure for a barrel reactor. (*Source:* From data in R.B. Herring, Applied Materials Tech. Rpt. HT–010, 1980.)

perfection will be severely degraded and the deposition rate reduced from that predicted from an extrapolation of an Arrhenius plot such as the one in Fig. 7.16.

As the chamber pressure is reduced, the deposition rate for a given feedstock concentration decreases first rather slowly and then quite rapidly when the pressure decreases below about 100 torr. Fig. 7.17 shows the general behavior for the case of $SiCl_4$ and SiH_2Cl_2

(62). The major effect of reducing pressure is to reduce the width of the stagnant layer next to the wafer surface. This reduction can, in turn, reduce autodoping (as discussed in a later section) and often can provide better thickness uniformity.

7.3.4 Effect of Orientation on Deposition Rate

Under some conditions, the deposition rate depends on the crystallographic orientation. Thus, flats or "facets" perpendicular to the slow-growth directions form on curved surfaces. The study of orientation effects therefore includes both a measurement of deposition rates on various chosen planes and an examination of the orientation of facet faces, their relative size, and their frequency of occurrence. The experimental data are rather sparse, but data based on growing from $SiCl_4$ at about 1200°C onto wafers cut to given orientations indicate that the deposition rate varies in the order (111) < (100) < (110), with the total variation being about 1:2 (10, 63). Differences in growth rates on variously oriented wafer surfaces would be expected to be a maximum when deposition rates are surface controlled (in region A of Fig. 7.14a). Unfortunately, most of the rate–orientation data was taken during the time period when high temperatures were deemed necessary for good-quality growth, and it is not clear whether depositions were in the diffusion-controlled or kinetically controlled regions. However, a more recent plot of log growth rate versus $1/T$ for SiH_2Cl_2 + HCl at reduced pressure in the 900°C–1000°C temperature range indicates the same activation energy for deposition on both (100) and (111) planes, with a rate difference of about 2:1 (59). If the activation energy is indeed independent of the plane of deposition, then the orientation effect is contained in the pre-exponential term of Eq. 7.15.

As an alternative to directly measuring rate, the comparative width of facets that form on round rods or hemispheres can be studied (64–66). During facet formation, the slow-growing faces become the largest. However, after prolonged growth, orientation of the long dimension of the rod or the base of the hemisphere affect which faces remain. For example, if crystal growth is on a rod running in a <111> direction, the rod first becomes roughly hexagonal, with alternate (110) and (211) planes running the length of the rod. With further growth, however, the (110) planes disappear, and the rod becomes bounded completely by (211) planes. If a (211) rod is used, two of the opposing faces will be (111), but to form a closed figure, two pairs of (311) planes are also required.

Facet formation is a commonly studied feature of crystal growth, whether it is from the melt, a solution, or a vapor. Column 1 of Table 7.7 lists facets that have been reported for silicon-melt growth along with the growth rate order in directions perpendicular to the facets. Column 2 lists the facets observed during vapor phase

TABLE 7.7

Silicon Facet Occurrence and Order of Growth Rate for Various Orientations

References					
(1)	*(2)*	*(3)*	*(4)*	*(5)*	*(2)*
110	211	110	110	110	100
311	100	211			311
100	110		100	100	110
	311	311	211		
111	111	111	111	111	111

Note: Rate increases from bottom to top.

References: (1) G.A. Wolff, *Am. Mineralogist 41,* p. 60, 1956. (2) C.H.J. van den Brekel, *J. Crystal Growth 23,* p. 259, 1974. (3) J.E. Allegretti et al., pp. 255–270, in Ralph O. Grubel, ed., *Metallurgy of Elemental and Compound Semiconductors,* Interscience Publishers, New York, 1961. (4) S. Mendelson, *J. Appl. Phys. 35,* p. 1570, 1964. (5) S.K. Tung, *J. Electrochem. Soc. 112,* p. 436, 1965.

growth onto a silicon hemisphere; column 3, the facets observed for growth onto variously oriented silicon rods. The other columns show data taken after growth onto wafers cut to specific orientations. It can be seen from columns 1–3 that the same facets occur during growth from the melt as during growth from the vapor, and all belong to planes of a (110) zone.[9] The fact that the same facets are found in both melt and vapor phase grown material indicates that it is crystal structure and not a surface reaction that determines which facets are likely. However, it is quite possible for the kinds of atoms adsorbed on the surface to change the relative growth rates of the facets.

Crystal habit changes (a change in the slow-growing planes) can be observed when crystals are grown from water solution if appropriate impurities are added to the growth solution (67). A similar phenomenon is observed when silicon is grown by vapor phase disproportionating reactions containing tellurium in that the slowest-growing planes become (100) and cubes of silicon grow instead of octahedrons (68). Some studies of faceting of small depressions in (100) oriented wafers have shown that the faceting becomes more pronounced as the temperature decreases and less pronounced as the operating pressure is decreased (69). It also appears that window faceting decreases as the number of chlorine atoms in the feedstock decreases (although deposition from SiH_4 still shows some faceting) (70). Decreased faceting with decreasing pressure is consistent with some reaction species adsorbed on the growing surface enhancing

[9]See Appendix A for a discussion of crystallographic zones.

the growth of a particularly oriented facet, probably a (311) in this case.

7.3.5 Layer Doping

To provide doping of the epitaxial layer requires co-depositing a suitable dopant with the silicon. The dopant concentration in the layer necessary for a given resistivity is well defined and is available in graphic form. (See Chapter 8 for a discussion and curve.) The first doping procedure used liquid $SiCl_4$ as the silicon feed material and mixed a dopant halide such as boron trichloride, antimony pentachloride, or phosphorus trichloride with it in the supply vessel. Thus, as the carrier gas picked up $SiCl_4$ in the bubbler,[10] it also picked up some of the dopant and simultaneously carried it into the reactor where it too was reduced by the hydrogen and co-deposited with the silicon. Since the two vapor pressures are usually not the same, the relative concentration changes as the volume of $SiCl_4$ is reduced during deposition.[11] Thus, in turn, the relative concentration in the gas stream changes and thus the resistivity of the layer.

In order to provide more uniformity of doping and to improve the ease of changing resistivity, the dopant and the silicon-bearing material are now added separately to the carrier gas stream. Liquid sources can still be used, but it is much more common to use gaseous sources such as phosphine (PH_3), diborane (B_2H_6), and arsine (AsH_3). Since these gases are very toxic (see Chapter 4), they are generally supplied diluted with either hydrogen or an inert gas. An alternate method of supplying small quantities of doping hydride (but one that has seldom been used) is to manufacture it on demand from a high-frequency discharge (71).

Over the range of temperatures and doping levels normally encountered in epitaxy, the silicon growth rate can be considered independent of the amount or kind of doping. However, in the 600°C temperature range commonly used for polysilicon deposition, some dopants do affect the silicon deposition rate. Phosphorus, for example, can severely reduce the rate apparently because, at low temperatures, the phosphorus successfully competes for surface sites and reduces silicon compound adsorption (72).

---

[10] A bubbler is a common contrivance for saturating a gas with vapor from a liquid. The carrier gas is allowed to bubble up through a column of liquid, during which time, if the liquid path is long enough, the bubbles become saturated with vapor from the liquid. By changing the temperature of the liquid, and hence its vapor pressure, the amount of liquid transferred to the carrier gas stream can be changed. For a further discussion, see Chapter 8.

[11] As pointed out by H.C. Theuerer of Bell Laboratories, by blowing the carrier gas over a frit (wick) constantly wet by the liquid, this problem can be eliminated since the wick provides isolation from the reservoir.

For a fixed dopant concentration, the amount of antimony, arsenic, and phosphorus incorporated into the layer decreases with increasing deposition temperature (73, 74), while boron increases (75). When growth is in the diffusion-controlled region, increasing the deposition rate causes the amount of antimony, arsenic, and phosphorus to increase so that the resistivity drops (74).

The behavior of phosphorus has been studied more extensively than that of other dopants, and the phosphorus-bearing species present in the gas stream as a function of temperature has been calculated for the case of a phosphine doping source (76, 77). As long as the doping is low enough for the layer to be intrinsic at the deposition temperature, the doping level is proportional to the concentration of atomic P in the gas stream. Over the lower range of PH_3 concentrations, the atomic P concentration is proportional to that of PH_3 in the input gas stream. Above a point that corresponds to a doping level of about 10^{18} atoms/cc, the atomic P concentration is proportional to the square root of the PH_3 concentration. When the conditions are such that the layer is extrinsic at the deposition temperature, the charge of the dopant ion affects its incorporation, and the doping level becomes proportional to the square root of the atomic P concentration (and thus generally to the fourth root of the PH_3 concentration).

EXAMPLE ☐ If the doping level is 10^{18} atoms/cc, for what deposition temperatures will the layer be extrinsic at deposition temperature?

As the criterion for intrinsic material, let the number of ionized dopant atoms be equal to $n_i/2$. From the curve of n_i versus temperature given in Chapter 8, it is found that $n_i = 2 \times 10^{18}$ carriers/cc at 800°C. ☐

Based on the preceding example, since normal doping levels for epitaxy are below 10^{18} atoms/cc and the temperatures are generally above 1000°C, phosphorus doping can be expected to be proportional to the phosphine concentration.

7.3.6 Impurity Redistribution during Epitaxy

When abrupt concentration profiles, such as a high-resistivity layer on a low-resistivity substrate, are attempted, the actual impurity profile, instead of being abrupt as shown in Fig. 7.18a, appears as shown in Fig. 7.18b. Not only is the transition region more graded than would be expected from diffusion alone, but also the high-resistivity layer may not be as high as anticipated. The gradation is due in part to diffusion from the substrate that occurs during epitaxial growth and in part from gas phase transport of dopant from the substrate (autodoping). The procedure for calculating the amount of

FIGURE 7.18

Effect of autodoping on concentration profile near interface between an epitaxial layer and a heavily doped substrate.

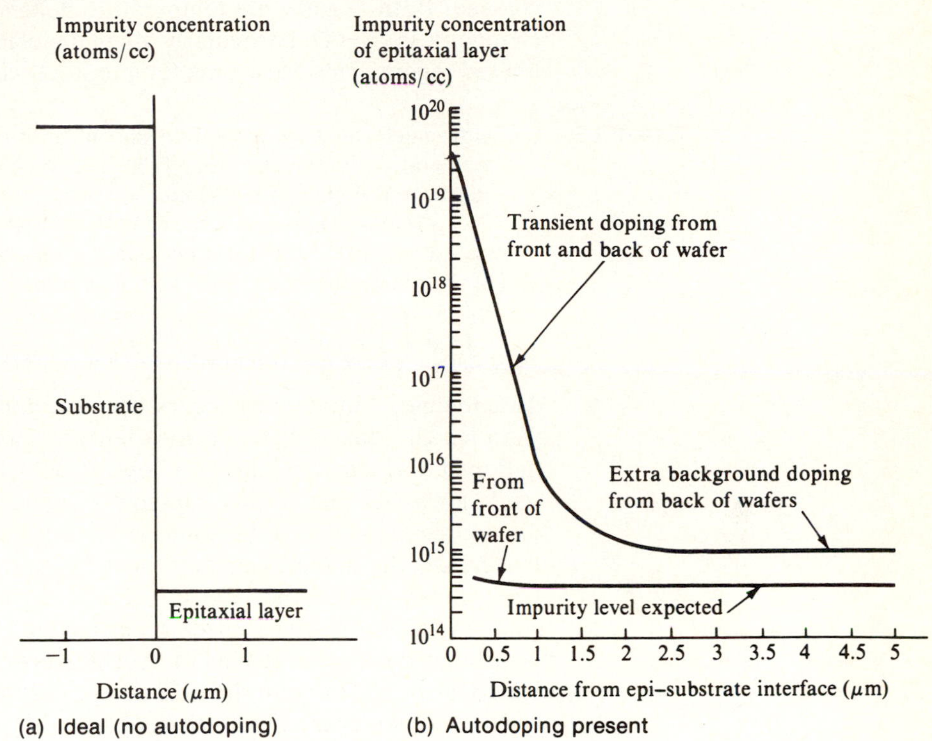

(a) Ideal (no autodoping)

(b) Autodoping present

diffusion is discussed in Chapter 8. Ordinarily, the effect of autodoping overshadows that of diffusion. However, when diffusion is dominant, there are some guidelines for reducing its effect.

Diffusion To a first approximation, when diffusion is dominant, the distance X from a concentration step to the point where the concentration is reduced by $1/e$ is given by

$$X = \sqrt{Dt} \tag{7.16}$$

where D is the solid-state diffusion coefficient and t is the diffusion time. X can be used as a measure of diffusion; smaller X's imply less diffusion. The time t is equal to the time of epitaxial growth and is d/r where d is the thickness of the layer and r is the epitaxial growth rate. Thus,

$$X = \sqrt{\frac{Dd}{r}} = \sqrt{d}\sqrt{\frac{D}{r}} \qquad 7.17$$

At a given temperature, X is decreased if the deposition rate is increased. Both D and r are temperature dependent, but, generally, r changes less than D, particularly when deposition is in the diffusion-limited region. Thus, lowering temperature will also reduce X.

EXAMPLE ☐ How much will X decrease for a silicon deposition if the deposition temperature is reduced from 1100°C to 900°C? Assume that the substrate is doped with arsenic.

A typical deposition rate at 1100°C is 1 μm per minute. A typical rate at 900°C is 0.1 μm per minute. The diffusion coefficients of arsenic in silicon at these two temperatures are $\sim 2 \times 10^{-14}$ and $\sim 1 \times 10^{-16}$ cm²/s, respectively. Substituting these values in Eq. 7.17 gives a reduction factor of ~4.5. ☐

Autodoping Autodoping occurs because impurities find their way from heavily doped wafer regions into the gas ambient and/or onto undoped surfaces and then co-deposit with the growing film. Initially, it was speculated that the growth mechanism using silicon halide sources would involve some etching as well as growth and by that means the dopant would enter the gas stream. However, it has been demonstrated that the effect is still present when a silane source is used, in which case no co-etching is possible. It now appears that the dopant leaves the surface primarily by evaporation. It may also diffuse from the bulk to the surface and then move laterally without ever leaving the surface. Fig. 7.19 shows possible paths for the two cases of a heavily doped wafer (appropriate for discrete devices or MOS) and a lightly doped wafer with heavily doped local areas (the usual bipolar case). During any time at temperature before deposition begins, dopant can evaporate, enter the

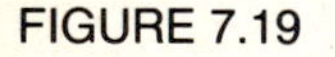

FIGURE 7.19

Sources of autodoping.

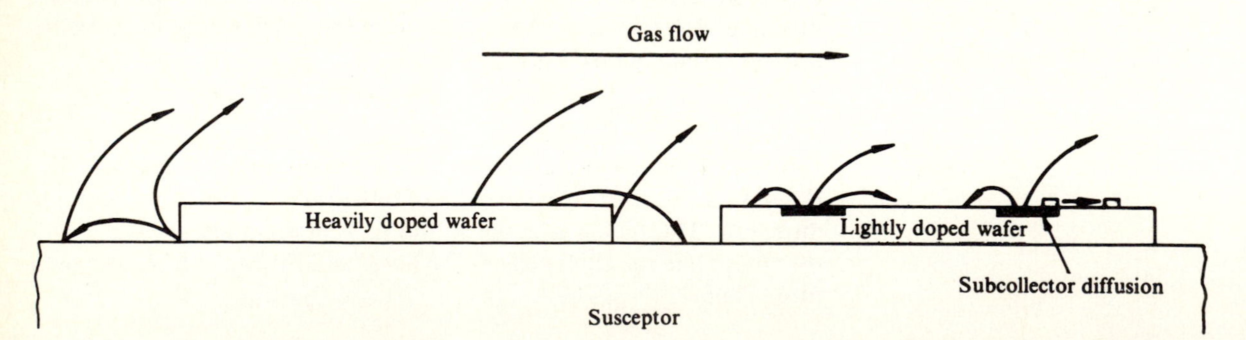

gas stream, and can either deposit on adjacent uncovered susceptor surfaces in the case of heavily doped wafers or can deposit adjacent to diffused areas in bipolar wafers. The edges and back of heavily doped wafers may not be totally covered during deposition and may continue to outgas throughout the deposition cycle. In this case, the outgassing dopant appears as extra background doping, as shown in Fig. 7.18b. For growth over small areas of local high doping, not only will autodoping occur above the heavily doped region as indicated in Fig. 7.18b (vertical autodoping), but also lateral doping will take place out from the region boundaries (lateral autodoping).

Autodoping is an effect that can materially affect the performance of devices made in epitaxial material,[12] and consequently numerous studies have been published since the effect was reported in 1961 (78). Early models were concerned with the effect from heavily doped wafers and considered that the dopant was transferred first to the ambient gas and then to the growing layer. One model considered only the front side as a source (79); others considered that the major contribution was from the back and that it was never completely covered and sealed (80–82). A model was next developed that considered front- and back-side sources as well as diffusion (83). Since bipolar wafers with subcollector diffusions have no heavily doped back-side, and since wafers that do have such doping are generally sealed with an oxide or polysilicon, a strictly front-side source model is required in these cases. The salient experimental observations are as follows:

1. In a horizontal reactor, the effect increases for downstream wafers (84).
2. A doping tail occurs both normal to the surface (axially) (78) and laterally out from individual heavily doped subcollector regions (85).
3. On an individual wafer, the effect is more pronounced the larger the fractional area covered by subepitaxial diffusions and varies approximately as the square root of the fraction (86).
4. Both chlorine-bearing silicon source and silane source depositions show autodoping (87).
5. Of the common silicon dopants, Sb produces the least autodoping (84).

---

[12]As examples, excess lateral doping can reduce packing density, and axial (vertical) autodoping, which has the same effect as excessive substrate up-diffusion, can cause reduced collector–base breakdown voltages unless the layer thickness is increased.

6. A high-temperature prebake in H_2 will minimize the effect (84).
7. For some reactor configurations, higher deposition rates increase autodoping (88).
8. For As and Sb, decreased deposition temperature increases autodoping (89). For B, it decreases autodoping (90). For P, it has been reported to decrease (90) and increase autodoping (44).
9. For As and Sb, decreased deposition pressure decreases autodoping (91). For B and P, the reverse is true (90).

Models now assume that the majority of the dopant comes from out-diffusion during preepitaxial heating and that it is temporarily stored by surface adsorption (89, 92–94). For some reactor designs and operating conditions, autodoping profile modeling is now included in the Stanford University computer modeling program SUPREM (95). The reduction in autodoping when Sb is used occurs because of its reduced efficiency in being transferred into the growing silicon (81). However, because the low segregation coefficient of Sb prevents growing heavily Sb-doped Si crystals and because its low solid solubility in Si prevents heavy doping by diffusion, the resistivity of Sb-doped wafers and Sb-diffused regions in high-resistivity wafers cannot be made as low as those of As. Hence, As doping is sometimes used. The increase in autodoping for As and Sb as the temperature is lowered and the decrease for B and P are apparently due to different temperature behaviors for the co-deposition processes of the different dopants. Note that the spread in the reported magnitude of this effect is considerable and that some data show little difference in behavior between dopants (96). The increase in autodoping with growth rate has been explained both in terms of a decreased time for the gas stream to carry away the dopant (88) and in terms of a decreased time for the concentration of adsorbed impurities on the surface to equilibrate with the gas ambient and the growing interface (89). Because it increases the gas diffusion coefficients, a reduction in operating pressure would seem to lead to a more efficient transfer of dopant to the part of the gas stream being swept from the reactor and thus decrease autodoping. Experimentally, autodoping for Sb and As is reported to be reduced at low pressures, while B and P autodoping is increased (90). It is also observed that autodoping behavior as a function of both pressure and growth rate is dependent on the deposition temperature (44, 96).

7.3.7 Pattern Shift and Distortion

When a shallow, flat-bottomed depression in a silicon wafer is overgrown by an epitaxial layer on the wafer surface, the edges of the

FIGURE 7.20

Diagrams showing various ways in which a shallow depression in a silicon surface changes in shape and position after epitaxial deposition.

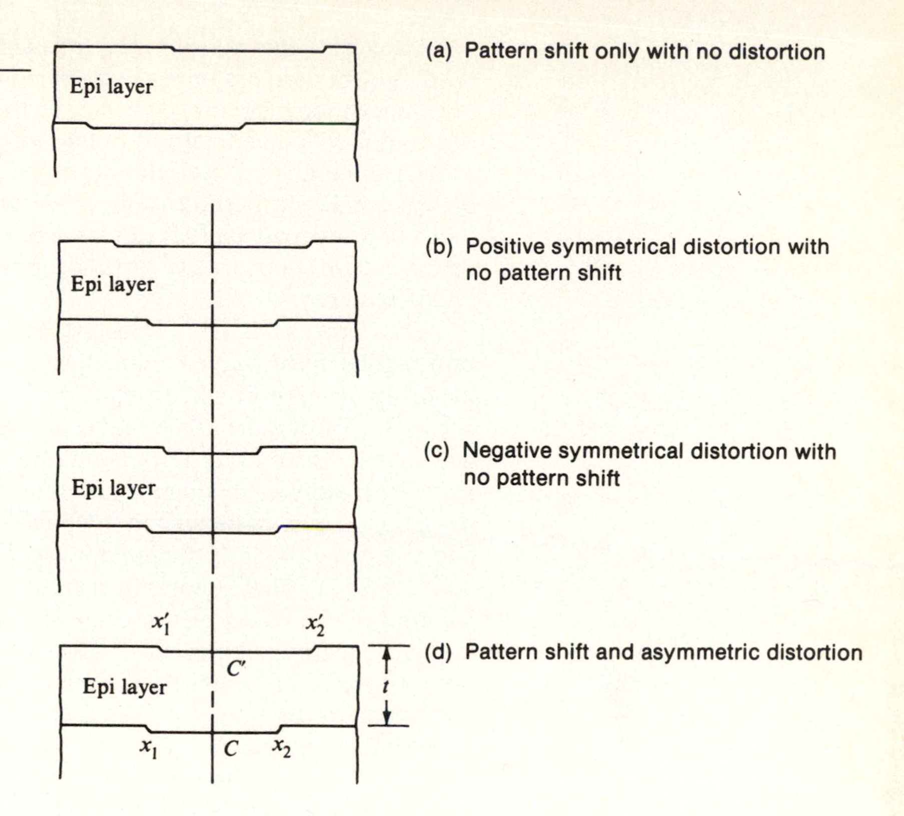

feature showing after growth will usually not be directly over the boundary of the original depression. The after-epi depression is ordinarily larger (pattern distortion) than the original, and the centers of the two depressions may not lay on top of each other (pattern shift). Examples of pattern distortion and shift are shown in Fig. 7.20. These depressions generally have a depth of about 1000 Å, and the epitaxial film thickness is on the order of 10 μm so that the step height to thickness ratio is actually much less than shown in the figure. The pattern shift is defined as the shift of the center of the pattern ($C' - C$ in Fig. 7.20d). The relative pattern shift ζ is $(C' - C)/t$, or

$$\zeta = \frac{(x_1' - x_1) + (x_2' - x_2)}{2t} \tag{7.18}$$

where t is the thickness of the epitaxial layer (70). The relative distortion δ is given by

$$\delta = \frac{(x_2' - x_1') - (x_2 - x_1)}{t} \tag{7.19}$$

Measuring the values x_1, x_2, x_1', and x_2' and thus determining δ is straightforward assuming that the boundaries can be seen. However, under some circumstances, referred to as pattern washout, the depression fills up enough to obliterate one or more sides. In order to determine $C' - C$ and thus ζ, cross sectioning and staining (97) or some indirect method such as electrically measuring the resistance of a test pattern (98) or else looking for stacking faults in the epitaxial layer deliberately introduced from the buried layer doping (99) is required.

The depressions just discussed typically arise in the production of a bipolar buried-layer (subcollector) epitaxial wafer, which is a major application of silicon epitaxy. The process flow is shown in Fig. 7.21. During the buried-layer diffusion or ion implant and anneal, the oxidation rate in the window is greater than that over the rest of the wafer since there is no initial oxide in the window to slow oxidation. Thus, more silicon is converted to oxide in the window, and when the oxide is stripped away prior to adding the epitaxial layer, there is a slight depression in the silicon surface. Knowing the location of the edges of the original depression is important since the subsequent isolation mask must be aligned to the buried-layer diffused region. In order to minimize space, the isolation should be very close to the transistor being isolated and hence to the buried-layer diffusion. However, if it actually contacts the buried layer, the breakdown voltage between the transistor collector and the p-isolation diffusion will be too low. Thus, if appreciable pattern shift and/or distortion occurs and the isolation mask is aligned to the top surface pattern without any compensation, serious misalignment and a subsequent low yield will result.

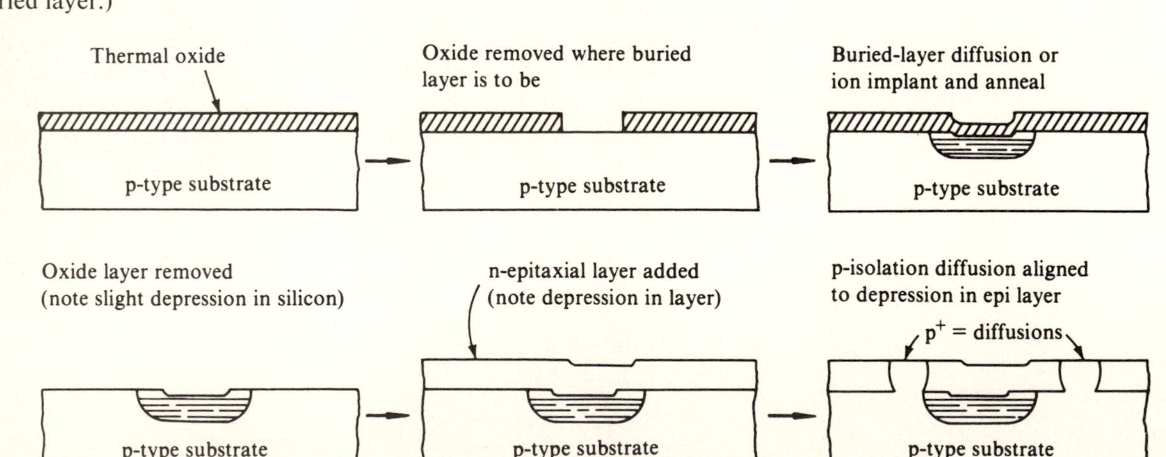
Thermal oxide
p-type substrate
Oxide removed where buried layer is to be
p-type substrate
Buried-layer diffusion or ion implant and anneal
p-type substrate
Oxide layer removed (note slight depression in silicon)
p-type substrate
n-epitaxial layer added (note depression in layer)
p-type substrate
p-isolation diffusion aligned to depression in epi layer
p+ = diffusions
p-type substrate

FIGURE 7.21

Process flow for producing buried-layer diffusions. (Note that if there is substantial pattern shift—shift in the position of the depression—the subsequent p-isolation diffusion will not register properly with the buried layer.)

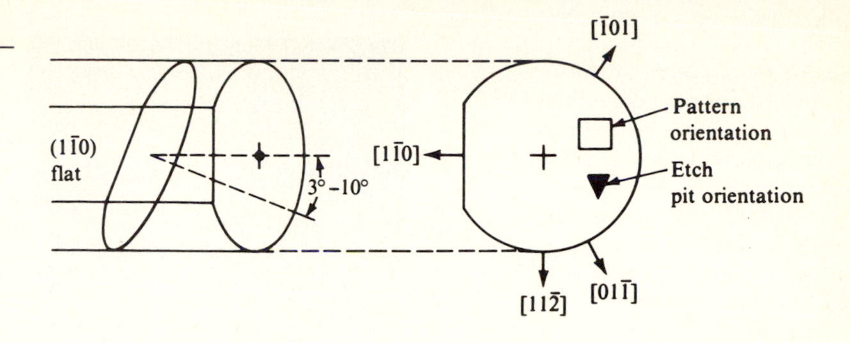

FIGURE 7.22

Method of orienting off-oriented (111) slices for minimum pattern shift. (The direction of tilt corresponding to "toward the nearest (110)" is shown. The orientation of etch pits is also shown, as well as the usual pattern orientation.) The flat is shown as (1$\bar{1}$0), but it, or the ($\bar{1}$01), or the (01$\bar{1}$) are all equivalent. (*Source:* Adapted from Duane O. Townley, *Solid State Technology,* p. 43, January 1973.)

Shift and distortion occur because of differences in nucleation and growth on the sides of the depression and on the large, flat areas. These differences depend on crystal orientation as well as on various growth parameters. It was recognized very early that cutting (111) slices a few degrees off the (111) plane would minimize the pattern shift (100, 101). The optimum amount is on the order of 4°, and the slice must be tilted toward the nearest (110) plane[13] (102) as shown in Fig. 7.22. For this kind of off-oriented wafer and rectangular depressions oriented with two sides parallel to (1$\bar{1}$0) traces and two sides parallel to (11$\bar{2}$) traces (the normal orientation) as shown in Fig. 7.22, there will be a small pattern shift and widening of the pattern in the [11$\bar{2}$] direction and noticeably more widening but no shift in the [1$\bar{1}$0] and [$\bar{1}$10] directions. An example of this effect is shown in Fig. 7.23. There is pattern distortion, but no pattern shift when growing on (100) surfaces, and deliberately misoriented (100) slices are seldom used. However, when small amounts of pattern shift are a problem, it may be necessary to examine the amount of shift to be expected by virtue of the normal manufacturing orientation tolerance.

The effect of processing conditions on pattern shift and orientation has also been studied extensively (70, 103, 104). Typical values of ζ are between 0 and ~2, while δ ranges from ~−1 to ~1. Deposition variables known to affect ζ and δ are rate, temperature, pressure, layer thickness, and silicon source material. Some general observations concerning the data are as follows:

1. There is little difference in the behavior of horizontal, pancake, or barrel reactors (105).
2. Increasing growth rate causes off-oriented (111) pattern

[13]For a discussion of this terminology, see Appendix A.

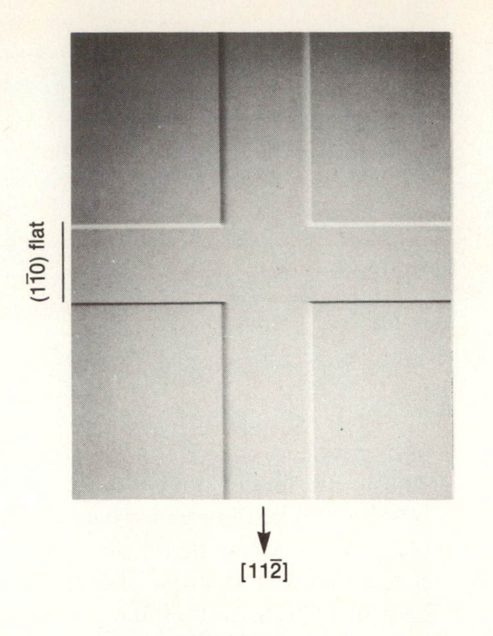

FIGURE 7.23

Example of pattern distortion. (The cross is the depression, although because of the interference contrast used during photography, it may appear raised. There is also pattern shift in the [11$\bar{2}$] direction, but it can be seen only by cross sectioning the wafer. The pattern distortion is primarily in the [1$\bar{1}$0] and [$\bar{1}$10] directions and shows up in this photograph as the vertical stripe being wider than the horizontal one.)

shift to increase and distortion to decrease for both (100) and off-oriented (111) (70, 103, 104).

3. Increasing temperature causes pattern shift to decrease and distortion to increase for both orientations (70, 103, 104). These trends for off-oriented (111) pattern shift are shown in Fig. 7.24.
4. As pressure is decreased from atmospheric to less than 100 torr, off-oriented (111) pattern shift changes from + to −, while (100) distortion changes from − to + (69, 106).
5. Both pattern shift and distortion are functions of thickness, with shift increasing and relative distortion decreasing as thickness is increased (70).
6. Changing the silicon source gas from silane (Cl atoms/molecule = 0) to SiH_2Cl_2 to $SiHCl_3$ to $SiCl_4$ (Cl atoms/molecule = 4) causes both ζ and δ to increase (70).

7.3.8 Selective Epitaxial Growth (SEG)

As has been discussed, to nucleate on an amorphous substrate is generally more difficult than to nucleate on silicon. It is thus, in principle, possible to use an amorphous layer, usually SiO_2, as a mask to prevent epitaxial deposition in selected regions of a wafer (107). Even if some polycrystalline deposition occurs over the mask, the epitaxial deposition will have been restricted to the openings in the mask, and a discrete device can be made in each of these regions (108). An alternative but seldom used approach to deposition only in the windows is to allow deposition everywhere and then either

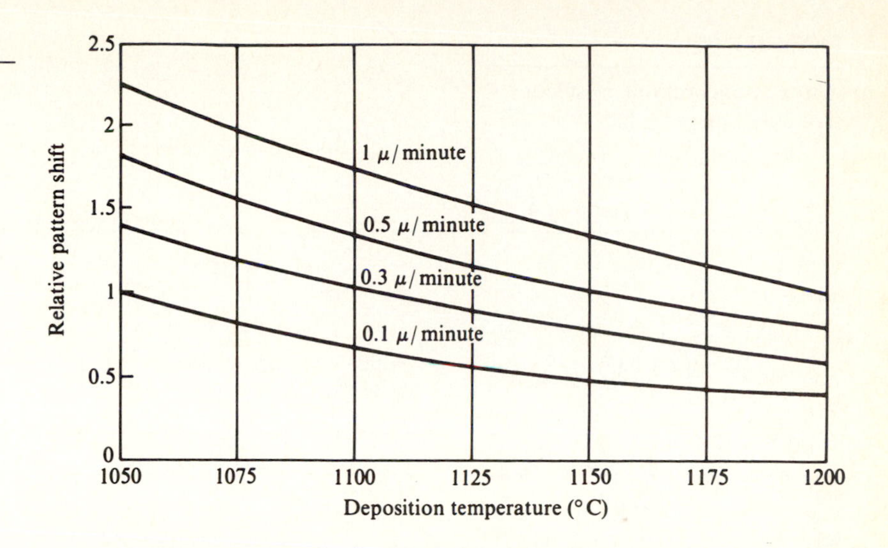

FIGURE 7.24

Relative pattern shift versus deposition temperature for deposition onto a (111) wafer from $SiCl_4$. (The wafer is off-oriented a nominal 4° toward the nearest (110) plane.) (*Source:* P.H. Lee et al., *J. Electrochem. Soc. 124,* p. 1824, 1977. Reprinted by permission of the publisher, The Electrochemical Society, Inc.)

strip off the mask material and the overlaying polycrystalline layer with it (109) or else polish away the polycrystalline layer.

The driving force for the use of selective depositions was originally the desire to produce local regions with either uniform or retrograde[14] doping normal to the wafer surface. A major perceived application was in the formation of IC complementary transistors (110). It was also hoped that all-epitaxial transistors, allowing separate tailoring of the base, collector, and emitter, would provide better performance than diffused transistors (111, 112). Fig. 7.25a is the oldest configuration of selective epitaxy and was used to fabricate a variety of devices such as diodes and bipolar transistors that required a thick epitaxial layer. However, except for some dielectrically isolated ICs, it is not clear that any commercial devices using this configuration were ever built.

For MOS transistors and some very shallow-base high-frequency bipolar transistors, the configuration shown in Fig. 7.25b or 7.25c has been proposed (24, 113, 114). Both of these configurations allow sidewall oxide isolation without the birdbeak that occurs in LOCOS-type sidewall isolation (see "Selective Oxidation" in Chapter 3 for details of this type of process). Further, since epitaxial growth generally proceeds more rapidly than diffusion, SEG can sometimes be used in place of diffusions—for example, for CMOS wells—to reduce the time a wafer is held at elevated temperature. The configuration in Fig. 7.25b and selective polycrystalline metal

[14]Retrograde doping means a profile that is the reverse of that obtainable by diffusion—that is, a lower impurity concentration at the surface.

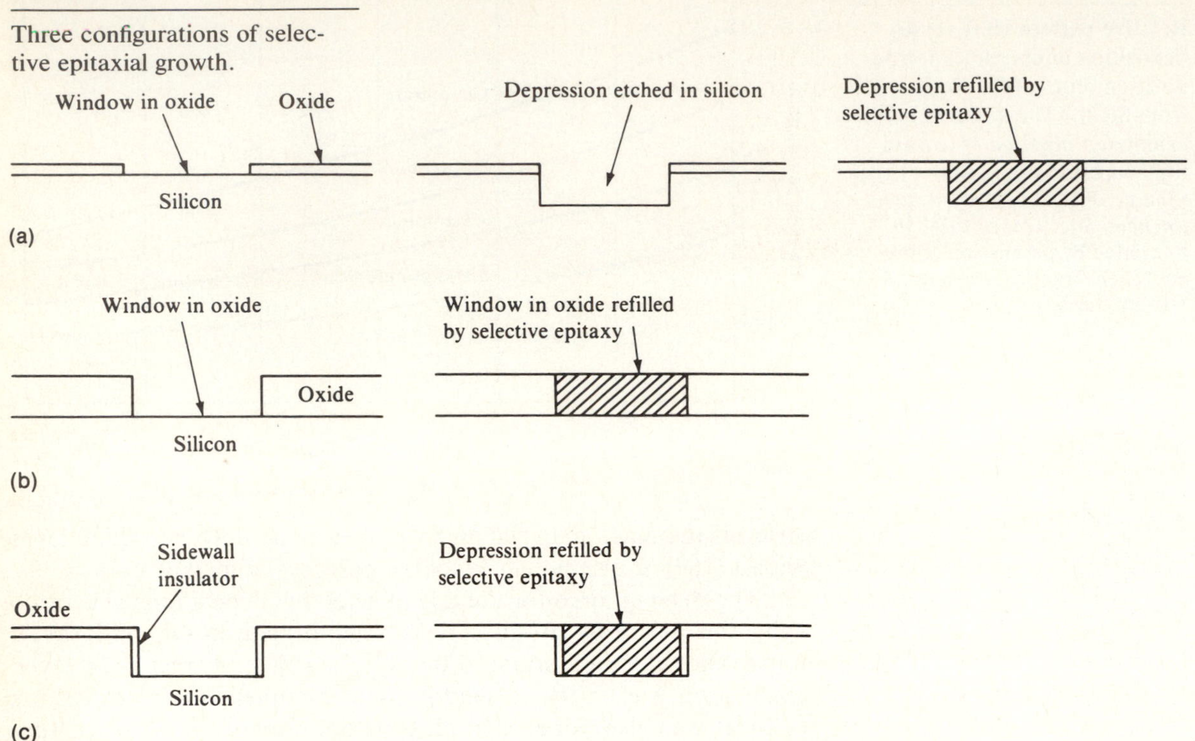

FIGURE 7.25

Three configurations of selective epitaxial growth.

CVD are used to raise contact surfaces from the bottom of the oxide to the top surface (planarization). The same configuration along with expanded single-crystal growth out over the top of the oxide have been used in silicon imagers (115).

Both thermal silicon dioxide and CVD silicon nitride have been used for selective epitaxial masking of silicon growth (107, 116–118). Nucleation occurs much more easily on CVD silicon dioxide than on either silicon nitride or thermal silicon oxide. Thermal oxide allows the most processing latitude but may degrade during high-temperature (~1200°C) depositions. Therefore, silicon nitride has sometimes been used but, with the move to lower temperatures, is now seldom considered for silicon. Along with the choice of mask material, other variables are mask layout, feedstock, deposition temperature, deposition pressure, and deposition rate.

It was observed by 1963 that even when substantial nucleation occurred on the mask, there was usually a band of oxide around the silicon window where no nucleation occurred (112). An example of this effect is shown in Fig. 7.26. The bare silicon acts as a sink and

FIGURE 7.26

Banding around windows in an SiO_2 mask. (In the band, there is no polycrystalline silicon buildup on the oxide during selective deposition.)

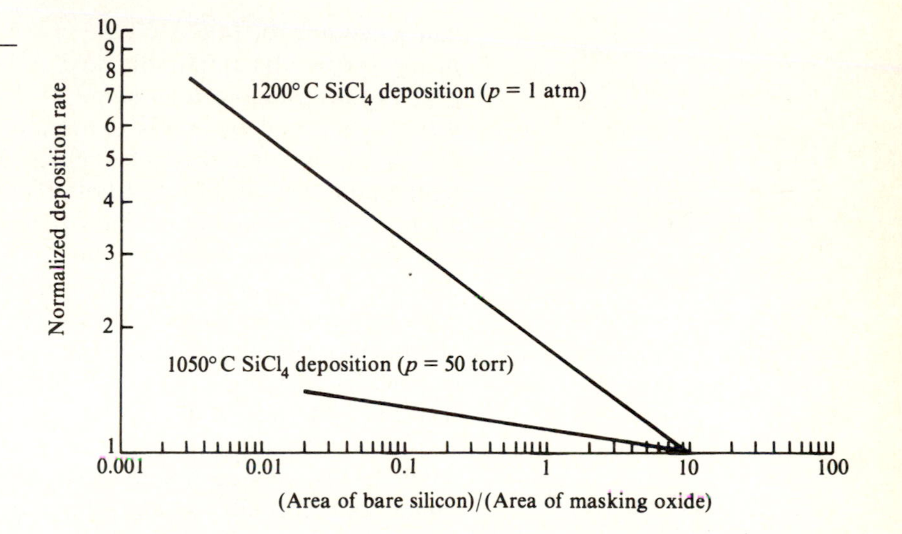

FIGURE 7.27

Trend lines showing effect of ratio of exposed silicon area to masking oxide area.
(*Source:* Based on data from Earl G. Alexander and W.R. Runyan, *Trans. Metall. Soc. AIME 236,* p. 284, 1966, and H.M. Liaw et al., p. 463, in *CVD 84,* Electrochemical Society, Pennington, N.J., 1984.)

keeps the concentration of any silicon-bearing species adsorbed on the oxide nearby below the critical concentration for nucleation. Thus, if the mask layout is such that there is no wider layer of oxide than the unnucleated strip just discussed, then there will be no nucleation on the oxide. This approach has the disadvantage of supplying extra nutrient to the edges of the silicon window so that a pronounced increased growth rate will occur at the edge ("rabbit ears") and a more gentle dish-shaped depression will extend to the middle (119). The additional silicon source from the oxide also has the effect of producing an overall growth rate that increases as the fraction of area covered by masking increases, as indicated in Fig. 7.27 (120). When the surface is mostly covered by bare silicon, there

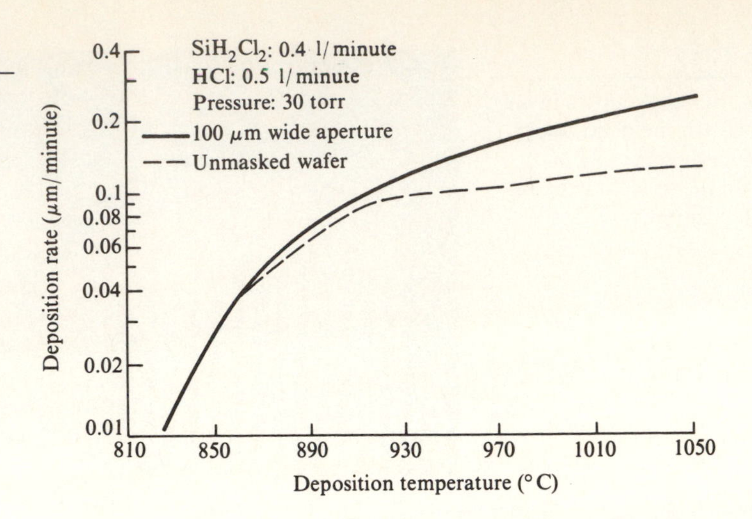

FIGURE 7.28

Curves showing effect of lower deposition temperature on relative rates of deposition over a whole wafer and in small areas. (In this case, the small area is a 100 μm wide slit. Deposition is from a mixture of SiH_2Cl_2 and HCl.) (*Source:* R. Pagliaro et al., *J. Electrochem. Soc. 134,* p. 1235, 1987. Reprinted by permission of the publisher, The Electrochemical Society, Inc.)

will, of course, be little increase in the rate from silicon species diffusing in from the surrounding oxide. However, when the oxide begins to occupy more than about 10% of the total area, and if the distance across the oxide is no more the width of the bare band just discussed, the effect becomes noticeable. Lowering the deposition temperature has the effect of both reducing surface diffusion and of moving the deposition reaction from one that is diffusion limited to one that is surface reaction limited. Thus, as the temperature is reduced, the effect of mask coverage on the deposition rate and excess growth at the edges both decrease (121). Fig. 7.28 shows the same diminishing difference between the two rates as temperature is decreased by comparing the deposition rate in a 100 μm wide window with that of a full wafer.

Since the early silicon epitaxial work used $SiCl_4$ as feedstock, the first selective depositions also used it. As previously mentioned, silicon tetrachloride requires high deposition temperatures, and when an SiO_2 mask is used for selective deposition, it will be substantially eroded. Thus, there was a move first to $SiHCl_3$ and SiH_4 (115, 122, 123) and later to SiH_2Cl_2 (115, 124, 125). $SiCl_4$ offers the best selectivity; silane, the least. It was observed, first for $SiCl_4$ (126) and later for the others, that the addition of HCl improves selectivity. $SiHCl_3$ has sufficient selectivity to be used without HCl; however, when silane or SiH_2Cl_2 is used, the addition of HCl is almost a necessity. The mechanism by which the HCl (or HBr) increases selectivity is not clear, but it does reduce the deposition rate for a given silicon species concentration and move the reaction closer to equilibrium (116). The possibility also exists that the halogen is strongly adsorbed on the mask surface and reduces the number of nucleation sites available. Grooves are sometimes observed at the boundary between the mask and selective deposition, and it has been reported that if the angle between the mask wall and the horizontal surface is greater than about 85° (difficult to maintain with a wet etch), the grooves are minimized (124). For a fixed set of geometries and mask edge profiles, it is often possible to substantially improve planarity by relatively small variations in processing conditions (127). When equilibrium conditions are approached, faceting of the silicon islands often occurs. Such faceting can be minimized on (100) wafers by orienting the sides of the oxide mask openings along (100) traces rather than along (111) traces as is normal (125, 128). Such reorientation could not, however, be satisfactorily done merely by rotating the printing mask since that would orient the scribe lines in a manner that would make chip separation more difficult. Stacking faults and dislocations tend to originate at the mask–silicon interface, and it has been reported that these defects are minimized by orienting the mask along (100) traces (129). The reason for the propensity for more defects when the mask sides are parallel with (111) traces has been ascribed both to the manner in which nucleation of each layer occurs (130) and to stresses at the oxide–silicon interface (131).

The use of reduced pressure generally improves selectivity (59). In the case of SiH_2Cl_2, without reduced pressure, no region of selectivity is found. Fig. 7.29 shows, at 25 torr, how the region of selectivity varies with HCl concentration and temperature. Note that even at this low pressure, there is no selective region without HCl.

The selectivity discussion has centered about chlorine–silicon feed material compounds, but silicon iodides can also be used for

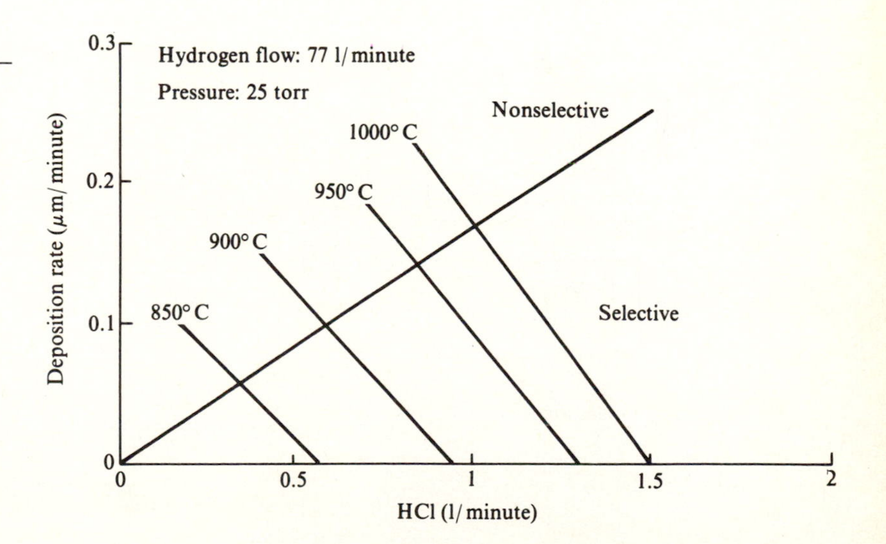

FIGURE 7.29

Effect of HCl in gas stream on selectivity of silicon deposition from dichlorosilane.
(*Source:* Adapted from John O. Borland and Clifford I. Drowley, *Solid State Technology,* pp. 141–148, August 1985.)

epitaxial deposition and, based on limited data, appear to afford greater selectivity (132).

7.3.9 Equipment for Silicon Vapor Phase Epitaxy

The major elements of an epitaxial reactor system are shown in Fig. 7.30. Reactors are generally described by the mode of heating (usually radiant or induction), the operating pressure (low or atmospheric), and the deposition chamber configuration. Many names have been used to describe the chambers, including such descriptive terms as dome, waffle iron, pancake, barrel, and carousel. The reason for such a proliferation of names is that the chamber has been the subject of almost as many redesigns as there have been epitaxial investigators. The original germanium and silicon epitaxial processes used a simple one-wafer vertical flow reactor chamber as shown in Fig. 7.31a. Shortly thereafter, multiwafer horizontal flow chambers as shown in Fig. 7.31b were introduced. Currently, the pancake and barrel configurations, shown schematically in Fig. 7.32, are primarily used. Two recent introductions are the carousel (Fig. 7.32d) for increased capacity and a single-wafer horizontal flow reactor for better parameter control. Silicon CVD epitaxial equipment is now purchased from a few specialized vendors, but during the first decade after the introduction of epitaxy, most of it was constructed in-house. Thus, it is difficult to trace the developmental stages that led to the current configurations. However, Fig. 7.33 lists some of the known designs.

The gas control system is generally quite straightforward, consisting of mass flow meters to monitor and control the inert purge gas (usually N_2), H_2, and the dopant gas. When silane or dichlorosilane is used as the silicon source, it can also be controlled by a mass flow meter, although the SiH_2Cl_2 cylinder may need tempera-

FIGURE 7.30

Major elements of an epitaxial CVD system. (If the system operates at reduced pressure, the reactor exhaust will include a vacuum pump before the burnoff scrubber.)

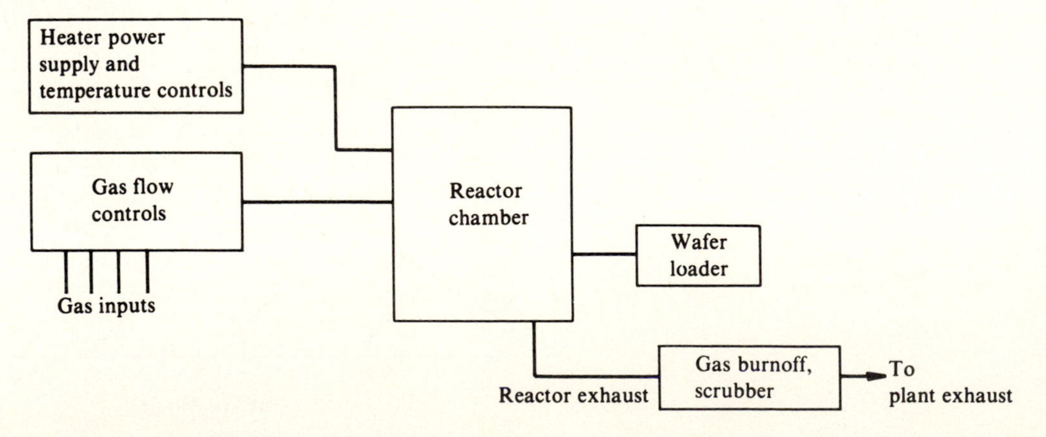

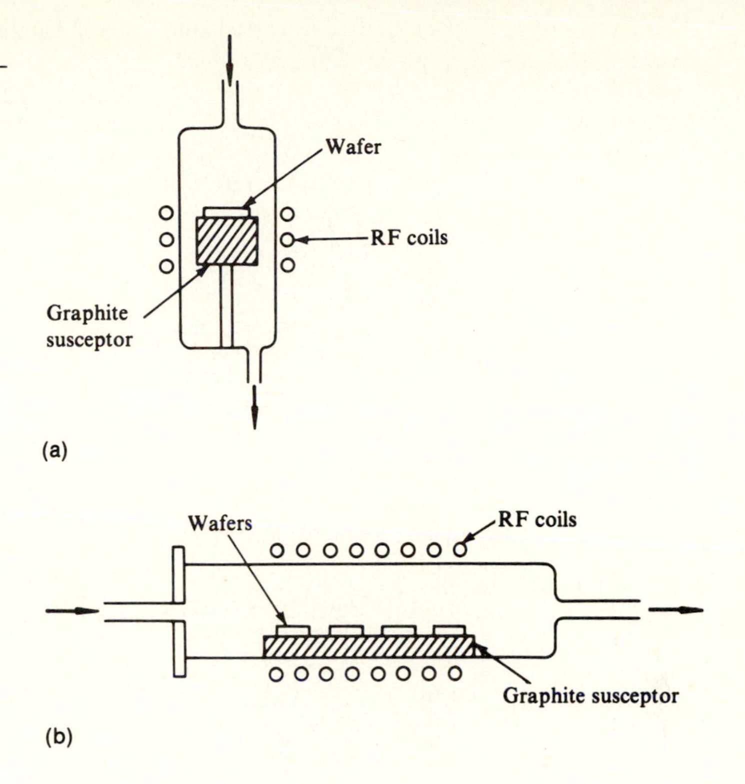

FIGURE 7.31

Schematic of (a) single-wafer vertical flow reactor and (b) multiwafer horizontal flow reactor.

ture controlling and either external heating or cooling to maintain a satisfactory vapor pressure above the liquid. Since silicon tetrachloride and trichlorosilane are liquids at room temperature, a bubbler is normally used to produce a vapor. With bubblers, the amount of material delivered depends on the vapor pressure of the liquid and the flow of carrier gas. Vapor pressure curves for $SiCl_4$, $SiHCl_3$, and SiH_2Cl_2 are given in Fig. 7.34. The mole fraction f of halide picked up by the gas through the bubbler is given by

$$f = \frac{p_h}{p_o} \qquad 7.20$$

where p_h is the vapor pressure of the halide and p_o is the pressure of the gas mixture flowing through the bubbler (usually approximately atmospheric). If the stream through the bubbler is then diluted by mixing with another flow (bypass gas) before it enters the reaction chamber (as is normal), then the mole percent $M_{\%}$ of halide entering is given by

$$M_{\%} = \frac{100fF}{F_b + F + Ff} \qquad 7.21$$

FIGURE 7.32

Schematic of epitaxial reactors: (a) pancake, (b) and (c) barrel, and (d) carousel.

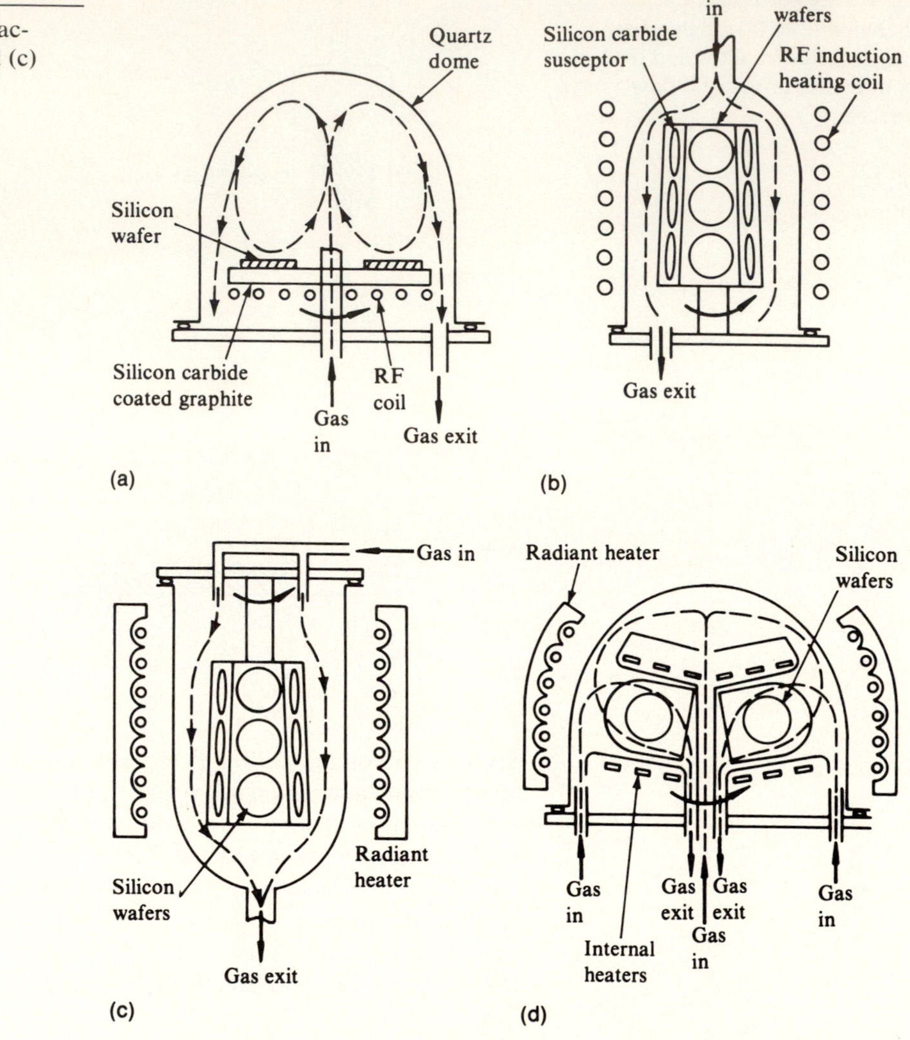

where F_b is the bypass gas flow rate and F is the flow rate of gas entering the bubbler. Note that since a molecular volume is the same for any gas, the mole percent equals the volume percent.

EXAMPLE ☐ If the flow of hydrogen through the bypass is 60 liters/minute and the bubbler contains $SiCl_4$ at 0°C, what gas flow through the bubbler is required to give 1 $M_{\%}$ of $SiCl_4$ in the total gas stream?

From Fig. 7.34, p_h is ~76 mm. Thus, f of Eq. 7.20 equals 0.1. From the problem, $F_b = 60$. Substituting these values into Eq. 7.21 gives $F = 6.7$ liters/minute. ☐

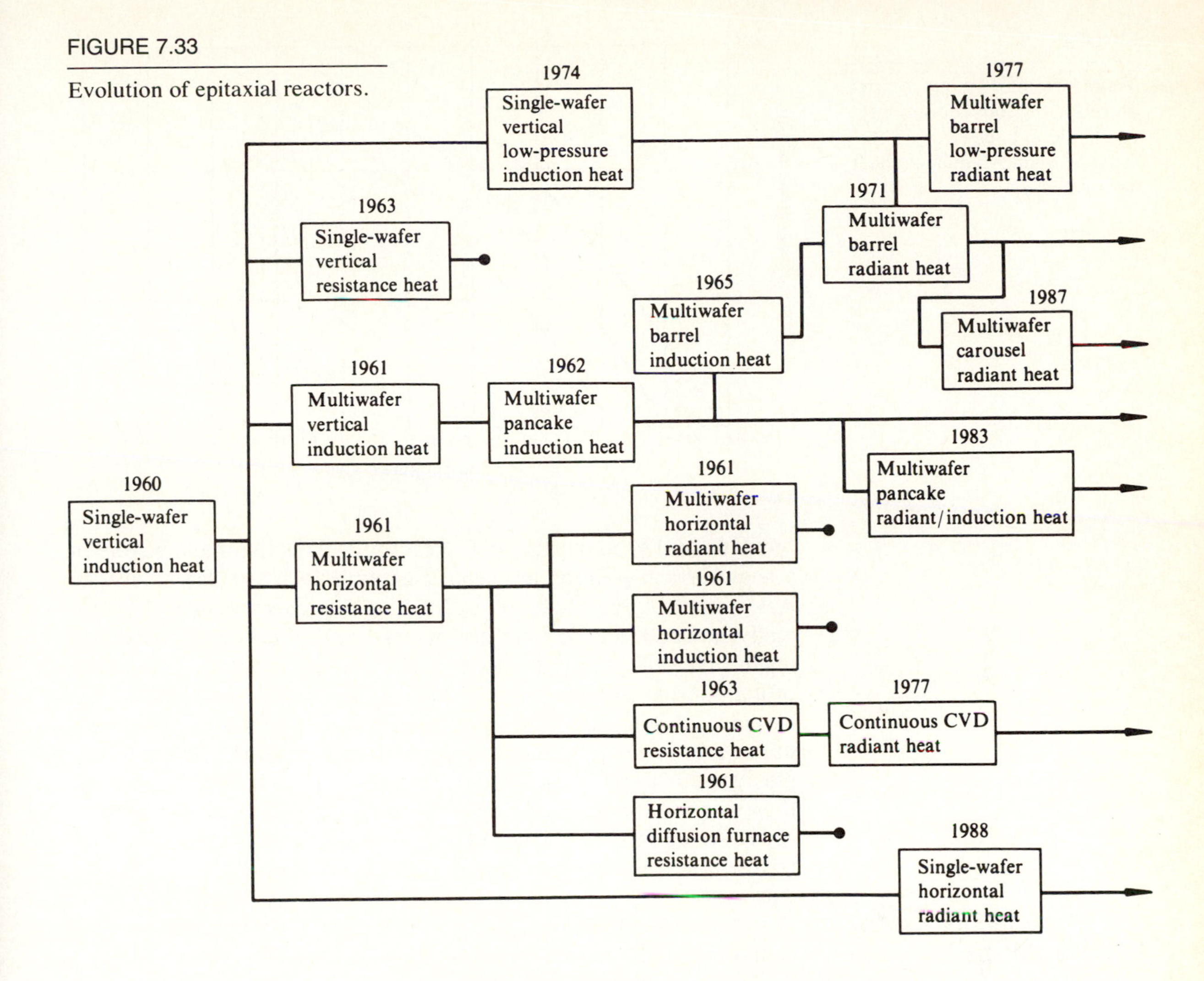

FIGURE 7.33

Evolution of epitaxial reactors.

Numerical calculations of deposition rate have largely centered on the diffusion- (transport-) limited region (51, 54, 133–139) where most atmospheric pressure reactors operate. A better understanding of the effect of flow, temperature gradients, and so on has led to the development of processes yielding better thickness uniformity.

Other equipment-related considerations are the maintenance of a low particulate count in the reactor, composition of the susceptors, and disposal of the toxic gases after they leave the reactor. Reactors are almost always operated with the outer walls cool in order to minimize deposition on them. However, where deposition does occur, the repeated temperature cycling causes the material to

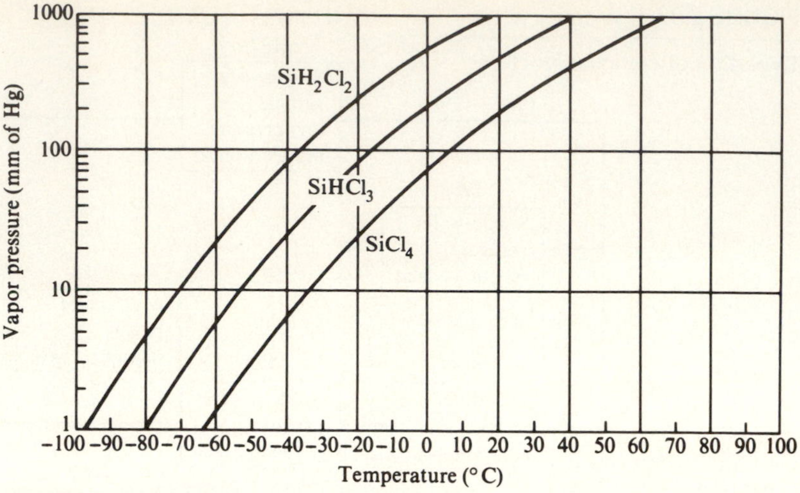

FIGURE 7.34

Vapor pressure versus temperature for the most commonly used silicon halides.
(*Source:* Union Carbide product information.)

break up and form particulates. Reactor designs that have wafers in a near-vertical position are less susceptible to particulates because less projected area exists to catch falling particles and because thermally induced convection currents can sweep particles away that might otherwise strike the wafer surface. Susceptors are commonly of high-purity graphite that has been coated with a layer of silicon carbide. Careful attention must be given to the coating thickness uniformity in order to minimize cracking. The large volume of explosive and often toxic gaseous effluent is ordinarily first burned and then water-scrubbed before it is discharged into the air. The scrubber residue must be monitored and, if required, chemically neutralized before it is put into a drain.

7.3.10 Molecular Beam Epitaxy (MBE)

Silicon molecular beam epitaxy (MBE) provides silicon atoms by evaporating silicon in a high vacuum and allowing the evaporating beam to strike the wafer. It is, in principle, like the evaporation method described in Chapter 4 for metallization. By using evaporation, no surface chemical reactions need to be considered and thus no reduction in rate as the deposition temperature is lowered. However, even though no reduction in the evaporation rate occurs as the wafer temperature decreases, the problem of reduced surface mobility exists. Thus, if single-crystal material is to be produced, the rate used must decrease with decreasing temperature in order to allow time for the atoms striking the surface at random locations to diffuse to proper positions. The same difficulty also exists at low temperatures in providing a clean surface on which to initiate epitaxy as was discussed earlier.

MBE was applied to silicon by 1962[15] (140). With the relatively poor vacuums used initially, the surface rapidly became contaminated, and only a poor-quality film was produced. By increasing the deposition rate so that the silicon impingement rate was large compared to that of contaminating gases in the chamber, the quality could be improved (142) but not enough to match that of CVD epitaxy in a hydrogen atmosphere. During the ensuing 25 years, vacuum systems became much better and MBE equipment much more sophisticated and expensive. However, for ordinary purposes, it still does not compete with conventional epitaxy. Nonetheless, it is capable of producing epitaxy at temperatures of a few hundred degrees and can provide for rapid changes in dopant concentration. It is also well suited for very thin epitaxial layers.

Fig. 7.35 is a schematic of a silicon molecular beam epitaxial system (143). The source is an electron-beam-heated piece of silicon. For cases where doping is to be constant throughout the layer,

FIGURE 7.35

Schematic of silicon MBE system. (*Source:* Adapted from J.C. Bean and E.A. Sadowiski, *J. Vac. Sci. Technol. 20,* p. 137, 1982.)

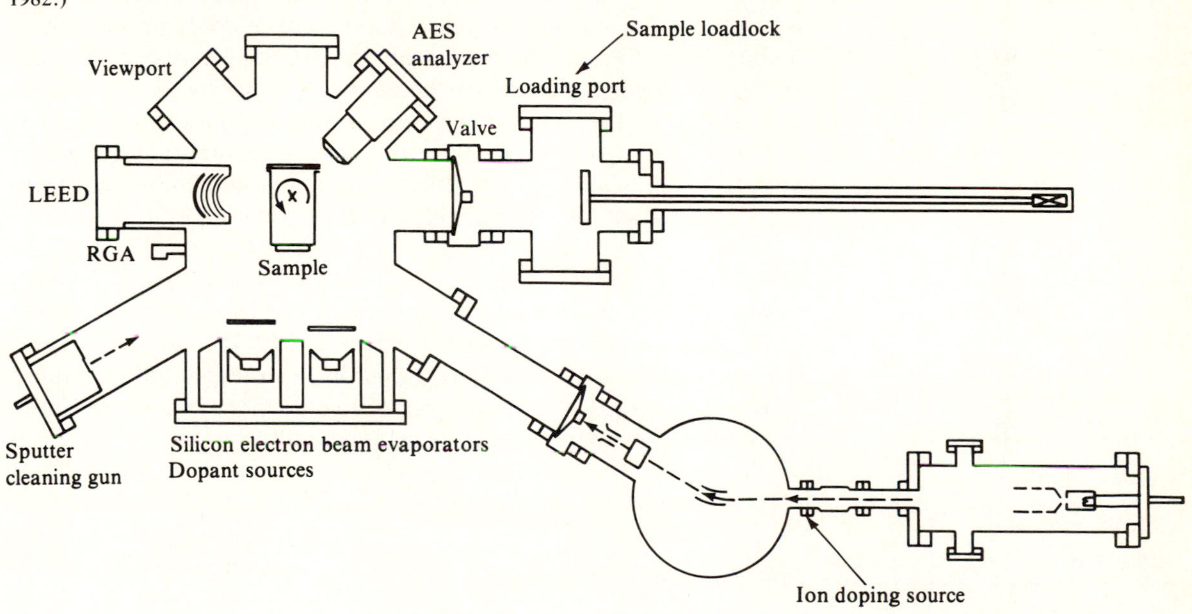

[15]At that time, it was simply referred to as "vacuum evaporation." In 1966, a high-vacuum deposition system using a silane beam decomposing on the hot silicon surface was referred to as "molecular beam epitaxy" (141). By the mid-1970s, MBE was routinely being used to describe the use of an evaporating silicon source in high vacuum.

the silicon source can be predoped. When doping is to be changed during deposition, separate doping source ovens and an ion implanter are also included. A residual gas analyzer (RGA) is attached to monitor the chamber composition. In order to increase throughput (a very-high-vacuum system usually requires overnight pumping after it has been open to the atmosphere), a loadlock station for loading and unloading wafers without breaking vacuum in the main chamber is included. To check layer crystal quality, a high-energy electron diffraction system is attached, as is an Auger electron scattering (AES) analyzer for examining surface contamination. An ion sputter gun is included in the system to clean the surface before deposition.

Most dopants have a very low probability of sticking to the growing silicon surface. However, the average residency time before desorption is sometimes hundreds of seconds. Therefore, unless special precautions are taken, abrupt changes in doping cannot be made. For example, at 700°C, antimony has an average time of over two hours. The adsorbed atoms on the surface provide the doping, and the doping level cannot be changed until their concentration is changed. When abrupt changes in doping are desired, growth can be stopped until the surface dopant adsorbed layer has re-equilibrated. When the dopant changes are accompanied by changes in growth rate and growth temperature, a more rapid response in the layer is possible (144). An alternative procedure is to incorporate a low-voltage ion implanter into the MBE equipment and implant the dopant as deposition takes place (145, 146).

7.3.11 Defects Introduced during Growth

Table 7.8 lists defects that may be found in silicon epitaxial layers (106). The first column lists defects that appear on the surface and are visible without any surface treatment. The second column lists defects that are more crystallographically oriented and often require etching to delineate. (A few appropriate etchants are given in Chapter 6, but for further details, see reference 97.)

TABLE 7.8

Silicon Epitaxial Layer Defects

Surface-Related Defects	*Bulk-Related Defects*
Haze	Dislocations
Pits	Misfit dislocations
Orange peel	Slip
Spikes	Stacking faults
Edge crown	Fog (visible after heat treatment)
Faceting	

Haze Generally, this high density of fine pits can be caused by oxygen in the incoming gas stream, air leaks, or moisture adsorbed on the reactor walls. It is different from the fog defect described later. As little as 1 ppm of oxygen present during the heatup cycle can cause a noticeable increase in haze. However, up to 10 ppm can be tolerated during the etch and deposition cycles. Further, 30 ppm can cause a gross degradation of surface quality (147). If a low concentration of HCl is present during the heatup and cooldown cycles, as from a leaky valve, a hazy surface will be formed.

Pits Widely separated individual pits probably originated during etching of the substrate and remained because enough material was not removed during etching.

Orange Peel This rough rippled surface can be caused by too high a growth rate. Orange peel also sometimes occurs during chemical etching of slices and, if it were present on a starting slice, would be visible after deposition.

Spikes The most common cause of spikes is particulates left on the surface. The spikes usually grow much more rapidly than the flat surface since they reach up through the boundary layer to a higher concentration of feed material. As reactor and slice cleanup procedures have improved, the spike density has declined dramatically. At one time, spikes were such a problem that spike crunchers, shavers, and etchers were used to remove them before the first contact print mask was applied. A less common cause of spikes is a localized high metal concentration on the surface leading to VLS (vapor–liquid–solid) growth, which is much more rapid than normal vapor phase growth. VLS growth occurs by vapor transport of the silicon to a molten metal–silicon alloy (148). Growth takes place at the alloy–silicon interface and keeps the alloy pushed to the end of the growth such that distinctive spikes with a ball at the end form as shown in Fig. 7.36. The high metal concentrations needed for VLS growth have, for example, occurred when metallic tips were substituted for plastic tips on vacuum slice pickups.

Edge Crown This raised region at the outer periphery of the slice is due to a more rapid growth near the edge than over the rest of the wafer, which is caused by the gas flow over the edge of the slice.

Faceting Two kinds of faceting can be seen. One kind occurs on the surface of (111) slices cut very close to orientation. The other

FIGURE 7.36

Example of unintentional VLS growth during silicon epitaxy.

kind of facets sometimes grow at the edge of a slice where chipping occurred after polishing but before epitaxial growth.

Dislocations Dislocations in the starting slice that intersect the top surface will propagate on into the epitaxial layer. The incomplete removal of slice surface mechanical damage will cause additional dislocations to form in the layer.

Misfit Dislocations When there is a lattice mismatch, as between a lightly doped layer and a very heavily doped substrate, misfit dislocations will be generated (see Chapter 8 and Appendix A). They are also sometimes deliberately introduced by growing a thin layer of germanium-rich silicon before the desired epitaxial layer and are used for gettering (149). They lie at the interface between the two regions of differing lattice spacing and seldom intersect the surface. They do, however, often cause a very slight rippling of the surface that can be seen under interference contrast. Fig. 7.37 (top) is an etched cross section showing misfit dislocations formed by a thin layer of co-deposited Ge and Si. Fig. 7.37 (bottom) is a top view of a similar slice and shows the plaid surface caused by the lattice distortion from the dislocations. Remember, however, that the dislocations themselves are parallel to the surface and lie well below it.

Slip Discussed in considerable detail in Chapter 8, slip occurs during the epitaxial cycle when there is enough temperature differential over the slice to cause stresses large enough to cause plastic flow of the silicon. Rapid nonuniform heating or cooling can cause the yield stress to be exceeded, but since wafers lay on flat, highly conductive susceptors, this is usually not the cause of slip. More often, it is caused either because the wafer is initially bowed, and thus only

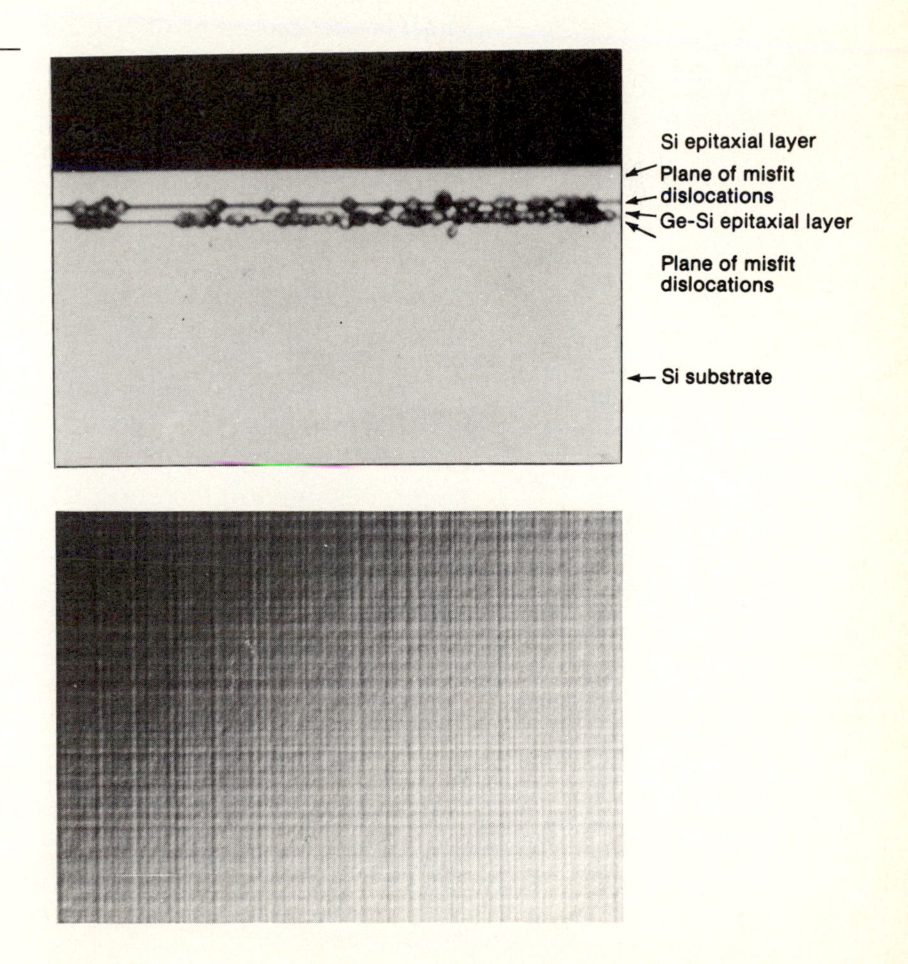

FIGURE 7.37

(Top): Cross section showing misfit dislocations between Si and a Ge–Si layer (magnified 1650X). (Bottom): Top surface of a layer above a plane of misfit dislocations as viewed with Nomarski interference microscopy (magnified).

touches the susceptor in the middle,[16] or because the thermal differential that normally exists axially through the slice causes it to bow and lose peripheral contact with the susceptor (150–152). The axial thermal gradient can be reduced by recesses in the susceptor if heating is by RF (153–155) or by radiant heating either by itself (106) or in conjunction with an RF-heated susceptor (155). Chipped slice edges will increase the likelihood of slip. The strength of silicon increases as the interstitial oxygen content increases. Thus, slices from float-zoned crystals are particularly prone to high-temperature slip. Precipitated oxygen weakens silicon so that if, for example, the oxygen content in the center of the slice is very high and during previous heat treatments has precipitated, then slip in that region is likely.

Stacking Faults Described in Appendix A and shown in Fig. 7.38, stacking faults can originate at the substrate–layer interface because either crystal damage or surface contamination was not completely removed before deposition began. Those that are initiated at the substrate surface will all have grown to the same size at the top surface since their size is a linear function of the layer thickness. For (111), one side of the fault is given by $t/0.816$, and for (100), by $t/0.707$, where t is the layer thickness. When many sizes of stacking faults

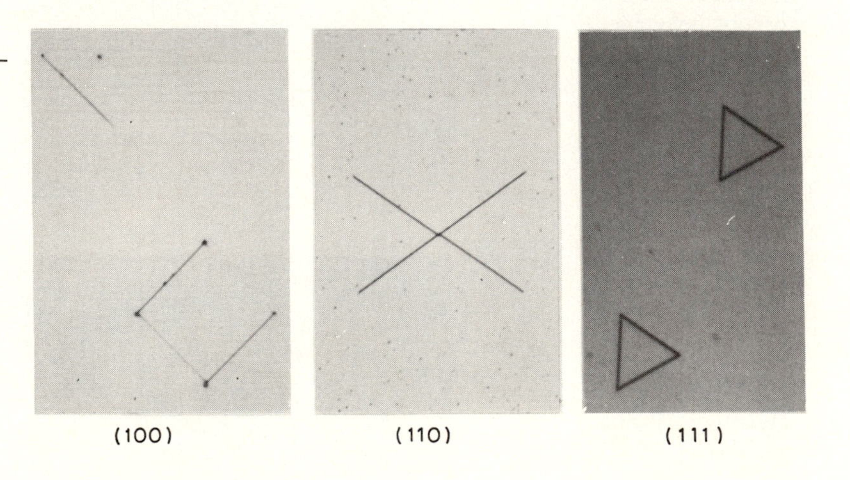

FIGURE 7.38

Silicon stacking faults that nucleated at substrate–layer interfaces. (The tops of the layers were treated with Sirtl etch to delineate the faults.)

[16]If a bowed wafer is placed with the concave side down (rim touching), the wafer will initially be hotter at the periphery, and thermal expansion will cause it to flatten and make better contact. If it is initially placed so that only the center of the wafer contacts the susceptor, the center will initially be hotter, causing the wafer to bow even more away from the susceptor and further increase the radial thermal gradient.

exist, they have been nucleated at varying times during growth, probably by particulates in the reactor falling on the slice.

Fog Also referred to as haze or S-pits, fog is not seen until a subsequent oxidation, followed by stripping the oxide and giving the silicon surface a light etch. S-pits, as discussed in Chapter 3, are caused, not by errors in epitaxial deposition, but by contamination of the slice by heavy metals. A subsequent oxidation causes precipitates to form, which, in turn, cause the S-pits. A likely source of contamination in an epitaxial reactor is moisture in the HCl line, which causes corrosion of stainless steel tubing.

7.4 VAPOR PHASE GALLIUM ARSENIDE EPITAXY

Vapor phase gallium arsenide epitaxial layers are primarily grown by either halide transport (the oldest method), metalorganic decomposition (MOCVD), or molecular beam epitaxy (MBE). Conventional halide or hydride CVD requires, unlike silicon halide transport, a hot-walled reactor, with its increased chance of particulate contamination from growth on the walls. Further, it is very difficult to transport aluminum, in the event that gallium aluminum arsenide growth is also desired. These shortcomings have led to increasing emphasis on the use of organometallic compounds or MBE. Because of the increased density of gallium vacancies that occur at higher temperatures, satisfactory growth is done only below about 850°C. As for silicon, except for MBE, vapor phase in situ etching of the wafers just before epitaxy is generally used. Also as for silicon, HCl is the common etchant, and it may either be fed into the reactor (156) or be produced in the reactor from the reaction of hydrogen with $AsCl_3$ (157). Temperatures of around 900°C are required to produce specular surfaces, so the wafer must be heated above deposition temperature for the etching cycle.

7.4.1 Halide Transport

By 1959, it was established that crystals of various III–V compounds, including GaAs, could be grown by iodine disproportionation in a closed tube (158). However, the first vapor phase epitaxial growth of GaAs used a chlorine disproportionation reaction (159). A GaAs source, the substrate, and a transport agent such as $AlCl_3$ were sealed in a tube. The source, which was in one end of the tube, was held in the 1000°C range, while the substrate at the other end of the tube was about 200°C cooler. As in such reactions, at the high-temperature end, the GaAs reacted with the transport gas; at the low-temperature end, GaAs deposited on the substrate. The current halide processes are similar except that they are done in a two-temperature-zoned horizontal open-tube reaction chamber as shown schematically in Fig. 7.39 (160). Typically, $AsCl_3$ from a bubbler reacts with hydrogen carrier gas in the front section of the heated

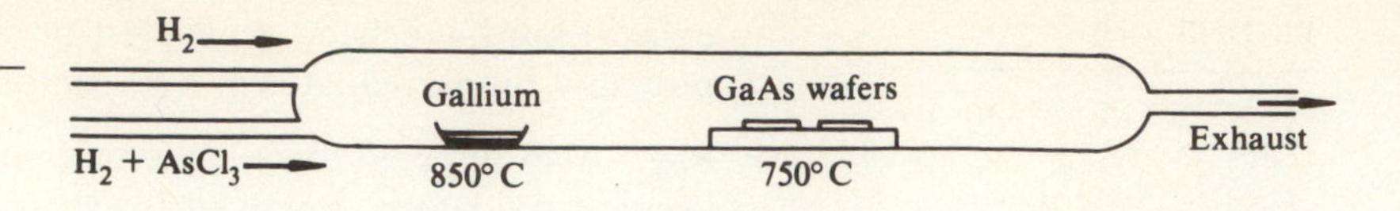

FIGURE 7.39

GaAs two-temperature-zone epitaxial deposition tube.

chamber to give HCl plus molecular arsenic (As_4). The HCl then reacts with either a GaAs or an arsenic-saturated gallium source held at about 850°C to give GaCl or some other Ga–Cl species. Downstream, at a lower temperature, such as 750°C, the GaCl, H_2, and As_4 react at the GaAs slice surface to give GaAs + HCl. The remaining reactants then leave the reactor chamber where they are burned, scrubbed, and exhausted. Because of the very low vapor pressure of the gallium compounds, all parts of the chamber from the point of gallium-compound production past the deposition region must be kept heated.

The exact reactions are somewhat speculative and apparently depend on the amount of $AsCl_3$ remaining in the gas stream at the gallium or GaAs source position (161). Experimentally, it is observed that the $AsCl_3$ begins reacting with H_2 near 550°C and, for temperatures in excess of 750°C, appears to be completely dissociated. Instead of using $AsCl_3$ to supply both the arsenic and the transport halide, HCl can be used to supply chlorine, and arsine can be used for the arsenic source (162). Regardless of which method is used, the final species in the reactor are quite similar, and the deposition behavior is much the same. Growth rates have been extensively studied, and various models have been proposed (163–168). The deposition rate goes through a peak as temperature or GaCl partial pressure increases, but for As, it first increases and then becomes relatively insensitive to further increases in the As_4 partial pressure. Fig. 7.40 shows the behavior with temperature. The rate dependency on crystal orientation is much more pronounced than it is for silicon, and since the (111) and $(\bar{1}\bar{1}\bar{1})$ planes are not equivalent in the zinc blende structure, a pronounced difference occurs in growth on the two. The maximum rate is observed on the $(111)_A$ face; the minimum, on the $(111)_B$ face. The (100) face shows an intermediate rate (166).

7.4.2 Metalorganic Source CVD (MOCVD)

The halide transport system just described requires a two-zone furnace and does not provide easy transport of materials such as aluminum, which is needed if GaAlAs is to be overgrown on GaAs. The use of organometallic[17] compounds for supplying the gallium (or

[17]Note that while chemistry books refer to these compounds as "organometallics," the common name used for the semiconductor process is "metalorganic."

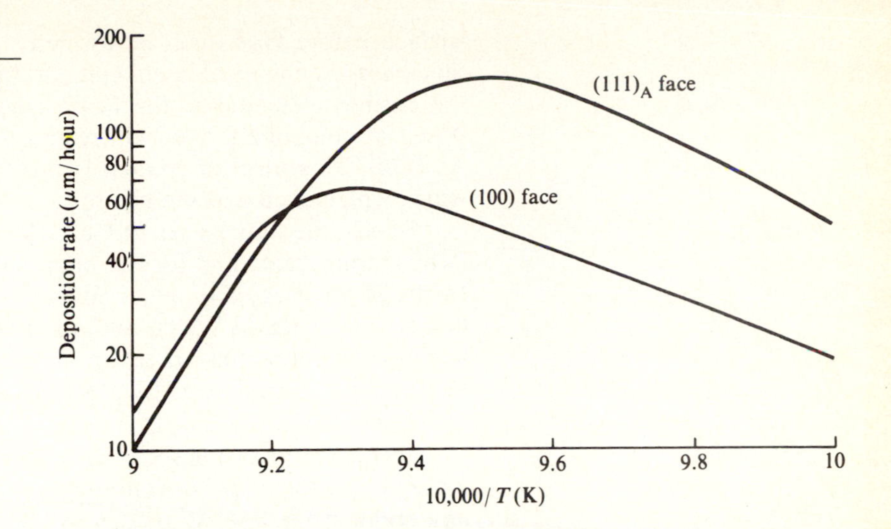

FIGURE 7.40

Effect of wafer temperature on GaAs deposition rate for a chlorine transport system. (*Source:* Data from Don W. Shaw, *J. Electrochem. Soc. 115*, p. 406, 1968.)

other group III metal) simplifies the process since several organometallic sources have sufficient vapor pressures to be vaporized in a bubbler. They can then be transported to the epitaxial reaction chamber in a carrier gas (normally H_2) for reaction with arsine (169–172). Thus, this process is particularly applicable to GaAs heterojunction growth applications.

$(C_2H_5)_3Ga$ (triethylgallium) and $(CH_3)_3Ga$ (trimethylgallium) are commonly used. The overall reaction is, for example,

$$(CH_3)_3Ga + AsH_3 \rightarrow GaAs + 3CH_4 \qquad 7.22$$

Unlike its effect on silicon growth, the presence of a carbon compound in this case does not prevent single-crystal GaAs growth. Group II, IV, and VI compounds are also available that can be used for shallow-level doping (172), and one compound, hexacarbonyl chromium, has been used for deep-level doping (173).

For deposition temperatures in the 600°C–850°C range, the growth rate is insensitive to temperature. As in the case of halide transport, an excess of arsenic is maintained, and the growth rate is proportional to the organometallic concentration. Crystallographic orientation effects also appear minimal. In some reactor and piping configurations, the reactants may form unwanted compounds before the hot reaction zone is reached and thus cause excessive variations in either growth rate or doping level.

7.4.3 GaAs Molecular Beam Epitaxy

Using equipment much like that discussed for silicon, GaAs layers can be grown by MBE. One difficulty is that arsenic generally evaporates as As_4 and then must dissociate into atomic arsenic at the surface before GaAs can be formed. An alternative that is sometimes used is chemical beam epitaxy. In this case, rather than using the desired elements as the beam constituents, the beam is composed of compounds that will react on the surface to give Ga and As (174). This form of epitaxy is much like the early silicon MBE work, which used a silane beam.

Two of the major uses of GaAs MBE are for the production of lattice strain fields and for the heteroepitaxial growth onto silicon. In the latter case, one major difficulty is in minimizing antiphase defects (see section 7.2.1), and another is in minimizing threading misfit dislocations that travel up into the epitaxial layer.[18] Both difficulties can be minimized by growing on a slightly misoriented (100) wafer (175, 176).

Surface preparation for MBE consists of first removing crystallographic damage by conventional wet etches and then vaporizing any residual Ga and As oxides in the MBE chamber. A potential disadvantage of such a procedure is that any carbon (from hydrocarbons) remaining on the wafer surface from the wet etch will not be removed in the chamber. (In this case, the presence of carbon on the surface does interfere with crystal growth.) However, it has been demonstrated that an HCL–ethanol wafer rinse in an inert atmosphere immediately before the wafer is inserted into the MBE equipment is a very effective cleanup (177).

7.5 LIQUID PHASE EPITAXY

Thin epitaxial layers can be grown from various melts in which the semiconductor to be grown is soluble. This process is known as liquid phase epitaxy (LPE). Silicon can be grown by LPE but has not been widely used; of the applications proposed, none have been IC oriented. GaAs liquid phase epitaxy is widely used, but mostly for light-emitting applications. LPE entails dissolving some of the material to be grown in a molten metal—for example, silicon in gallium or tin—inserting the substrate wafer, and then cooling the melt until the required amount of material has frozen onto the substrate surface. Growth of various materials from lower-temperature melts (fluxes) has long been used for producing various high-melting-point refractory material crystals, and it was noted in 1953 that silicon single crystals could be grown from gallium, indium, or tin melts (178, 179). The LPE growth of thin layers of GaAs onto a wafer and a method of doing it were proposed in 1963 (180), and LPE silicon was produced by 1965 (84).

---

[18]Two kinds of misfit dislocation form in the GaAs layer. When the layer is (100) oriented, one is in the (100) interface plane and is quite immobile. The other can move in (110) planes and thus can travel up into the layer.

The early silicon applications were primarily directed to those requiring very heavily doped layers on heavily doped substrates such as tunnel diodes. Suggested applications since have included ohmic contacts (181), gallium-doped infrared detectors (182), thin-film solar cells (183), vertical-junction solar cells (184), field effect transistors (184), and field-controlled thyristors (184). Gallium arsenide (185), gallium phosphide (186), indium phosphide (187), gallium aluminum arsenide (188), indium gallium arsenide on indium phosphide (189), and indium gallium arsenic phosphide (180) have all been grown by LPE, and as mentioned earlier, such material is widely used in light-emitting diodes and other electro-optical devices. LPE has also been applied to the growth of mercury cadmium telluride for infrared applications (181). Some advantages of LPE, when it can be used, are the simplicity and economy of equipment, the rapidity with which thick layers can be grown, and the ease of growing heavily doped layers. A major problem in LPE application to GaAs ICs is a lack of thickness control when thin layers are grown.

7.5.1 Silicon LPE

Silicon has enough solubility in several metals to allow it to be grown from them at rather modest temperatures. As an example, Fig. 7.41 shows silicon solubility in gallium, from which it can be

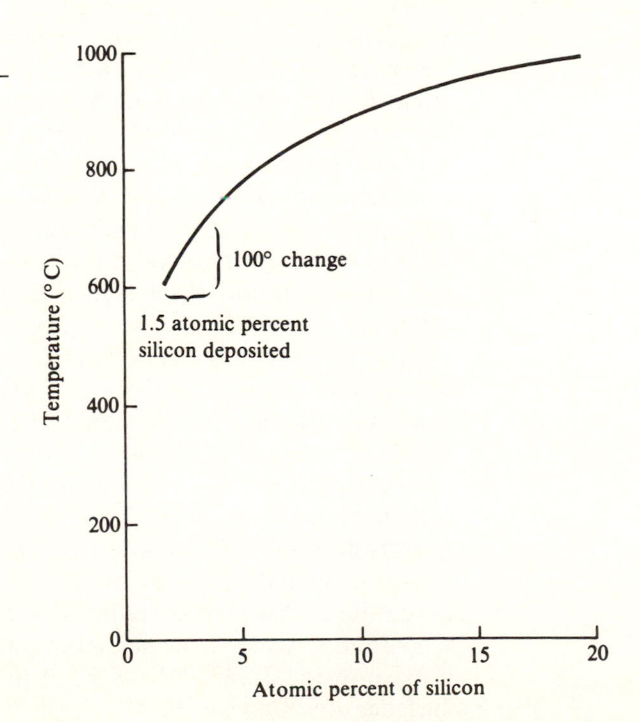

FIGURE 7.41

Portion of gallium–silicon liquidus line showing solubility of silicon in molten gallium.

FIGURE 7.42

Solubility of silicon in various molten metals versus temperature. (*Source:* Plotted from data in C.D. Thurmond and M. Kowalchik, *Bell Syst. Tech. J. 39*, p. 169, 1960; P.H. Keck and J. Broder, *Phys. Rev. 90*, p. 521, 1953; and M. Hansen, *Constitution of Binary Alloys*, McGraw-Hill Book Co., New York, 1958.)

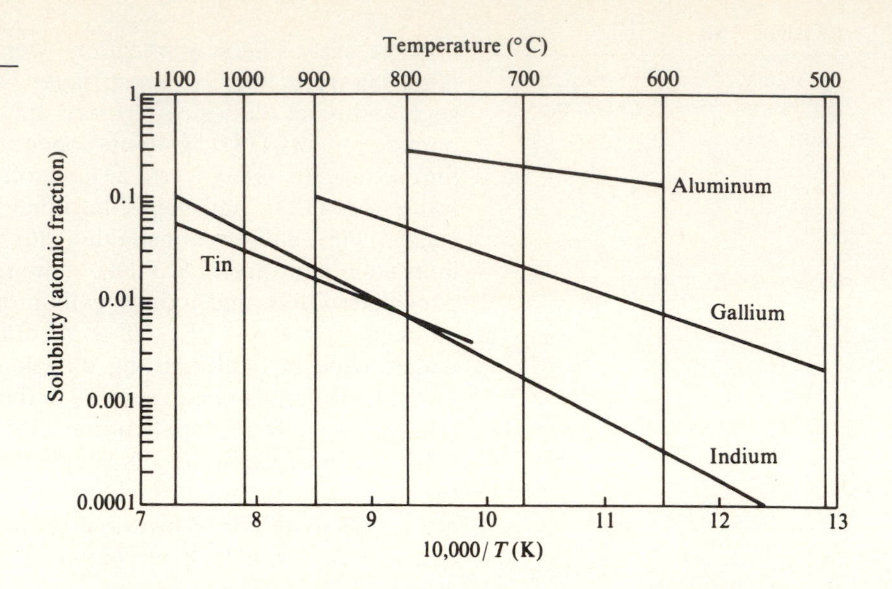

seen that if a gallium melt at 700°C were saturated with silicon and then reduced in temperature to 600°C, approximately 1.5 atomic percent (0.015 atomic fraction) of silicon could deposit (freeze) onto wafers immersed in the gallium. Fig. 7.42 gives solubilities for other materials that have been used as silicon solvents. The curves in this figure, rather than being plotted in the conventional phase diagram manner, are plotted as solubility versus $1/T$ (K) because, for small percentages dissolved, theory predicts (and it is experimentally observed in many cases) that such curves should be straight lines (192). Regardless of which solvent is used, the grown silicon layer will be saturated with it. At the lower temperatures, solubility may be reduced, and, except for tin, all of the metals shown produce p-doping. Tin is a group IV element and apparently does not affect the electrical properties of silicon. Therefore, it has been extensively studied as a silicon solvent (184, 193–195), although gallium (84, 182) and gallium aluminum (196) have also been considered. (A figure showing the solid solubility versus temperature for a variety of materials is given in Chapter 8.)

Potential advantages of silicon LPE are relatively rapid growth at temperatures in the 600°C or less range, fewer carbon-induced growth defects (197), reduced concentration of heavy metals because of limited solubility at the reduced temperature, and more abrupt junctions through reduced autodoping (195). Some difficulties are that, except for aluminum, none of the solvents listed will remove residual SiO_2; it is difficult to produce thin layers; and the surfaces are often rough.

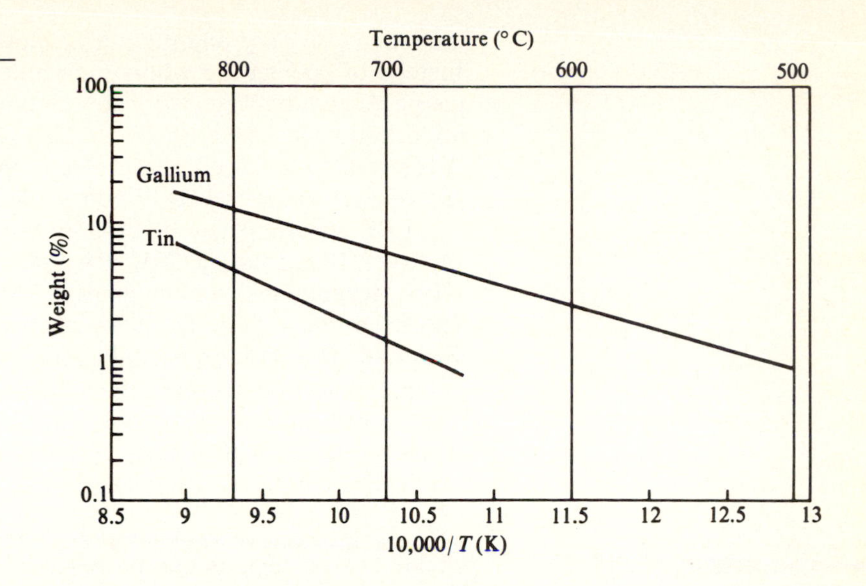

FIGURE 7.43

Solubility of GaAs in molten gallium and tin versus temperature. (*Source:* Plotted from data in B.R. Pamplin, *Crystal Growth,* Pergamon Press, New York, 1980.)

7.5.2 GaAs LPE

Two common solvents for GaAs LPE are gallium and tin. Fig. 7.43 gives the solubility of GaAs in these two solvents. Tin is incorporated into the GaAs layer as an n-type dopant. Gallium, however, being a constituent of GaAs, does no doping, and further, near-stoichiometric GaAs can be grown from a low-temperature gallium melt. When they are grown from a GaAs melt during conventional crystal pulling, GaAs crystals have a gallium deficiency (vacancies on gallium sublattice sites).

The growth rate of LPE layers is determined either by the rate at which the solute can diffuse to the surface or by the nucleation rate on the surface. Unlike vapor phase epitaxy, chemical reaction rates are not involved. Since GaAs LPE has been so widely used, various growth rate modeling studies have been made (198–201). It appears that solute transport to the surface (arsenic when growing in gallium or both gallium and arsenic when growing in tin) is generally the limiting step. If the possibility of transport by convection is neglected, and only diffusion from the main body of solvent through a boundary layer to the surface is assumed, an equation for the growth rate R of the form

$$R \cong \frac{D(C_l - C_e)}{\delta} \qquad 7.23$$

results, where D is the diffusion coefficient of the solute in the melt, C_l is the concentration of solute—for example, arsenic—in the main body of liquid, C_e is the concentration in the liquid adjacent to the growing surface, and δ is the width of the solute-depleted layer. The

ered stepwise once and growth is allowed to proceed isothermally. Instead of cooling the whole melt and wafer, it is also possible to pass a current between the wafer and melt and produce Peltier cooling at the wafer–melt interface. This technique, still experimental, is referred to as liquid phase electroepitaxy and has been used to grow layers of several III–V compounds (202).

LPE apparatus can be broadly broken into tipping (the earliest method) (180), rotating (203, 204, 186), direct immersion, and sliding (205), depending on the manner in which the melt is transferred to and from the wafers. These variations are shown schematically in Fig. 7.44. The sliding system has the advantage of wiping off excess solvent that would otherwise solidify on the wafer surface. The joints need only be reasonably tight to prevent leakage since the melts used have high surface tensions and the material of construction will generally be graphite and not wet by the molten metal.

7.6 EPITAXIAL CRYSTAL GROWTH FROM A SOLID

Single-crystal regions can be enlarged at temperatures well below the melting point as is demonstrable by the grain growth during annealing that is often observed. In some cases, it is postulated that the driving force is just the reduction in interfacial energy that occurs when some grains enlarge and others disappear. In other cases, the internal stresses in the grains left after cold-working apparently provide the force. By stressing materials and then annealing, large metal crystals can be grown. The first intentional growth of crystals by this method was apparently in 1912 (206). More directly related to IC fabrication are the presumption that the damaged surfaces left after slice polishing regrow as single crystal in this manner and the observation that silicon amorphized by ion implant will regrow from the solid phase as single crystal. The latter is a very important feature since the ion implant doses often used produce an amorphized layer on the surface. Afterward, if the layer had to be regrown by surface melting, substantial redistribution of dopant would occur during the time the surface was molten. The rate of regrowth of silicon layers amorphized by silicon implantation has been measured as varying from ~100 Å/s near 600°C to ~1.4μm/s at 745°C (207). It has also been reported that amorphous silicon layers evaporated onto well-cleaned (100) silicon surfaces can be regrown by annealing at temperatures above 550°C (208). However, the use of a series of subsequent ion implants to first ensure a completely amorphous layer and then to produce ion mixing[19] at the substrate–layer interface appears to be more reproducible (209). The use of such a combination of low-temperature amorphous or near-amorphous deposition, followed by an implant and relatively low-temperature

[19] For a discussion of ion mixing, see Chapter 9.

FIGURE 7.44

Various geometries for LPE. (All of them are enclosed in furnaces that are not shown.)

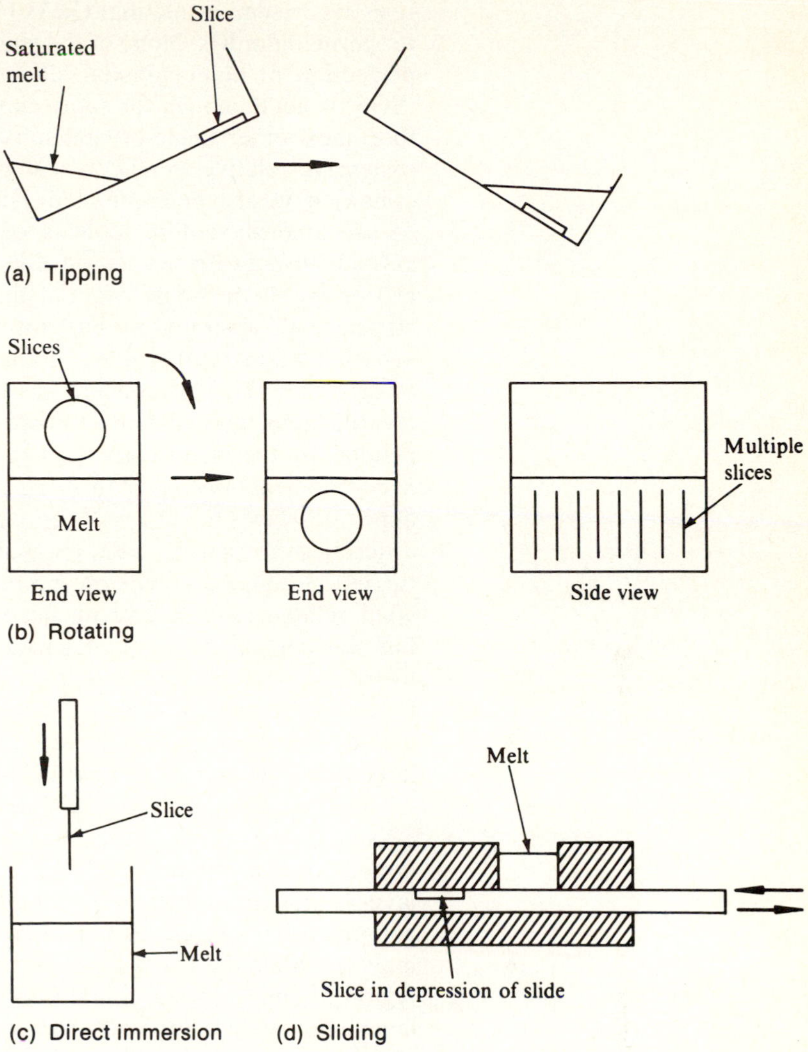

(a) Tipping

(b) Rotating

(c) Direct immersion

(d) Sliding

boundary conditions, and hence the final form of R, will depend on whether the temperature is continually lowered or whether it is low-solid-state epitaxial regrowth (anneal), is an alternative to direct low-temperature chemical vapor phase or molecular beam epitaxy.

7.7 CRYSTAL GROWTH OVER INSULATING SUBSTRATES*

Almost since the inception of the silicon IC, true epitaxy over single-crystal insulating substrates such as sapphire, spinel, and quartz has been studied (211–213). In addition, germanium epitaxy on high-

*See reference (210).

resistivity (semi-insulating) GaAs (13) was considered as a material for germanium ICs. None of the silicon on foreign substrates has the perfection of homoepitaxial silicon, although silicon on sapphire (SOS) is good enough for some classes of devices to be made. Further, most other single-crystal substrates such as sapphire are quite expensive relative to silicon wafers. Thus, alternatives to heteroepitaxy have also been pursued. One of the oldest is the suggestion to use amorphous-film replicas of the steps occurring on single-crystal surfaces for nucleation. Since nucleation begins most easily at steps, growth should begin at the replicated steps, and since the steps are all separated by integral multiples of lattice spacings, the separately nucleated regions should grow together with a minimum of defects (214). This concept has been expanded to use artificially created steps (grooves) on amorphous substrates (215, 216) and, depending on the investigators, has been called *artificial epitaxy, diataxy*, and *graphoepitaxy*. In some cases, recrystallization following deposition gives better results apparently because even if there is no ordering over the total area, enough material is oriented adjacent to the ledges to act as seeds during recrystallization (217). In a somewhat related (and largely unsuccessful) approach (rheotaxy), the film was deposited on a liquid surface that was subsequently allowed to freeze. The concept was that small single-crystal nuclei would form on the molten surface and, upon contact with one another, would have enough mobility to shift to the proper orientation for large-area single-crystal growth (218).

Several methods have also been proposed for producing single-crystal layers on insulating films that do not depend on the initial growth of a single-crystal layer. A polycrystalline or amorphous layer is first deposited and then melted and allowed to freeze in such a manner that single-crystal regrowth occurs from a seed. The seed may be either the first part of the remelted layer that freezes or a region of single crystal initially in contact with the polycrystalline layer. The second approach has the advantage of providing a seed of known and reproducible orientation, while the first has the advantage of not requiring external seeding.

If there is no initial seed, the area melted at any time must be small, and either a spot traveling over the film in raster fashion or a line moving laterally over the film is generally used. Spot heating was originally by electron beam (219) but may also be by laser beam. When a narrow line is to be melted, strip heaters can be used (220). An alternative approach has been to restrict the deposition to a small area at any one time and to sweep it over the surface (221). A more successful nonseed method for silicon is to pattern the polycrystalline layer so that only small areas remain. During laser annealing, islands of up to 2 μm × 20 μm have been observed to regrow as single crystals (222).

When seeds are used, they are most often exposed regions of a single-crystal wafer otherwise covered with an insulating layer as shown in Fig. 7.45 (223). Not shown in this figure is a coating often deposited over the silicon layer. Such a coating is usually desirable for any of the melt regrowth methods, whether external seeding is used or not. Surface tension can cause the molten surface to be undulating and sometimes to ball up and loose contact with the remainder of the film unless the film is first covered with a coating such as SiO_2. The technique of growth shown in Fig. 7.45, along with the one described in the next paragraph, is referred to as lateral overgrowth. Melting has been by E-beam (224), laser (223), strip heater (220), or halogen lamps (225). Scanning can be used, or the film over a whole wafer can be melted from the top, after which, if cooling is from the bottom of the wafer, the single-crystal seed areas will freeze first and seed the lateral growth over the oxide. This occurs because of the higher thermal conductivity of the semiconductor in the windows relative to the oxide over the rest of the wafer.

There are conditions for selectively growing GaAs (226) or silicon that produces some overgrowth of single-crystal material onto the mask. Typical overhangs for silicon are about equal to the thickness grown above the mask, as is shown in the cross section of Fig. 7.46. However, by changing growth conditions, larger lateral-to-vertical growth ratios have been reported (227). For GaAs, the lateral spread is typically much more than for silicon. An early use for

Oxide
Single-crystal silicon
CVD polycrystalline
layer added
Polycrystalline layer melted
and regrown as a single-crystal film
Exposed regions act
as seeds during regrowth

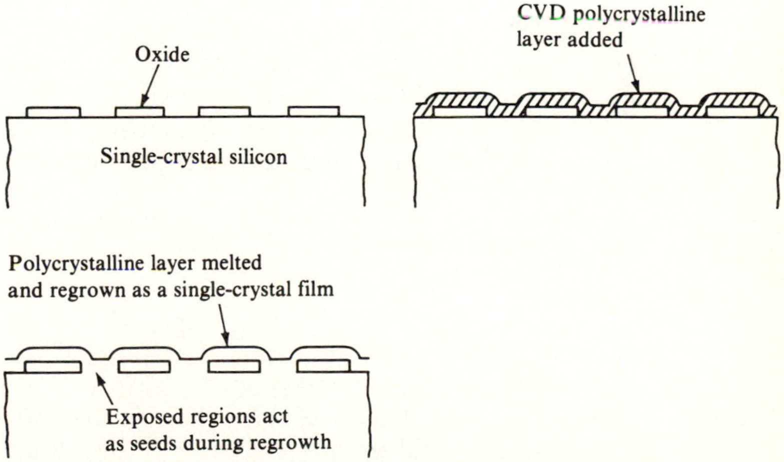

FIGURE 7.45

Lateral overgrowth process for preparing single-crystal films over amorphous layers.

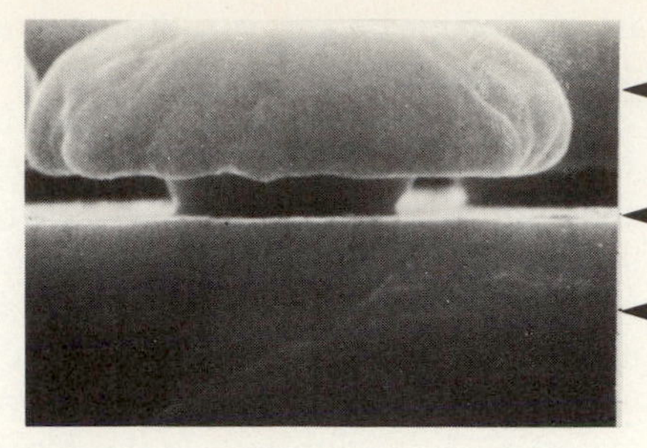

FIGURE 7.46

Cross section of a silicon pillar selectively grown through a hole in thermal oxide. (The thermal oxide has been etched away so that the amount of lateral spreading that occurred after the hole was filled can be seen.)

this spreading was to increase the collection area of photodiodes (115), but more recently it has been proposed as an alternative to the seeded recrystallization discussed previously (227). In order to obtain reasonable lateral spreading distance and good crystal quality, the film thickness may be excessive, in which case it has been proposed to thin it by a series of oxidation–etch steps (228).

CHAPTER 7 SAFETY

Safety is probably more of an issue in the epitaxial operation than in any other, and very careful procedures must be followed to prevent leaks in the epitaxial reactor or any of the supply lines feeding it. Outward leaks can bring highly toxic or flammable materials into the atmosphere, and inward leaks may allow explosions. The explosion and safety risks in epitaxial deposition are similar to those encountered during thin-film CVD and discussed in Chapter 4, although the gas volumes and flow rates are usually much higher in epitaxial operations. The increased volumes of explosive gases found in most large-scale production epitaxial reactors offer the potential of much more damaging explosions than might be expected of the typical LPCVD reactor. Because of the higher flows of hazardous gases encountered, more care must be taken in the design of ventilation systems and in ensuring that there are no stagnant pockets where either a poisonous or an explosive gas can collect.

CHAPTER 7 KEY IDEAS

- The original concept of epitaxy was that of oriented overgrowth of some material onto a naturally occurring crystallographic face of a mineral.
- The first semiconductor industry application of oriented overgrowth, the vapor phase growth of a germanium layer onto a germanium wafer, was also referred to as epitaxy.

- □ Growth onto a foreign substrate—for example, silicon on sapphire—is called heteroepitaxy to distinguish it from growth onto a substrate of the same material (epitaxy, isoepitaxy, autoepitaxy).
- □ Epitaxial layers can be doped by co-depositing a dopant with the semiconductor.
- □ The epitaxial process allows a layer to be overgrown onto a substrate of radically different impurity concentration.
- □ Silicon epitaxial growth is almost always from the vapor. GaAs epitaxy for discrete devices is sometimes from the liquid, but GaAs for ICs is from the vapor phase.
- □ The hydrogen reduction of SiH_2Cl_2, $SiHCl_3$, and $SiCl_4$ and the thermal decomposition of SiH_4 are the common sources of epitaxial silicon.
- □ AsH_3 + GaCl manufactured in the deposition chamber or AsH_3 + gallium from the decomposition of an organometallic such as trimethylgallium is commonly used for GaAs epitaxial layers.
- □ Molecular beam epitaxy can be used for either silicon or GaAs, but it is too slow for most applications.
- □ In situ etching of wafer surfaces before epitaxial deposition is widely used to remove native grown surface oxide and residual polish damage.
- □ Because of autodoping and, to a lesser degree, because of solid-state diffusion, the doping concentration transition from the substrate to the epitaxial layer is not as abrupt as the substrate–layer transition itself.
- □ Any shallow depressions originally in the substrate wafer surface are usually transferred to the epitaxial layer surface, but the boundaries may be displaced laterally. The effect is referred to as pattern shift and distortion.
- □ By growing a layer heavily doped with an impurity with a large misfit ratio, misfit dislocations (useful for gettering) can be deliberately produced in the layer.

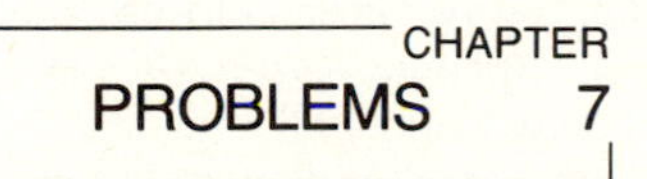

1. If $SiCl_4$ + H_2 is to be used for etching silicon, give a set of conditions that might work, and explain your choice.
2. By a series of at least three sketches, indicate where a single-crystal, round rod of silicon with the long dimension in a <111> direction first shows (110) and (211) facets, and then demonstrate that the faster-growing (110) facets (planes) will eventually disappear.
3. If a 1% by volume $SiHCl_3$ concentration in hydrogen is desired, what should the temperature of the bubbler be if all of the hydrogen gas goes through it? What should the temperature be if 1 liter/minute goes through the bubbler and a total of 50 liters/minute goes through the reactor.
4. If a 1% by volume of $SiHCl_3$ concentration in hydrogen flows into a reactor at a rate of 50 liters/minute and if the deposition rate is 1 μm/minute onto twenty 100 mm diameter wafers, what is the deposition efficiency in grams of silicon onto a wafer divided by grams of silicon in inlet gas stream?
5. If a 20 μm thick silicon epitaxial layer over a buried layer is grown from $SiCl_4$ at 1125°C at a rate of 1 μm/minute, how much and in which direction should the isolation diffusion mask be offset from the surface pattern in order to center the isolation diffusion over the buried collector diffusion?
6. Sketch a top view of a stacking fault occurring in a (100) silicon epitaxial layer, and calculate the thickness of the layer if the length of one side of the fault is 3 μm. If the length of one side of

a (111) stacking fault is also 3 μm, how thick is the layer?

7. If misfit dislocations are to be minimized when growing a lightly doped silicon layer onto a heavily n-doped silicon substrate, what would be the best substrate dopant to use?

CHAPTER 7 REFERENCES

1. M.L. Royer, "Recherches Experimentales Sur L'epitaxie ou Orientation Mutuelle de Cristaux D'especes Differentes," *Bull. Soc. Franc. Mineral 51*, pp. 7–159, 1928.
2. H. Christiansen and G.K. Teal, U.S. Patent 2,692,839, October 26, 1954 (filed April 7, 1951).
3. H.C. Theuerer et al., "Epitaxial Diffused Transistor," *Proc. IRE 48*, pp. 1642–1643 (L), 1960.
4. B.T. Murphy, "Monolithic Semiconductor Devices," U.S. Patent 3,237,026, February 22, 1966 (filed October 20, 1961).
5. J.P. Hirth and G.M. Pound, *Condensation and Evaporation*, The Macmillan Co., New York, 1963.
6. W.K. Burton, N. Cabrera, and F.C. Frank, "The Growth of Crystals and the Equilibrium Structure of Their Surfaces," *Phil. Trans. Royal Soc. 243*, pp. 299–358, 1951.
7. Tsunenori Sakamoto and Gen Hashiguchi, "Si(001)—2×1 Single Domain Structure Obtained by High Temperature Annealing," *Jap. J. Appl. Phys. 25*, pp. L78–L80, 1986.
8. D. Walton, "Nucleation of Vapor Deposits," *J. Chem. Phys. 37*, pp. 2182–2188, 1962.
9. H.P. Zeindl et al., "Influence of Substrate Misorientation and Temperature on MBE-Grown Si," *J. Crystal Growth 81*, pp. 231–236, 1987.
10. S.K. Tung, "The Effects of Substrate Orientation on Epitaxial Growth," *J. Electrochem. Soc. 112*, pp. 436–438, 1965.
11. C.H.J. van den Brekel, "Growth Rate Anisotropy and Morphology of Autoepitaxial Silicon Films from $SiCl_4$," *J. Crystal Growth 23*, pp. 259–266, 1974.
12. B.A. Joyce et al., "The Influence of Substrate Surface Conditions on the Nucleation and Growth of Epitaxial Silicon Films," *Surface Science 15*, pp. 1–13, 1969.
13. G.O. Krause and E.C. Teague, "Observations of Misfit Dislocations in GaAs–Ge Heterojunctions," *Appl. Phys. Lett. 10*, pp. 251–253, 1967.
14. Herbert Kroemer, "Polar-on-Nonpolar Epitaxy," *J. Crystal Growth 81*, pp. 193–204, 1987.
15. References in J. Bloem, "Nucleation and Growth of Silicon by CVD," *J. Crystal Growth 50*, pp. 581–604, 1980.
16. B.S. Meyerson and M.L. Yu, "In-Situ Doped Silicon Via LPCVD; Interactions of the Dopant/Silane/Silicon Surface System," pp. 287–294, in *CVD 84*, Electrochemical Society, Pennington, N.J., 1984.
17. A.A. Chernov and M.P. Rusaikin, "Theoretical Analysis of Equilibrium Adsorption Layers in CVD Systems (Si–H–Cl, Ga–As–H–Cl)," *J. Crystal Growth 45*, pp. 73–81, 1978.
18. See, for example, Georg-Maria Schwab, Hugh S. Taylor, and R. Spence, *Catalysis*, D. Van Nostrand Co., New York, 1937.
19. E.G. Alexander, "A Surface Reaction Approach to the Growth Kinetics of Epitaxial Silicon from $SiCl_4$," abst. no. 105, Electrochemical Society Meeting, Spring 1967.
20. John C. Bean, "Silicon Molecular Beam Epitaxy: 1984–1986," *J. Crystal Growth 81*, pp. 411–420, 1987.
21. Wilford J. Corrigan, "Gas Etching," U.S. Patent 3,243,323, March 29, 1966 (continuation of a patent filed June 11, 1962).
22. Kenneth Bean and Paul Gleim, "Vapor Etching Prior to Epitaxial Deposition of Silicon," late news paper, Electrochemical Society Meeting, Fall 1963.
23. G.A. Lang and T. Stavish, "Chemical Polishing of Silicon with Anhydrous Hydrogen Chloride," *RCA Review 24*, pp. 488–499, 1963.
24. P. Rai-Choudhury and D.K. Schroder, "Selective Growth of Epitaxial Silicon and Gallium Arse-

nide," *J. Electrochem. Soc. 118*, pp. 106–110, 1971.

25. F. Jona, "Preparation of Atomically Clean Surfaces of Si and Ge by Heating in Vacuum," *Appl. Phys. Lett. 6*, pp. 205–206, 1965.
26. E. Kasper and K. Worner, "Applications of Si-MBE for Integrated Circuits," pp. 429–447, in K.E. Bean and G.A. Rozgonyi, eds., *VLSI Science and Technology/84*, Electrochemical Society, Pennington, N.J., 1984.
27. Koichi Kugimiya et al., "Si-Beam Radiation Cleaning in Mole-Beam Epitaxy," *Jap. J. Appl. Phys. 24*, pp. 564–567, 1985.
28. Yusuke Ota, "Si Molecular Beam Epitaxy (n on n^+) with Wide Range Doping Control," *J. Electrochem. Soc. 124*, pp. 1795–1802, 1977.
29. Steve Wright and Herbert Kroemer, "Reduction of Oxides on Silicon by Heating in a Gallium Molecular Beam at 800°C," *Appl. Phys. Lett. 36*, pp. 210–211, 1980.
30. E.R. Levin, J.P. Dismukes, and M.D. Coutts, "Electron Microscope Characterization of Defects on Gaseous-Etched Silicon Surfaces," *J. Electrochem. Soc. 118*, pp. 1171–1175, 1971.
31. H.C. Theuerer, "Epitaxial Silicon Films by Hydrogen Reduction of $SiCl_4$," *J. Electrochem. Soc. 108*, pp. 649–653, 1961.
32. S.B. Kulkarni and A.K. Gaind, "'Hydrogen Etching' of Silicon at High Temperatures," abst. no. 546, Electrochemical Society Pub. 79–2, 1979.
33. T.L. Chu, "A Low-Temperature Nonpreferential Gaseous Etchant for Silicon," *J. Electrochem. Soc. 115*, p. 1207, 1968.
34. P. Rai-Choudhury and A.J. Noreika, "Hydrogen Sulfide as an Etchant for Silicon," *J. Electrochem. Soc. 116*, pp. 539–541, 1969.
35. J.P. Dismukes and R. Ulmer, "Gas-Phase Etching of Silicon with Chlorine," *J. Electrochem. Soc. 118*, pp. 634–636, 1971.
36. L.D. Dyer, HI–H_2 Vapor Etch for Low Temperature Silicon Epitaxial Manufacturing," *AIChE J. 18*, pp. 728–734, 1972.
37. P. Rai-Choudhury, "Sulfur Hexafluoride as an Etchant for Silicon," *J. Electrochem. Soc. 118*, pp. 266–269, 1971.
38. L.J. Stinson et al., "Sulfur Hexafluoride Etching Effects in Silicon," *J. Electrochem. Soc. 123*, pp. 551–555, 1976.
39. P. Van Der Putte et al., "Surface Morphology of HCl Etched Silicon Wafers," *J. Crystal Growth 41*, pp. 133–145, 1977.
40. T.L. Chu and R.L. Tallman, "Water Vapor as an Etchant for Silicon," *J. Electrochem. Soc. 111*, pp. 1306–1307, 1964.
41. J.G. Gualtieri, M.J. Katz, and G.A. Wolff, "Gas Etching and Its Effect on Semiconductor Surfaces," *Z. Fur Krist. 114*, pp. 9–22, 1960.
42. L.V. Gregor, P. Balk, and F.J. Campagna, "Vapor-Phase Polishing of Silicon with H_2–HBr Gas Mixtures," *IBM J. Res. Develop. 9*, pp. 327–332, 1965.
43. J.C. Bean et al., "Dependence of Residual Damage on Temperature during Ar^+ Sputter Cleaning of Silicon," *J. Appl. Phys. 48*, pp. 907–913, 1977.
44. John Borland et al., "Silicon Epitaxial Growth for Advanced Device Structures," *Solid State Technology*, pp. 111–119, January 1988.
45. E.A.V. Ebsworth, *Volatile Silicon Compounds*, The Macmillan Co., New York, 1963.
46. F. Mieno et al., "Low Temperature Silicon Epitaxy Using Si_2H_6," *J. Electrochem. Soc. 134*, pp. 2320–2323, 1987.
47. Harold P. Klug and Robert C. Brasted, *Comprehensive Inorganic Chemistry*, Vol. 7, D. Van Nostrand Co., New York, 1958.
48. E. Sirtl and H. Seiter, "Vapor Deposited Microcrystalline Silicon," *J. Electrochem. Soc. 113*, pp. 506–508, 1966.
49. Lee P. Hunt, "Thermodynamic Equilibrium in the Si–H–Cl and Si–H–Br Systems," *J. Electrochem. Soc. 135*, pp. 206–209 and references therein, 1988.
50. J. Nishizawa and M. Saito, "Mechanism of Chemical Vapor Deposition of Silicon," *J. Crystal Growth 52*, pp. 213–218, 1981.
51. R.F.C. Farrow, "The Kinetics of Silicon Deposition on Silicon by Pyrolysis of Silane," *J. Electrochem. Soc. 121*, pp. 899–907, 1974.
52. S.E. Bradshaw, "The Effects of Gas Pressure and Velocity on Epitaxial Silicon Deposition by the Hydrogen Reduction of Chlorosilanes," *Int. J. Electronics 23*, pp. 381-391, 1967.
53. E.G. Bylander, "Kinetics of Silicon Crystal Growth from $SiCl_4$ Decomposition," *J. Electrochem. Soc. 109*, pp. 1171–1175, 1962.
54. J. Bloem, "Trends in the Chemical Vapor Depo-

sition of Silicon," pp. 180–190, in Howard R. Huff and Ronald R. Burgess, eds., *Semiconductor Silicon/73*, Electrochemical Society, Princeton, N.J., 1973.

55. W.H. Shepherd, "Vapor Phase Deposition and Etching of Silicon," *J. Electrochem. Soc. 112*, pp. 988–994, 1962.
56. J. Nishizawa et al., "Silicon Epitaxial Growth," *J. Crystal Growth 17*, pp. 241–248, 1972.
57. J.M. Charig and B.A. Joyce, "Epitaxial Growth of Silicon by Hydrogen Reduction of $SiHCl_3$ onto Silicon Substrates," *J. Electrochem. Soc. 109*, pp. 957–962, 1962.
58. Anders Lekholm, "Epitaxial Growth of Silicon from Dichlorosilane," *J. Electrochem. Soc. 119*, pp. 1122–1123, 1972.
59. Kohetsu Tanno et al., "Selective Silicon Epitaxy Using Reduced Pressure Techniques," *Jap. J. Appl. Phys. 21*, pp. L564–L566, 1982.
60. B.A. Joyce and R.R. Bradley, "Epitaxial Deposition of Silicon from the Pyrolysis of Monosilane on Silicon Substrates," *J. Electrochem. Soc. 110*, pp. 1235–1240, 1963.
61. J. Bloem and W.A.P. Claassen, "Rate-Determining Reactions and Surface Species in CVD of Silicon," *J. Crystal Growth 49*, pp. 435–444, 1980.
62. R.B. Herring, "Silicon Epitaxy at Reduced Pressure," Applied Materials Tech. Rpt. HT–010, 1980.
63. S. Mendelson, "Stacking Fault Nucleation in Epitaxial Silicon on Variously Oriented Silicon Substrates," *J. Appl. Phys. 35*, pp. 1570–1581, 1964.
64. C.H.J. van den Brekel, "Growth Rate Anisotropy and Morphology of Autoepitaxial Silicon Films from $SiCl_4$," *J. Crystal Growth 23*, pp. 259–266, 1974.
65. J.E. Allegretti et al., pp. 255–270, in Ralph O. Grubel, ed., *Metallurgy of Elemental and Compound Semiconductors*, Interscience Publishers, New York, 1961.
66. J. Nishizawa et al., "Layer Growth in Silicon Epitaxy," *J. Crystal Growth 13/14*, pp. 297–301, 1972.
67. H.E. Buckley, *Crystal Growth*, John Wiley & Sons, New York, 1951.
68. C. Holm and E. Sirtl, "Large-Area Epitaxial Growth of Chalcogen-Doped Silicon," pp. 303–309, in *CVD 84*, Electrochemical Society, Pennington, N.J., 1984.
69. J.F. Carboy et al., "Pattern Distortion during CVD Epitaxy on (100) Silicon at Atmospheric and Reduced Pressure," pp. 434–453, in *CVD 84*, Electrochemical Society, Pennington, N.J., 1984.
70. S.P. Weeks, "Pattern Shift and Pattern Distortion during CVD Epitaxy on (111) and (100) Silicon," *Solid State Technology*, pp. 111–117, November 1981.
71. J.H. Alexander and H.F. Sterling, "Semiconductor Epitaxy–Gas Phase Doping by Electric Discharge," *Solid-State Electronics 10*, pp. 485–490, 1967.
72. B.S. Meyerson and M.L. Yu, "In-Situ Doped Silicon Via LPCVD; Interactions of the Dopant/Silane/Silicon Surface System," pp. 287–294, in, *CVD 84,* Electrochemical Society, Pennington, N.J., 1984.
73. R. Nuttall, "The Dependence on Deposition Conditions of the Dopant Concentration of Epitaxial Layers," *J. Electrochem. Soc. 111*, pp. 317–323, 1964.
74. W.H. Shepherd, "Doping of Epitaxial Silicon," *J. Electrochem. Soc. 115*, pp. 541–545, 1968.
75. P. Rai-Choudhury and E.I. Salkovitz, "Doping of Epitaxial Silicon," *J. Crystal Growth 7*, pp. 353–360, 1970.
76. J. Bloem et al., "The Incorporation of Phosphorus in Silicon Epitaxial Layer Growth," *J. Electrochem. Soc. 121*, pp. 1354–1357, 1974.
77. L.J. Giling and J. Bloem, "The Incorporation of Phosphorus in Silicon: The Temperature Dependence of the Segregation Coefficient," *J. Crystal Growth 31*, pp. 317–322, 1975.
78. H. Basseches et al., "Factors Affecting the Resistivity of Epitaxial Silicon Layers," pp. 69–86, in John B. Schroeder, ed., *Metallurgy of Semiconductor Materials*, Interscience Publishers, New York, 1962 (proc. of meeting held in 1961).
79. C.O. Thomas et al., "Impurity Distribution in Epitaxial Silicon Films," *J. Electrochem. Soc. 109*, pp. 1055–1061, 1962.
80. B.A. Joyce et al., "Impurity Redistribution Processes in Epitaxial Silicon Layers," *J. Electrochem Soc. 112*, pp. 1100–1106, 1965.

81. W.H. Shepherd, "Autodoping of Epitaxial Silicon," *J. Electrochem. Soc. 115*, pp. 652–656, 1968.
82. G. Skelly and A.C. Adams, "Impurity Atom Transfer during Epitaxial Deposition of Silicon," *J. Electrochem. Soc. 120*, pp. 116–122, 1973.
83. Paul H. Langer and Joseph I. Goldstein, "Impurity Redistribution during Silicon Epitaxial Growth and Semiconductor Device Processing," *J. Electrochem. Soc. 121*, pp. 563–571, 1974.
84. W.R. Runyan, *Silicon Semiconductor Technology*, McGraw-Hill Book Co., New York, 1965.
85. H.B. Poage et al., in Fifth International Conference on Chemical Vapor Deposition, Electrochemical Society, Princeton, N.J., 1970.
86. G.R. Srinivasan, "A Flow Model for Autodoping in VLSI Substrates," *J. Electrochem. Soc. 127*, pp. 2305–2306, 1980.
87. S.E. Meyer and D.E. Shea, "Epitaxial Deposition of Silicon Layers by Pyrolysis of Silane," *J. Electrochem. Soc. 111*, pp. 550–556, 1964.
88. Carl O. Bozler, "Reduction of Autodoping," *J. Electrochem. Soc. 122*, pp. 1705–1709, 1975.
89. G.R. Srinivasan, "Kinetics of Autodoping in Silicon Epitaxy," *J. Electrochem. Soc. 125*, pp. 146–151, 1978.
90. M.W.M. Graef and B.J.H. Leunissen, "Autodoping in Silicon Epitaxy," *Electrochem. Soc. Ext. Abst. 84–2*, abst. no. 456, pp. 655–656, 1984.
91. J.L. Deines and A. Spiro, "Low Pressure Silicon Epitaxy," *Electrochem. Soc. Ext. Abst. 74–1*, abst. no. 62, 1974.
92. Michiharu Tabe and Hiroaki Nakamura, "Adsorbed Layer Model for Autodoping Mechanism in Silicon Epitaxial Growth," *J. Electrochem. Soc. 126*, pp. 822–826, 1979.
93. G.R. Srinivasan, "Autodoping Effects in Silicon Epitaxy," *J. Electrochem. Soc. 127*, pp. 1334–1342, 1980.
94. G.K. Ackermann and E. Ebert, "Autodoping Phenomena in Epitaxial Silicon," *J. Electrochem. Soc. 130*, pp. 1910–1915, 1983.
95. A. Reif and R.W. Dutton, "Computer Simulation in Silicon Epitaxy," *J. Electrochem. Soc. 128*, pp. 909–918, 1981.
96. Hseuh-Rong Chang, "Autodoping in Silicon Epitaxy," *J. Electrochem. Soc. 132*, pp. 219–224, 1985.
97. For lapping and staining techniques, see, for example, W.R. Runyan, *Semiconductor Measurements and Instrumentation*, McGraw-Hill Book Co., New York, 1975.
98. Hiroyasu Kubota and Shigeo Kotani, "A New Method of Pattern Shift Measurement during Silicon Epitaxy," *Electrochem. Soc. Ext. Abst. 84–2*, abst. no. 458, pp. 659–660, 1984.
99. J. Wang and P. Howell, "Pattern Distortion and Shift Measurement Technique for 1 Micron Epitaxial Layers," *Electrochem. Soc. Ext. Abst. 84–2*, abst. no. 500, pp. 728–729, 1984.
100. Kenneth E. Bean and Walter R. Runyan, "Semiconductor Wafer with at Least One Epitaxial Layer and Methods of Making Same," U.S. Patent 3,379,584, April 23, 1968 (filed September 4, 1964).
101. C.E. Benjamin and E.J. Patzner, "(111) Faceting in Epitaxy Grown over Stepped Surfaces," recent news paper, Electrochemical Society Meeting, May 1965.
102. D.W. Boss and V.Y. Doo, "Pattern Washout and Displacement in (111) Epitaxy," *Electrochem. Soc. Electronics Div. Ext. Abst. 15–2*, abst. no. 186, pp. 59–65, 1966 (Philadelphia meeting).
103. C.M. Drum and C.A. Clark, "Geometric Stability of Shallow Surface Depressions during Growth of (111) and (100) Epitaxial Silicon," *J. Electrochem. Soc. 115*, pp. 664–669, 1968.
104. C.M. Drum and C.A. Clark, "Anisotropy of Macrostep Motion and Pattern Edge-Displacement during Growth of Epitaxial Silicon on Silicon near {100}," *J. Electrochem. Soc. 117*, pp. 1401–1405, 1970.
105. P.H. Lee et al., "Epitaxial Pattern Shift Comparison in Vertical, Horizontal, and Cylindrical Reactor Geometries," *J. Electrochem. Soc. 124*, pp. 1824–1826, 1977.
106. Walter Benzing et al., Applied Materials Seminar, Dallas, 1982.
107. B.D. Joyce and J.A. Baldrey, "Selective Epitaxial Deposition of Silicon," *Nature 195*, pp. 485–486, 1962.
108. Jack M. Hirshon, "Silicon Epitaxial Junctions with Compatible Masking," *Electrochem. Soc.*

Electronics Div. Ext. Abst., abst. no. 101, pp. 242–243, Spring 1962.

109. R. Glang and E.S. Wajda, "Status of Vapor Growth in Semiconductor Technology," pp. 27–47, in John B. Schroeder, ed., *Metallurgy of Semiconductor Materials*, Interscience Publishers, New York, 1962.
110. Hiroshi Katsumura et al., "Isolation in Integrated Circuits Using Preferential Epitaxial Growth Technique," *Electronics and Communications in Japan 49*, pp. 180–186, 1966.
111. "Localized Epitaxial Growth for Wristwatch, Space MOS," *Electronics*, January 4, 1971.
112. G.L. Schnable et al., "Preferential Silicon Epitaxy with Oxide Masking," *Electrochem. Soc. Electronics Div. Ext. Abst.*, abst. no. 129, pp. 36–39, Fall 1963; expanded in *Electrochem. Technol. 4*, pp. 485–491, 1966.
113. Andrew F. McKelvey, "The Use of Preferential Etching and Preferential Epitaxy in Microelectronics," pp. II-B-1 to II-B-6, in *Proc. 2nd Annual Microelectronics Symposium*, St. Louis, April 1964.
114. Juliana Manoliu, "CMOS Device Isolation and Latch-Up Prevention," *Semiconductor International*, pp. 90–92 and references therein, April 1988.
115. Kenneth E. Bean, "Chemical Vapor Deposition Applications in Microelectronics Processing," *Thin Solid Films 83*, pp. 173–186, 1981.
116. Erhard Sirtl and Hartmut Seiter, "Selective Epitaxy of Silicon under Quasi-Equilibrium Conditions," pp. 189–199, in Rolf R. Haberecht and Edward L. Kern, eds., *Semiconductor Silicon*, Electrochemical Society, New York, 1969.
117. M. Nomura, "Silicon Refill Epitaxial Growth with Si_3N_4 Masking," abst. no. 181, Electrochemical Society Meeting, Fall 1969.
118. Hiroshi Ogawa et al., "The Selective Epitaxial Growth of Silicon by Using Silicon Nitride Films as a Mask," *Jap. J. Appl. Phys. 10*, pp. 1675–1679, 1971.
119. W.G. Oldham and R. Holmstrom, "The Growth and Etching of Si through Windows in SiO_2," *J. Electrochem. Soc. 114*, pp. 381–388, 1967.
120. Earl. G. Alexander and W.R. Runyan, "A Study of Factors Affecting Silicon Growth on Amorphous SiO_2 Surfaces," *Trans. Metall. Soc. AIME 236*, pp. 284–290, 1966.
121. R. Pagliaro et al., "Uniformly Thick Selective Epitaxial Silicon," *J. Electrochem. Soc. 134*, pp. 1235–1238, 1987.
122. D.J. Dumin, "Selective Epitaxy Using Silane and Germane," *J. Crystal Growth 8*, pp. 33–36, 1971.
123. Manfred Druminski and Roland Gessner, "Selective Etching and Epitaxial Refilling of Silicon Wells in the System $SiH_4/HCl/H_2$," *J. Crystal Growth 31*, pp. 312–316, 1975.
124. H.M. Liaw et al., "Effect of Substrate Preparation and Growth Ambient on Silicon Selective Epitaxy," pp. 463–475, in *CVD 84*, Electrochemical Society, Pennington, N.J., 1984.
125. John O. Boreland and Clifford I. Drowley, "Advanced Dielectric Isolation through Selective Epitaxial Growth Techniques," *Solid State Technology*, pp. 141–148, 1985.
126. Don M. Jackson, Jr., "Advanced Epitaxial Processes for Monolithic Integrated-Circuit Applications," *Trans. Metall. Soc. AIME 233*, pp. 596–602, 1965.
127. Alan R. Stivers et al., "Growth Condition Dependence of SEG Planarity and Electrical Characteristics," pp. 389–395, in *CVD 87*, Electrochemical Society, Pennington, N.J., 1987.
128. Akihiko Ishitani et al., "Facet Formation in Selective Silicon Epitaxial Growth," *Jap. J. Appl. Phys. 24*, pp. 1267–1269, 1985.
129. Alfred C. Ipri et al., "Selective Epitaxial Growth for the Fabrication of CMOS Integrated Circuits," *IEEE Trans. on Electron Dev. ED-31*, pp. 1741–1748 and earlier references, 1984.
130. A. Ishitani et al., "Silicon Selective Epitaxial Growth for CMOS Technology," pp. 355–365, in *CVD 87*, Electrochemical Society, Pennington, N.J., 1987.
131. L. Jasterzebeski et al., "Issues and Problems Involved in Selective Epitaxial Growth of Silicon for SOI Fabrication," pp. 334–353, in *CVD 87*, Electrochemical Society, Pennington, N.J., 1987.
132. "Epitaxial Process Is Highly Selective in Depositing Silicon," *Electronics*, January 31, 1980.
133. P.C. Rundle, "The Epitaxial Growth of Silicon in Horizontal Reactors," *Int. J. Electronics 24*, pp. 405–413, 1968.
134. F.C. Eversteyn et al., "A Stagnant Layer Model for the Epitaxial Growth of Silicon from Silane

in a Horizontal Reactor," *J. Electrochem. Soc. 117*, pp. 925–931, 1970.

135. Vladimir S. Ban and Stephen L. Gilbert, "The Chemistry and Transport Phenomena of Chemical Vapor Deposition of Silicon from $SiCl_4$," *J. Crystal Growth 31*, pp. 284–289, 1975.
136. J.C. Gillis et al., "Fluid Mechanical Model for CVD in a Horizontal RF Reactor," pp. 21–29, in *CVD 84*, Electrochemical Society, Pennington, N.J., 1984.
137. E. Fujii et al., "A Quantitative Calculation of the Growth Rate of Epitaxial Silicon from $SiCl_4$ in a Barrel Reactor," *J. Electrochem. Soc. 119*, pp. 1106–1113, 1972.
138. C.W. Manke and L.F. Donaghey, "Analysis of Transport Processes in Vertical Cylinder Epitaxy Reactors," *J. Electrochem. Soc. 124*, pp. 561–569, 1977.
139. H.A. Lord, "Convective Transport in Silicon Epitaxial Deposition in a Barrel Reactor," *J. Electrochem. Soc. 134*, pp. 1227–1235, 1987.
140. B.A. Unvala, "Epitaxial Growth of Silicon by Vacuum Evaporation," *Nature 194*, pp. 966–967, 1962.
141. B.A. Joyce and R.R. Bradley, "A Study of Nucleation in Chemically Grown Epitaxial Silicon Films Using Molecular Beam Techniques. I: Experimental Methods," *Phil. Mag. Series 8, Vol. 14*, pp. 289–299, 1966.
142. G.R. Booker and B.A. Unvala, "Vacuum Evaporated Silicon Layers Free from Stacking Faults," *Phil. Mag. Series 8, Vol. 8*, pp. 1597–1598, 1963.
143. J.C. Bean and E.A. Sadowiski, "Silicon MBE Apparatus for Uniform High-Rate Deposition on Standard Format Wafers," *J. Vac. Sci. Technol. 20*, pp. 137–142, 1982.
144. Subramanian S. Iyer et al., "Dopant Incorporation Processes in Silicon Grown by Molecular Beam Epitaxy," pp. 473–488, in K.E. Bean and G.A. Rozgonyi, eds., *VSLI Science and Technology/84*, Electrochemical Society, Pennington, N.J., 1984.
145. Yusuke Ota, "n-Type Doping Techniques in Silicon Molecular Beam Epitaxy by Simultaneous Arsenic Ion Implantation and by Antimony Evaporation," *J. Electrochem. Soc. 126*, pp. 1761–1765, 1979.
146. References in John C. Bean and Sigrid R. McAfee, "Silicon Molecular Beam Epitaxy: A Comprehensive Bibliography 1962–82," Proc. Int. Meeting on the Relationship between Epitaxial Growth Conditions and the Properties of Semiconducting Epitaxial Layers, Perpignan, France, August 1982.
147. Raymond P. Roberge et al., "Gaseous Impurity Effects in Silicon Epitaxy," *Semiconductor International*, pp. 77–81, 1987.
148. R.S. Wagner and W.C. Ellis, "Vapor–Liquid–Solid Mechanism of Single Crystal Growth," *Appl. Phys. Lett. 4*, pp. 89–90, 1964.
149. G.A. Rozgonyi et al., "Defect Engineering for VLSI Epitaxial Silicon," *J. Crystal Growth 85*, pp. 300–307, 1987.
150. H.R. Huff et al., "Influence of Silicon Slice Curvature on Thermally Induced Stresses," *J. Electrochem. Soc. 118*, pp. 143–145, 1971.
151. L.D. Dyer et al., "Plastic Deformation in Central Regions of Epitaxial Silicon Slices," *J. Appl. Phys. 42*, pp. 5680–5688, 1971.
152. J. Bloem and A.H. Goemans, "Slip in Silicon Epitaxy," *J. Appl. Phys. 43*, pp. 1281–1283, 1972.
153. A.H. Goemans and L.J. Van Ruyven, "Control of Slip in Horizontal Silicon Epitaxy with Profiled Susceptors," *J. Crystal Growth 31*, pp. 308–311, 1975.
154. McD. Robinson et al., "Low Dislocation Density RF-Heated Epitaxial Silicon Depositions," *J. Electrochem. Soc. 129*, pp. 2858–2860, 1982.
155. J. McDiarmid et al., "Slip-Free Epitaxial Silicon Deposition in an Induction-Heated Vertical Reactor," pp. 351–355, in *CVD 84*, Electrochemical Society, Pennington, N.J., 1984.
156. Rajaram Bhat et al., "Vapor-Phase Etching and Polishing of Gallium Arsenide Using Hydrogen Chloride Gas," *J. Electrochem. Soc. 122*, pp. 1378–1382, 1975.
157. Rajaram Bhat and Sorab K. Ghandhi, "Vapor-Phase Etching and Polishing of GaAs Using Arsenic Trichloride," *J. Electrochem. Soc. 124*, pp. 1447–1448, 1977.
158. N. Holonyak et al., "Halogen Vapor Transport and Growth of Epitaxial Layers of Intermetallic Compounds and Compound Mixtures," pp. 49–59, in John B. Schroeder, ed., *Metallurgy of Semiconductor Materials*, Interscience Publish-

ers, New York, 1962 (proc. of meeting held in 1961).

159. G.R. Antell and D. Effer, "Preparation of Crystals of InAs, InP, GaAs, and GaP by a Vapor Phase Reaction," *J. Electrochem. Soc. 106*, pp. 509–511, 1959.

160. J.R. Knight et al., "The Preparation of High Purity Gallium Arsenide by Vapor Phase Epitaxial Growth," *Solid-State Electronics 8*, pp. 178–180, 1965.

161. J. Nishizawa et al., "Reaction Mechanism of GaAs Vapor-Phase Epitaxy," *J. Electrochem. Soc. 133*, pp. 2567–2575, 1986.

162. James J. Tietjen and James A. Amick, "The Preparation and Properties of Vapor Deposited Epitaxial $GaAs_{1-x}P_x$ Using Arsine and Phosphine," *J. Electrochem. Soc. 113*, pp. 724–728, 1966.

163. Don W. Shaw, "Epitaxial GaAs Kinetic Studies: {001} Orientation," *J. Electrochem. Soc. 117*, pp. 683–687, 1970.

164. A. Boucher and L. Hollan, "Thermodynamic and Experimental Aspects of Gallium Arsenide Vapor Growth," *J. Electrochem. Soc. 117*, pp. 932–936, 1970.

165. D.J. Kirwan, "Reaction Equilibria in the Growth of GaAs and GaP by the Chloride Transport Process," *J. Electrochem. Soc. 117*, pp. 1572–1577, 1970.

166. Don W. Shaw, "Influence of Substrate Temperature on GaAs Epitaxial Deposition Rates," *J. Electrochem. Soc. 115*, pp. 405–408, 1968.

167. Don W. Shaw, "Kinetic Aspects in the Vapor Phase Epitaxy of III–V Compounds," *J. Crystal Growth 31*, pp. 130–141, 1975.

168. J.C. Hong and H.H. Lee, "Epitaxial Growth Rate of GaAs: Chloride Transport Process," *J. Electrochem. Soc. 132*, pp. 427–432, 1985.

169. H.M. Manasevit and W.I. Simpson, "The Use of Metal-Organics in the Preparation of Semiconductor Materials," *J. Electrochem. Soc. 116*, pp. 1725–1732, 1969.

170. P. Rai-Choudhury, "Epitaxial Gallium Arsenide from Trimethyl Gallium and Arsine," *J. Electrochem. Soc. 116*, pp. 1745–1746, 1969.

171. S.J. Bass, "Device Quality Epitaxial Gallium Arsenide Grown by the Metal Alkyl-Hydride Technique," *J. Crystal Growth 31*, pp. 172–178, 1975.

172. M.J. Ludowise, "Metalorganic Chemical Vapor Deposition of III–V Semiconductors," *J. Appl. Phys. 58*, pp. R31–R55 and extensive list of included references, October 1985.

173. S.J. Bass, "Growth of Semi-Insulating Epitaxial Gallium Arsenide by Chromium Doping in the Metal-Alkyl + Hydride System," *J. Crystal Growth 44*, pp. 29–33, 1978.

174. W.T. Tsang, "Chemical Beam Epitaxy of InP and GaAs," *Appl. Phys. Lett. 45*, pp. 1234–1236, 1984.

175. R. Fischer et al., "Growth and Properties of GaAs/AlGaAs on Nonpolar Substrates Using Molecular Beam Epitaxy," *J. Appl. Phys. 58*, pp. 374–383, 1985.

176. N. Otsuka et al., "Study of Heteroepitaxial Interfaces by Atomic Resolution Electron Microscopy," *J. Vac. Sci. Technol. B4*, pp. 896–899, 1986.

177. R.P. Vasquez, "Cleaning Chemistry of GaAs(100) and InSb(100) Substrates for Molecular Beam Epitaxy," *J. Vac. Sci. Technol. B1*, pp. 791–794, 1983.

178. Paul H. Keck and Jacob Broder, "The Solubility of Silicon and Germanium in Gallium and Indium," *Phys. Rev. 90*, pp. 521–522, 1953.

179. A.J. Goss, "Crystallization of Silicon from Solution in Tin," *J. of Metals 5*, part 2, p. 1085, 1953.

180. H. Nelson, "Epitaxial Growth from the Liquid State and Its Applications to the Fabrication of Tunnel and Laser Diodes," *RCA Review 24*, pp. 603–615, 1963.

181. H.J. Kim, "Liquid Phase Epitaxial Growth of Silicon in Selected Areas," *J. Electrochem. Soc. 119*, pp. 1394–1398, 1972.

182. B.E. Sumner and R.T. Foley, "Liquid-Phase Epitaxial Growth of Gallium-Doped Silicon," *J. Electrochem. Soc. 125*, pp. 1817–1824, 1978.

183. W.R. Runyan, "Melt and Solution Growth of Silicon," pp. 26–39 in *Low Cost Polycrystalline Silicon Solar Cells*, Southern Methodist University, Dallas, 1976.

184. B. Jayant Baliga, "Silicon Liquid Phase Epitaxy—A Review," *J. Electrochem. Soc. 133*, pp. 5C–14C, January 1986.

185. *J. Crystal Growth 27*, pp. 1–332, 1974. (This complete issue is devoted to liquid phase epitaxy.)

186. Tatsuhiko Niini, "GaP Red Light Emitting Diodes Produced by a Rotating Boat System of

Liquid Phase Epitaxy," *J. Electrochem. Soc. 124*, pp. 1285–1289, 1977.

187. R.B. Wilson et al., "Investigation of Melt Carry-Over during Liquid Phase Epitaxy: I. Growth of Indium Phosphide," *J. Electrochem. Soc. 132*, pp. 172–176, 1985.

188. J.M. Woodall et al., "Liquid Phase Epitaxial Growth of $Ga_{(1-x)}Al_xAs$," *J. Electrochem. Soc. 116*, pp. 899–903, 1969.

189. Kazuo Nakajima and Jiro Okazaki, "Substrate Orientation Dependence of the In–Ga–As Phase Diagram for Liquid Phase Epitaxial Growth of $In_{0.53}Ga_{0.47}As$ on InP," *J. Electrochem. Soc. 132*, pp. 1424–1432 and references therein, 1985.

190. P. Besome et al., "Investigation of Melt Carry-Over during Liquid Phase Epitaxy: II. Growth of Indium Gallium Arsenic Phosphide Double Heterostructure Material Lattice Matched to Indium Phosphide," *J. Electrochem. Soc. 132*, pp. 176–179, 1985.

191. J.G. Fleming and D.A. Stevenson, "Isothermal Liquid Phase Epitaxial Growth of Mercury Cadmium Telluride," *J. Electrochem. Soc. 134*, pp. 1225–1227, 1987.

192. C.D. Thurmond and M. Kowalchik, "Germanium and Silicon Liquidus Curves," *Bell Syst. Tech. J. 39*, pp. 169–204, 1960.

193. B. Jayant Baliga, "Kinetics of the Epitaxial Growth of Silicon from a Tin Melt," *J. Electrochem. Soc. 124*, pp. 1627–1631, 1977.

194. B. Jayant Baliga, "Isothermal Silicon Liquid Phase Epitaxy from Supersaturated Tin," *J. Electrochem. Soc. 125*, pp. 598–600, 1978.

195. B. Jayant Baliga, "Boron Autodoping during Silicon Liquid Phase Epitaxy," *J. Electrochem. Soc. 128*, pp. 161–165, 1981.

196. B. Girault et al., "Liquid Phase Epitaxy of Silicon at Very Low Temperatures," *J. Crystal Growth 37*, pp. 169–177, 1977.

197. L.A. D'Asaro et al., "Low Defect Silicon by Liquid Phase Epitaxy," pp. 233–242, in Rolf R. Haberecht and Edward L. Kern, eds., *Semiconductor Silicon*, Electrochemical Society, New York, 1969.

198. I. Crossley and M.B. Small, "Computer Simulations of Liquid Phase Epitaxy of GaAs in Ga Solutions," *J. Crystal Growth 11*, pp. 157–165, 1971.

199. R. Ghez, "An Exact Solution of Crystal Growth Rates under Conditions of Constant Cooling Rate," *J. Crystal Growth 19*, p. 153, 1973.

200. D.L. Rode, "Isothermal Diffusion Theory of LPE: GaAs, GaP, Bubble Garnet," *J. Crystal Growth 20*, pp. 13–23, 1973.

201. T. Bryskiewicz, "Investigation of the Mechanism and Kinetics of Growth of LPE GaAs," *J. Crystal Growth 43*, pp. 101–114, 1978.

202. T. Bryskiewicz et al., "Bulk GaAs Crystal Growth by Liquid Phase Electroepitaxy," *J. Crystal Growth 82*, pp. 279–288 and references therein, 1987.

203. J.A. Donahue and H.T. Minden, "A New Technique for Liquid Phase Epitaxy," *J. Crystal Growth 7*, pp. 221–226, 1970.

204. O.G. Lorimor et al., "High Capacity Liquid Phase Epitaxy Apparatus Utilizing Thin Melts," *Solid-State Electronics 16*, pp. 1289–1295, 1973.

205. L.R. Dawson, "Near-Equilibrium LPE Growth of GaAs-$Ga_{1-x}Al_xAs$ Double Heterostructures," *J. Crystal Growth 27*, pp. 86–96, 1974.

206. H.E. Buckley, p. 93, in *Crystal Growth*, John Wiley & Sons, New York, 1956.

207. A. Lietoila et al., "The Rate of CW Laser Induced Solid Phase Epitaxial Regrowth of Amorphous Silicon," *Appl. Phys. Lett. 39*, pp. 810–812, 1981.

208. M. von Allmen et al., "Solid-State Epitaxial Growth of Deposited Si Films," *Appl. Phys. Lett. 35*, p. 280, 1979.

209. R.C. Cole et al., "Thin Epitaxial Silicon Regrowth Using Ion Implantation Amorphization Techniques," *J. Electrochem. Soc. 135*, pp. 974–979, 1988.

210. For background reading on this topic, see H.W. Lam et al., Chap. 1, "Silicon-on-Insulator for VLSI and VHSIC," in *VLSI Electronics: Microstructure Science 4*, Academic Press, New York, 1982; and complete issue of *J. Crystal Growth*, October 1983.

211. H.M. Manasevit and W.I. Simpson, "Single-Crystal Silicon on a Sapphire Substrate," American Physical Society Meeting, August 1963; and *J. Appl. Phys. 35*, pp. 1349–1351, 1964.

212. H.M. Manasevit and D.H. Forbes, "Single-Crystal Silicon on Spinel," American Physical Society Meeting, January 1965; and H. Seiter and C. Zaminer, "Epitaktische Siliziumschichten auf Mg–Al–Spinell," *Zeit. angew. Phys. 20*, pp. 158–161, 1965.

213. B.A. Joyce et al., "Epitaxial Deposition of Sili-

con on Quartz," *Solid State Comm. 1*, pp. 107–108, 1963; and R.W. Bicknell et al., "The Epitaxial Deposition of Silicon on Quartz," *Phil. Mag. Series 8, Vol. 9*, pp. 965–978, 1964.

214. J.D. Filby and S. Nielsen, "Single Crystal Films of Silicon on Insulators," *Brit. J. Appl. Phys. 18*, pp. 1357–1382, 1967.

215. Henry I. Smith et al., "Silicon-on-Insulator by Graphoepitaxy and Zone-Melting Recrystallization of Patterned Films," *J. Crystal Growth 63*, pp. 527–546, 1983.

216. V.I. Klykov and N.N. Sheftal, "Diataxial Growth of Silicon and Germanium," *J. Crystal Growth 52*, pp. 687–691, 1981.

217. Henry I. Smith et al., "The Mechanism of Orientation in Si Graphoepitaxy by Laser or Strip Heater Recrystallization," *J. Electrochem. Soc. 130*, pp. 2050–2053, 1983.

218. E. Rasmanis, "Thin Film p–n Junction Silicon Devices," *Semiconductor Products*, pp. 30–33, July 1963.

219. J. Maserjian, "Single Crystal Germanium Films by Micro Zone-Melting," *Solid-State Electronics 6*, pp. 477–484, 1963.

220. M.W. Geis et al., "Zone-Melting Recrystallization of Si Films with a Moveable-Strip-Heater Oven," *J. Electrochem. Soc. 129*, pp. 2812–2818, 1982.

221. See H.F. Matare, "Heteroepitaxy of Silicon on Insulator Crystal Substrates," pp. 249–290, in Rolf R. Haberecht and Edward L. Kern, eds., *Semiconductor Silicon*, Electrochemical Society, New York and included references 52–54, 1969.

222. J.F. Gibbons et al., "CW Laser Recrystallization of <100> Silicon on Amorphous Substrates," *Appl. Phys. Lett. 34*, pp. 831–833, 1979.

223. M. Tamura et al., "Si Bridging Epitaxy from Si Windows onto SiO_2 by Q-Switched Ruby Laser," *Jap. J. Appl. Phys. 19*, pp. L23–L26, 1980.

224. J.R. Davis et al., "Characterization of the Dual E-Beam Technique for Recrystallizing Polysilicon Films," *J. Electrochem. Soc. 132*, pp. 1919–1924, 1985.

225. G.K. Celler et al., "Dielectrically Isolated Thick Si Films by Lateral Epitaxy from the Melt," *J. Electrochem. Soc. 132*, pp. 211–219, 1985.

226. A.G. Lapierre et al., "Anomalous Unconstrained Crystal Growth of GaAs," pp. 301–304, in H. Steffen Peiser, ed., *Crystal Growth*, Pergamon Press, London, 1967.

227. References in L. Jastrzebski et al., "Issues and Growth of Silicon for SOI Fabrication," pp. 334–353, in *CVD 87*, Electrochemical Society, Pennington, N.J., 1987.

228. L. Jastrzebski et al., "Preparation of Thin (0.6 μm) Continuous Monocrystalline Silicon over SiO_2," *J. Electrochem. Soc. 132*, pp. 3056–3057, 1985.

CHAPTER

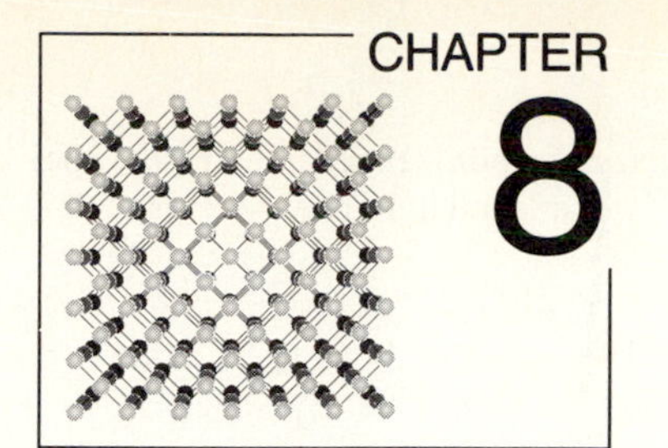

8

Impurity Diffusion

8.1 INTRODUCTION

Thermal diffusion is a thermally activated process in which atoms or molecules in a material (solid, liquid, or gas) move from high- to low-concentration regions. The phenomenon is encountered in a number of semiconductor processes and has already been alluded to in several chapters. In silicon thermal oxidation, the oxygen travels through the growing oxide to the silicon surface by diffusion. In the CVD of thin films and in epitaxial growth, the gaseous reactants must diffuse through a relatively stagnant gas layer adjacent to the growing surface. Similarly, in liquid etching, the reacting species must generally diffuse through stagnant liquid layers. However, when diffusion is mentioned in the context of semiconductor processing, the motion of doping impurities in the semiconductor itself is generally assumed. Such diffusion will occur whenever there is an impurity gradient and the temperature is high enough to allow appreciable motion. Thus, while a high-temperature diffusion is widely used intentionally to move doping impurities in the semiconductor lattice, motion by diffusion can also take place during high-temperature processing steps not related to the diffusion step.

While appreciable diffusion in gases and liquids can occur at room temperature, the solid-state diffusion required for semiconductor device fabrication generally requires temperatures in the 1000°C range. For silicon, two primary groups of diffusing impurities are to be considered. One is the group IIIA and VA substitutional shallow-level impurities that diffuse rather slowly. The other is comprised mainly of gold and transition elements such as Cu, Ni, and Fe, which are deep-level lifetime killers and which diffuse orders of magnitude faster than the group IIIA and VA atoms. The groupings are similar for GaAs, except that the normal dopant atoms of interest are groups II and VI.

After diffusing for some time from an initially abrupt concentration step, the concentration–distance profile will have the general characteristics shown in Fig. 8.1. $N(x)$, the impurity concentration,

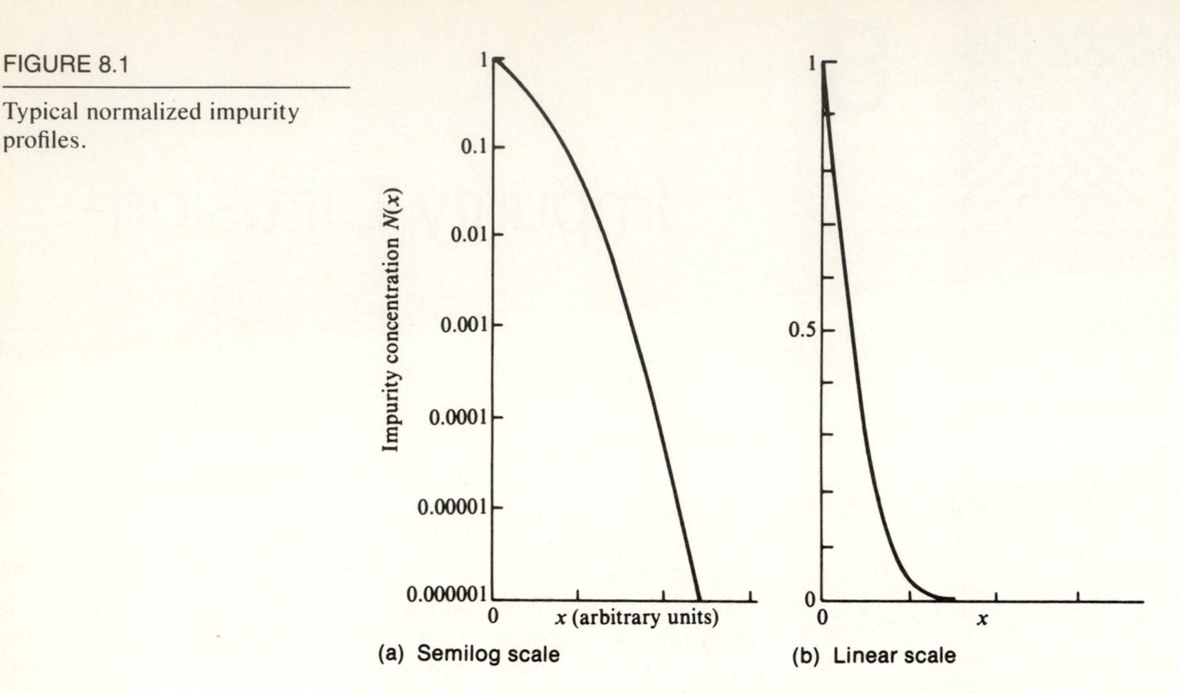

FIGURE 8.1

Typical normalized impurity profiles.

is shown plotted both on a linear and on a semilogarithmic scale. The latter scale is more useful since several decades of concentration must generally be considered. Increasing the temperature or extending the time will cause the curve to spread out along the x axis. In a typical diffusion step, impurities are desired in localized areas, such as MOS sources and drains or bipolar bases and emitters. To accomplish this diffusion, a wafer is coated with a masking material (something relatively impervious to the desired diffusant) and openings are cut in the mask where diffusions are required. The wafer is then heated to the required temperature in an atmosphere containing the impurities to be diffused and held there long enough for the necessary diffusion to take place.

Ideal device impurity profiles are usually abrupt,[1] as shown in Fig. 8.2a for a simple bipolar transistor. One of the process engineering challenges is to use a mechanism that inherently provides a sloped profile (Fig. 8.1) to approximate the ideal abrupt profiles closely enough for acceptable device operation. Fig. 8.2b shows the general manner in which this is done. In the illustration and in most practical cases, all diffusions are from the front surface of the wafer. Thus, the wafer background doping is present in the base, along with

[1]Sometimes, a concentration gradient with its attendant electric field is desirable in, for example, a base region in order to decrease the carrier transit time.

FIGURE 8.2

Bipolar diffused transistor doping profiles.

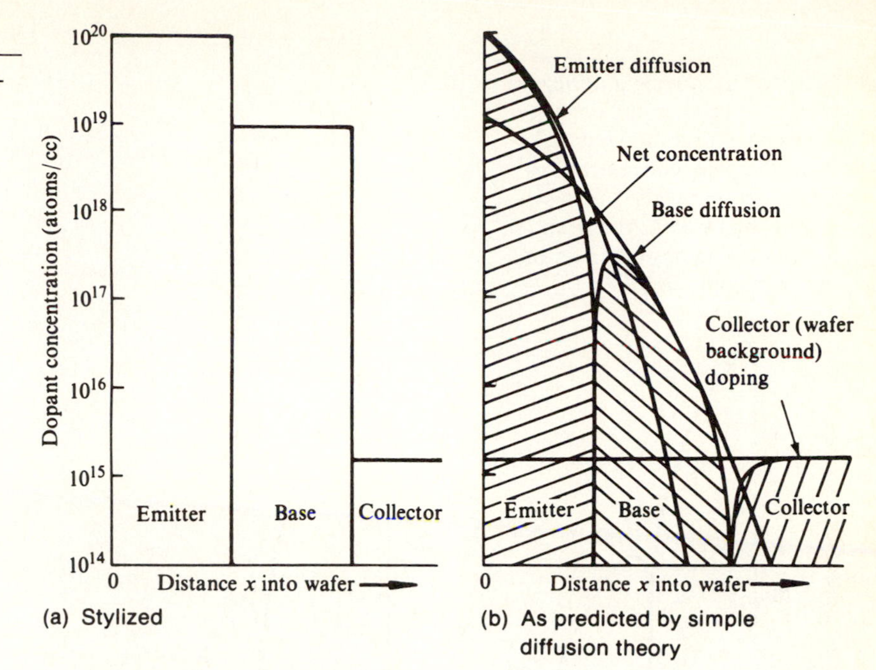

doping from the base diffusion. In the emitter region, the wafer background doping and the base doping are both present in the emitter, along with the emitter doping. The profile of most interest, and one also shown in Fig. 8.2b, is not that of the individual impurities, but rather that of the *net* concentration $|N_A - N_D|$, where N_A is the concentration of acceptors and N_D is the concentration of donors. Efforts to match diffusion profiles to the desired ideal impurity profiles of various device structures have required a variety of diffusion boundary conditions that will be discussed in the following sections. Fig. 8.3 illustrates, again using a bipolar transistor as an example, how interrelated the various properties become when diffusion is used to place impurities. If it is desired to reduce the base width, diffusing the emitter in deeper is a logical approach.[2] However, as the sequence of parts a, b, and c shows, while the base width is reduced by 35%, the peak base impurity concentration is reduced by a factor of 10. Thus, maintaining an appropriately high Gummel number[3] while producing very thin bipolar bases is a very real problem indeed.

[2]There is also a phenomenon referred to as emitter push, which causes the base impurities to rapidly diffuse ahead of the emitter and maintain a much wider base than expected. This phenomenon will be discussed in more detail later.

[3]The Gummel number is the net number of impurities per cm^2 in the base.

FIGURE 8.3

Effect of narrowing a diffused transistor base on the peak base concentration. (Asterisks mark the peak concentrations and show that while the base is narrowing by 35%, the peak concentration has fallen an order of magnitude.)

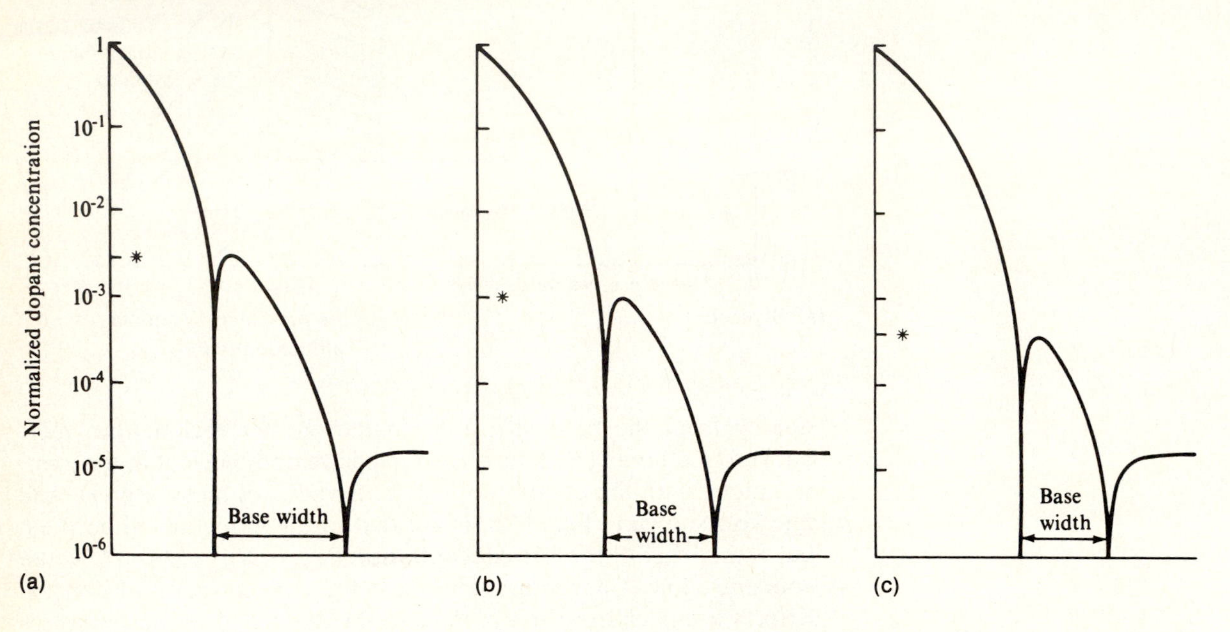

Diffusion can be discussed in terms of solutions to a set of diffusion equations that will predict impurity profiles such as those just shown for a wide variety of boundary conditions. It can also be discussed in terms of the atomic motion necessary to provide the observed transport. The first view is more useful in developing semiconductor processes, but the second is required in order to offer explanations of the many deviations from simple theory that are actually observed and to provide the necessary background for computer modeling of the diffusion process.

8.2 ATOMIC DIFFUSION MECHANISMS

Even though there are relatively fixed locations for each atom of a solid, the thermal vibration of the atoms will occasionally be of sufficient magnitude to allow an atom to surmount its potential barrier and move to an adjacent location. This frequency ν_b is given by

$$\nu_b = \nu_e^{-E_b/kT} \qquad 8.1$$

where E_b is the energy of the barrier, k is Boltzmann's constant, and ν is the frequency with which the atom is vibrating in the direction of the jump. ν is normally assumed to be the same as the Debye frequency.[4] The frequency with which a jump will actually occur will also depend on the availability of empty sites and the number of directions in which a given atom can jump.

In order to relate the atomic jumps to a net transfer of impurities, consider first the one-dimensional case shown in Fig. 8.4 (1). Let two adjacent atomic planes be located at x_1 and x_2, and suppose that the density of diffusing atoms residing in plane 1 is n_1 atoms/cm^2 and in plane 2 is n_2 atoms/cm^2. Atoms can jump only in the $+x$ or the $-x$ direction and have equal probability of going in either direction. If the jump frequency is ν_b, then there will be $\nu_b n_1/2$ atoms/cm^2 jumping to the right per unit time from x_1 and $\nu_b n_2/2$ atoms/cm^2 jumping to the left per unit time from x_2. The net flux is then given by

$$J = -\frac{\nu_b(n_1 - n_2)}{2} \tag{8.2}$$

However, it is customary to express the concentration in terms of N atoms/cm^3, not n atoms/cm^2. If, as in an actual crystal lattice, all atoms are located on planes separated by a distance Δx, then N is given by $n/\Delta x$ atoms/cm^3. The concentration gradient $\partial N/\partial x$ is given by $\Delta N/\Delta x$ where $\Delta N = N_1 - N_2 = (n_1 - n_2)/\Delta x$ and Δx (the jump distance) $= x_1 - x_2$. Substituting these values into Eq. 8.2 gives

$$J = -\left(\frac{\nu_b}{2}\right)(\Delta x)^2 \frac{\partial N}{\partial x} \tag{8.3}$$

Defining the diffusion coefficient D as $\Delta x^2(\nu_b/2)$ gives

$$J = -D\frac{\partial N}{\partial x} \tag{8.4}$$

which is often referred to as Fick's first law of diffusion.

To relate D to the cubic diamond lattice constant, consider jumps from two adjacent (400) planes, for which $\Delta x = a_0/4$ where a_0 is the lattice constant.[5] An examination of the bond directions shows that for this choice of planes, all jumps must be to adjacent planes and none to sites within the same plane. Since there are 4 bonds per atom, there are 2 possible jumps in each direction. However, even

FIGURE 8.4

Jumping of atoms from plane to plane. (With atoms jumping in both the $-x$ and the $+x$ directions with equal probability, net flow will only result when there is a difference in atomic densities between planes 1 and 2.)

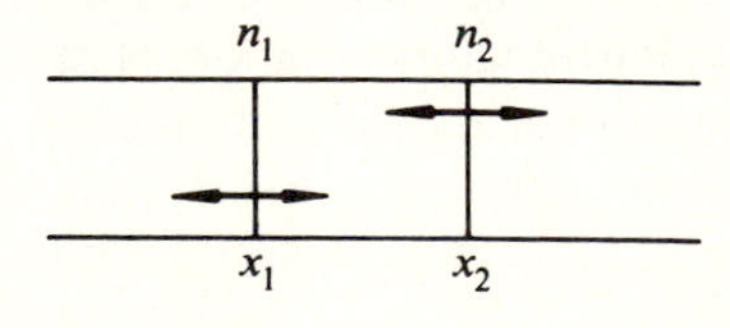

[4]The Debye frequency is given by $(k/h)\Theta_D$ where h is Planck's constant and Θ_D is the Debye temperature. ν for silicon is approximately 10^{13}/s.

[5]For a discussion of crystallographic planes, see Appendix A.

if an atom has the energy to surmount the barrier, there is no assurance that the site will be vacant. Therefore, a probability P of a site availability must be assigned. Eq. 8.3 then becomes

$$J = -P\nu_b\left(\frac{a_0^2}{8}\right)\frac{\partial N}{\partial x} \tag{8.5}$$

P may, for example, be the probability of a vacancy being adjacent to the diffusing atom and will be given by the density of vacancies divided by the number of atomic sites. The general form of P is $P = P_0e^{-E_p/kT}$. If this expression is combined with the value of ν_b from Eq. 8.1,

$$D = \nu e^{-E_b/kT} P_0e^{-E_p/kt} = D_0e^{-E_d/kT} \tag{8.6}$$

where D_0 is a constant and E_d is the diffusion activation energy. In practice, the D values used are experimentally measured quantities. To calculate them accurately from first principles requires a more elegant treatment than that used to arrive at the general form given by Eq. 8.6.

If jumps between adjacent (220) planes are considered, $\Delta x = a_0/2\sqrt{2}$, but only half of the jumps will be to adjacent planes. Thus, again,

$$J = -P\nu_b\left(\frac{a_0^2}{8}\right)\frac{\partial N}{\partial x} \tag{8.7}$$

Regardless of the set of planes chosen (equivalent to direction of diffusion), the value for D will remain the same.

If a vacancy is diffusing, it can go to any filled site. Therefore, if the number of vacancies is much less than the number of possible sites, the probability of a jump direction being toward an acceptable site is 1 and $D_V = (a_0/8)\nu e^{-E/kT}$ where E is the free energy required for a vacancy to surmount its barrier.

In the case of atoms, the problem is more complex. Examples of the ways in which motion can occur are shown in Fig. 8.5 (1, 2). Part a shows a simple exchange of sites by a rotation of adjacent atoms. The energy required for this process is substantial and is not believed[6] to occur in either Si or GaAs. Part b shows motion by an atom moving into an empty adjacent site (a vacancy). Several dopants apparently move by this mechanism. For those that do, any part of the diffusion process that generates more than an equilibrium

[6]Even though diffusion mechanisms in semiconductors have now been studied for about 40 years, no experiments have been devised that conclusively establish the mode of motion of most of the impurities in common use.

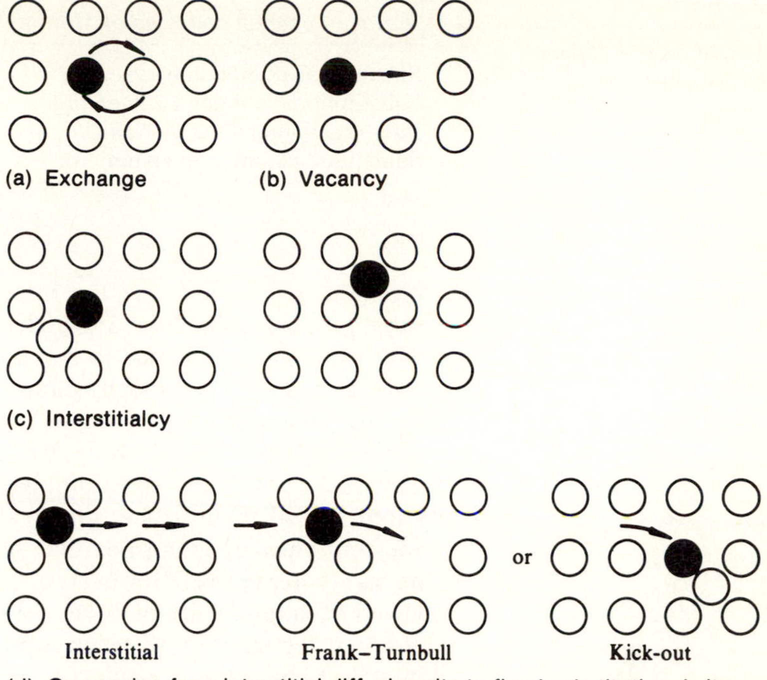

FIGURE 8.5

Atomic motion involved in various types of diffusion.

number of vacancies[7] will enhance the jump rate. Similarly, if the vacancy density is depressed, the jump rate will decrease. Part c shows motion in which an interstitial atom kicks a substitutional atom out of place and occupies its space. The kicked-out atom then kicks out an adjacent atom, and the process repeats itself. Note that in this case the interstitial atom does not move through the lattice via interstices but only via the kick-out mechanism. If both of the atoms involved belong to the host, then self-diffusion occurs; if the kicked-out atom is a substitutional impurity, then impurity diffusion occurs. In this mechanism, no impurity atom will move until an interstitial moves adjacent to it. Thus, any process that increases interstitials over equilibrium will enhance the jump rate, and a reduction will reduce it. Some impurities ordinarily occupy substitutional sites but can move into an interstitial site, travel rapidly via the interstices, and then return to a substitutional site. This motion is shown in part d. The return can be via a vacancy (Frank–Turnbull mechanism) or via a kick-out as discussed previously. Such impur-

[7]Both vacancies and interstitials are present in small quantities and are thermodynamically stable.

TABLE 8.1

Atomic Motion in Silicon Versus Impurity

Lattice Location and Kind of Motion	*Impurity*
Substitutional; motion by vacancy	P, Sb, Si, Al, Ga, As
Substitutional; motion by interstitialcy	Si, B, P, As
Interstitial; interstitial motion	O
Substitutional; interstitial motion	Au

ities generally have a very low interstitial solubility relative to their substitutional solubility. Table 8.1 lists various mechanisms by which atoms are assumed to move in silicon. As indicated, a given atomic species may move by more than one mechanism.

In gallium arsenide, group II impurities are substitutional in the gallium sublattice, and group VI are substitutional in the arsenic sublattice. Since any substitutional atom's four nearest neighbors are on the other sublattice, simple atomic motion by moving into an adjacent vacancy is not possible, and a rotational exchange would be highly disruptive. However, a substitutional–interstitial step followed by interstitial motion would allow the longer travel distance required for an atom to reach a proper site and is the mechanism thought to be active in gallium arsenide.

8.2.1 Diffusion by Vacancies

Diffusion by vacancy must include terms for each of the vacancy charge states (neutrals, $-$, $=$, $+$) so that the diffusion coefficient is (3)

$$D_V^* = D^{*0} + D^{*-} + D^{*=} + D^{*+} \tag{8.8}$$

where an $*$ indicates the value when a component is at its equilibrium value. When the doping level increases to the point where the semiconductor is no longer intrinsic at diffusion temperature, the distribution of charged vacancies will change. Thus, the relative contribution to diffusion from each of the vacancy charge states will change, and D_V becomes doping-level dependent. A further implication is that since the concentration decreases from the wafer surface inward, D_V will change with position. The D value will vary linearly with the number of vacancies, and the variation of vacancy concentration with doping level will obey the mass action law:

$$V^0 + h \rightarrow V^+, V^0 + e \rightarrow V^-, \ldots$$

so that

$$K_e = \frac{N_V^+}{pN_V^0} \cdots$$

where V^0 is a neutral vacancy, V^+ is a vacancy with a single plus charge, V^- is a vacancy with a single minus charge, N_V is the concentration of vacancies, h is a hole, p is the hole concentration, e is an electron, and K_e is the mass action equilibrium constant. K_e may be evaluated under intrinsic conditions when $p = n = n_i$ and $N_V^+ = N_{Vi}^+$; n is the electron concentration:

$$K_e = \frac{N_{Vi}^+}{n_i N_V^0} = \frac{N_V^+}{p N_V^0}$$

Substituting $p = n_i^2/n$ gives

$$\frac{N_{Vi}^+}{N_V^+} = \frac{n}{n_i}$$

Performing similar arithmetic on the other terms gives

$$D_V = D^{*0} + \left(\frac{n}{n_i}\right) D^{*-} + \left(\frac{n}{n_i}\right)^2 D^{*=} + \left(\frac{n_i}{n}\right) D^{*+} \qquad 8.9$$

The expression for D_V just derived is in terms of electron concentration but, for p-type diffusions, could have been derived with hole concentration used instead. Remember that, for intrinsic conditions, $n = n_i$, and thus there is no reduction in the diffusion constant for low carrier concentrations. Depending on the experimental conditions, different terms in Eq. 8.9 will dominate. Therefore, it is necessary to know the values of all of the D^*'s in order to adequately describe a material that diffuses in this manner.

8.2.2 Diffusion by Interstitialcy

Apparently, interstitialcy diffusion is never observed by itself, and most interstitials are neutral so that their concentration will be unaffected by doping level. They are, however, produced in nonequilibrium numbers by the thermal oxidation of silicon, and hence any diffusion that proceeds partially by interstitialcy will be enhanced near an oxidizing surface. When an impurity diffuses by both vacancy and interstitialcy, the total diffusivity D^* is the sum of the two components. That is,

$$D^* = D_V^* + D_I^*$$

and the fractional vacancy diffusivities are

$$f_V = \frac{D_V^*}{D^*}$$

and

$$f_I = \frac{D_I^*}{D^*}$$

When the concentration N_V or N_I is changed from its equilibrium value, each of the fractional diffusivities is assumed to change by the same amount so that (4)

$$D = D^*\left(\frac{f_V N_V}{N_V^*} + \frac{f_I N_I}{N_I^*}\right) \tag{8.10}$$

8.2.3 Interstitial Diffusion

Some materials, such as gold in silicon, have a very low interstitial solubility and a reasonably high substitutional solubility. However, they have a high interstitial diffusivity and a low substitutional diffusivity. Hence, appreciable motion can occur only by moving from a substitutional to an interstitial site, diffusing, and then becoming substitutional again. For the general case, with combined vacancy and interstitial motion, the effective diffusion coefficient D_{eff} is given by

$$D_{\text{eff}} = \frac{D_I N_I}{N_I + N_S} + \frac{D_S N_S}{N_I + N_S} \tag{8.11}$$

where the subscripts I and S refer to the interstitial and substitutional components, respectively. When $D_I >> D_S$ and $N_S >> N_I$ and there are no limitations by either vacancies or self-interstitials,

$$D_{\text{eff}} \equiv D_1 \cong \frac{D_I N_I}{N_S} \tag{8.12}$$

When the Frank–Turnbull interstitial–vacancy reaction occurs,

$$\text{Interstitials} + \text{Vacancies} \rightarrow \text{Substitutionals} \tag{8.13}$$

and a restricted source of vacancies—for example, dislocation-free material—may limit diffusion. As an example, consider that before an interstitial–substitutional conversion can occur, a vacancy has to diffuse to the site. Further, suppose that D_I is much greater that D_V where D_V is the vacancy diffusivity. D_{eff} then becomes

$$D_{\text{eff}} = \frac{D_V N_V}{N_V + N_S} + \frac{D_S N_S}{N_V + N_S} \tag{8.14a}$$

When N_V is much less than N_S and D_V is less than D_S,

$$D_{\text{eff}} \equiv D_2 \cong \frac{D_V N_V}{N_S} \tag{8.14b}$$

If the kick-out mechanism is dominant,

$$\text{Interstitials} \rightarrow \text{Substitutionals} + \text{Silicon Interstitials} \tag{8.15}$$

and a buildup of silicon interstitials can be the limiting factor. In this case, the diffusion profile will be determined by silicon self-diffusivity. Early studies of gold diffusion in silicon indicated that the vacancy mechanism played the dominant role (5), but later work has favored Eq. 8.15 as being more likely (6, 7).

As another example, consider the specific case of diffusion of zinc in gallium arsenide, which is assumed to be covered by the conditions of Eq. 8.12 (8). Zinc will move from an interstitial position to a gallium site vacancy. Substitutional zinc has a negative charge; interstitial zinc, a positive charge; and the gallium vacancy is neutral. The concentration of gallium vacancies is assumed to be independent of the doping level N_S. Thus, in equilibrium,

$$[\mathrm{Zn}]_S^- = [\mathrm{Zn}]_I^+ + 2e \quad \text{or} \quad N_S = N_I + 2n$$

where N_S is the concentration of substitutional zinc and N_I is the concentration of interstitial zinc. The hole density p arises from the negatively charged substitutional zinc ions and equals N_S. n can be calculated from $pn = n_i^2$. Then, by applying the mass action law,

$$K = \frac{n^2[\mathrm{Zn}]_I}{[\mathrm{Zn}]_S}$$

where K is an equilibrium constant. Since $n = n_i^2/p = n_i^2 N_S$,

$$K = \frac{N_I}{N_S^3} \quad \text{or} \quad N_I = KN_S^3$$

Substituting this value into Eq. 8.12 gives $D_1 = KD_I N_S^2$; that is, the diffusion coefficient depends on the concentration of substitutional zinc. Expressed somewhat differently,

$$D_1(N_{S2}) = D_1(N_{S1}) \times \left(\frac{N_{S2}}{N_{S1}}\right)^2$$

where the subscripts 1 and 2 denote two different values of N_S.

8.2.4 Electric-Field-Aided Diffusion

If an electric field is applied to the semiconductor, a velocity v will be imparted to any mobile ions present. When the field is parallel to the direction of diffusion, Eq. 8.4, Fick's first law, becomes

$$J = -D\left(\frac{\partial N}{\partial x}\right) + vN \tag{8.16}$$

v is given by μE where μ is the ionic mobility and E is the electric field. D and μ are related as $D/\mu = kT/qZ$. Thus, Eq. 8.16 becomes

$$J = -D\left(\frac{\partial N}{\partial x}\right) + \frac{qZEN}{kT} \tag{8.17}$$

where qZ is the ionic charge.[8] Some ions, such as lithium, have a high enough mobility that they can be moved at room temperature (drifted) by the field developed in a space charge region. At the high temperatures required for most diffusing species, a small electric field can be produced by the IR drop of an electric current flow. An internal field will also exist due to the diffusion gradient itself if the concentration is heavy enough for the semiconductor to be extrinsic at diffusion temperature (9). This field is given by

$$E = -\left(\frac{kT}{q}\right)\left(\frac{1}{n}\right)\left(\frac{dn}{dx}\right) = -\left(\frac{kT}{q}\right)\left(\frac{1}{n}\right)\left(\frac{dn}{dN}\right)\left(\frac{\partial N}{\partial x}\right) \tag{8.18}$$

where n is the free-carrier concentration. Only when $N >> n_i$ does $n = N$, and this condition is often not met during diffusion. For the general case,

$$n = \left(\frac{N^2}{4} + n_i^2\right)^{1/2} + \frac{N}{2} \tag{8.19}$$

A combination of Eqs. 8.16–8.19 gives[9]

$$J = -gD\frac{\partial N}{\partial x} \tag{8.20}$$

where g is the enhancement factor due the field:

$$g = 1 + \frac{ZN}{(N^2 + 4n_i^2)^{1/2}} \tag{8.21}$$

From Eq. 8.21, it can be seen that the maximum internal field enhancement for a singly charged ion occurs when N is considerably greater than n_i and is $2x$. If a p-dopant is being diffused into a region where a gradient due to an n-diffusion already exists, then a retardation rather than an enhancement is to be expected (10) and can cause a dip in base profiles that will be discussed later.

---

[8]Remember that some species diffuse as neutrals and thus will not be affected by the field. This could occur either because the atom is un-ionized or because it diffuses as a neutral complex.

[9]For refinements to this simple derivation, see reference 10.

8.2.5 Effect of Thermal Oxidation on the Diffusion Coefficient*

The thermal oxidation of silicon produces a higher-than-equilibrium number of silicon interstitials that will in turn increase the diffusivity of any atom that has an interstitialcy component. Conversely, if an atom diffuses solely by vacancies, extra interstitials will depress the number of vacancies present and decrease the diffusivity. Boron, phosphorus, and arsenic all show enhancements, while antimony shows a retardation. The enhancement effect is seen in both wet and dry oxidations and is more pronounced at lower oxidation/diffusion temperatures. Early work reported a retarding effect above 1150°C–1200°C, which corresponds approximately to the temperature where oxidation-induced stacking faults begin to shrink, but more recent data merely show the effect disappearing at high temperature.

The number of interstitials generated at the interface will depend on the oxidation rate, which decreases with time (thickness). Plus, some of the interstitials diffuse into the oxide and recombine with incoming oxygen, while the rest diffuse into the silicon and eventually recombine. If the additional interstitials affect only the interstitial component of D, then, from Eq. 8.10, $D = D^*(1 + f_I N_I / N_I^*)$, and the average diffusivity $\langle D \rangle$ over a diffusion/oxidation time t is given by

$$\langle D \rangle = \frac{1}{t}\int_0^t D t = D^*\left(1 + \frac{f_I}{N_I^*}\right)\int_0^t (N_I - N_I^*)dt \qquad 8.22$$

N_I is the concentration of interstitials where diffusion is occurring, and N_I^* is their equilibrium concentration. Various complex expressions for the interstitial supersaturation have been derived (16–20). It is predicted that for low temperatures and dry oxygen, the supersaturation starts at zero and increases with time. At longer times, it is experimentally deduced to be given by

$$N_I - N_I^* = K\left(\frac{dx}{dt}\right)^p \qquad 8.23$$

where p is in the 0.2–0.4 range and K is a constant. Since dx/dt decreases as the oxide layer thickness increases, the supersaturation decreases with time in this region. The supersaturation also decreases with depth into the wafer. A simple analysis gives

$$N_I - N_I^* \sim e^{x/L} \qquad 8.24$$

where L is a characteristic length, measured experimentally to be 30 μm at 1000°C and 25 μm at 1100°C (21). Fig. 8.6 shows the magnitude of the enhancement of D for boron, arsenic, and phosphorus.

---

*See references 11–20.

FIGURE 8.6

Effect of oxidation during diffusion on diffusion coefficient. (*Source:* D.A. Antoniadis et al., *Appl. Phys. Lett. 33*, p. 1030, 1978, and D.A. Antoniadis et al., *J. Electrochem. Soc. 125*, p. 813, 1978. Reprinted by permission of the publisher, The Electrochemical Society, Inc.)

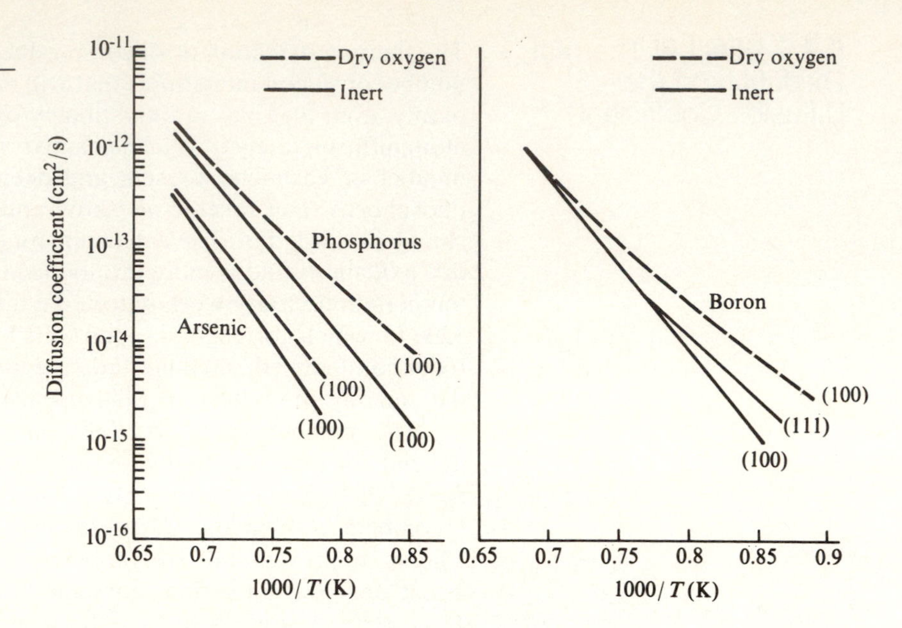

8.2.6 Effect of Diffusing Direction (Orientation)

If the crystal in which diffusion is occurring is anisotropic, the value of D will depend on direction. In the general case, Eq. 8.4 becomes

$$J_i = D_{ij}\, \partial N \partial x_j \tag{8.25}$$

from which it is seen that the diffusion coefficient is a second-rank tensor. In cubic crystals, examples of which are Si and GaAs, properties that are second-rank tensors are independent of crystallographic orientation (22) (which is the reason that Eq. 8.5 is identical to Eq. 8.7). However, recalling that, depending on the diffusion mechanism, a deviation of the density of vacancies and/or interstitials may affect D, it can be surmised that a nonisotropic flow of either of them could cause a nonisotropic effective D. It is, in fact, experimentally observed that in silicon, orientation-dependent diffusions are often encountered (23–26) during simultaneous thermal oxidation and diffusion. For diffusion in an inert atmosphere, there is no evidence of anisotropic diffusion coefficients (14, 27). This orientation-dependent phenomenon can be explained by the different rates at which nonequilibrium numbers of interstitials are generated by oxidation on different crystallographic surfaces. The effect becomes greater as the temperature is lowered, and its magnitude at any temperature is in the order of (100) > (110) > (111). Impurities showing the effect will be the same ones showing oxidation-enhanced diffusion (OED), and the magnitude of the effect can be judged by the curves of Fig. 8.6.

8.2.7 Effect of Heavy Doping

As was shown in the discussion on vacancies, the density of charged point defects can change with doping concentration when the level approaches n_i. Most silicon dopants have a diffusion component depending on charged vacancies and will thus show a doping-level dependency. In addition, the field enhancement discussed in the previous section (and which becomes noticeable only when $N > n_i$) must also be included where appropriate. In the case of boron, and perhaps the other acceptors as well, diffusion is apparently by a neutral complex—for example, $B^- V^+$. Arsenic and antimony diffuse by V^0 and V^-; phosphorus, by V^0 and $V^=$; and boron and gallium, by V^0 and V^+. Thus, the expressions for D^* are as follows (28):

$$\text{As, Sb} \qquad D = g\left(D^{*0} + D^{*-}\frac{n}{n_i}\right)$$

$$\text{P} \qquad D = g\left[D^{*0} + D^{*=}\left(\frac{n}{n_i}\right)^2\right]$$

$$\text{B, Ga, Al} \qquad D = \left(D^{*0} + D^{*+}\frac{p}{n_i}\right)$$

When $N > n_i$, the neutral vacancy contribution to the total D generally becomes small. If, at the same time, the diffusing profile is such that an appreciable electric field occurs and g (from Eq. 8.21) approaches 2, then, if D^* is the diffusivity in intrinsic material, for donors,

$$\text{As, Sb} \qquad D \to 2D^*\frac{n}{n_i}$$

$$\text{P} \qquad D \to 2D^*\left(\frac{n}{n_i}\right)^2$$

For acceptors, field-aided diffusion appears to be negligible, so

$$\text{B, Ga, Al} \qquad D \to D^*\,\frac{p}{n_i}$$

In intermediate doping ranges and at some temperatures, contributions from charge states other than those just listed may also become important. Fig. 8.7 is a curve of n_i versus temperature for silicon and can be used to tell when the doping concentrations are in the region where these effects may occur.

When the concentration of dopants in silicon approaches the 10^{21} atoms/cc range, precipitates, complexes, and atomic misfit strain begin to play a role in diffusion. For arsenic, complexes begin to form at slightly above 10^{20} atoms/cc, have a reduced diffusivity,

FIGURE 8.7

n_i versus temperature for silicon. (*Source:* From data in F.J. Morin and J.P. Maita, *Phys. Rev.* *96*, p. 28, 1955.)

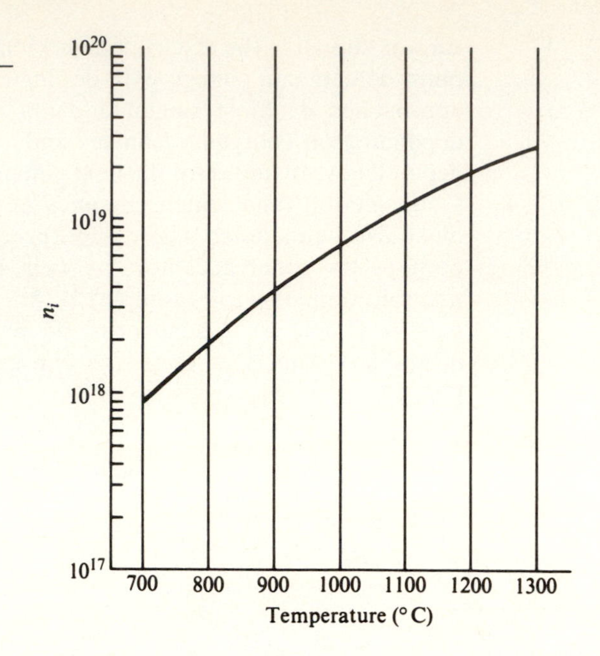

and are not fully ionized. The result is that D_{As} versus concentration has a maximum as shown in Fig. 8.8 (29). This figure also shows the increase in diffusivity with concentration up to the point where complexes form. In the case of phosphorus, the strain effect (see next section) becomes important for concentrations above about 4×10^{20} atoms/cc.

8.2.8 Effect of Temperature

From Eq. 8.6, the simple expression for D is of the form $D = D_0 e^{-E/kT}$. When the doping level is below n_i over the whole temperature range being covered and when there are no oxidation enhancement effects—that is, in the intrinsic diffusivity range—Eq. 8.6 is appropriate. E will be different for each element, and in cases where differing diffusing conditions can favor a particular charge state for the point defect involved, different E values for each state are required. Diffusion data are sometimes presented in graphic form, in which case D versus $1/T$ gives a straight line if Eq. 8.6 is appropriate. When the data give a straight line, an alternative is to present D_0 and E in tabular form, as is done in Table 8.2 for silicon. Unfortunately, most gallium arsenide diffusants are not well enough behaved to be depicted in this manner. However, the table still lists various n- and p-dopants.

8.2.9 Effect of Stress/Strain

Lattice strain will change the bandgap (30, 31) and thus can affect the distribution of vacancies and the diffusivity (28, 31). It has also been suggested that, in the case of GaAs, it affects the jump fre-

FIGURE 8.8

Effect of concentration on arsenic diffusivity. (The peak position at intermediate temperatures follows the line.)
(*Source:* Adapted from R.B. Fair, *Semiconductor Silicon/81*, p. 963, Electrochemical Society, Pennington, N.J., 1981.)

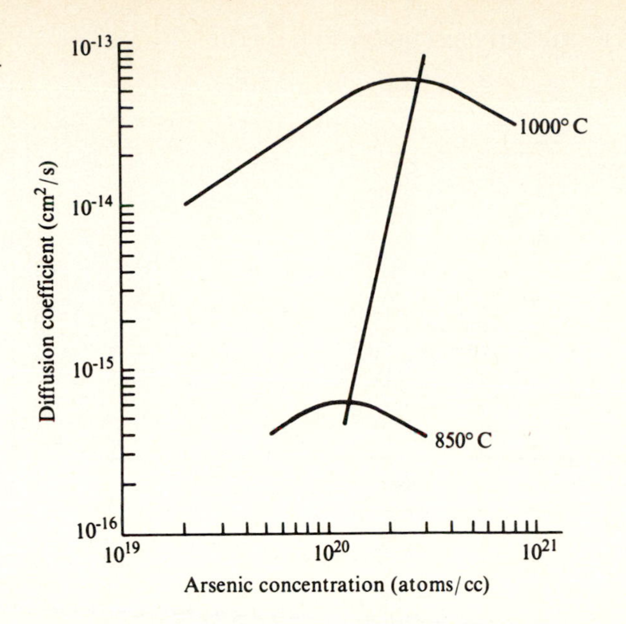

TABLE 8.2

Diffusion Coefficients

	Silicon			Gallium Arsenide	
	D_0 (cm^2/s)	E (eV)		D_0 (cm^2/s)	E (eV)
Donors			*Donors*		
Phosphorus			Sulfur	0.0185	2.6
V^0	3.85	3.66	Selenium	3000	4.16
V^-	4.44	4.0	Tellurium		
$V^=$	44.2	4.37			
Arsenic					
V^0	0.066	3.44			
V^-	12	4.05			
Antimony					
V^0	0.214	3.65			
V^-	15	4.08			
Acceptors			*Acceptors*		
Boron			Beryllium	7.3×10^{-6}	1.2
V^0	0.037	3.46			
V^+	0.41	3.46			
Aluminum			Magnesium	0.026	
V^0	1.385	3.41			
V^+	2480	4.20	Zinc		
Gallium					
V^0	0.374	3.39	Cadmium		
V^+	28.5	3.92	Mercury		
Self-Interstitial					
			Gallium	0.1	3.2
			Arsenic	0.7	5.6

Source: From references 28 and 29.

FIGURE 8.9

Diffusion interactions.

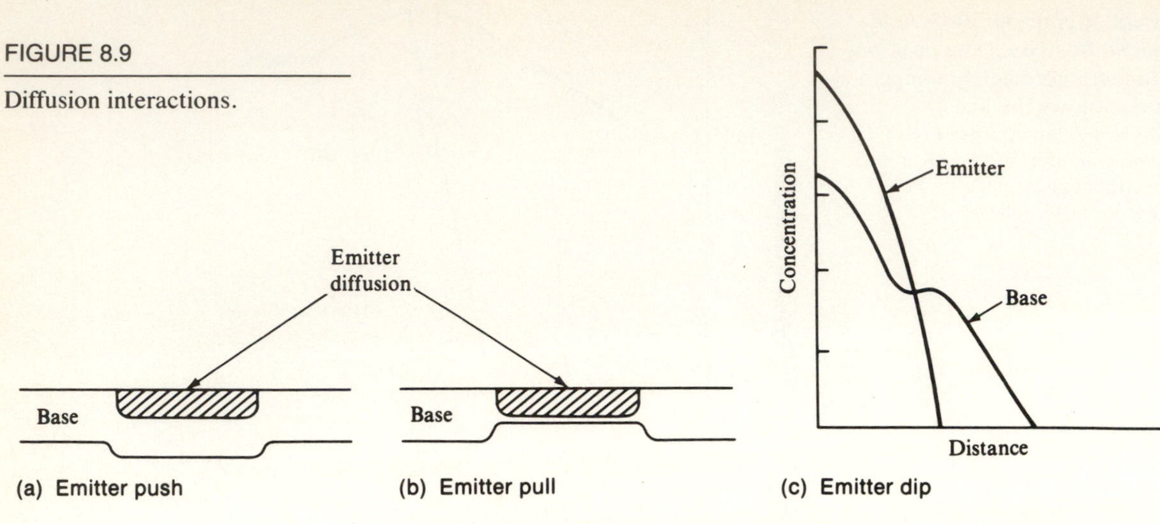

quency directly (32). Strain can arise from atomic misfit of diffused or implanted atoms (see section 8.7) or from layers such as SiO_2 or silicon nitride deposited on the surface. In particular, the stress around windows cut in oxide and nitride produces noticeable changes in D values (32, 33).

8.2.10 Emitter Push, Pull, and Dip

Simple theory predicts that each impurity will diffuse quite independently of others that may be present. There are, however, at least three phenomena—emitter push, pull, and dip—where this is not true (10, 34). These interactions are shown in Fig. 8.9. The earliest one reported (in 1959) was emitter push, which caused a gallium base to move deeper under a diffused phosphorus emitter (35). The effect was also present with boron base diffusions and indeed, in some cases, made it difficult to obtain the desired narrow bipolar base widths. Sometime later, when gallium bases and arsenic emitters were used, the reverse situation, emitter pull, was reported. Emitter push and pull can be explained in terms of the concentration of vacancies associated with the emitter diffusion (3, 31). The other interactive effect shown in Fig. 8.9 is a dip in the concentration of base impurity that is sometimes seen at or near the intersection of the emitter profile with the base profile. In this case, internal field retardation appears to cause the dip (34, 36).

8.3 SOLUTIONS TO THE DIFFUSION EQUATIONS

Impurity diffusion mathematics parallels that for carriers, except that the diffusion coefficient has more causes of variability in the region of interest. Combining Fick's first law with the continuity equation and assuming that D is independent of concentration give Fick's second law:

$$\frac{\partial N}{\partial t} = D\frac{\partial^2 N}{\partial x^2} \tag{8.26}$$

where t is the diffusion time. In three dimensions,

$$\frac{\partial N}{\partial t} = D\nabla^2 N \tag{8.27}$$

When D is a function of the concentration N, Eq. 8.26 becomes

$$\frac{\partial N}{\partial t} = \frac{\partial(D\partial N/\partial x)}{\partial x} = \frac{\partial D}{\partial x}\frac{\partial N}{\partial x} + D\frac{\partial^2 N}{\partial x^2} \tag{8.28}$$

In regions where the diffusion coefficients are concentration dependent, closed-form analytical solutions for the boundary conditions of interest are usually not available. In those cases, numerical integration and computer modeling are used to obtain solutions. Large programs are available that are designed to run on mainframe computers—for example, SUPREM, the Stanford University process engineering model for diffusion and oxidation. Nevertheless, it is still very instructive to first look at solutions of Eq. 8.26 for various boundary conditions applicable to semiconductor processing. To a first approximation, such solutions are valid. However, of more importance, the use of the simple closed-form solutions in studying a process does not screen the engineer from the physical processes as the massive computer modeling programs do.

Eq. 8.26 may be solved by separation of the variables (37)—that is, by considering that

$$N(x,t) = X(x)\Upsilon(t) \tag{8.29}$$

where X is a function only of x and Υ is a function only of the time t. The general solution to Eq. 8.29 is

$$N(x,t) = \int_0^\infty [A(\lambda) \cos \lambda x + B(\lambda) \sin \lambda x]e^{-\lambda^2 Dt} d\lambda \tag{8.30}$$

Solutions with boundary conditions appropriate for semiconductor processing are given in the following sections. Where multiple impurities are present, it is assumed that each impurity species diffuses independently of all others. Thus, separate, independent solutions to the diffusion equation can be obtained for each impurity. The total impurity concentration $N'(x,t)$ can be determined by summing the individual distributions. The net impurity concentration is given by $|\Sigma N_A - \Sigma N_D|$.

8.3.1 Diffusion from Infinite Source on Surface

This condition is one of the more common conditions used and gives the profiles that were shown in Figs. 8.2 and 8.3. The impurity concentration at the surface is set by forming a layer of a doping source on the surface. The layer either is thick enough initially or else is continually replenished so that the concentration N_0 is maintained at the solid solubility limit of the impurity in the semiconductor during the entire diffusion time. Assuming that the semiconductor is infinitely thick, the solution is as follows (37):

$$N(x,t) = N_0[1 - \left(\frac{2}{\sqrt{\pi}}\right) \int_0^z e^{-z^2} dz \tag{8.31}$$

where $z = x/2\sqrt{Dt}$. The integral is a converging infinite series referred to as the error function, or erf(z), and occurs in the solution of many diffusion problems. Toward the end of this chapter, some properties of the error function and a short table of values are given (see Table 8.12). Using the erf abbreviation, Eq. 8.31 can be written as

$$N(x,t) = N_0\left[1 - \operatorname{erf}\left(\frac{x}{2\sqrt{Dt}}\right)\right] \tag{8.32}$$

The expression $1 - \operatorname{erf}(z)$ is often referred to as the complementary error function erfc(z). If there is a background concentration N_1 of the same species at the beginning of diffusion, Eq. 8.32 becomes

$$N(x,t) = N_1 + (N_0 - N_1)\left[1 - \operatorname{erf}\left(\frac{x}{2\sqrt{Dt}}\right)\right] \tag{8.33}$$

If there is a background concentration N_1 of a different species, then they act independently, and

$$N(x,t) = N_1 + N_0\left[1 - \operatorname{erf}\left(\frac{x}{2\sqrt{Dt}}\right)\right] \tag{8.34}$$

If the species are of opposite type, a junction will occur when

$$N_0\left[1 - \operatorname{erf}\left(\frac{x}{2\sqrt{Dt}}\right)\right] = N_1 \tag{8.35}$$

The value of x where this occurs is given by

$$x_j = 2\sqrt{Dt}\, \operatorname{erf}^{-1}\left(1 - \frac{N_1}{N_0}\right) \tag{8.36}$$

which can be rewritten as

$$x_j = A\sqrt{t} \tag{8.37}$$

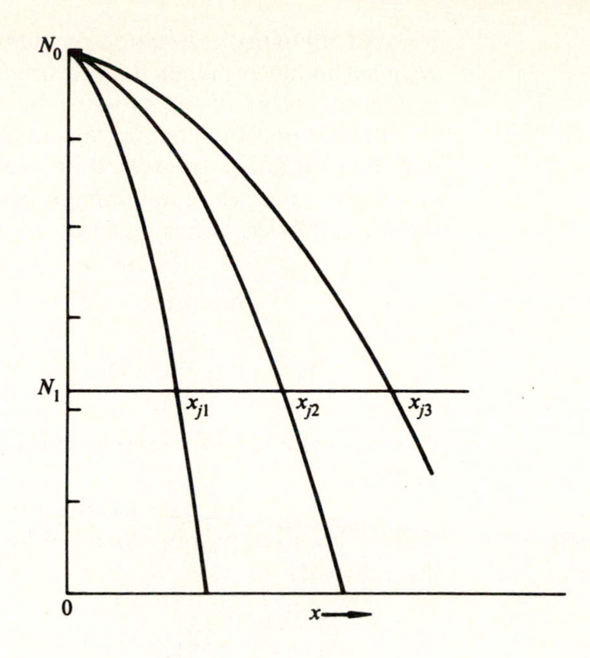

FIGURE 8.10

Position of junction after diffusion times in ratio of 1:4:9. (x_j is linear with $\sqrt{t}$ regardless of N_0/N_1 ratio.)

where A is a constant given by $A = 2[\text{erf}^{-1}(1 - N_1/N_0)]\sqrt{D}$. Thus, the junction depth x_j increases as the square root of the diffusion time, as illustrated in Fig. 8.10 where the diffusion times are in the ratio of 1:4:9. A linear plot of x_j versus $\sqrt{t}$ provides assurance that D is remaining constant over the range of concentrations covered. Then, a best-fit curve of several values of x_j and t can be used to determine a particular x_j and t that can be substituted into Eq. 8.36 to determine D.

EXAMPLE ☐ If N_0 is 5×10^{19} atoms/cc and a diffusion is made into an opposite-type background of 10^{16} atoms/cc, how long will it take to diffuse to the point where the junction is 1 μm deep if the temperature is such that $D = 10^{-14}$ cm²/s?

Substituting into Eq. 8.36 (cm and s are a consistent set of units) gives

$$10^{-4}\text{ cm} = (2 \times 10^{-7}\text{ cm}/\sqrt{\text{s}}) \times \sqrt{t}\,\text{erf}^{-1}(1 - 10^{16}/5 \times 10^{19})5 \times 10^{2}\sqrt{\text{s}} = \sqrt{t}\,\text{erf}^{-1}(0.998)$$

From Table 8.12, erf(0.998) ≅ 2.18. Thus, t = 52,900 s, or 14.7 hours. ☐

Because N_0 is set by the solubility limit of the dopant being used, at a given temperature, only one N_0 is available per dopant per semiconductor. The use of a diffusion with the concentration set by solid solubility gives good control, but the surface concentration N_0 may be much higher than desired. To provide greater flexibility in the choice of N_0, a two-step diffusion process comprised of a short diffusion from an infinite source followed by its removal and a continued diffusion from the thin layer of impurities already diffused in is often used. The total number of impurities S available is set by the first diffusion and is given by

$$\begin{aligned} S &= \int_0^\infty N(x)dx \\ &= \int_0^\infty N_0\left[1 - \operatorname{erf}\left(\frac{x}{2\sqrt{Dt}}\right)\right]dx = \frac{2N_0}{\sqrt{\pi}}\sqrt{Dt} \end{aligned} \qquad 8.38$$

The behavior of the second diffusion is described in the next section.

8.3.2 Diffusion from Limited Source on Surface

At $t = 0$, a fixed number S/cm^2 of impurities is on the surface. N is given by (37)

$$N(x,t) = \frac{S}{\sqrt{\pi Dt}}\, e^{-x^2/4Dt} \qquad 8.39$$

In this case, the distribution is Gaussian[10] and not erf. The surface concentration $N(0,t)$ continually diminishes with time as shown in Fig. 8.11 and is given by

$$N(0,t) = \frac{S}{\sqrt{\pi Dt}} \qquad 8.40$$

Thus, in the limited-source case, the surface concentration decreases linearly with $\sqrt{t}$. The junction depth, unlike that of the previous case, varies in a more complex manner:

$$e^{-x_j^2/4Dt} = \frac{N_B}{S}\sqrt{\pi Dt} \qquad 8.41a$$

or

$$x_j = \left[4Dt \log\left(\frac{S}{N_B\sqrt{\pi Dt}}\right)\right]^{1/2} \qquad 8.41b$$

Eq. 8.39 was derived based on S being located at $x = 0$, but if S actually is in a layer from a previous short diffusion of the type described by Eq. 8.32, and if the Dt product of the second diffusion

[10] Functions of the form $y = e^{-x^2/a^2}$ are referred to as Gaussian.

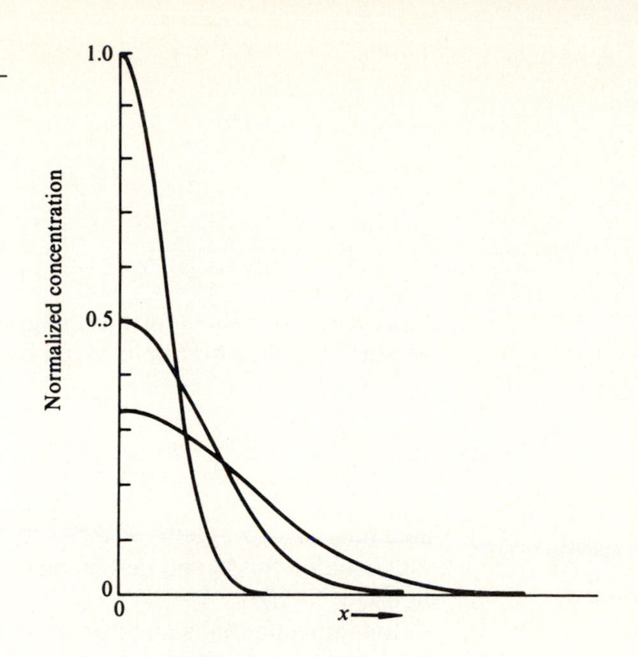

FIGURE 8.11

Diffusion from limited source plotted on linear scale for diffusion times in ratio of 1:4:9. (Note that S decreases as square root of time.)

is as much as four or five times that of the first one, the impurities from the first can still be considered as all lying at $x = 0$. S for the second diffusion will be the amount of impurities introduced during the first diffusion. Substituting the value for S from Eq. 8.38 into Eq. 8.39 gives

$$N(x',t,t') = \left(\frac{2N_0}{\pi}\right)\left(\frac{Dt}{D't'}\right)^{1/2} e^{-x'^2/4D't'} \qquad 8.42$$

and the surface concentration is

$$N_0(t,t') = \frac{2N_0}{\pi}\sqrt{\frac{Dt}{D't'}} \qquad 8.43$$

where the prime values indicate values for the second diffusion. When the second Dt product is not large compared to the first, as is likely during short, low-temperature diffusions, the exact solution instead of the approximations of Eqs. 8.42 and 8.43 can be used (38, 39):

$$N(x,t,t') = N_0\frac{2}{\sqrt{\pi}}\int_{\sqrt{\beta}}^{\infty} e^{-m^2}\operatorname{erf}(\alpha m)dm \qquad 8.44$$

where

$$\alpha = \sqrt{\frac{Dt}{D't'}}$$

$$\beta = \frac{x^2}{4(Dt + D't')}$$

(Representative values for the integral are tabulated in Table 8.13.) The surface concentration is given by

$$N_{sur} = N_0 \frac{2}{\pi} \tan^{-1} \sqrt{\frac{Dt}{D't'}} \qquad 8.45$$

Since $\tan^{-1} x = x - x^3/3 + x^5/5 - \ldots$, the difference between the exact solution for N_0' and that given by Eq. 8.43 is only a few percent when $D't' > 5Dt$.

Ion implantation is an alternative to the first diffusion. The ions implanted can be counted very accurately, and they are quite close to the surface (see Chapter 9). They will not have the atoms all located on one plane, however, and the boundary conditions are much more like those discussed in the next section.

8.3.3 Diffusion from Interior Limited Source

This condition is like the case just described, except that the sheet of impurity atoms is located at $x = X_0$ where X_0 is in the interior of the semiconductor so that the source is depleted by diffusion in both directions. In this case (37), assuming that X_0 is far removed from a free surface,

$$N(x,t) = \frac{S}{2\sqrt{\pi Dt}} e^{-(x - X_0)^2/4Dt} \qquad 8.46$$

Fig. 8.12 shows two profiles with their respective diffusion times differing by a factor of 4. This exponential profile has the same shape as the Gaussian or normal distribution often encountered in probability theory. For that application, it is written as

$$\frac{N_i}{\sigma\sqrt{2\pi}} e^{-(x - X_0)^2/2\sigma^2} \qquad 8.47$$

where N_i is the number of observations and σ is the standard deviation.[11] A comparison of Eqs. 8.46 and 8.47 shows that N_i

[11] 68.3% of the area under the curve is contained in the portion of the curve between $x - X_0 = 1\sigma$ and $x - X_0 = -1\sigma$.

FIGURE 8.12

Profile for diffusion from source S of width $= 0$ located at $x = X_0$.

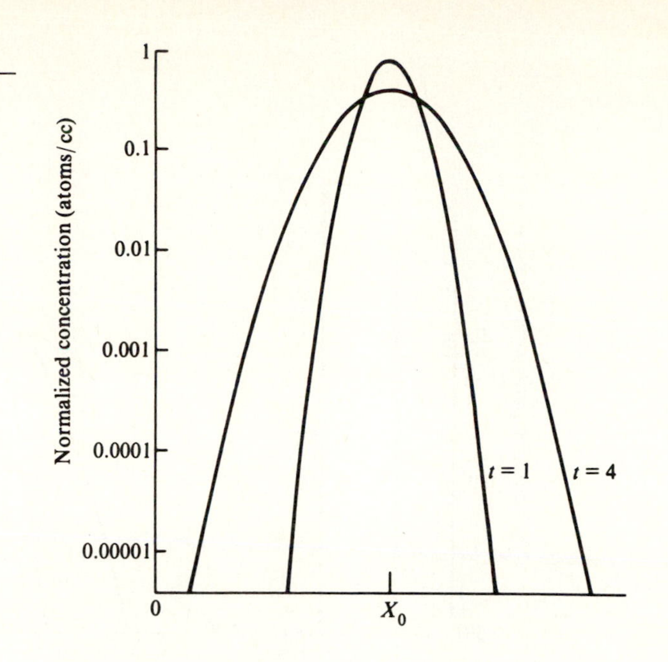

corresponds to S and $\sigma = \sqrt{2Dt}$. This correlation is useful when considering subsequent diffusions after an ion implant since the initial ion implant profile is often described by a total flux ϕ (S), a range R_p (X_0), and a standard deviation ΔR_p (σ or $\sqrt{2Dt}$).

EXAMPLE ☐ Suppose that an ion implant is characterized by a total dose ϕ of 10^{13} atoms/cm^2, a range of 0.2 μm, and a standard deviation of 0.1 μm. What will the profile look like after a 20 minute heat cycle at a temperature where $D = 10^{-14}$ cm^2/s? Neglect the fact that because of the shallowness of the implant, there might be a surface boundary effect.

ϕ is the total number of impurities and hence corresponds to S. The scatter during implant is equivalent to a first diffusion from an initial sheet source. $\Delta R_p = \sigma = 0.1$ μm $= \sqrt{2Dt} = 10^{-5}$ cm. Solving for Dt gives 5×10^{-11} cm^2. The subsequent diffusion has a Dt product of 10^{-14} cm^2/s $\times$ 1800 s $= 1.8 \times 10^{-11}$cm^2, which is much less than the equivalent Dt of the implant. Hence, it would be expected to change the profile very little. However, as will be discussed in section 8.3.13, in this case the final distribution can be calculated by substituting a $(Dt)_{\text{eff}} = \Sigma D_1t_1 + D_2t_2 = 5 \times 10^{-11} + 1.8 \times 10^{-11} = 6.8 \times 10^{-11}$ cm^2 for the Dt of Eq. 8.46. ☐

8.3.4 Diffusion from Layer of Finite Thickness

This condition is a case similar to the one just described, except that initially a rectangular rather than a sheet source distribution is located internally to the body of the semiconductor. As is shown in Fig. 8.13a, if the initial thickness 2ℓ is much greater than $\sqrt{Dt}$, it is

FIGURE 8.13

Profile for diffusion from interior layer of width $= 2\ell$ located at $x = X_0$.

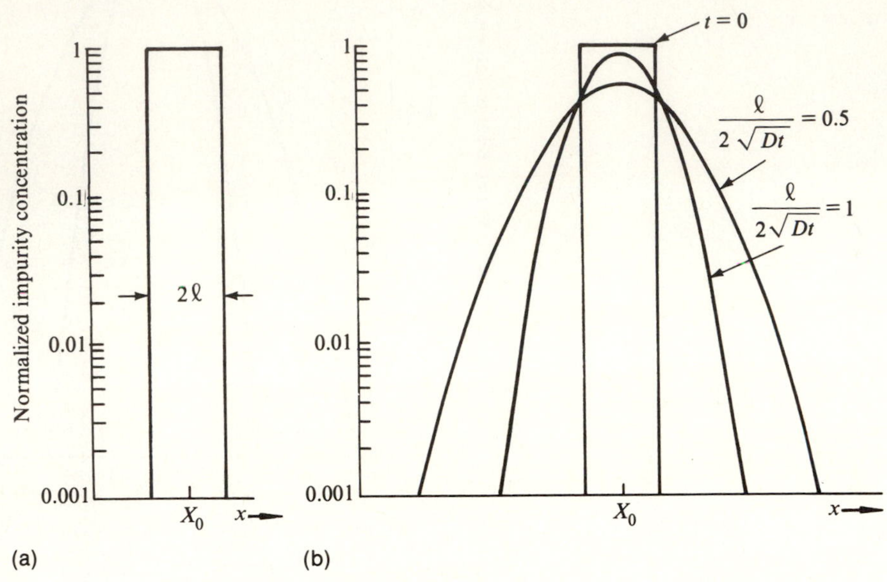

best treated as the two separate step distributions to be described in the next section. Of more interest is the case where $\sqrt{Dt}$ is greater than ℓ, where the peak distribution N_0 decreases with increasing diffusion time.

Application for this set of boundary conditions can occur if the dopant redistribution of multiple layers of molecular beam epitaxy after heat treatments is being considered. It is less applicable to conventional epitaxial processing since autodoping will usually overshadow the diffusion.

To simplify the algebra, consider that a new origin is chosen with its 0 at the X_0 of Fig. 8.13a. If the width of the doped layer is 2ℓ, $N(x,t)$ is given by (37)

$$N(x,t) = \frac{N_0}{2}\left[\operatorname{erf}\left(\frac{\ell + x}{2\sqrt{Dt}}\right) + \operatorname{erf}\left(\frac{\ell - x}{2\sqrt{Dt}}\right)\right] \tag{8.48}$$

where 2ℓ is the width of the layer. As diffusion progresses, impurities will diffuse out in both directions as shown in Fig. 8.13b. The peak concentration N_0 will decrease as $\operatorname{erf}(\ell/2\sqrt{Dt})$ and thus will change imperceptibly until $\ell/2\sqrt{Dt}$ becomes less than about 2.

8.3.5 Diffusion from Concentration Step

The initial boundary conditions for this case, illustrated in Fig. 8.14 as the abrupt steps of $t = 0$, approximate the conditions just after a conventional epitaxial deposition and very closely match those after molecular beam epitaxy (MBE). If there are a series of thin MBE layers, the previous case may be more appropriate. This case also provides an approximation of a grown junction crystal distribution. However, grown junction technology has now been obsolete for 20 years. Diffusion from each side of the step can be considered to proceed independently as described by Eqs. 8.49 and 8.50 (37):

$$N(x,t) = \frac{N_1}{2}\left[1 - \text{erf}\left(\frac{x}{2\sqrt{D_1 t}}\right)\right] \tag{8.49}$$

$$N(x,t) = \frac{N_2}{2}\left[1 + \text{erf}\left(\frac{x}{2\sqrt{D_2 t}}\right)\right] \tag{8.50}$$

N_1 is the doping level for $-x$ and $t = 0$; N_2, for $+x$ and $t = 0$. D_1 is the diffusion coefficient for the N_1 species; D_2, for the N_2 species. Eqs. 8.49 and 8.50 may be added together to give a single expression covering diffusion from both sides of the step:

$$N(x,t) = \frac{N_1}{2}\left[1 - \text{erf}\left(\frac{x}{2\sqrt{Dt}}\right)\right] + \frac{N_2}{2}\left[1 + \text{erf}\left(\frac{x}{2\sqrt{Dt}}\right)\right] \tag{8.51}$$

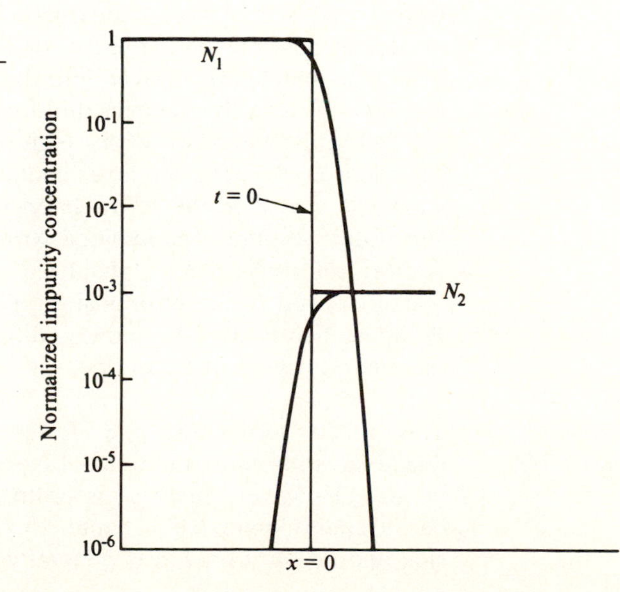

FIGURE 8.14

Profile for diffusion occurring at a concentration step. (In this example, N_1 and N_2 have the same Dt product.)

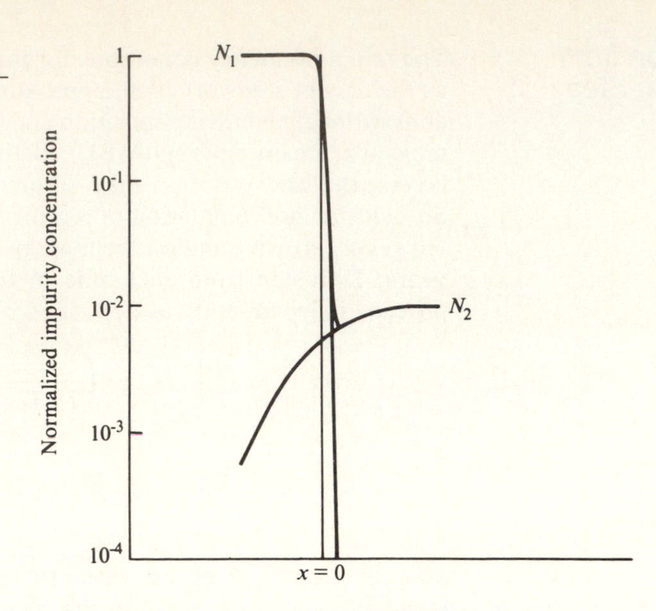

FIGURE 8.15

Dip in net concentration occurring near original high/low boundary because of a more rapid diffusant in the high-resistivity layer.

Fig. 8.14, along with the initial conditions, also shows a profile for each separate diffusion (Eqs. 8.49 and 8.50) and their sum (Eq. 8.51) for the case of the same impurity on each side of the step ($D_1t = D_2t$). When the two Dt's are not equal, substantially different distributions can sometimes occur. For example, if $N_1 >> N_2$ and $D_2 > D_1$, it is possible to get a dip in the concentration curve as shown in Fig. 8.15 (40). This dip might occur, for example, if an epi substrate were doped with antimony and overgrown with an arsenic layer. The maximum possible dip is a factor of 2 and occurs before the N_1 diffusion has perceptibly moved into the N_2 region. If a heavily doped n-substrate partially compensated by having some spurious p-dopant present is overgrown with a high-resistivity n-layer, then if $N_2 < N_{1p}$ and if $D_{1n} << D_{2p}$, a p-layer could occur in the epitaxial layer as shown in Fig. 8.16 (41). This happens if low-resistivity antimony substrates become contaminated with boron. The resulting p-layer is often referred to as a "phantom" p-layer, although it is indeed real. If N_1 and N_2 are of opposite type, the junction will move into the more lightly doped region as diffusion proceeds, regardless of the relative values of D_1 and D_2.

8.3.6 Diffusion from Concentration Step into Moving Layer

This case, as shown by Fig. 8.17, describes diffusion from a heavily doped substrate into an epitaxial layer being grown at a velocity v in cm/s. However, this case is ordinarily of little importance both because autodoping (see Chapter 7) will usually overshadow diffusion and because growth is generally much more rapid than diffu-

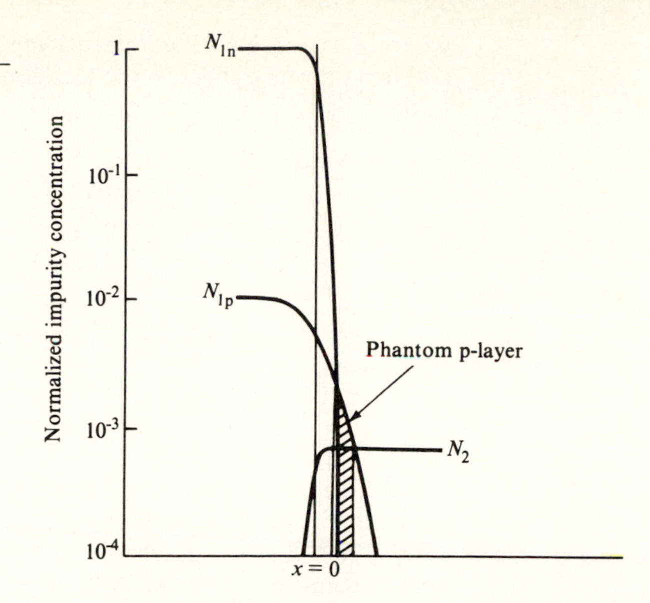

FIGURE 8.16

Origin of phantom p-layer that occurs in n-epi layers because of n-substrate contamination by a faster diffusing p-dopant.

sion. In the event that the epitaxial growth temperature is reduced to the point where autodoping is negligible, diffusion will be reduced to the point that the solution for diffusion into an infinite thickness (Eq. 8.50) can be used. Even at higher temperatures, the use of Eq. 8.50 causes little error in most cases.

To account for the moving boundary, instead of Eq. 8.26, the equation to be solved is as follows (42, 43):

$$D\frac{\partial^2 N}{\partial x^2} = \frac{\partial N}{\partial t} + v\frac{\partial N}{\partial x} \qquad 8.52$$

where v is the epitaxial growth velocity. In the solutions to follow, v is assumed to remain constant. One solution represents diffusion from an infinitely thick substrate into the growing layer and is generally the one of interest. In the event that the "substrate" is a thin epitaxial layer just grown or a thin high-concentration layer diffused into a lightly doped wafer, then it is not infinite in extent, and its concentration will decrease with growth (diffusion) time. The other represents diffusion from the growing layer into the substrate. For the case of out-diffusion from an infinite substrate, the boundary conditions are as follows:

FIGURE 8.17

Geometry for diffusion during epitaxial growth.

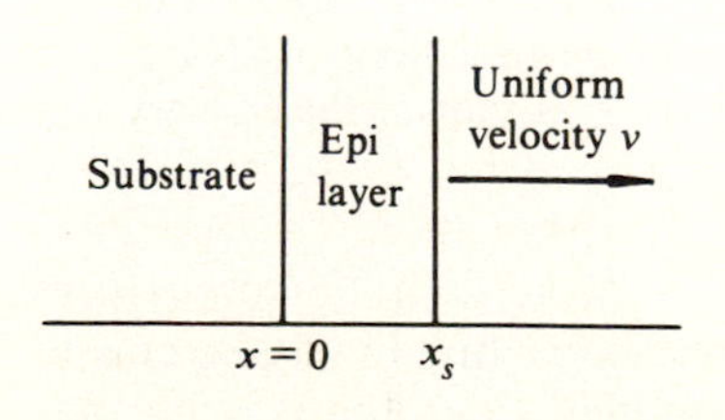

$$N(-x,0) = N_1$$
$$N(-\infty,t) = N_1$$
$$J = -D\frac{\partial N}{\partial x} = (K + v)N \quad \text{at } x_s$$

where K is a rate constant describing the loss of dopant at the epi–ambient interface and x_s is the location of the interface. The solution is given by

$$\begin{aligned}\frac{N(x,t)}{N_1} &= \frac{1}{2}\,\mathrm{erfc}\left(\frac{x}{2\sqrt{Dt}}\right) \\ &\quad - \left(\frac{K+v}{2K}\right)e^{(v/D)(vt-x)} \times \mathrm{erfc}\left(\frac{2vt-x}{2\sqrt{Dt}}\right) \\ &\quad + \left(\frac{2K+v}{2K}\right)e^{[(K+v)/D][(K+v)t-x]} \\ &\quad \times \mathrm{erfc}\left(\frac{2(K+v)t-x}{2\sqrt{Dt}}\right)\end{aligned} \qquad 8.53$$

When the rate constant K approaches infinity (no impediment at the surface), Eq. 8.53 will predict a smaller N for a given x than will Eq. 8.50. When K goes to zero (no loss at the surface), Eq. 8.53 will predict a higher value for N at a given x than will Eq. 8.50. For v^2t/D greater than about 10, Eqs. 8.53 and 8.50 will very closely match, regardless of the K value. The reason for the insensitivity to K in this case is that the layer appears infinitely thick and no impurity gets to the boundary. Little data exist on K values, but data in reference 43 suggest values in the range of 2×10^{-7}cm/s. In the event that the substrate is not infinitely thick—for example, for one epitaxial layer deposited on top of another or on top of a diffused layer—then the solution is more related to diffusion from the finite thickness sources discussed earlier. (See reference 42 for further details.)

EXAMPLE □ For the case of a 1200°C silicon epitaxial deposition, calculate the v^2t/D values for a layer 0.05 μm thick if the deposition rate is 0.5 μm/minute and the substrate is doped with antimony.

The D value for antimony, found from Fig. 8.26, which appears in a later section, is about 10^{-13} cm²/s. The time t to grow 0.05 μm at 0.5 μm/minute is 0.1 minute, or 6 s. The rate of 0.5 μm per minute equals 8.3×10^{-7} cm/s. Thus, $v^2t/D \cong 42$. As the film gets thicker, the expression gets larger. It also gets larger as v increases and as D decreases. D, as will be seen later, decreases rapidly as the temperature decreases. Typical silicon epitaxial deposition conditions are 1100°C at 1 μm/minute where v^2t/D will be much larger than 42. Thus, in most epitaxial depositions, $v^2t/D >> 10$, and Eq. 8.50 is applicable for thicknesses greater than a very small fraction of a micron. □

The other half of the diffusion, that from the moving layer into the initial material, is considerably simpler since there is no out-diffusion problem. When the initial material is infinitely thick, the boundary conditions are as follows:

$$N(-x,0) = 0$$
$$N(x_s,t) = N_2$$
$$N(-\infty,t) = 0$$

and when applied to Eq. 8.52 give

$$N(x,t) = \frac{N_2}{2}\left[1 + \text{erf}\left(\frac{x}{2\sqrt{Dt}}\right) + e^{(v/D)(vt-x)}\text{erfc}\left(\frac{2vt - x}{2\sqrt{Dt}}\right)\right] \quad 8.54$$

When v^2t/D is large, this equation reduces to the nonmoving boundary solution of Eq. 8.50.

8.3.7 Out-Diffusion with Rate Limiting at Surface

This set of conditions would describe the motion of impurities in an initially uniformly doped semiconductor heated to a high temperature. The impurity flux J_s across the semiconductor–ambient interface at $x = 0$ is given by

$$J_s = K(N_e - N_s)$$

where K is the same rate constant discussed in the preceding case, N_s is the concentration of impurity in the semiconductor at the surface, and N_e is the equilibrium concentration of impurity in the ambient adjacent to the semiconductor. Note that if $N_e > N_s$, there will be in-diffusion rather than a loss of impurities by out-diffusion. If, as is common, N_e is assumed to be much less than N_s, then $J_s = -KN_s$, and, by setting the growth velocity v equal to zero, Eq. 8.53 can be used to calculate the impurity profile following out-diffusion (43). Thus,

$$\frac{N(x,t)}{N_1} = \text{erf}\left(\frac{-x}{2\sqrt{Dt}}\right) + e^{(K/D)(Kt-x)}\text{erfc}\left(\frac{2Kt - x}{2\sqrt{Dt}}\right) \quad 8.55$$

For the limit of $K = 0$, nothing will escape; for the other limit of K very large so that there is no surface barrier, Eq. 8.55 reduces to $N(x,t) = N_1\text{erf}(-x/2\sqrt{Dt})$. The solution for the case of $N_e \neq 0$ is available (44) but is more complex and not included. Graphical aids to its use are found in reference 45.

8.3.8 Diffusion from a Fixed Concentration into Moving Layer

These conditions could arise from a simultaneous diffusion and surface etching or evaporation. During some silicon diffusions, the initial diffusion source is a mixed oxide layer formed on the surface.

When the diffusion is done in an oxidizing atmosphere, as is normal, oxide will grow at the glass–silicon interface. However, the mixed oxide layer dissolves it as it forms so that the source of fixed concentration N_0 stays in contact with the moving wafer surface.

For the conditions of the surface moving in the $+x$ direction with a velocity v,

$$N(vt,t) = N_0 \qquad \text{for all } t\text{'s}$$
$$N(x,0) = 0 \qquad \text{for } x > 0$$

By changing to a new variable $x' = x - vt$ (x' is measured from the interface and not from the original position), Eq. 8.26 becomes

$$D\frac{\partial^2 N}{\partial x'^2} + v\frac{\partial N}{\partial x} = \frac{\partial N}{\partial t} \qquad 8.56$$

The solution is as follows (46):

$$N(x - vt,t) \equiv N(x',t) = \frac{N_0}{2}\left[\left(\text{erfc}\frac{x' + vt}{2\sqrt{Dt}}\right) + e^{-vx'/D}\text{erfc}\frac{x' - vt}{2\sqrt{Dt}}\right] \qquad 8.57$$

where x is measured from the original surface and x' from the actual boundary at a time t.

8.3.9 Diffusion through Thin Layer

One way in which these conditions could occur would be if diffusion were through a layer of polysilicon on top of single-crystal silicon or through a layer of a mixed III–V compound on top of gallium arsenide. Another configuration is an oxide or other masking layer over silicon or gallium arsenide. Since the solutions to be given here do not consider a moving boundary, they are an approximation to diffusion through a thermal oxide growing on silicon. Fig. 8.18 shows the two circumstances considered. In Fig. 8.18a, there is no segregation coefficient between the two media, and thus the concentration is continuous across the boundary. This situation is applicable to the polysilicon/Si and mixed III–V/GaAs cases. For an oxide layer, the concentration is usually not continuous, as is shown in Fig. 8.18b. The equations for case (a) are as follows (47):

$$D_1\frac{\partial^2 N_1}{\partial x^2} = \frac{\partial N_1}{\partial t} \qquad \text{for} -a < x < 0 \qquad 8.58a$$

$$D_2\frac{\partial^2 N_2}{\partial x^2} = \frac{\partial N_2}{\partial t} \qquad \text{for } x > 0 \qquad 8.58b$$

The conditions are

$$J_1 = J_2 \qquad \text{at } x = 0$$

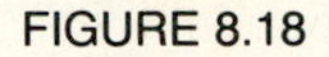

FIGURE 8.18

Diffusion through thin layer into infinitely thick layer of different material.

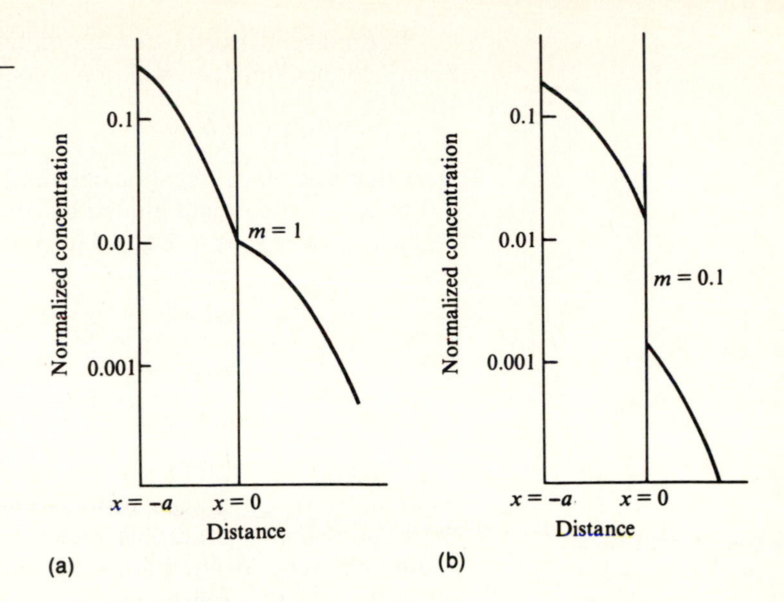

$$N_1(-a,t) = N_0 \qquad \text{where } a = \text{thickness of thin layer}$$

$$N_1(0,t) = N_2(0,t)$$

$$N_2(x,t) \to 0 \text{ as } x \to \infty$$

The solutions are

$$N_1(x,t) = N_0 \sum_{j=0}^{\infty} \left(\frac{1-\mu}{1+\mu}\right)^j \times \left[\operatorname{erfc}\frac{a(2j+1)+x}{2\sqrt{D_1 t}} - \frac{1-\mu}{1+\mu}\operatorname{erfc}\frac{a(2j+1)-x}{2\sqrt{D_1 t}}\right] \qquad 8.59$$

$$N_2(x,t) = \frac{2\mu N_0}{1+\mu} \sum_{j=0}^{\infty} \left(\frac{1-\mu}{1+\mu}\right)^j \times \operatorname{erfc}\left[\frac{a(2j+1)}{2\sqrt{D_1 t}} + \frac{x}{2\sqrt{D_2 t}}\right] \qquad 8.60$$

where $\mu = D_1/D_2$. If $a/2\sqrt{Dt} > 1$, the first term in each series is a good approximation. That is,

$$N_1(x,t) \approx N_0\left(\operatorname{erfc}\frac{a+x}{2\sqrt{D_1 t}} - \frac{1-\mu}{1+\mu}\operatorname{erfc}\frac{a-x}{2\sqrt{D_1 t}}\right) \qquad 8.61a$$

$$N_2(x,t) \approx \frac{2\mu N_0}{1+\mu}\operatorname{erfc}\left(\frac{a}{2\sqrt{D_1 t}} + \frac{x}{2\sqrt{D_2 t}}\right) \qquad 8.61b$$

In the event that the concentration is not continuous from medium 1 to medium 2, an additional boundary condition is needed:

$$N_2(0,t) = mN_1(0,t)$$

Note that m is the segregation coefficient and is discussed in Chapter 3. For $m < 1$ and N_2 not limited by solid solubility, the solutions for N_1 and N_2 analogous to Eq. 8.61 become (48)

$$N_1(x,t) \approx N_0\left(\operatorname{erfc}\frac{a + x}{2\sqrt{D_1 t}} - \frac{m - \mu}{m + \mu}\operatorname{erfc}\frac{a - x}{2\sqrt{D_1 t}}\right) \qquad 8.62a$$

$$N_2(x,t) \approx \frac{2m\mu}{m + \mu}N_0 \operatorname{erfc}\left(\frac{a}{2\sqrt{D_1 t}} + \frac{x}{2\sqrt{D_2 t}}\right) \qquad 8.62b$$

The problem of diffusion through a growing rather than a static oxide (a moving boundary problem) has also been solved (49), but the requirement for a profile under such conditions is seldom encountered. What is often needed is the amount of dopant diffusing through a simultaneously growing thermal oxide mask in order to see whether it will cause a surface problem. The likelihood of a problem can be determined by calculating $N_2(0)$ for any desired time from Eq. 8.62a and comparing it with the background doping. However, the source material reacting with the masking oxide may reduce its thickness by a substantial amount and thus increase the amount of dopant that diffuses through it, as is particularly true in the case of phosphorus diffusions.

8.3.10 Diffusion during Thermal Oxidation

This solution concerns the redistribution during oxidation of impurities already in a silicon wafer. Fig. 8.19 shows the geometry. During oxidation, the silicon wafer surface moves inward. Simultaneously,

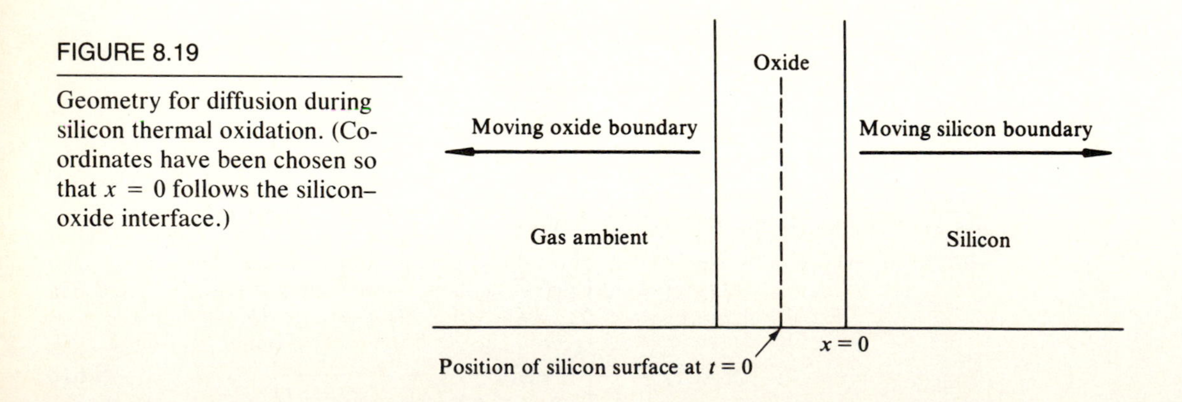

FIGURE 8.19

Geometry for diffusion during silicon thermal oxidation. (Coordinates have been chosen so that $x = 0$ follows the silicon–oxide interface.)

the oxide increases in thickness. The impurities originally in the silicon consumed by the oxide may be rejected and diffuse ahead of the silicon or be incorporated into the growing oxide, depending on whether the oxide–silicon segregation coefficient m is >1 or <1. (As defined earlier, m is the impurity concentration in silicon at the interface divided by the concentration in the oxide at the interface.) Solutions to the various boundary conditions that can arise are very complex (50), and most have been solved by numerical integration (51, 52).

Thus, if the silicon is initially uniformly doped, the oxidation will either pile up some of those impurities in the silicon next to the interface or else deplete the silicon and cause a concentration dip at its surface. Fig. 8.20 shows the kind of profile to be expected for the case of $m > 1$, where there is a depletion of impurity from the silicon surface. When there is a nonuniform distribution, as, for example, if an earlier diffusion or implant step had taken place, the same pileup or depletion occurs. A profile comparable to that of Fig. 8.20 is shown in Fig. 8.21 but represents a boron dopant after a predep and subsequent drive-in.[12] In the case of a very shallow predep or when an implant is used, care must be taken to ensure that the oxide layer is not grown thick enough to consume all of the impurities.

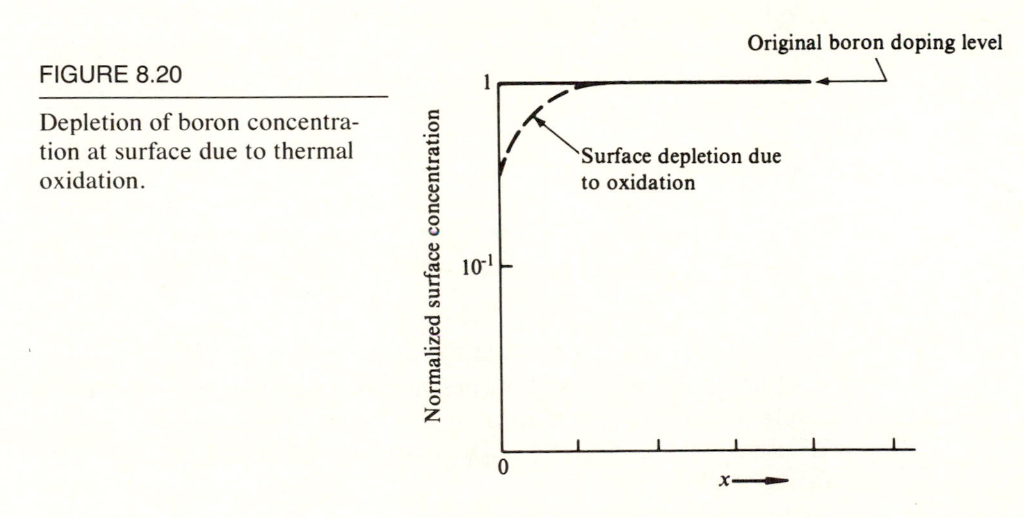

FIGURE 8.20

Depletion of boron concentration at surface due to thermal oxidation.

[12]A drive-in is used to intentionally lower the surface concentration after an initial predeposition diffusion. Such a lowering can occur in two ways. Since the drive-in diffusion source is limited to only those impurities introduced during the predep, continued diffusion causes the surface concentration to decrease as was shown in Fig. 8.11. In addition, in the case of boron, the growing oxide not only retains all of the boron in the silicon consumed but also acts as a sink so that boron from silicon adjacent to the oxide diffuses into it.

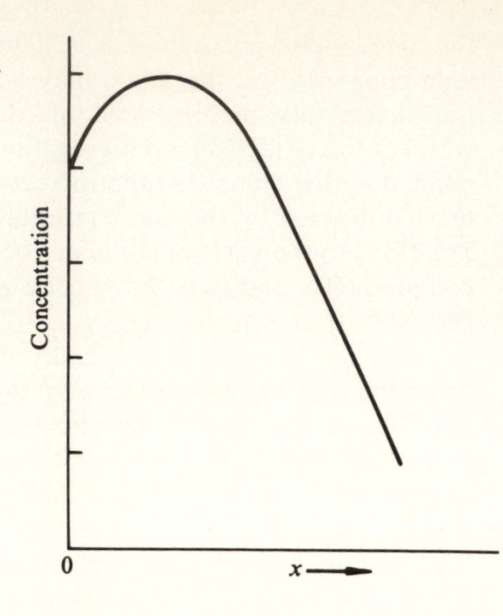

FIGURE 8.21

Boron diffusion profile after combined oxidation and drive-in.

8.3.11 Diffusion from Infinite Source into Finite Thickness

Most diffusions are intended to extend only a short distance into the wafer and thus the case described in section 8.3.1 is appropriate. However, in the case of fast-diffusing impurities such as the heavy metals or for long diffusion times and thin wafers, it is quite possible for a diffusion originating from one side to travel completely through the wafer. If the back of the wafer is considered to be impermeable (37),

$$N(x,t) = N_0\left[1 - \frac{4}{\pi}\left(e^{-y}\sin\frac{\pi x}{2a} + \frac{1}{3}e^{-9y}\sin\frac{3\pi x}{2a} + \ldots\right)\right] \quad 8.63$$

where the thickness of the wafer is a and $y = \pi^2 Dt/4a^2$.

Should the back of the wafer not be impermeable, as, for example, if it were coated with an oxide that acted as a sink for the diffusant, then a diffusion current would exist across the back surface given by

$$J = K[N(x = a,t) - N(+)] \quad 8.64$$

where K is a rate constant and $N(+)$ is the concentration just outside the wafer boundary. For solutions of this case, see reference 53.

Occasionally, diffusion into a wafer from both sides is of interest. In this case, Eq. 8.63 is slightly modified:

$$N(x,t) = N_0\left[1 - \frac{4}{\pi}\left(e^{-y'}\sin\frac{\pi x}{a} + \frac{1}{3}e^{-9y'}\sin\frac{3\pi x}{a} + \ldots\right)\right] \quad 8.65$$

where y' is given by $y' = \pi^2 Dt/a^2$. These distributions are shown in Fig. 8.22.

8.3.12 Two-Dimensional Diffusion from High-Diffusivity Path

The idealized geometry for this case is shown in Fig. 8.23a. Solutions to this case were derived to cover diffusion along and out from metal grain boundaries and dislocations (54–56). When semiconductor processing was less developed, there was concern about enhanced diffusion along grain boundaries, twin boundaries, stacking faults, and dislocations, and the effect was studied in germanium and silicon (57–59). Grain boundaries exhibit a pronounced effect,

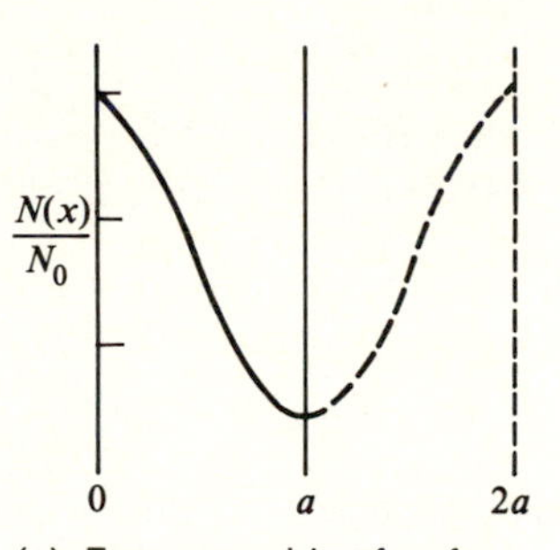

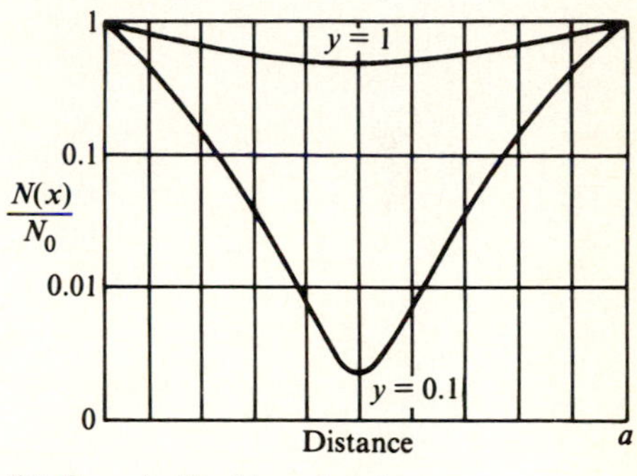

(a) From one side of wafer (b) From both sides of wafer

FIGURE 8.22

Diffusion profile when $x/2\sqrt{Dt}$ is comparable to wafer thickness.

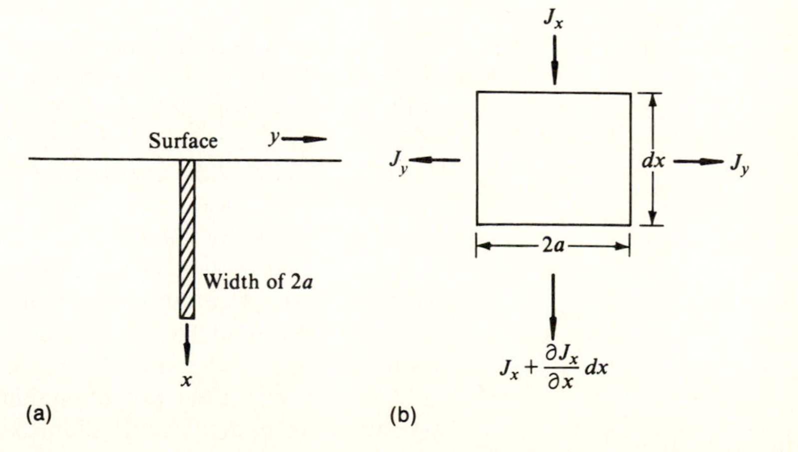

(a) (b)

FIGURE 8.23

(a) Geometry for diffusion from a high-diffusivity layer. (b) Fluxes into and out of an element of the high-diffusivity layer that are needed to develop continuity equation.

but now grain boundaries are not tolerated in the starting wafers used in IC processing, and any incoming starting slices having grain boundaries are rejected. No enhancement has been observed in either first- or second-order twins (also not tolerated in IC processing) or in stacking faults. Enhanced diffusion can apparently occur at a single dislocation, but not a high enough density of dislocations exists to materially affect overall diffusivity or the shape of the diffusion front. Enhanced diffusion occurs in polycrystalline material, and some of the newer structures using polysilicon-filled trenches may have trenches thin enough that solutions to be given here are useful. It is also observed in the defect network adjacent to the sapphire substrate in silicon-on-sapphire wafers (60).

It is assumed that the segregation at the boundary is 1 and that the surface concentration N_0 of each medium is the same. Let N and D be the concentration and diffusivity in the single crystal; N' and D', that in the high-diffusivity layer. At the boundary between the two media ($y \pm a$), $N = N'$ and $J(y) = J'(y)$. Outside the layer,

$$D\nabla^2 N = \frac{\partial N}{\partial t} \qquad 8.66$$

Using the fluxes for an element of the layer, as shown in Fig. 8.23b, and the continuity equation within the layer gives as an additional boundary at $y = \pm a$

$$D'\frac{\partial^2 N}{\partial x^2} - \frac{D}{a}\frac{\partial N}{\partial x} = \left(\frac{D'}{D} - 1\right)\frac{\partial N}{\partial t}$$

Eq. 8.66 has been solved analytically, but the expression is very complex and unwieldy (55). To simplify its use, tables of values for various parametric ratios have been computed (56). Calculated plots of $N(x,y)$ are shown in Fig. 8.24a for differing values of the parameter $\beta = [(D'/D) - 1]a/\sqrt{Dt}$. β can change by either changing D'/D or the diffusion time. In Fig. 8.24a, β changed because of a change in the D ratio since a change in time would have changed the depth of the diffusion front at large y. This change is shown in Fig. 8.24b, which is actual data from a silicon grain boundary diffusion (57). In this case, the time was changed but not the ratio of grain boundary to single-crystal diffusivities. The D'/D ratio for phosphorus has been experimentally measured as $\sim 10^4$–10^5 (57, 61). This data leads to the conclusion that with a close spacing of boundaries, a large increase in effective diffusivity would be expected. Depending on diffusion conditions, the interstitial sinks at closely spaced grain boundaries could also increase the overall diffusivity. Experimentally, it is observed that polycrystalline silicon does show enhanced D values. The effect is modeled in some programs by assuming styl-

FIGURE 8.24

Grain boundary isoconcentration profiles. (*Source:* (a) Adapted from R.T.P. Whipple, *Phil. Mag. Series 7, Vol. 45*, pp. 1225–1236, 1954. (b) Adapted from Van E. Wood et al., *J. Appl. Phys. 33*, pp. 3574–3579, 1962.)

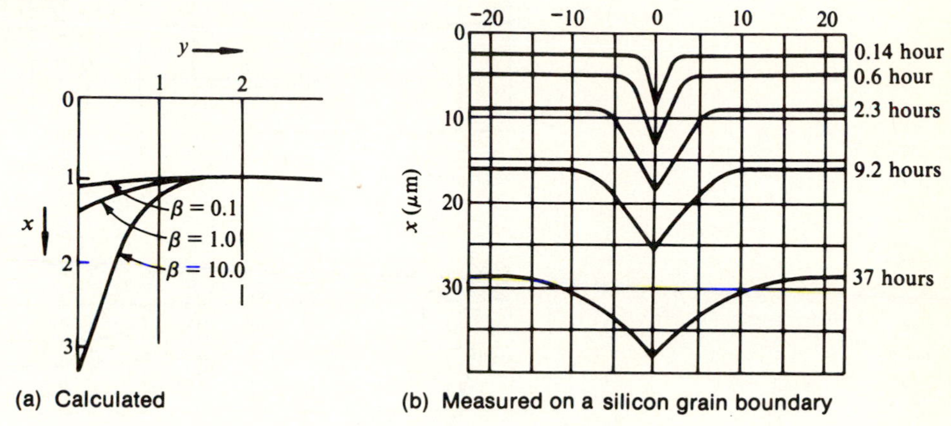

ized grains, bulk and grain boundary diffusion coefficients, and simultaneous flow through grains and along boundaries (62). Heavily doped amorphous silicon also shows enhanced diffusivity in the 500°C–600°C range (63) (at higher temperatures, the amorphous silicon recrystallizes).

8.3.13 Two-Dimensional Solutions

Except for diffusion from the high-diffusivity layer, all of the solutions described thus far have only considered a planar diffusion front of infinite extent. In actual device fabrication, the diffusions do not extend to infinity but terminate near the edge of the mask used to define the diffusions. To calculate the behavior at the edge, two-dimensional solutions are required. The display method in this case is generally one of a series of isoconcentration lines as shown in Fig. 8.25 rather than a concentration–distance profile as used for one-dimensional cases. Fig. 8.25a is the case of diffusion from a limited source with the two mask edges infinitely close together, which gives a line source of S' atoms/cm. Fick's second law in two-dimensional spherical coordinates with axial symmetry is

$$D\frac{\partial^2 N}{\partial r^2} + \frac{D}{r}\frac{\partial N}{\partial r} = \frac{\partial N}{\partial t} \tag{8.67}$$

FIGURE 8.25

Isoconcentration contours at edge of diffusion mask.
(*Source:* Curves b and c from D.P. Kennedy and R.R. O'Brien, *IBM J. Res. Develop. 9*, p. 179, 1965.)

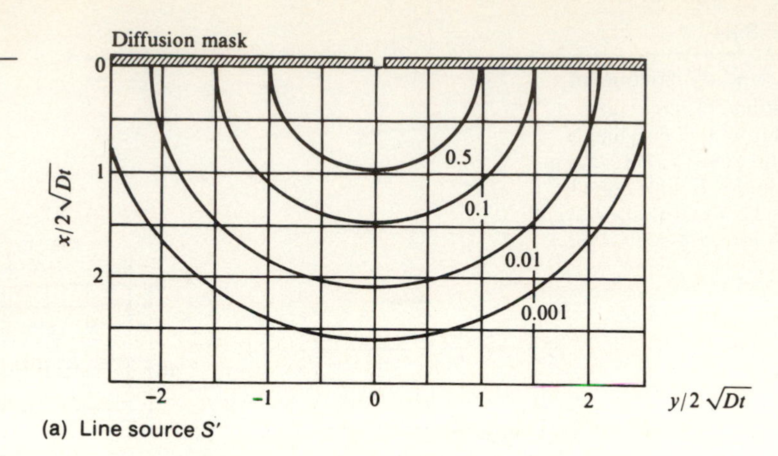

(a) Line source S'

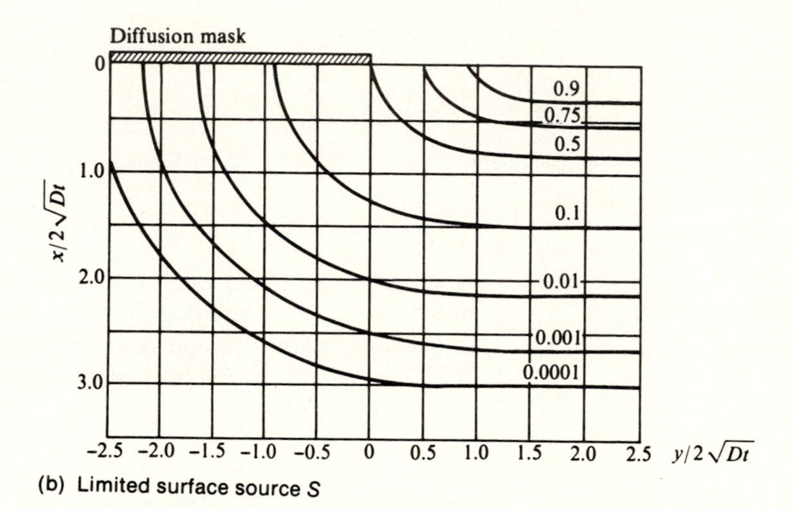

(b) Limited surface source S

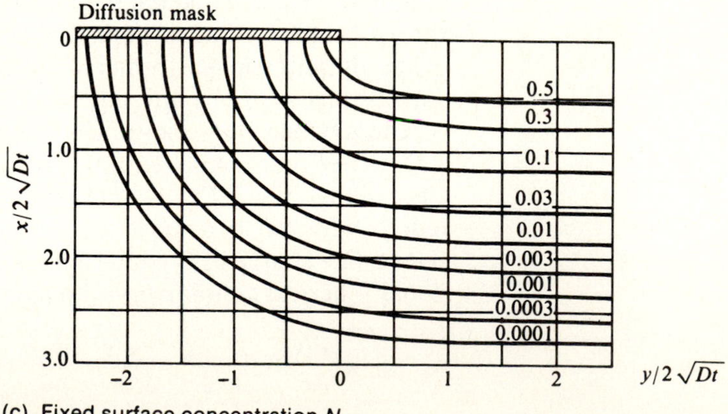

Diffusion mask

0.5
0.3
0.1
0.03
0.01
0.003
0.001
0.0003
0.0001

$x/2\sqrt{Dt}$: 0, 1.0, 2.0, 3.0

$y/2\sqrt{Dt}$: −2, −1, 0, 1, 2

(c) Fixed surface concentration N_0

and it has as a solution (37)

$$N(r,t) = \frac{S'}{2\pi Dt} e^{-r^2/4Dt} \tag{8.68}$$

This case is not actually encountered, and it is the only case that has radial symmetry about $r = 0$ (where diffusion along the surface equals that normal to the surface).

For the other extreme, the case of a limited surface concentration of S atoms/cm^2 and the two edges of the mask separated by an infinite distance rather than by an infinitely narrow strip, the solution near one mask edge in x,y coordinates is (64)

$$N(x,y,t) = \frac{S}{2\sqrt{\pi Dt}} e^{-x^2/4Dt}\left[1 + \operatorname{erf}\left(\frac{y}{2\sqrt{Dt}}\right)\right] \tag{8.69}$$

where the coordinate axes are as shown in Fig. 8.25b. Note that diffusion out under the mask causes increased depletion of the surface near the mask edge and that any particular isoconcentration line extends further down beneath the open window than laterally beneath the mask. For intermediate cases where the mask opening (source) has a finite width w but the isoconcentration line of interest is at a depth much greater than w, Eq. 8.68 can be used with S' given by Sw (29).

The solution for a fixed surface concentration (not given) is considerably more complex and requires the use of hypergeometric series (64). A plot is shown in Fig. 8.25c. When the diffusion coefficient D is concentration dependent, it appears, both experimentally and from modeling (65), that the ratio of lateral to vertical motion is somewhat reduced from that predicted by Eq. 8.69.

8.3.14 Effect of Temperature Varying with Time

Often, the determination of a diffusion profile after several heat-treatment steps at different times and temperatures is necessary. Since only the Dt product occurs in the various diffusion equations, if the initial conditions do not change from one cycle to the next, then, even though a series of discrete times and temperatures are actually used, a single $(Dt)_{\text{eff}}$ is equivalent:

$$(Dt)_{\text{eff}} = \Sigma D_1t_1 + D_2t_2 + D_3t_3 + \ldots \tag{8.70}$$

If the temperature varies continuously, the $\int D(T)dt$ is required. Examples of cases where this approach is applicable are diffusion from a step and diffusion from a limited source. If a diffusion were being made from an infinite source, and the surface concentration varied with temperature as is sometimes true, then Eq. 8.70 is not sufficient. However, that case can be treated by a series solution involving the various Dt's and their respective surface concentrations (66).

8.4 DIFFUSION PROFILE CALCULATIONS

In order to calculate a diffusion profile, a D value, the appropriate $N(x,t)$ function, and a time–temperature sequence are required. As was mentioned earlier, if the concentration N remains below n_i, then D for substitutional diffusion can generally be considered as independent of N. Fig. 8.7 showed n_i versus temperature for silicon, and Fig. 8.26 shows typical low concentration values of D for the common silicon substitutional dopants. These values may be substituted into the expressions given in the previous section to give profiles for the desired boundary conditions. For dopant concentrations high enough for the semiconductor to be extrinsic at the diffusion temperature, the enhancement described in section 8.2.7 is observed for most group IIIA and VA impurities. Experimental data are available for As, Sb, P, Ga, and B (3, 28, 67). Typically, they all look much like the profile shown in Fig. 8.27 for arsenic. It should be noted that while this profile has the enhancement plotted versus the concentration of arsenic atoms, some data are plotted versus the number of carriers. The effect of an enhanced diffusivity at high concentrations is to make the diffusion profile fuller at high concentrations as shown in Fig. 8.28. To account for this effect, the diffusion equations may be solved numerically, or a polynomial approximation may be used for the portion of the profile in the extrinsic region, or else an "average"[13] value of D may be used that will give the correct junction depth but an incorrect profile.

The arsenic profile in the extrinsic region can be reasonably approximated by (68)

$$\frac{N}{N_0} = 1 - 0.87Y - 0.45Y^2 \qquad 8.71$$

where $Y = (8N_0D^*t/n_i)^{-1/2}x$. While the boron profile has the same general features, it is best fitted by an expression of the form (69)

$$\frac{N}{N_0} = 1 - Y^{2/3} \qquad 8.72$$

This expression gives a very good fit to the experimental data, but no simple relation exists between Y and the diffusivity. Y is approximately given by

---

[13]This approach was widely used before a better understanding of the diffusion mechanisms was developed and led to D values that were dependent on background doping.

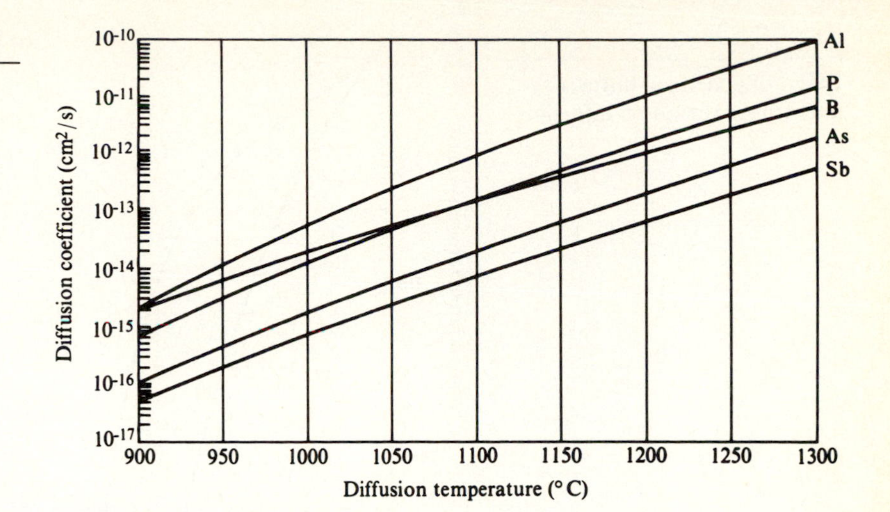

FIGURE 8.26

Diffusion coefficients for common dopants diffusing in intrinsic silicon. (*Source:* From data in Richard B. Fair, *Impurity Doping Processes in Silicon,* F.F. Wang, ed., North-Holland, New York, 1981.)

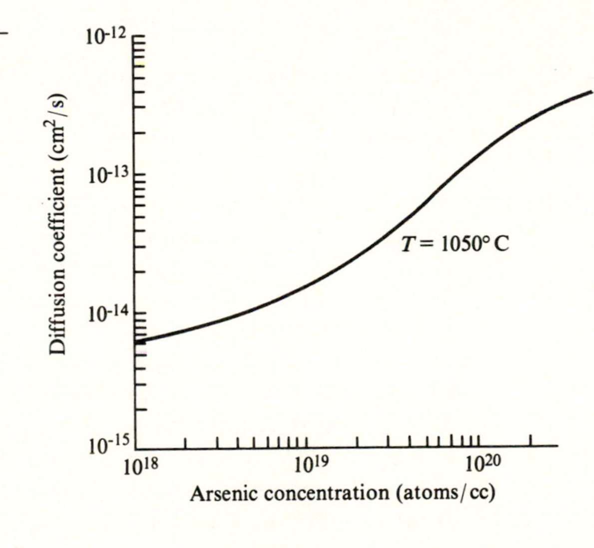

FIGURE 8.27

Arsenic diffusivity enhancement versus arsenic concentration. (*Source:* Adapted from Richard B. Fair and Joseph C.C. Tsai, *J. Electrochem. Soc. 122,* p. 1689, 1975.)

$$Y = \left(\frac{x^2}{6D_s t}\right)^{3/2} \qquad 8.73$$

where $D_s = D^*N_0/n_i$.

Diffusion in polycrystalline silicon has not been nearly as well characterized as diffusion in single-crystal material. In general, however, diffusivity is much higher in polycrystalline material. Data for arsenic show a D value with an activation energy approximately the same as that of single-crystal silicon diffusivity, but with a mag-

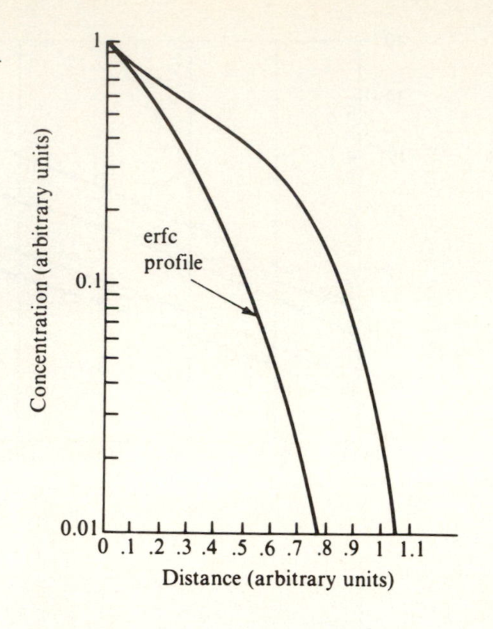

FIGURE 8.28

Effect of enhanced diffusivity in extrinsic region on diffusion profile.

nitude about 5 orders of magnitude greater (70). Phosphorus appears about 2 orders of magnitude higher (71); and boron, about 1 order higher (72).

8.4.1 Phosphorus Profile

The phosphorus profile (3), shown in Fig. 8.29, has peculiarities that do not allow its modeling by the polynomial approach just described. The flat top, shown by a dashed line, is sometimes found in other profiles and is really an artifact of measurement. It represents the maximum amount of electrically active phosphorus present. The solid line shows the total phosphorus content as determined by SIMS. In the region from maximum concentration down to approximately $N = 10^{20}$ atoms/cc, D is given reasonably well by $D^{*-}\ (1 + N/n_i)^2$. That is, the double-charged vacancy term of Eq. 8.9 predominates. However, when the impurity concentration drops enough, the P-double vacancy complex dissociates, giving an excess of V^- vacancies, which then diffuse away. At the same time, the excess V^- provide an enhanced number of P^+V^- complexes and increase the diffusivity over that ordinarily observed during intrinsic diffusion by increasing D^{*-}.

8.4.2 Silicon Interstitial Diffusants

Fig. 8.30 gives D values for a number of fast interstitial diffusants in silicon. However, since the diffusivity of most of these impurities has not been examined in depth, profiles calculated from these numbers may be grossly in error. As an example, consider gold, which has been studied extensively (5, 73). Gold is thought to diffuse both

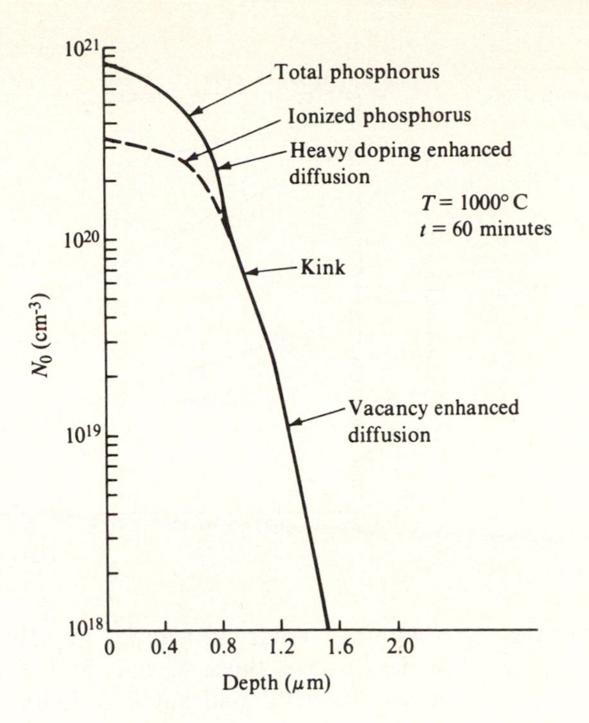

FIGURE 8.29

Profile of phosphorus diffusion with high surface concentration. (*Source:* Adapted from R.B. Fair and J.C. Tsai, *J. Electrochem. Soc. 124*, p. 1107, 1977.)

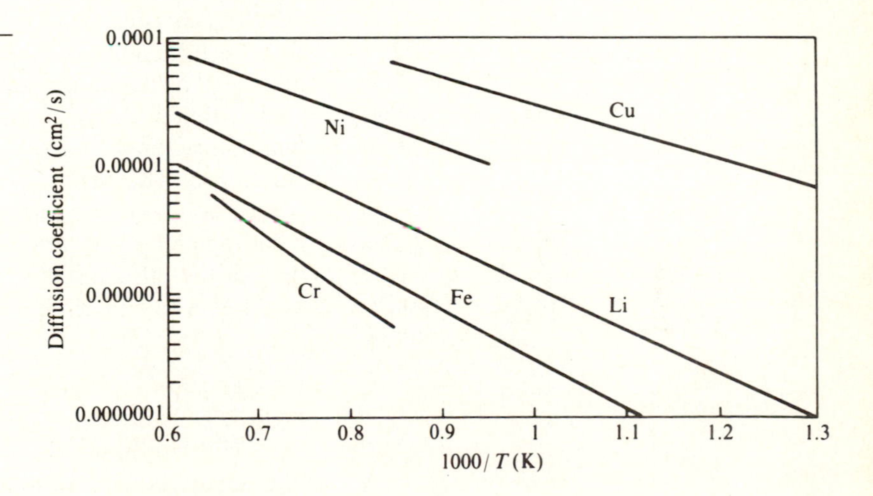

FIGURE 8.30

Diffusion coefficients of some fast diffusers in silicon. (*Source:* From data in Eicke R. Weber, *Appl. Phys. A30*, p. 1, 1983.)

substitutionally and interstitially, with $D_{int} >> D_{sub}$. D_{int} and D_{sub} are shown in Fig. 8.31 as the upper and lower curves. Experimentally, much of the reported data has followed one of the three lower curves (5) and has been interpreted as following either Eq. 8.14b (curve II) or Eq. 8.12 (curve I). Other data appear to be best explained by assuming that silicon self-interstitial diffusion is the limiting factor in gold diffusion as is described by Eq. 8.15 (6, 7).

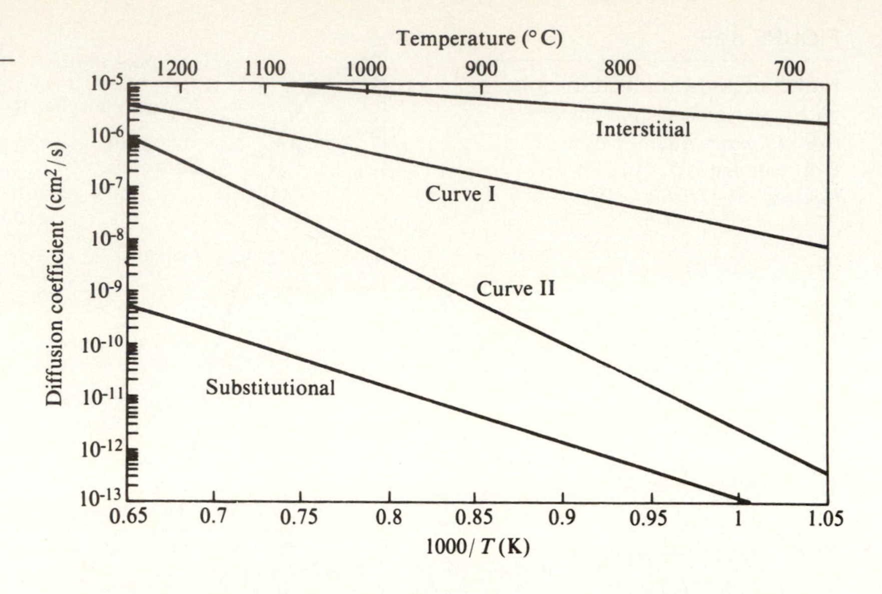

FIGURE 8.31

Diffusion coefficients of gold in silicon. (*Source:* Adapted from W.R. Wilcox and T.J. LaChapelle, *J. Appl. Phys. 35*, p. 240, 1964.)

8.4.3 Gallium Arsenide Profile

Gallium arsenide diffusivities are neither as well behaved nor as well understood as those of silicon. The donors are thought to diffuse substitutionally, and junction depths are linear with $\sqrt{t}$ (29). The acceptors apparently diffuse by an interstitial–substitutional mechanism (74) (see the discussion of zinc in gallium arsenide in section 8.2.3). The effective diffusion coefficients are generally concentration dependent as well as dependent on such things as the wafer gas environment, wafer defect density, and wafer capping (74, 75). Because of this variability, only two D values and activation energies are given in Table 8.2. The behavior of zinc (an acceptor) under carefully controlled conditions is shown in Fig. 8.32 (76). It is characterized by a flat top and a very abrupt falloff. Such curves are not described by any of the diffusion cases discussed, but a plot of the depth x_j versus $\sqrt{t}$ gives a straight line with a slope of D_{sur}, which is given by

$$D_{sur} = \frac{KN_{sur}^2}{(P_{As4})^{1/4}} \tag{8.74}$$

where N_{sur} is the dopant concentration at the surface, K is an equilibrium constant, and P_{As4} is the partial pressure of arsenic over the wafer.

8.4.4 Solid Solubility Data

In order to calculate profiles when using a fixed surface concentration, solubility versus temperature data are required as shown in Fig. 8.33 for the common substitutional dopants in silicon. Experimental gold solubility data are for substitutional atoms and mostly

FIGURE 8.32

(a) Diffusion profiles for zinc in gallium arsenide. (*Source:* R. Jett Field and Sorab K. Ghandhi, *J. Electrochem. Soc. 129,* p. 1567, 1982. Reprinted by permission of the publisher, The Electrochemical Society, Inc.), (b) x_j from part a plotted versus square root of diffusion time.

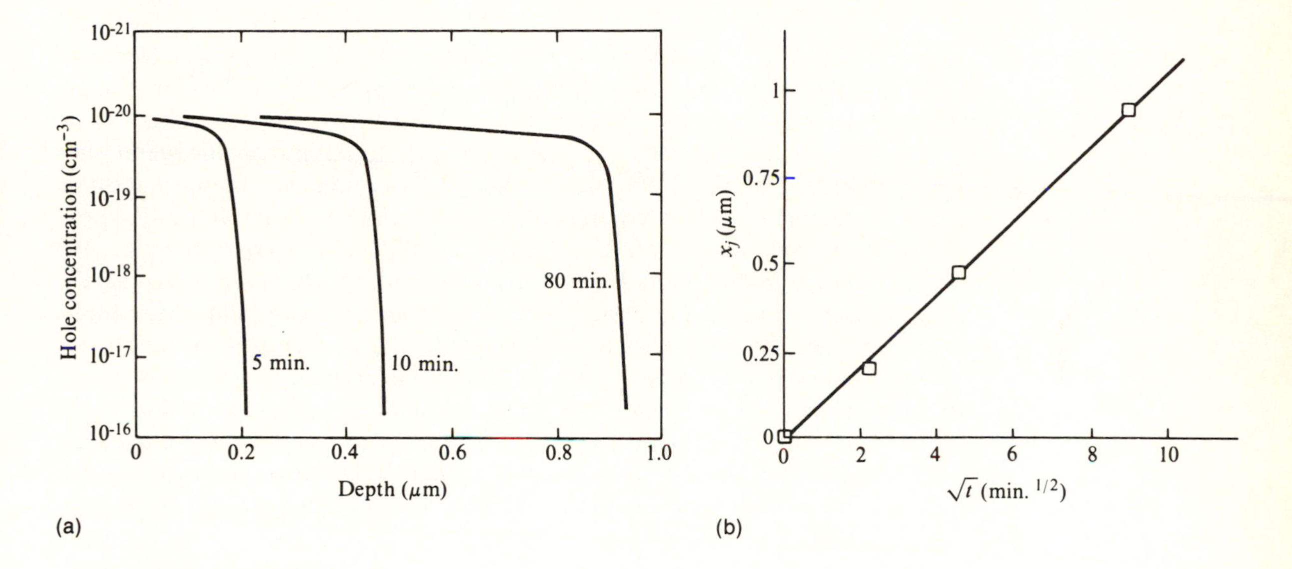

FIGURE 8.33

Solid solubility limits of various dopants in silicon. (*Source:* Adapted from F.A. Trumbore, *Bell Syst. Tech. J. 39,* p. 205, 1960, and G.L. Vick and K.M. Whittle, *J. Electrochem. Soc. 116,* p. 1142, 1969.)

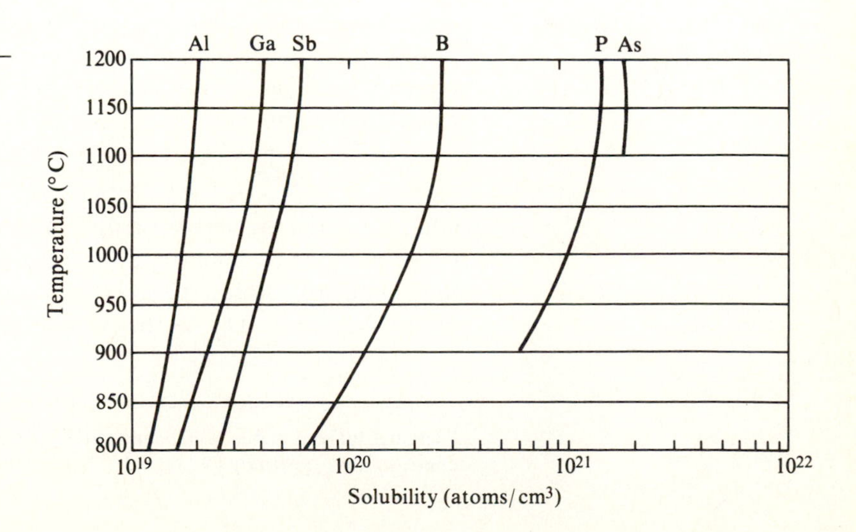

for temperatures above 1200°C, but both interstitial and substitutional solubilities for lower temperatures have been calculated (73). Log solubility is linear in $1/T$ (K) and, for substitutional atoms, is $\sim 10^{17}$ atoms/cc at 1200°C and $\sim 5 \times 10^{13}$ atoms/cc at 700°C. Interstitial solubilities are $\sim 10^{16}$ atoms/cc at 1200°C and $\sim 10^{13}$ atoms/cc at 800°C. (The substitutional solubility is shown later in Fig. 8.39.)

In diffusion processes using ion implanting to put a fixed number of atoms near the surface as a source, solid solubility data are still required to make sure that the implant dose does not lead to precipitation.

8.5 DIFFUSION CHARACTERIZATION

The full characterization of a diffusion usually means a determination of the profile—that is, the determination of the impurity concentration versus distance from the surface—and an evaluation of the uniformity of the profile from point to point over the wafer surface. Unfortunately, such profiling is very time consuming and often requires complex equipment. Consequently, various alternative procedures have evolved. The two most common measurements are junction depth and sheet resistance, both of which are more closely related to device design and performance than the impurity profile itself. From these two measurements, the surface concentration can be inferred if N as a function of depth is known. When one is trying to control or change the junction depth x_j and the diffusion sheet resistance, the value of surface concentration becomes important and thus must be periodically evaluated. Diffusion uniformity over the wafer surface is generally specified in terms of sheet resistance uniformity.

8.5.1 Conversion from Dopant Concentration to Resistivity

During the course of characterization, it is sometimes the dopant concentration and sometimes the resistivity due to that dopant that is measured. Consequently, it is convenient to be able to quickly convert from one value to the other. The conductivity σ may be calculated from

$$\sigma = q\mu N_{ne} \tag{8.75}$$

where q is the electronic charge, μ is the carrier mobility, and N_{ne} is the net ionized impurity concentration.[14] For concentrations less than about 10^{17} atoms/cc, all of the dopant atoms will be ionized and $N_{ne} = N_{\text{net}}$. However, as the concentration increases, there will not

[14]The net ionized impurity concentration is determined from the net impurity concentration N_{net} given by $|N_A - N_D|$.

TABLE 8.3

Percent Ionization Versus Dopant Concentration

Concentration	10^{16}	10^{17}	10^{18}	10^{19}	10^{20}	10^{21}
Phosphorus (1, 2)	100		90	100*	90*	30*
Boron (3)	100	93	75			

*Experimental data.

1. S.S. Li and W.R. Thurber, *Solid-State Electronics 20,* p. 609, 1977.
2. R.B. Fair and J.C.C. Tsai, *J. Electrochem. Soc. 124,* p. 1107, 1977.
3. Sheng S. Li, *Solid-State Electronics 21,* p. 1109, 1978.

be complete ionization, and, in addition, there may be dopant–defect pairing that prevents part of the dopant from even being electrically active. Ionization data for boron and phosphorus are given in Table 8.3. For conversion of dopant concentration to resistivity, a set of curves as shown in Fig. 8.34 can be used instead of Eq. 8.75. Their nonlinearity is due to a combination of concentration-dependent mobility and the incomplete ionization shown in Table 8.3.

8.5.2 Sheet Resistance Measurement

Sheet resistance (R_s) of the diffused layer can be measured directly by a four-point probe (77) if a junction exists between the sheet being measured and the main body of the wafer. Such measurements are destructive only insofar as the wafer surface is either damaged or contaminated by the probe points. R_s is related to the doping of the diffused layer by

$$R_s = \frac{1}{q}\int_0^\infty \mu N_{ne}(x)dx \quad \Omega/\text{sq} \tag{8.76}$$

The average resistivity of a layer of thickness x_j is given by

$$\rho_{av} = R_s x_j \quad \Omega\text{-cm} \tag{8.77}$$

and the average conductivity is given by

$$\sigma_{av} = \frac{1}{R_s x_j} \tag{8.78}$$

When large quantities of data are desired, as, for example, when the uniformity of sheet resistance over a whole wafer is mapped, it may be more expeditious to put an array of metal patterns on the wafer surface and then use automatic wafer test equipment to collect the data. Specialized equipment is also available that steps probes over the wafer in a predetermined pattern and then displays the data in various formats such as contour maps, single-dimensional cross-sectional profiles, and percent deviation (78).

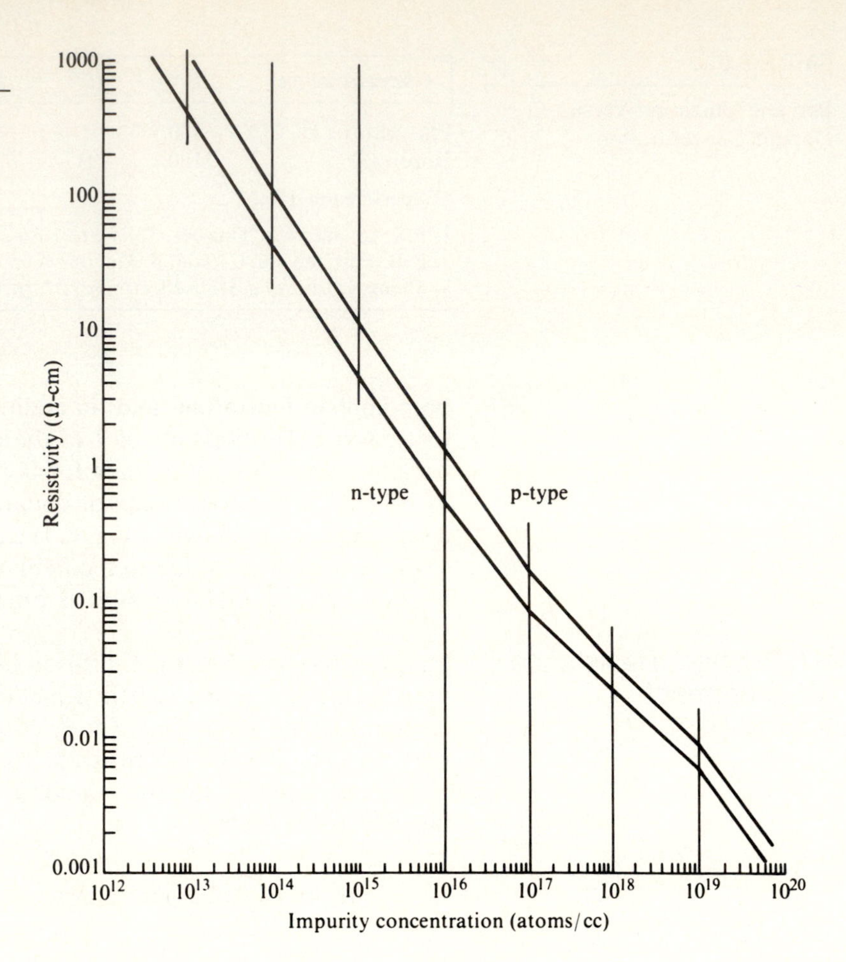

FIGURE 8.34

Resistivity versus impurity concentration for silicon. From data in S.M. Sze and J.C. Irving, *Solid State Electronics 11*, p. 599, 1968; W.R. Thurber et al., *J. Electrochem. Soc. 127*, p. 1807, 1980; and L.C. Linares and S.S. Li, *J. Electrochem. Soc. 128*. p. 601, 1981.

8.5.3 Determination of N_0

When the N_{ne} of Eq. 8.76 is known—for example,

$$N_{ne} \cong N_{net} = \left|N_0 \operatorname{erfc}\left(\frac{x}{2\sqrt{Dt}}\right) - N_B\right| \tag{8.79}$$

the sheet resistance can be calculated as a function of N_0 and N_B by using Eqs. 8.76 and 8.79, providing the mobility as a function of N_{ne} is known (see, for example, references 1 and 2 in Table 8.3 for data). By having made such calculations, curves such as those shown in Fig. 8.35 can be plotted for the range of values of interest and then N_0 read from them (79, 80). References 79 and 80 both have a selection of curves for error and Gaussian function distributions in silicon. However, reference 79 has used more recent electrical data. The Hall constant of the diffused layer can be measured, and, by using an analogous set of curves, the surface concentration can

FIGURE 8.35

Relation between average conductivity ($1/R_s x_j$) of n-type layers diffused into p-type wafer and surface concentration N_0. (Curves for four wafer background doping levels N_B are included.) (*Source:* Adapted from J.C. Irving, *Bell Syst. Tech. J. 41*, p. 387, 1962.)

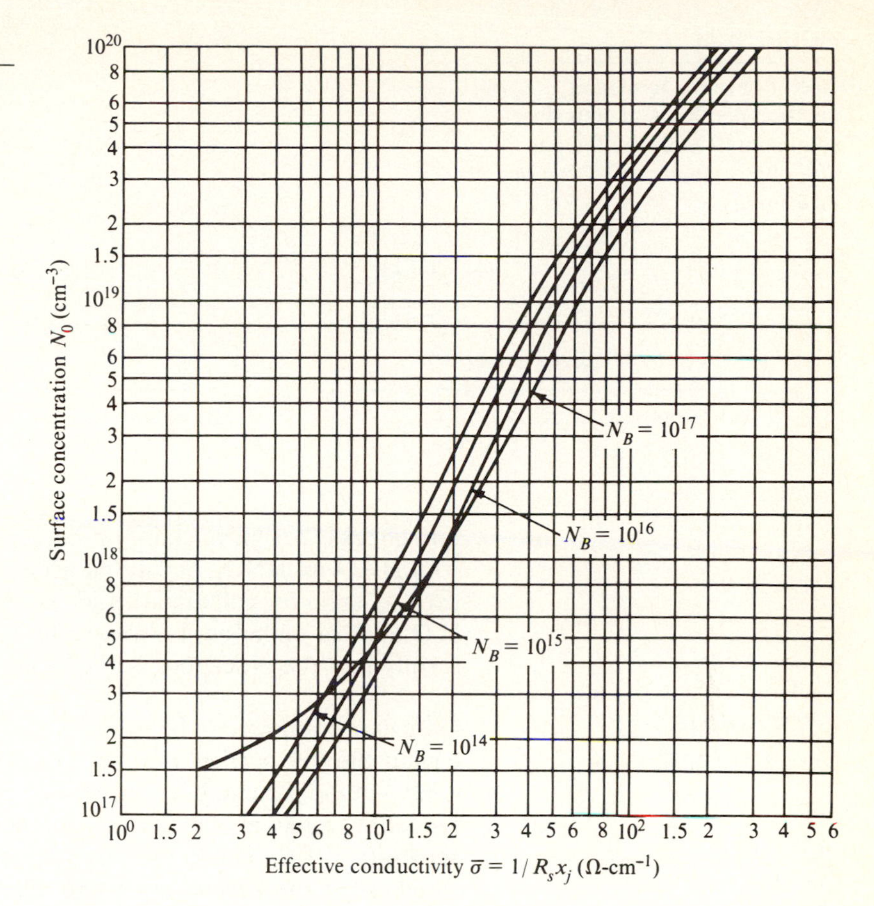

be determined (81). Since a Hall measurement, like the resistance measurement just discussed, includes only ionized impurities, conversion to total impurities requires a consideration of nonionized and precipitated impurities. In addition to these two rather specialized procedures, any of the profiling methods will give N_0 when extrapolated to $x = 0$.

8.5.4 Junction Depth Measurement

Junction depth measurement is ordinarily done by lapping and staining and is destructive. It is, however, quick and simple and requires only a small portion of a wafer. The general procedure is to expose a section through the wafer by some sort of bevel lapping in order to get mechanical magnification and then stain with a chemical solution that differentiates between p- and n-material. Fig. 8.36 shows two types of bevel that are often used. The beveling shown in part a is more appropriate for low angles and high magnification; that shown in part b is quicker and is normally used for in-line produc-

FIGURE 8.36

Use of beveling to magnify junction depth. (A chemical stain is used to define the position of x_j so that x and y can be measured.)

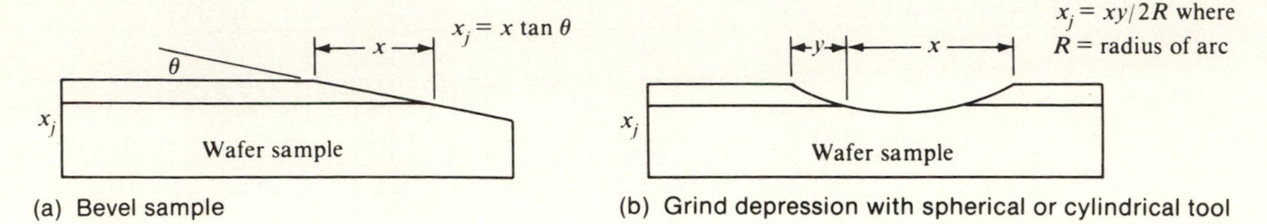

(a) Bevel sample

(b) Grind depression with spherical or cylindrical tool

tion checks. A typical stain for silicon is 1–3–10 etch, comprised of 1 ml HF, 3 ml HNO_3, and 10 ml acetic acid. The p-region will stain dark. A stain for gallium arsenide is 1 ml HF, 1 ml H_2O_2, and 10 ml H_2O applied under an intense white light. (For a more detailed description of the procedure, see Chapter 7 of reference 82.)

8.5.5 Profile Measurement

Diffusion profiling can be done by counting either the impurity atoms or the free carriers produced by the impurities. The oldest profiling method is a combination of sheet resistance measurements and sequential removal of material from the wafer surface (66). It can be shown that if R_1 and R_2 are respectively the sheet resistances measured when the surface is at two closely spaced depths x_1 and x_2 below the original surface, then the bulk resistivity of the thin layer Δx between x_1 and x_2 is given by

$$\rho = \frac{R_1 R_2 \Delta x}{R_2 - R_1} \qquad 8.80$$

The thin layers can be removed by etching the semiconductor (84) and, in the case of silicon, also by anodically oxidizing the surface and then stripping the oxide (85).

The value of $\rho(x)$ from Eq. 8.80 depends linearly on the value of Δx, which is difficult to measure accurately. It also depends on the difference between successive values of sheet resistance. To simplify the task of smoothing the data, the log of the surface conductance G (which is $1/R$) versus x can be plotted and smoothed as desired, without having the possibility of gross errors from an incorrect Δx or $R_n - R_{n+1}$ (86). The slope s of that curve for any value of x can be determined from the smoothed curve and is also given by

$$s = d(\log G)dx = \frac{1}{G}\frac{dG}{dx} \tag{8.81}$$

The bulk conductivity σ in terms of the measured sheet conductance G is given by

$$\sigma(x) = \frac{dG}{dx} \tag{8.82}$$

Combining these two equations gives

$$\sigma(x) = G(x)s(x) = \frac{1}{\rho(x)} \tag{8.83}$$

Rather than successively removing layers, the wafer can be beveled, and either a 4-point probe or a spreading resistance probe can be stepped down the bevel (86–88).

Capacitance–voltage profiling, using a metal Schottky barrier contact or an MIS capacitor, can profile from near the surface to the depth where avalanche breakdown occurs. By using a MOSFET transistor with back-gate bias, a depletion zone can be moved out to where avalanche occurs, and the profile can be determined to that point by measuring the MOS transistor voltages (89–91).

Secondary ion mass spectrometry (SIMS) is widely used to directly measure the dopant atom concentration as a function of depth. It is most applicable to shallow profiling, but the industry trend for several years has been toward shallower junctions. The SIMS technique consists of using an ion beam (usually oxygen or cesium) to sputter away the semiconductor and produce ions that can then be mass-analyzed (92). Table 8.4 shows the sensitivities that can be expected. The ease of ionization is ordinarily the largest factor in determining sensitivity. However, interference between the desired ion and ion complexes causes reduced sensitivity in some cases. This problem can sometimes be alleviated by measuring a different isotope. An example of the problem is Si–H, with masses

TABLE 8.4

SIMS Sensitivity to Dopant Atoms

Dopant	*Semiconductor*	*Sensitivity (atoms/cc)*
Boron	Silicon	10^{14}
Phosphorus	Silicon	2×10^{16}
Antimony	Silicon	10^{16}
Arsenic	Silicon	10^{16}
Gold	Silicon	10^{15}
Chromium	Gallium arsenide	2×10^{15}

of 29, 30, and 31, which interferes with the only stable isotope of phosphorus, ^{31}P. To eliminate the problem with ion complexes being formed during sputtering, an accelerator mass spectrometer has been substituted for the conventional mass spectrometer used in a normal SIMS instrument. The result is a lowering of the detection limits of phosphorus and arsenic by over an order of magnitude (93).

Rutherford backscattering (RBS) can be used for profiling and, in addition has the potential for separating interstitial and substitutional atoms. It is more applicable to atoms of higher mass number than the semiconductor they are in and has, for example, been used to study antimony profiles in silicon (92).

Before the advent of the newer techniques of SIMS and RBS, sectioning by etching, combined with radioactivity counting, was used to obtain profiles independently of the electrical properties. Either the dopant could be a radioactive isotope (radiotracer analysis), or else after diffusion, the sample could be subjected to neutron irradiation in order to produce a radioactive species (neutron activation analysis) (94). The radiotracer method also allowed profiles after an isoconcentration diffusion to be determined. In this kind of study, the radioactive source vapor pressure is the same as the equilibrium vapor pressure due to the nonradioactive dopant of the same species already present so that there is no net flow of impurity into or out of the wafer surface.

8.5.6 Determination of Diffusion Coefficient

When the diffusion coefficient is independent of concentration (Eq. 8.26 is applicable) and the solution for $N(x)$ is known, there are several ways to rather easily determine D (95). If a series of diffusions with differing t's are made from a fixed concentration source into wafers of opposite conductivity type with a concentration N_B, a plot of x_j versus $\sqrt{t}$ will give a straight line (Eq. 8.36). From the slope of the line, D can be determined. In principle, only one t and x_j need to be measured, but plotting a series of values allows data smoothing. If N_0 cannot be found by some method such as the use of initial sheet resistance and curves like those of Fig. 8.35, then two wafers with widely differing N_B's can be simultaneously diffused so that the two surface concentrations are the same. Dividing the expression for N_{B1} by that for N_{B2} (Eq. 8.32) eliminates N_0 and gives

$$\frac{N_{B1}}{N_{B2}} = \frac{1 - \mathrm{erf}(x_{j1}/2\sqrt{Dt})}{1 - \mathrm{erf}(x_{j2}/2\sqrt{Dt})} \qquad 8.84$$

from which D may be determined by successive approximations (96). Alternatively, the approximate expression for erfc(z) for large z from section 8.10 can be used to solve Eq. 8.84 (97):

$$D \cong \frac{(1/4t)(x_{j2}^2 - x_{j1}^2)}{\ln(N_{B1}x_2/N_{B2}x_1)} \qquad 8.85$$

It must be emphasized that if the profile is not an error function, interpreting the data in this fashion will lead to erroneous results. If, for example, the actual profile is like the one of Fig. 8.37, which is typical of high-concentration diffusions, not only will an N_0 determination from a sheet resistance and an x_j measurement be wrong, but the calculated D value will appear to depend on the background doping level (10). This kind of interpretation apparently was the reason for some early reports of such D value dependency.

If diffusion is from a limited source instead of from an infinite source, if S is known—for example, by having ion implanted a known density of diffusant—and if the diffusion is for a long enough time for x_j to be much greater than the implant range R, then D can be determined from Eq. 8.41b. Diffusing into wafers with two different doping levels allows S to be eliminated in the same manner that N_0 was in Eq. 8.84. In this case, D is given by

$$D = \frac{(1/4t)(x_{j2}^2 - x_{j1}^2)}{\ln(N_{B1}/N_{B2})} \qquad 8.86$$

It should be noted that in Eqs. 8.85 and 8.86, D depends on the difference between the squares of the junction depth and thus is quite sensitive to errors in measurement of the x_j's. If a profile of N versus x over a wide range of x is made for a diffusion time t, a plot

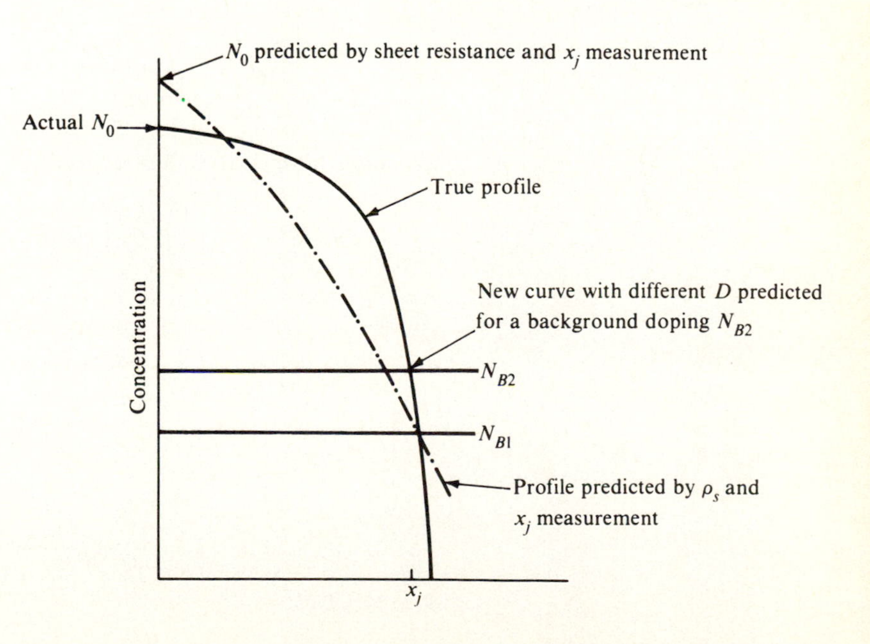

FIGURE 8.37

Errors introduced during attempt to interpret a diffusion profile due to a concentration-dependent dopant in terms of a constant D profile.
(*Source:* Adapted from S.M. Hu in D. Shaw, ed., *Atomic Diffusion in Semiconductors,* Plenum Publishing Co., London, 1973.)

of log N versus x^2 should be a straight line with a slope of $-1/4Dt$. That this is true can be seen by taking the log of both sides of Eq. 8.39, which gives

$$\log N = \log\left(\frac{S}{\sqrt{\pi Dt}}\right) - \frac{x^2}{4Dt} \tag{8.87}$$

When D is concentration dependent, the methods just described are not appropriate for determining its value. For the conditions of diffusion from an infinite source and from a concentration step, the Boltzmann–Matano method is ordinarily used (1). This method is based on substituting a new variable $\zeta = x/\sqrt{t}$ into Fick's second law as written for a concentration-dependent D. This substitution gives an ordinary differential equation that can be solved analytically.[15] First, relate $\partial N/\partial t$ and $\partial N/\partial x$ to the new variable ζ. That is,

$$\frac{\partial N}{\partial t} = \left(\frac{\partial \zeta}{\partial t}\right)\frac{\partial N}{\partial \zeta} = -\left(\frac{x}{2t^{3/2}}\right)\frac{dN}{d\zeta} \tag{8.88a}$$

$$\frac{\partial N}{\partial x} = \left(\frac{\partial \zeta}{\partial x}\right)\frac{\partial N}{\partial \zeta} = \left(\frac{1}{t^{1/2}}\right)\frac{dN}{d\zeta} \tag{8.88b}$$

Substituting these values into

$$\frac{\partial N}{\partial t} = \frac{\partial D}{\partial x}\frac{\partial N}{\partial x} + D\frac{\partial^2 N}{\partial x^2} \qquad \text{(Fick's second law)}$$

gives

$$-\left(\frac{\zeta}{2}\right)\frac{dN}{d\zeta} = \frac{d(D\,dN/d\zeta)}{d\zeta} \tag{8.89}$$

By integrating Eq. 8.89 once and rearranging,

$$D(N) = -\left(\frac{1}{2}\right)\frac{d\zeta}{dN}\int_0^N \zeta\, dN \tag{8.90}$$

or

$$D(N) = -\left(\frac{1}{2t}\right)\left(\frac{dx}{dN}\right)_N \int_0^N x\, dN \tag{8.91}$$

[15] Boltzmann used the new variable as a means of solving the diffusion equation (*Wied. Ann. 53*, p. 959, 1894). Later, Matano used it to determine D (*Jap. J. Phys. 8*, p. 109, 1933).

Thus, if an N versus x profile is determined, dx/dN at a desired point on the graph can be determined, and the integration can be performed graphically. When the variable $\zeta = x/\sqrt{t}$ is not appropriate, for example, in diffusion from a limited source, often other relations will allow D to be determined in a similar fashion (98–101).

EXAMPLE ☐ Using the curves of Fig. 8.38, determine the diffusion coefficient using the Boltzmann–Mantano method. Since these curves have been plotted assuming a constant D, any choice of position should give the same value. The diffusion time was 10 hours.

Choose point A for evaluation. The value under the horizontal curve is the integral required and is about 6.2×10^{12} atoms/cm². The slope of the profile dx/dN at point A (measured from the linear plot) is about -3.3×10^{-22} cm⁴/atom. The time is 7.2×10^4 s. Substituting these values into Eq. 8.91 gives $D = 2.8 \times 10^{-14}$ cm²/s. The curve was plotted using a value of $D = 2.5 \times 10^{-14}$ cm²/s, so the agreement is satisfactory. ☐

When one is examining the effects of variables such as crystallographic orientation and oxidation conditions on diffusion results, it is often difficult to compare results obtained from separate wafers. Consequently, devising structures and/or flows that enable comparisons to be made on the same wafer is often helpful. A simple example is the use of a nitride mask over part of a wafer during drive-in to study the effect of oxidation-produced defects on diffusion rate. More complex examples are the use of orientation-dependent

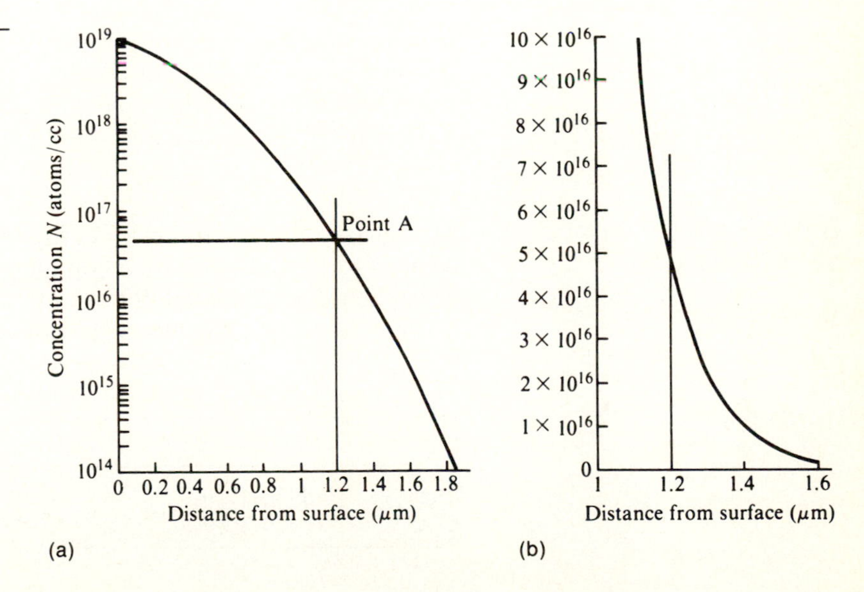

FIGURE 8.38

Profile to use with Boltzmann–Matano method of determining D. (Curve b is a linear plot to simplify the determination of dx/dN and the numerical integration of xdN.)

etch to expose both (111) and (100) faces on the same wafer (102) or the use of twinned or polycrystalline wafers to provide two or more orientations on the same planar surface.

8.6 DIFFUSION PROCESSES

Diffusion processes can be characterized in terms of the function of the diffused structure such as a base or a source/drain diffusion. From a processing standpoint, however, they can be better described in terms of high or low surface concentration, high or low diffusion temperature, and deep or shallow diffusion depths. The approximate limits of each of these factors as defined by common usage are shown in Table 8.5.

8.6.1 Depth

MOS circuits are characterized by shallow diffusions. MOS transistors require only a very thin layer for the channel, and any source–drain depth below it usually only adds unwanted capacitance. Thus, source/drain diffusions can be 1 μm or less in depth. (With very thin diffusions, great care must be exercised to ensure that the contacts do not alloy completely through the diffused layer.) The exception to shallowness is the formation of diffused wells in CMOS, but even then the well depth is usually only 5–6 μm. The collector isolation diffusion is the only deep bipolar diffusion. It must extend from the surface down to the epitaxial substrate, which may be up to 20 μm away for older designs. Truly deep diffusions are found in discrete devices like triacs, SCRs, and power rectifiers.

Deep diffusion processes will normally be at high temperature in order to reduce diffusion time and to provide the high surface concentration that is also usually required. For very deep power device p-diffusions, aluminum is often used because of its high diffusion coefficient. In order to control shallow diffusions, low temperatures are usually required. Since the total amount of impurity is small, an ion implantation predep, followed by a thermal drive-in, is usually used. In many cases, the implantation is followed only by an anneal step.

8.6.2 Surface Concentration

The surface concentration is normally set by the device requirements and is controlled by a combination of diffusion temperature or implant dose and total diffusion time. When a one-step diffusion is used, the solid solubility of the dopant at that temperature can be

TABLE 8.5

Diffusion Process Ranges

	Temperature (°C)	*Surface Concentration*	*Depth (μm)*
High	1150–1250	$>5 \times 10^{19}$	>15
Low	850–1000	$<5 \times 10^{19}$	<2

used to hold the surface concentration constant. As Fig. 8.33 showed, the solid solubilities versus temperature for common dopants in silicon change no more than a factor of 3 over the 800°C–1200°C temperature range. Thus, while the concentration can be fixed by solid solubility, it cannot be varied over a very wide range of values by changing temperature. To obtain a wider range, other methods must be used (none of which will provide higher concentrations). A direct approach, but one that does not give very good control, is just to reduce the dopant source flow if a gas or liquid source is being used. Another approach is to first grow a thin thermal oxide and let the glassy source layer form on top. If the diffusion time is very short, as for an emitter diffusion, the dopant can diffuse through the thin oxide, with the result that the concentration at the silicon surface is reduced (103). With longer times, the glassy source layer will dissolve the thin oxide and then again be in contact with the silicon surface. When a two-step diffusion (predep or ion implant followed by a drive-in) is used, the initial surface concentration is set by temperature or implant dose, but the final lower concentration is determined by the time and temperature of the subsequent drive-in. Alternatively, CVD diffusion source layers with lesser dopant concentrations can be deposited on the surface to obtain a given surface concentration even though the diffusion is still made at a higher temperature. (See section 8.9 on diffusion sources for more details on various diffusion sources.)

8.6.3 Lifetime Control

In most cases, a high semiconductor bulk lifetime is desired for reasons such as minimizing reverse junction leakage, improving bipolar transistor gain, and increasing refresh time in dynamic MOS memories. Some circuits, however, such as TTL logic, depend on a low lifetime in the collector regions of the transistors to improve switching time. Substitutional gold, whose deep levels provide for carrier recombination, is most often used (73, 104), although platinum is a possible alternative. For a given concentration, platinum lowers the lifetime more than gold does and, for a given lifetime, produces fewer deep-level centers and hence less generation leakage current (105, 106).

Both gold and platinum are introduced by diffusion, and both are fast diffusers. Thus, the length of time required to diffuse to the required region is generally quite short. The diffusion is complicated by the fact that the highest solubility regions for the killer impurity are the heavily doped ones, while the regions where the lifetime reduction is required are the lightly doped ones. Thus, heavy doping acts as a sink and "getters" the gold (and other lifetime killing impurities). Data indicate that for phosphorus concentrations over about 2×10^{19}, substantial gettering occurs (107, 108). Curves such as the one in Fig. 8.39 can be used to relate the amount of gold required in a region to give the desired device parameters (transistor switching time or diode recovery time). It is, however, difficult to estimate the amount that may be trapped in adjacent regions. Therefore, gold or platinum concentrations are difficult to predict. Usually, the procedure is to diffuse at a temperature at which the solid solubility of the substitutional gold or platinum is adequate to control lifetime and for a time adequate to ensure that the region of interest is saturated. Fig. 8.39 also relates gold solubility to processing temperature. After diffusion, the wafer is quenched rapidly enough to prevent precipitation. As will be discussed in a later section, since rapid cooldowns can generate excessive crystallographic damage, process trade-offs may be required.

Since diffusion times required are relatively short, the gold (or platinum) is introduced at the end of the diffusion cycles, and because of gettering by heavily doped regions, introduction is almost always directly into the collector region. The simplest procedure is to introduce the gold from the back of the wafer. If, however, a heavily doped layer exists between the back of the wafer and the collector–base junction, as for an n^+-epi substrate or a subcollector diffusion, that layer will interfere with the gold's diffusing to the required region. Sometimes, diffusion is through collector contact openings in the oxide, but if the concentration of gold is too high, the collector near the surface will be compensated, causing high contact resistance and poor device performance. To eliminate the problem, some designs open up special lifetime doping holes in the oxide.

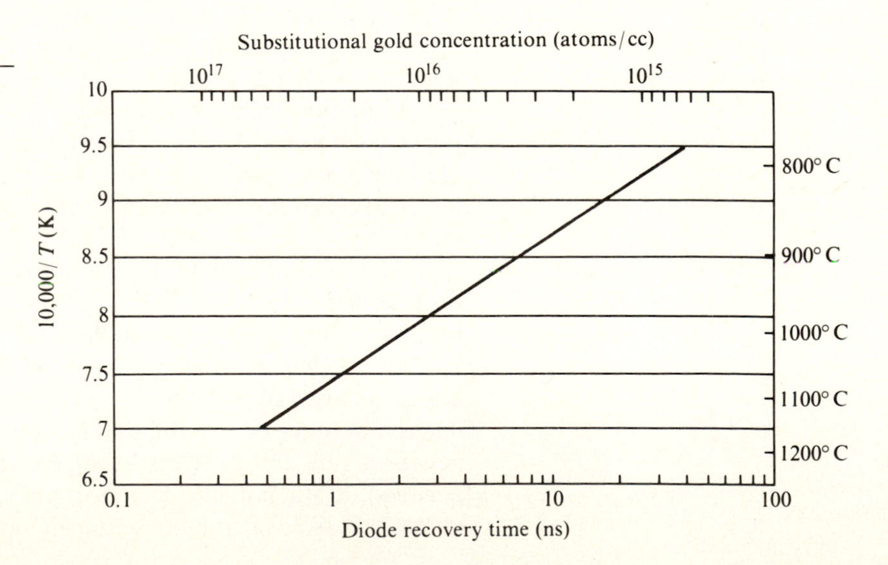

FIGURE 8.39

Relations between diode recovery time, gold concentration, and gold diffusion temperature. (*Source:* Adapted from W.M. Bullis, *Solid-State Electronics* 9, p. 143, 1966.)

Lifetime can also be controlled by nondiffusion processes such as high-energy particle irradiation (electrons and protons). Irradiation effects can usually be rather easily annealed out and hence are seldom used in production devices. Localized ion implanted argon can be used for selective lifetime control and is reported to provide a useful range of lifetimes even after 1200°C processing (109).

8.6.4 Gettering of Silicon

Gettering steps are used to remove lifetime reducing dopants (usually some of the heavy metals) from regions of the circuit where their presence would degrade performance. The impurities most likely to be found are gold, nickel, copper, and iron. However, all of the transition metals (Ti, V, Cr, Mn, Fe, Co, Ni) are reported to be deleterious (110). The most common effect is that of lifetime reduction, but gold, for example, reduces mobility in MOS structures (108, 110–113) and increases the resistivity for both n- and p-type as the substitutional gold concentration approaches that of the IIIA/VA dopant. In addition, any of the metals with high solubilities at processing temperature and very low solubility at room temperature are prone to form precipitates and cause excessive leakage or even direct shorts between transistor elements. It is also reported that metallic precipitates near the surface can reduce the thickness of thermal oxide grown over them (114). As noted in Chapter 3, the presence of these metals will cause S-pits after oxidation and lead to the subsequent formation of stacking faults. The stacking fault generation is particularly troublesome in the oxidation–diffusion–epitaxy sequence used to produce bipolar buried-collector layers. Much effort is directed toward maintaining a clean process (see Chapter 3), but, in most instances, some gettering in conjunction with the diffusion steps is still required during silicon IC fabrication. Gold is the earliest lifetime killer identified (115), the most common contaminant, and the one most studied.

Gettering processes depend either on enhanced solubility of the metal in heavily doped n-material or on providing sites for enhanced precipitation in regions where they will do no harm. Various gettering methods are listed in Table 8.6 and shown schematically in Fig. 8.40. While lifetime killers are characterized as fast diffusers, as both diffusion times and temperatures are reduced, it becomes more difficult for back-side gettering sinks to function. Consequently, those methods that provide gettering locations closer to the front surface and the active devices become more desirable.

Heavy n-layers on the back of the wafer (116–121) were the first gettering sinks used and are probably still the most common. They are the natural consequence of an npn transistor emitter diffusion or an n-channel source/drain diffusion and an unprotected back surface. Hence, their use for gettering requires no additional processing

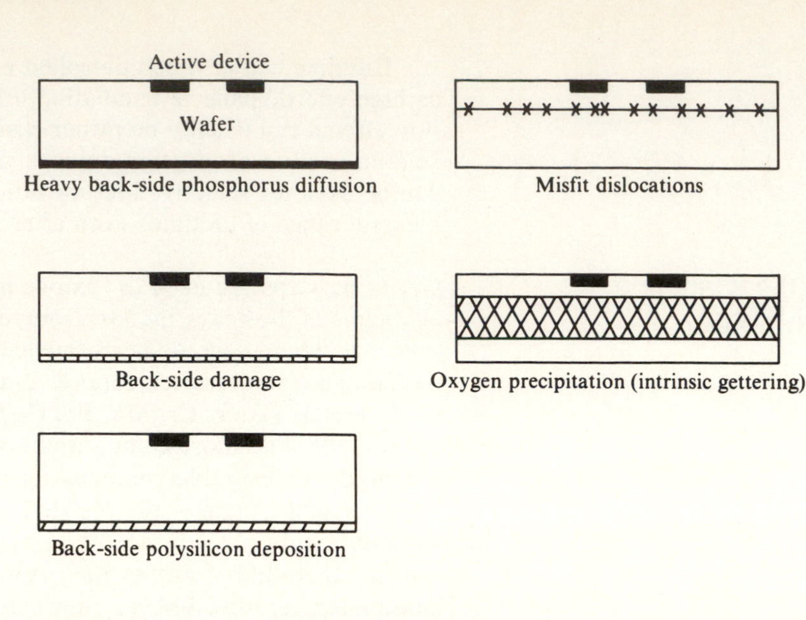

FIGURE 8.40

Location on wafer of gettering region for various gettering methods. (Note that only misfit dislocations and oxygen precipitation allow gettering close to the active device junctions.)

steps. Heavily n-doped regions have increased substitutional solubility for those lifetime killers that behave as acceptors when substitutional (for example, Au and Cu), so, in principle, heavily doped phosphorus, arsenic, or antimony layers should perform equally as well. Experimentally, however, it is found that phosphorus is much more effective than the others (118). This fact has been explained by the formation of phosphorus–metal pairs (118) and by the extra crystal defects formed when silicon phosphide precipitates (119). Regions of misfit dislocations induced into (100) wafers have also been shown to be effective gettering sites (122). Heavy boron doping should increase the solubility of interstitial gold and hence also getter. However, interstitial solubility is much less than substitutional solubility, and it is experimentally observed that p^+-gettering is not nearly as effective as n^+. Some evidence exists that, unlike phosphorus, the boron glassy source associated with p^+-diffusions collects some of the gold (100).

Mechanical back-side damage gettering has been studied at least since 1965 (123–126). By using controlled sandblasting or other forms of light abrasion, the results are reasonably reproducible, and indeed wafers with such damage are now standard items of commerce. This sort of damage is applicable to processes where the formation of a heavy phosphorus layer presents problems. Within limits, the more abrasion, the more pronounced the gettering effect. For process control of such mechanical damage, light scattering from the abraded surface as measured by an instrument such as a paint gloss meter can be used.

Localized laser melting can be used to generate the back-side damage and does not involve the dirty processing required of abrasion (127, 128). A pulsed laser with a wavelength absorbed by the wafer, such as Nd:YAG, is used to scan the surface, generally producing nonoverlapping rows of overlapping melted regions. The thermal shock associated with the rapid heating and cooling produces a network of dislocations for the gettering. Some minimum power per pulse is required to ensure that the damage is severe enough that it does not anneal out and is on the order of 15–20 J/cm^2 (128).

Back-side ion implantation gettering is now used with some regularity. By selective masking, it can also be used locally on the front of the wafer and possibly be closer to the volume needing gettering than if it were on the back-side. A wide range of procedures and results have been reported (129–135). Methods of annealing and the point in the process at which the implant is done affect the final results, as does the implant species and the wafer orientation. Argon is one of the more effective ion implant species, and some correlation exists between gettering efficiency and the misfit ratio of the implanted ion to silicon (130). Much more residual damage remains in (111) oriented silicon than in (100), and the gettering efficiency is correspondingly higher. Dosages for implanting into bare silicon range from 10^{14} and 10^{15} ions/cm^2, with an implant energy of 100–200 keV. When implanting through a thin oxide layer, 10^{16} atoms/cm^2 have been used (135).

A silicon nitride layer deposited on the back of a wafer will provide enough stress during high-temperature processing to effectively getter (136, 137). Both the low-temperature plasma-deposited nitride and the higher-temperature CVD nitride appear effective. Thicknesses required are in the 1000–4000 Å range. Gettering is less for the thinner films, and wafer bowing can occur if the film is much above 4000 Å.

A thin layer of low-temperature polycrystalline silicon deposited on the back of the slice will provide for gettering sinks (138). The polycrystalline layers are deposited at low temperature, such as 650°C, to a thickness of about 0.5 μm. Even if the layers are thin enough to be fully oxidized during processing, stacking faults that continue to getter are propagated into the single-crystal silicon (138, 139). The gettering efficiency of a polysilicon layer is reported to be less easily annealed out during high-temperature cycling than the surface damage processes discussed earlier. The relative efficiency of back-side damage and a polysilicon layer in reducing S-pits as a function of the number of heat cycles is illustrated in Fig. 8.41.

A different kind of gettering, and one not included in Table 8.6, is the use of an atmosphere in the furnace tube during diffusion (or

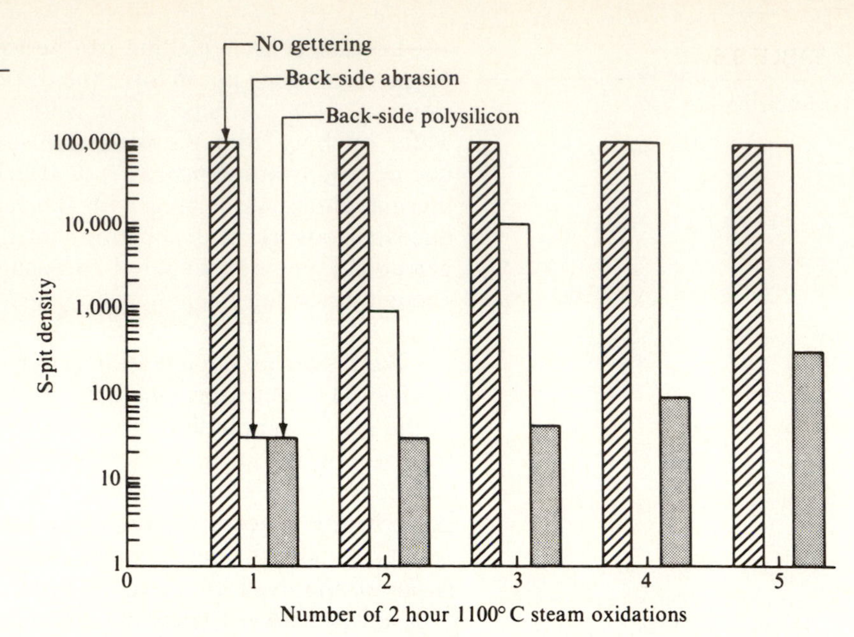

FIGURE 8.41

General trend of back-side gettering efficiency versus amount of processing as measured by S-pit reduction. (The final S-pit value will depend not only on the gettering process but also on the initial contamination level.) (*Source:* Adapted from data in *Monsanto Applications Note,* AN9-7/82, revised January 1, 1983.)

oxidation) that will assist in transferring unwanted impurities from the wafer surface to the gas stream. It has been known since 1960 that the use of chlorine or a chlorine-bearing species such as PCl_3 or BCl_3 would improve lifetime (140). In the case of phosphorus and boron, the relative impact of the glassy layer, the heavy doping in the silicon, and the chlorine species in the tube ambient were not resolved. Since then, the use of HCl to clean tubes has become standard practice. The mechanism is one of forming chlorides volatile at the tube temperature, and the same mechanism can be used to clean wafers if etching of the surface can be prevented. Such conditions prevail during HCl oxidation, and substantial lifetime improvement is sometimes observed (141–143). In this case, the effectiveness will also depend on how easily the impurity to be gettered can diffuse through the protective thermal oxide and whether or not it is trapped at the interface. In the case of gold, it appears that there is a substantial pileup at the Si–SiO_2 interface (144) and that gold is not effectively gettered from wafers during an HCl oxidation (145).

Another gettering method not included in Table 8.6 and seldom used is the application of metallic layers to the silicon surface to either trap impurities or prevent them from entering the wafer (146, 147).

The misfit dislocations shown near the surface in Fig. 8.40 are normally introduced during an epitaxial operation and were dis-

TABLE 8.6

Gettering Methods

Item	Comments
Back-side phosphorus diffusion	
Back-side abrasion	Usually uses sandblasting.
Laser damage	Crystal damage introduced by local thermal shock.
Ion implant damage	Typically uses argon ions.
Silicon nitride deposition	~4000 Å layer deposited on the back of the wafer.
Back-side polysilicon deposition	Combination of polysilicon, residual oxide, and diffusion temperature introduces crystal damage.
Misfit dislocations	Must be introduced during an epitaxial step (see Chapter 7).
Oxygen precipitates	Requires correct oxygen level in crystal and a prescribed heat-treat cycle to be effective.

cussed in Chapter 7. They could also, in principle, be introduced by the high-energy implant of an appropriate impurity atom.

Intrinsic gettering generally refers to gettering by oxygen precipitates in the bulk of a silicon wafer. The term could, however, equally as well be applied to some other gettering mechanism originating with grown-in defects. The majority of silicon crystals are grown by the Teal–Little pulling process (often called CZ, or Czochralski) from a fused silica container. The molten silicon dissolves a small part of the fused silica, and some of the oxygen from it is incorporated into the single-crystal silicon. The amount depends on specific growth procedures but is in the 10^{18} atoms/cc range. As the crystal comes from the puller, most of the oxygen is dispersed as interstitial atoms and is electrically inactive. However, heat treatments such as are involved in diffusions can cause it to aggregate and eventually form rather large precipitates. The small aggregates are donors and can be sufficient to reduce high-resistivity material down to a few tenths Ω-cm n-type (148). Treatment in the 1300°C temperature range will cause a redissolving of the oxygen structures, but any crystallographic defects, such as stacking faults that formed because of the precipitates, will remain.

The oxygen-precipitation-induced defects have been shown to act as heavy-metal gettering sites (149), but if such sites are within the active volume of the IC devices, a yield degradation results. The idealized approach to intrinsic oxygen gettering is to first form a region from the wafer surface toward the wafer interior that is thick enough to contain the devices made in that wafer and has so little oxygen that it will not precipitate. Later, during subsequent processing, this region will remain free of oxygen-induced defects and thus not adversely affect devices made in it. The high-temperature operation to allow oxygen near the surface to out-diffuse and also dissolve any oxygen aggregates that may be in the wafer is referred to as a "denuding" step. After denuding, the wafer is subjected to a low-temperature heat cycle in order to allow precipitate nuclei to form. With this preparation, the dissolved oxygen will form large precipitates during the heat cycles associated with subsequent oxidations and diffusions. These large precipitates, in turn, produce crystal defects that act as gettering sites. This temperature sequence is often referred to as "high–low–high." In the interest of process time minimization, all or part of the required temperature steps may be combined with various regular processing steps so that an optimized high–low–high sequence may not exist. The approximate temperature ranges over which these various effects occur are shown in Fig. 8.42.

As will be discussed in section 8.7.4 on wafer warpage, interstitial oxygen improves silicon yield strength, while oxygen precipitates reduce it and sometimes lead to wafer slip during heatup and cooldown cycles. Thus, trade-offs between gettering and strength may be required in order to obtain maximum yield.

Under ordinary circumstances, an as-grown crystal is supersaturated with oxygen. It can be seen by noting that typical oxygen concentration specifications range from 25–40 ppma[16] and then com-

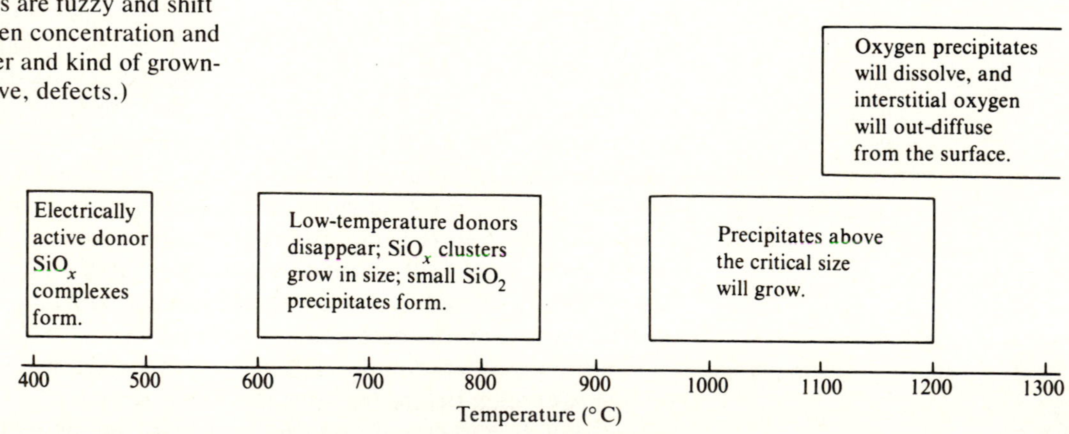

FIGURE 8.42

Behavior of oxygen in silicon as a function of processing temperature. (The temperature boundaries are fuzzy and shift with oxygen concentration and the number and kind of grown-in, or native, defects.)

[16]Since there are 5×10^{22} atoms/cc of silicon, 25–40 ppma converts to $1.25\text{–}2 \times 10^{18}$ atoms/cc.

FIGURE 8.43

Oxygen interstitial solubility in silicon versus temperature. (*Source:* From data in R.A. Craven, in H.R. Huff and R.J. Kriegler, eds., *Semiconductor Silicon/ 81,* p. 254, Electrochemical Society, Pennington, N.J., 1981.)

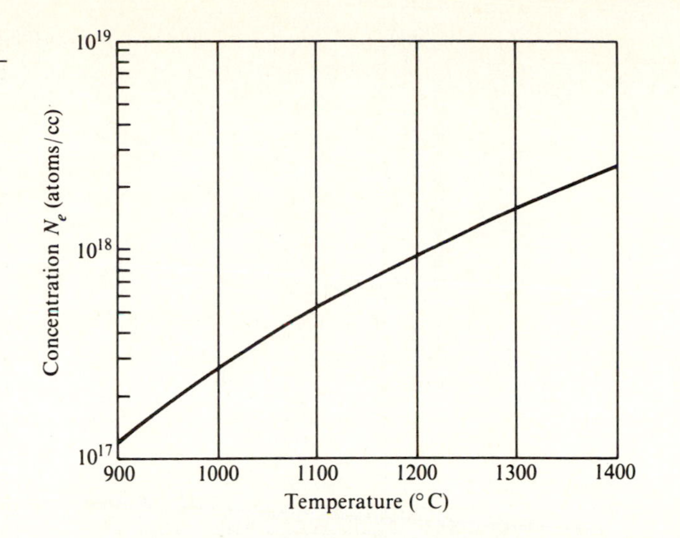

paring those numbers with Fig. 8.43, which gives the equilibrium concentration N_e versus temperature (150). When cooled to room temperature, the oxygen is immobile and will remain dispersed so that the silicon is supersaturated. However, during high-temperature processing, part of the oxygen can diffuse out of the wafer and into the ambient, and part of that remaining can precipitate. If it is assumed that there is no barrier to oxygen leaving the wafer surface, the oxygen profile $N(x,t)$ is given by (151)

$$N = N_e + (N_0 - N_e)\operatorname{erf}\left(\frac{x}{2\sqrt{Dt}}\right) \quad 8.92$$

where N_0 is the initial concentration of oxygen in the wafer[17] and N_e is the equilibrium concentration at the heat-treat temperature. One reported value of the diffusion coefficient D is (152)

$$D(T) = 0.17e^{-2.54/kT} = 0.17e^{-29449/T} \quad \text{cm}^2/\text{s} \quad 8.93$$

where T is the diffusion temperature in K.

Before Eq. 8.92 can be used to determine the width of a denuded zone, the concentration below which no precipitation will occur at the processing temperature must be determined. Unfortunately, determining this concentration is not easy since the value depends not only on oxygen supersaturation but also on the amount and kind of grown-in defects, the amount of impurities such as carbon that are present, and the prior heat-treat history. For a given wafer, however, the width should vary as the square root of denude time and exponentially with temperature. Typical plots are shown in Fig. 8.44.

The rate of nucleation depends in a complex way on interstitial

FIGURE 8.44

General behavior of oxygen-denuded region width.

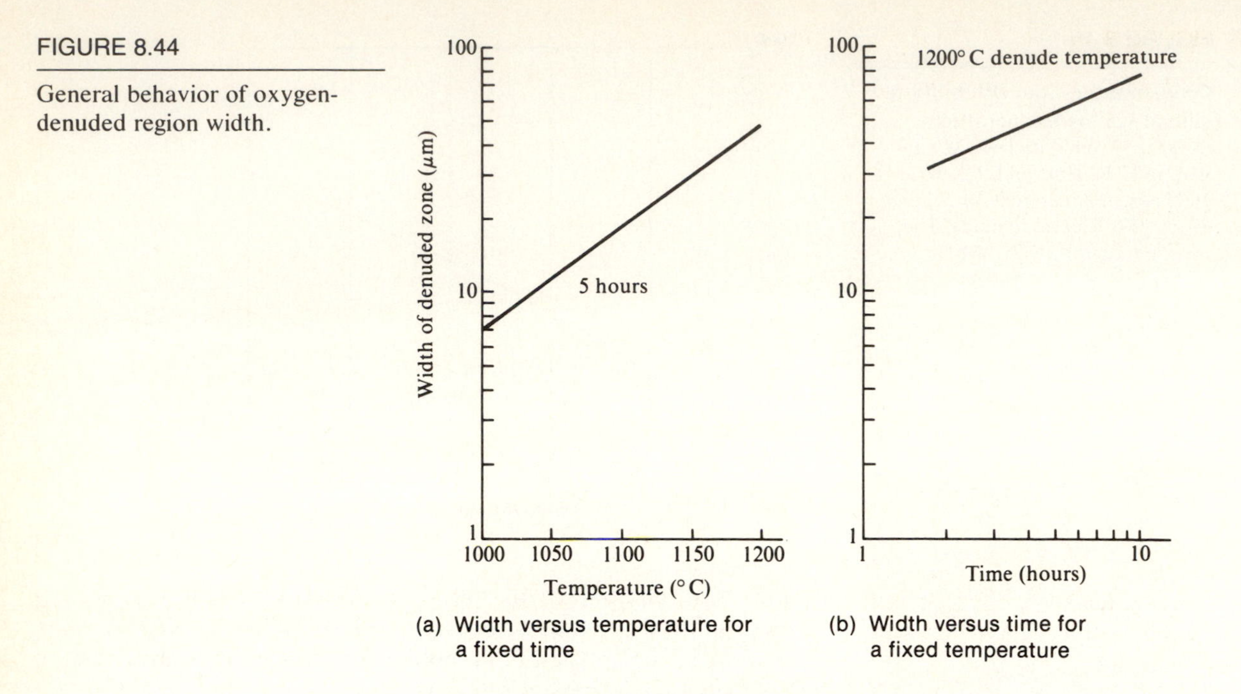

(a) Width versus temperature for a fixed time

(b) Width versus time for a fixed temperature

oxygen concentration, temperature, oxygen diffusivity, dissolution enthalpy, and precipitate interfacial energy (153). A plot of nucleation rate versus anneal temperature for a given oxygen concentration is shown in Fig. 8.45.

In the intermediate temperature range, between low-temperature nucleation and high-temperature dissolution, the nuclei that are above a critical size will grow comparatively rapidly, while the smaller ones will dissolve. Fig. 8.46 shows the trend of critical size versus temperature for two different oxygen levels. Experimentally, it is observed that no optically observable precipitates form for $N_0 < \sim 6 \times 10^{17}$ atoms/cc (154). Apparently, below this concentration, the precipitate nuclei do not reach a size sufficiently large to continue growth when the temperature is increased to the point where

[17]The interstitial oxygen concentration can be measured nondestructively by IR absorption in the 1106 cm^{-1} band. (See, for example, B. Pajot, "Characterization of Oxygen in Silicon by Infrared Absorption: A Review," *Analusis 5*, p. 293, 1977.) Precipitated oxygen absorbs weakly in a different band, so the amount of precipitation can be inferred by the change in interstitial absorption. However, after the first diffusion, the IR absorption due to free carriers becomes so high that no further measurements can be made. By stopping the spectrometer aperture down to a few millimeters, the oxygen profile across starting slices and heat-treated wafers can be determined. Either selective etching combined with optical microscopy or X-ray topography can be used to observe precipitates directly.

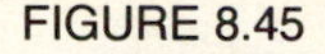

FIGURE 8.45

Nucleation rate versus anneal temperature. (*Source:* Adapted from N. Inoue et al., in H.R. Huff and R.J. Kriegler, eds., *Semiconductor Silicon/81*, p. 282, Electrochemical Society, Pennington, N.J., 1981.)

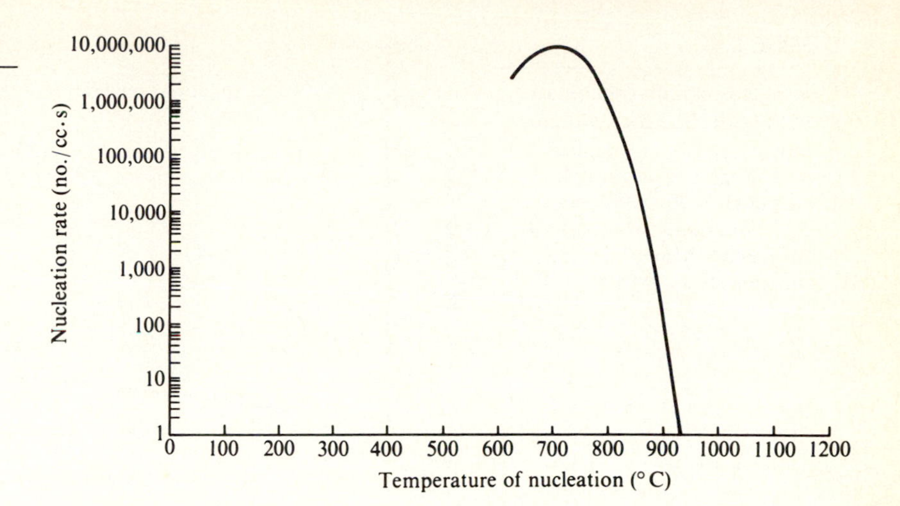

FIGURE 8.46

Effect of temperature on critical radius of oxygen precipitate nuclei. (*Source:* Adapted from R.A. Craven, in H.R. Huff and R.J. Kriegler, eds., *Semiconductor Silicon/81*, p. 254, Electrochemical Society, Pennington, N.J., 1981.)

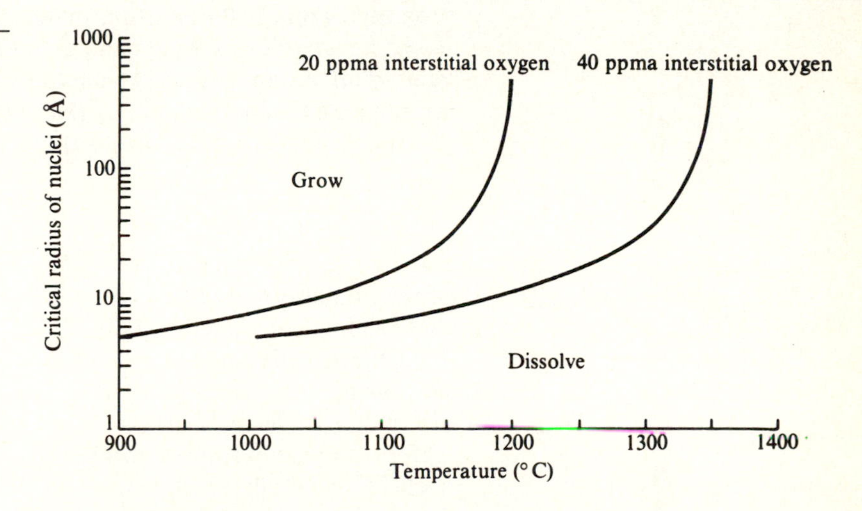

oxygen diffusion is large enough to allow growth. Even if there is sufficient oxygen, if the nucleation cycle time is insufficient for critically sized nuclei to form, or if the temperature is too high to nucleate at all, no precipitate growth will occur. When there are critically sized nuclei, the amount of oxygen precipitated will increase with the number of nuclei, the growth time and temperature, and the oxygen concentration. Fig. 8.47 shows how the precipitate size is thought to increase with time for a fixed temperature.

Because of the complex interactions between the variables involved in intrinsic gettering, it is difficult to experimentally devise an optimum process. In attempts to solve this problem, various computer modeling programs have been devised (155, 156). These

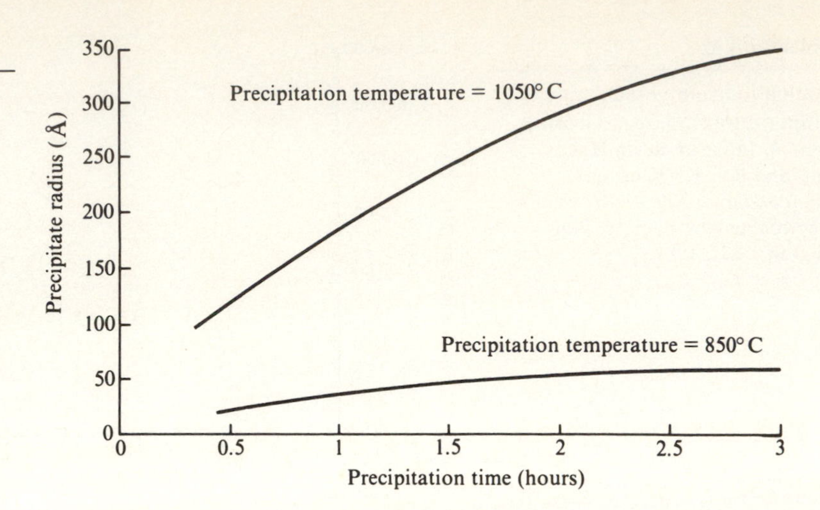

FIGURE 8.47

Calculated oxygen precipitate growth with time for two temperatures. (*Source:* Adapted from B. Rogers et al., in K.E. Bean and G.A. Rozgonyi, eds., *VLSI Science and Technology/84*, p. 74, Electrochemical Society, Pennington, N.J., 1984.)

programs require the starting material's oxygen content and proposed heat cycles as inputs and give the amount of precipitated oxygen as an output. It should be noted that if processing temperatures are reduced to about 800°C, it appears that there will be insuffficent precipitate formation for intrinsic oxygen gettering to be viable.

8.7 DIFFUSION-INDUCED DEFECTS

During the diffusion steps required to produce the desired impurity profile, a number of diffusion-related defects often appear. Some of these defects are listed in Table 8.7.

Chipped edges are likely to occur whenever wafers are handled, but during most processing, the wafers are either lying flat or else are sitting in plastic holders. During diffusion (and oxidation), the wafers are held in slots or grooves in carriers made of hard materials such as fused silica, silicon, or silicon carbide and are thus much more likely to be damaged without additional handling care. Further, the diffusion cycle sometimes produces a dopant-rich glass that flows at diffusion temperatures and collects where the wafers contact the carrier. The wafers may then stick to the carrier after cooldown and be chipped as they are removed from the carrier. Reducing the concentration of dopant reactant in the gas stream will generally prevent the formation of excessive glass.

Many of the gaseous doping compounds used for silicon will react with a clean silicon surface at high temperatures and cause pitting. To prevent pitting, an oxidizing ambient is generally used, and a thin layer of oxide is grown before the doping gases are admitted to the diffusion furnace tube.

TABLE 8.7

Diffusion Process Induced Defects

Item	*Comments*
Chipped wafer edges	Boat slots too small or source buildup on boat causing wafer sticking.
Wafer surface pitting	Flow of Cl- or Br-containing gas across hot wafer before protective oxide formed.
Localized unwanted diffusions	From initial pinholes in masking oxide or from oxide failure during diffusion.
Diffusion pipes	Filamentary diffusion paths connecting active elements (see Chapter 11 for a discussion of their effects).
Lifetime-killing impurities	May be brought in on wafer surface or be transported to hot wafer from contaminated diffusion boat or tube.
Precipitates	Due to choice of cooling cycle combined with an excess of impurity.
Strain-induced dislocations	From excessive concentration of diffusant.
Slip	From thermally induced stresses exceeding yield point of material.
Wafer warpage	Due to massive slip from stresses introduced during diffusion cycle.

8.7.1 Diffusion Flaws and Pipes

Either pinholes already in the masking oxide due to lithography problems or local failure of the mask during diffusion will cause small regions of unwanted impurities. In addition, large-scale mask failure could result from an incorrect choice of oxide thickness. The two most likely causes of local area flaws are phase separation (oxide flowers), which sometimes occurs in spin-on sources, and particulates containing a high concentration of dopant alighting on the wafer surface. Particulates high in dopant concentration may be brought in from the outside[18] or form in the diffusion tube. Phosphorus diffusions are particularly prone to generating products in the tube that cause oxide failures. The end-product of most phosphorus sources is P_2O_5, which tends to coat not only the wafers but the tube as well. The coating forms most heavily on cooler portions of the tube (near the end). Moisture from the air will interact with it while the wafers are being removed and gradually form a syrupy layer that

[18]According to legend, one of the earliest identified sources of unwanted phosphorus diffusions was phosphorus-rich lawn fertilizer tracked into the diffusion room.

can drop off onto the next set of wafers being loaded and subsequently dissolve the masking oxide during the diffusion cycle. Thin spots in the initial oxide due to surface contamination are also potential failure spots. For the case of phosphorus, $POCl_3$ is reported to produce a higher surface concentration and more defects than PBr_3 (157).

When a surface particle contains a higher concentration of dopant than the diffusion provides over the rest of the surface, the diffusion front[19] will be deeper under the particle and thus produce a diffusion "pipe" extending past the rest of the diffusion. If the diffusion covers a crystallographic defect such as a grain boundary, enhanced diffusion will occur along the boundary, as was discussed in section 8.3.12, and will also produce a diffusion pipe extending out from the main diffusion front. (The effect of pipes on devices is discussed in Chapter 11.)

8.7.2 Lifetime-Reducing Impurities and Precipitates

Contamination of the furnace by lifetime-reducing impurities (sometimes referred to as heavy metals or as transition metals) and methods of minimizing it were discussed in Chapter 3. Also discussed were various cleanup procedures designed to ensure that a clean wafer surface enters the diffusion or oxidation tube. Unfortunately, such extensive precautions do not always provide the required low level of contamination. Thus, various gettering steps are usually used during high-temperature processing. These steps were discussed in detail in the previous section. In some cases, however, the deliberate use of lifetime "killers" is necessary. The lifetime-reducing impurities usually have a much higher solid solubility at diffusion temperature than at room temperature so that if their concentration is high at diffusion temperature, either from intentional or unintentional doping, precipitation may occur during cooldown. Precipitates that are in space charge regions will cause excessive leakage (see Chapter 11). Precipitates near the surface can cause a thinning of thermal oxide grown over that region, and long conductive precipitates may even short elements together. Precipitate prevention is done either by having previously reduced the concentration below the precipitation level or by cooling the wafer so rapidly that there is not enough time for the impurities to diffuse to precipitate nuclei and cause growth. This method is used, for example, with the gold doping of a TTL circuit. Unfortunately, however, too rapid a cooldown can cause slip in the wafers, as will be discussed in section 8.7.4.

---

[19]Some chosen isoconcentration line is often referred to as a "front."

8.7.3 Strain-Induced Dislocations

Since dopant atoms are almost never close in size to the atoms they displace, in high concentrations, they can produce enough strain to generate misfit dislocations (158–161). If Vegard's law is followed,[20] the volume strain β is given by $(V_i - V)/V$, where V_i is the new volume after introduction of impurities. This can be rewritten as

$$\beta = \frac{\Delta V}{V} = f(\Gamma^3 - 1) \qquad 8.94$$

where ΔV is the change in volume, V is the original volume, and f is the atomic fraction of impurity present (N/N'), with N as the concentration of impurity atoms/cc and with N' as the number of host atoms/cc. Γ, the misfit ratio, is the ratio of the impurity covalent radius to that of the host. The linear strain ε is given by[21] the expression $[(V + \Delta V)/V]^{1/3} - 1$, or

$$\varepsilon = \frac{\Delta \ell}{\ell} = [1 + f(\Gamma^3 - 1)]^{1/3} - 1 \qquad 8.95$$

where ℓ is a linear dimension. Eq. 8.95 can be approximated by $f(\Gamma^3 - 1)/3 = \beta/3$.

If the new atom is smaller than the host, then the lattice will contract and ε will be negative. If the atom is larger, then the lattice will expand. When an impurity atom enters the lattice interstitially rather than substitutionally, the lattice will expand regardless of the relative radii, and Eq. 8.95 is not applicable. Table 8.8 lists some values of covalent radii and misfit ratios. Fig. 8.48 shows experimental strain data for boron, antimony, and phosphorus in silicon and compares them with the predictions of Eq. 8.95. It should be noted that as the strain increases, the bandgap will also change, causing n_i to change and, in turn, affecting D for cases where $N > n_i$ (162). Since the strain is in a thin layer on one surface, it will cause the wafer to bow, and indeed one method of estimating the amount of misfit-induced strain is to measure the amount of bowing on wafers whose elastic limits have not been exceeded. Bowing will not occur if a balanced diffusion occurs on the opposite side of the wafer, but most processes keep the back side protected by a thick oxide during diffusion.

TABLE 8.8

Covalent Radii

Element	*Radius (Å)*	*Misfit Ratio*
C	0.77	0.66
B	0.88	0.75
P	1.10	0.94
Si	**1.17**	**1.00**
As	1.18	1.01
Ge	1.22	1.04
Ga	1.26	1.08
Al	1.26	1.08
Sb	1.36	1.16
Sn	1.40	1.23

Source: Linus Pauling, *The Nature of the Chemical Bond,* Table 7–13, Cornell University Press, Ithaca, New York, 1960.

[20] Originally proposed for ionic crystals, Vegard's law is almost never followed for metallic solutions but, for at least some dopants in silicon, appears to apply reasonably well (see Fig. 8.48).

[21] The symbols σ and ε or T and S are commonly used for stress and strain, respectively. In this discussion, σ and ε will be used even though a conflict exists with earlier symbol definitions.

FIGURE 8.48

Linear strain versus amount of impurity. (*Source:* Adapted from G. Celotti et al., *J. Mat. Science* *9*, p. 821, 1974; and F.H. Horn, *Phys. Rev. 97*, p. 1521, 1955.)

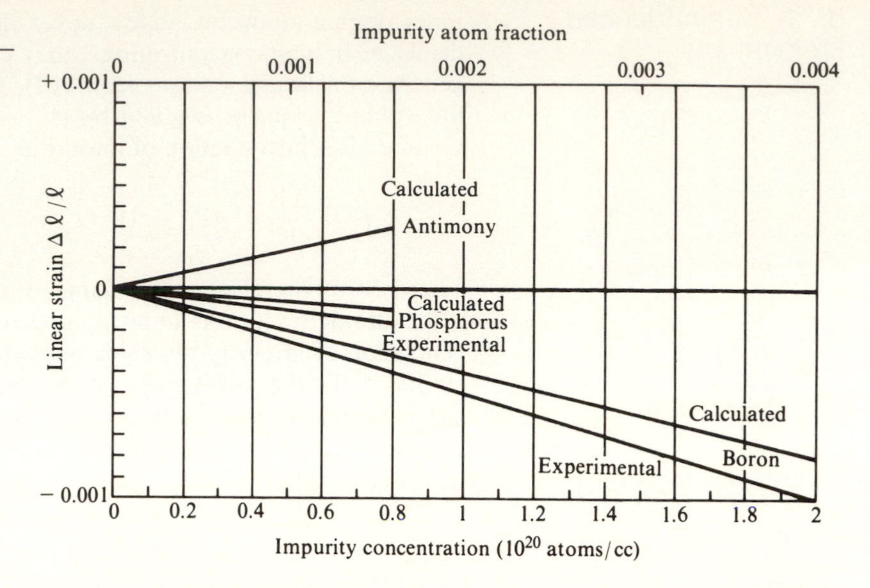

Analyses of the stresses involved as a function of depth and impurity profile are available (159), with the maximum stress σ_s occurring at the surface just as diffusion starts and given by

$$\sigma = \frac{Y\varepsilon}{1 - \nu} \qquad 8.96$$

or

$$\sigma = \frac{\beta Y}{3(1 - \nu)} \qquad 8.97$$

where Y is Young's modulus and ν is Poisson's ratio. The fraction of impurity at the surface and the f to be used in calculating β for Eq. 8.97, are given by $f = N_0/N'$ where N_0 is the surface concentration. Because of the stress relief afforded by dislocation production, the highest density of dislocations generally occurs not at the surface but somewhat below it.

EXAMPLE ☐ Estimate the value of σ, and decide whether or not dislocations will be induced from a 1100°C boron diffusion with a surface concentration of 10^{19} atoms/cc.

From Fig. 8.48, the strain induced by a boron diffusion with a concentration of 10^{19} atoms/cc is about 5×10^{-5}. Young's modulus and Poisson's ratio are both orientation and temperature dependent. As typical values, use $Y = 10^{12}$ dynes/cm^2 and $\nu = 0.3$. Thus, the stress σ is approximately 7×10^7 dynes/cm^2. Fig. 8.54, shown in the next section, gives the stress required for dislocation generation to become noticeable in silicon, and from it, one concludes that dislocations will probably not be generated. However, if a surface concentration of 10^{20} is used, ε becomes 0.0005, T increases to 7×10^8 dynes/cm^2, and from Fig. 8.54, at 1100°C, generation dislocation is very likely. □

High-concentration phosphorus diffusions, such as might be used for bipolar emitters, can cause widespread slip to occur in wafers. Its occurrence is apparently associated with the onset of precipitates (163). Data are shown in Fig. 8.49 for a $POCl_3$ source diffusion into (111) silicon at 1100°C. Background doping is 2 Ω-cm boron, and the junction depth is approximately 2 μm. Experimentally, it is observed that diffusion pipe losses increase drastically when there is heavy dislocation damage in the emitter. Hence, from the data of Fig. 8.49, the phosphorus emitter diffusion sheet resistance for the described set of diffusion conditions should be kept above about 4 Ω/sq. Since higher concentrations of phosphorus afford better gettering and usually give better emitter efficiency, the process engineer must balance these gains against losses from increased leakage due to the presence of excessive dislocations.

The strain introduced by a small atom can be compensated by simultaneously adding a larger one. Thus, germanium or tin, which would not be expected to appreciably affect the electrical properties of silicon, can be used to eliminate strain during a boron or phosphorus diffusion (164–167). Such a procedure will not, however, solve the problem of precipitates forming and producing additional mechanical damage. Arsenic atoms are closely matched to silicon

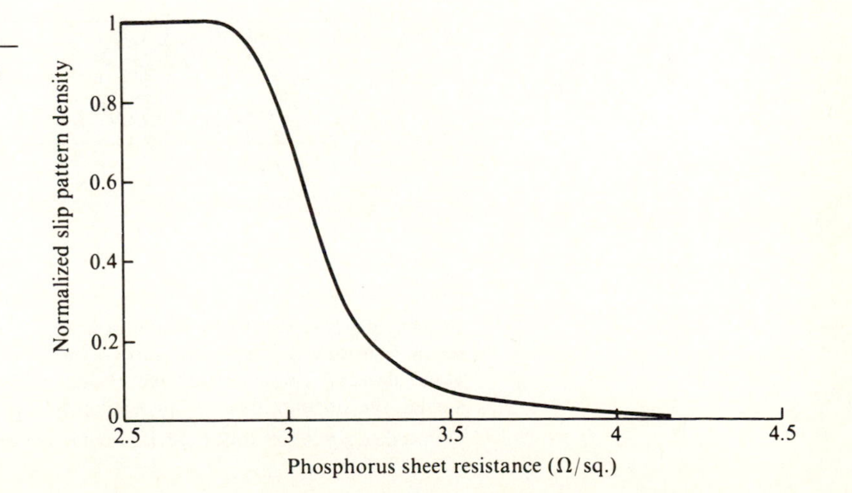

FIGURE 8.49

Slip density versus phosphorus diffusion sheet resistance for an 1100°C diffusion.
(*Source:* Adapted from R.A. McDonald et al., *Solid-State Electronics 9*, p. 807, 1966.)

and hence produce few misfit dislocations, but like the others, they do form precipitates at high doping levels.

8.7.4 Permanent Wafer Warpage

The most common cause of wafer warpage is excessive thermal stresses occurring during wafer insertion and removal from diffusion or oxidation tube hot zones. The high thermal stresses are often enough to produce plastic flow and thus permanent deformation[22] of the wafer (168–172). These thermal stresses arise primarily because as the wafers are inserted or removed from the hot furnace, heating or cooling must be produced from the outside in as shown in Fig. 8.50. This effect can produce a transient temperature differential between the center and edge of the wafer of up to a few hundred degrees. In addition, the thermal mass of the boat[23] will prevent it from changing temperature as rapidly as the wafers and will thus keep the portion of a wafer in contact with it from heating or cooling as fast as the rest of the wafer. Fig. 8.51 shows some experimental data for 100 mm diameter wafers, taken radially along a direction not influenced by boat contact. The temperature depends on the wafer diameter and thickness, the spacing between wafers, and the boat design. The slower the insertion or withdrawal rate, the less the temperature differential, so that by decreasing the rate, ΔT can be reduced to any desired level. Since it is difficult to reproducibly withdraw wafers manually, programmable "push–pull" mechanisms are used. As can be seen from Fig. 8.51, lower furnace temperatures produce less temperature difference; therefore, as an alternative or sometimes as an adjunct, the furnace temperature can be gradually reduced or increased (ramped).

Aspects other than thermally induced damage may need to be

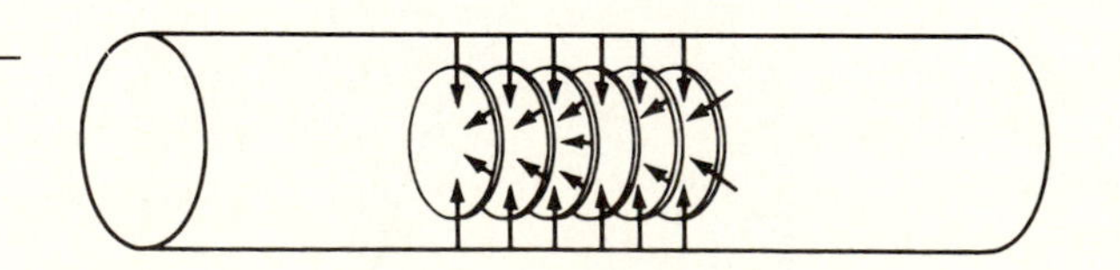

FIGURE 8.50

Heating from periphery because of external cylindrical heat source.

[22]This defect is separate from wafer bowing caused by mechanical surface damage, strain from low-concentration diffusions, or stresses from interconnect layers. Those defects seldom exceed the elastic limits of the silicon, so if they are removed, the silicon returns to its original shape.

[23]For a description of wafer carrier boats, see section 8.8.

considered in choosing push–pull or ramp rates, and the final choice must be based on the total effect. Slow withdrawal will allow precipitation if there is a high concentration of an impurity such as gold, which has a high diffusion coefficient and a low room-temperature solubility. When the gold is needed to kill lifetime, a slow withdrawal may be harmful in that it will allow the gold to precipitate rather than remain in the lattice interstitially. Even if a large amount of such an impurity is not present, a slow cool may allow it to diffuse to crystallographic defect sites, decorate them, and cause device failures. The processing window becomes progressively narrower as the wafer diameter increases, and it becomes quite difficult to choose a withdrawal rate for 150 mm wafers that will prevent thermally induced slip and yet not allow small haze-like precipitates to form on the surface. The electrical characteristics of a silicon–oxide interface depend on the kind of ambient used during the cooling cycle as well as on the rate. When a push–pull system is used, maintaining the desired ambient may be difficult, so the combination of a ramp-down to a low enough temperature to minimize slip, followed by a rapid withdrawal, may be the best compromise.

In order to estimate the temperature differential that can be tolerated, the strain due to the thermal expansion can be converted into mechanical stress, and the stress can be compared with the stress required for plastic flow at the temperature in question. With a radially symmetric temperature gradient from the center to the edge of a wafer, both radial and tangential stress will develop. The tangential stress will change sign in going from center to edge and will have a substantially larger absolute value at the edge than at the center. The radial component, equal to the tangential component in the center of the wafer, decreases to zero at the edge (169–172). An example of calculated stress, using the radial temperature distribution of Fig. 8.51, is shown in Fig. 8.52 (172). Since the maximum stress is at the periphery, that is where slip would be expected to begin. However, if the yield point of the silicon is less near the center than near the edge, then slip could occur elsewhere.

To evaluate the likelihood of plastic flow (slip), the applied stress must be resolved into shear stress lying in the slip plane and directed in the slip direction. For silicon, slip predominantly occurs between (111) planes in a [110] direction. The resolved shear stress is given by

$$\sigma_{\text{shear}} = \sigma_0 \cos \alpha \cos \beta \tag{8.98}$$

where σ_0 is the stress being resolved, α is the angle between σ_0 and the normal to the slip plane, and β is the angle between σ_0 and the slip direction.

EXAMPLE ☐ When σ_0 lies in the (100) plane with a [012] direction, what is the shear stress in the (111) plane in the [110] shear direction?

Using the expression

$$\cos\theta = \frac{hh' + kk' + \ell'}{\sqrt{(h^2 + k^2 + \ell^2)(h'^2 + k'^2 + \ell'^2)}}$$

where $hk\ell$ and $h'k'\ell'$ are the indices of any two planes, let $hk\ell$ = 012 and $h'k'\ell'$ = 111 to solve for cos α and $hk\ell$ = 012 and $h'k''$ = 110 to solve for cos β. Note that in the cubic system, the direction normal to a $(hk\ell)$ plane is $[hk\ell]$. Alternatively, spherical coordinates of each of the direction vectors could be calculated with respect to the wafer surface (168). This allows for an easier choice of azimuthal angle of the applied stress. ☐

If the resolved shear stress is greater than the critical shear stress for the wafer material, then slip will occur. The radial dependence of the resolved stress (Eq. 8.98) is such that for (111) oriented silicon, 12 stress maxima occur. It is at those points on the periphery that slip is normally first seen, although, often, every other one will be much more pronounced and thus give the familiar six-point star pattern. On (100) silicon, there are 4 maxima. Fig. 8.53 shows a typical (111) slip pattern after being delineated by etching.

FIGURE 8.51

Temperature differential between center and various points along radius r of 100 mm wafer while load of wafers is inserted and withdrawn from hot diffusion tube. (T_1 is the temperature of the tube. The curves in part a are for insertion/removal rates of 100 mm/minute; the curves in part b, for 500 mm/minute.)
(*Source:* A.E. Widmer and W. Rehwald, *J. Electrochem. Soc. 133*, p. 2402, 1986. Reprinted by permission of the publisher, The Electrochemical Society, Inc.)

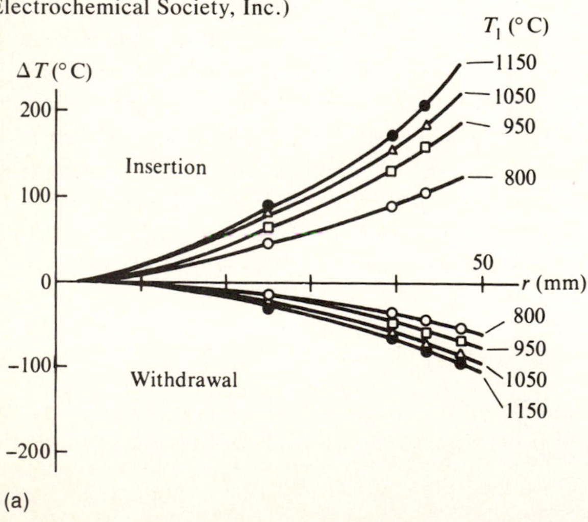

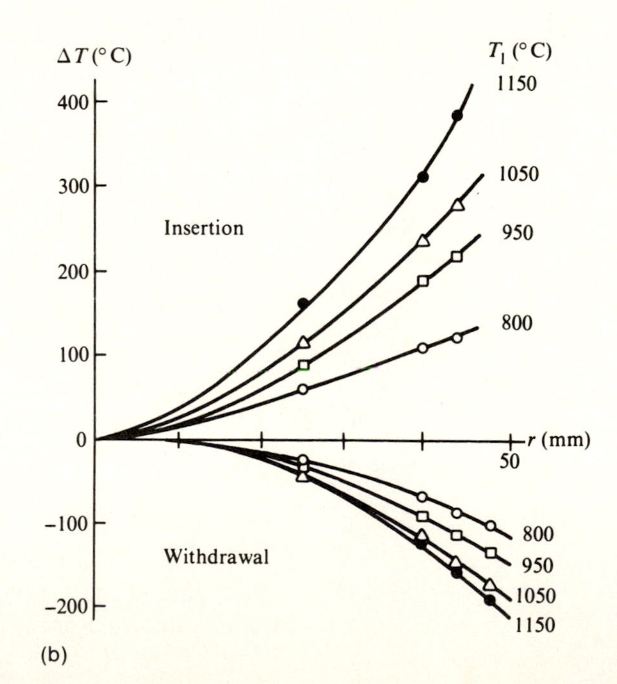

FIGURE 8.52

Calculated radial and tangential stresses induced in 100 mm silicon wafer when withdrawn from 1050°C furnace at rate of 50 cm/minute. (*Source:* A.E. Widmer and W. Rehwald, *J. Electrochem. Soc. 133*, p. 2403, 1986. Reprinted by permission of the publisher, The Electrochemical Society, Inc.)

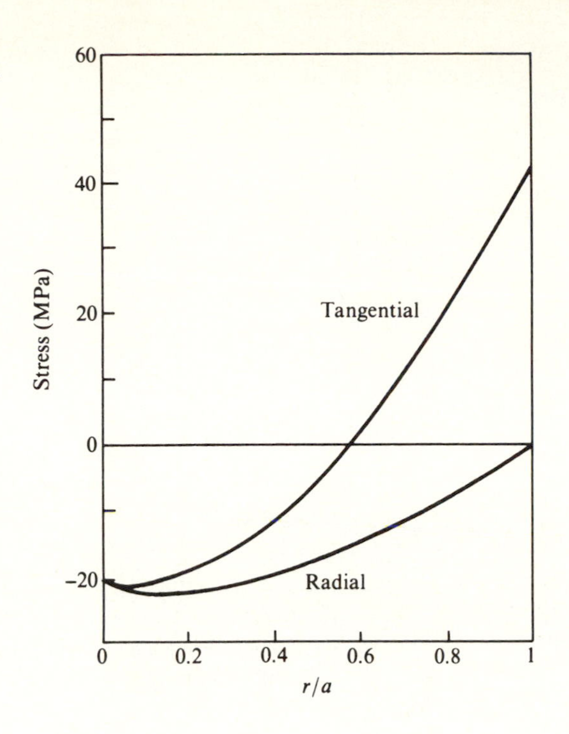

FIGURE 8.53

A (111) oriented silicon wafer with severe slip introduced by a large thermal gradient. (The slip has been delineated by etching to show the emergence of dislocations.)

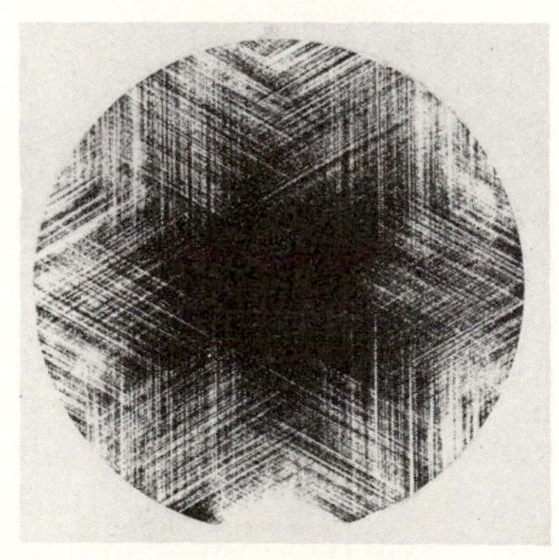

The value of the critical shear stress for a given material is quite variable and is still well described by a statement in A.H. Cottrell's book *Dislocations and Plastic Flow in Crystals*[24] written about 40 years ago: "Precise values for individual crystals have little significance apart from their order of magnitude because the critical stress not only depends upon the temperature and speed of straining but is also very structure sensitive. Variations in the purity of the material, the conditions of growth and the state of the surface of a crystal all greatly affect its shear strength." Higher temperatures reduce silicon yield strength in the general manner shown in Fig. 8.54 (173–175). Increasing the strain rate somewhat increases yield strength at a given temperature, thus making values dependent on the strain rate used by the measuring instrument. Interstitial oxygen increases silicon strength, while precipitated oxygen decreases it (176–180). Fig. 8.55 shows the general behavior. It should be noted that while the higher concentrations of interstitial oxygen enhance strength, they also have the potential to precipitate more during oxidation and diffusion cycles and thus reduce the strength below the level observed with no oxygen. Nitrogen, in concentrations of about 10^{15}

[24] Oxford University Press, London, 1953.

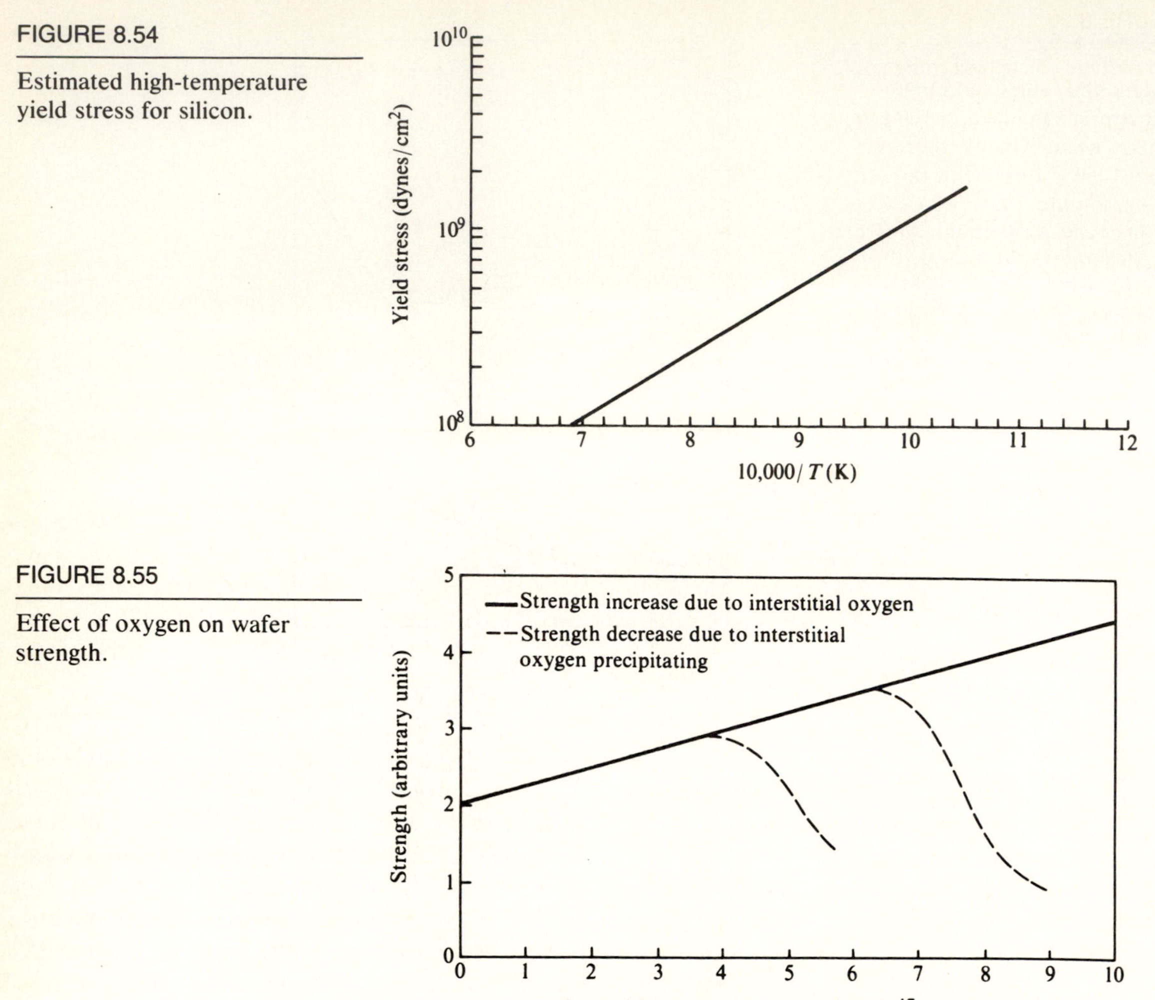

FIGURE 8.54

Estimated high-temperature yield stress for silicon.

FIGURE 8.55

Effect of oxygen on wafer strength.

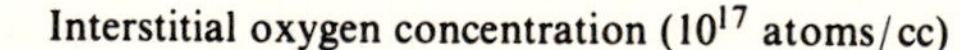

atoms/cc, and apparently interstitial, will increase yield strength and seems less likely to precipitate than oxygen (181, 182).

The wafer photographed for Fig. 8.53 was intentionally subjected to very severe thermal stresses in order to produce an exaggerated effect. In good IC processing, no discernible slip occurs. Marginal processing, however, produces a number of distinctive slip region patterns. Fig. 8.56 shows the shape of these patterns. They range from slip over the whole wafer (very rapid removal) to no slip. Reject chips very closely follow the pattern of slip, and the relation between a reject map and the slip pattern was used in early experiments to establish the correlation between slip and yield loss. The

FIGURE 8.56

Slip patterns observed in processed wafers. (The cross-hatching denotes area of high slip.)

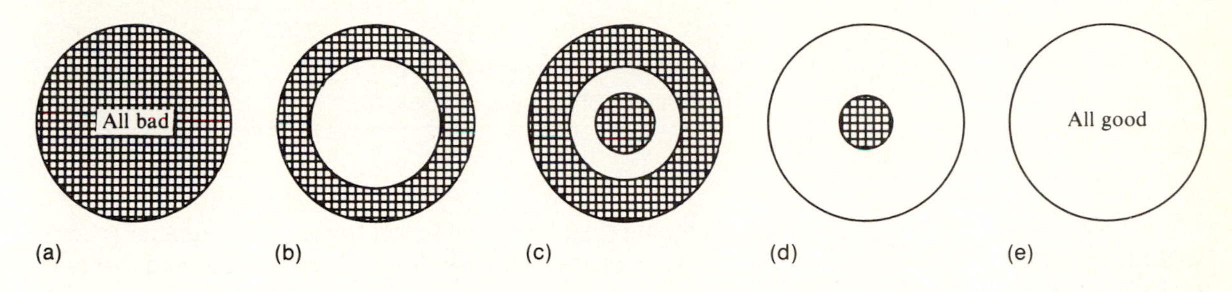

pattern in part b occurs when the thermal differential is marginal. The ring of unslipped material in the pattern in part c apparently occurs from a combination of boron and oxygen precipitation in the center of the wafer, which reduces the yield point. The pattern in part d is on a wafer that did not have enough temperature differential to cause peripheral slip but that had enough oxygen precipitation in the middle to cause a reduction in yield strength and subsequent slip. Such central precipitation is usually associated with oxygen coring, in which a much higher than normal concentration of oxygen is incorporated into the central region of the crystal during growth.

8.8 DIFFUSION EQUIPMENT

Diffusions are carried out in diffusion furnaces that are comprised of a fused silica tube surrounded by a heater element. Temperatures are in the 900°C–1300°C range, with the trend being to lower temperatures. Historically, the tubes have been horizontal with the wafers held vertically. However, vertical-tube furnaces with the wafers held horizontally are also available. The diameter of the tube is a few centimeters larger in diameter than the wafers to be processed. Table 8.9 shows typical values. The length of the flat part of the furnace hot zone is about 100 cm and is heated by elements broken into at least three segments so that the two end zones can operate at higher power to compensate for heat losses out the ends of the tube. Most of the temperature controllers use microprocessors with appropriate control algorithms to match the thermal properties of the furnace. Some trade-offs must be made in the thermal design. Too little insulation will require excessive power, while too much insulation will cause cooldown time to be excessive. Attached to one end of the furnace proper is a gas cabinet to house controls for the gases (source cabinet).

TABLE 8.9

Diffusion Tube Diameters

Wafer Diameter (mm)	*Tube Diameter (mm)*
100	170/176*
125	184/190
150	215/224 or 225/235
200	250/275

*Inside diameter/outside diameter (ID/OD).

Source: Data courtesy of Thermco Systems, Inc.

8.8.1 Gas Flow Measurement and Control

In older plumbing systems and in laboratories, the rotameter is commonly used to measure gas flow rates. It is a simple and virtually foolproof instrument but is not very amenable to remote reading. It consists of a vertical glass tube with a tapered bore that linearly increases in diameter from bottom to top. Inside is a round ball that is free to move up and down and that fits rather tightly at the bottom of the tube. As the gas flow rate through the tube increases, the ball (float) will move up the tube, and its vertical position is a measure of flow.

Electronic mass flow meters depend on the fact that the temperature differential up and downstream from a heat exchanger supplying a constant amount of heat per unit time to the gas stream is proportional to the mass of gas per unit time flowing through the exchanger. The temperatures are sensed by resistance thermometers in a bridge network, and the difference is expressed in terms of a DC voltage. After multiplying by constants appropriate for the gas being measured, the voltage swing corresponding to zero to full flow is usually 0–5 V. This voltage can be read directly on a voltmeter, used to control the gas flow, and/or fed into an A/D converter for a computer input.

The gas flow can be varied by simple, manually operated needle valves, which are usually the valves used in conjunction with rotameter flow measurement. However, for automatic control, more sophisticated valves are required. They are still based on the needle valve concept but use piezoelectric elements or electrically heated thermal expansion elements to move the needle in and out of its seat. Combining the output of a mass flow meter with the input to an electrically operated valve provides an automatic mass flow controller that is widely used to control diffusion gas flows. On/off control is either by manual cutoffs or by air- or solenoid-operated valves. Solenoid valves do not supply as much closure force as those that are air operated. Thus, more problems with leakage may occur when they are used. When a valve is chosen, care must be taken to ensure that the valve seal is compatible with the gas being controlled.

8.8.2 Diffusion Tubes

Historically, the diffusion tube (the inside tube in a furnace, and the one containing the wafers and carrying the process gases) has been made of clear fused quartz.[25] Sometimes, between the diffusion tube and the furnace winding is a thick-walled tube, or liner, which is generally made of opaque fused SiO_2. The liner offers protection for the more expensive inner tube, provides additional thermal mass, and minimizes the propensity of the inner tube to sag at temperatures above about 1150°C. Problems with the fused quartz that have led to a search for alternative materials are the high-temperature sagging, the gradual devitrification with use at high temperature, and the relative ease with which some impurities can penetrate the quartz. Vapor-deposited polycrystalline silicon and sintered silicon carbide are two alternative materials that are partially replacing fused quartz. Both are now available in high purity, sag less at elevated temperature, and do not devitrify. They are, however, more expensive. Since both Si and SiC will react with metals at diffusion temperature, care must be taken to ensure that the heating coils do not become distorted and touch either of them; otherwise, alloying will take place, and both the heater and the tube will be ruined. Si and SiC are quite conductive at diffusion temperatures; thus, even if no reaction occurs, arcing between turns will occur if the heater and tube touch. In the case of SiC, this possibility can be minimized by using a tube coated with a nonconductor such as zirconia.

The inside walls of fused quartz tubes will react enough with the doping atmosphere to become a source. Thus, to prevent cross contamination, different tubes should be reserved for each of the dopants being used. The buildup of dopant on the tube walls will act as an additional source and may affect the diffusion.[26] A few dummy runs may be required to saturate the tube before reproducible results are obtained. Therefore, a tube also may not provide reproducible diffusions if it is used at two widely differing temperatures.

8.8.3 Wafer Diffusion Boats

The materials used for tubes—that is, SiO_2, Si, and SiC—are also used for the boats. The same care must be used in keeping separate

---

[25]Some usage refers to synthetically prepared SiO_2 that has been melted and fabricated as fused silica and to naturally occurring quartz that has been melted and fabricated as fused quartz. The terms are also sometimes used interchangeably. There are some differences in physical properties, with fused naturally occurring quartz generally having a slightly higher viscosity than the synthetic material.

[26]Sometimes the buildup in a single run is enough to cause the wafer boats to stick to the tube.

boats for each process as was used for tubes. In addition, cross contamination can also occur during the boat cleaning operation if boats from separate diffusion steps are etched in the same cleaning solution. Diffusion boats are designed to impede flow as little as possible and to have minimal thermal inertia.

8.8.4 Wafer Insertion

The original manual method for placing wafers in a conventional horizontal-tube furnace was to load each wafer individually into a slotted fused silica boat by tweezers and then slide the boatload of wafers into the furnace with a push rod. Difficulties with this procedure are that it is slow, tweezers introduce lifetime-killing contaminants and mechanical damage, the sliding boat generates particles, and the insertion and withdrawal rates are ill-defined. The cleanup step that precedes diffusion is almost always done with a number of wafers (25 or less, depending on diameter) in a carrier. Usually, the carrier is designed to withstand the cleaning solutions and is plastic. If the spacing of wafers in the cleanup boat matches the spacing for diffusion boats, conventional "flip–transfer" (dump–transfer) can be used to load the diffusion boat and eliminate the tweezer step. When spacings of the two boats are different, a slice-by-slice unload and reload machine can be used, but the slice pickups used can introduce damage or contamination. To provide control over insertion and withdrawal rates, paddles of fused silica or silicon carbide that either slide or ride on rollers driven by a "boat pusher" were introduced to carry the diffusion boats into the furnace. Since such sliding or rolling produces many defect-generating particles, cantilever boat loaders were developed (183, 184). The boat is supported on the end of a long arm that moves the wafers and boat(s) into the furnace without touching the walls. The arm may then set down the boat and withdraw, or it may remain in the furnace for the duration of the diffusion.

Vertical-tube furnace loading varies somewhat from one manufacturer to the next, but wafers are generally automatically transferred from cleanup carriers to a special fused quartz carrier that is part of the furnace.

8.8.5 Computer Control of Tube Functions

Studies made before the introduction of computer-controlled furnaces showed that most yield loss came from human error and that the numerical losses were several yield points. Some of these errors are listed in Table 8.10. The industrywide desire to eliminate such errors became the driving force for computer control. Several levels of control can be exercised, as shown in Fig. 8.57. The first level, usually done by an on-board microprocessor, is used to control the time sequencing of the various operations, the gas flows, temperature, temperature ramping, and wafer insertion and withdrawal rates. The next level, also usually done by the local microprocessor,

TABLE 8.10

Major Causes of Diffusion Process Associated Yield Loss

1. Lot to wrong diffusion tube.
2. Error in selection of sequencing.
3. Wrong temperature or flow set.
4. Push–pull rates in error.
5. Boat loaded into wrong part of heat zone.
6. Lot skipped cycle or double cycled.
7. Tubes and boats not cleaned on proper schedule.
8. Failure to find plumbing leaks or cracked tubes.
9. Liquid sources run dry during processing.

includes things like tube diagnostics and data collection. Some examples of tube diagnostics are given in Table 8.11. Data collection includes records of specified and actual flow rates and temperatures.

The remainder of the functions shown in Fig. 8.57 are usually done by a higher-level (host) computer. Without computer control, one of the main causes of misprocessed diffusion lots is either the skipping of a diffusion or giving a lot the same diffusion twice. Thus, a system that keeps track of a given lot's correct location and gives a warning when it is logged into an incorrect step is very useful and is a natural adjunct to a lot tracking system. If control is totally local, a number of recipes in memory can be called up by an operator as required when a lot is started into a particular tube. With a host computer, the specifications for each device type at each furnace operation can be kept on file so that when a particular lot number is entered, the proper settings are automatically made. Automatic checks are also available to ensure that the lot is put in a tube compatible with the lot requirements. Finally, as part of an overall line-balancing program, lots can be automatically scheduled to particular tubes in order to maximize utilization and minimize cycle time.

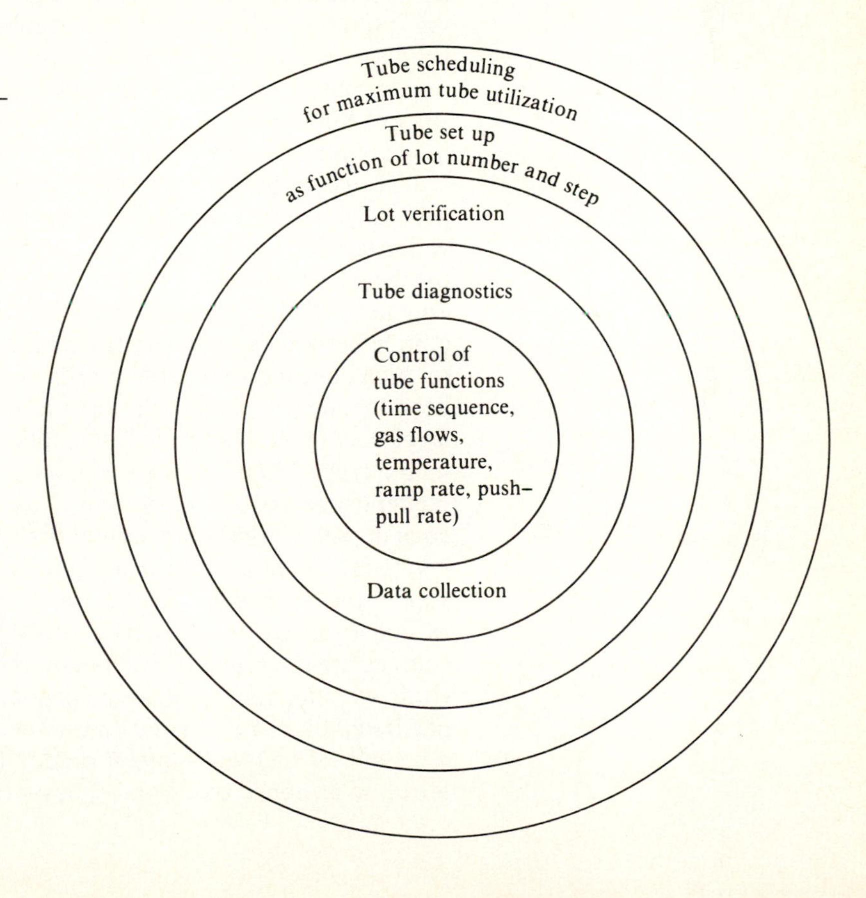

FIGURE 8.57

Levels of computer control of diffusion tubes. (The functions in the two inner circles can be handled locally; the outer two levels require a host, or supervisory, computer.)

TABLE 8.11

Tube Diagnostics

Item	*Approach*
Plumbing leak	Close appropriate valves; measure residual gas flow with sensitive flow meter.
Furnace health	Check status of cooling water flow and temperature, air flow, line voltage, and so on.
Tube preconditioning	Check time since tube last used; cycle source if required.
Tube cleaning	Check time since last tube cleaning; give warning; prevent start-up if warning time exceeded.
Thermocouple drift	Will vary with control system used.
Mass flow meter clogging	Sequence valves to put two meters in series; compare readings.

8.9 DIFFUSION SOURCES

Diffusion processes can be broadly categorized as open tube or closed tube. In the first catagory, the diffusion tube is operated at atmospheric pressure and a carrier gas flows through the tube during the diffusion operation. In closed-tube diffusion, the wafers and a solid that will provide impurity atoms for the diffusion are sealed in an evacuated ampule. Closed-tube diffusions are seldom used for silicon, but since they allow the vapor pressure to be controlled, they are often used for gallium arsenide.

In open-tube systems, the final source for silicon diffusions is generally a mixed oxide layer of silicon and the diffusant on the wafer surface. As shown in Fig. 8.58, as diffusion progresses, the oxide–silicon interface moves to the right. If the concentration of dopant in the mixed oxide is relatively low, the thermal oxide will remain between silicon and the mixed oxide (Fig. 8.58c). However, for high concentrations, the thermal oxide will be dissolved, and the doped oxide will contact the silicon surface (Fig. 8.58b). The doping layer may be derived from the reaction of a gaseous dopant species with oxygen and the thermal silicon oxide grown during diffusion or predeposited on the wafers before loading into the furnace. The gaseous dopant can be from an external gaseous source such as diborane (BH_3), from a vaporized liquid source such as BBr_3, or from a vaporizing solid in the furnace, such as an upstream boat of P_2O_5 or an adjacent wafer of boron nitride. Examples of predeposited sources are the separate low-temperature CVD deposition of silicon oxide with doping impurities incorporated into the oxide and the use of a liquid that when coated on the surface and allowed to dry leaves a residue of SiO_2 and dopant oxide. These sources are usually referred to as doped oxide and spin-on (sometimes paint-on) sources,

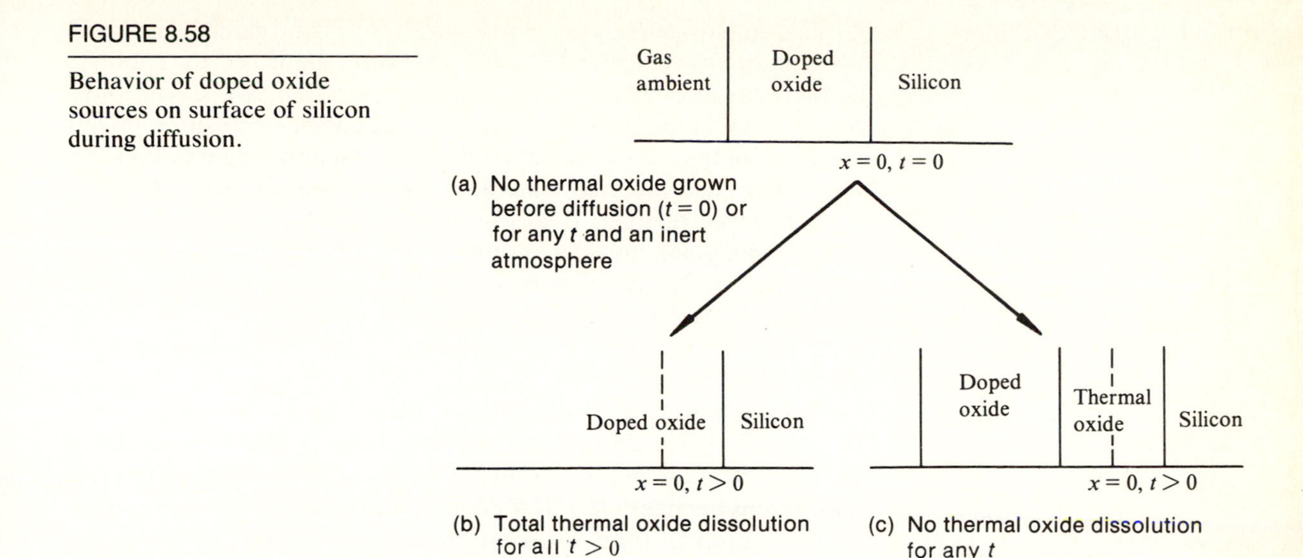

FIGURE 8.58

Behavior of doped oxide sources on surface of silicon during diffusion.

respectively. Since the growth of an additional oxide on the wafer surface is not required with these sources, they can, in principle, be applied equally well to either silicon or gallium arsenide. Indeed, the doped oxide concept was first applied to gallium arsenide.[27]

The choice of the kind of source depends somewhat on personal preference, but important factors are the availability and purity of source material, the ease of use, the doping uniformity, the amount of damage introduced, and the surface concentration control. $POCl_3$ and BBr_3 are reasonably satisfactory liquid sources and are widely used. Liquid As and Sb sources have generally not worked well, and doped oxide sources are more common. For cases where maximum concentration is not desired, doped oxide sources are more applicable. Solid wafer sources are probably the most convenient to use but are not available for all dopants.

The source cabinet alluded to earlier that is part of the furnace console usually contains controls for all of the gases used in the furnace. In addition, if the source of the dopant is a liquid that must be vaporized prior to being introduced into the furnace, the liquid reservoir and vaporizer will be in it as well.

[27]There is little application for diffusion sources in the fabrication of gallium arsenide ICs. Ion implantation instead of a predep diffusion is normally used. Diffused junctions are used in discrete devices, however.

8.9.1 Liquid Sources

Typical liquid sources for silicon are boron tribromide (BBr_3) and phosphorus oxychloride ($POCl_3$). They are vaporized by bubbling an inert gas through the liquid. The concentration can be controlled by the liquid's temperature, which sets the vapor pressure, flow of gas through the liquid, and amount of dilution after vaporization. An alternative, and one that eliminates the problem of incomplete saturation in the bubbler, is to use mass flow meters to measure the ratio of gas going into the bubbler to the amount of dopant picked up. The input flow can then be varied as necessary to give the proper amount of dopant in the final input stream. Fig. 8.59 shows a diagram of a typical liquid source using the mass flow control (MFC). MFC–2 controls the total flow of nitrogen through the tube. The combination of MFC–3 and MFC–4 allows the amount of vapor to be controlled. The dopant can be vented until its flow has stabilized. Valves isolate the bubbler or bypass it if gas flow is needed to clear liquid from the lines. The flow constrictor is used to provide enough pressure drop to force nitrogen through the bubbler. The oxygen input is required to react with the incoming source gas and prevent pitting of the silicon surface.

The reactions of BBr_3, which has a boiling point of 90°C, are assumed to be

$$2BBr_3 + \text{heat} \rightarrow 2B + 3Br_2$$

$$4B + 3O_2 \rightarrow 2B_2O_3$$

$$B_2O_3 + SiO_2 \text{ on Si surface} \rightarrow \text{Mixed oxide on surface}$$

$$2B_2O_3 \text{ in mixed oxide} + 3Si \rightarrow 4B + 3SiO_2$$

$$\text{Excess B} + Si \rightarrow SiB_4 \quad \text{or} \quad SiB_6$$

If the silicon surface does not have some protective oxide, the bromine can etch it at diffusion temperature; if too much free boron is formed, it reacts with the silicon wafer to give a mixture of SiB_4 and SiB_6. The boron silicon compounds are difficult to remove and usually must be oxidized away. With a lower BBr_3 concentration in the gas stream, the insoluble layer can be prevented and the doped oxide removed with HF. At an intermediate set of conditions, a nitric acid soluble layer results. In addition, the presence of SiB_4 and SiB_6 increases the surface concentration and gives erratic and nonuniform surface concentration and sheet resistance. Increasing the oxygen decreases the likelihood of forming the unwanted layers, but more than a few percent will produce such thick oxide that surface concentration will begin to decrease. Spacing the wafers too closely will cause an increase in sheet resistance toward the center of the wafer because of restricted gas flow (185–187).

Methyl borate (188, 189), made by saturating methyl alcohol

FIGURE 8.59

Piping diagram for liquid diffusion source.

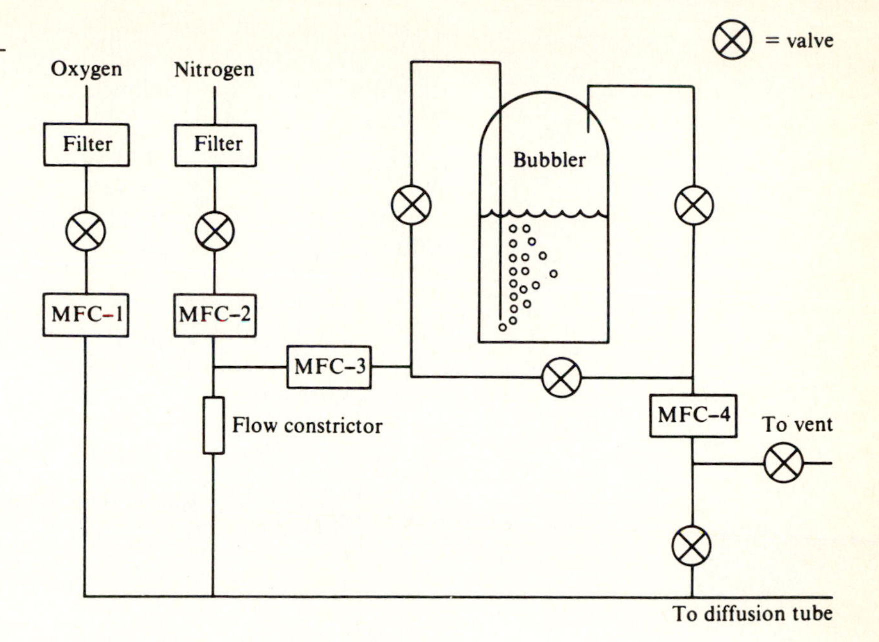

with boric acid, was examined as a liquid source by 1961; however, it has never been widely used. In principle, the carbon in the compound should exit the tube as CO_2, but carbon does deposit on the cooler walls of the entry side of the diffusion tube.

The phosphorus liquid source in common use is $POCl_3$ (having a boiling point of 107°C), although some years ago, PBr_3 was also used (163, 190). As in the case of BBr_3, oxygen must also be included in the gas stream. In the case of PBr_3, the reaction products are presumably analogous to those occurring during the use of BBr_3, with P_2O_5 being the compound deposited on the surface. However, no evidence exists of the formation of a hard-to-remove surface phosphorus compound. It has been reported that with heavy-concentration phosphorus diffusions, silicon phosphide precipitates form in the bulk (191). $POCl_3$ apparently dissociates into several components, such as Cl_2, PCl_3, P, and P_2O_4. The P and P_2O_4 react with oxygen to form P_2O_5 (192). The mixed oxide that forms on the silicon surface prevents attack by the chlorine.

8.9.2 Gaseous Sources

Diborane (B_2H_6), phosphine (PH_3), and arsine (AsH_3) are gas phase dopants that have been used for silicon (189). They are particularly hazardous to health, and extreme caution should be exercised in their use. Like the vaporized liquid sources just discussed, they depend on a high-temperature reaction with oxygen to form an oxide deposit on the wafer surface. Oxygen is also needed to form an initial oxide on the wafer and prevent pitting. In principle, pitting should not occur, but evidence does exist that if free phosphorus is formed on the surface, it will occur. The plumbing for gas phase doping is very simple, requiring only a path for oxygen like the one shown in Fig. 8.59 and one each for nitrogen and the dopant.

Some potential chemical reactions involving phosphine are as follows:

Without oxygen

$$2PH_3 + 440°C \text{ heat} \rightarrow 3H_2 + 2 \text{ red phosphorus}$$

With oxygen

$$PH_3 + 2O_2 + 150°C \text{ heat} \rightarrow H_3PO_4$$

The H_3PO_4 can decompose through a series of steps and thus give $P_2O_5 + H_2O$. Alternatively,

$$2PH_3 + 4O_2 + 300°C \text{ heat} \rightarrow P_2O_5 + 3H_2O$$

However, in an experimental study of reaction products from the oxidation of phosphine, no water, only hydrogen, was found (193). In the case of the oxidation of diborane, both were observed. Regardless of the path followed, the final result is that with phosphine, B_2H_6 or AsH_3, an oxide of the dopant is formed on the surface. Stibine (SbH_3) could also, in principle, be used, but it is even more unstable than arsine and often partially decomposes in its storage cylinder.

Unwanted gaseous sources must also be considered. Since the incoming gases may be contaminated with very fine metal particles that are carried into the furnace, point-of-use gas filters should always be used. Some kinds of furnace tube materials may out-gas. An example is the sintered silicon carbide diffusion tubes introduced several years ago. Vapors from high-temperature heater components such as the heating element, thermocouples, and various impurities in the refractory insulation may penetrate quartz diffusion tubes, particularly after long usage and partial devitrification.

8.9.3 Planar Sources

Instead of bringing the dopant into the furnace via a stream, doping-wafers that are the same size as silicon wafers can be placed adjacent to them in carriers in the diffusion furnace. From the planar source wafers, a suitable dopant such as B_2O_3 can be transferred across the narrow spacing to the silicon. The earliest planar sources were slices of boron nitride (BN) (194). In a high-temperature oxygen ambient preoxidation step, their surfaces are oxidized to B_2O_3. The B_2O_3 will volatilize, diffuse across the narrow spacing between source and wafer, slightly react with the silicon, and form a coating on the wafer that has a lower vapor pressure than that of the B_2O_3 and that will act as a diffusion source. However, a little moisture either in the diffusion ambient or absorbed in the source will lead to the formation of HBO_2, which has a much higher vapor pressure. It will then react with the silicon wafer surface such that (195)

$$2HBO_2 + 2Si \rightarrow 2SiO_2 + 2B + H_2$$

forming a mixed oxide on the surface. Variable moisture will lead to sheet resistance variability, but by keeping the BN dry and then deliberately adding water, generally formed by reacting hydrogen with oxygen in the tube ("hydrogen injection"), reproducibility is improved (196). The B_2O_3 formed on the BN is gradually depleted so that the doping wafers must be periodically reoxidized. Also, since the B_2O_3 layer is hygroscopic, the wafers must be stored between runs at about 350°C in dry nitrogen. To minimize the chance of contamination from impurities in the BN, sources made of B_2O_3 mixed with SiO_2 and other oxides are also available (197). A similar approach has been used to make planar sources for phosphorus (198, 199) and arsenic (200).

8.9.4 Doped Oxide Sources

Instead of forming a doping layer on the wafer during the diffusion cycle, the layer can be predeposited. Either doped spin-on glasses (201–205) or CVD oxides have been used (201). Methods of producing these films were discussed in Chapter 4. Spin-on sources are commercially available for both silicon and gallium arsenide dopants and can be applied with modified photoresist spinners. This type of source is widely used for antimony buried-layer diffusions in silicon.

CVD oxides are not as convenient as spin-on glasses since a CVD reactor and substantially more processing are required and thus are seldom used. They do, however, allow more leeway in the choice of dopants. Typically, a mixed oxide consisting of SiO_2 and an oxide of the dopant are co-deposited. Examples of doping oxides are P_2O_5, B_2O_3, As_2O_3, SnO, and ZnO (206–208, 76).

8.9.5 Closed-Tube Sources

Closed-tube diffusions seal both the wafer to be diffused and the source in a capsule, usually of fused silica (209–212). The tube is evacuated at room temperature, but at operating temperature, substantial pressure due to the partial pressures of the doping agent constituents can exist. Generally, no buildup of a doping layer on the wafer surface will occur; that is, the dopant transfer is directly from the gas phase into the wafer and allows control of surface concentration over a wide range. For diffusion into wafers that tend to dissociate at diffusion temperature, such as GaAs, a sealed tube containing arsenic as well as a doping source affords a way of maintaining a high enough pressure to prevent wafer degradation. By using a long tube and a two-zone furnace, relatively independent control of diffusion temperature and vapor pressure from the additive can be maintained.

If chips are diffused individually, as was sometimes done in the early 1960s, large quantities can be sealed and diffused economically. However, when wafers are used rather than chips, as is now done, and when wafer diameters are quite large, closed-tube diffusions are inconvenient and very expensive.

8.10 THE ERROR FUNCTION

8.10.1 Error Function Algebra

The error function erf(z) (1), given by the integral $(2/\sqrt{\pi})\int_0^z e^{-u^2}du$, can also be evaluated from the series

$$\frac{2}{\sqrt{\pi}}\left[z - \frac{z^3}{3 \cdot 1!} + \frac{z^5}{5 \cdot 2!} \cdots \frac{(-1)^n z^{2n+1}}{(2n+1) \cdot n!}\right.$$

$$\text{erfc}(z) = \frac{2}{\sqrt{\pi}}\int_z^\infty e^{-u^2}du = 1 - \text{erf}(z)$$

$$\text{erf}(-z) = -\text{erf}(z)$$

$$\text{erf}(0) = 0$$

$$\text{erf}(\infty) = 1$$

$$\text{erf}(z) \cong \frac{2z}{\sqrt{\pi}} \qquad \text{for } z << 1$$

$$\text{erf}(z) \cong 1 - \left(\frac{1}{z\sqrt{\pi}}\right)e^{-z^2} \qquad \text{for } z >> 1^{28}$$

$$\frac{d[\text{erf}(z)]}{dz} = \left(\frac{2}{\sqrt{\pi}}\right)e^{-z^2}$$

$$\int_0^\infty \text{erfc}(z)dz = \frac{1}{\sqrt{\pi}}$$

8.10.2 Calculation of Error Function Values

Abbreviated sets of error function and Smith function values are given in Tables 8.12 and 8.13. The highest surface concentration encountered in diffusion problems will be in the range of 10^{20} atoms/cm^3, and a typical wafer background level will be 5×10^{14} atoms/

---

[28]Functional form only; not accurate enough for most calculations (see expression in the next section).

TABLE 8.12

Error Function erf(z)

z	erf(z)	z	erf(z)	z	erf(z)	z	erf(z)
0.00	0.000 000	0.88	0.786 687	1.76	0.987 190	2.64	0.999 811
0.02	0.022 565	0.90	0.796 908	1.78	0.988 174	2.66	0.999 831
0.04	0.045 111	0.92	0.806 768	1.80	0.989 091	2.68	0.999 849
0.06	0.067 622	0.94	0.816 271	1.82	0.989 943	2.70	0.999 866
0.08	0.090 078	0.96	0.825 424	1.84	0.990 736	2.72	0.999 880
0.10	0.112 463	0.98	0.834 232	1.86	0.991 472	2.74	0.999 893
0.12	0.134 758	1.00	0.842 701	1.88	0.992 156	2.76	0.999 905
0.14	0.156 947	1.02	0.850 838	1.90	0.992 790	2.78	0.999 916
0.16	0.179 012	1.04	0.858 650	1.92	0.993 378	2.80	0.999 925
0.18	0.200 936	1.06	0.866 144	1.94	0.993 923	2.82	0.999 933
0.20	0.222 703	1.08	0.873 326	1.96	0.994 426	2.84	0.999 941
0.22	0.244 296	1.10	0.880 205	1.98	0.994 892	2.86	0.999 948
0.24	0.265 700	1.12	0.886 788	2.00	0.995 322	2.88	0.999 954
0.26	0.286 900	1.14	0.893 082	2.02	0.995 719	2.90	0.999 959
0.28	0.307 880	1.16	0.899 096	2.04	0.996 086	2.92	0.999 964
0.30	0.328 627	1.18	0.904 837	2.06	0.996 423	2.94	0.999 968
0.32	0.349 126	1.20	0.910 314	2.08	0.996 734	2.96	0.999 972
0.34	0.369 365	1.22	0.915 534	2.10	0.997 021	2.98	0.999 975
0.36	0.389 330	1.24	0.920 505	2.12	0.997 284	3.00	0.999 977 91
0.38	0.409 009	1.26	0.925 236	2.14	0.997 525	3.02	0.999 980 53
0.40	0.428 392	1.28	0.929 734	2.16	0.997 747	3.04	0.999 982 86
0.42	0.447 468	1.30	0.934 008	2.18	0.997 951	3.06	0.999 984 92
0.44	0.466 225	1.32	0.938 065	2.20	0.998 137	3.08	0.999 986 74
0.46	0.484 655	1.34	0.941 914	2.22	0.998 308	3.10	0.999 988 35
0.48	0.502 750	1.36	0.945 561	2.24	0.998 464	3.12	0.999 989 77
0.50	0.520 500	1.38	0.949 016	2.26	0.998 607	3.14	0.999 991 03
0.52	0.537 899	1.40	0.952 285	2.28	0.998 738	3.16	0.999 992 14
0.54	0.554 939	1.42	0.955 376	2.30	0.998 857	3.18	0.999 993 11
0.56	0.571 616	1.44	0.958 297	2.32	0.998 966	3.20	0.999 993 97
0.58	0.587 923	1.46	0.961 054	2.34	0.999 065	3.22	0.999 994 73
0.60	0.603 856	1.48	0.963 654	2.36	0.999 155	3.24	0.999 995 40
0.62	0.619 411	1.50	0.966 105	2.38	0.999 237	3.26	0.999 995 98
0.64	0.634 586	1.52	0.968 413	2.40	0.999 311	3.28	0.999 996 49
0.66	0.649 377	1.54	0.970 586	2.42	0.999 379	3.30	0.999 996 94
0.68	0.663 782	1.56	0.972 628	2.44	0.999 441	3.32	0.999 977 34
0.70	0.677 801	1.58	0.974 547	2.46	0.999 497	3.34	0.999 997 68
0.72	0.691 433	1.60	0.976 348	2.48	0.999 547	3.36	0.999 997 983
0.74	0.704 678	1.62	0.978 038	2.50	0.999 593	3.38	0.999 998 247
0.76	0.717 537	1.64	0.979 622	2.52	0.999 634	3.40	0.999 998 478
0.78	0.730 010	1.66	0.981 105	2.54	0.999 672	3.42	0.999 998 679
0.80	0.742 101	1.68	0.982 493	2.56	0.999 706	3.44	0.999 998 855
0.82	0.753 811	1.70	0.983 790	2.58	0.999 736	3.46	0.999 999 008
0.84	0.765 143	1.72	0.985 003	2.60	0.999 764	3.48	0.999 999 141
0.86	0.776 100	1.74	0.986 135	2.62	0.999 789	3.50	0.999 999 257

(*continues*)

TABLE 8.12 (*continued*)

z	erf(z)	z	erf(z)	z	erf(z)	z	erf(z)
3.52	0.999 999 358	3.68	0.999 999 805	3.84	0.999 999 944		
3.54	0.999 999 445	3.70	0.999 999 833	3.86	0.999 999 952		
3.56	0.999 999 521	3.72	0.999 999 857	3.88	0.999 999 959		
3.58	0.999 999 587	3.74	0.999 999 877	3.90	0.999 999 965		
3.60	0.999 999 644	3.76	0.999 999 895	3.92	0.999 999 970		
3.62	0.999 999 694	3.78	0.999 999 910	3.94	0.999 999 975		
3.64	0.999 999 736	3.80	0.999 999 923	3.96	0.999 999 979		
3.66	0.999 999 773	3.82	0.999 999 934	3.98	0.999 999 982		

Note: For a more complete table, see L.J. Comrie, "Chambers Six Figure Mathematical Tables," vol. 2, W. & R. Chambers, Ltd., Edinburgh, 1949, or "Tables of the Error Function and Its Derivative," National Bureau of Standards Applied Mathematical Series, no. 41, Oct. 22, 1954.

TABLE 8.13

Smith Function

α \ β	0.1	0.2	0.4	0.6	0.8	1.0	1.2	1.4	1.6	1.8	2.0	2.5	3.0	4.0	5.0	β / α
0.1	0.09015	0.08155	0.06672	0.05459	0.04467	0.03655	0.02990	0.02446	0.02002	0.01638	0.01340	0.00811	0.00491	0.00180	0.00066	0.1
0.2	0.17838	0.16119	0.13162	0.10748	0.08777	0.07167	0.05853	0.04779	0.03903	0.03187	0.02603	0.01568	0.00945	0.00343	0.00125	0.2
0.4	0.34254	0.30837	0.24993	0.20259	0.16422	0.13314	0.10794	0.08752	0.07097	0.05756	0.04668	0.02766	0.01640	0.00577	0.00204	0.4
0.6	0.48366	0.43290	0.34692	0.27814	0.22308	0.17900	0.14368	0.11538	0.09268	0.07448	0.05988	0.03475	0.02021	0.00688	0.00236	0.6
0.8	0.59940	0.53264	0.42100	0.33317	0.26398	0.20940	0.16628	0.13219	0.10519	0.08379	0.06680	0.03806	0.02180	0.00724	0.00244	0.8
1.0	0.69176	0.60975	0.47475	0.37066	0.29013	0.22765	0.17903	0.14109	0.11141	0.08814	0.06985	0.03931	0.02231	0.00733	0.00246	1.0
1.2	0.76448	0.66808	0.51232	0.39486	0.30574	0.23772	0.18553	0.14529	0.11412	0.08989	0.07098	0.03969	0.02244	0.00735	0.00246	1.2
1.4	0.82144	0.71164	0.53781	0.40979	0.31449	0.24286	0.18855	0.14706	0.11517	0.09051	0.07134	0.03979	0.02247	0.00735	0.00246	1.4
1.6	0.86601	0.74388	0.55469	0.41865	0.31914	0.24530	0.18983	0.14774	0.11552	0.09070	0.07144	0.03981	0.02247	0.00735	0.00246	1.6
1.8	0.90095	0.76759	0.56562	0.42369	0.32147	0.24638	0.19033	0.14797	0.11563	0.09075	0.07147	0.03982	0.02247	0.00735	0.00246	1.8
2.0	0.92838	0.78491	0.57254	0.42646	0.32258	0.24682	0.19051	0.14804	0.11566	0.09076	0.07147	0.03982	0.02247	0.00735	0.00246	2.0
2.5	0.97404	0.81009	0.58029	0.42887	0.32335	0.24707	0.19059	0.14807	0.11567	0.09076	0.07147	.....	.....	.....	.....	2.5
3.0	0.99920	0.82094	0.58234	0.42928	0.32343	0.24708	0.19059	0.14807	0.11567	0.09076	0.07147	.....	.....	.....		3.0
∞	1.02843	0.82795	0.58291	0.42933	0.32343	0.24709	0.19059	0.14807	0.11567	0.09076	0.07147	0.03982	0.02247	0.00735	0.00246	∞

Source: From R.C.T. Smith, "Conduction of Heat in the Semi-Infinite Solid with a Short Table of an Important Integral," *Australian J. Phys. 6*, pp. 127–130, 1953.

cm^3. To plot a diffusion profile over this range will require an N variation of about 6 orders of magnitude. Thus, from Eq. 8.32, for example, the value of the error function must range from 1 to 10^{-6}. The z range corresponding to this is 0.0 to 3.5 and is covered in Table 8.12. If the series expansion from section 8.10.1 is used to calculate erf(z), it should be remembered that it is given by small differences of large numbers and that single-precision BASIC calculations will not provide the required accuracy. A better way, if errors of no more than $\pm 1.5 \times 10^{-7}$ are acceptable, is, for positive z, to use the approximate expression[29]

$$\text{erf}(z) = 1 - (a_1T + a_2T^2 + a_3T^3 + a_4T^4 + a_5T^5)e^{-z^2}$$

where $T = 1/(1 + Pz)$. P and the a_i values are as follows:

$$P = 0.3275911$$

$$a_1 = 0.254829592$$

$$a_2 = -0.284496736$$

$$a_3 = 1.421413741$$

$$a_4 = -1.453152027$$

$$a_5 = 1.061405429$$

SAFETY — CHAPTER 8

Several materials used for diffusion sources are toxic and should therefore be handled with caution. In addition, some safeguards must be included in the facility. Some hazardous materials currently in common use are listed in Table 8.14. In addition, the quantities of nitrogen used in a diffusion area, if vented into an unexhausted room, as, for example, during a power failure, can cause asphyxiation. Of the materials listed in the table, diborane and phosphine are the most toxic. They have, respectively, a TLV of 0.1 ppm and 0.3 ppm and an IDLH of 40 ppm and 200 ppm.[30] Once applied, spin-on

TABLE 8.14

Toxic Materials Used in Diffusion

Item	*Use*
Arsine	Gaseous arsenic source
Diborane	Gaseous boron source
Phosphine	Gaseous phosphorus source
Phosphorus oxychloride	Liquid phosphorus source
Boron tribromide	Liquid boron source
Arsenic spin-on	Liquid spin-on source
Antimony spin-on	Liquid spin-on source

[29] Cecil Hastings, Jr., *Approximations for Digital Computers*, Princeton University Press, Princeton, N.J., 1955.

[30] Recommended threshold limit value (TLV) and immediately dangerous to life or health (IDLH).

materials should present little hazard, but the spin applicator should be carefully shielded and ventilated in order to remove toxic vapors. (See also the safety discussion given earlier in Chapter 4.)

KEY IDEAS CHAPTER 8

- ☐ Diffusion is a mechanism by which impurities migrate from high- to low-concentration regions.
- ☐ If the species is diffusing in silicon and normally resides at a substitutional position in the crystal lattice, it generally moves by jumping into an adjacent vacant lattice space.
- ☐ If the diffusing species normally resides in space between lattice sites (interstices), it generally moves by jumping to an adjacent interstice.
- ☐ The frequency of jumps from position to position is thermally activated.
- ☐ Simple theory describes diffusion in terms of Fick's two laws:

$$J = -D\frac{\partial N}{\partial x}$$

$$\frac{\partial N}{\partial t} = D\frac{\partial^2 N}{\partial x^2}$$

where J is the diffusing atoms' flux, D is the diffusion coefficient, N is the concentration of diffusant, t is the time, and x is the distance.

- ☐ D usually increases exponentially with temperature.
- ☐ When D is not independent of concentration, Fick's laws must be modified to include $\partial D/\partial x$.
- ☐ Solutions of Fick's laws using the appropriate boundary conditions are used to describe the diffusion profiles observed after semiconductor diffusions.
- ☐ The solutions generally involve either erf or Gaussian distributions.
- ☐ Diffusion coefficients are experimentally measured and are available for dopants of interest.
- ☐ Interstitial atoms diffuse much more rapidly than substitutional atoms.
- ☐ Diffusion processes usually are in two steps, with a shallow predep using a diffusion source followed by a drive-in. Particularly in the case of gallium arsenide, the diffused predep step has generally been superseded by an ion implant step.
- ☐ The same formalism used to describe a specific diffusion process can also be used to describe impurity redistributions (diffusion that occurs during other high-temperature processing steps such as oxidation).
- ☐ The most common measurements used to characterize diffusions are the sheet resistance and the junction depth.
- ☐ Gettering is a process for collecting lifetime-degrading impurities present in wafers and storing them in designated regions of the wafer.
- ☐ Gettering steps are often combined with diffusion.
- ☐ Common gettering processes use a heavy back-side concentration of phosphorus, mechanically damaged wafer back side, oxygen precipitation in the middle of the wafer, or misfit dislocations at epi–substrate interfaces.

- If wafers are heated or cooled too rapidly, slip will be generated. Very heavy surface concentrations of impurities will also generate slip.

- Diffusion sources are described as liquid, gaseous, doped oxide, or closed-tube, with liquid as the one most commonly used with silicon.

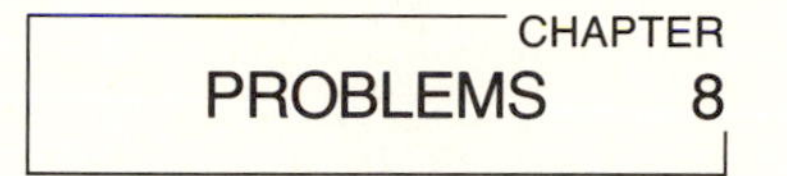

1. Express Fick's first law in words. Show how Eq. 8.5 can be derived from Eq. 8.3.
2. Describe the mechanism of diffusion by interstitialcy and compare it to diffusion by interstitial motion. Will an increase in the number of vacancies affect either type of diffusion?
3. What is the distinction between the concentration N_D of an n-type impurity in a wafer and the concentration n of majority carriers in the same wafer? Using the n_i versus temperature curve of Fig. 8.7, calculate the carrier concentration n versus temperature over the range from room temperature to 1100°C when $N_D = 10^{18}$ atoms/cc.
4. Assuming a constant phosphorus surface concentration of 10^{19} atoms/cc and a concentration-independent diffusion constant, calculate the time to produce a junction at a depth of 15 μm in 1 Ω-cm p-type silicon when (a) the diffusion temperature is 1250°C and when (b) it is 1050°C.
5. A sheet of atoms of density 10^{14} atoms/cm^2 is diffused at a temperature that gives a D value of 5×10^{-14} cm^2/s. Plot the impurity profile after diffusing for 30 minutes. Plot the surface concentration versus time for a period of 1 hour.
6. Show that for diffusion from a step, regardless of the relative magnitude of the diffusion coefficients of the impurities on each side of the step, the position where the two concentrations are equal moves more and more into the lightly doped side of the step as the diffusion time increases.
7. A 5μm thick, 1 Ω–cm n-type epitaxial layer is grown on a 10 Ω-cm p-type substrate. A p-type ring is to be diffused into the layer so that the ring will electrically isolate the n-region inside the ring from the n-region outside it. Show what the cross section will look like after diffusion. If the p-diffusant is boron, what is the minimum diffusion time at 1150°C required for isolation, assuming no substrate up-diffusion and no autodoping? Use D values from Table 8.2.
8. The material on one side of an abrupt silicon junction is 1 Ω-cm n-type. On the other side, it is 1 Ω-cm p-type. What is the impurity concentration in atoms/cc on each side? After a 2 hour diffusion, will the junction location have moved into the region initially n-type or into the region initially p-type? Justify your answer.
9. A p-type diffusion profile has the shape given by the following points. When the diffusion was into 1 Ω-cm material for 1 hour, the junction depth was 0.98 μm. When an identical diffusion was made in 10 Ω-cm material, the junction depth was 1.14 μm. Determine the diffusion coefficient.

Normalized Concentration	*Depth (μm)*
1.0	0
4.6×10^{-1}	0.2
6.6×10^{-2}	0.5
3.3×10^{-3}	0.8
2.5×10^{-4}	1.0
5.2×10^{-5}	1.1

10. What is the primary advantage of using a heavy phosphorus back-side wafer diffusion for gettering? What is the primary advantage of using misfit dislocations at an epi–substrate interface for gettering? What is a possible major disadvantage of the misfit dislocation approach?

11. Given a (001) oriented wafer, consider the quadrant included between the [010] and $[\bar{1}00]$ directions. If slip occurs between $(\bar{1}11)$ planes in the $[\bar{1}\bar{1}0]$ direction, plot the normalized resolved shear stress as a constant magnitude stress changes direction from $[\bar{1}\bar{1}0]$ to $[0\bar{1}0]$. Based on this plot, sketch where slip might first be expected to appear on a (100) wafer.

REFERENCES CHAPTER 8

1. Paul G. Shewmon, *Diffusion in Solids*, McGraw-Hill Book Co., New York, 1963.
2. John R. Manning, *Diffusion Kinetics for Atoms in Crystals*, D. Van Nostrand Co., Princeton, N.J., 1968.
3. R.B. Fair and J.C. Tsai, "A Quantitative Model for the Diffusion of Phosphorus in Silicon and the Emitter Push Effect," *J. Electrochem. Soc. 124*, pp. 1107–1118, 1977.
4. Paul Fahey, "New Models of Dopant Diffusion Appropriate for VLSI Fabrication Processes," pp. 216–229, in Kenneth E. Bean and George A. Rozgonyi, eds., *VLSI Science and Technology/84*, Electrochemical Society, Pennington, N.J., 1984.
5. W.R. Wilcox and T.J. LaChapelle, "Mechanism of Gold Diffusion into Silicon," *J. Appl. Phys. 35*, pp. 240–246 and references therein, 1964.
6. U. Gosele et al., "Mechanism and Kinetics of the Diffusion of Gold in Silicon," *Appl. Phys. 23*, pp. 361–368, 1980.
7. M. Hill et al., "Diffusion of Silicon in Gold," *J. Electrochem. Soc. 129*, pp. 1579–1587, 1982.
8. L.R. Weisberg and J. Blanc, "Diffusion with Interstitial–Substitutional Equilibrium. Zinc in GaAs," *Phys. Rev. 131*, pp. 1548–1552, 1963.
9. F.M. Smits, "Formation of Junction Structures by Solid State Diffusion," *Proc. IRE 46*, pp. 1049–1061, 1958.
10. S.M. Hu, "Diffusion in Silicon and Germanium," pp. 217–350, in D. Shaw, ed., *Atomic Diffusion in Semiconductors*, Plenum Publishing Co., London, 1973. (This is an extensive review of diffusion processes and includes almost 400 references.)
11. S.M. Hu, "Formation of Stacking Faults and Enhanced Diffusion in the Oxidation of Silicon," *J. Appl. Phys. 45*, pp. 1567–1573, 1974.
12. Dimitri A. Antoniadis et al., "Boron in Near-Intrinsic <100> and <111> Silicon under Inert and Oxidizing Ambients—Diffusion and Segregation," *J. Electrochem. Soc. 125*, pp. 813–819, 1978.
13. D.A. Antoniadis et al., "Oxidation-Enhanced Diffusion of Arsenic and Phosphorus in Near-Intrinsic <100> Silicon," *Appl. Phys. Lett. 33*, pp. 1030–1033, 1978.
14. R. Francis and P.S. Dobson, "The Effect of Oxidation on the Diffusion of Phosphorus in Silicon," *J. Appl. Phys. 50*, pp. 280–284, 1979.
15. A. Miin-Ron Lin et al., "The Oxidation Rate Dependence of Oxidation-Enhanced Diffusion of Boron and Phosphorus in Silicon," *J. Electrochem. Soc. 128*, pp. 1131–1137, 1981.
16. S.M. Hu, "Kinetics of Interstitial Supersaturation during Oxidation of Silicon," *Appl. Phys. Lett. 43*, pp. 449–451, 1983.
17. R.M. Harris and D.A. Antoniadis, "Silicon Self-Interstitial Supersaturation during Phosphorus Diffusion," *Appl. Phys. Lett. 43*, pp. 937–939, 1983.
18. S.M. Hu, "Kinetics of Interstitial Supersaturation and Enhanced Diffusion in Short-Time/Low-Temperature Oxidation of Silicon," *J. Appl. Phys. 57*, pp. 4527–4533, 1985.
19. Scott T. Dunham and James D. Plummer, "Point-Defect Generation during Oxidation of Silicon in Dry Oxygen. I. Theory," *J. Appl. Phys. 59*, pp. 2541–2550, 1986.
20. Scott T. Dunham and James D. Plummer, "Point-Defect Generation during Oxidation of Silicon in Dry Oxygen. II. Comparison to Experiment," *J. Appl. Phys. 59*, pp. 2551–2561, 1986.
21. K. Taninguchi et al., "Oxidation Enhanced Diffusion of Boron and Phosphorus in (100) Silicon," *J. Electrochem. Soc. 127*, pp. 2243–2248, 1980.
22. J.F. Nye, *Physical Properties of Crystals*, Oxford University Press, London, 1960.
23. J.F. Shepard et al., "Study of a Liquid Boron Dif-

fusion Source for Silicon," *Electrochem. Soc. Ext. Abst.*, abst. no. 196, pp. 87–89, Fall 1966.

24. G.N. Wills, "The Orientation Dependent Diffusion of Boron in Silicon under Oxidizing Conditions," *Solid-State Electronics 12*, pp. 133–134, 1969.
25. L.E. Katz, "Orientation Dependent Diffusion Phenomena," in Charles P. Marsden, ed., *Silicon Device Processing*, NBS Special Pub. 337, pp. 192–199, 1970.
26. W.G. Allen, "Effect of Oxidation on Orientation Dependent Boron Diffusion in Silicon," *Solid-State Electronics 16*, pp. 709–717, 1973.
27. C. Hill, "Measurement of Local Diffusion Coefficients in Planar Device Structures," pp. 988–998, in Howard R. Huff and Rudolph J. Kriegler, eds., *Semiconductor Silicon/81*, Electrochemical Society, Pennington, N.J., 1981.
28. Richard B. Fair, Chap. 7, "Concentration Profiles of Diffused Dopants in Silicon," in F.F. Wang, ed., *Impurity Doping Processes in Silicon*, North-Holland, New York, 1981.
29. Sorab K. Ghandhi, *VLSI Fabrication Principles*, John Wiley & Sons, New York, 1983.
30. J.J. Wortman et al., "Effect of Mechanical Stress on p-n Junction Device Characteristics," *J. Appl. Phys. 35*, pp. 2122–2131, 1964.
31. R.B. Fair, "Modeling Anomalous Phenomena in Arsenic Diffusion in Silicon," pp. 963–987, in H.R. Huff and R.J. Kriegler, eds., *Semiconductor Silicon/81*, Electrochemical Society, Pennington, N.J., 1981.
32. J. Kasahara et al., "The Effect of Stress on the Redistribution of Implanted Impurities in GaAs," *J. Electrochem. Soc. 130*, pp. 2275–2279, 1983.
33. Y. Todokoro and I. Teramoto, "The Stress Enhanced Diffusion of Boron in Silicon," *J. Appl. Phys. 49*, pp. 3527–3529, 1978.
34. A.F.W. Willoughby, "Interactions between Sequential Dopant Diffusions in Silicon—A Review," *J. Phys. D: Appl. Phys. 10*, pp. 455–480, 1977.
35. L.E. Miller, "Uniformity of Junctions in Diffused Silicon Devices," pp. 303–322, in Harry C. Gatos, ed., *Properties of Elemental and Compound Semiconductors*, Interscience Publishers, New York, 1960.
36. S.M. Hu and S. Schmidt, "Interactions in Sequential Diffusion Processes in Semiconductors," *J. Appl. Phys. 39*, pp. 4272–4283, 1968.
37. W. Jost, *Diffusion in Solids, Liquids, and Gases*, Academic Press, New York, 1952.
38. Ron C. Wackwitz, Texas Instruments Incorporated, unpublished work.
39. R.C.T. Smith, "Conduction of Heat in the Semi-Infinite Solid with a Short Table of an Important Integral," *Australian J. Phys. 6*, pp. 127–130, 1953.
40. W.R. Runyan, *Silicon Semiconductor Technology*, McGraw-Hill Book Co., New York, 1965.
41. C.C. Allen and W.R. Runyan, "An Epitaxial Grown–Diffused Silicon Transistor," *IEEE Trans. on Electron Dev. ED-10*, pp. 289–290, 1963.
42. Warren R. Rice, "Diffusion of Impurities during Epitaxy," *Proc. IEEE 52*, pp. 284–295, 1964.
43. A.S. Grove, A. Roder, and C.T. Sah, "Impurity Distribution in Epitaxial Growth," *J. Appl. Phys. 36*, pp. 802–810, 1965
44. F.M. Smits and R.C. Miller, "Rate Limitation at the Surface for Impurity Diffusion in Semiconductors," *Phys. Rev. 104*, pp. 1242–1245, 1956.
45. R.C. Miller and F.M. Smits, "Diffusion of Antimony out of Germanium and Some Properties of the Antimony–Germanium System," *Phys. Rev. 107*, pp. 65–70, 1957.
46. T.I. Kucher, "The Problem of Diffusion in an Evaporating Solid Medium," *Soviet Phys.—Solid State 3*, pp. 401–404, 1961.
47. R.B. Allen et al., "Effect of Oxide Layers on the Diffusion of Phosphorus into Silicon," *J. Appl. Phys. 31*, pp. 334–337, 1960.
48. C.T. Sah et al., "Diffusion of Phosphorus in Silicon Oxide Film," *J. Phys. Chem. Solids 11*, pp. 288–298, 1959.
49. J.C.C. Tsai, "The Simultaneous Diffusion of Donor and Acceptor Impurities into Silicon," Ph.D. thesis, Ohio State University, 1962.
50. A.S. Grove et al., "Redistribution of Acceptor and Donor Impurities during Thermal Oxidation of Silicon," *J. Appl. Phys. 35*, pp. 2695–2701, 1964.
51. Taketoshi Kato and Yoshio Nishi, "Redistribution of Diffused Boron in Silicon by Thermal Oxidation," *Jap. J. Appl. Phys. 3*, pp. 377–383, 1964.
52. R.H. Krambeck, "Numerical Calculation of Im-

purity Redistribution during Thermal Oxidation of Semiconductors," *J. Electrochem. Soc. 121*, pp. 588–591, 1974.
53. E.D. Fabricius, "Diffusion of Impurities in a Thin Semiconductor Slab," *J. Appl. Phys. 33*, pp. 753–754, 1962.
54. J.C. Fisher, "Calculation of Diffusion Penetration Curves for Surface and Grain Boundary Diffusion," *J. Appl. Phys. 22*, pp. 74–77, 1951.
55. R.T.P. Whipple, "Concentration Contours in Grain Boundary Diffusion," *Phil. Mag. Series 7, Vol. 45*, pp. 1225–1236, 1954.
56. Van E. Wood et al., "Theoretical Solutions of Grain-Boundary Diffusion Problems," *J. Appl. Phys. 33*, pp. 3574–3579, 1962.
57. H.J. Queisser et al., "Diffusion along Small Angle Grain Boundaries in Silicon," *Phys. Rev. 123*, pp. 1245–1254, 1961.
58. A. Goetzberger and H. Queisser, "Structural Imperfections in Silicon p-n Junctions," Shockley Transistor Corp. Interim Rpt. No. 1, Contract AF19(604)8060, August 1961.
59. H. Queisser, "Failure Mechanisms in Silicon Semiconductors," Shockley Transistor Corp. Final Rpt., Contract AF30(602)2556, January 1963.
60. T.I. Kamins and S.Y. Chiang, "Lateral Dopant Diffusion in Implanted Buried-Oxide Structures," *Appl. Phys. Lett. 47*, pp. 1197–1199, 1985.
61. P.H. Holloway, "Grain Boundary Diffusion of Phosphorus in Polycrystalline Silicon," *J. Vac. Sci. Technol. 21*, pp. 19–22, 1982.
62. Kouichi Sakamoto et al., "Complete Process Modeling for VLSI Multilayer Structures," *J. Electrochem. Soc. 132*, pp. 2457–2462, 1985.
63. R.G. Elliman et al., "Diffusion and Precipitation in Amorphous Si," *Appl. Phys. Lett. 46*, pp. 478–480, 1985.
64. D.P. Kennedy and R.R. O'Brien, "Analysis of the Impurity Atom Distribution near the Diffusion Mask for a Planar p-n Junction," *IBM J. Res. Develop. 9*, pp. 179–186, 1965.
65. D.D. Warner and C.L. Wilson, "Two Dimensional Concentration Dependent Diffusion," *Bell Syst. Tech. J. 59*, pp. 1–41, 1980.
66. R.C. Wackwitz, "Analytical Solution of the Multiple-Diffusion Problem," *J. Appl. Phys. 33*, pp. 2909–2910, 1962.
67. R.B. Fair and J.C.C. Tsai, "The Diffusion of Ion-Implanted Arsenic in Silicon," *J. Electrochem. Soc. 122*, pp. 1689–1696, 1975.
68. Y. Nakajima et al., "Simplified Expression for the Distribution of Diffused Impurity," *Jap. J. Appl. Phys. 10*, pp. 162–163, 1971.
69. Richard B. Fair, "Boron Diffusion in Silicon—Concentration and Orientation Dependence, Background Effects, and Profile Estimation," *J. Electrochem. Soc. 122*, pp. 800–805, 1975.
70. B. Swaminathan et al., "Diffusion of Arsenic in Polycrystalline Silicon," *Appl. Phys. Lett. 40*, pp. 795–798, 1982.
71. T.I. Kamins et al., "Diffusion of Impurities in Polycrystalline Silicon," *J. Appl. Phys. 43*, pp. 81–91, 1972.
72. C.J. Coe, "The Lateral Diffusion of Boron in Polycrystalline Silicon and Its Influence on the Fabrication of Sub-Micron MOSTs," *Solid-State Electronics 20*, pp. 985–992, 1977.
73. W.M. Bullis, "Properties of Gold in Silicon," *Solid-State Electronics 9*, pp. 143–168 and various included references, 1966.
74. C.W. Farley and B.G. Streetman, "Simulation of Anomalous Acceptor Diffusion in Compound Semiconductors," *J. Electrochem. Soc. 134*, pp. 453–458 and references therein, 1987.
75. I.K. Naik, "Annealing Behavior of GaAs Ion Implanted with p-Type Dopants," *J. Electrochem. Soc. 134*, pp. 1270–1275, 1987.
76. R. Jett Field and Sorab K. Ghandhi, "An Open-Tube Method for Diffusion of Zinc into GaAs," *J. Electrochem. Soc. 129*, pp. 1567–1579, 1982.
77. F.M. Smits, "Measurement of Sheet Resistivity with the Four-Point Probe," *Bell Syst. Tech. J. 37*, pp. 711–718, 1958.
78. David S. Perloff et al., "Four-Point Resistance Measurements of Semiconductor Doping Uniformity," *J. Electrochem. Soc. 124*, pp. 582–590, 1977.
79. J.C. Irving, "Resistivity of Bulk Silicon and Diffused Layers in Silicon," *Bell Syst. Tech. J. 41*, pp. 387–410, 1962.
80. Gerhard Backenstoss, "Evaluation of the Surface Concentration of Diffused Layers in Silicon," *Bell Syst. Tech. J. 37*, pp. 699–710, 1958.
81. O.N. Tufte, "The Average Conductivity and Hall Effect of Diffused Layers in Silicon," *J. Electrochem. Soc. 109*, pp. 235–238, 1962.

82. W.R. Runyan, *Semiconductor Measurements and Instrumentation*, McGraw-Hill Book Co., New York, 1975.
83. C.S. Fuller and J.A. Ditzenberger, "Diffusion of Donor and Acceptor Elements in Silicon," *J. Appl. Phys. 27*, pp. 544–553, 1956.
84. P.A. Illes and B. Leibenhaut, "Diffusant Impurity Concentration Profiles in Thin Layers on Silicon," *Solid-State Electronics 5*, pp. 331–339, 1962.
85. E. Tannenbaum, "Detailed Analysis of Thin Phosphorus Diffused in p-Type Silicon," *Solid-State Electronics 2*, pp. 123–132, 1961.
86. Stacy B. Watelski et al., "A Concentration Gradient Profiling Method," *J. Electrochem. Soc. 112*, pp. 1051–1053, 1965.
87. T.H. Yeh, "Current Status of the Spreading Resistance Probe and Its Application," pp. 111–122, in Charles P. Marsden, ed., *Silicon Device Processing*, NBS Special Pub. 337, 1970.
88. O. Kudoh et al., "Impurity Profiles within a Shallow p-n Junction by a New Differential Spreading Resistance Method," *J. Electrochem. Soc. 123*, pp. 1751–1754, 1976.
89. Masami Konaka et al., "Non-Destructive Determination of Impurity Concentration in Silicon Epitaxial Layer Using Metal–Silicon Schottky Barrier," *Jap. J. Appl. Phys. 7*, pp. 790–791, 1968.
90. K.H. Zaininger and F.P. Heiman, "The *C–V* Technique as an Analytical Tool—Parts I and II," *Solid State Technology*, pp. 49–56 (May), pp. 46–55 (June), 1970.
91. M.G. Buehler, "The D.C. MOSFET Dopant Profile Method," *J. Electrochem. Soc. 127*, pp. 701–704, 1980.
92. C.W. White and W.H. Christie, "The Use of RBS and SIMS to Measure Dopant Profile Changes in Silicon by Pulsed Laser Annealing," *Solid State Technology*, pp. 109–116, September 1980.
93. J.M. Anthony et al., "Super SIMS for Ultrasensitive Impurity Analysis," *Proc. Materials Research Society Symposium 69*, pp. 311–316, 1986.
94. Philip F. Kane and Graydon B. Larrabee, *Characterization of Semiconductor Materials*, McGraw-Hill Book Co., New York, 1970.
95. T.H. Yeh, "Experimental Methods for Determining Diffusion Coefficients in Semiconductors," pp. 155–215, in D. Shaw, ed., *Atomic Diffusion in Semiconductors*, Plenum Publishing Co., London, 1973.
96. Howard Reiss and C.S. Fuller, "Diffusion Processes in Germanium and Silicon," pp. 222–268, in N.B. Hannay, ed., *Semiconductors*, Reinhold Publishing Corp., New York, 1959.
97. B.I. Boltaks, *Diffusion in Semiconductors*, Academic Press, New York, p. 149, 1963.
98. M. Ghezzo, "Diffusion from a Thin Layer into a Semi-Infinite Medium with Concentration Dependent Diffusion Coefficient," *J. Electrochem. Soc. 119*, pp. 977–979, 1972.
99. D. Anderson and K.O. Jeppson, "Nonlinear Two-Step Diffusion in Semiconductors," *J. Electrochem. Soc. 131*, pp. 2675–2679, 1984.
100. Dan Anderson and Kjell O. Jeppson, "Evaluation of Diffusion Coefficients from Non-Linear Impurity Profiles," *J. Electrochem. Soc. 132*, pp. 1409–1412, 1985.
101. R. Ghez et al., "The Analysis of Diffusion Data by a Method of Moments," *J. Electrochem. Soc. 132*, pp. 2759–2761, 1985.
102. C. Hill, "Measurement of Local Diffusion Coefficients in Planar Device Structures," pp. 988–998, in Howard R. Huff et al., eds., *Semiconductor Silicon/81*, Electrochemical Society, Pennington, N.J., 1981.
103. B.L. Morris and L.E. Katz, "Reduction of Excess Phosphorus and Elimination of Defects in Phosphorus Emitter Diffusions," *J. Electrochem. Soc. 125*, pp. 762–765, 1978.
104. W.R. Thurber and W.M. Bullis, "Resistivity and Carrier Lifetime in Gold-Doped Silicon," Final Rpt., PRO Y–71–71–906, AFCRL–72–0076 and the extensive references, January 1972.
105. K.P. Lisiak and A.G. Milnes, "Platinum as a Lifetime-Control Deep Level Impurity in Silicon," *J. Appl. Phys. 46*, pp. 5229–5235, 1975.
106. R. Saito et al., "Dual Diffusion of Gold and Platinum into Silicon," *J. Electrochem. Soc. 132*, pp. 225–229, 1985.
107. S.F. Cagnina, "Enhanced Gold Solubility Effect in Heavily n-Type Silicon," *J. Electrochem. Soc. 116*, pp. 498–502, 1969.
108. P.C. Parekh, "Gettering of Gold and Its Influence on Some Transistor Parameters," *Solid-State Electronics 13*, pp. 1401–1406, 1970.
109. A. Mogro-Campero and R.P. Love, "Localized

Lifetime Control by Argon Ion Implantation into Silicon," *Electrochem. Soc. Ext. Abst. 83–1*, abst. no. 319, pp. 502–503, 1983.

110. Eicke R. Weber, "Transition Metals in Silicon," *Appl. Phys. A 30*, pp. 1–22, 1983.

111. Paul Richmond, "The Effect of Gold Doping upon the Characteristics of MOS Field-Effect Transistors with Applied Substrate Voltage," *Proc. IEEE 56*, pp. 774–775, 1968.

112. D.R. Lamb et al., "The Effect of Gold Doping on the Threshold Voltage, Hall Mobility, Gain, and Current Noise of MOS Transistors," *Int. J. Electronics 30*, pp. 141–147, 1971.

113. Takashi Nishioka et al., "MOSFET's on Au-Diffused High Resistivity Si Substrates," *IEEE Trans. on Electron Dev. ED-29*, pp. 1507–1510, 1982.

114. Kouichirou Hondo et al., "Breakdown in Silicon Oxides—Correlation with Cu Precipitates," *Appl. Phys. Lett. 45*, pp. 270–271, 1984.

115. A. Goetzberger and W. Shockley, "Metal Precipitates in Silicon p-n Junctions," *J. Appl. Phys. 31*, pp. 1821–1824, 1960.

116. J.W. Adamic and J.F. McNamara, "A Study of the Removal of Gold from Silicon Using Phosphorus and Boron Glass Getters," *Electrochem. Soc. Ext. Abst. 13–2*, abst. no. 153, pp. 94–95, October 1964.

117. M. Ghezzo, "Vapor Deposition of Phosphosilicate Glasses from Mixtures of SiH_4, O_2 and $POCl_3$," *J. Electrochem. Soc. 119*, pp. 1428–1430, 1972.

118. R.L. Meek et al., "Diffusion Gettering of Au and Cu in Silicon," *J. Electrochem. Soc. 122*, pp. 786–796, 1975.

119. O. Paz et al., "$POCl_3$ and Boron Gettering of LSI Si Devices: Similarities and Differences," *J. Electrochem. Soc. 126*, pp. 1754–1761, 1979.

120. Livio Balsi et al., "Heavy Metal Gettering in Silicon-Device Processing," *J. Electrochem. Soc. 127*, pp. 164–169, 1980.

121. A. Ourmaszd and W. Schroter, "Phosphorus Gettering and Intrinsic Gettering of Nickel in Silicon," *Appl. Phys. Lett. 45*, pp. 781–783, 1984.

122. G.A. Rozgonyi et al., "Elimination of Oxidation-Induced Stacking Faults by Preoxidation Gettering of Silicon Wafers," *J. Electrochem. Soc. 122*, pp. 1725–1729, 1975.

123. E.J. Mets, "Poisoning and Gettering Effects in Silicon Junctions," *J. Electrochem. Soc. 112*, pp. 420–425, 1965.

124. D.I. Pomerantz, "A Cause and Cure of Stacking Faults in Silicon Epitaxial Layers," *J. Appl. Phys. 38*, pp. 5020–5026, 1967.

125. G.A. Rozgonyi et al., "The Identification, Annihilation, and Suppression of Nucleation Sites Responsible for Epitaxial Stacking Faults," *J. Electrochem. Soc. 123*, pp. 1910–1915, 1976.

126. C.L. Reed and K.M. Mar, "The Effects of Abrasion Gettering on Silicon Material with Swirl Defects," *J. Electrochem. Soc. 127*, pp. 2058–2062, 1980.

127. C.W. Pearce and V.J. Zalackas, "A New Approach to Lattice Damage Gettering," *J. Electrochem. Soc. 126*, pp. 1436–1437, 1979.

128. Y. Hayafuji et al.,"Laser Damage Gettering and Its Application to Lifetime Improvement in Silicon," *J. Electrochem. Soc. 128*, pp. 1975–1980, 1981.

129. T.M. Buck et al., "Gettering Rates of Various Fast-Diffusing Metal Impurities at Ion-Damaged Layers on Silicon," *Appl. Phys. Lett. 21*, pp. 485-487, 1972.

130. T.E. Seidel et al., "Direct Comparison of Ion-Damage Gettering and Phosphorus-Diffusion Gettering of Au in Si," *J. Appl. Phys. 46*, pp. 600–609, 1975.

131. T.W. Sigmon et al., "Ion Implant Gettering of Gold in Silicon," *J. Electrochem. Soc. 123*, pp. 1116–1117, 1976.

132. M.R. Poponiak et al., "Argon Implantation Gettering of Bipolar Devices," *J. Electrochem. Soc. 124*, pp. 1802–1805, 1977.

133. H.J. Geipel and W.K. Tice, "Reduction of Leakage by Implantation Gettering in VLSI Circuits," *IBM J. Res. Develop. 24*, pp. 310–317, 1980.

134. James A. Topich, "Reduction of Defects in Ion Implanted Bipolar Transistors by Argon Backside Damage," *J. Electrochem. Soc. 128*, pp. 866–870, 1981.

135. K.D. Beyer and T.H. Yeh, "Impurity Gettering of Silicon Damage Generated by Ion Implantation through SiO_2 Layers," *J. Electrochem. Soc. 129*, pp. 2527–2530, 1982.

136. P.M. Petroff et al., "Elimination of Process-Induced Stacking Faults by Preoxidation Gettering

of Si Wafers," *J. Electrochem. Soc. 123*, pp. 565–570, 1976.

137. M.C. Chen and V.J. Silvestri, "Post-Epitaxial Polysilicon and Si_3N_4 Gettering in Silicon," *J. Electrochem. Soc. 129*, pp. 1294–1299, 1982.

138. W.T. Stacy et al., "The Microstructure of Polysilicon Backsurface Gettering," *Electrochem. Soc. Ext. Abst. 83–1*, abst. no. 310, pp. 484–485, 1983.

139. W.J.M.J. Jousquin and M.J.E. Ulenaers, "Oxidation-Induced Defects at the Poly/Mono Silicon Interface," *J. Electrochem. Soc. 131*, pp. 2380–2386, 1984.

140. M. Waldner and L. Sivo, "Lifetime Preservation in Diffused Silicon," *J. Electrochem. Soc. 107*, pp. 298–301, 1960.

141. R.S. Ronen and P.H. Robinson, "Hydrogen Chloride and Chlorine Gettering as an Effective Technique for Improving Performance of Silicon Devices," *J. Electrochem. Soc. 119*, pp. 747–752, 1972.

142. D.R. Young and C.M. Osburn, "Minority Carrier Generation Studies in MOS Capacitors on n-Type Silicon," *J. Electrochem. Soc. 120*, pp. 1578–1581 and references therein, 1973.

143. P.D. Esqueda and M.B. Das, "Dependence of Minority Carrier Bulk Generation in Silicon MOS Structures on HCl Concentration in an Oxidizing Ambient," *Solid-State Electronics 23*, pp. 741–746, 1980.

144. S.D. Brotherton, "Electrical Properties of Gold at the Silicon–Dielectric Interface," *J. Appl. Phys. 42*, pp. 2085–2094, 1971.

145. Thomas A. Baginski and Joseph R. Monkowski, "The Role of Chlorine in the Gettering of Metallic Impurities from Silicon," *J. Electrochem. Soc. 132*, pp. 2031–2033, 1985.

146. S.J. Silverman and J.B. Singleton, "Technique for Preserving Lifetime in Diffused Silicon," *J. Electrochem. Soc. 105*, pp. 591–594, 1958.

147. N. Momma et al., "Gettering of Gold and Copper in Silicon during Gallium Diffusion," *J. Electrochem. Soc. 125*, pp. 963–968, 1978.

148. P. Rava et al., "Thermally Activated Oxygen Donors in Si," *J. Electrochem. Soc. 129*, pp. 2844–2849 and references therein, 1982.

149. C.Y. Tan et al., "Intrinsic Gettering by Oxide Precipitation Induced Dislocations in Czochralski Si," *Appl. Phys. Lett. 30*, pp. 175–176, 1977.

150. Robert A. Craven, "Oxygen Precipitation in Czochralski Silicon," pp. 254–271, in Howard R. Huff and Rudolph J. Kriegler, eds., *Semiconductor Silicon/81*, Electrochemical Society, Pennington, N.J., 1981.

151. John Andrews, "Oxygen Out-Diffusion Model for Denuded Zone Formation in Czochralski-Grown Silicon with High Interstitial Oxygen Content," *Electrochem. Soc. Ext. Abst. 83–1*, abst. no. 271, pp. 415–416, 1983.

152. M. Stavola et al., "Diffusivity of Oxygen at the Donor Formation Temperature," *Appl. Phys. Lett. 42*, pp. 73–75, 1983.

153. N. Inoue et al., "Oxygen Precipitation in Czochralski Silicon—Mechanism and Application," pp. 382–393, in Howard R. Huff and Rudolph J. Kriegler, eds., *Semiconductor Silicon/83*, Electrochemical Society, Pennington, N.J., 1981.

154. Charles W. Pearce, "Defect Contamination Control in Silicon Wafer Processing," *Proc. Third Annual Microelectronics Measurement Techniques*, pp. v27–v52 and references therein, 1981.

155. R.A. Hartzell et al., "A Model That Describes the Role of Oxygen, Carbon, and Silicon Interstitials in Silicon Wafers during Device Processing," *Proc. Materials Research Society Symposium 36*, pp. 217–222, 1985.

156. B. Rogers et al., "Computer Simulation of Oxygen Precipitation and Denuded Zone Formation," pp. 74–84, in Kenneth E. Bean and George Rozgonyi, eds., *VLSI Science and Technology/84*, Electrochemical Society, Pennington, N.J., 1984.

157. S. Blackstone et al., "Microdefects during Phosphorus Diffusion," *J. Electrochem. Soc. 129*, pp. 667–668, 1982.

158. H.J. Queisser, "Slip Patterns on Boron-Doped Silicon Surfaces," *J. Appl. Phys. 32*, pp. 1776–1780, 1961.

159. S. Prussin, "Generation and Distribution of Dislocations by Solute Diffusion," *J. Appl. Phys. 32*, pp. 1876–1881, 1961.

160. J.E. Lawrence, "Diffusion Induced Stress and Lattice Disorders in Silicon," *J. Electrochem. Soc. 113*, pp. 819–824, 1966.

161. K.G. McQuhae and A.S. Brown, "The Lattice Contraction Coefficient of Boron and Phospho-

rus in Silicon," *Solid-State Electronics 15*, pp. 259–264, 1972.

162. R.B. Fair, "The Effect of Strain-Induced Band-Gap Narrowing on High Concentration Phosphorus Diffusion in Silicon," *J. Appl. Phys. 50*, pp. 860–868, 1979.
163. R.A. McDonald et al., "Control of Diffusion Induced Dislocations in Phosphorus Diffused Silicon," *Solid-State Electronics 9*, pp. 807–812, 1966.
164. T.H. Yeh and M.L. Joshi, "Strain Compensation in Silicon by Diffused Impurities," *J. Electrochem. Soc. 116*, pp. 73–77, 1969.
165. T.H. Yeh et al., "Diffusion of Tin into Silicon," *J. Appl. Phys. 39*, pp. 4266–4271, 1968.
166. Y. Yukimoto et al., "Effect of Tin on the Diffusion of Impurities in Transistor Structure," pp. 692–697, in Howard R. Huff and Ronald R. Burgess, eds., *Semiconductor Silicon/73*, Electrochemical Society, Pennington, N.J., 1973.
167. Satoru Matsumoto et al., "Effects of Diffusion-Induced Strain and Dislocation on Phosphorus Diffusion in Silicon," *J. Electrochem. Soc. 125*, pp. 1840–1845, 1978.
168. Kenji Morizane and Paul S. Gleim, "Thermal Stress and Plastic Deformation in Thin Silicon Slices," *J. Appl. Phys. 40*, pp. 4104–4107, 1969.
169. S.M. Hu, "Temperature Distribution and Stresses in Circular Wafers in a Row during Radiative Cooling," *J. Appl. Phys. 40*, pp. 4413–4423, 1969.
170. B. Leroy and C. Plougonven, "Warpage of Silicon Wafers," *J. Electrochem. Soc. 127*, pp. 961–970, 1980.
171. D. Thebault and L. Jastrzebski, "Review of Factors Affecting Warpage of Silicon Wafers," *RCA Review 41*, pp. 592–611, 1980.
172. A.E. Widmer and W. Rehwald, "Thermoplastic Deformation of Silicon Wafers," *J. Electrochem. Soc. 133*, pp. 2403–2409, 1986.
173. G.L. Pearson et al., "Deformation and Fracture of Small Silicon Crystals," *Acta Metallurgica 5*, pp. 181–191, 1957.
174. W.D. Sylwestrowicz, "Mechanical Properties of Single Crystals of Silicon," *Phil. Mag. Series 8, Vol. 7*, pp. 1825–1845, 1962.
175. Masato Imai and Koji Sumino, "In Situ X-Ray Topograph Study of the Dislocation Mobility in High-Purity and Impurity Doped Silicon Crystals," *Phil. Mag. Series A, Vol. 47*, pp. 599–621, 1983.
176. J.R. Patel and A.R. Chaudhuri, "Oxygen Precipitation Effects on the Deformation of Dislocation-Free Silicon," *J. Appl. Phys. 33*, pp. 2223–2224, 1962.
177. S.M. Hu and W.J. Patrick, "Effect of Oxygen on Dislocation Movement in Silicon," *J. Appl. Phys. 46*, pp. 1869–1883, 1975.
178. K. Yasutake et al., "Mechanical Properties of Heat-Treated Czochralski-Grown Silicon Crystals," *Appl. Phys. Lett. 73*, pp. 789–791, 1980.
179. Yojiro Kondo, "Plastic Deformation and Preheat Treatment Effects in CZ and FZ Silicon Crystals," pp. 220–231, in Howard R. Huff et al., eds., *Semiconductor Silicon/81*, Electrochemical Society, Pennington, N.J., 1981.
180. Ichiro Yonenga and Koji Sumino, "Mechanical Strength of Silicon Crystals as a Function of the Oxygen Concentration," *J. Appl. Phys. 56*, pp. 2346–2350, 1984.
181. T. Abe et al., "Impurities in Silicon Single Crystals," pp. 54–71, in Howard R. Huff et al., eds., *Semiconductor Silicon/81*, Electrochemical Society, Pennington, N.J., 1981.
182. Koji Sumino et al., "Effects of Nitrogen on Dislocation Behavior and Mechanical Strength in Silicon Crystals," *J. Appl. Phys. 54*, pp. 5016–5020, 1983.
183. Joe Lambert and Chris Bayne, "A Suspended Boat Loader Based on the Cantilever Principle," *Semiconductor International*, pp. 150–155, 1983.
184. Arthur Waugh and Bryan D. Foster, "Design and Performance of Silicon Carbide Cantilever Paddles in Semiconductor Diffusion Furnaces," *American Ceramic Society Bul. 64*, pp. 550–554, 1985.
185. Pravin C. Parekh and David R. Goldstein, "The Influence of Reaction Kinetics between BBr_3 and O_2 on the Uniformity of Base Diffusion," *Proc. IEEE 57*, pp. 1507–1512, 1969.
186. G.M. Oleszek and W.M. Whittemore, "The Effect of Process Variables on the Open-Tube Diffusion of Boron into Silicon from Boron Tribromide," pp. 490–501, in Rolf R. Haberecht and Edward L. Kern, eds., *Semiconductor Silicon/69*, Electrochemical Society, New York, 1969.

187. P. Negrini et al., "Boron Predeposition in Silicon Using BBr_3," *J. Electrochem. Soc. 125*, pp. 609–613, 1978.
188. Gary Calson, Texas Instruments Incorporated, unpublished work.
189. R.M. Burger and R.P. Donovan, eds., *Fundamentals of Silicon Integrated Device Technology*, Prentice-Hall, Englewood Cliffs, N.J., 1967.
190. W Greig et al., "Diffusion Technology for Advanced Microelectronics Processing," pp. 168–174, in Charles P. Marsden, ed., *Silicon Device Processing*, NBS Special Pub. 337, 1970.
191. P.F. Schmidt and R. Stickler, "Silicon Phosphide Precipitates in Diffused Silicon," *J. Electrochem. Soc. 111*, pp. 1188–1189, 1964.
192. P.C. Parekh, "On the Uniformity of Phosphorus Emitter Concentration for Shallow Diffused Transistors," *J. Electrochem. Soc. 119*, pp. 173–177, 1972.
193. K. Strater and A. Mayer, "The Oxidation of Silane, Phosphine and Diborane during Deposition of Doped Oxide Diffusion Sources," pp. 469–480, in Rolf R. Haberecht and Edward L. Kern, eds., *Semiconductor Silicon/69*, Electrochemical Society, New York, 1969.
194. N. Goldsmith et al., "Boron Nitride as a Diffusion Source for Silicon," *RCA Review 28*, pp. 344–350, 1967.
195. David Rupprecht and Joseph Stach, "Oxidized Boron Nitride Wafers as an In-Situ Boron Dopant for Silicon Diffusions," *J. Electrochem. Soc. 120*, pp. 1266–1271, 1973.
196. J. Stach and J. Kruest, "A Versatile Boron Diffusion Process," *Solid State Technology*, pp. 60–67, October 1976.
197. J.J. Steslow et al., "Advances in Solid Planar Dopant Sources for Silicon," *Solid State Technology*, pp. 31–34, January 1975.
198. N. Jones et al., "A Solid Planar Source for Phosphorus Diffusion," *J. Electrochem. Soc. 123*, pp. 1565–1569, 1976.
199. R. Wheeler and J.E. Rapp, "Improved Transistor Characteristics Using Solid Planar Diffusion Sources for Emitter Diffusion," *Solid State Technology*, pp. 203–205, August 1985.
200. R.E. Tressler et al., "Present Status of Arsenic Planar Diffusion Sources," *Solid State Technology*, pp. 165–171, October 1984.
201. M.L. Barry, "Diffusion from Doped Oxide Sources," pp. 175–181, in Charles P. Marsden, ed., *Silicon Device Processing*, NBS Special Pub. 337 and references therein, 1970.
202. K. Reindl, "Spun-On Arsenosilica Films as Sources for Shallow Arsenic Diffusions with High Surface Concentration," *Solid-State Electronics 16*, pp. 181–189, 1973.
203. B.H. Justice et al., "Diffusion Processing of Arsenic Spin-On Diffusion Sources," *Solid State Technology*, pp. 39–42, July 1978.
204. T.C. Chandler et al., "Debris-Induced Effects from Spin-On Diffusion Sources," *J. Electrochem. Soc. 126*, pp. 2216–2220, 1979.
205. B.H. Justice et al., "A Novel Spin-On Dopant," *Solid State Technology*, pp. 153–159, October 1984.
206. D.M. Brown et al., "Characteristics of Doped Oxides and Their Use in Silicon Device Fabrication," *J. Crystal Growth 17*, pp. 276–287, 1972.
207. M. Gezzo and D.M. Brown, "Arsenic Glass Source Diffusion in Si and SiO_2," *J. Electrochem. Soc. 120*, pp. 110–116, 1973.
208. B. Jayant Baliga and Sorab K. Ghandhi, "Planar Diffusion in Gallium Arsenide from Tin-Doped Oxides," *J. Electrochem. Soc. 126*, pp. 135–138, 1979.
209. R. Gereth et al., "Solid–Solid Vacuum Diffusion Processes in Silicon," *J. Electrochem. Soc. 120*, pp. 966–971, 1973.
210. A. Kostka et al., "A Physical and Mathematical Approach to Mass Transport in Capsule Diffusion Processes," *J. Electrochem. Soc. 120*, pp. 971–974, 1973.
211. K.K. Shih, "High Surface Concentration Zn Diffusion in GaAs," *J. Electrochem. Soc. 123*, pp. 1737–1740, 1976.
212. O. Hasegawa and R. Namazu, "Zn Diffusion into GaAs by a Two-Temperature Method," *Appl. Phys. Lett. 36*, pp. 203–205, 1980.

CHAPTER 9

Ion Implantation

9.1 INTRODUCTION

Ions that have been accelerated by several kilovolts have enough energy to penetrate a solid surface and, unlike the ions of a typical diffusion process, can do so even when the solid is at room temperature. Such an operation is referred to as ion implantation and offers an alternative to thermal diffusion for introducing impurities into a semiconductor surface. The penetration depth is quite small, and unlike diffusion, ion implantation produces the maximum concentration beneath the wafer surface, as is illustrated in Fig. 9.1. The depth of the concentration peak increases as the accelerating voltage increases, and the total number of ions injected is proportional to the beam current and implant time. The depth spread depends inversely on the ratio of the mass of the host atom to the implanted ion.

As-implanted dopants are generally not in the proper lattice position and are mostly electrically inactive. In addition, the implant operation can generate substantial damage to the host crystal lattice. Most of the crystal damage and the electrical inactivity can be corrected by appropriate high-temperature anneals. Thus, while an implant can be, and generally is, done near room temperature, before it is useful, the semiconductor must be subjected to a high temperature heat treatment.

Applications of ion implanting to the semiconductor industry can grouped into the following categories:

1. A source of doping atoms.
2. A method of introducing gettering damage into the semiconductor.
3. The introduction of a layer of different composition into the wafer.
4. A means of supplying a known quantity of atoms for subsequent study.

Use as a source of doping atoms is the oldest and the most important implant application. As discussed in Chapter 8, in most cases, im-

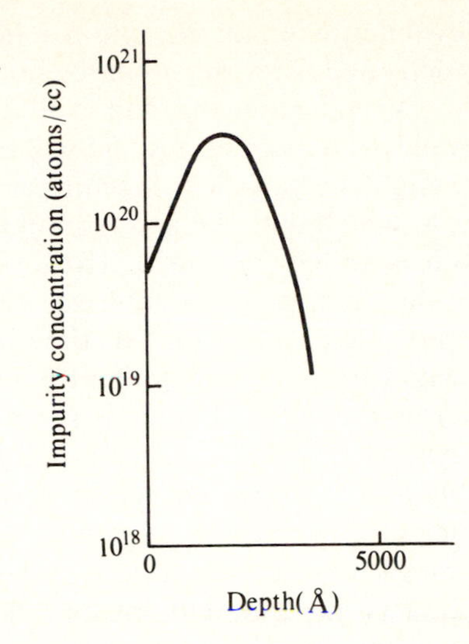

FIGURE 9.1

Typical implanted impurity profile showing that, unlike a diffused profile, the concentration peak can be well below surface.

plantation can be used directly as an alternative to a diffusion predeposition. However, if no clear-cut advantage exists, diffusion is preferred since it is a less-expensive step. Indeed, for several years, this was the case for all diffusions, and high-volume ion implant operations were restricted to MOS transistor threshold adjusts, in which there is no reasonable alternative. Then, as shallower diffusions and lower-temperature processing became necessary, ion implantation became more appropriate than diffusion for some kinds of bipolar and MOS transistors. In the case of gallium arsenide, where substantial decomposition may occur during a diffusion predep, ion implantation is used almost exclusively.

Implantation can introduce an appreciable amount of damage, even to the point of producing amorphous layers in many cases. In conventional applications, the damage presents a problem and must be removed. However, some applications, such as gettering and the enhanced mixing of atoms at interfaces, depend on the damage.

By implanting a high concentration of oxygen or nitrogen in a silicon wafer, a discrete silicon oxide or nitride layer can be formed beneath the surface. If the implanting is done in a manner that does not amorphize the single-crystal silicon above the oxide or nitride, the result is a buried insulating layer surrounded with single-crystal silicon.

Ion implantation affords a relatively simple means of placing a known number of atoms in a wafer. Such a capability is very useful in determining, for example, the effect of a specific amount of impurity on lifetime.

9.2 IMPLANT DEPTH (RANGE)

As a low-energy ion moves into a solid, it will follow a zigzag path as it is deflected from one atom to another by nuclear collisions (see Fig. 9.2). Very-high-energy ions will initially lose much of their energy by electron interactions and will travel a relatively straight path until enough energy is lost for nuclear collisions to begin absorbing energy, at which time the zigzag path will begin. The total length of this path is called the range. However, it is the depth below the surface that is of practical interest, and that distance is referred to as the projected range—that is, the range projected onto the x axis as shown in Fig. 9.2. The amount of lateral travel (spreading of the beam) from the point of entry is given by projecting the range onto the y axis. Unlike that of diffusion, lateral penetration here is considerably less than penetration in the direction of the beam. Since each ion is subjected to a somewhat different set of conditions, the actual range will be different for each ion. The projected range just referred to is the arithmetic mean (average) of the projected range of a large number of implanted ions. To a first approximation, the scatter can be described by the standard deviation of the projected ranges.

9.2.1 Range Calculations

The range can be expressed in terms of the distance of travel required for the energy of the incoming ion to be reduced from its initial value of E_0 to 0. The incremental energy loss $dE/d\zeta$, where ζ is a coordinate along the path of travel (and whose direction will change with each collision), is given by

$$\frac{dE}{d\zeta} = -N[S_n(E) + S_e(E)] \tag{9.1}$$

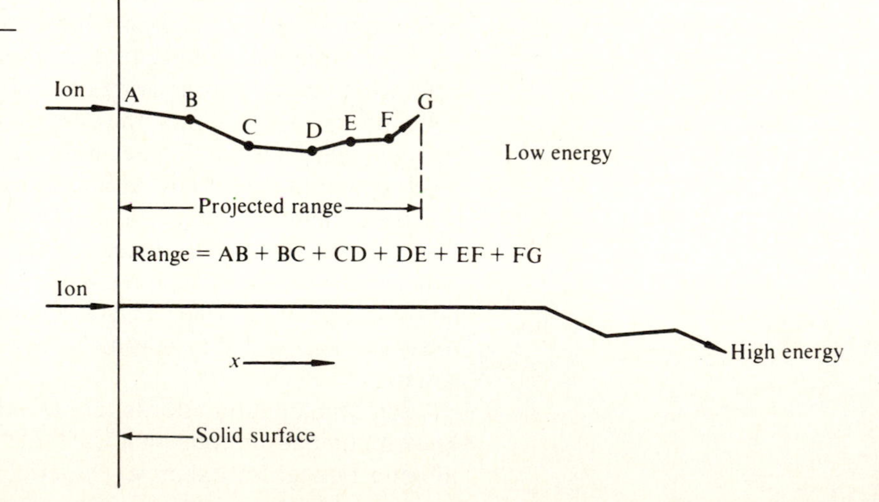

FIGURE 9.2

Paths of low- and high-energy ions. (The high-energy ion initially follows a straight path and loses energy by electron interactions. When it slows down sufficiently, nuclear collisions become most important.)

where N is the number of atoms per unit volume in the material being implanted, S_n is the nuclear stopping power, and S_e is the electronic stopping power. From Eq. 9.1,

$$\int d\zeta = -\frac{1}{N}\int \frac{dE}{S_n + S_e} \qquad 9.2$$

where the integral of $d\zeta$ is just the range R. Thus, if expressions for S_n and S_e are available, Eq. 9.2 can be integrated to give the range. To a first approximation (1, 2),

$$S_n = \frac{2.8 \times 10^{-15} M_1 Z_1 Z_2}{(M_1 + M_2)(Z_1^{2/3} + Z_2^{2/3})^{1/2}} \quad \text{eV-cm}^2 \qquad 9.3$$

where Z_1 is the atomic number of the ion, Z_2 is the atomic number of the target material, M_1 is the atomic mass of the ion, and M_2 is the atomic mass of the target. In this approximation, S_n is independent of the ion energy. In a similar approximation,

$$S_e = KE^{1/2} \quad \text{eV}-\text{cm}^2 \qquad 9.4$$

where K is proportional to $Z_1 + Z_2$ and is, for a silicon target, $0.2 \times 10^{-15}\sqrt{\text{eV}}\ \text{cm}^2$. Experimentally, as shown in Fig. 9.3, S_e oscillates with Z (3, 4), at least for lower Z numbers. By assuming a somewhat different electronic screening function than was assumed in the original derivation, the shape of the curve of Fig. 9.3 can be reasonably well predicted theoretically (5).

What constitutes high-energy ions in relation to determining which loss mechanism dominates depends on the mass and charge

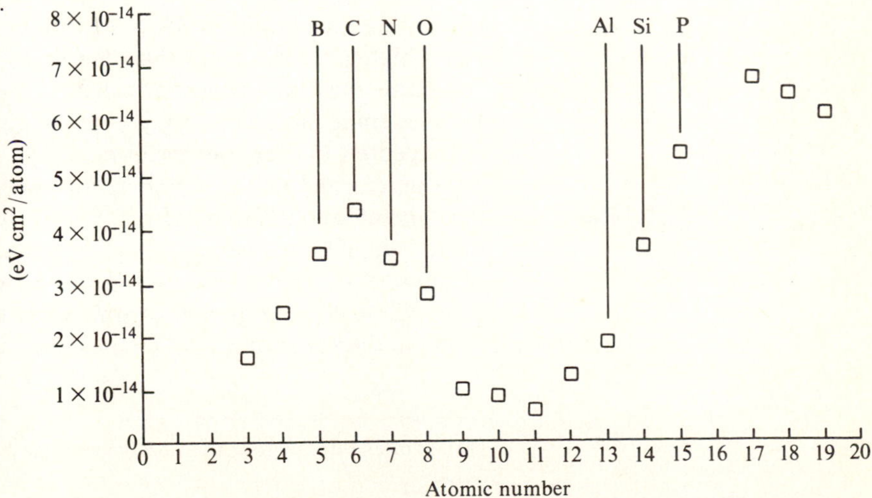

FIGURE 9.3

Electronic stopping power S_e versus atomic number for ions traveling with initial velocity of 1.5×10^8 cm/s in a [110] direction of silicon. (*Source:* From data in F.H. Eisen, *Can. J. Phys.* *46*, p. 561, 1968, and James Comas and Robert G. Wilson, *J. Appl. Phys.* *51*, p. 3697, 1980.)

FIGURE 9.4

Approximate energy of ion being implanted into silicon for which nuclear and electronic stopping powers are equal.

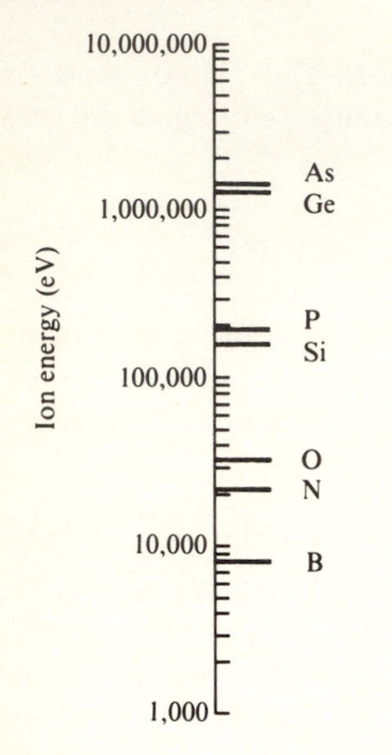

of the ion. As an approximation, for energies of E_c, given by Eq. 9.5, losses from nuclear collisions equal those from electron interactions (1):

$$E_c = \frac{14Z_1Z_2M_1}{(M_1 + M_2)(Z_1^{2/3} + Z_2^{2/3})^{1/2}} \qquad 9.5$$

As the energy increases above E_c, the electronic contribution becomes dominant. Fig. 9.4 shows E_c for several materials used in silicon IC processing. Thus, except for boron, nitrogen, and perhaps oxygen, nuclear interactions will dominate since most implants use single-charge-state ions and energies of less than 200 keV.

In order to calculate the projected range R_p with reasonable accuracy, an integral much more complex than that of Eq. 9.2 must be used (1, 6).[1] In amorphous materials, the projected range can be roughly characterized by a mean depth and a standard deviation. These two parameters have been calculated for a wide variety of ions, targets, and energies and are available in book form (7, 8). Also, simplified approaches can be run on programmable calculators, and somewhat more sophisticated approaches can be used with relatively simple FORTRAN programs (9).

To a first approximation, the depth depends only on the implant ion mass and its energy. The standard deviation of the depth, however, depends on the ratio of the mass of the implant ion to the host atom. The larger the ratio, the smaller the standard deviation. As a reasonable approximation in the region where S_n dominates, the projected range is given by (1)

$$R_p = \frac{3M_1R}{3M_1 + M_2} \qquad 9.6$$

and is more accurate for $M_1 > M_2$, in which case the implanted ion has a higher atomic mass than the matrix. Calculated values (7) of R_p versus implant energy for selected impurities in silicon and gallium arsenide are given in Figs. 9.5 and 9.6. From Eq. 9.3, it would be expected that at low beam energies, the range would decrease with increasing atomic number, and that is what is shown in Fig. 9.5. The atomic numbers of most of the ions of interest for implanting in silicon lie between those of boron and antimony, so the ranges of interest are included between these two curves. At higher energies, where S_e is of greater significance, the range will then depend more on the value of S_e, which oscillates as Z increases. Thus, while

[1]This is often referred to as the LSS theory after the authors Linhard, Scharff, and Schiøtt.

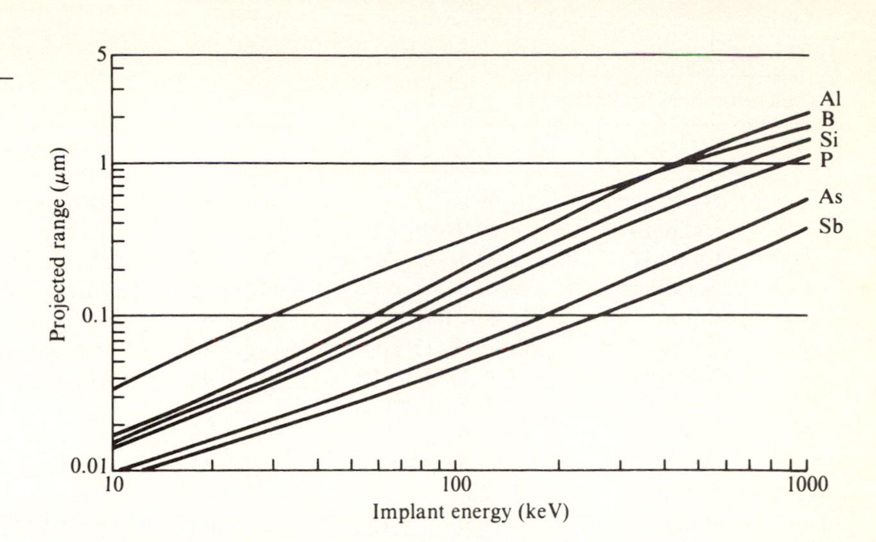

FIGURE 9.5

Projected range versus implant energy for ions in silicon. (*Source:* From data in James F. Gibbons et al., *Projected Range Statistics,* 2d ed., Dowden, Hutchinson and Ross, Stroudsburg, Pa., 1975.)

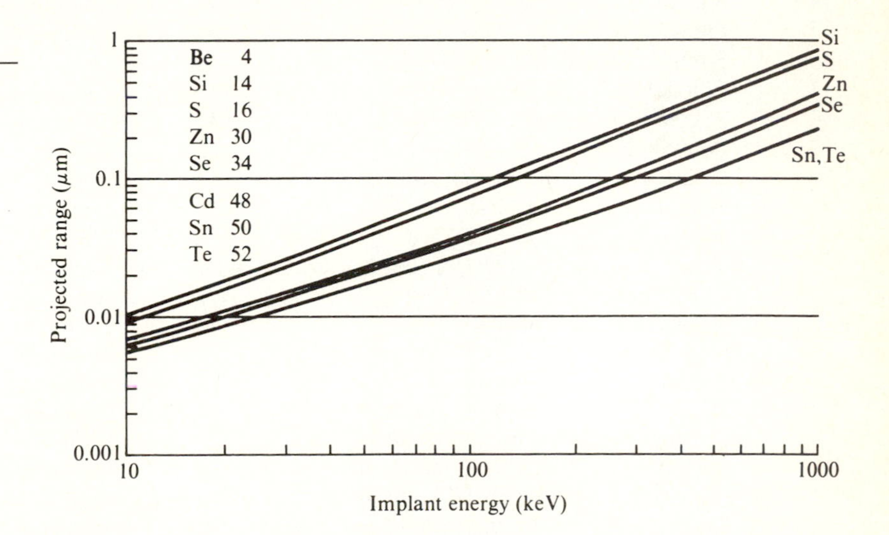

FIGURE 9.6

Projected range versus implant energy for ions in gallium arsenide. (*Source:* From data in James F. Gibbons et al., *Projected Range Statistics,* 2d ed., Dowden, Hutchinson and Ross, Stroudsburg, Pa., 1975.)

the ranges of aluminum, silicon, and phosphorus (atomic numbers of 13, 14, and 15) are very close together for an implant energy of 10 keV, because of the S_e variation shown in Fig. 9.3, a much wider separation occurs at 1 MeV. For reference, Table 9.1 lists for many of the elements the atomic number, mass, and relative abundance of the two major isotopes.

9.2.2 Range Dispersion

The projected range R_p is defined as $R_p = \Sigma r_p/N$ where r_p is the projected range of an individual ion and N is the total number of ions. If the ranges are distributed in a "normal" fashion about R_p,

TABLE 9.1

Atomic Numbers and Masses

No.	*Element*	*Mass(% Abundance)**	*No.*	*Element*	*Mass(% Abundance)*
1	Hydrogen	1(99.985)	35	Bromine	79(50.7), 81(49.3)
2	Helium	4(99.99986)	36	Krypton	84(57.0), 86(17.3)
3	Lithium	7(92.5), 6(7.5)	37	Rubidium	85(72.2), 87(27.8)
4	Beryllium	9(100)	38	Strontium	88(82.6), 86(9.9)
5	Boron	11(80), 10(20)	39	Yttrium	89(100)
6	Carbon	12(98.9), 13(1.1)	40	Zirconium	90(51.5), 94(17.3)
7	Nitrogen	14(99.6), 15(0.4)	41	Niobium	93(100)
8	Oxygen	16(99.76), 18(0.2)	42	Molybdenum	98(24.1), 96(16.7)
9	Fluorine	19(100)	43	—	—
10	Neon	20(90.5), 22(9.2)	44	Ruthenium	102(31.6), 104(18.7)
11	Sodium	23(100)	45	Rhodium	103(100)
12	Magnesium	24(79.0), 25(11.0)	46	Palladium	106(27.3), 108(26.5)
13	Aluminum	30(100)	47	Silver	107(51.8), 109(48.2)
14	Silicon	28(92.2), 29(4.67)	48	Cadmium	114(28.7), 112(24.1)
15	Phosphorus	31(100)	49	Indium	115(95.5), 113(4.5)
16	Sulfur	32(95), 34(4.2)	50	Tin	120(32.4), 118(24.3)
17	Chlorine	35(76), 37(24)	51	Antimony	121(57.3), 123(42.7)
18	Argon	40(99.0), 38(0.6)	52	Tellurium	130(33.8), 128(31.7)
19	Potassium	39(93.3), 41(6.7)	53	Iodine	127(100)
20	Calcium	40(96.94), 44(2.1)	54	Xenon	132(26.9), 129(26.4)
21	Scandium	45(100)	55	Cesium	133(100)
22	Titanium	48(73.8), 47(8.0)	56	Barium	138(71.7), 137(11.2)
23	Vanadium	51(99.75), 50(0.25)	57–71	Rare earths	—
24	Chromium	52(83.8), 53(9.5)	72	Hafnium	180(35.2), 178(27.1)
25	Manganese	55(100)	73	Tantalum	181(99.98)
26	Iron	56(91.7), 54(5.8)	74	Tungsten	184(30.7), 186(28.6)
27	Cobalt	59(100)	75	Rhenium	187(62.6), 185(37.4)
28	Nickel	58(68.3), 60(26.1)	76	Osmium	192(41.0), 190(26.4)
29	Copper	63(69.2), 65(30.8)	77	Iridium	193(62.7), 191(37.3)
30	Zinc	64(48.6), 66(27.9)	78	Platinum	195(33.8), 194(32.9)
31	Gallium	69(60.1), 71(39.9)	79	Gold	197(100)
32	Germanium	74(36.5), 72(27.4)	80	Mercury	202(29.6), 200(23.1)
33	Arsenic	75(100)	81	Thallium	205(70.5), 203(29.5)
34	Selenium	80(49.6), 78(23.5)	82	Lead	208(52.4), 206(24.1)
			83	Bismuth	209(100)

*The two most abundant isotopes are listed.

Source: Data from *Handbook of Chemistry and Physics*, 68th ed. (1987–1988), CRC Press, Boca Raton, Fla.

the distribution can be characterized in terms of the standard deviation σ where

$$\sigma = \frac{\sqrt{\Sigma R^2}}{N}$$

and $R = R_p - r_p$. The distribution of the number of ions $N(x)$ or $N(r_p)$ is given by

$$N(x) = \left(\frac{\phi}{\sigma\sqrt{2\pi}}\right)e^{-(x-R_p)^2/2\sigma^2} \qquad 9.8$$

where ϕ is the fluence per cm^2 (usually referred to as "dose"). $N(x)$ is in ions/cm^3. In implantation literature, the symbol ΔR_p is used instead of σ and is generally termed "ion straggle." Eq. 9.8 thus becomes

$$N(x) = \left(\frac{\phi}{\Delta R_p\sqrt{2\pi}}\right)e^{-(x-R_p)^2/2\Delta R_p^2} \qquad 9.9$$

At the same time as the r_p's are calculated and an R_p is determined, a σ can also be calculated and is usually tabulated along with R_p. Unfortunately, there is nothing a priori that requires the distribution of r_p to actually be normal, and there are other curves that would meet the requirements of a given R_p and σ. Experimentally, it has been observed that the distributions do appear somewhat normal, and, for some time, Eq. 9.9 was a sufficient approximation.[2] However, more sophisticated uses and the widespread implementation of computer modeling have required a better description of the distribution. When a normal distribution as in Fig. 9.7a is skewed as in Fig. 9.7b, a measure of the skewness is given by the third moment Π_3. The curve might also have either a sharper or a blunter peak as in Figs. 9.7c or 9.7d. This distribution is referred to as kurtosic. The fourth moment[3] Π_4 is a measure of kurtosis. When Π_4 is normalized by dividing by σ^2, a value of 3 gives a normal distribution, values less than 3 indicate a curve with a flatter top (pla-

---

[2]When implanting is done into crystalline rather than amorphous solids, channeling, to be described in a later section, may occur. In such cases, major deviations from a normal distribution can occur. The present discussion pertains only to amorphous implanting.

[3]The first moment Π_1 is defined as $(\Sigma R)/N$ and is zero. The second moment Π_2 is given by $(\Sigma R^2)/N$ and is just σ^2. The third moment Π_3 is $(\Sigma R^3)/N$ and will be nonzero only when there is skew. The fourth moment Π_4 is $(\Sigma R^4)/N$ and is a measure of kurtosis. A normalized skewness γ_1 is sometimes defined by Π_3/σ; a normalized kurtosis β, by Π_4/σ. In implant literature, a first moment μ_1 is sometimes defined as R_p. (For more information, see a standard text on statistics.)

FIGURE 9.7

(a) Normal (Gaussian), (b) skewed, (c) leptokurtic, and (d) platykurtic distributions.

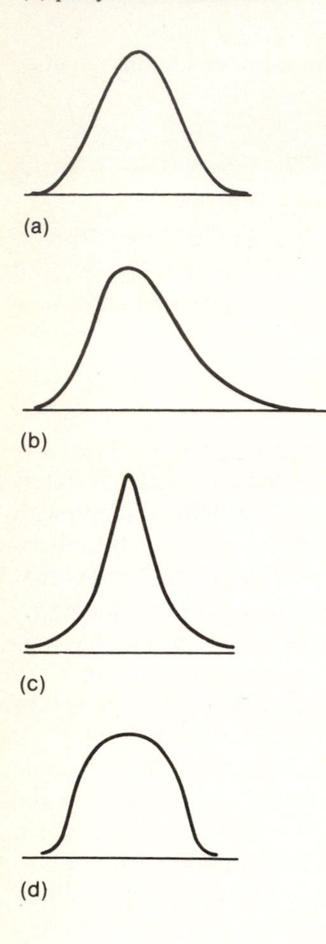

tykurtic), and values greater than 3 indicate a more peaked curve (leptokurtic).

The approach to finding an analytical expression for the implant profile has been to search for variations (usually algebraically complex) of Eq. 9.9 that will, with the tabulated inputs of R_p, σ, and Π_3, predict reasonably well the experimentally observed curves. One method of modifying Eq. 9.9 is by multiplying it by a polynomial. An early example of this approach, used long before the advent of ion implantation, was to use the first two terms of the Gram–Charlier series (10). Applied to Eq. 9.8, it gives

$$N(x) = \left(\frac{\phi}{\sigma\sqrt{2\pi}}\right)e^{-(x-R_p)^2/2\sigma^2}\left[1 - \frac{\alpha_3}{2}\left(\frac{x}{\sigma} - \frac{x^3}{3\sigma^3}\right)\right] \qquad 9.10$$

α_3 is the normalized skewness factor given by

$$\alpha_3^2 = \beta_1 = \frac{\Pi_3^2}{\Pi_2^3} \qquad 9.11$$

How the third moment and Eq. 9.10 can change a normal (Gaussian) profile is illustrated in Fig. 9.8. This particular equation gives a very poor fit with experimental implant data. However, an extension of Eq. 9.10 with more terms has been used, as has the joining of halves of two separate Gaussian curves with different σ's (11). To get a good fit of some ion profiles, the fourth moment must be also included, although it is often calculated from the third moment and hence is not an independent variable. The most successful equation is that of the Pearson IV distribution[4] (12, 13), but it is sometimes combined with an exponential function to further tailor the distribution tails. A FORTRAN program for profile calculations using Pearson IV is given in reference 13. SUPREM modeling of implant profiles is based on Pearson IV plus an exponential for each half of the distribution.

In general, the standard deviation σ decreases as the implant energy increases and as the atomic number of the implanted species increases. Fig. 9.9 shows the relative dispersion (σ/R_p) versus implant energy for several ions implanted into silicon. As Z increases, the nature of the curve changes from that of boron ($Z = 5$) to that of phosphorus ($Z = 15$) and then to that of As ($Z = 33$) and Sb ($Z = 51$). Typically, the lighter ions have a larger relative dispersion at low energies and a lower one at high energies than the heavy ions.

[4]Karl Pearson was a professor of applied mathematics at University College, London. He published tables of distribution function values that were determined uniquely by not just the second moment (normal curve), but by the second and third, and second, third, and fourth moments.

EXAMPLE ☐ Using the simple distribution equation (Eq. 9.9), determine the total implant flux ϕ required to give a peak concentration of 10^{18} atoms/cc if boron is implanted at 200 keV into silicon with negligible background doping.

Inspection of Eq. 9.9 shows that the peak $N(x)$ will occur when $x = R_p$ and that its value is given by $N(x) = (\phi/\Delta R_p\sqrt{2\pi})e^0$. ΔR_p can be found in a set of range tables, or it can be inferred from Figs. 9.5 and 9.9, which give R_p and the ratio σ/R_p (remember that $\sigma \equiv \Delta R_p$). From the figures, $R_p = 0.53$ μm and $\sigma/R_p = 0.175$ so that $\sigma = 0.093$ μm. Solving for ϕ gives

$$\phi = (10^{18}\ \text{atoms/cm}^3)(2\pi)^{0.5}(0.09\ \mu\text{m} \times 10^{-4}\ \text{cm}/\mu\text{m})$$
$$= 2.3 \times 10^{14}\ \text{atoms/cm}^2$$

☐

The impact of the change of standard deviation on the implant profile is shown in Fig. 9.10 for the specific case of boron in silicon.

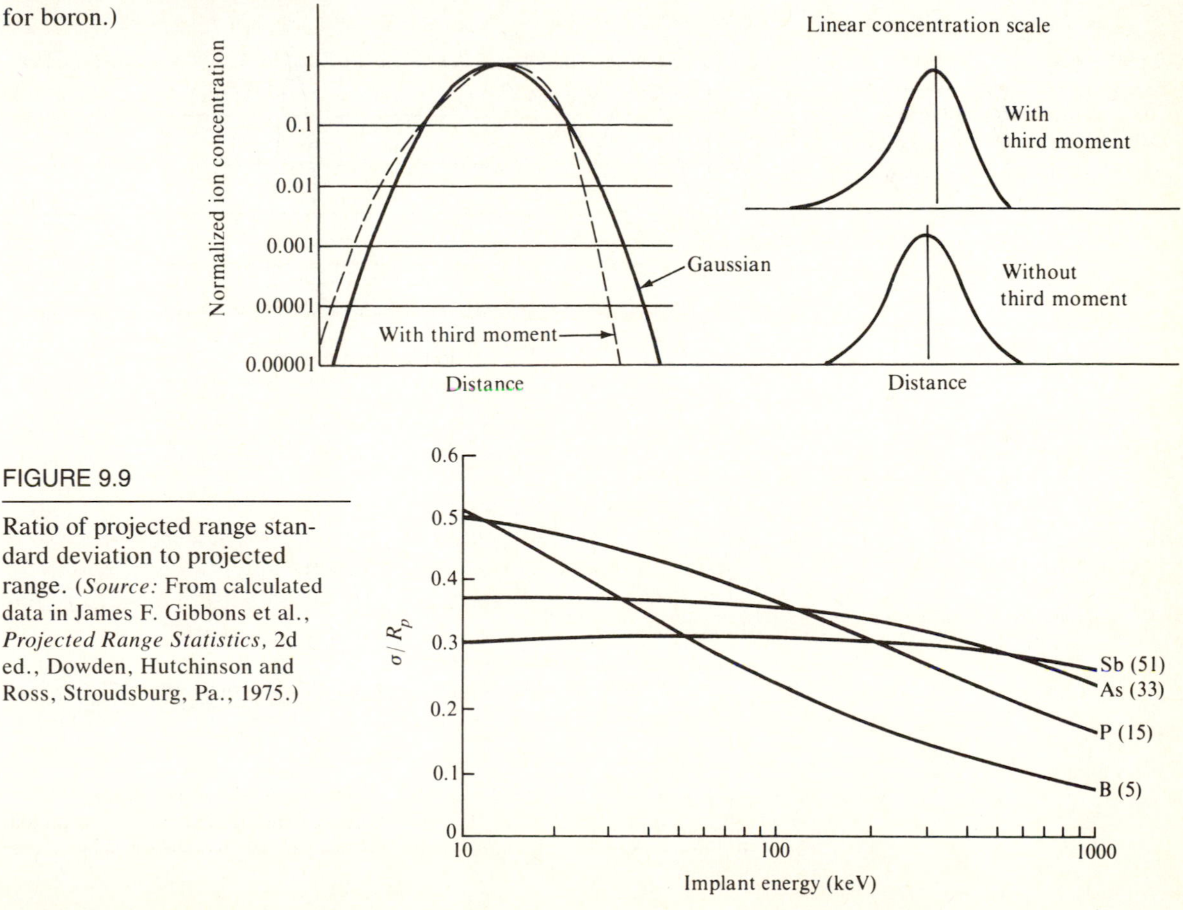

FIGURE 9.8

Example of Gaussian curve and one using a third moment to skew curve to left. (Gaussian, based on range and second moment of 150 keV boron implant in silicon; third moment, one third that calculated for boron.)

FIGURE 9.9

Ratio of projected range standard deviation to projected range. (*Source:* From calculated data in James F. Gibbons et al., *Projected Range Statistics,* 2d ed., Dowden, Hutchinson and Ross, Stroudsburg, Pa., 1975.)

The curves are plotted from the simple expression of Eq. 9.9 and make use of a high and a low implant energy to provide differing dispersions.

9.2.3 Lateral Range Dispersion

Because of the same random path travel that leads to ion straggle in the forward direction, there is also a scatter of ions perpendicular to the path of the incident beam. This number, like the standard deviation of the range, is also calculable and is listed in some range tables, such as the one in reference 7. It is referred to as lateral standard deviation $R_\perp$. $R_\perp$ is calculated based on cylindrical coordinates so that in Cartesian coordinates with the beam normal to the x axis, the lateral standard deviations σ_y and σ_z (also referred to as ΔY and ΔZ) are given by $\sigma_y = \sigma_z = R_\perp/\sqrt{2}$. $R_\perp$ is generally larger than σ and, for lighter atoms, may be twice σ at high implant energies. The values come closer together as the implant energy decreases, and as the ion mass increases, the two values are close together over the whole energy range. These trends are shown in Fig. 9.11.

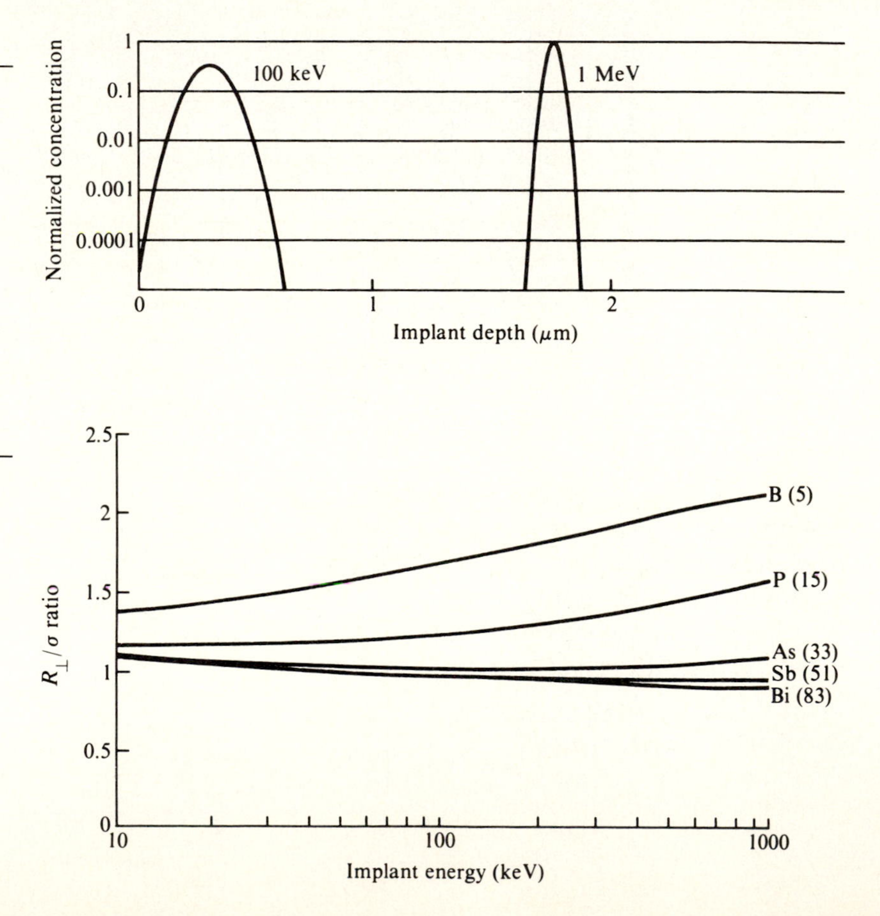

FIGURE 9.10

Effect of implant voltage on implant depth and on spread of ions, plotted using Gaussian distribution.

FIGURE 9.11

Ratio of $R_\perp/\sigma$ versus implant energy for ions of progressively higher atomic weight. (*Source:* From data in James F. Gibbons et al., *Projected Range Statistics,* 2d ed., Dowden, Hutchinson and Ross, Stroudsburg, Pa., 1975.)

The edge profile due to this scatter has been calculated for the specific case of implanting through a slit mask with the length much greater than the width and is given by (14)

$$N(x,y) = \left(\frac{\phi}{2\pi\sigma^2}\right)e^{(x-R_p)^2/2\sigma^2}\left[\left(\frac{1}{\sqrt{\pi}}\right) erf\left(\frac{x-a}{\sigma_y\sqrt{\pi}}\right)\right] \qquad 9.12$$

The erf expression does not arise from the solution to a diffusion equation but rather is an approximation when the slit length to width ratio is large. Eq. 9.12 is based on the assumption that the mask is straight walled and allows no transmission for $y > \pm a$. This gives impurity profiles like the one shown in Fig. 9.12. If the walls are tapered, then some ions will penetrate the mask and give the impression of an extralarge lateral spreading (15). Some assumptions lead to the conclusion that the lateral spreading can even be as much as that found during diffusion—that is, an amount comparable to R_p. While calculated R_p and σ values can and have been checked experimentally and found to agree very well, it is much more difficult to experimentally check Eq. 9.12 because of the masking difficulties. However, some data using implanted krypton indicate good agreement (16).

9.2.4 Channeling

The range and its standard deviation as discussed in the previous section assume that the implanted material is amorphous so that, regardless of the direction of travel of an ion, there will, on the average, be the same number of atoms per unit length of path. In single-crystal material, there are directions in which no nuclei will be encountered, and as long as the ion travels in one of those directions, its only stopping force will be due to electronic interactions (S_e) and the range will be considerably increased (channeled). The maximum channeling range for phosphorus implantation in <111> and <110> directions in single-crystal silicon, along with the cal-

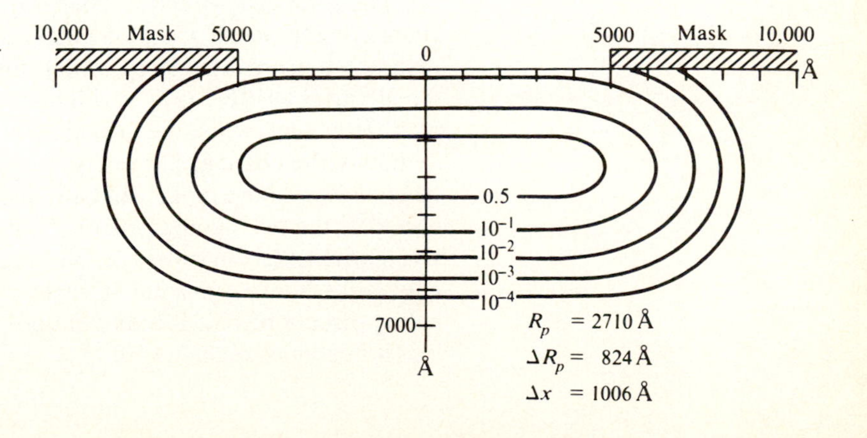

FIGURE 9.12

Calculated ion isoconcentration lines for use of implanting 70 keV boron ions into silicon through an infinitely long, 1 μm wide opening.
(*Source:* Seijiro Furukawa et al., *Jap. J. Appl. Phys. 11*, p. 134, 1972.)

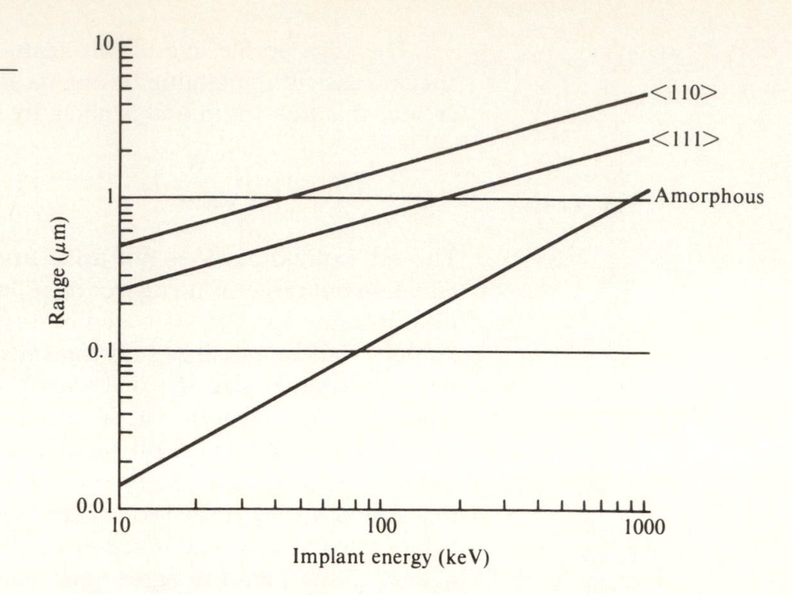

FIGURE 9.13

Maximum channeling range and amorphous range versus implant energy for phosphorus implanted in silicon.
(*Source:* From data in Goode et al., *Radiation Effects* 6, p. 237, 1970, and James F. Gibbons et al., *Projected Range Statistics, Semiconductors and Related Materials,* 2d ed., Dowden, Hutchinson, and Ross, Stroudsburg, Pa., 1975.)

culated range in amorphous silicon, is shown in Fig. 9.13. This figure shows that the maximum range varies as the square root of implant energy. It also shows that since the range for amorphous materials has a different energy dependency at low energies, the separation between the two ranges decreases as energy increases. When the energy is high enough for S_e to be the major loss mechanism, the amorphous curve will become parallel to the single-crystal channel curves but will remain somewhat below them. It is for this reason that channeling becomes less important as implant energy increases.

As larger and larger fractions of the ions are channeled, the observed profile will change as shown in Fig. 9.14. For the case of the diamond and zinc blende structures, the most open directions are the <110> and <100> (see Appendix A), but many other directions offer some degree of freedom.

The ion beam direction does not have to be exactly aligned for channeling to occur. If an ion enters an open path at an angle of less than some critical angle ψ_c, which for most implant energies of interest varies as $1/E^{1/4}$ (17), it will be trapped and channeled along the path. However, when the direction of travel is not straight down the middle of the channel, the energy loss per unit distance of travel will be greater, and the range will be reduced. There is an increased likelihood of the ion's being deflected out of the channel (dechanneled). Even when the ion enters at an angle greater than ψ_c, channeling may still occur if subsequent nuclear scattering places it in a channeling direction. It will then continue to travel in that direction unless it becomes dechanneled.

FIGURE 9.14

Change in implant profile from amorphous to maximum channeling. (*Source:* Adapted from Robert G. Wilson et al., *NBS Special Pub. 400–49,* November 1978.)

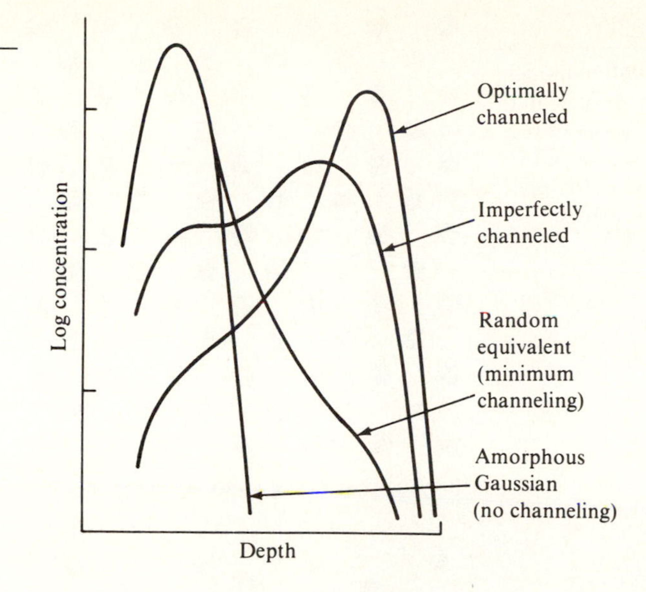

In most cases, one goal of implanting procedures is to keep the implant profile as near Gaussian as possible, which means minimizing or eliminating channeling. This can be done by preamorphizing the front surface of the wafer (18), by implanting through an amorphous layer such as thermal silicon oxide, or by changing the direction of the ion beam so that it does not enter in a direction that favors channeling. A qualitative picture of the effect of changing beam direction can be obtained from viewing a crystal model from different directions. Fig. 9.15a shows how the lattice appears when looking directly into a (100) face. A regular network of openings exists in which an ion might travel into the crystal in a <100> direction and not collide with a lattice atom. Contrast this view with the view of Fig. 9.15b, in which the crystal has been rotated 10° about the <001> zone axis and 20° about the <010> zone axis. In this case, the lattice atom positions appear to be nearly random, and no obvious paths exist down which an ion could travel without collision. Typically, misorientations of 7°–10° are used, but often little care is taken to ensure that the tilt is always about the optimum axis.

Theoretical modeling indicates that <100> and <110> axial channeling and (111) and (220) planar channeling are quite likely for boron implanted into silicon in a <100> direction (19). Planar channels are those in which the ions are constrained to move only between sets of parallel planes and not down given axial directions. Experimentally, over 30 different directions within 20° of the <100> direction (microchannels) have been observed that allow enhanced range (20). Based on both theoretical and experimental data (21), it

FIGURE 9.15

View of diamond lattice model showing position of atoms and bonds. (In part (b) some of the open spaces occur because the computer program producing the view used only a small number of atoms.)
(*Source:* Program courtesy of Dr. Anthony Stephens, Texas Instruments Incorporated.)

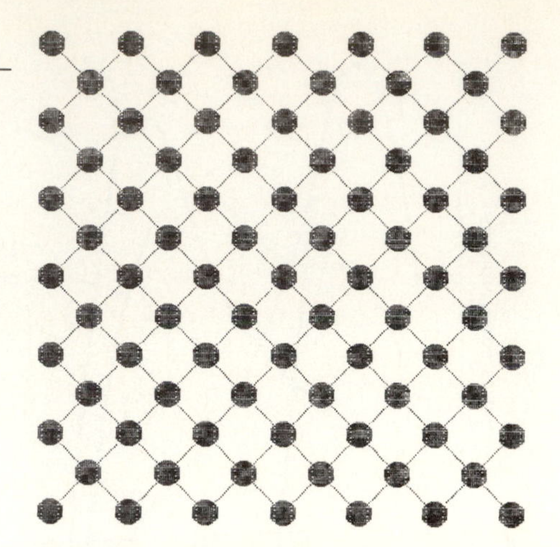

(a) Looking directly into a (100) face

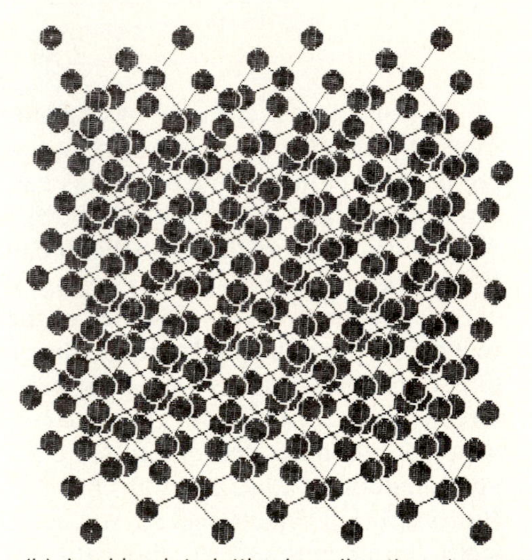

(b) Looking into lattice in a direction chosen to impede motion of an ion (see text for direction)

appears that for (100) silicon wafers, a 7° tilt about an axis making an angle of 30° with the flat (a "thirty degree twist") should minimize channeling. Views of a crystal model (similar to that of Fig. 9.15b) also indicate a high degree of randomizing when that orientation is used (22). As the implant energy decreases, progressively more tilt is required. For example, if a 4° critical angle is calculated for arsenic implanted at 80 keV, then at 10 keV, the angle would have increased to 7° and, from a practical standpoint, the tilt angle should be about twice the critical angle in order to ensure that most ions are not channeled (23).

9.2.5 Implanting through Layers

Implanting through a surface layer occurs, for example, when ion implanting is being used for threshold adjust, ion mixing, or the shifting of Schottky barrier heights. These applications are discussed in section 9.4. In addition, when a layer of some material is used as a mask to localize the implant area, implanting through the layer will occur if it is not thick enough. The more commonly encountered layer materials are photoresist, silicon oxide, and silicon nitride. The range and moments for them have also been calculated (7, 8) and can be used in estimating a composite profile. In the case of CVD layers, the range may change with deposition conditions. The tabulations indicate substantial differences in resist range values as the formulation changes. For the specific cases of AZ111 (positive resist) and KTFR (negative resist), the range and ΔR_p in μm for 200 keV implants are as follows (7, 8):

	AZ111		*KTFR*	
	R_p	ΔR_p	R_p	ΔR_p
Boron	1.84	0.15	0.67	0.072
Phosphorus	0.85	0.13	0.40	0.097
Arsenic	0.44	0.073	0.17	0.038

As a very rough approximation, the implant profile can be calculated by considering the layer and substrate as one material. A better approximation for the case of the layer having a thickness d greater than the range R_{p1} in the layer is (13, 24)

$$N(x) = \left(\frac{\phi}{\Delta R_{p1}\sqrt{2\pi}}\right) e^{-(x-R_{p1})^2/2\Delta R_{p1}^2} \quad \text{for } x < d \qquad 9.13a$$

$$N(x) = \left(\frac{\phi}{\Delta R_{p2}\sqrt{2\pi}}\right) e^{-[\{d+(R_{p1}-d)\Delta R_{p2}/\Delta R_{p1}-x\}^2/2\Delta R_{p1}^2} \quad \text{for } x > d \qquad 9.13b$$

where R_{p2} is the range in the substrate. For the case in which the layer thickness is considerably less than the range R_{p1}, an approximation that can be used with higher moment distributions (such as Pearson IV) is (13)

$$N(x) = \left(\frac{\Delta R_{p2}}{\Delta R_{p1}}\right) N(x') \quad \text{for } x < d \qquad 9.14a$$

$$N(x) = N(x'') \quad \text{for } x > d \qquad 9.14b$$

where $x' = (\Delta R_{p2}/\Delta R_{p1})x$ and $x'' = x - d(1 - \Delta R_{p2}/\Delta R_{p1})$.

In specifying masking, either the thickness of masking required to absorb a given fraction of the implanted dose or the thick-

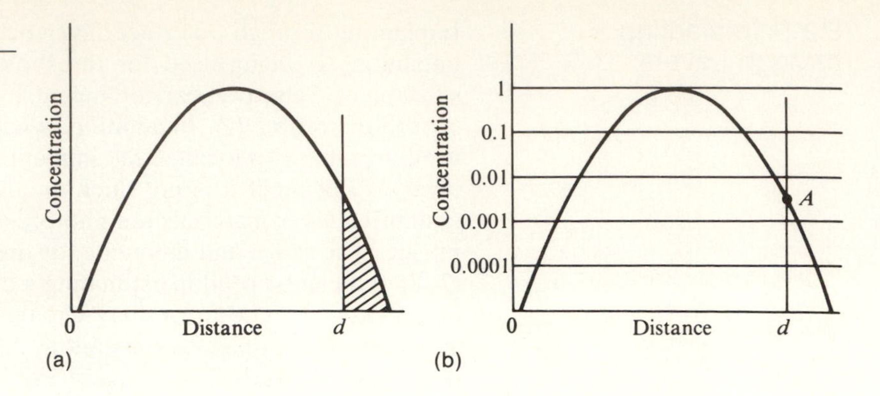

FIGURE 9.16

Masking efficiency required as expressed either in terms of allowed flux leakage (crosshatched portion of part a), or in terms of maximum concentration in material being masked (approximately at *A* in part b; see text). (The mask–semiconductor interface is at *d*.)

ness required to ensure that $N(x)$ in the masked material is less than some value may be specified. These two choices are shown in Fig. 9.16, which has been plotted from Eq. 9.9 rather than from the better approximation of Eq. 9.13. In the example shown (using a range of 0.5 μm and a ΔR_p of 0.1 μm), $\phi_e/\phi = 0.0003$, while N/N_0 is only 0.0035. ϕ_e is the number of ions per cm^2 that escape the masking layer (crosshatched portion of Fig. 9.16a), ϕ is the total dose, N is the concentration at the substrate surface (at *A* in Fig. 9.16b), and N_0 is the maximum concentration. This example illustrates the need for an almost complete blocking of ions in order to prevent problems with high surface concentration. N can be calculated[5] from Eqs. 9.9, 9.13, or 9.14. ϕ_e is given by $\int_d^\infty N(x)dx$, and when Eq. 9.9 is used,

$$\frac{\phi_e}{\phi} = \int_d^\infty \left(\frac{1}{\Delta R_p\sqrt{2\pi}}\right) e^{-[(x-R_p)^2/2\Delta R_p^2]} dx \qquad 9.15$$

Since

$$\text{erfc}(z) = \left(\frac{2}{\sqrt{\pi}}\right)\int_z^\infty e^{-z^2}dz$$

Eq. 9.15 becomes

$$\frac{\phi_e}{\phi} = \frac{1}{2}\,\text{erfc}\left(\frac{d - R_p}{\Delta R_p\sqrt{2}}\right) \qquad 9.16$$

Fig. 9.17 shows some required thicknesses, based on Eq. 9.16.

[5]It must be remembered that, like many of the other calculated values discussed in this book, substantial errors can occur in these numbers and experimental data must be taken to verify the predictions.

FIGURE 9.17

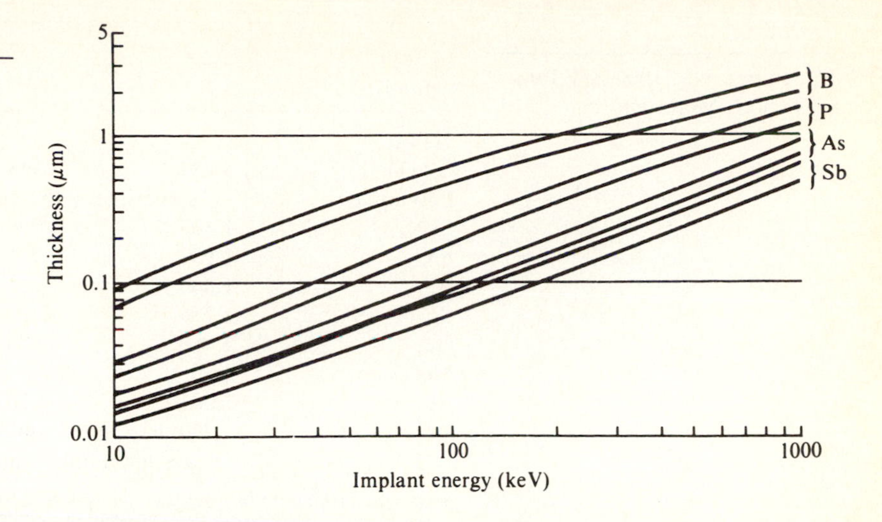

Calculated thickness of layer required to stop 99.999% of implanted ions. (For each pair, the top curve is for silicon dioxide and the bottom one is for silicon nitride.)
(*Source:* From Eq. 9.9 and James F. Gibbons et al., *Projected Range Statistics, Semiconductors and Related Materials,* 2d ed., Dowden, Hutchinson and Ross, Stroudsburg, Pa., 1975.)

EXAMPLE ☐ When a boron implant at 200 keV is used, what thickness of SiO_2 will be required to mask 99.999% of the implanted ions?

$\phi_e/\phi = 10^{-5}$ and, from the previous example, $R_p = 0.53$ μm and $\Delta R_p = 0.093$ μm. Substituting values into Eq. 9.16 gives

$$2 \times 10^{-5} = \text{erfc}\left(\frac{d \times 10^{-4} - 0.53 \times 10^{-4}}{0.93 \times 1.414}\right) = \text{erfc}(z)$$

where d is the desired thickness in μm. So $\text{erfc}(z) = 2 \times 10^{-5} = 1 - \text{erf}(z)$. Thus, $\text{erf}(z) = 0.99998$ and, from the error function table in Chapter 8, $z_c = 3.02$. Since, by definition,

$$z = \frac{d \times 10^{-4} - 0.53 \times 10^{-4}}{0.93 \times 1.414}$$

$d = 0.93$ μm, which matches the data of Fig. 9.17. ☐

When Eq. 9.16 is used, the thickness will be overestimated if the masking material has a negative third moment. If the third moment is positive and/or if there is substantial tailing (fourth moment), the thickness will be underestimated. Also, during high-dose implants, the edges of photoresist may round (25) so that because of the effect discussed in section. 9.2.3, implanted geometries may be larger than anticipated. As can be seen from Fig. 9.17 and Table 9.2, high-energy implants require an inordinate mask thickness for all of the materials normally used in IC fabrication unless implants are restricted to the higher atomic weight materials such as arsenic and antimony. Of the materials listed, gold has the shortest range but would still need to be over a micron thick to mask boron at 1000 keV implant energy. Polyimide, even though it does not have great stop-

TABLE 9.2

Range in μm of 1000 keV Ions in Various Materials

Material	Ion B	P	As	Sb
KTFR			0.849	
AZ111	5.889	3.718	2.200	1.548
Polyimide	Experimentally, appears comparable to KTFR.			
Si_3N_4	1.500	0.720	0.373	0.242
SiO_2	1.939	0.931	0.483	0.313
Al	1.802	1.025	0.518	0.332
Si	1.756	1.123	0.585	0.375
Ge	2.066	0.783	0.366	0.230
Mo			0.214	
Au	1.004	0.320	0.145	0.093
W	No data, but should be close to Au since density, atomic number, and atomic weight are similar.			

Sources: From data in reference 7, with information on KTFR from data in reference 8 and information on polyimide from T.O. Herndon et al., *Solid State Technology 179,* November 1984.

ping power, has potential because of its ability to have relatively straight-walled features etched in thick layers. Unfortunately, thick layers of organic materials sometimes out-gas enough during high-dose implants to affect implanter performance (26).

9.3 IMPLANT-INDUCED DEFECTS

As the ions being implanted move through a crystal lattice, they displace atoms in their path and thus cause a substantial amount of crystallographic damage. For most device applications, it is necessary to remove as much of the damage as possible. Further, the implanted ions generally do not come to rest in substitutional lattice sites. In order for them to exhibit the proper electrical properties, those implanted ions that would normally occupy substitutional sites must be induced to move to their proper positions. In the case of silicon (and germanium), all substitutional sites are equivalent. In the case of gallium arsenide (and other III–V compounds), there are two sites, corresponding to the normal locations of each constituent. Sometimes, even though the implanted ions move to substitutional sites, they move to the wrong ones and produce antistructure crystal defects. Thermal annealing is used both to repair the crystallographic damage produced by the passage of the implanted ions and to allow the implanted ions to move to their proper sites. Such annealing can be done slowly in a conventional diffusion tube or quickly by a rapid thermal anneal (RTA). The latter may be by scanned laser heating or by a quick exposure to high-intensity radiant heat.

9.3.1 Lattice Damage*

The kind of lattice damage initially produced depends somewhat on whether the ion is dissipating energy primarily by nuclear collisions or by inelastic electronic interactions. When nuclear collisions occur, atoms will be displaced from their substitutional sites and become interstitials, leaving vacancies behind. The energy imparted to atoms in the path is often enough to cause them to displace additional atoms so that a broad region of damage results. In some cases, each ion will leave a trail of amorphous material.

When there is enough overlapping of damage, an amorphous layer is produced in silicon. In the case of gallium arsenide, heavy lattice damage but no true amorphous regions are observed (28). A substantial difference exists between the annealing behavior of amorphous silicon regions and those that are heavily damaged but still crystalline. Ordinarily, the amorphous region is easier to regrow into damage-free material than is the region that has not quite been amorphized. Thus, the conditions for producing it have been studied extensively. For a given ion, the lower the implant temperature, the smaller the dose required for amorphizing. For a given temperature, the higher the atomic number of the ion, the smaller the dose. This behavior is shown in Fig. 9.18. The curves, which fit the experiment reasonably well, are based on a simple theory that assumes that the amorphous region formed by each ion is a right circular cylinder of length R_p and radius $r(T)$ that is temperature sensitive. The projection of the cross-sectional area of the amorphous cylinder onto the wafer surface is assumed to be A_i. The growth of the amorphized fractional area A_a is given by (27, 29)

$$\frac{dA_a}{d\phi} = A_i P_a = A_i(1 - A_a) \qquad 9.17$$

where P_a is the probability that an unamorphized area will be hit by the next ion. The solution of Eq. 9.17 is

$$A_a = 1 - e^{-A_i\phi} \qquad 9.18$$

A_i is a function of temperature through the radius r of the damage cluster. It is assumed that the outlying damage will be repaired by defects diffusing away and thus reduce the final value of r by an amount $\Delta r(T)$. The amount of diffusion will increase as the implanted wafer temperature increases. As diffusion increases, r and consequently A_i will decrease and slow down the rate of amorphization. A critical implant dose ϕ^* can be defined as the amount required to amorphize a large fraction of the total area—for example,

*See reference 27.

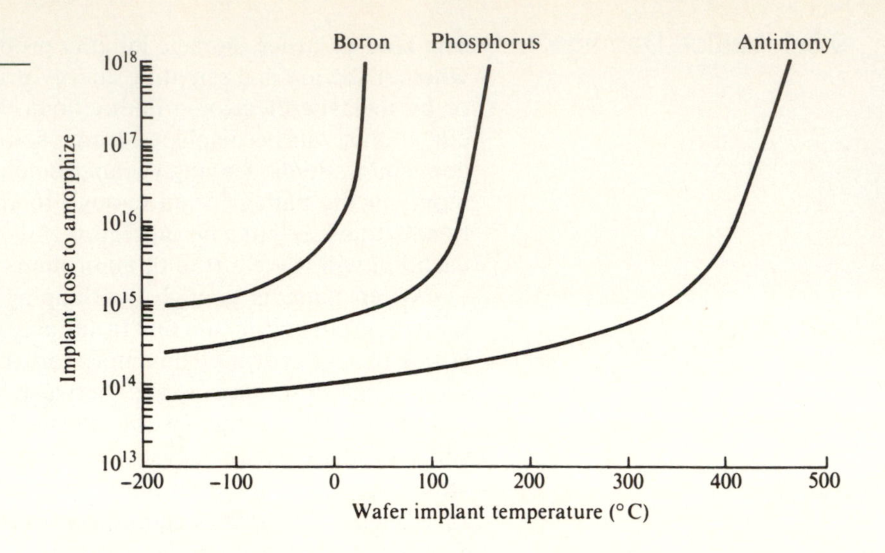

FIGURE 9.18

Effect of wafer temperature during implant on dose required to produce an amorphous layer. (*Source:* From curves in James F. Gibbons, *Proc. IEEE 60*, p. 1062, 1972.)

90%. Combining this criterion with a variation of r with T of the form $r = r_0 - \Delta r(T) = r_0 - ae^{-b/kT}$ an appropriate relation between R_p and S_n, and Eq. 9.18 gives a ϕ^* of the form

$$\phi^* = \frac{A}{[1 - B(S_n^{-1/2})e^{-C/kT}]^{1/2}} \qquad 9.19$$

where A, B, and C are constants that depend on the implant species and energy.

As the total implant dose increases, the critical value will be reached first at the peak of the distribution (at a depth R_p), and a narrow amorphous band will form. Further implanting will cause the band to increase in width, and eventually it will reach the surface. On either side of the amorphous band, a thin band of relatively undamaged material is found, and next to it is a band containing heavy damage (30).

An amorphous layer can be detected by any of the usual methods such as X-ray and electron diffraction and Rutherford backscattering. In addition, the optical properties of amorphous silicon are different enough from those of single-crystal silicon that an amorphous layer will appear cloudy. Enough reflection even occurs at the amorphous–crystalline interface to sometimes produce observable interference color (31). In visually determining amorphization, it should be kept in mind that the onset of cloudiness does not necessarily mean that the whole layer is amorphous. It has been suggested that full amorphization requires nearly a 10× larger dose (32).

When light doses such as those required for threshold adjust are implanted, the lattice damage is easily removed by annealing. When very heavy doses are used and amorphization occurs, the amorphous layer can be epitaxially regrown to give good crystalline quality silicon. For intermediate dose ranges, it is sometimes difficult to remove the damage, and often a preamorphizing step using, for example, implanted silicon, will be used.

9.3.2 Annealing

Annealing requires temperatures of above ~600°C for silicon and above 800°C or 900°C for gallium arsenide. Silicon is quite stable at high temperatures, but gallium arsenide begins to dissociate above ~600°C and thus requires a cap layer to prevent arsenic loss (33). Layers of silicon nitride and CVD silicon oxide can be used, but because of differences in their expansivity and that of gallium arsenide, they sometimes crack. Annealing can also be done in enclosed compartments with an increased partial pressure of arsenic. By laying a slice of gallium arsenide directly on top of the wafer being annealed, much of the vaporization can be prevented without the inconvenience of either a closed chamber or an added CVD cap film. The partial pressure of arsenic over a molten layer of tin on a gallium arsenide slice is higher than it is over uncoated gallium arsenide, and it is thus possible to use such coated slices in close proximity to gallium arsenide wafers to further increase the arsenic pressure (34).

Two heating regimes are of interest for annealing. In one regime, the heating is slow enough that annealing is substantially isothermal. The two ways used to accomplish this isothermal annealing are by furnace annealing over a period of many minutes or by a rapid thermal anneal (35, 36) lasting for a few seconds. In either case, however, the time is considerably longer than the thermal time constant τ of the semiconductor[6] (37). In the other regime, the heating time is comparable to or less than τ and only a thin layer is heated.

---

[6]The thermal time constant τ is somewhat analogous to the electrical *RC* time constant and arises, for example, from the approximate solution of the equation for heat flow from each face of a slab (a wafer) of thickness d to the middle of the slab. If the temperature of the faces is suddenly increased by an amount T, the first-order approximation of the temperature difference ΔT between the faces and the middle of the slab is given by $\Delta T = Ae^{-t/\tau}$, where A is a constant and $\tau = d^2/\pi^2 D_T$. D_T is not the diffusivity discussed in Chapter 8, but rather the *thermal* diffusivity, which equals $k/\rho c$ where k is the thermal conductivity of the slab, ρ is its density, and c is its specific heat. In some literature, the π^2 is dropped and τ is defined as d^2/D_T.

EXAMPLE ☐ Calculate τ for a 75 mm diameter silicon wafer that is 20 mils thick.

As long as the diameter is much greater than the thickness, its value plays no part in the calculation. The thickness is 0.051 cm and is much less than the 7.5 cm wafer diameter. $D_T = k/\rho c$ and $\rho = 2.33$ grams/cc. Since k and c are functions of temperature, the temperature should have been specified. Assume 1000°C, in which case $k = 0.29$ W/(cm·K) and $c = 1.8$ cal/(gram·°C). Since these units are not compatible, express k in terms of cal/(s·cm·°C), or $k = .07$ cal/(s·cm·°C). D_T then equals 0.17 cm²/s. Using the expression $\tau = d^2/\pi^2 D_T$, $\tau = 1.5$ ms. Thus, a rapid anneal source lasting a few seconds will heat the wafer throughout. However, laser pulses lasting a microsecond or less will heat only a thin surface layer. ☐

In some cases, surface melting may be allowed so that regrowth is by liquid phase rather than solid phase epitaxy. Generally, the activation energy for dopant diffusion is less than for implant defect annealing, so when it is necessary to minimize diffusion, higher-temperature, shorter-time anneals are preferred. However, in the extreme case where annealing is done by rapid melting and cooling of a thin layer, the impurities will be redistributed throughout the melted region. To adequately remove defects when this sort of annealing is used, melting well past the peak impurity distribution and down into the inner band of defects may be necessary. This procedure, unfortunately, will allow the impurities to redistribute to a much greater depth than that of the original implant. Even when a nonmelting, rapid anneal cycle is used, more diffusion than predicted by the standard diffusion equations is sometimes observed and is apparently due to diffusion enhancement from point defects liberated during the anneal (38).

Depending on the anneal sequence, it is possible to grow large extended defects from the implant-induced point defects during the annealing process. In particular, high-concentration implants, followed by a high-temperature anneal coupled with simultaneous oxidation, can lead to a high density of stacking faults and dislocation loops. A low-temperature—for example, 800°C—initial oxidation for a few minutes will reduce defects observed after a follow-on high-temperature oxidation (as required for many drive-ins after implantation) by orders of magnitude (39).

The conditions for optimum activation of dopant atoms and defect impurity will depend on the implant species, implant energy, and dose. In the case of silicon, low-dose boron implants can be properly annealed in the 900°C range, but heavier ion implants will require temperatures of over 1000°C. For a rapid thermal anneal time of 2 s, a temperature of 1100°C is required to fully activate and remove damage from a 10^{16} ion dose of arsenic implanted at 80 keV

(35). For gallium arsenide, a furnace anneal of 15–20 minutes at 850°C–900°C is typically used. For a given set of implant conditions, the amount of implanted impurity activated often goes through a peak, giving rise to an "optimum" anneal point. Optimum rapid thermal anneals, using incoherent heat sources, last a few seconds with maximum temperatures ranging from 800°C to over 1000°C, depending on the implant dose and species. The fraction of dopant activated is in the range of from 50% to 90%.

9.4 APPLICATIONS

Fig. 9.19 shows typical implant dose ranges for the various applications to be considered. The required dose varies over at least 7 orders of magnitude, and to satisfactorily cover this range, medium-current machines with beam currents up to about 1 mA and high-current implanters with currents up to about 20 mA are typically used. However, new requirements for implanting buried layers of oxygen and nitrogen have led to high-current versions designed to operate near 100 mA. How the beam current relates to the dose requirements shown in Fig. 9.19 depends on the time allowed for implanting each wafer. For singly charged ions (the usual case), each

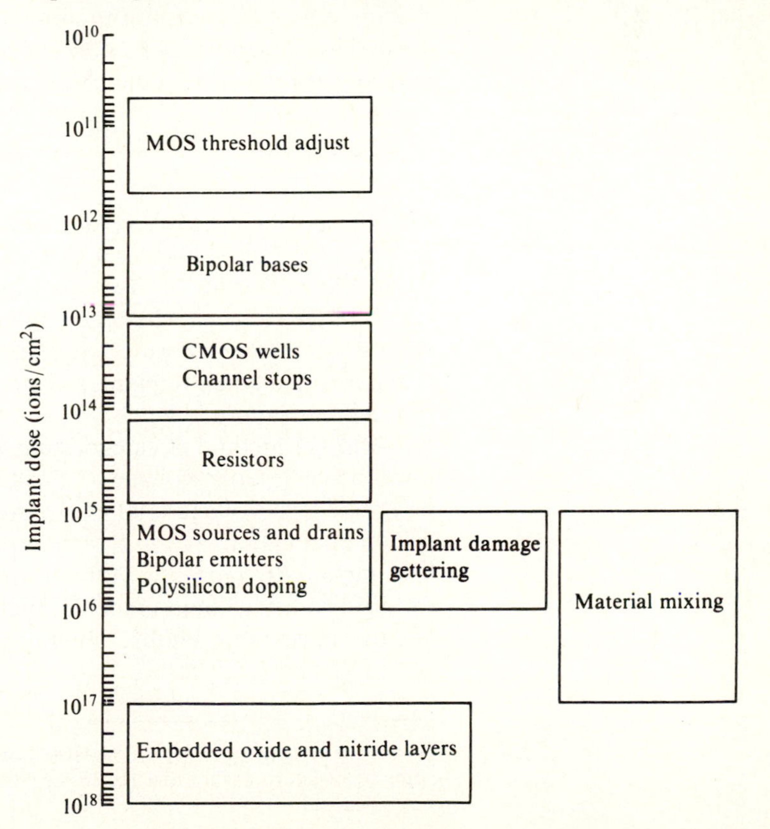

FIGURE 9.19

Typical implant dose ranges for various silicon IC applications.

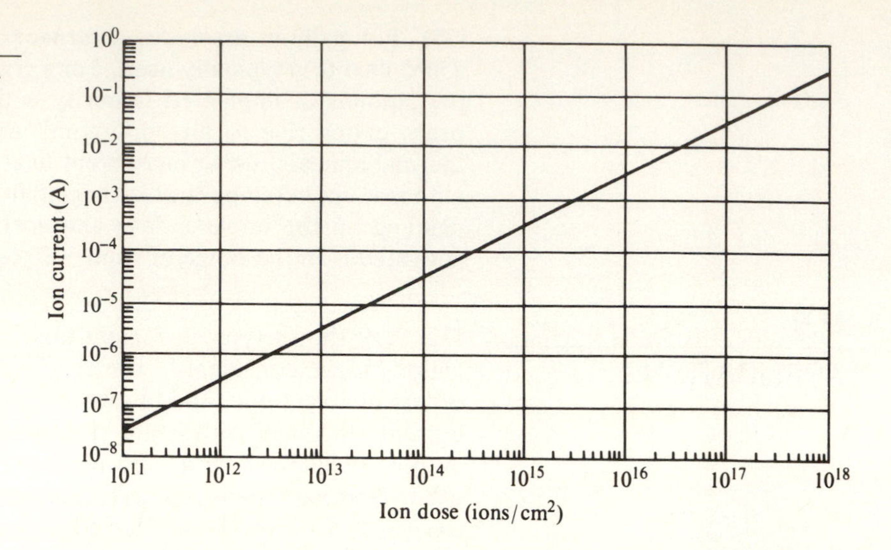

FIGURE 9.20

Ion current versus ion dose for implanting a single 125 mm diameter wafer in 1 minute with a singly charged ion.

ion has the charge q of one electron—that is, 1.6×10^{-19} coulombs (C). Thus, if ϕ ions per cm^2 are to be implanted, for a wafer of area A cm^2, $qA\phi$ coulombs are required. If the implant is to be accomplished in t seconds, $qA\phi/t$ coulombs per second are required. Since current in amperes is coulombs per second, the beam current I is

$$I \text{ in mA} = \frac{1.6 \times 10^{-16} A\phi}{t} \tag{9.20}$$

For the specific case of implanting a single 125 mm diameter wafer in 1 minute,[7] Fig. 9.20 gives a plot of dose versus beam current. This figure shows that for the low flux (dose) range, very little beam current is required, even when the implant time is reduced substantially from one minute per wafer. It also shows that for the high doses required for oxygen implants, even with 100 mA of beam current, throughput is quite low.

Another method of categorizing applications and machines is by low ion accelerating voltage[8] (~5–50 kV), medium voltage (~50–200 kV), and high voltage (300 kV to greater than 1 MV). The bulk of silicon applications fall in the medium-voltage range, but very shallow implants require low voltage. In order to substitute implanting for diffusion in gallium arsenide, high implant energy is required to get the necessary depths. High-energy beams also allow deeper

[7]This does not equate to implanting 60 wafers an hour since load time and perhaps some overscan to ensure that the wafer edges are completely covered must also be considered.

[8]Alternatively, by ion energy—for example, 50 keV.

amorphization in silicon, as well as the direct implantation of other structures such as deep retrograde CMOS wells.

9.4.1 MOS Threshold Adjust

Despite the fact that semiconductor device manufacturing applications for ion implanting have been discussed at least since 1952 (40), the first commercial application seems to have been that of MOS transistor threshold adjust begun by Mostek[9] in 1970. The threshold voltage V_t is given by

$$V_t = \phi_{MS} + \phi_B - \frac{Q_f}{C_{ox}} - \frac{Q_B}{C_{ox}} \quad 9.21$$

where ϕ_{MS} is the metal–semiconductor work function,[10] ϕ_B is the potential barrier, Q_f is the fixed positive oxide charge, Q_B is a function of the substrate doping, and C_{ox} is a parallel-plate capacitance per unit area given by $C_{ox} = \varepsilon\varepsilon_0/w$ where w is the oxide thickness, ε is the dielectric constant of the oxide (~3.9), and ε_0 is the permittivity of free space (8.85×10^{-14} F/cm). If an additional charge Q_I is added just below the oxide–semiconductor surface by ion implantation (41–44), V_t will change by an amount

$$\Delta V_t = \frac{Q_I}{C_{ox}} \quad 9.22$$

The implanted impurities that do not travel all the way through the oxide will not be ionized. Those going through the oxide and far enough into the semiconductor to stop beyond the edge of the space charge region will be in the neutral portion of the semiconductor and will also not be ionized. The rest will be ionized and are the part that causes a threshold shift. Eq. 9.22 is rigorously true only when the implanted charge is at the semiconductor–oxide interface. The farther the charge is from the interface, the less effective it is (45) so that all of the charge implanted beneath the oxide but still within the channel and space charge region cannot be equally weighted in determining a threshold shift.[11] When the peak of the distribution is between the oxide–semiconductor interface and the edge of the

---

[9]The earliest patent on threshold adjust by ion implantation appears to be held by someone other than any of the early authors of scientific papers on the subject (as is often the case). See David P. Robinson et al., "Method of Making Semiconductor Devices," U.S. Patent 3,653,978, March 4, 1972, claiming priority to March 11, 1968.

[10]Note that the symbol ϕ is used here to denote something other than implant dose. Both meanings conform to commonly used terminology.

[11]The range suggested in order to simultaneously minimize implant dose and provide the maximum localization of impurities near the silicon surface is somewhat greater than the oxide thickness (24).

space charge region, calculations of the shift based on the total charge and the location of its centroid can be made.[12] However, the implant dose actually needed is generally determined experimentally, with the initial trial based on Eq. 9.22.

For silicon, boron, which is a p-dopant and produces negative ions and a negative threshold shift in p-channel accumulation devices, is normally used. If an n-type dopant were used instead, it would increase the threshold of a p-channel transistor since its positive charge would add to the already positive Q_f.

EXAMPLE ☐ Estimate the implant dose required to reduce a p-channel threshold voltage by 1 V if the gate oxide is 400 Å thick.

Assume that the implant voltage is adjusted so that the peak of the distribution occurs at the oxide–silicon interface. Thus, half of the implant goes into the silicon. Further assume that 90% of the implanted ions in the silicon are electrically activated by the annealing process used. These assumptions allow 45% of the implanted ions to be used for threshold adjusting. Also assume that all of the charge in the silicon is effectively at the silicon–oxide interface so that Eq. 9.22 can be used. Thus,

$$\Delta V_t = 1\ \text{V} = \frac{Q_I}{C_{ox}}$$

$$Q_I\ (\text{in C/cm}^2) = \frac{1\ \text{V} \times 3.9 \times (8.85 \times 10^{-14}\ \text{F/cm})}{4 \times 10^{-6}\ \text{cm}}$$

$$Q_I = 86.5 \times 10^{-9}\ \text{C/cm}^2$$

The number of singly charged ions required for that amount of charge is Q_I/q, where q is the electronic charge (1.6×10^{-19} C per electron). Thus, $N_I = 5.4 \times 10^{11}$ ions, and since this represents 45% of the ions implanted, the total dose required is 1.2×10^{12} ions per cm^2. ☐

9.4.2 Self-Aligned MOS Gate

When the gate electrode appreciably overlaps source and drain, the extra capacitance degrades transistor performance. However, if the electrode does not reach to the source and drain, the uncovered portion of the channel cannot be turned on, and the transistor will not work. When the source and drain are defined in one step and the gate electrode in a later one, with the normal lithographic tolerances available, it is very difficult to exactly position the gate electrode so that no overlap or uncovered channel occur. To prevent this, the electrode can be deliberately made longer than the channel so that even with poor alignment no uncovered channel will occur. A better

---

[12]For details of these calculations, see, for example, S.M. Sze, *Physics of Semiconductor Physics*, 2d ed., John Wiley & Sons, New York, 1981.

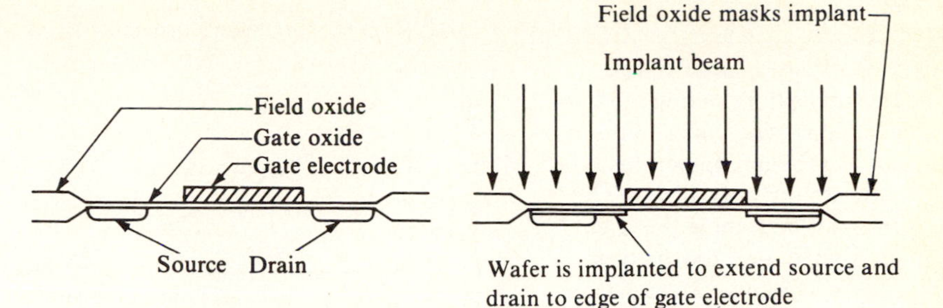

FIGURE 9.21

MOS self-aligned gate process.

solution is the use of the self-aligned gate (SAG) process. In it, the source and drain regions are separated by more than the intended channel length so that the gate electrode positioning tolerance will allow it to be placed between them as shown in Fig. 9.21. The source and drain can then be extended up to the edge of the gate by using the gate electrode as the implant mask and implanting through the gate oxide that extends out from it (46). Alternatively, a somewhat heavier implant dose can be used, and the initial source–drain doping step can be eliminated. When polysilicon gates are used, the implant step can be used to dope the polysilicon as well.

9.4.3 Polysilicon Doping

Since heavy doping of polycrystalline silicon during CVD is difficult, the usual procedure is to dope it later by either diffusion or ion implantation. The implant dose required depends on the dopant used, the thickness of the silicon layer, and the desired sheet resistance. To a lesser extent, there may be some dependency on the grain structure of the polysilicon and on the anneal temperature. If enough doping ions are implanted to meet or exceed the solubility limit for active doping, the sheet resistance R_s will decrease as the thickness is increased as shown in Fig. 9.22. While it would be expected that R_s would vary as $1/w$, where w is the thickness, experimentally it is observed (47) that $R_s \cong 1/(w - w_0)$ where w_0 is a constant dependent on the process and is in the range of 100–200 Å when the polysilicon is on an oxide. It is thought that w_0 represents a thin layer adjacent to the oxide that has a very low conductivity. If a fixed quantity of dopant is used and the polysilicon thickness is increased to the point that there is less dopant concentration than is required for electrical active saturation, the sheet resistance will stop decreasing, as shown by the straight lines in Fig. 9.22.

During thermal annealing after implanting, the dopant is activated and distributed uniformly through the thickness of the poly film by diffusion. In the case of impurities with high vapor pressures, such as arsenic, an oxide cap layer is required during annealing to prevent substantial loss of arsenic. Losses are higher for implants

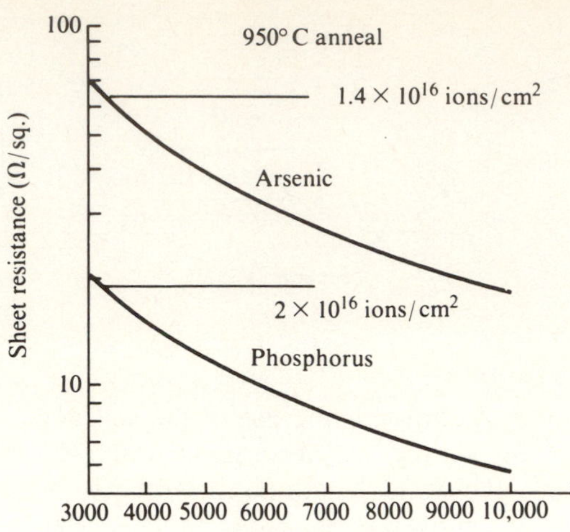

FIGURE 9.22

Sheet resistance of arsenic- and phosphorus-implanted polysilicon. (The curved lines assume enough implant dose to saturate the polysilicon. The straight lines show the sheet resistance limit for the doses shown.) (*Source:* From data in N. Lifshitz, *J. Electrochem. Soc. 130*, p. 2464, 1983.)

into polysilicon than for those into single crystal, presumably because of the increased diffusivity in polycrystalline material, which allows a more rapid arsenic diffusion to the surface (48). The anneal time required depends on implant species and the wafer temperature–time profile. As an example, with a 5×10^{15} arsenic implant into silicon, an anneal in a rapid anneal furnace set at 1200°C requires 10 s to reduce R_s to about 200 Ω/sq. and a further 15 s anneal to stabilize R_s at about 90 Ω/sq.

9.4.4 Alternative to Diffusion Predep

One of the most obvious uses of ion implantation is as an alternative to the diffusion predeposition step. However, ion implanting is usually more expensive than thermal diffusion, so unless there are clear-cut advantages in yield, throughput, or final device performance, there is no incentive to use it. Two advantages that have emerged are greater lateral doping uniformity over the wafer and lower-temperature processing. As discussed in Chapter 8, most predeps are done at a high temperature, and the solid solubility of the dopant is at that temperature, with diffusion time used to set the amount of dopant. By ion implanting, the total dopant added to the wafer can be controlled by the implant machine, and thus diffusion at high temperatures is not necessary. With the move to shallower diffusions, the lower temperatures are a distinct advantage.

9.4.5 Ion Beam Mixing and Ion Beam Damage

By irradiating an interface between two materials with a beam of ions before annealing, the interface can be smeared, and often the reaction between the components on either side of the interface will proceed at a lower temperature (49, 50). A major potential application is the reduced-temperature formation of various silicides useful as lead and contact materials.

An ion beam can also produce damage that may be used to advantage. As already discussed in Chapter 8, it can be used to provide gettering damage. Another use is amorphization, as discussed in this chapter. Finally, it can be used to increase the etch rate of, for example, silicon, silicon dioxide, and silicon nitride (51). Etch rate enhancements of from 2 to 5 can be obtained.

9.4.6 Formation of Buried Layers

Because of the high concentrations of ions that can be injected below the surface, it is possible, in principle, to form buried layers of a completely different composition. Examples are layers of silicon oxide and silicon nitride formed, respectively, by implanting oxygen and nitrogen into silicon and then annealing (52, 53). Such structures are possible alternatives to a thin silicon layer epitaxially deposited on single-crystal sapphire (SOS). By using SOS, devices built in the silicon layer are already partially electrically isolated because of the nonconductive sapphire substrate. Complete isolation is obtained by etching away the silicon layer as required between devices. Such total isolation allows very-high-speed, high-performance ICs to be built, but sapphire substrates are substantially more expensive than silicon slices.

Silicon dioxide requires an implant dose of about 2×10^{18} atoms of oxygen per cm^2. Silicon nitride requires somewhat less ($\leq 7 \times 10^{17}$ atoms of nitrogen per cm^2). In either case, care must be taken to ensure that the amorphous silicon damage region that forms does not reach to the surface. If it does, no single-crystal silicon will be left above the implant region to seed regrowth during the annealing step. Without such single-crystal regrowth, only polycrystalline silicon will remain between the surface and the buried oxide or nitride. To minimize damage, implanting at temperatures around 500°C are used. Generally, the implant depth will not be enough to allow devices to be made in the thin single-crystal layer above the oxide or nitride, but an epitaxial layer can then be added to provide additional layer thickness (54).

9.5 ION IMPLANT EQUIPMENT

Fig. 9.23 is a simplified schematic that shows the major elements of a medium-current ion implanter. The ion source starts with an appropriate molecular species and converts it into ions. The ions are accelerated and then enter the mass analyzer for ion selection. The exit beam of desired implant ions is chosen based on the charge-to-mass ratio of the ions, and the analyzer is generally sensitive enough to discriminate against adjacent mass numbers. The ions are then given a final acceleration, after which the ion beam will be slightly

FIGURE 9.23

Schematic of a medium-current ion implanter.

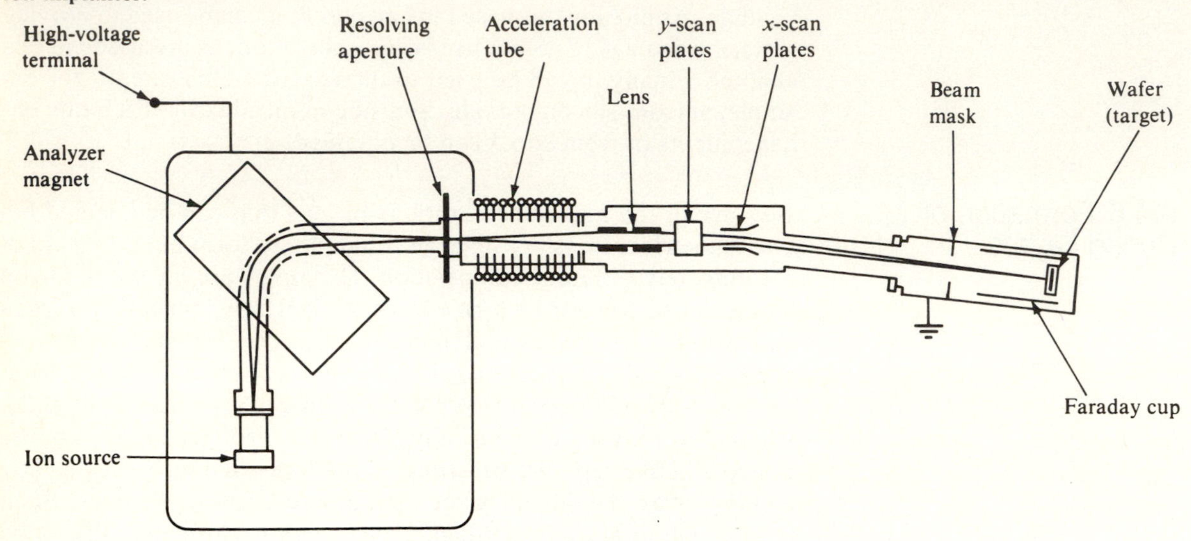

electrostatically deflected to separate it from any neutral atoms that may have formed. The beam is then scanned over the wafer surface, either electrostatically, or mechanically, or by a combination of the two. In addition, an electron source may be near the wafers to "flood" the surface with electrons and prevent a charge buildup on insulating surfaces such as SiO_2 or silicon nitride (55). This kind of charge buildup can also occur on isolated conductors such as gate electrodes and can be severe enough to cause gate oxide failure because of electrical breakdown from the gate to the substrate through the oxide.

The ion beam is focused and deflected in the same manner as electron beams used in electron microscopy. Ion beam lenses, both convergent and divergent, can be made by appropriately shaping electron or magnetic fields. For example, as shown in Fig. 9.24, either a pair of closely spaced metal plates with holes in them or a pair of hollow metal cylinders placed end to end with a small gap between them can be used to focus ions. Acceleration of the ions is accomplished electrostatically while they travel down the high-voltage column. From a space standpoint, it is desirable to keep the column as short as possible, but the necessity of preventing electric arcing requires a minimum spacing between each electrode. Several electrodes are required in order to produce an electric field that will

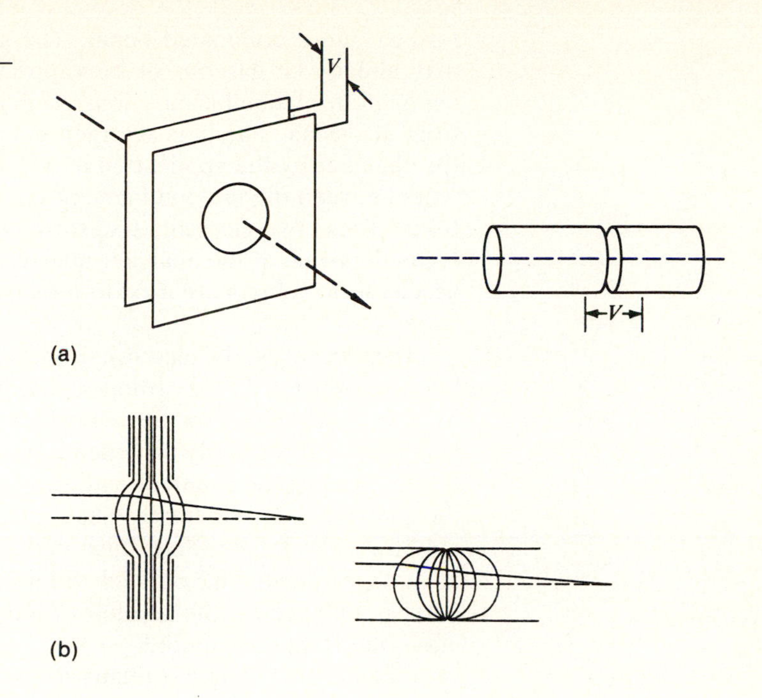

FIGURE 9.24

(a) Schematic of parallel-plate and hollow-tube electrostatic ion beam lenses. (b) Ion path through the two lenses.

not only provide ion acceleration but also keep the beam from substantially diverging while in the column.

There are two common ways to arrange the order in which the beam traverses the various elements just described. In both of them, however, the ions must have an initial accelerating voltage of at least a few kilovolts before entering the mass analyzer. In the preanalysis configuration (Fig. 9.23), most of the ion acceleration takes place after beam analysis. In the postanalysis configuration, the beam is accelerated to its final velocity and then analyzed. For very-low-voltage machines, preanalysis combined with a post-deaccelerating voltage may be necessary. A preanalysis machine can use a smaller magnet for analysis, but the analyzer section must operate at high voltage. With postanalysis, a larger magnet is required because of the higher ion velocity, but the analyzer can operate at ground potential (56, 57).

9.5.1 Ion Sources

The ion source supplies ions, usually singly positively charged, in enough quantity to provide beam currents of from 10 μA to 100 mA, depending on the rating of the implanter. Further, the source must be constructed so that the ions produced can be extracted and formed into a collimated beam. The species to be ionized, which may already be gaseous or be vaporized in or near the source, is confined in a chamber and ionized in a gaseous plasma by impact from electrons. The ions are then extracted through an opening in the chamber by the application of a negative voltage of several kilovolts between the ion confinement chamber and an extraction electrode. These two elements also form an electrostatic lens that helps focus the beam at the analyzer aperture. Additional lenses after the acceleration column are used to refocus the beam at the plane of the wafers.

In most cases, the electrons come from a hot filament (cathode), but occasionally, a cold cathode is used, and electrons are generated by secondary emission from positive ion bombardment. Also, a magnetic field is usually provided to cause spiraling of the electrons, thus increasing the length of path traveled before the electrons reach the anode and thereby improving ionization efficiency. Many different physical configurations have been used (58), with their relative advantages depending on such things as the physical form of the desired ion species, the amount of ions needed, and the ion generation rate stability required.

The Penning source is relatively simple and is often used for low beam currents. The anode is cylindrical and inside a chamber held at cathode potential. One end of the chamber is the cathode, and ions are extracted from a circular hole in it. The Freeman source is widely used in medium- and high-current machines. The chamber is at anode potential, and the ions are extracted from a slot in one wall. The cathode is a straight wire filament running parallel with and quite close to the slot. A magnetic field is directed along the length of the filament and slot. The Bernas source (based on the calutron source) also has the container at anode potential and extracts from a slot. However, the filament is located at one end of the container and near one end of the slot. As in the Freeman source, a magnetic field is parallel with the slot.

If the desired species is available as a gas, it can be fed directly into the source chamber. If it is available as a solid[13] with a reasonably high vapor pressure, a heated vaporizer immediately adjacent to the source chamber can be used to generate a source vapor. Since the source chamber pressure is $\sim 10^{-4}$ torr, it is feasible to have a sputtering capability built into the chamber. Thus, when the implan-

---

[13]Remember that elemental sources are not required since the implanter mass analyzer will separate out the desired atomic weight species from the rest of those present. Care must be taken, however, to ensure that a spurious complex ion of the same ionic weight as the desired ion is not formed—for example, $^{31}(CF)^+$ versus $^{31}P^+$.

tation of a material with a very low vapor pressure is required, a sputtering target of the desired material can be installed inside the source chamber, and sputtering can be used for vaporization. Table 9.3 lists some typical source materials.

9.5.2 Ion Analyzer

The ion analyzer works by injecting the ion beam into a magnetic field so that the beam is perpendicular to the field. In this case, the ions will travel in a circular path while remaining perpendicular to the field. The force exerted by the magnetic field is given by

$$f = nq(\boldsymbol{v} \times \boldsymbol{B}) \qquad 9.23a$$

where q is the charge of an electron, n is the charge state of the ion, v is the velocity of the ion, and B is the magnetic field. For ion travel perpendicular to the field, the vector cross product of Eq. 9.23a can be expressed in scalar form as

$$f = nqvB \qquad 9.23b$$

This force, tending to make the ion travel in an inward spiral, is counteracted by a centrifugal force f_c given by

$$f_c = \frac{mv^2}{r} \qquad 9.24$$

TABLE 9.3

Possible Implant Source Materials

Implant Species	*Source*
Antimony	Sb_2O_3(s)*
Arsenic	AsF_3, AsH_3, GaAs(s)
Beryllium	Be(s), $BeCl_2$(s), BeF_2(s)
Boron	BCl_3, BF_3, B_2H_6
Cadmium	Cd(s), CdS(s)
Phosphorus	Red P(s), PCl_3, PF_3, PH_3, PF_5
Selenium	Se(s), CdSe(s), SeO_2(s)
Silicon	$SiCl_4$, SiF_4, SiH_4
Sulfur	S(s), SO_2, CdS(s), H_2S
Tellurium	Te(s), CdSe(s)
Tin	Sn(s), $SnCl_2$(s)
Zinc	$ZnCl_2$(s)

*Solid material. Unmarked entries are gaseous.
Note: Chlorine and fluorine compounds may cause source deterioration. Hydrogen may find its way into the rest of the vacuum system and prevent cryopumping.

Source: From information in reference 58 and in A. Axmann, "Ionizable Materials To Produce Ions for Implantation," *Solid State Technology,* p. 47, February 1975.

where m is the mass of the ion and r is the radius of the path. These two forces will be equal so that

$$nqB = \frac{mv}{r} \qquad 9.25$$

The velocity of the ion is given by

$$\frac{mv^2}{2} = nqV \qquad 9.26$$

where V is the voltage through which the ion was accelerated before reaching the analyzer. Note that the use of a doubly charged ion is equivalent to doubling the acceleration voltage. Substituting Eq. 9.26 into Eq. 9.25 and using a consistent set of units give

$$r = \frac{1}{B}\sqrt{\frac{2mV}{nq}} = \frac{144}{B}\sqrt{\frac{MV}{n}} \qquad 9.27$$

where r is in centimeters, M is the ion mass in atomic mass units (as given in Table 9.1), B is in gauss units, and V is in volts.

EXAMPLE ☐ If the apertures in the analyzer are set for a 10 cm radius and the ions are extracted from the source with 30 kV, what magnetic field is required to pass singly charged arsenic?

From Table 9.1, $M_{As} = 75$. Substituting this value of M, an r of 10, and an n of 1 (singly charged) into Eq. 9.27 gives

$$B = \frac{144}{10}\sqrt{75 \times 3 \times 10^4} = 21.6 \text{ kgauss}$$

Under these conditions, how far from the exit slit would germanium ions hit (how effective will the system be in separating the two ions)?

Assume the mass 74 Ge isotope. In this case, modifying Eq. 9.27 gives

$$\Delta r = \frac{144}{21.6 \times 10^3} [(75 - 74) \times 3 \times 10^4]^{0.5} = 1.2 \text{ cm}$$

To a first approximation, the separation of the two beams will equal Δr. Thus, if the slits were 5 mm wide, the two beams would just be separated. ☐

9.5.3 Wafer Handling

Wafer handling includes transporting the wafer into and out of the chamber, controlling the temperature of the wafer during implanting, and moving the wafer as necessary during implanting to ensure implant uniformity. In addition, provision must be made for adjusting the angle of implant in order to minimize channeling. Wafer scanning is usually done by a combination of a rotating multiple-slice wafer holder and radial beam deflection. The wafers can generally be automatically transferred from multiple-wafer cassette holders to the wafer holder. In some designs, the wafers enter the implant chamber individually and are then placed on the holder. In other designs, a holder is loaded with wafers in normal room atmosphere and then is transferred to the implant chamber. This design has the advantage of easily allowing dedicated wafer holders in order to minimize cross-contamination (see the following section). In either case, great care must be taken to minimize particulate contamination during the load/transfer operation. Even particles that do not initially land on wafers may be blown onto them later during venting.

9.5.4 Wafer Contamination

Wafer contamination arises from hydrocarbons that are adsorbed on the wafer and polymerized or carbonized by the beam, from material sputtered from fixturing by the ion beam redepositing on the wafer, and from an impure ion beam. Organic-based pump oils used in an improperly trapped vacuum system are a possible source of hydrocarbons. The use of perfluorinated polyether pump oil will minimize this problem (59). Cryopumping is an alternative, but only if the ion source feedstock does not produce appreciable hydrogen, as occurs, for example, when phosphine is used (60).

Some ion beam sputtering will occur wherever the ion beam strikes, and when the striking surface is other than the wafer itself, the sputtered material can be a source of wafer contamination (60, 61). Because some overtravel of the ion beam is necessary to ensure that the whole wafer surface is implanted, the beam will strike the metal fixturing a portion of the time. Therefore, metals such as stainless steel are usually avoided, and aluminum is used instead. In addition, dopant atoms are deposited on the fixturing by the implant beam and then subsequently sputtered away. Thus, cross-contamination from run to run is possible. However, by using dedicated fixturing for each dopant species, this effect can be eliminated. Dopant buildup can also be reduced by periodically making dummy cleanup runs. If implanting is always done through a layer that will be removed before heat treatment, then the layer will act as a shield for sputter contamination.

To determine the purity of the beam, the degree to which ions can be separated must be considered. In principle, ions with a single mass number (atomic weight) may be selected. However, most elements have several isotopes, and for those with higher atomic num-

TABLE 9.4

Atomic Weights for Natural Isotopes of Tin, Antimony, and Tellurium

Tin	*Antimony*	*Tellurium*
112(1.0)*		
114(0.7)		
115(0.4)		
116(14.7)		
117(7.3)		
118(24.3)		
119(8.6)		
120(32.4)		**120**(0.1)
	121(57.3)	
122(4.6)		**122**(2.5)
	123(42.7)	**123**(0.9)
124(5.6)		**124**(4.6)
		125(7.0)
		126(18.7)
		130(31.7)

*Numbers in parentheses give the percentage abundance of the isotope.

Source: Data from *Handbook of Chemistry and Physics*, 68th ed. (1987–1988), CRC Press, Boca Raton, Fla.

bers, some may overlap. For example, Table 9.4 shows the naturally occurring isotopes of tin, antimony, and tellurium. Thus, if a source were contaminated, beam contamination could also occur. The species ionized can also affect the implanted ion. For example, if molecular nitrogen were ionized to a +1 charge ($^{14}N_2^+$), it could be implanted with the same instrument setting as that for implanting $^{28}Si^+$

9.5.5 Implant Uniformity Measurements

The total number of ions implanted is tracked during implantation by integrating the beam current, but the uniformity over the wafer depends on the scan uniformity and is not readily amenable to in situ monitoring. One common way of measuring uniformity is by profiling the sheet resistance of an implanted layer after it has been annealed for activation (62). The results of a 1985 round robin test involving various semiconductor manufacturers indicated that sheet resistance values measured in the various facilities varied by no more than about 1% (63). It also showed that the actual sheet resistance, based on the measured implant dose, varied by several percent from facility to facility, thereby highlighting the fact that substantial difficulty can be encountered in accurately and reproducibly measuring dose. Another technique, applicable to low-value implants where the resulting sheet resistance is very high and difficult to measure, is to measure the threshold voltage of MOS transistors that have had their threshold shifted by the implant. Capacitance–voltage measurements on an array of MOS capacitors can also be used (64).

Rather than measuring the electrical properties of ions implanted in the semiconductor wafer, changes in the properties of the silicon itself can be used. The implant damage causes changes in optical properties and in thermal properties, both of which can be sensed. The change in optical transmissivity of silicon deposited on sapphire test wafers can be correlated with ion implant dose (65). In principle, ellipsometry and Raman spectroscopy (66) are also applicable, and both can be used to make observations from the top side of a wafer.

When thermal waves are introduced in the wafer by laser beam heating, ion-implant-induced damage can be detected (67, 68). The signal level can be related to implant dose and can be used to produce uniformity contours similar to those obtained from spreading resistance measurements. Such a technique is advantageous in that no annealing to electrically activate the impurities is required.

The change in optical transmissivity of some ion-sensitive coating other than silicon on a transparent test wafer can also be measured. Materials that have been used are photoresist (65) and nylon impregnated with a radiation-sensitive dye (69).

SAFETY — CHAPTER 9

Since ion implanters use high voltages, care against electrocution is a major consideration. Hence, safety interlocks should never be circumvented unless required during machine servicing, and then only by qualified repair and maintenance personnel. Also, because of the high operating voltage, X rays are generated; therefore, shields must not be indiscriminately removed. The major source of X rays is electrons that become trapped in the ion accelerator column and are accelerated in the opposite direction so that they strike various metal parts in the vicinity of the source (or the analyzer in post-acceleration machines). As part of the initial installation of an implanter, a check of X-ray radiation should be required. However, even after that check, X-ray dosimetry badges should be worn by all personnel working near ion implant machines.

Highly toxic ion source gases are often used, and provision should be made to continuously monitor for leaks. The supply is usually limited to small lecture bottles, but even so, a leak could be very serious. (Refer to Chapter 4 for additional discussion on the handling of toxic gases.)

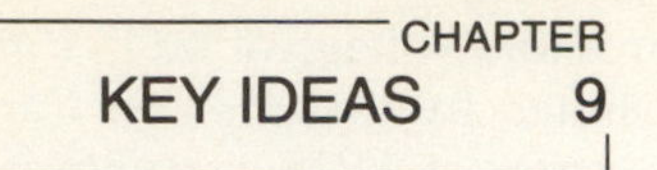

KEY IDEAS — CHAPTER 9

- ☐ Ion implantation introduces impurities into the wafer by accelerating ions to a high velocity and directing them toward the wafer surface.
- ☐ The depth distribution of implanted impurities in an amorphous material is roughly Gaussian, with the center of the distribution at a depth (range) determined by the implant energy.
- ☐ The range is generally less than 1 μm, but with light ions and implanters operating at several MeV, it can be a few μm.
- ☐ Implantation can be done at room temperature.
- ☐ Implantation produces crystallographic damage. Very high doses cause enough damage to convert crystalline material to amorphous material. To remove damage and electrically activate the impurities, an anneal between 500°C and 900°C is generally required.
- ☐ During implantation into crystalline materials, if the implant direction is along a major crystallographic direction, ion channeling may occur and substantially increase the depth of some of the ions.
- ☐ Ion implantation can be substituted for shallow diffusions in most cases and can also be used for MOS transistor threshold adjusting by implanting through the gate oxide.

PROBLEMS — CHAPTER 9

1. If a 400 keV implanter is available and an n-type dopant is to be implanted into silicon, which dopant will give the most range? What will be the range if the ion used is doubly charged? (Assume no channeling.)
2. What will the straggle be when implanting arsenic at 300 keV into silicon? Assuming no moment other than the second, plot the impurity distribution.
3. If a p-type silicon wafer of 1 Ω-cm resistivity is to have localized low-resistivity n-regions (10^{19} atoms/cc peak concentration) implanted, how thick should a resist masking layer be if the implant is phosphorus done at 200 keV? List the assumptions made.
4. If a silicon wafer is held at 100°C, estimate the dose of implanted silicon required for amorphization? What is a major advantage of using silicon as the implant species?
5. A 150 mm wafer is to be implanted in 1 minute with a dose of 10^{13} ions/cm^2 of a doubly charged ion. What beam current will be required?
6. An n-channel silicon MOS transistor needs to have its V_t adjusted downward by 2 V. The oxide is 500 Å thick. What is a suitable ion for implanting, and what will be the approximate dose required for V_t correction?
7. If a 3000 Å layer of SiO_2 is to be formed inside a silicon wafer, what is the minimum number of oxygen ions that must be implanted? List the assumptions made.
8. If the magnetic field used for deflection is the same in each case, how much greater will the radius of curvature in an analyzer be when 1000 keV rather than 100 keV ions are used?

CHAPTER 9 REFERENCES

1. J. Linhard, M. Scharff, and H. Schiøtt, "Range Concepts and Heavy Ion Ranges," *Mat. Fys. Medd. Dan. Vid. Sclsk. 33*, pp. 1–39, 1963.
2. James F. Gibbons, "Ion Implantation in Semiconductors—Part I, Range Distribution Theory and Experiments," *Proc. IEEE 56*, pp. 295–319, 1968.
3. F.H. Eisen, "Channeling of Medium Mass Ions through Silicon," *Can. J. Phys. 46*, pp. 561–572, 1968.
4. James Comas and Robert G. Wilson, "Channeling and Random Equivalent Depth Distributions in 150 keV Li, Be, and B Implanted in Si," *J. Appl. Phys. 51*, pp. 3697–3701, 1980.
5. I.M. Cheshire et al., "The Z_1 Dependence of Electronic Stopping," *Physics Letters 27A*, pp. 304–305, 1968.
6. J.P. Biersack and J.F. Ziegler, "The Stopping and Range of Ions in Solids," pp. 122–156, in H. Ryssel and H. Glawischnig, eds., *Ion Implantation Techniques*, Springer-Verlag, New York, 1982.
7. James F. Gibbons, William S. Johnson, and Steven W. Mylroie, *Projected Range Statistics, Semiconductors and Related Materials*, 2d ed., Dowden, Hutchinson and Ross, Stroudsburg, Pa., 1975.
8. B. Smith, *Ion Implantation Range Data for Silicon and Germanium Device Technologies*, Research Studies, Forest Grove, Oreg., 1977.
9. J.P. Biersack and J.F. Ziegler, "The Calculation of Ion Ranges in Solids with Analytic Solutions," in H. Ryssel and H. Glawischnig, eds., *Ion Implantation Techniques*, Springer-Verlag, New York, 1982.
10. W.A. Shewhart, *Economic Control of Quality of Manufactured Product*, D. Van Nostrand Co., New York, 1931.
11. James F. Gibbons and Steven F. Mylroie, "Estimation of Impurity Profiles in Ion Implanted Amorphous Targets Using Joined Half-Gaussian Distributions," *Appl. Phys. Lett. 22*, pp. 568–569, 1973.
12. W.K. Hofker, *Philips Research Rpts., Suppl. 8*, 1975.
13. Heiner Ryssel, "Range Distributions," pp. 177–205, in H. Ryssel and H. Glawischnig, eds., *Ion Implantation Techniques*, Springer-Verlag, New York, 1982.
14. Seijiro Furukawa et al., "Theoretical Considerations on Lateral Spread of Implanted Layers," *Jap. J. Appl. Phys. 11*, pp. 134–142, 1972.
15. H. Runge, "Distribution of Implanted Ions under Arbitrary Shaped Mask Edges," *Phys. Stat. Sol. (a) 39*, pp. 595–599, 1977.
16. Seijiro Furukawa and Hideka Matsumura, "Backscattering Study on Lateral Spread of Implanted Ions," *Appl. Phys. Lett. 22*, pp. 97–98, 1973.
17. James W. Mayer et al., *Ion Implantation in Semiconductors*, Academic Press, New York, 1970, p. 40.
18. I. Suni et al., "Effect of Preamorphization Depth on Channeling Tails in B^+ and As^+ Implanted Silicon," in B.R. Appleton et al., eds., *Ion Beam Processes in Advanced Electronic Materials and Device Technology*, Vol. 45, Materials Research Society, 1985.
19. Karen W. Brannon and R.F. Lever, "Computational Investigation of Channeling of Boron in Silicon," *Electrochem. Soc. Ext. Abst. 86–2*, pp. 813–814, October 1986.
20. J.F. Ziegler, "The Channeling of Ions near the <100> Axis," in B.R. Appleton et al., eds., *Ion Beam Processes in Advanced Electronic Materials and Device Technology*, Vol. 45, Materials Research Society, 1985.
21. Masayasu Miyake et al., "Incidence Angle Dependence of Planar Channeling in Boron Ion Implantation in Silicon," *J. Electrochem. Soc. 130*, pp. 716–719, 1983.
22. Norman L. Turner et al., "Effects of Planar Channeling Using Modern Ion Implantation Equipment," *Solid State Technology*, pp. 163–172, February 1985.
23. Robert G. Wilson et al., "Angular Sensitivity of Controlled Implanted Doping Profiles," *NBS Special Pub. 400–49*, November 1978.
24. H. Ishiwara et al., "Projected Range Distribution of Implanted Ions in a Double-Layer Sub-

strate," pp. 423–428, in S. Namba, ed., *Ion Implantation*, Plenum Press, New York, 1975.
25. T.O. Herndon et al., "Polyimide for High Resolution Ion Implantation Masking," *Solid State Technology*, pp. 179–183, November 1984.
26. D. Roche, "Outgassing of Photoresist during Ion Implantation," *Proc. Materials Research Society Symposium 45*, pp. 203–210, 1985.
27. James F. Gibbons, "Ion Implantation in Semiconductors—Part II, Damage Production and Annealing," *Proc. IEEE 60*, pp. 1062–1096, 1972.
28. Sorab K. Ghandhi, *VLSI Fabrication Principles*, John Wiley & Sons, New York, p. 324, 1983.
29. F.F. Morehead, Jr., and B.L. Crowder, "A Model for the Formation of Amorphous Si by Ion Bombardment," pp. 25–30, in L. Chadderton and F. Eisen, eds., *Proc. 1st International Conference on Ion Implantation*, Gordon and Breach, New York, 1971.
30. J. Narayan and O.W. Holland, "Characteristics of Ion-Implantation Damage and Annealing Phenomena in Semiconductors," *J. Electrochem. Soc. 131*, pp. 2651–2662, 1984.
31. D.G. Beanland and D.J. Chivers, "Color-Band Generation during High Dose Ion Implantation of Silicon Wafers," *J. Electrochem. Soc. 125*, pp. 1331–1338, 1978.
32. James W. Mayer, Lennart Eriksson, and John A. Davies, *Ion Implantation in Semiconductors*, Academic Press, New York, p. 111, 1970.
33. F. Eisen et al., "Implantation into GaAs," pp. 117–144, in James V. DiLorenzo and Deen D. Khandelwal, eds., *GaAs FET Principles and Technology*, Artech House, Inc., Dedham, Mass., 1982.
34. C.A. Armiento et al., "Capless Rapid Thermal Annealing of GaAs Implanted with Si Using an Enhanced Overpressure Proximity Method," *J. Electrochem. Soc. 134*, pp. 2010–2016, 1987.
35. R. Kwor et al., "Effect of Furnace Preanneal and Rapid Thermal Anneal on Arsenic-Implanted Silicon," *J. Electrochem. Soc. 132*, pp. 1201–1206, 1985.
36. S.K. Tiku and W.M. Duncan, "Self-Compensation in Rapid Thermal Annealed Silicon-Implanted Gallium Arsenide," *J. Electrochem. Soc. 132*, pp. 2237–2239, 1985.
37. T.O. Sedgwick, "Short Time Annealing," *J. Electrochem. Soc. 130*, pp. 484–492, 1983.
38. K. Cho et al., "Transient Enhanced Diffusion during Rapid Thermal Annealing of Boron Implanted Silicon," *Appl. Phys. Lett. 47*, pp. 1321–1323, 1985.
39. D. Hagmann et al., "A Method To Impede the Formation of Crystal Defects after High Dose Arsenic Implants," *J. Electrochem. Soc. 133*, pp. 2597–2600, 1986.
40. R.S. Ohl, "Properties of Ionic Bombarded Silicon," *Bell Syst. Tech. J. 31*, pp. 104–122, 1952.
41. K.G. Aubuchon, "The Use of Ion-Implantation To Set the Threshold Voltage of MOS Transistors," pp. 575–593, in *Proc. International Conference on Properties and Use of MIS Structures*, Grenoble, France, 1969.
42. J. MacDougall, K. Manchester, and R. Palmer, "Ion Implantation Offers a Bagfull of Benefits for MOS," *Electronics 43*, pp. 86–90, June 1970.
43. M.R. MacPherson, "The Adjustment of MOS Transistor Threshold Voltage by Ion Implantation," *Appl. Phys. Lett. 18*, pp. 502–504, 1971.
44. R.B. Palmer et al., "The Effect of Oxide Thickness on Threshold Voltage of Boron Ion Implanted MOSFET," *J. Electrochem. Soc. 120*, pp. 999–1001, 1973.
45. S.D. Brotherton and P. Burton, "The Influence of Non-Uniformly Doped Substrates on MOS CV Curves," *Solid-State Electronics 13*, pp. 1591–1595, 1970.
46. Robert W. Bower et al., "MOS Field Effect Transistors Formed by Gate Masked Ion Implantation," *IEEE Trans. on Electron Dev. ED-15*, pp. 757–761, 1968.
47. N. Lifshitz, "Solubility of Implanted Dopants in Polysilicon: Phosphorus and Arsenic," *J. Electrochem. Soc. 130*, pp. 2464–2467, 1983.
48. S.R. Wilson et al., "Properties of Ion-Implanted Polycrystalline Si Layers Subjected to Rapid Thermal Anneal," *J. Electrochem. Soc. 132*, pp. 922–929, 1985.
49. B.Y. Tsaur et al., "Ion-Beam Induced Metastable Pt_2Si_3 Phase: I. Formation, Structure, and Properties," *J. Appl. Phys. 51*, pp. 5326–5333, 1980.
50. T.W. Orent et al., "Effects of Ion Implantation on the Thermal Growth of Pt and NiPt Silicides," *J. Electrochem. Soc. 130*, pp. 687–691, 1983.

51. A.H. van Ommen et al., "Etch Rate Modification of Si_3N_4 Layers by Ion Bombardment and Annealing," *J. Electrochem. Soc. 133*, pp. 2140–2147 and included references, 1986.
52. C.G. Tuppen and G.J. Davies, "An AES Investigation into the Phase Distribution of Ion-Implanted Oxygen in Silicon N-Channel Devices," *J. Electrochem. Soc. 131*, pp. 1423–1427 and included references, 1984.
53. L. Nesbit et al., "Microstructure of Silicon Implanted with High Doses of Nitrogen and Oxygen," *J. Electrochem. Soc. 133*, pp. 1186–1190 and included references, 1986.
54. W.W. Lloyd and R. Dexter, "Ion Implanted and Conventional Epitaxy to Produce Dielectrically Isolated Silicon Layers," U.S. Patent 3,855,009, December 17, 1974.
55. N. White et al., "Wafer Charging and Beam Interactions in Ion Implantation," *Solid State Technology*, pp. 151–158, February 1985.
56. Hans Glawischnig, "Ion Implantation System Concepts, pp. 3–21, in H. Ryssel and H. Glawischnig, eds., *Ion Implantation Techniques*, Springer-Verlag, New York, 1982.
57. Pieter Burggraaf, "Ion Implanters: Major 1986 Trends," *Semiconductor International*, pp. 78–89, April 1986.
58. D. Aitken, "Ion Sources," pp. 23–71, in H. Ryssel and H. Glawischnig, eds., *Ion Implantation Techniques*, Springer-Verlag, New York, 1982.
59. M.Y. Tsai et al., "Study of Surface Contamination Produced during High Dose Ion Implantation," *J. Electrochem. Soc. 126*, pp. 98–102, 1979.
60. G. Ryding, "Evolution and Performance of the Nova NV–10 Predep™ Implanter," pp. 319–342, in H. Ryssel and H. Glawischnig, eds., *Ion Implantation Techniques*, Springer-Verlag, New York, 1982.
61. L.A. Larson and M.I. Current, "Metallic Impurities and Dopant Cross-Contamination Effects in Ion Implanted Surfaces," *Proc. Materials Research Society Symposium 45*, pp. 381–388, 1985.
62. D.S. Perloff et al., "Dose Accuracy and Doping Uniformity of Ion Implantation Equipment," *Solid State Technology*, pp. 112–120, February 1981.
63. M.I. Current and W.A. Keenan, "A Performance Survey of Production Ion Implanters," *Solid State Technology*, pp. 139–146, February 1985.
64. R.O. Demming and W.A. Keenan, "Low Dose Ion Implant Monitoring," *Solid State Technology*, pp. 163–167, September 1985.
65. J.R. Golin et al., "Advanced Methods of Ion Implant Monitoring Using Optical Dosimetry," *Solid State Technology*, pp. 155–163, June 1985.
66. A.C. deWilton et al., "Raman Spectroscopy for Nondestructive Depth Profile Studies of Ion Implantation in Silicon," *J. Electrochem. Soc. 133*, pp. 988–995, 1986.
67. W. Lee Smith et al., "Ion Implant Monitoring with Thermal Wave Technology," *Appl. Phys. Lett. 47*, pp. 584–586, 1985.
68. W.L. Smith et al., "Ion Implant Monitoring with Thermal Wave Technology," *Solid State Technology*, pp. 85–92, January 1986.
69. Kranti V. Anand and Myron Cagin, "Fluence (Dose) Monitoring of Energetic H^+, B^+, N^+, P^+, and As^+ Ions Using Ionization in a Radiachromic Film," *J. Electrochem. Soc. 132*, pp. 1206–1208, 1985.

CHAPTER 10

Ohmic Contacts, Schottky Barriers, and Interconnects

10.1 INTRODUCTION

Before a circuit element formed in silicon or gallium arsenide can perform a useful function, it generally must be electrically connected with other elements on the same chip and most certainly must be connected with circuitry not on the same chip. An integrated circuit is by definition a number of electrically interconnected circuit elements on the same chip. Some of the interconnections are done in the silicon itself, but most are done by means of thin conductive stripes running across the top surface of the wafer. The connection to off-chip circuitry is included in packaging technology and will not be considered here.

Fig. 10.1 is a cross section of a small portion of a silicon wafer, showing a contact, a Schottky diode, and two levels of interconnections. The usual definition of a contact is metal contacting the semiconductor and providing a low-resistance ohmic electrical connection. What constitutes a low-resistance connection depends on the particular circuit it is used in, but generally it is a few microohms per square centimeter ($\mu\Omega/cm^2$) of contact area. When two levels of conductors contact each other at a via (hole in the separating insulator), a contact resistance will also be present. Its magnitude will primarily depend on how clean the first surface was before the second layer was applied. A Schottky barrier metal is one that contacts a semiconductor surface and forms one element of a Schottky diode. The interconnects are the electrical conductors connecting various contacts and Schottky diodes together so that they perform a useful function. The conductors must also be of low resistance, and, again, what constitutes low resistance depends on where they are used. In general, however, the resistance will be in the range of 25–100 Ω per centimeter of length.

In order to reliably fabricate a structure such as that of Fig. 10.1, interactions between the layers must be considered. In many cases, some of the functional layers shown in the figure will be comprised of layers of two or more different materials because no single

FIGURE 10.1

Cross section of a silicon IC interconnect system with two levels of metallization (not to scale).

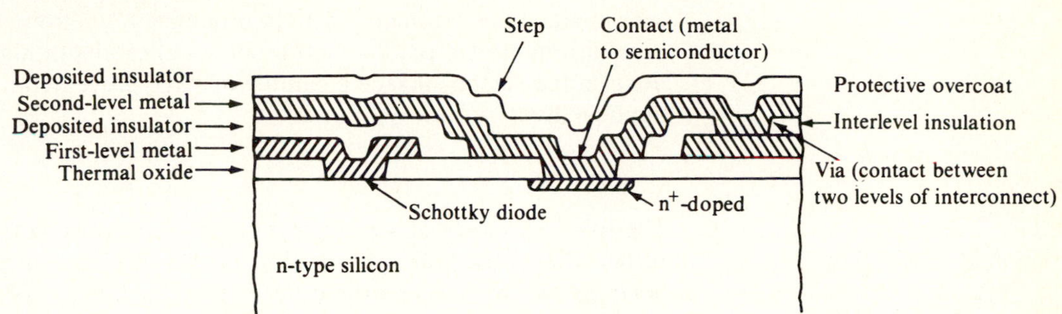

universal contact and interconnect material exists that will work properly in all applications. As soon as more than one material is involved, interactions between materials must be considered, and the likelihood of deleterious interactions grows. Table 10.1 lists properties that must be considered in such a multiple-material interconnect system. The first three electrical properties are dictated by the required IC performance. All the other properties are related to reliability of the chip.

Aluminum was used almost from the beginning of silicon device fabrication for low-resistivity contacts,[1] and was extended to the

TABLE 10.1

Material Properties of Importance to an Interconnect System

Electrical	*Physical*	*Chemical*	*Purity*
Resistivity Contact resistance Schottky barrier height Amount of electromigration	Adhesion to SiO_2 or other insulators Adhesion to adjacent conductive layers Diffusion barrier efficacy Thermal expansivity Stress induced during deposition Surface morphology	Resistance to oxidation Resistance to corrosion Reactivity with adjacent interconnect materials	Level of sodium Level of radioactive impurities

[1]Early grown junction silicon transistors (the kind first commercially available) used either electroless nickel plating or baked-on silver paste. (For a discussion of electroless plating on silicon, see, for example, Miles V. Sullivan and John H. Eigler, *J. Electrochem. Soc. 104*, p. 226, 1957.)

interconnect material as well when planar ICs were introduced. Aluminum has high electrical conductivity, makes good contact to silicon, is easy to deposit by evaporation or sputtering, and unlike some metals such as gold, adheres well to silicon dioxide. Unfortunately, some of its other properties, such as having only a modest resistance to electromigration, a propensity to grow spikes and puncture interlevel oxides, and a silicon alloy depth highly dependent on the silicon surface cleanliness, have prevented its complete applicability.

10.2 OHMIC CONTACTS

The total resistance of a contact is comprised of two components in series. One is the bulk spreading resistance r_s of the semiconductor beneath the contact, and the other is the resistance r_c of the metal–semiconductor interface. Thus, the total resistance R_T of a contact is given by $r_s + r_c$. By convention, "contact resistance" means only r_c. The spreading resistance is always ohmic,[2] but the contact resistance often is not.

10.2.1 Spreading Resistance

For flat, circular contacts on a semi-infinite body of uniform resistivity ρ (see Fig. 10.2), the spreading resistance r_s is given by

$$r_s = \frac{\rho}{4a} \qquad 10.1$$

where a is the radius of the contact.

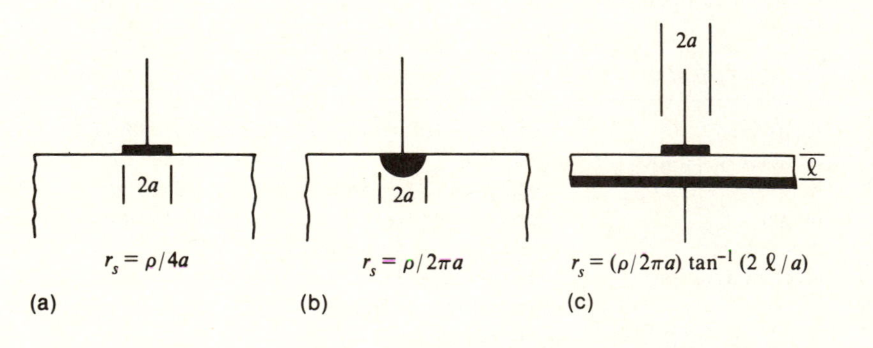

FIGURE 10.2

Geometries used in spreading resistance calculations.

[2]An ohmic contact is one in which the I–V curve is linear and symmetric about the origin.

If the contact is an imbedded hemisphere, then

$$r_s = \frac{\rho}{2\pi a} \tag{10.2}$$

For the more general case of the flat contact on a layer of finite thickness ℓ and resistivity ρ, Eq. 10.1 becomes (1)

$$r_s = \frac{\rho}{2\pi a} \tan^{-1} \frac{2\ell}{a} \tag{10.3}$$

For the case of $\ell << a$,

$$r_s = \frac{\rho\ell}{\pi a^2} \tag{10.4}$$

or the familiar $R = \rho\ell/A$ expression.

These cases are not exactly applicable to most IC geometries. In an IC, contact is generally made to a thin semiconductor layer isolated from the rest of the wafer by a pn junction. Thus, as shown in Fig. 10.3, the current must often turn and flow laterally in the thin section. This current will generally then become crowded near the leading edge of the contact (Fig. 10.3a). In such cases, increasing the length d of the contact past some value d' will not noticeably decrease the resistance. For the case in which the contact resistance r_c is negligible compared to the resistance of the layer (2), the dis-

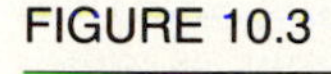

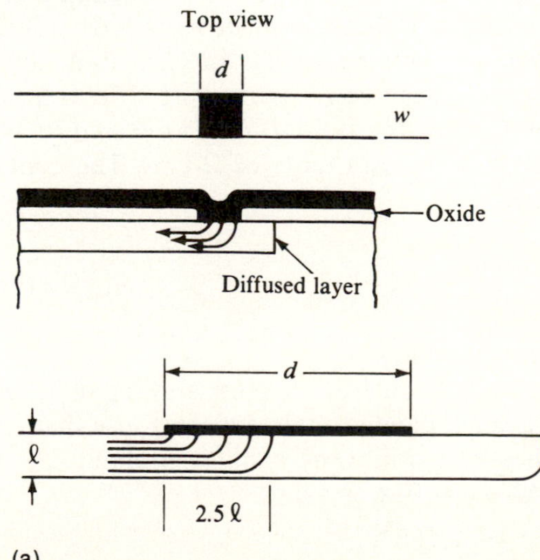

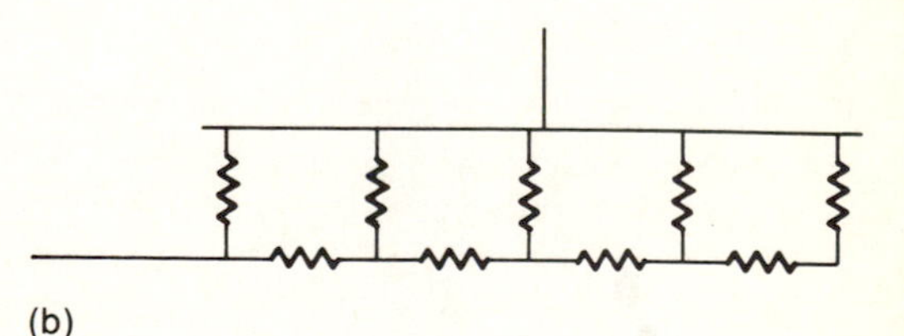

FIGURE 10.3

(a) Current flow from a contact to a diffused lead when no contact resistance is present. (The layer thickness ℓ is given as $2\sqrt{Dt}$ and represents the distance not to the isolating junction but to the point where the doping concentration has decreased by a factor of ~1000.) (b) Transmission line equivalent circuit when contact resistance is present.

tance d' can be approximated by $5\sqrt{Dt}$, where Dt is the diffusivity–time product of the diffusion used to produce the diffused layer being contacted. In most cases, $2\sqrt{Dt}$ is a little less than the distance from the surface to the isolating junction, so d' is about 2.5 times the layer thickness (ℓ).

A derivation for the resistance, based on the assumption that r_c is not negligible, gives rise to a transmission line equivalent circuit as shown in Fig. 10.3b. For the case in which the contact covers the full width w of a thin layer, as was shown in Fig. 10.3a, the total resistance of the contact can be approximated by

$$R_T = \frac{1}{w}\sqrt{R_s A r_c}\coth\left(d\sqrt{\frac{R_s}{A r_c}}\right) \qquad 10.5$$

where R_s is the sheet resistance of the layer and A is the area (wd) (3, 4). (For more expressions obtained by making various assumptions, see reference 4.)

10.2.2 Contact Resistance

Contact resistance arises from the fact that a metal in contact with a semiconductor surface forms a Schottky diode current barrier. The I–V relationship, like that of a pn junction, is, in the ideal case, given by

$$I = I_o(e^{qV/kT} - 1) \qquad 10.6$$

where I_o is a constant (the diode saturation current), q is the electronic charge, and k is Boltzmann's constant. The curve is nonlinear, but a diode differential resistance for any voltage can be defined as dV/dI. To be useful, an ohmic contact must have a low voltage drop and be usable for voltage swings in each direction. Hence, dV/dI evaluated at zero voltage is appropriate. If current density J is used instead of current, the ratio dV/dJ is referred to as the specific contact resistance R_c with units of $\Omega\text{-cm}^2$. The contact resistance then becomes

$$r_c = \frac{R_c}{A} \qquad 10.7$$

where A is the contact area.

Fig. 10.4a shows a diode I–V trace for a swing of several volts, while Fig. 10.4b is an expanded scale covering less than ± 100 mV and shows that near the origin the contact is ohmic. $J(V)$ has the same form as $I(V)$ in Eq. 10.6 so that

$$R_c = \frac{dV}{dJ} = \frac{kT}{qJ_o} \quad \text{at } V = 0 \qquad 10.8$$

FIGURE 10.4

(a) Forward and reverse I–V curve of an ideal diode. (b) Middle portion of same curve with both scales expanded 50×. (The curve is quite linear for $-3kT/q < V < 3kT/q$.)

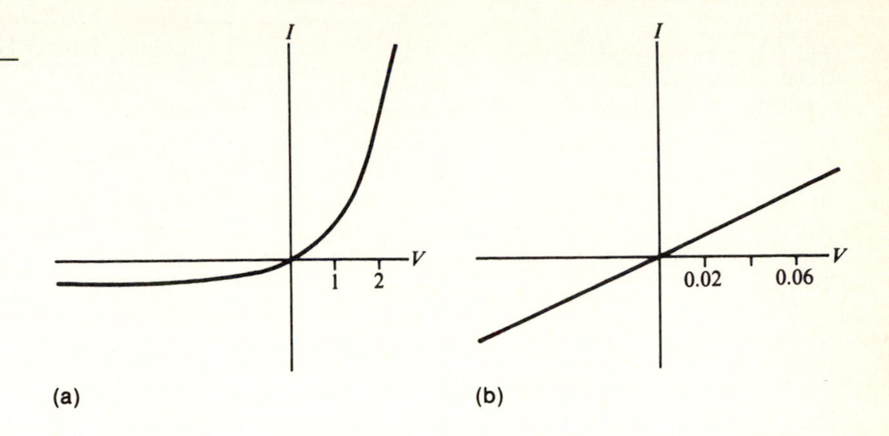

Current flow across a metal–semiconductor interface is primarily by thermionic emission when the doping level is in the 10^{17} atoms/cc and less range and J is given by

$$J = A^{**}T^2e^{-q\phi_B/kT} \times (e^{qV/kT} - 1) \qquad 10.9$$

where A^{**} is the Richardson constant and ϕ_B is the Schottky barrier height in volts. The specific contact resistance near zero voltage is given by

$$R_c = \frac{dV}{dJ} = \left(\frac{k}{qA^{**}T}\right)e^{q\phi_B/kT} \qquad 10.10$$

Thus, the larger the barrier height, the larger R_c. Table 10.2 gives barrier heights for several metals on silicon and gallium arsenide. Table 10.3 gives values of R_c as calculated for silicon from Eq. 10.10 (5) and shows a range of 10 orders of magnitude as ϕ_B goes from 0.25 V to 0.85 V. As Table 10.2 shows, heights for p-type material are generally much lower than those for n-type, and it is therefore easier to make acceptable contacts to p-type material. It is clear from these tables that finding a metal for n-type silicon with a low enough barrier height to give an R_c in the useful range of $1\text{–}2 \times 10^{-6}$ $\Omega\text{-cm}^2$ is unlikely, although the lanthanides are close. From a practical standpoint, however, the R_c values can be much lower than those predicted from Eq. 10.10. Without considerable care in manufacturing, the diode saturation current I_o will be several orders of magnitude larger than theoretical and will reduce R_c. In the early days of the transistor, satisfactory contacts to etched 2 Ω-cm silicon were made with baked-on silver paste.

By increasing the doping level of the semiconductor, the space charge region width can be reduced to a thickness that allows sub-

TABLE 10.2

Metal–Semiconductor Schottky Barrier Heights

	Barrier Height in Volts			
Metal	*n-Si*	*p-Si*	*n-GaAs*	*p-GaAs*
Ag			0.93	0.44
Al	0.72	0.58	0.8	0.63
Au	0.8	0.35	0.95	0.48
Cr	0.61			
La	0.4	0.7		
Mo	0.68	0.42		
Pt	0.9		0.94	0.48
Ti	0.50	0.61		
W	0.67	0.45	0.77	
Y	0.4	0.7		

Note: A substantial spread in the values reported in the literature occurs, depending on the method of measurement, the surface treatment given the semiconductor before the metal was applied, and the specific heat treatment given the contact. There is also a small variation due to the semiconductor doping level.

TABLE 10.3

Calculated R_c for an n-Type Silicon Schottky Diode near $V = 0$

ϕ_B (V)	R_c (Ω-cm^2)
0.85	4.7×10^5
0.70	1.4×10^3
0.55	4.3
0.40	1.3×10^{-2}
0.25	4×10^{-5}

Source: From data in L.P. Lepselter and J.M. Andrews, "Ohmic Contact to Silicon," in Bertram Schwartz, ed., *Ohmic Contacts to Semiconductors*, Electrochemical Society, New York, 1969.

stantial electron tunneling (as opposed to the electrons surmounting the barrier by thermal emission). This reduction in thickness provides for a much increased current flow and drop in specific contact resistance.

When the doping level is greater than about 10^{19} atoms/cc (6), rather than $R_c :: e^{q\phi_B/kT}$,

$$R_c :: e^{B\phi_B/\sqrt{N}_D} \tag{10.11}$$

$B = (4\pi/h)\sqrt{\varepsilon m^*}$ where h is Planck's constant, ε is the dielectric permittivity, and m^* is the effective mass of the charge carrier. The trend of R_c versus doping for silicon and gallium arsenide over both the thermionic emission and the tunneling regimes is shown in Figs. 10.5 and 10.6.

Experimentally, n-type silicon follows the theoretical prediction rather well. In the case of n-type gallium arsenide with alloyed germanium contacts, the data scatter is substantial, and the contact resistance decreases over a wide range of initial doping levels approximately as $1/N_D$ (7). For p-type silicon with alloyed aluminum contacts, a similar behavior is observed (8). The trends are shown in Fig. 10.7. The failure to show an $e^{q\phi_B/kT}$ dependence at low N_D values is presumably because the contact systems dope the material, regardless of its starting resistivity, to the point where tunneling dominates. For the case of gallium arsenide, it has been postulated that the $1/N_D$ rather than the $1/\sqrt{N_D}$ variation is the outgrowth of an uneven and spiked contact–GaAs interface (6). In such cases, the

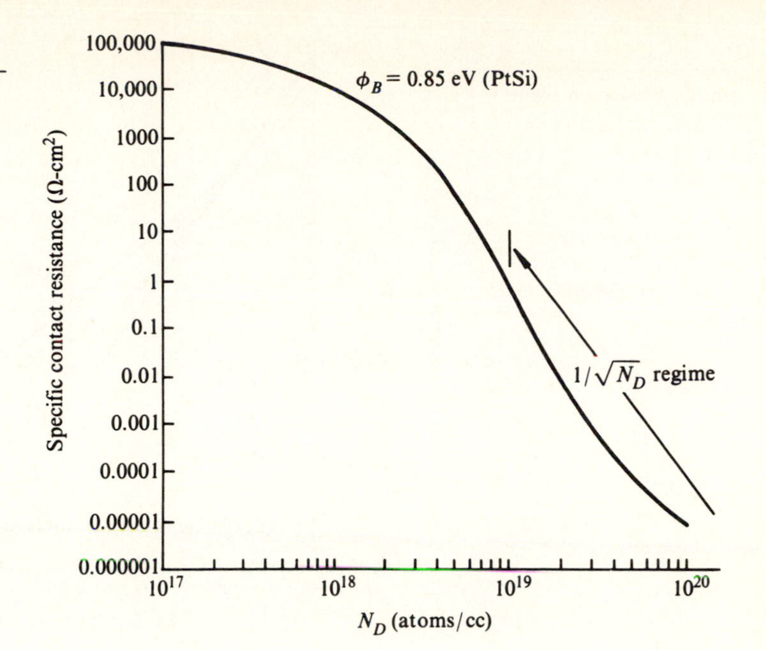

FIGURE 10.5

Theoretical specific contact resistance as a function of silicon donor doping level. (*Source:* From data in C.Y. Chang et al., *Solid-State Electronics 14*, p. 541, 1971.)

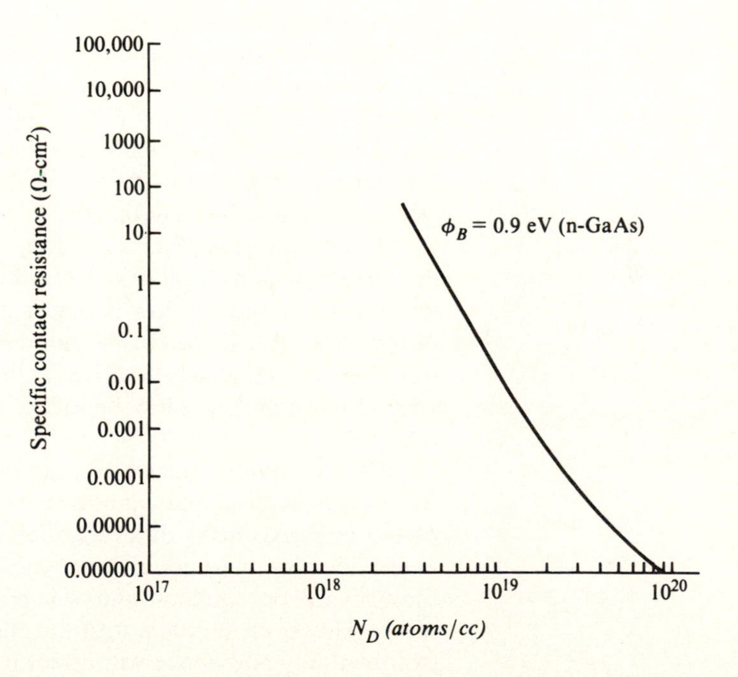

FIGURE 10.6

Theoretical specific contact resistance as a function of gallium arsenide donor doping level. (*Source:* From data in Gary Robinson, CEI, September 1986.)

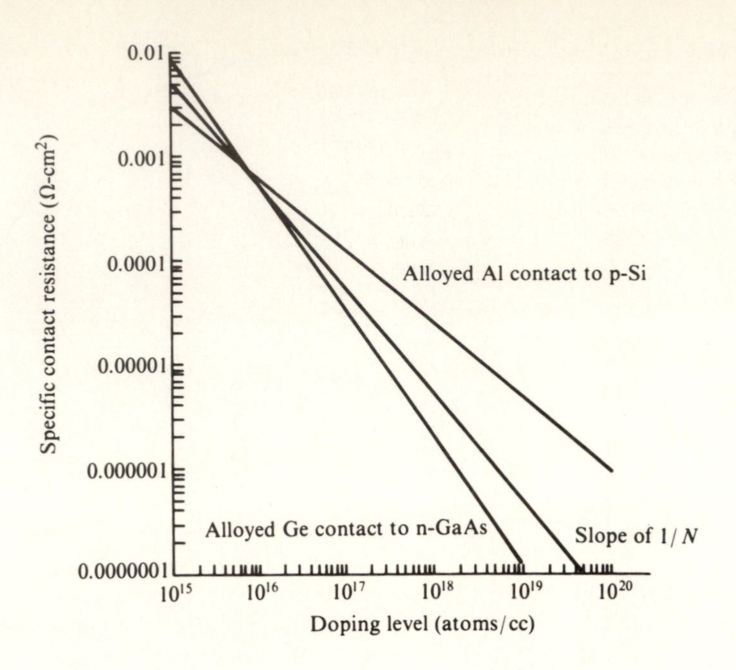

FIGURE 10.7

Experimentally observed trend of specific contact resistance versus doping level for alloyed contacts to silicon and gallium arsenide. (*Source:* Data from H.H. Berger, *J. Electrochem. Soc. 119,* p. 507, 1972, and N. Braslau, *J. Vac. Sci. Technol. 19,* p. 803, 1981.)

contact resistance would then be determined by the series of point contacts, which according to Eq. 10.2 would be proportional to ρ, which in turn is proportional to $1/N_D$.

Table 10.3 showed how R_c varies with barrier height when thermal emission dominates. However, since heavy doping and tunneling are actually used, it is more helpful to examine the effect of ϕ_B in the region where R_c varies as $1/\sqrt{N_D}$. For silicon, as shown in Fig. 10.8, R_c still decreases with decreasing barrier height, but the effect is much less pronounced.

An alternate approach to making ohmic contacts to a wide bandgap material such as gallium arsenide has been proposed in which a thin layer of a second, low bandgap material is added to the surface of the wafer (9). If the two doping levels are properly adjusted, the heterojunction formed should offer little resistance, and because the second material has a low bandgap, making ohmic contacts to it is relatively easy.

The theoretical specific contact resistance curves just shown are for n-type material and cannot be used for p-type material because of the effective mass difference between holes and electrons. m^* enters into the thermionic emission expression (Eq. 10.10) through the Richardson equation and will have only a multiplicative effect on R_c. However, it enters into the tunneling expression (Eq. 10.11) exponentially and hence will affect it much more.

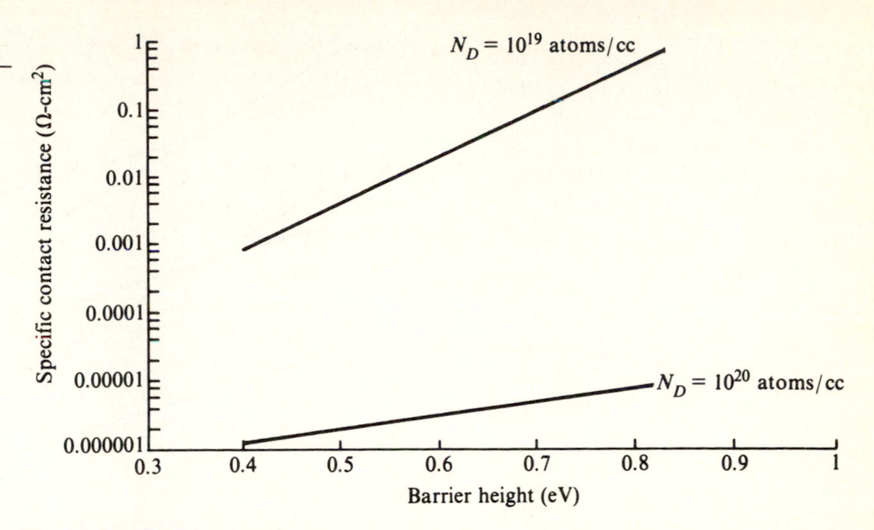

FIGURE 10.8

Theoretical trend of specific contact resistance as a function of Schottky barrier height for two silicon doping levels. (*Source:* Based on data in C.Y. Chang et al., *Solid-State Electronics 14,* p. 541, 1971.)

10.2.3 Total Contact Resistance

The value of total contact resistance R_T required in an integrated circuit varies with the specific device but is generally in the 10–100 Ω range. The R_c value that can be tolerated varies from 10^{-6} Ω-cm² to about 10^{-5} Ω-cm². Of concern as devices continue to decrease in size is how contact resistance must scale in order to not degrade the device. Bipolar behavior seems to dictate that the current density of emitters remain near 400 A/cm² regardless of emitter area. Since allowable circuit voltage drops will remain constant, independent of current and geometry, as the current decreases, the resistance can increase commensurately. Thus, the move to smaller bipolar geometries should not put additional restraints on contact resistance. In MOS circuits, however, reduced geometries may lead to reduced voltages as well as currents. In such cases, a constant resistance value during size reduction may be required, and that must come from a reduction of the contact resistance r_c.

To see how R_c and bulk resistivity ρ interact, consider Fig. 10.9. It has curves of contact resistance versus contact area for representative values of R_c and curves of spreading resistance versus area for different silicon doping levels. These curves show that for the total resistance to be in the <100 Ω range for the small-diameter contacts, the silicon doping must be 0.01 Ω-cm or less. The heavy doping is required for r_s as well as for R_c reduction. Current technology primarily uses locally diffused regions (p^+ or n^+) under the contact to secure the high doping necessary. However, the doping that occurs during regrowth from a metal-alloyed region can also be used. For example, regrowth from aluminum will give p-type aluminum-doped silicon, as will regrowth from gold to which small

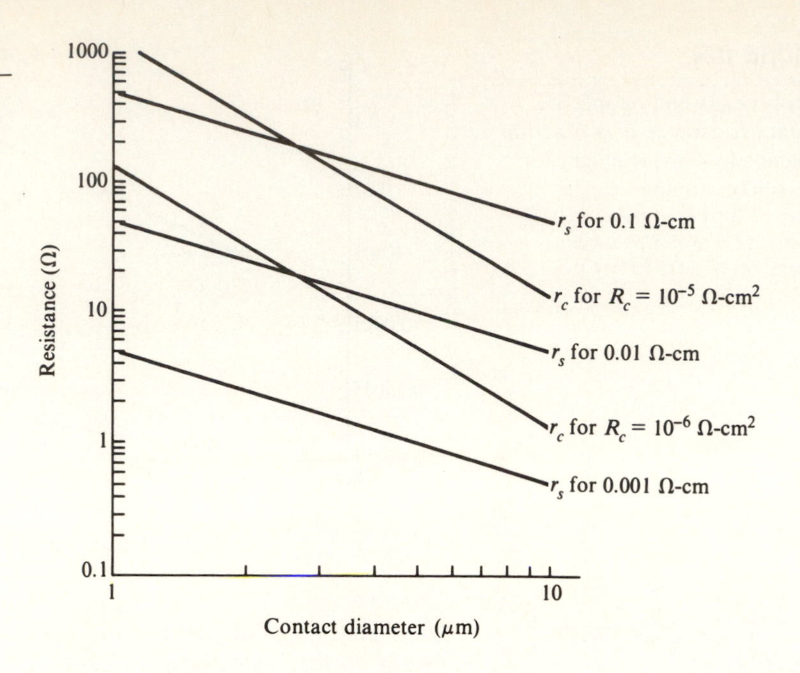

FIGURE 10.9

Spreading resistance of circular contacts for various resistivities of material and contact resistance for two values of specific contact resistance.

amounts of gallium have been added. Similarly, gold with small amounts of antimony will regrow an antimony-doped n-region. Sandblasting has also been used to provide a damaged surface layer whose electrical properties are similar to those of heavily doped surfaces.

Perhaps more directly applicable to the understanding of these relations as they pertain to integrated circuits is a reconsideration of Fig. 10.3, which showed contact to a thin isolated diffused layer. The total contact resistance R_T for that geometry can be approximated by Eq. 10.5. Using that same equation, R_T versus specific contact resistance is plotted for representative sheet resistances of 5, 50, and 500 Ω/sq. and shown in Fig. 10.10. For example, with a contact 1 μm long by 3 μm wide, a contact resistance of less than 50 Ω can, for any of the three sheet resistance values shown, only be achieved if R_c is less than about 10^{-6} Ω-cm^2. However, with a 5 μm contact, 50 Ω can be obtained with a 5 Ω/sq. sheet and an R_c of just a little less than 10^{-5} Ω-cm^2. For this reason, as contacts have shrunk, more emphasis has been placed on reducing specific contact resistance.

10.2.4 The Aluminum–Silicon Ohmic Contact

Silicon has an appreciable solubility at elevated temperatures in several of the possible contact materials. Aluminum, which has been used since the inception of the planar silicon IC for contacts and metallization, is an example. Its ability to dissolve thin layers of

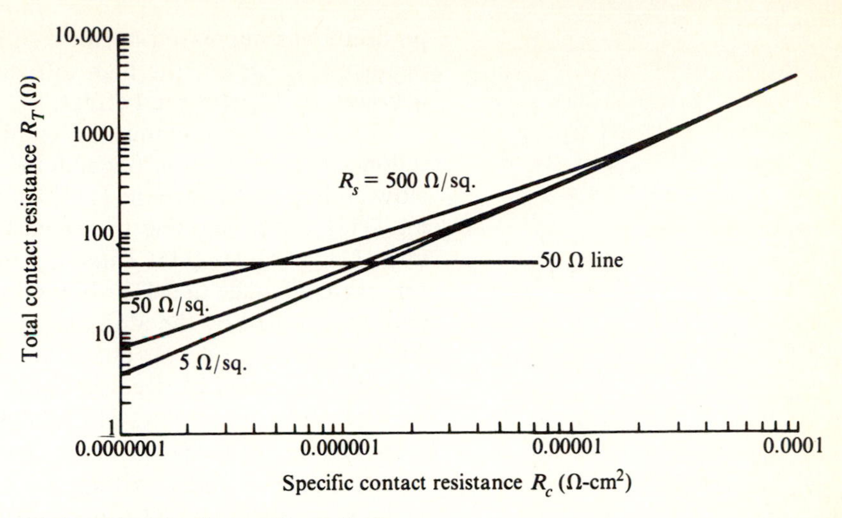

(a) 1 μm long by 3 μm wide contact

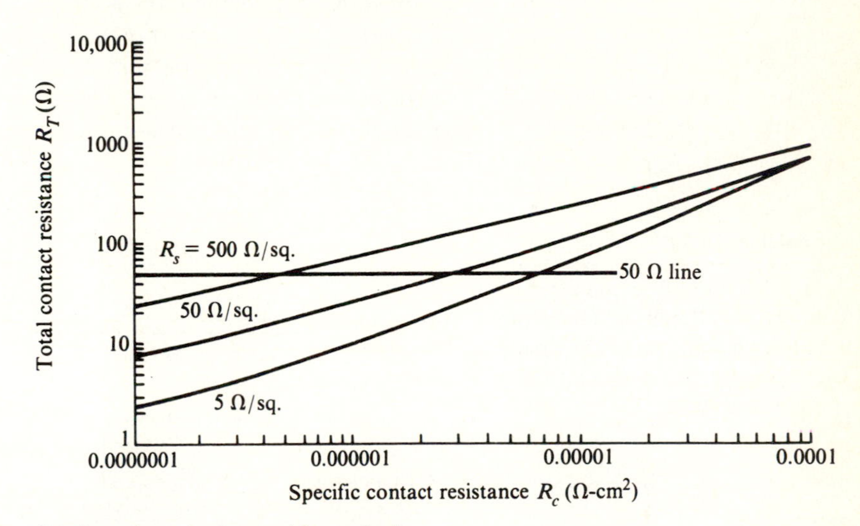

(b) 5 μm long by 3 μm wide contact

FIGURE 10.10

Specific contact resistance versus total contact resistance for various sheet resistance values.

SiO_2 helps ensure good physical contact to the silicon even if surface cleaning is not complete. However, enough silicon will dissolve in the aluminum during the contacting operation to form pits in the silicon surface. The pits are filled with aluminum, and the phenomenon is often referred to as aluminum spiking. Fig. 10.11 shows the way the dissolution (alloying) takes place. This pitting is seen well below the aluminum–silicon eutectic temperature of 577°C. (See Appendix B for a discussion of solubility, eutectic temperature, and phase diagrams.) Some early work postulated that the eutectic temperature was depressed because of stress and that melting actually occurred. However, the high diffusivity of silicon in aluminum can account for the observed effect.

Like aqueous etching, the etching of silicon by metals is crystallographic orientation dependent, and as in aqueous etching, the slow etching planes are the {111}'s (10). Because of the slow dissolving (111)'s, it is possible to get very flat alloy fronts when alloying into (111) surfaces (11). Thus, germanium and silicon alloy transistors were all made in (111) oriented wafers. The equilibrium form of patterns formed by the aluminum–silicon interaction will appear as flat-bottomed, triangular pits in (111) and square-topped, pyramidal pits in (100) material. Occasionally, hexagonal shapes will be seen on (111) material because at early stages the $(\bar{1}11)$, $(1\bar{1}1)$, and $(11\bar{1})$ planes as well as the $(\bar{1}\bar{1}1)$, $(1\bar{1}\bar{1})$, and $(\bar{1}1\bar{1})$ planes limit the reaction.[3] In cross section, they will appear as in Fig. 10.11. Since no (111) plane is parallel to a (100) surface to act as an etch stop, the aluminum–silicon interface of contacts on (100) wafers will not be smooth. Of more importance, because the thin layer of interfacial oxide that must first be dissolved will not fail uniformly, different regions will begin reacting at different times. The first ones can allow enough silicon into the aluminum to saturate much of the contact

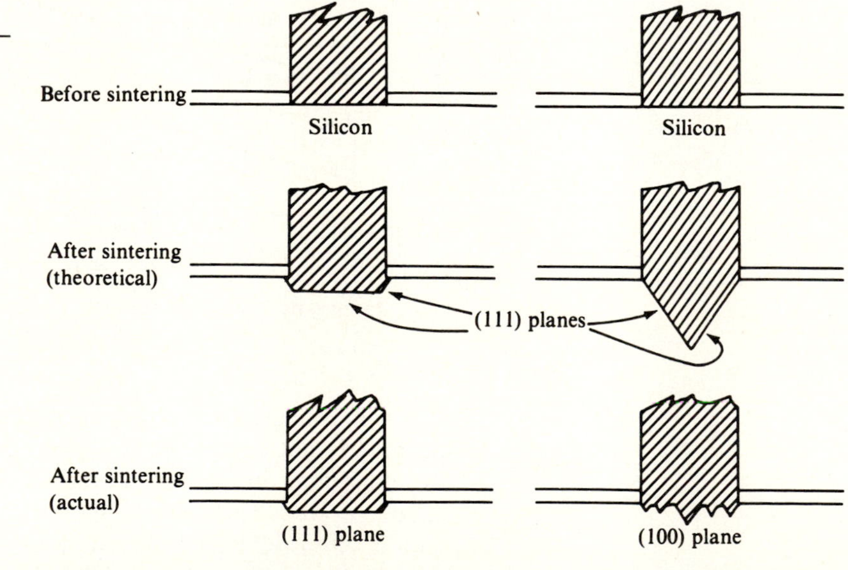

FIGURE 10.11

Metal front before and after sintering (alloying). (The actual depth depends on sinter time, temperature, and amount of metal present.)

[3]It is arguable that the equilibrium shapes formed by solid-state diffusion might be different from those occurring during liquid dissolution, but experimentally, they are observed to be the same.

and thus limit dissolution over the remainder of the contact area. The unfortunate result is that the first pits will be much deeper than the rest and, in shallow junction devices, may extend through a junction and cause shorting. Fig. 10.12 shows scanning electron microscope photographs of (111) and (100) silicon surfaces after the aluminum contacts have been removed. The excessive pitting of the (100) surface is clearly visible.

A planar transistor or an integrated circuit has rather thin metallization, and the sintering temperature is intentionally held below the metal–silicon eutectic. Nevertheless, the solubility of silicon in aluminum is high enough for appreciable dissolution to occur (12–18). The total amount of silicon dissolved can be calculated from the

FIGURE 10.12

SEM views of (111) and (100) silicon surfaces showing pits remaining after removing aluminum metallization from contact windows.
(*Source:* Photographs courtesy of Dr. P.B. Ghate, Texas Instruments Incorporated.)

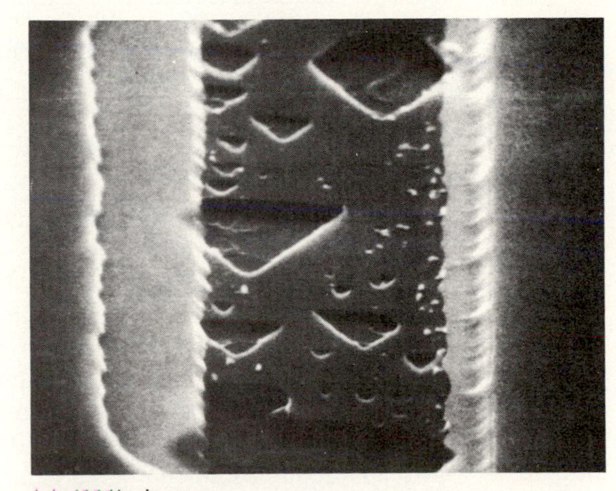

(a) (111) plane

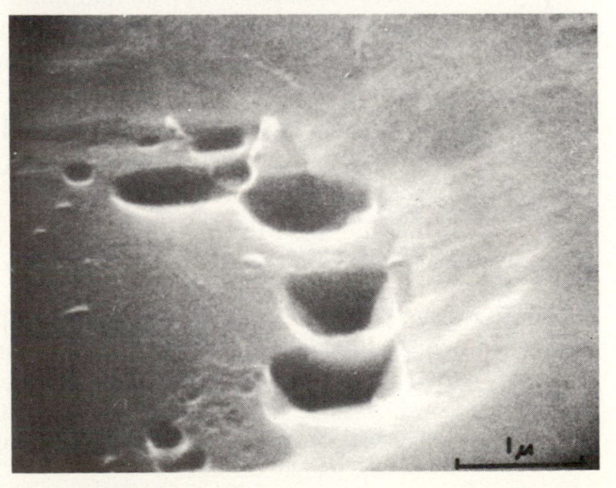

(b) (100) plane

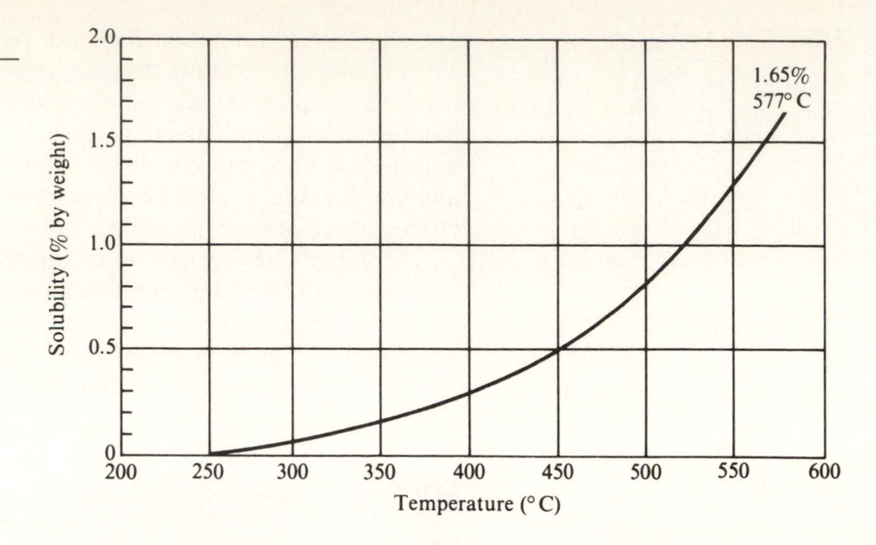

FIGURE 10.13

Solubility of silicon in aluminum. (*Source:* From data in Max Hansen, *Constitution of Binary Alloys,* McGraw-Hill Book Co., New York, 1958.)

solid solubility of silicon in aluminum and the amount of aluminum available. Fig. 10.13 gives solubility data up to 577°C. Determining the amount of solvent aluminum is not completely straightforward, however. In principle, given enough time for the silicon to diffuse along its length, the entire aluminum lead system could become saturated. In practice, the time at high temperature is not enough for this to happen. Substantial diffusion distances are possible, however, because of the high silicon diffusivity (Fig. 10.14). The effect of leads acting as silicon sinks can clearly be seen by noting that when a contact has a lead extending out from it in one direction only, more dissolution generally occurs on that side of the contact. Since the removal of the silicon does not produce a void under the aluminum, it must be assumed that a simultaneous motion of the aluminum occurs (aluminum self-diffusion) that keeps voids from forming. In order to use the data of Fig. 10.14, it must be further assumed that it is the silicon diffusion, and not that of aluminum, that limits motion of silicon in the aluminum.

EXAMPLE ☐ If an aluminum lead the same width as the contact (10 μm) and 1 μm thick extends out from the contact to infinity in one direction only, how much silicon will diffuse into it in 20 minutes at 450°C? Assume that the aluminum over the contact is already saturated with aluminum and that it will remain saturated by diffusion from the wafer.

FIGURE 10.14

Diffusivity of silicon in aluminum. (The upper curve is representive of IC metallization; the lower curve for wrought aluminum is intended only as a trend line. The spread of reported data is substantial.) (*Source:* Upper curve data from J.O. McCaldin and H. Sankur, *Appl. Phys. Lett. 19,* p. 524, 1971; lower curve based on old data tabulated in McCaldin and Sankur.)

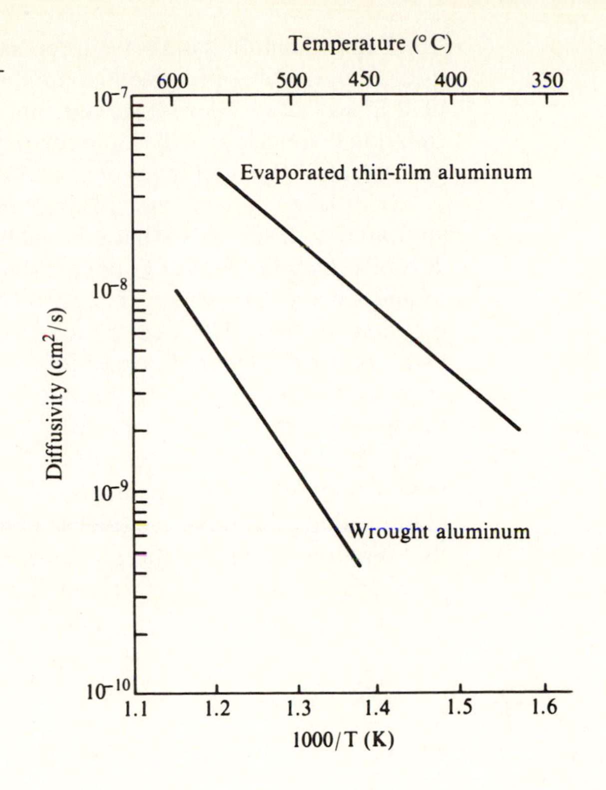

From Fig. 10.13, at 450°C, the solubility limit of silicon in aluminum is ~0.5% by weight, or 0.014 gram/cc. Let that equal N_0. $N(x) = N_0 \operatorname{erfc}(x/2\sqrt{Dt})$. (For details on diffusion, see Chapter 8.) The total amount M of silicon that diffused in is given by $\int_0^\infty N(x)dx$, or

$$M = \int_0^\infty \operatorname{erfc}\left(\frac{x}{2\sqrt{Dt}}\right) = 2N_0\sqrt{\frac{Dt}{\pi}}$$

From Fig. 10.14, at 450°C, $D \cong 10^{-8}$ cm²/s. Substituting these numbers into the equation above gives $M = 5.5 \times 10^{-5}$ grams/cm². The cross section of the lead is 10^{-7} cm², so 5.5×10^{-12} grams of silicon diffused down the lead. Since silicon has a density of 2.3 grams/cc, this represents a volume of $\sim 2.4 \times 10^{-12}$ cc. If the contact were square, its area would be 10^{-6} cm²; if the silicon were uniformly removed from beneath it, the surface would move down 2×10^{-7} cm. While this number may seem very small, if all the silicon came from one 3 μm square area, as might happen in a (100) wafer, the pit would be 0.8 μm deep. □

Experimentally,[4] it has been reported that sintering at 300°C produces discernible pitting with some depths to 0.2 μm. At 350°C, pits of 0.75 μm have been observed, and at 450°C, pits of 2 μm (19). Individual behavior will, of course, depend on the amount and placement of the aluminum over and around the contact window.

In order to prevent such pitting, the aluminum may be deposited saturated with silicon so that it is unable to absorb any more (12, 13, 20), or it may be deposited over a thin sacrificial layer of polycrystalline silicon (21). When this problem was first recognized, there was considerable reticence to the use of silicon-doped aluminum because of the difficulty of deposition. A number of other alternatives, such as the deposition of a thin layer of aluminum followed by sintering and then a subsequent thick aluminum deposition, were also tried (18). With the later development of improved sputtering equipment came the routine deposition of aluminum–silicon of controlled composition.[5] However, oxide radiation damage may be induced that requires additional thermal annealing for removal.

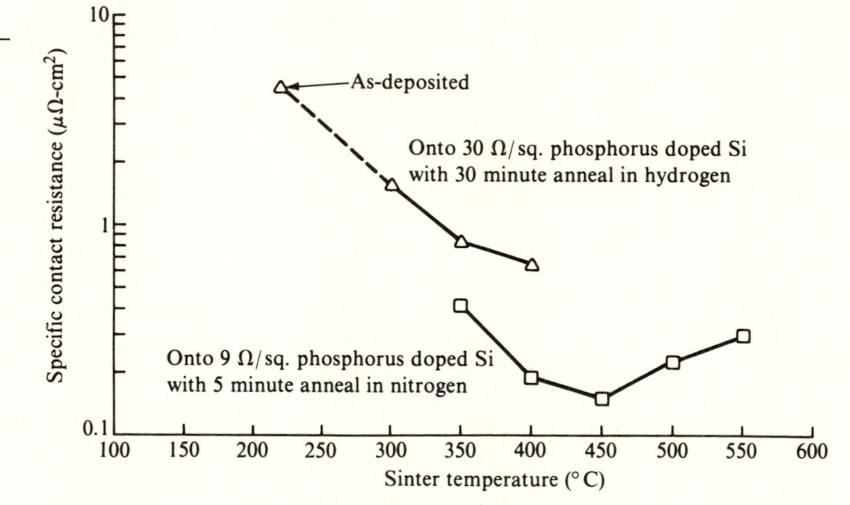

FIGURE 10.15

Specific contact resistance versus sinter temperature of aluminum on heavily doped phosphorus-diffused silicon. (The two sets of data were obtained from two different test patterns that may not give comparable results.) (*Source:* Upper curve data from R.A. Levy et al., *J. Electrochem. Soc. 132,* p. 159, 1985; lower curve calculated from data in H.M. Naguib and L.H. Hobbs, *J. Electrochem. Soc. 124,* p. 573, 1977.)

[4]Some idea of the penetration depth can be obtained from cross sections, but it is difficult to find a section that stops just at the bottom of a pit. Referenced data were obtained by replicating the contact surface after etching away the aluminum and examining the replica with a SEM.

[5]The addition of silicon requires some modification of the procedures used to bond connecting wires to the chip bonding pads. Because of the silicon precipitates, somewhat different etching techniques are needed to prevent silicon from being left on the wafer surface after lead pattern definition and to prevent jagged edge definition.

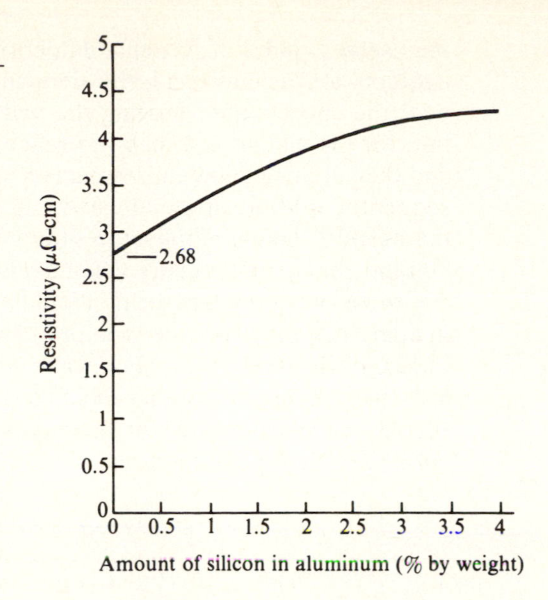

FIGURE 10.16

Aluminum resistivity at room temperature versus amount of silicon present. (*Source:* From data in L.W. Kempf, "Properties of Aluminum–Silicon Alloys," in Taylor Lyman, ed., *Metals Handbook,* American Society for Metals, 1948.)

To assess the minimum temperature that can be used for sintering, and thus to reduce pitting, several studies of specific contact resistance versus sinter temperature have been made (19, 22, 23). Fig. 10.15 shows the general behavior for n-type silicon and illustrates the need for sinter temperatures in the 400°C–450°C range. The reason for the upturn near 450°C for n-type silicon is not due to an increase in the bulk resistance of aluminum because of increased silicon uptake since, as shown in Fig. 10.16, the effect of silicon is considerably less than that of the sintering. The increase, not observed in p-type, apparently occurs because enough aluminum-doped silicon (p-type) regrows in the contact window to affect the contact resistance.

10.2.5 The Silicide–Silicon Ohmic Contact

One of the requirements of contact materials is that they withstand heating to a few hundred degrees while in contact with the silicon. Often, such heating is required for breaking through the residual surface oxide so that good electrical contact can be made. In addition, annealing near 500°C is sometimes required to adjust the SiO_2–Si interface states, and in multilevel metal systems, the interlevel insulator may require a 300°C–500°C deposition temperature. The result is that many potential metal–silicon contacts cannot exist because the silicon and the metal react to form a silicide. However, the silicides generally have very high electrical conductivity and in themselves make very dependable silicon contacts.[6] Silicides for contacts are usually made by depositing a thin layer of the metal over the entire wafer, heating the wafer to a high enough temperature for the silicon and metal to react in the contact window areas, and then etching away the unreacted metal on top of the oxide. Subsequently, additional metallization is added that makes contact to the silicide. This multiple-step operation prevents voids due to the volume change that occurs when a silicide is formed.

Most of the metals form silicides, as shown in Table 10.4, and in many cases, form more than one. Examples are PtSi, Pt_2Si, Pt_3Si, Pt_5Si_2, Pt_7Si_3, and $Pt_{12}Si_5$. For some of the compositions, high- and low-temperature crystallographic modifications may also occur. Although the list of silicides is formidable, when they are grouped according to their position in the periodic chart, some commonality of

TABLE 10.4

Metals Reported to Form Silicides

IA	*IB**	*IIA*	*IIB**	*IIIA*	*IIIB**	*IVB†*	*VB†*	*VIB†*	*VIIB*	*VIII‡*		
	Li			None reported								
	Na	Mg										
Cu	K		Ca		Sc	Ti	V	Cr	Mn	Fe	Co	Ni
	Rb		Sr		Y	Zr	Nb	Mo		Ru	Rh	Pd
	Cs		Ba		Rare earths§	Hf	Ta	W	Re	Os	Ir	Pt
						Th‖		U‖		Np‖	Pu‖	

*Group IB contains the alkali metals; group IIB contains the alkali earth metals; and group IIIB contains rare earth metals, also referred to as the lanthanides.

†The first three rows of groups IVB, VB, and VIB are sometimes referred to as transition and sometimes as refractory metals.

‡The third column of group VIII is sometimes referred to as near-noble elements. (The noble elements are gold, silver, mercury, and copper, so named because they are found free in nature.)

§La, Ce, Pr, Nd, Sm, Eu, Gd, Tb, Dy, Ho, Er, Tm, Yb, Lu.

‖The elements thorium, uranium, neptunium, plutonium, and the others from atomic No. 89–103 are the actinides.

Sources: From compilations in Harold P. Klug and Robert C. Brasted, "The Elements and Compounds of Group IVA," in M. Cannon Sneed and Robert C. Brasted, eds., *Comprehensive Inorganic Chemistry,* Vol. 7, D. Van Nostrand Co., New York, 1958; in Bertil Aronsson, Torsten Lundstrom, and Stig Rundqvist, *Borides, Silicides, and Phosphides,* Meuthen & Co. Ltd., London, 1965; and in Marc-A. Nicolet and S.S. Lau, "Formation and Characterization of Transition-Metal Silicides," in Norman G. Einspruch and Graydon B. Larrabee, eds., *VLSI Electronics Microstructure Science 6,* Academic Press, New York, 1983. (For an extensive listing of the properties of silicides, see tables in the third reference just given.)

[6]Some silicides also make good Schottky diodes when contacting high-resistivity silicon (see section 10.3 on Schottky diodes). Thus, if windows are opened for both the Schottky diodes and the ohmic contacts, both structures can be made at the same time. Because of their low resistivity and high-temperature stability, silicides are also sometimes used as lead conductors in parallel with the higher-resistivity polycrystalline silicon.

the silicides and their siliciding properties can be obtained (24). These properties can be used to initially screen the silicides for potential semiconductor use. The reasons for elimination vary from reactivity with water to radioactivity.

Groups I and II silicides are generally considered inappropriate for semiconductor use. Cu, group IA, is a very fast diffusing lifetime killer in silicon. The group IB silicides that have been studied all spontaneously ignite in air. Most of the group II silicides slowly decompose from moisture in the air (25). The group IIIB rare earth metals form disilicides at very low temperatures and have n-Si barrier heights in the 0.3–0.4 eV range. The metals themselves oxidize rather easily and hence, from a manufacturing standpoint, are rather intractable (24). The actinide series (thorium, uranium, neptunium, and plutonium) are radioactive, and their use would build in a source of radiation damage. Thus, about half the known silicides do not appear to be candidates for semiconductor metallization at this time.

The transition (refractory) metals (groups IVB, VB, and VIB) require temperatures near 600°C for silicide formation. Silicides of the form $MoSi_2$ are generally the final form and generally have the lowest resistivity. In most cases, the species diffusing during silicide formation is silicon. When the silicide is formed from a co-deposition of metal and silicon, if there is a deficiency of metal, silicon will be rejected. If there is an excess of metal, it will be corrected by silicon diffusing in from the contact window. If no free silicon is available and the silicide is contacting silicon dioxide, the oxide is sometimes reduced.

TABLE 10.5

Silicide Barrier Heights on n-Type Silicon

Material	ϕ_B *(V)*
$CoSi_2$	0.65
$CrSi_2$	0.57
$HfSi_2$	0.55
$MoSi_2$	0.55
$NbSi_2$	0.62
$NiSi_2$	0.7
Pd_2Si	0.75
PtSi	0.84
$TaSi_2$	0.6
$TiSi_2$	0.6
VSi_2	0.55
WSi_2	0.65
$ZrSi_2$	0.55

Source: From data in Marc-A. Nicolet and S.S. Lau, "Formation and Characterization of Transition-Metal Silicides," in Norman G. Einspruch and Graydon B. Larrabee, eds., *VLSI Electronics Microstructure Science 6*, Academic Press, New York, 1983.

The near-noble elements Pt and Pd form metal-rich silicides or monosilicide at temperatures near 400°C. When produced from a Pt or Pd layer on Si, the siliciding is by metal rather than silicon diffusion. PtSi is widely used, both for contacts and for Schottky diodes. It was the first silicide commercially used for silicon contacts and was chosen because of its metallurgical stability, corrosion resistance, ability to be formed in the solid phase, and a visual appearance distinctly different from either Pt or Si (26). The latter property allowed optical examination of the contact areas after siliciding. Palladium has been used occasionally as an alternative since it can be deposited by evaporation and thus not produce radiation damage.

The rest of the group VIII elements, as well as those of group VIIB, are apparently potentially usable. Except for those with very low Schottky barrier heights, the silicides, like aluminum, make good ohmic contact only when contacting highly doped silicon. Contact resistance values have been measured for a few of the materials, but their behavior can be judged from the barrier height (see Table 10.5) and Fig. 10.8. It should be remembered, however, that just because a silicide is relatively stable and has a low resistivity does not necessarily mean that it will make a satisfactory contact material. Properties such as thermal expansivity, adhesion, and internal stresses must also be considered. When used only as a contact, the silicide can be very thin, and stresses may cause no problem. When contacts are combined with leads, much thicker layers are required. Then, if the silicide is made in situ, where lateral growth is confined, as in contact windows, stresses may be high enough to cause the silicide to break away from the silicon (27).

10.2.6 Alloyed GaAs Ohmic Contacts

As was shown in Fig. 10.6, the doping level of n-type GaAs required for good ohmic contact is in excess of 10^{19} atoms/cc. Since the making of n^+-regions by conventional diffusion is difficult and generally avoided if possible, the n^+-silicon contact has no commercial GaAs equivalent. Instead, alloyed contact metallization that includes a dopant source is generally used. Since such contacts present problems of metal balling and nonuniform dissolution of the GaAs, however, nonalloyed contacts continue to be examined.

One of the earliest contact metals was tin, but it has significant motion at device operating temperature when it is under the influence of high electric fields. As a result, it can be transported through a device and produce conducting channels (28). Silver, which does not dope, mixed with germanium for n-type and zinc for p-type contacts has also been used (29). Currently, gold with germanium for doping is used almost universally (7, 30). A eutectic mixture of germanium and gold (88% Au, 12% Ge) is now the most commonly used contact (31). The eutectic provides a liquid alloy source at the low temperature of 356°C. Germanium can be either a p- or an n-dopant but apparently in this case resides on a gallium site and dopes the GaAs n-type. Even with this doping, however, the specific contact resistance declines with increasing doping as was shown in Fig. 10.7 so that n^+-doping is still desirable. The surface melting and regrowth can leave a very rough surface, but the addition of an overlaying nickel layer will minimize that roughness. The ternary combination (Au–Ge–Ni) interacts with the GaAs and provides for much deeper penetration of the germanium than expected. Substantial lattice disruption also occurs, and there are some reports of a thin, high-resistivity layer underlying the contact (32).

10.2.7 Nonalloyed GaAs Ohmic Contacts

Three different nonalloyed contact systems have been reported. One uses ion implantation to introduce the heavy doping in the contact windows and then rapid thermal annealing to activate the dopant (33–35). The second forms a germanide on the GaAs surface and then depends on germanium diffusing out of it and into the GaAs for the high-concentration doping (36, 37). The third uses molecular beam epitaxy either to deposit highly doped GaAs in the windows

(38–40) or to deposit a layer of material with a lower bandgap (heterojunction) (17, 41).

10.2.8 Measurement of Contact Resistance

One of the earliest methods used for measuring semiconductor contact resistance was to use an epitaxial structure as shown in Fig. 10.17a and measure R_T for several contact sizes (29). For any given contact,

$$R_T = r_c + r_s + r_N \qquad 10.12$$

where r_N is the resistance of the n^+-layer and lower contact and is assumed to be approximately the same for all contact sizes. The spreading resistance r_s of the contact due to the epitaxial layer can be calculated from Eq. 10.3. If $R_T - r_s$ is then plotted versus 1/area A, the slope of the curve is R_c (or Ar_c), and the y axis intercept is r_N.

The Kelvin bridge is applicable to metal–metal contact resistance (as in a via) and to metal–diffused-layer contact resistance. Three versions are shown in Fig. 10.17b (42, 43). The first is appropriate for metal-to-metal measurements, and the two layers may be separated by an oxide except for the desired contact area. The other two depend on a diffused pattern and pn junction isolation to restrict current flow in the bottom half of the test pattern. They further assume that the current density is uniform over the whole contact area, in which case r_c is given directly by the expression $\Delta V/I$, where $\Delta V = |V_1 - V_2|$.

For contacts to a diffused stripe, such as a resistor, as discussed in section 10.2.1, a transmission line model is more appropriate, and Eq. 10.5 must be solved (8). If $d\sqrt{R_s/R_c} > 2$, then the coth term becomes 1, and

$$R_c = \frac{w^2 R_T^2}{R_s} \qquad 10.13$$

where R_c, the specific contact resistance, has been substituted for Ar_c. $d\sqrt{R_s/R_c} > 2$ is not a very stringent requirement since typically R_s is greater than 10 and R_c is less than 10^{-5}. Thus, a $d > 20$ μm will suffice. An alternate analysis, which eliminates the requirement for knowing the contact area, is shown in Fig. 10.17c (44). A series of contacts are made, and the voltage along the resistor is measured and plotted as shown. The curve is then extrapolated back to L_0, and R_c is given by

$$R_c = R_s L_0^2 \qquad 10.14$$

Other schemes as well have been used on occasion. For example, the behavior of a series of concentric contacts to a thin conducting layer has been interpreted in terms of the transmission line

FIGURE 10.17

Contact resistance measuring methods.

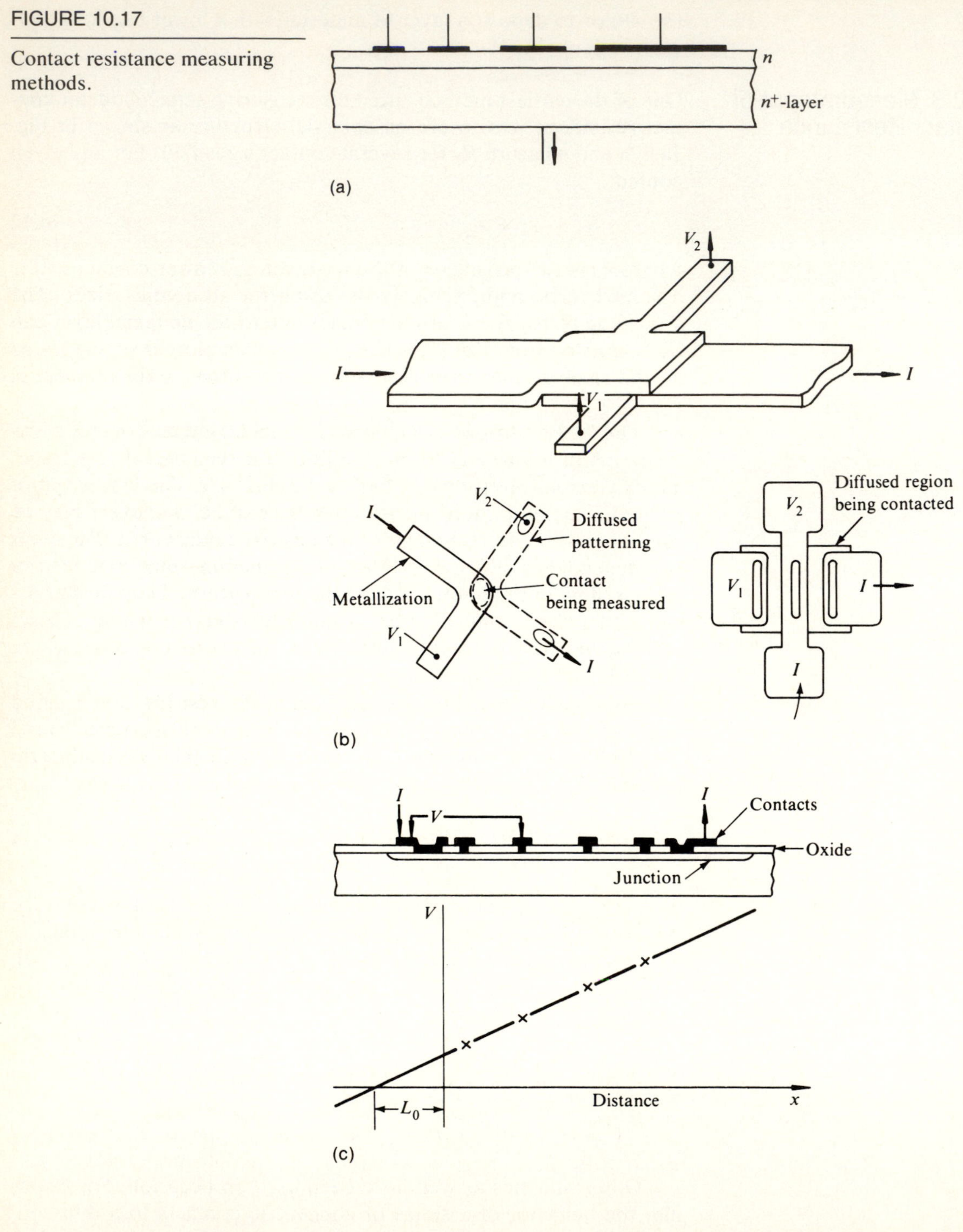

model (45). The voltage distribution under one of the current contacts of a transmission line structure, combined with the sheet resistance, can also be used to calculate R_c (46). In this case, an additional very thin diffusion is required to bring out test points that can be contacted. (For a discussion of errors associated with various methods, see reference 8.)

10.3 SCHOTTKY DIODES

10.3.1 Rectifying Schottky Barriers

In previous sections, Schottky barriers were discussed in terms of a metal or metal-like material contacting a highly doped semiconductor so that the *I–V* behavior was linear. When the semiconductor is n-type and the doping level is less than perhaps 10^{17} atoms/cc, the *I–V* characteristics have the same general characteristics as those of a pn junction. There are, however, several basic differences. The two that make it a desirable circuit element in bipolar circuits are the fact that substantially all current is majority carrier current and the fact that its forward voltage drop is less than that of a pn junction. Thus, when polarity is changed, there is no long delay while minority carriers are recombining. A Schottky diode can be put in parallel with the collector–base of a bipolar switching transistor and act as a short to keep the collector–base junction from being driven into saturation. This allows the transistor to switch much more rapidly and is the basis for the high-speed Schottky TTL ICs. Fig. 10.18a shows how the diode is placed in the circuit. Fig. 10.18b shows how it is integrated on the chip and demonstrates that some integration can be done without resorting to additional leads. The Schottky barrier makes ohmic contact to the base but, where it overlaps the collector, forms a rectifying junction. In MOS circuitry, Schottky barriers are used as gates in GaAs MESFET transistors as shown in Fig. 10.18c.

In the fabrication of Schottky diodes, the same parameters as those applicable to pn junction diodes are important—namely, the reverse breakdown voltage, the reverse leakage well below breakdown, and the forward current. (For a discussion of both pn junction and Schottky diode diagnostics, see Chapter 11.) The breakdown voltage, as in pn junctions, is determined by avalanching in the semiconductor. The field at the edge of the Schottky metallization will be higher, however, and breakdown will appear to be premature. In addition, space charge effects at the barrier periphery can cause substantial excess current, both in the forward and the reverse directions (47). In particular, an accumulated surface will cause excess forward current, and because of the properties of silicon thermal oxide (see Chapter 3), oxidized n-type silicon surfaces show accumulation. To solve these problems, either pn junction guard rings (48) or MOS field plates (49) can be used. These structures are

FIGURE 10.18

Schottky diode applications to integrated circuits.

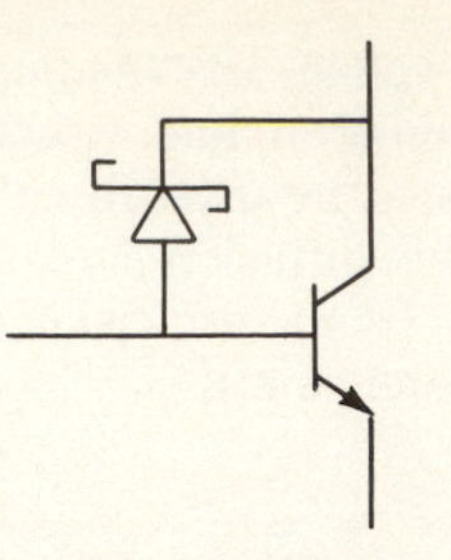

(a) Schottky clamped silicon transistor

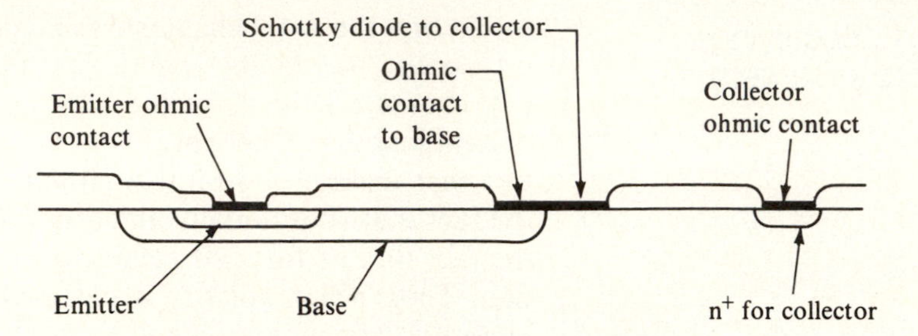

(b) Implementation of Si Schottky clamped transistor

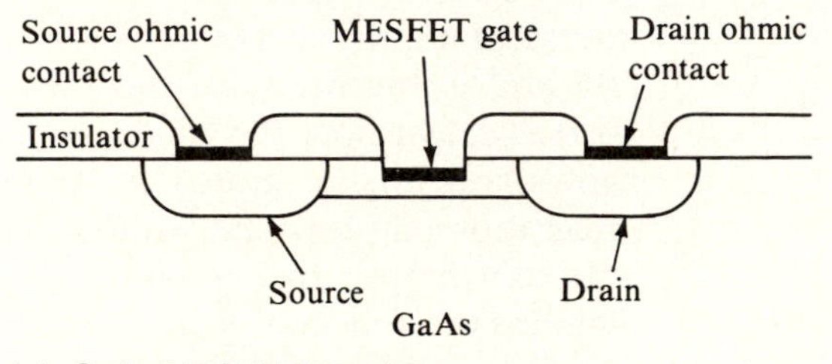

(c) GaAs MESFET transistor

shown in Fig. 10.19. The pn junction guard ring has the disadvantage of requiring additional fabrication steps and increasing the diode area, while the field plates are often not as effective as guard rings, particularly if the oxide is quite thick.

The theoretical Schottky forward current density J is given by

$$J = A^{**}T^2 e^{-q\phi_B/kT} e^{qV/nkT} \tag{10.15}$$

where n is slightly greater than 1 because of barrier lowering (50). Experimentally, depending on the cleanliness of the surface and how well edge effects are minimized by guard rings or field plates, n may range from about 1.03 to as much as 1.5. Fig. 10.20 shows experimental I–V plots for PtSi and Cr metallization on silicon. The

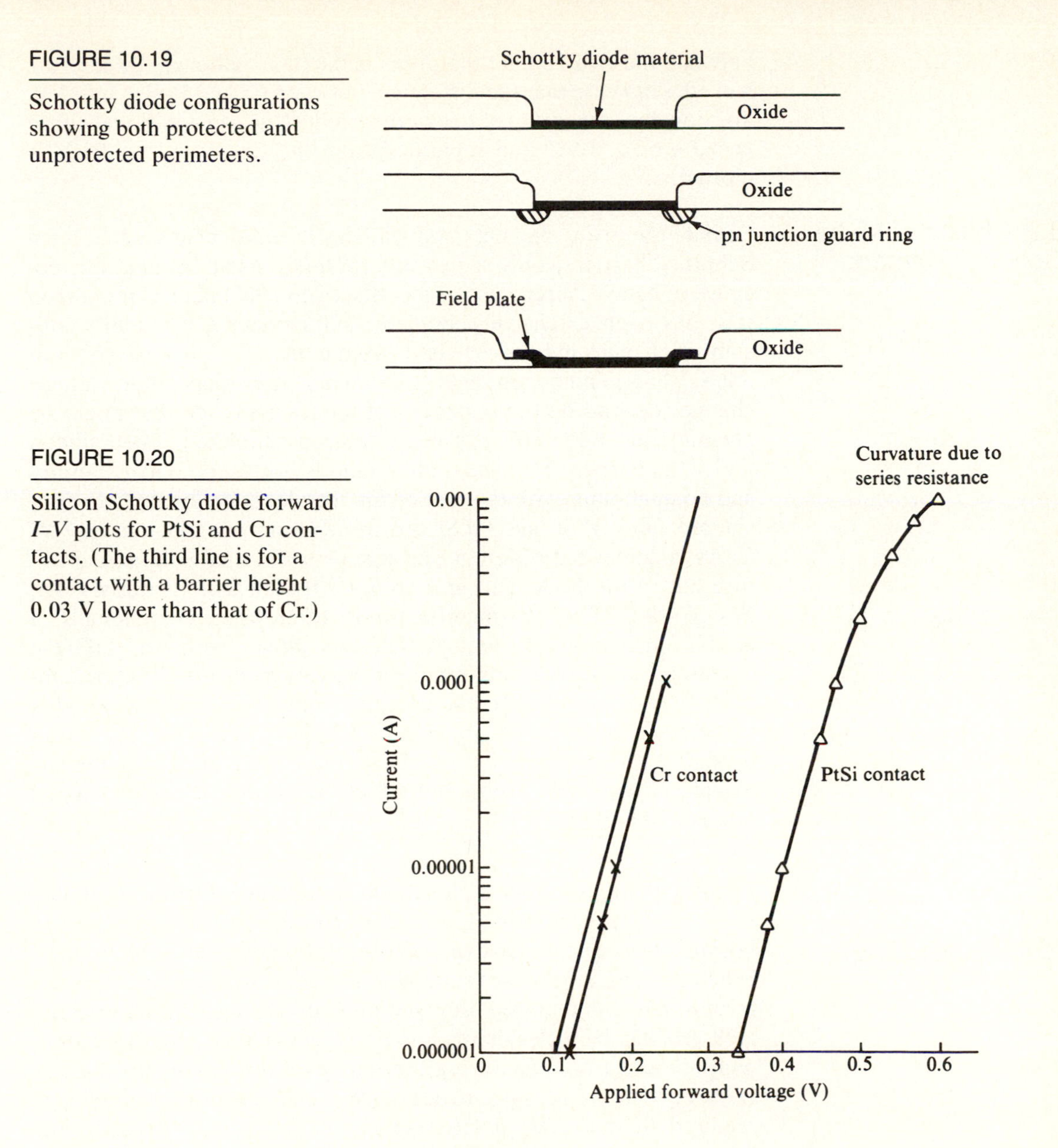

FIGURE 10.19

Schottky diode configurations showing both protected and unprotected perimeters.

FIGURE 10.20

Silicon Schottky diode forward *I–V* plots for PtSi and Cr contacts. (The third line is for a contact with a barrier height 0.03 V lower than that of Cr.)

areas are the same, and, in each case, the n value is about 1.04.[7] The figure also shows the effect of changing barrier height on the current for a given voltage. Based on Eq. 10.15, the barrier height difference

---

[7]The n value can be determined by multiplying the voltage change required to change the current by 2 orders of magnitude by 8.38.

between the two contact materials is 0.021 V, although Tables 10.2 and 10.5 indicate that the difference between Cr and PtSi is 0.023 V. To show the sensitivity of forward current to ϕ_B, an additional curve based on Eq. 10.15 and separated from the Cr curve by 0.03 V is plotted.

10.3.2 Choice of Metal for Schottky Barriers

Despite the many elements and silicides discussed earlier that form Schottky barriers, only a few are routinely used. Aluminum, because of its widespread use as a contact and lead material for silicon ICs, has been extensively studied. Unfortunately, it exhibits substantial changes in barrier height (from a low of ~0.6 V to a high of ~0.8 V) depending on heat treatments after deposition. These changes depend on the thickness of interfacial oxide that might be present and on the effect of recrystallized aluminum-doped silicon under the barrier (51). Thus, aluminum is not used. When stability and compatibility with the rest of the metallization system are considered, only PtSi and Pd_2Si are in common use. In the case of GaAs, a somewhat different problem exists in that gallium will diffuse out of the GaAs and into some of the possible materials. Al, Mo, W, and Ti are the materials most often used. Occasionally, a circuit application will require Schottky diodes with two different "turn-on" voltages.[8] This requirement can most easily be accomplished by using two different materials with differing barrier heights (as shown, for example, in Fig. 10.20). Pd_2Si and PtSi have been used, but it is also possible to change the doping directly under the barrier by ion implantation and affect the barrier height by several tenths of a volt (52).

10.4 LEADS (INTERCONNECTS)

There are several possible choices for interconnect leads, as shown in Fig. 10.21, where the complexity and cost increase from top to bottom. The least expensive and easiest-to-use metal is aluminum. When the geometries are reduced to the point where more than one level of metal is required, doped aluminum is used to minimize hillock growth. Hillock minimization is necessary to keep the aluminum hillocks from puncturing the interlevel insulation. Further, as leads become narrower, current density increases so that doping is required to minimize electromigration.[9] The same forces driving

[8]Even though the forward I–V curves of pn and Schottky diodes appear linear on a semilog plot, they turn up rather abruptly somewhere below 1 V when plotted linearly (the familiar forward junction curve). For a given current density, the "turn-on" voltage is less for Schottky than for pn junction diodes, and for Schottkys, is less, the smaller the barrier height.

[9]Electromigration will be discussed in section 10.8.1.

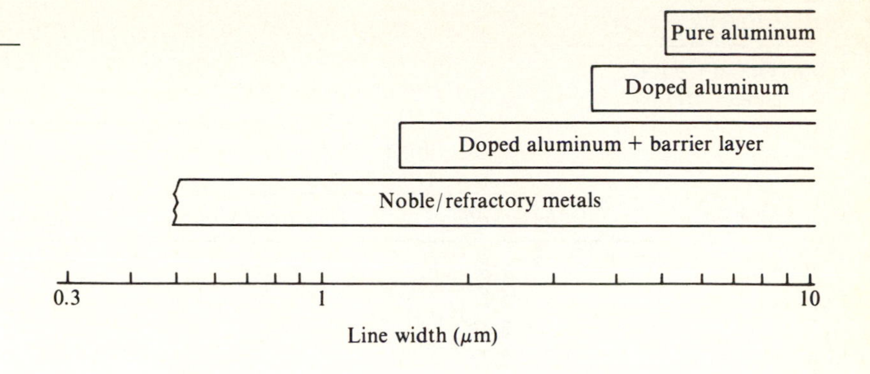

FIGURE 10.21

Applicability of various metal combinations to silicon IC device interconnections.

closer spacing and narrower leads also drive toward shallower junctions, which makes aluminum contact spiking more of a problem. Hence, the smaller geometries require a barrier layer between aluminum and silicon. Finally, some requirements are so demanding that aluminum is no longer applicable, and various noble/refractory metal combinations must be used. The various properties of potential lead materials is examined next.

10.4.1 Lead Resistance

When IC chips are very small and geometries are large, lead resistance is not a serious problem. However, as geometries shrink and operating frequency and chip size increase, the series resistance, coupled with lead-distributed capacitance, can cause undesirable pulse delays. New processing flows such as those using self-aligned gates and multilevel metal have required lead materials that can stand higher subsequent processing temperatures than aluminum and that often have electrical conductivities much lower than that of aluminum. Table 10.6 lists pertinent properties of several materials used in silicon contact/interconnect systems. Fig. 10.22 shows the range of lead lengths encountered in a particular VLSI chip (53). While the majority are less than 1 mm in length, a substantial number are considerably longer. One empirical rule is that the longest lead length is approximately half the square root of the chip area (54, 55). The impact of the available range of resistivities can be seen in Table 10.7, where line resistance and voltage drop for a representative set of conditions (including a lead length of 1 mm) are tabulated. From this table, it is clear that if the lead is of significant length and a few milliamperes of current are to be carried, neither polysilicon nor thin diffused layers will be satisfactory[10] because of their high resistance.

[10]In many parts of a circuit, a few millivolts drop in a lead is intolerable.

TABLE 10.6

Properties of Contact/ Interconnect Metals

Material	Resistivity* (10^{-6} Ω-cm) Bulk Value	Thin-Film Value	Melting Point (°C)	Adherence to SiO_2†
Aluminum	2.6	2.7–3	660	3
Copper	1.7		1083	2
Gold	2.4	3.4–5	1063	1
Molybdenum	5.8		2625	3
Palladium	11		1555	1
Platinum	10.5		1755	1
Polysilicon		1000	1420	3
Diffused silicon		1000	1420	NA
Silver	1.6		960	—
Tantalum	15.5		2850	—
Titanium	47.8		1800	3
Ti/W pseudo alloy				3
Tungsten	5.5	11–16	3370	2

*Thin-film values of resistivity are generally somewhat higher than those of bulk material because of additional crystallographic defects introduced during deposition. For a further discussion of this effect, see section 10.10.
†On a scale of 1 (poor), 2 (medium), and 3 (good).

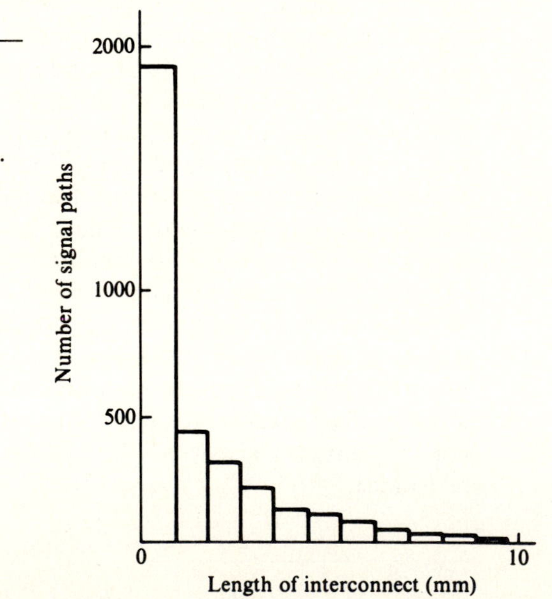

FIGURE 10.22

Distribution of signal path lengths for a combination of the first and second levels of metallization of a VLSI circuit. (*Source:* Adapted from P.B. Ghate, *Proc. 20th IEEE Reliability Physics,* 1982.)

TABLE 10.7

Resistance of an IC Lead

*Material**	*Resistance* *(Ω)*	*Voltage drop (1 mA)* *(mV)*
Aluminum	10	10
Molybdenum	30	30
Titanium	150	150
$MoSi_2$†	300	300
Polysilicon	3000	3000
Diffused silicon	3000	3000

*3 μm wide, 1 μm thick, and 1 mm long.
†The resistivity of various silicides is given in Table 10.8.

The combination of lead resistance and its associated capacitance will cause a pulse delay time that increases as the lead material resistivity increases (55, 56). The delay time will remain constant as the lead width is increased since the resistance decreases and the capacitance increases in step. Thus, the delay time cannot be decreased by making the leads wider. Making them thicker will help since in that case the resistance is decreased without increasing the capacitance, but then lead definition becomes more difficult. When the lead attaches to a separate capacitor, as, for example, a MOS gate, the gate capacitance and the lead resistance may determine the delay time. Then, increasing the lead width to lower its resistance will help until the lead becomes wide enough for its capacitance to dominate the gate capacitance. At that point, the delay time will begin to increase. This behavior is shown in Fig. 10.23, which illustrates that even a 10 μΩ-cm resistivity may not be adequate to allow top performance of very fine geometry circuits (57).

For short runs, such as crossunders to provide paths under conventional leads over oxide, diffused layers have been used since the first ICs were made and work well for most purposes. However, as operating frequencies increase, air-isolated bridges such as are now used in GaAs, may be required for silicon. Because of its ability to withstand subsequent high-temperature operations, polysilicon is commonly used as a combination MOS gate electrode material and partial first-level interconnect (58). In this application, current flow is small, but the high resistance can cause pulse rise-time and propagation delay problems. The problem of high resistance causing delay was recognized soon after the use of polysilicon was introduced and led to the early proposal for a refractory metal lead system (56). This approach has not been well accepted, but processes that add a layer of metal silicide to the top of the polysilicon are now being used. The silicides, as shown in Table 10.8, have between 10 and 100 times the conductivity of polysilicon and thus offer considerable

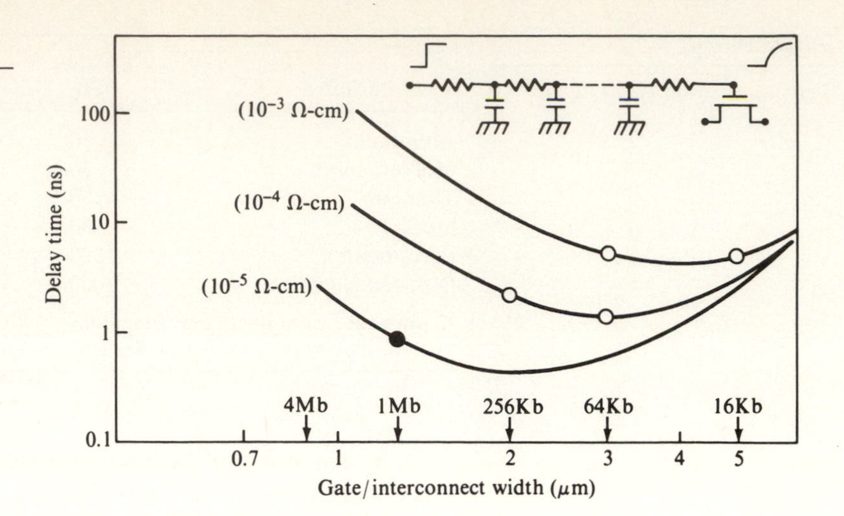

FIGURE 10.23

Delay time versus gate/lead width for various electrical conductivities. (*Source:* N. Yamamoto et al., *J. Electrochem. Soc. 133,* p. 401, 1986. Reprinted by permission of the publisher, The Electrochemical Society, Inc.)

TABLE 10.8

Silicide Resistivity at Room Temperature

Material	*Co-Sputter* (μΩ-cm)	*Metal–Polysilicon Reaction* (μΩ-cm)	*Reference*
$CoSi_2$	25	17–20	(1)
$HfSi_2$		45–50	(1)
$MoSi_2$	100		(1)
$NbSi_2$	70		(2)
$NiSi_2$	50–60	50	(1)
$PdSi_2$		30–35	(1)
PtSi		28–35	(1)
$TaSi_2$	50–55	35–45	(1)
$TiSi_2$	25	13–16	(1)
WSi_2	40*–70		(1, 3)
$ZrSi_2$		35–40	(1)

*Slightly lower values are reported for CVD co-deposition. See, for example, S. Sachdev and R. Castellano, *Semiconductor International,* May 1985.

1. S.P. Murarka, *J. Vac. Sci. Technol. 17,* p. 775, 1980.
2. T.P. Chow et al., *J. Electrochem. Soc. 133,* p. 175, 1986.
3. J. Kato et al., *J. Electrochem. Soc. 133,* p. 794, 1986.

improvement. The layer can be formed by depositing a layer of metal on the surface of polysilicon and then alloying it and the top portion of the polysilicon together. Alternatively, a silicide can be deposited by sputtering from a silicide target, by co-sputtering from silicon and metal targets, or by CVD. Of the materials listed, titanium silicide affords a good compromise of low resistivity, high adherence, ease of forming, and the ability to make acceptable

FIGURE 10.24

Effect of temperature on electrical resistivity of several metals. (*Source:* From data in Alexander Goldsmith et al., *Handbook of Thermophysical Properties of Solid Materials,* Vol. 1, Pergamon Press, New York, 1961.)

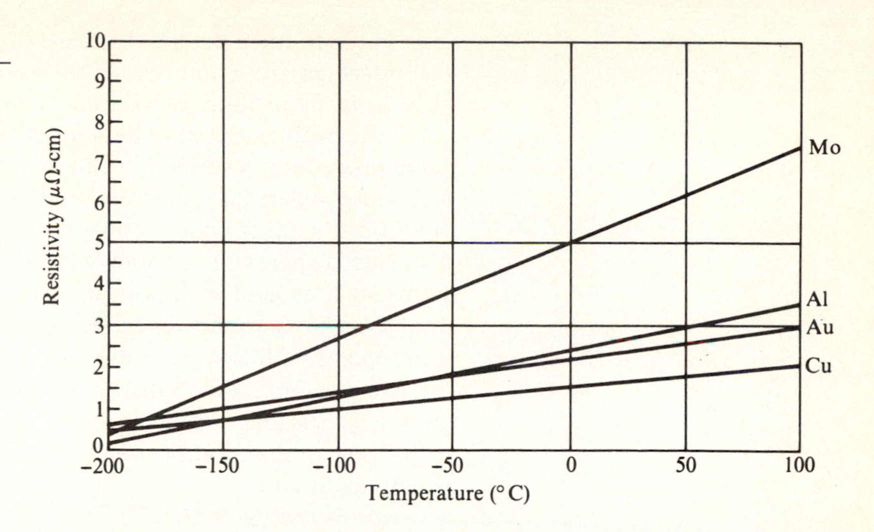

Schottky diodes so that one material can be used for both leads and Schottky contacts. Its conductivity is, however, 5 to 10 times less than that of the metals commonly used for leads. As can be seen from Fig. 10.23, and as has been projected, even the metals of Table 10.6 may become inadequate for very fine-line geometries. One method of further reducing the lead resistance is by cooling the leads (and the semiconductor device as well) (59). Fig. 10.24 shows how several potential lead materials behave at low temperatures. By going to liquid nitrogen temperature, the resistance could be reduced by a factor of ~5. Before such a step is taken, however, the device design must be optimized for this temperature. If higher-temperature superconducting materials become a reality, if they can remain superconducting at the high current densities at which ICs operate, and if they can be made compatible with IC processing, they would be a very appealing alternative (60). Another high-speed interconnect system, but one that is more specialized, is the use of light pipes and optical coupling directly on the chip (61, 62).

10.4.2 Adhesion to SiO_2

Most IC leads run over silicon dioxide, and thus it is very important that adhesion be good. Table 10.6 categorized the adhesive properties of the more common interconnects as good, medium, and poor. Aluminum and polysilicon adhere very well on clean oxide, and when a lead is torn away, it will usually take away the oxide and some silicon beneath. A case of peeling leads usually occurs because the oxide surface was not well cleaned but may occur because of a faulty deposition. In the case of Ti:W, adhesion is dependent on such deposition parameters as sputter target potential, system pressure, and system leaks (63). In the case of the medium adherers, it is possible to have peeling because of excessive shear developed at the interface from internal film stresses. In such cases, either a change in deposition conditions or a post-deposition anneal may solve the problem. It may be helpful to monitor the stress using the same procedure described in Chapter 3 for determining stress in oxide films. When the leads are multicomponent, adhesion may depend on the exact composition. For example, silicon-rich tungsten disilicide appears to adhere better than tungsten-rich disilicide. Films such as gold with poor adhesive qualities should not be used directly. They can, however, be used on top of a thin layer of some other metal with less desirable electrical properties. An example is the use of a very thin buffer layer of titanium, followed by a thick gold layer. In this particular example, yet another layer is required to prevent interaction between the gold and the titanium, and it is usually platinum.

10.4.3 Diffusion Barrier

Diffusion barriers may be used to prevent undesirable impurities from reaching some part of the circuit. An example is the shielding of a MOS gate oxide from impurities diffusing through the gate after a source–drain ion implantation. More often, however, diffusion barriers are thin layers inserted between two other layers of multilayer leads or between the main body of the lead and the semiconductor to prevent interdiffusion and chemical reaction (63, 64). A common application is between aluminum leads and silicon to prevent contact pitting. Another is the use of a layer between PtSi and aluminum to prevent the aluminum–PtSi reaction from changing the Schottky barrier height. Under some circumstances, layers may serve the dual role of improving adhesion over oxide and preventing interactions in the contact area.

Barriers have been categorized based on the manner in which they work as passive or sacrificial (65). Passive layers are quite inert to the layer on each side and thus will, in principle, always keep the two layers separated. Unfortunately, there may be substantial diffusion of one or both components through the passive barrier. It should be noted that high-temperature diffusion coefficient data extrapolated to low temperatures will usually substantially understate the low-temperature value. Sometimes, it is possible to add a component to the barrier to reduce diffusion. This combination is referred to as a stuffed barrier. An example of a stuffed barrier is thin-film TiN that has been sputtered in the presence of oxygen. Without the oxygen, substantial diffusion of silicon through the layer occurs when the layer is used as a barrier between silicon and aluminum. The diffusion apparently takes place along grain boundaries, where the inclusion of oxygen will inhibit diffusion (66). Sacrificial barriers will react, but at a slow enough rate that they are still useful.

Almost all barriers in common use are either sacrificial or else allow noticeable diffusion. However, some of the transition metal nitrides, borides, and carbides may be truly passive. As a class, they have high electrical conductivity and are stable in the temperature range of interest (67, 68). For example, TiN can be sputter deposited with a resistivity of about 20 $\mu\Omega$-cm (69) and will make good ohmic contact to low-resistivity silicon. When the behavior of the barrier is considered, the effect of additives to the films being separated should not be neglected. For example, it is common to add a few percent copper to aluminum to reduce electromigration, and the copper's presence retards the reaction between aluminum and titanium (70). Table 10.9 shows a few of the many combinations of leads and barriers now in use. It is to be emphasized that as new metallization systems are introduced, they must be carefully examined for latent failures caused by interdiffusion and unexpected compound formation.

Not listed in the table is a barrier for use between aluminum and gold, the lack of which produced the most famous case of premature failure in the history of the semiconductor industry. Gold wires are commonly used to connect from gold-plated package leads to aluminum bonding pads. At the upper end of the temperature range over which silicon transistors will operate, aluminum and gold slowly form a series of colored intermetallics (the "purple plague") that will ultimately cause failure. By somewhat restricting the temperature range (also necessary for plastic packages), the problem can be contained. For cases where higher temperatures are required (as for components for some military applications), an all-aluminum system using aluminum leads, aluminum wires, and aluminum-coated package connections is sometimes used. As an alternative, several gold metallization systems have been developed. Among these are the beam lead method (26) and a gold lead metallization that can be used directly with gold wire bonding.

TABLE 10.9

Examples of Diffusion Barrier Applications

Barrier Material	*Films Separated*	*Reason*
PtSi	Si and Al	Prevents aluminum contact spiking.
Ti:W	PtSi and Al	Prevents Al–PtSi reaction.
Pt	Ti and Au	Prevents Ti–Au reaction.
W	Au–Ge–GaAs and Au	Prevents Ga diffusing into Au.
Si (poly)	Gate oxide and poly n^+	Prevents dopant from reaching gate.

10.4.4 Interconnect Material Contamination

Care must be taken to ensure that the interconnect materials used are free of sodium contamination. High-purity materials processed especially for the semiconductor industry should pose no problem, but alternative materials being investigated may not have been refined to semiconductor-grade standards. Radioactive impurity levels, although generally quite low, are a potential problem in the case of high-density MOS memories, where soft errors can occur because of carrier generation from random radioactive particles.

10.4.5 Step Coverage

An important part of lead technology, and one that has little to do with lead composition, is step coverage. Leads are generally about 1 μm thick, and steps range from about 0.25 μm to greater than 1 μm high. The step profiles range from gentle, relatively easy-to-cover slopes (Fig. 10.25a) to virtually impossible-to-cover overhangs (Fig. 10.25d). (For a discussion of the causes of these various profiles, none of which are metal related, see Chapter 6.) In addition to etching procedures, some other process-related factors may affect the step profile. If a contact is being etched in a relatively thin oxide window nested inside a thicker oxide as shown in Fig. 10.26a, misalignment of the mask before etching can cause an unexpectedly high and abrupt step as shown in Fig. 10.26b. If steps are in CVD oxides, the oxide can be doped with phosphorus, arsenic, or boron to reduce the melting point and allow the oxide to be reflowed after etching, thus rounding the corners and allowing better coverage. For the specific case of leads covering holes in oxides leading down to a contact area, selective metal deposition can be used to deposit conductive plugs and fill the holes as shown in Fig. 10.27. In the following step, metal for the leads can be deposited without worrying about the step down into the contact opening.

FIGURE 10.25

Oxide step profiles encountered in IC manufacturing.

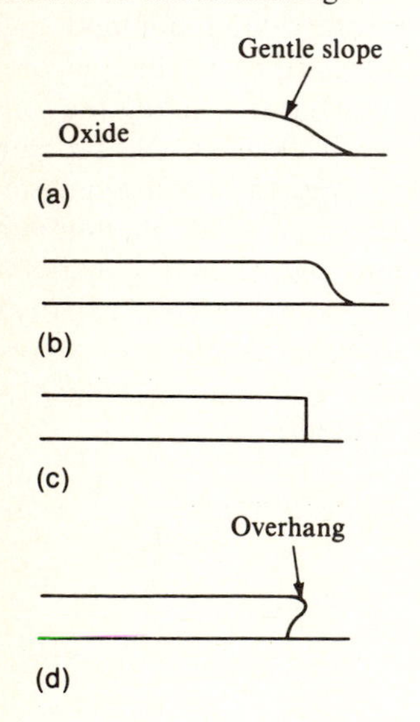

In order to produce a uniform (conformal) coating over steps, the incident atoms should arrive over a wide range of directions. However, most metallization sources allow arrivals from a very limited solid angle so that step coverage is determined by geometric shadowing. Fig. 10.28a shows a cross section of a lead that traverses a relatively modest slope and yet that has substantial thinning over the step. As the step approaches 90°, the profile becomes like that of Fig. 10.28b. The deep groove at the bottom causes a serious reliability problem and illustrates the effect of a profile changing from that of Fig. 10.25a to that of Fig. 10.25b. The low step of Fig. 10.25a,

FIGURE 10.26

Effect of contact window misregistration on oxide step height.

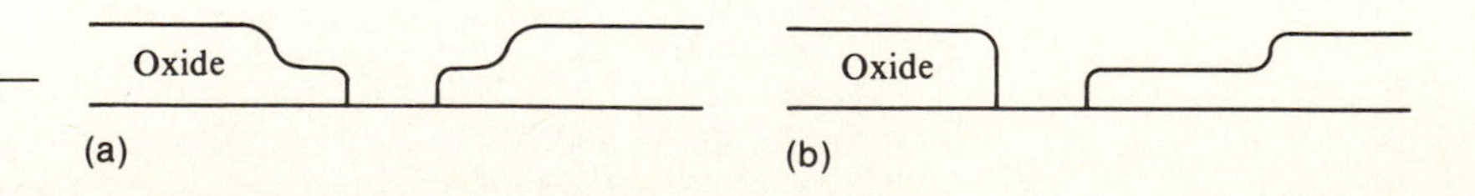

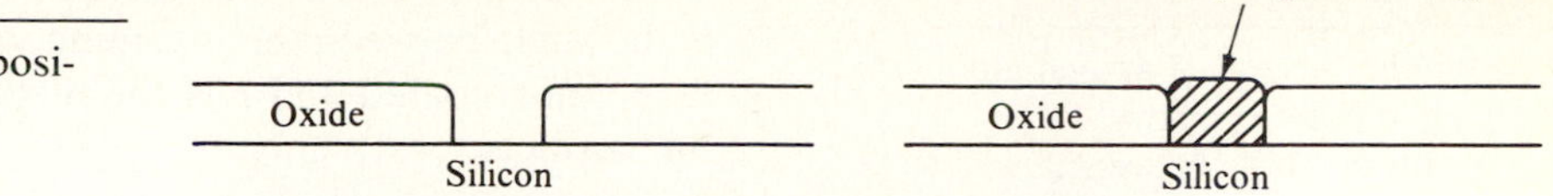

FIGURE 10.27

Use of selective metal deposition to eliminate a step.

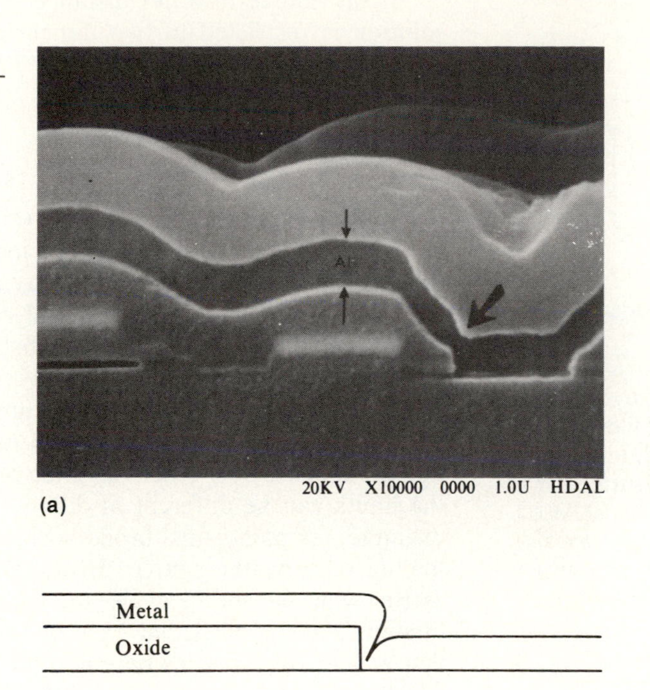

(a)

Metal

Oxide

(b)

FIGURE 10.28

(a) Metal and CVD oxide step coverage. (*Source:* Photograph courtesy of Dr. P.B. Ghate, Texas Instruments Incorporated.) (b) Metal coverage to be expected if the step were vertical.

combined with a thick metallization, can be acceptably covered. However, the high step of Fig. 10.25b leads to a metal coverage profile like that of Fig. 10.28b.

Calculations of the coverage to be expected from an evaporation or sputter source have been made (71, 72). Assumptions usually made are as follows:

1. In evaporation, the mean free path distance is greater than the source-to-wafer distance.
2. The source-to-wafer distance is large compared to the step height.
3. The arriving atomic sticking coefficient is 1, and there is no surface mobility.

FIGURE 10.29

Maximum spread of arrival angles to a point (x,y) on a vertical wall of a hole or trench.

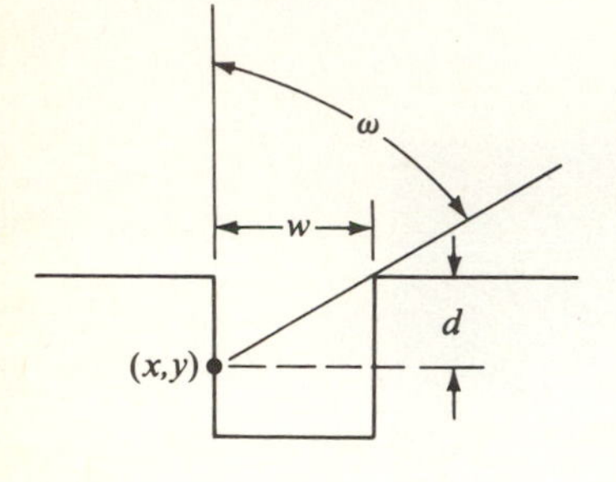

FIGURE 10.30

Calculated step coverage for various ratios of film thickness to step height. (A total arrival angle on the flat surface of 1π was assumed.) (*Source:* A.C. Adams, *Solid State Technology,* p. 135, April 1983. Reprinted with permission of *Solid State Technology*.)

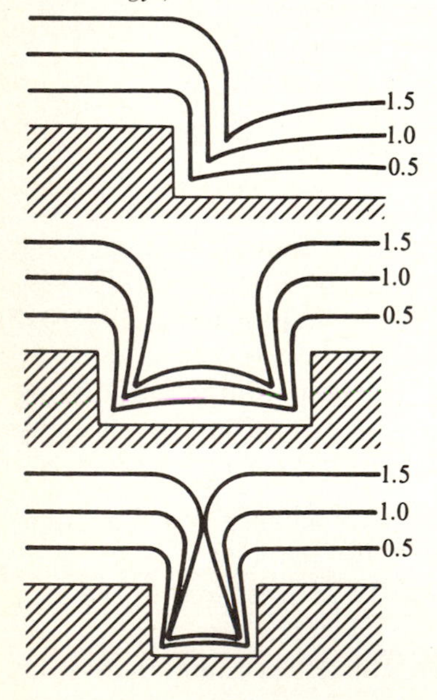

4. The layer grows at a rate proportional to cos ω/r, where ω is the angle between the incoming vapor stream and the surface being coated and r is the distance from the source to the surface.

If the source-to-wafer distance is considered constant and the solution is restricted to two dimensions, expressions of the form given by Eq. 10.16 result (71):

$$\Delta x = C(\cos \omega_1 - \cos \omega_2) \qquad 10.16a$$

$$\Delta y = C(\sin \omega_2 - \sin \omega_1) \qquad 10.16b$$

where Δx and Δy are the changes in the coordinates of a point (x,y) on the surface, C is a constant that includes the deposition rate and the time span over which Δx and Δy are calculated, and ω_1 and ω_2 are the angular limits of ω and will change as x and y change. To determine a final profile after a substantial thickness has been deposited, it is necessary to do many calculations, advancing points on the surface each time by a small amount of Δx and Δy. The limits of ω_1 and ω_2 can change with each new calculation since an additional thickness may block off some of the incoming beam. Further, the limits can be different in different regions of the surface. For example, for points on a broad horizontal surface, the angle will be limited only by the source. However, as shown in Fig. 10.29, no matter what the source angle, the incident angle along a side wall is limited. For the particular case of vertical walls of the well of the figure, $\omega \cong \tan^{-1}(w/d)$ where w is the width of the opening and d is the distance down to the point (x,y) (73).

Use of this kind of model gives predictions as shown in Fig. 10.30 and demonstrates how voids can be produced in holes such as could be encountered in small-geometry vias and contacts (73, 74). Fig. 10.31 shows the manner in which the thickness of a lead over an isolated step changes as the step angle varies (64). Computer programs are available for calculating step coverage, and SAMPLE is a very popular one (72).

Evaporation sources are very small. Thus, if the slices are stationary, there is only one angle of arrival, and shadowing would be completely unacceptable. To extend the angles of arrival, planetary slice holders are used that move the slices so that the source appears to be nearly hemispherical. Sputter sources are generally large flat plates, and the pressures are higher so that the mean free paths are shorter. However, under normal sputtering conditions, little difference exists between evaporation and sputter coverage (64). However, it has been reported that if the conditions can be changed so

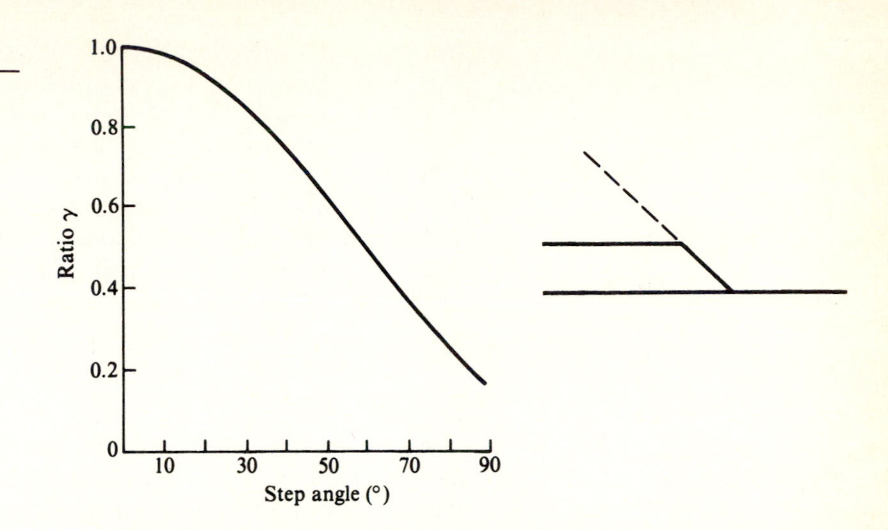

FIGURE 10.31

Calculated effect of step angle on ratio γ of metal thickness on step wall to that on an unshadowed horizontal surface. (Conventional planetary slice rotating tooling was assumed.) (*Source:* Adapted from P.B. Ghate, *Thin Solid Films 93,* p. 359, 1982.)

that resputtering of the depositing surface occurs, substantial redistribution and much better coverage occur (74). Plasma deposition atmospheres have short mean free paths and lead to an incoming angular spread very close to π. Chemical vapor deposition comes close to giving uniform coatings regardless of the profile. The mean free path of reacting atoms is very short, and atoms reach the surface, not by a line-of-sight path from the source, but by diffusion. Thus, even with overhangs, reacting atoms can reach the surface. In addition, there is usually a high degree of surface mobility of the reacting species, which further improves coverage. It is for this reason that the plug shown in Fig. 10.27 could be deposited. Not all materials and flows are amenable to CVD, however. Consequently, it is not a solution to all step coverage problems.

10.4.6 Protective Overcoats

Metal leads are generally thin, soft, and narrow and thus are subject to damage by scratching, which is most likely to occur either during multiprobe or during chip separation. To minimize scratch damage, a hard overcoat of SiO_2 had been added to the process flow by the late 1960s. The coating methods were originally E-beam evaporation or sputtering but later were expanded to include chemical vapor deposition of silicon dioxide and the plasma deposition of silicon nitride. All are low temperature in order to not damage aluminum leads, and, originally, most layers were cracked. The cracking was not particularly important for scratch protection nor for the subsequent use of protecting leads from being electrically shorted by loose particles in the package. However, with the realization that the overcoat could also afford protection from moisture,[11] emphasis was then placed on producing crackfree coatings. The primary way of producing such coatings is by co-depositing a small amount of phosphorus with CVD oxide and forming a phosphosilicate glass that has a lower internal stress than CVD SiO_2 alone. However, as will be discussed in the section on reliability, too much phosphorus will appreciably increase the conductivity of the glass and cause corrosion in the presence of small amounts of moisture. To minimize these effects, a maximum of about 7% phosphorus is now normally used.

10.5 MOS GATE MATERIALS

Silicon gate metallization choices are shown in Fig. 10.32. The earliest gate material was aluminum, which was superseded by polysilicon in order to have a self-aligned[12] gate. It is convenient to have the gate and its interconnection to the next circuit element of the same material, but where the geometries are very small, the polycrystalline silicon resistivity is too high. In this case, various silicides can be used. (Table 10.8 gave typical silicide resistivities.)

Polyicide is short for a composite gate composed of polycrystalline silicon with a layer of silicide on top to reduce lead resistance (75). Salicide (self-aligned silicide) is a name applied to the simultaneous application of a silicide-forming metal to the gate poly and

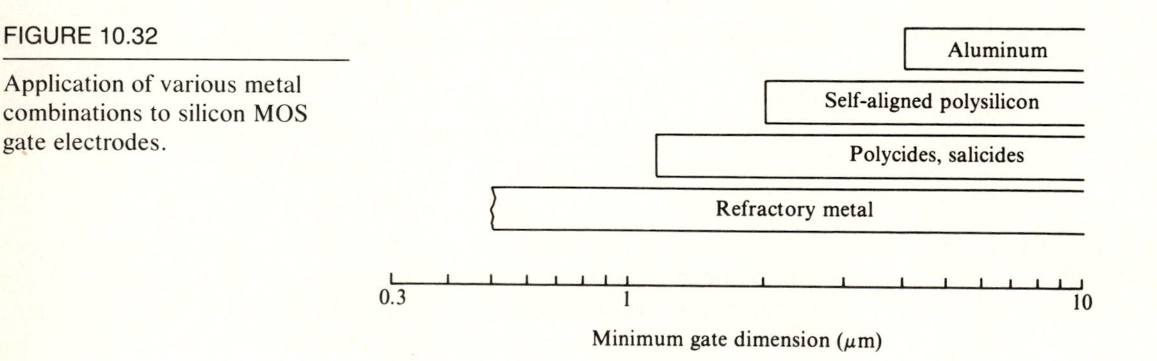

FIGURE 10.32

Application of various metal combinations to silicon MOS gate electrodes.

[11]The possibility of moisture arises because plastic packages are slightly permeable to moisture and because moisture can migrate along the plastic–lead interface into the package.

[12]Self-aligned means that the gate does not overlap part of the source and drain. This is accomplished by defining the gate in a high-temperature material such as polysilicon or a refractory metal and then using it as a mask during the source/drain implant step. The gate material must then withstand the implant anneal and any additional diffusion drive-in that may be required.

silicon contact windows (76, 77). After heat treatment and formation of the silicide where the metal touches bare silicon, the unreacted metal can be etched away, leaving the silicide self-aligned in contact openings and over the polysilicon gate.

For cases where aluminum leads and contacts are not desirable, refractory metals can be used. The processing is then simplified by using the same refractory metal for gates as well.

When one gate material is substituted for another in an existing circuit design, not only must possible deleterious chemical interactions be considered, but also the effect of the gate on threshold voltage. In principle, the threshold voltage V_t should change from gate material to gate material by the amount of the work function change since

$$V_t = \phi_{ms} + \cdots$$

where ϕ_{ms} is defined by

$$\phi_{ms} = \phi_g - \phi_{\text{Si}} \qquad 10.17$$

where ϕ_g and ϕ_{Si} are the work functions of the gate material and the silicon, respectively. Table 10.10 gives values of ϕ_g for selected materials. Unfortunately, for a given gate material, since ϕ_{ms} may vary with processing, tabular values inserted into Eq. 10.17 can predict results that differ from experiment. These changes with processing

TABLE 10.10

Approximate Work Function of Potential Gate Materials

Material (Metal)	ϕ_g *(eV)**†	*Material (Compound)*	ϕ_g *(eV)**†
n^+-Si	4		
Aluminum	4.25		
Chromium	4.5	$CrSi_2$	3.9
Cobalt	5.0	$CoSi_2$	4.4
Gold	4.8		
Molybdenum	4.3	$MoSi_2$	4.5
Platinum	5.7	PtSi	5.4
		Pt_2Si	5.6
Palladium	5.1	Pd_2Si	5.1
Tantalum	4.2	$TaSi_2$	4.2
Titanium	4.3	$TiSi_2$	4
Tungsten	4.6	WSi_2	4.7
Zirconium	4.0	$ZrSi_2$	3.9

*From a compilation in Farrokh Mohammadi, *Solid State Technology,* p. 65, January 1981.

†A range of values is often reported. A "mid-value" number is listed here.

have been studied extensively for the aluminum–silicon system and are thought to be associated with changing interface properties, both at the aluminum–oxide and the silicon–oxide interface (78). In any event, changes in ϕ_{ms} will generally be small enough so that, when thresholds are adjusted by ion implantation, a change of gate material will only require a change of implant dosage and thus be easily accommodated.

10.6 MULTILEVEL METALLIZATION

As circuits became more complex, larger and larger portions of the surface area must be dedicated to leads, and, in many cases, the chip size determining factor is lead area. Without provisions for the crossing of leads, the problem of lead routing is insurmountable for all but the very simplest of circuits. Diffused crossunders have been used from the beginning, but such diffused paths add circuit capacitance and resistance, use appreciable area themselves, and still only allow one level.

An example of multilevel interconnects was shown in Fig. 10.1. The first level of metal is placed on the thermal oxide (Si ICs). Next, the wafer surface (thermal oxide and first-level leads) is covered by an insulator, usually CVD silicon dioxide, and via holes are cut through the CVD oxide where the second-level leads must contact the first level. Another layer of metal is then added and patterned. In some cases, yet another level is used, but the manufacturing problems increase with each level. In MOS circuits, if a polysilicon gate and polysilicon leads are to be used along with aluminum leads, the polysilicon must be deposited before the aluminum. Otherwise, the temperature required for the silicon deposition will ruin the aluminum–silicon contacts.

Multilevel capability has allowed more complex chips to be built in the same area and, in some cases, provided enhanced performance. The main process problem encountered in multilevel metallization is the difficulty in running leads up and down over the surface contours that would ordinarily get worse as the number of levels increases. Additionally, when aluminum is used, there can be difficulty in ensuring that all oxide is removed in the via contact area before the next level is added.

10.6.1 Interlevel Shorts

Interlevel shorts can occur either because of flaws in the interlevel oxide or because of hillock growth from the metal surface after it and the oxide have been deposited. Oxide flaws can occur because of particulate contamination during the oxide deposition or can stem from unwanted holes etched in the oxide due to a lithographic failure or from poor oxide step coverage (usually not a problem). Hillock growth is a serious problem, particularly when aluminum leads are used. Grain growth occurs during temperature cycling—for example, during deposition of the interlevel oxide—and because of film constraints, some of the growth is in the form of hillocks rising above the rest of the surface. Hillocks also occur as the result of electromigration (see section 10.8.1) and because of thermally induced or deposition-induced stresses in the film. These hillocks can penetrate the interlevel insulation and contact the next level of metal. Aluminum is a particularly poor material to use in this regard, but the addition of a few percent copper reduces the problem substantially.

10.6.2 Planarization

To prevent the worsening of surface contours as more layers are added,[13] various planarization processes are used. One partial solution, already mentioned, is the use of selectively deposited metal plugs in vias. Another is the use of CVD silicon dioxide heavily doped with boron and phosphorus (BPSG, or borophosphosilicate glass), which reduces its softening temperature and allows it to be reflowed. Reflowing allows both the sharp edges of openings and the hills formed by the oxide going over lower-level leads to be smoothed. Unfortunately, reflowing requires temperatures higher than those applicable to the commonly used aluminum, but such coatings are suitable for use over polysilicon.

Processes used for planarization after aluminum leads are generally based either on a simultaneous deposition/etchback process for the interlevel dielectric or on a chemically vapor deposited dielectric followed by a sacrificial conformal coating of either spin-on glass[14] or photoresist (79). Combined deposition/etching methods include biased sputtering (80) and plasma-enhanced CVD combined with ion etching (81). If a LPCVD process is available that gives a smooth surface as thickness increases, a very thick coating can be added and then etched back to the desired thickness by an anisotropic plasma etch (82). The process is simpler than a combination etch/deposition, and the results may be adequate. The sacrificial conformal coating process is based on the premise that the coating will be thinner over the high spots and thicker in depressions and that its upper surface will be much smoother than that of the underlying CVD oxide. If a plasma etch with the same etch rate for the spin-on coating and underlying CVD oxide is then used to remove all of the coating, the surface of the remaining CVD oxide will mirror the original surface contour of the coating (83).

---

[13]That steps will become more troublesome with each layer added can be seen from Fig. 10.23, which shows that as the thickness increases (which could come from additional layers) steps tend to become more pronounced.

[14]Usually a water-soluble silicate that, when heated, will decompose into volatiles and silica. The familiar water glass is one example of a water-soluble silicate.

10.7 INTERCONNECT AND GATE METAL PATTERN DEFINITION

Metal definition is done by a combination of lithography and wet etching, dry etching (plasma), or lift-off. Wet etching requires the least expensive equipment, is conceptually the simplest process, and has been used the longest. Plasma etching has little undercutting and thus is more applicable for fine geometries. Lift-off definition is a specialized process ordinarily used in fine-line applications. Both wet and dry etching procedures are covered in detail in Chapter 6. Lift-off is described here next.

When multiple layers are etched, the problem of preventing undercutting of at least one of the layers can be quite severe. Plasma etching is the most uniform, but the problem still remains, when multilayer systems are used, of providing the necessary series of reaction gases. Further, the requirement of minimal undercutting implies straight walls, and straight walls are more difficult to cover by the follow-on insulating layer. The lift-off process makes use of the fact that it is difficult to get good coverage over steps when most deposition systems are used (84–86). Thick resist is patterned so that it remains where the leads are unwanted. When the resist is covered with lead material, there will be a thinning or, in some cases, a break at the step. Thus, when the resist is removed, the metal on top of it will also be removed, leaving the leads in the areas where there was no resist. By using a multilayer resist, the lower layer can be undercut so that a pronounced shadowing of the edge of the leads occurs. This has the effect of producing a lead with sloped edges that will be easy to cover. Lift-off processing of metal is more widely used on gallium arsenide than on silicon.

10.8 FAILURE MODES

Metallization failure modes can be broken into five broad categories:

1. Electromigration-induced open leads that result from high lead currents causing a mass transport of lead material and ultimately a break in the lead.
2. Time-induced break in leads over steps caused by a combination of thermal cycling and stress in the film.
3. Electrical-overstress-induced open leads that result most often from an electrostatic discharge vaporizing a section of the lead.
4. Corrosion-induced open lead or a short between adjacent leads caused by moisture and/or poor cleaning.
5. Interlevel shorts caused either by breaks in the interlevel oxide or by growth of metal hillocks.

In addition, it is possible to have failures at the lead–wire-bond junction due either to an improper match of wire and lead materials causing undesirable intermetallic compound formation or to faulty bonding procedures. Neither of these problems will be discussed further.

10.8.1 Electromigration

Electromigration is the movement of lead material in the direction of electron flow because of momentum transfer from electrons to the metal ions. It is noticeable only when current density is very high, as it often is in IC leads. For example, a 1 μm thick strip 5 μm wide carrying 10 mA of current has a current density of 2×10^5 A/cm^2. This value is to be compared to a typical house wiring code that specifies a maximum of 30 A for a #14 wire, or less than 10^3 A/cm^2. Because of the possibility of such large current densities, electromigration properties must be considered both in circuit design and in the choice of lead material.

The metal ion transport flux J_m in the ideal case is given by (87)

$$J_m = \frac{NDZ^*e\rho j}{kT} \qquad 10.18$$

where N is the density of metal ions, D is the self-diffusion coefficient of the metal, Z^*e is the effective charge of the ion, ρ is the resistivity of the metal, and j is the density of the current flowing through the metal. For application to thin films with many grain boundaries, the expression is modified somewhat, but the form is the same.

The effects of such migration are the growth of hillocks toward the positive end of the lead and the formation of voids and ultimately a break in the lead near the negative end. An example of a gap in a lead is shown in Fig. 10.33. That there can be enough transport to constitute a transistor reliability problem was recognized as early as 1965 (88) and has been studied extensively since (89, 90). Transport depends on the material and increases exponentially with temperature since D of Eq. 10.18 is of the form $D = D_o e^{-E/kT}$ where D_o is a constant and E is the diffusion activation energy. For a given material, the more crystal defects, the more pronounced the migration. However, the addition of small amounts of another metal, such as 1% or 2% copper in aluminum, reduces the effect. In the case of aluminum, an overcoating of silica glass also reduces transport (91).

FIGURE 10.33

Failure of an aluminum lead at a contact because of electromigration. (*Source:* Photograph courtesy of Dr. P.B. Ghate, Texas Instruments Incorporated.)

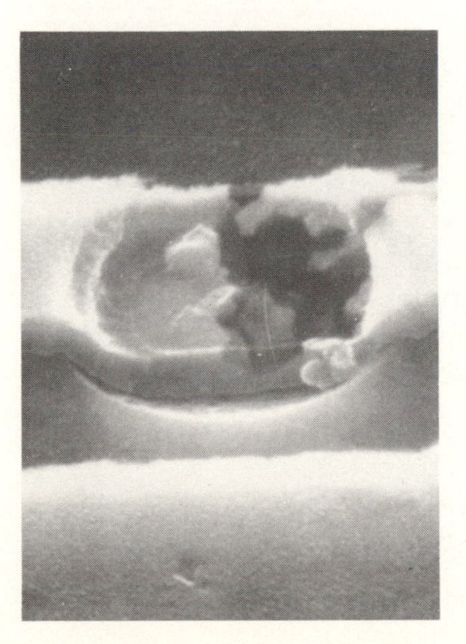

MTF, the median time to failure,[15] which in this case is the time to move enough material to cause a fatal flaw, is given by (90)

$$\text{MTF} = A'j^{-m}e^{E/kT} \qquad 10.19$$

[15]The median time to failure is the time at which half the units have failed.

where A' is a constant and m is between 1 and 3. Experimentally, the MTF of aluminum leads evaporated onto a cold substrate has been found to be approximated by (89)

$$\text{MTF} = 4.1 \times 10^{16}(ABj^{-2})e^{0.48/kT} \tag{10.20}$$

where MTF is in hours, T is the temperature in K, A is the cross section of the lead in cm^2, B is a units conversion constant and equals 1 (A^2·hour)/cm^6, and k is Boltzmann's constant expressed in eV/K. For aluminum evaporated onto a hot substrate so that large grains are produced, the activation energy typically increases to ~0.7 eV, and the MTF over the normal operating range is substantially increased. Other materials such as gold and tungsten show a pronounced increase in activation energy and in MTF.

EXAMPLE ☐ Compare MTF for leads 1 μm thick and 10 μm wide produced by evaporation onto a cold substrate if the IC chip is operating at 70°C and the lead is carrying first 30 mA and then 50 mA.

The cross-sectional area of the lead is $10^{-4} \times 10^{-3}$, or 10^{-7} cm^2. The current density j is either 3×10^5 A/cm^2 or 5×10^5 A/cm^2. The temperature is 343 K, and $k = 8.62 \times 10^{-5}$ eV/K. $ABj^{-2} = 1.1 \times 10^{-18}$ hours or 0.4×10^{-18} hours. Substituting these values and kT into Eq. 10.20 gives 46 and 17 years, respectively. Also, from Eq. 10.20, MTF varies as $(j_2/j_1)^2$, or as 25/9, which closely matches the ratio of 46/17. It should be remembered that this is median time to failure and that the actual time could be substantially less. Also, Eqs. 10.19 and 10.20 relate only to two particular sets of conditions. In actual manufacturing processes, a whole smear of activation energies ranging from 0.48 to almost 1.4 is observed. Also, the preexponential may vary over 6 or more orders of magnitude. ☐

In most cases, the failure rate $\lambda(t)$ rather than the median time to failure is of more importance to a systems designer. The distribution of failures versus time has been experimentally observed to obey, not a normal,[16] but a log normal distribution.[17] Log normal distributions show a pronounced skew with a long tail extending to longer time, as shown in Fig. 10.34a, which by replotting using a logarithmic horizontal scale, then appears as the normal curve in Fig. 10.34b. In this transformation, by using $x = \log t$, a particular skewed distribution in t becomes a normal distribution in x. In this

---

[16]For a discussion of normal distributions, see Chapter 9.

[17]This kind of distribution is not peculiar to transistor and integrated circuit failures. It has, for example, been observed in one locale for the kilowatt hours of electrical power used versus the number of homes.

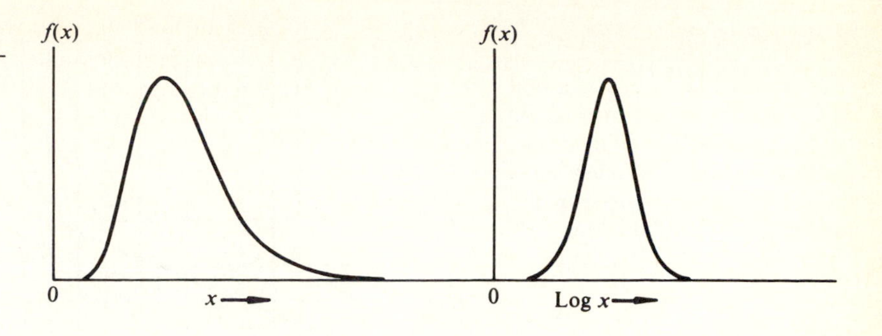

FIGURE 10.34

Effect of plotting $f(x)$ versus log x rather than $f(x)$ versus x when the distribution is log normal.

case, the fraction f that fails between time t and $t + dt$ as a function of time t is given by (92)

$$f(t) = \frac{0.4343}{\sigma t\sqrt{2\pi}} e^{-(\log t - \log t')^2/2\sigma^2} \qquad 10.21$$

where σ is the dispersion, not in time, but in log time and t' is the median life (MTF). When experimental data are used, log t' is given by $(\Sigma \log t_i)/N$ where the t_i's are the times associated with N experimental observations. This expression can be rewritten in the more conventional form as

$$\log [(t_1 \cdot t_2 \cdot t_3 \cdot \cdots t_N)^{1/N}] \qquad 10.22$$

where the bracketed term is referred to as the geometric mean. For the case of a log normal distribution, the geometric mean and the median are the same. Note that in Eq. 10.21 a $0.4343/t$ term appears that does not appear in the normal distribution. The t arises because during the transformation a $d(\log t)dt$ is involved. Since most experimental data are plotted on $\log_{10}$ graph paper, the 0.4343, or $\log_{10} e$, is needed to allow the direct use of $\log_{10}$ data.

The instantaneous fractional failure rate $\lambda(t)$ is the decrease in the fraction of surviving units per unit time at time t. That is (92),

$$\lambda(t) = \frac{f(t)}{1 - F(t)} \qquad 10.23$$

where $F(t)$ is the fraction of initial units that have failed at time t and is given by

$$F(t) = \int_0^t f(t)dt \qquad 10.24$$

The unit of λ from Eq. 10.23 is $1/t$, but typical terminology uses "device hours," with a common unit being FIT. One FIT is defined

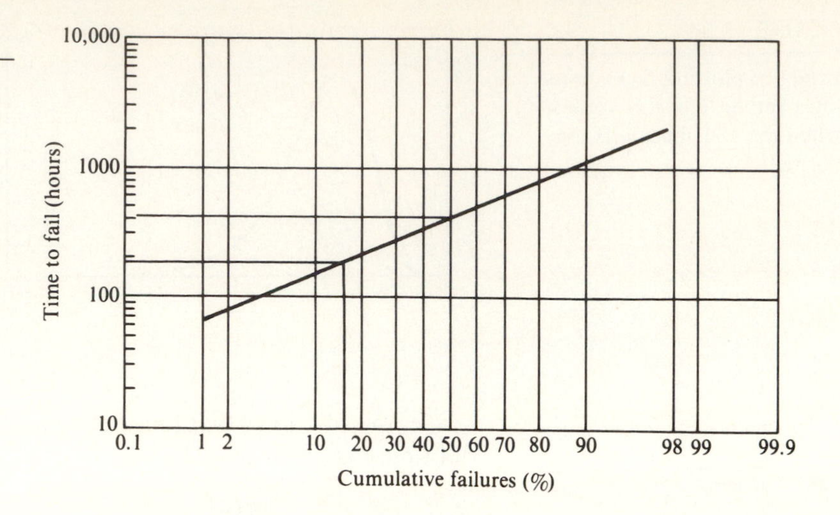

FIGURE 10.35

Log normal plot of accelerated lift test data, where the median life (time for 50% failures) is 410 hours and σ for a log normal distribution is given by ln (time for 50% fail/time for 15.9% fail) = ln (410/180) = 0.82.

as one failure in 10^9 device hours and is, for example, equivalent to 0.1% failures per million hours.

While t' (MTF) can be, in principle, determined from Eq. 10.19, varying experimental conditions are hard to account for, and σ is not very amenable to calculation. Thus, failure rate predictions are almost always made from experimental cumulative failure data plotted on log probability paper[18] as shown in Fig. 10.35, from which both t' and σ can be read.

10.8.2 Corrosion

In the presence of moisture, corrosion of many metals used for leads can occur during IC operation. Polysilicon leads are immune to such effects, and silicides appear to be. Most corrosion is caused by electrolysis arising from a combination of moisture leaking into the chip, a water-soluble ionizable material either on the chip surface or along the moisture path, and electric current flow. Moisture most commonly reaches the chip by traveling along the leads of plastic packages to the bonding pads. Then, if cracks exist in the passivation layer or if the layer is a heavily doped phosphorus oxide (93), moisture can penetrate the coating and provide a conduction path between leads. Ions can be from contamination not removed from the chips before assembly, from phosphorus leaching from a phosphosilicate (PSG) overcoat (94), or from the package itself.

[18]One scale is based on cumulative percent, which, from Eq. 10.24, involves an integral of the form

$$\int_0^w e^{-z^2} dz = erf(w)$$

To make the time-to-fail plot linearly, a scale based on erf^{-1} of cumulative percent is used.

Aluminum electrolytic corrosion can be either anodic (lead-biased positive) or cathodic (lead-biased negative), but it is generally much more pronounced on negative leads (95, 96). In either case, the result is conversion of the aluminum to aluminum hydroxide [$Al(OH)_3$], a nonconductor. The anodic rate is temperature independent over the current density range studied (0.32–3.5 mA/cm^2), and one atom of aluminum is oxidized for each three electrons supplied. As the temperature increases, the cathodic reaction rate first increases with an activation energy of 0.47 eV and is relatively independent of current density. At some point, depending on the current density, the rate rather abruptly becomes temperature independent and depends only on current density, with one aluminum metal atom being converted to hydroxide for each electron supplied. For a current density of 0.32 mA/cm^2, the break occurs just below 40°C (97). Various chemical reactions have been suggested (90, 97, 98), but they all involve forming OH^- ions that then react with the Al to form $Al(OH)_3$.

Continued corrosion (oxidation) will ultimately break lead continuity, and, in addition, the increased volume of $Al(OH)_3$ will cause additional cracks in the protective overcoat. It has even been observed that wider leads sometimes fail sooner because they cause more overcoat cracking than narrow leads (99). Even if there are initially no cracks or moisture permeability of the layer, if moisture reaches the bonding pads, their oxidation can cause the raising and cracking of the adjacent overcoat, thus exposing the much smaller leads to corrosion as well.

Both gold and platinum grow dendrites in the presence of moisture, which can cause shorting of leads (100), and the titanium used in silver–titanium solar cell contacts gradually converts to TiO_2 (101).

10.8.3 Internal Stresses

Most lead materials will have high internal stresses induced during the deposition process. In addition, the layered construction, with constituent expansion coefficients ranging from 5×10^{-7}/°C (SiO_2) to 2.3×10^{-5}/°C (Al) will generate additional thermal stresses as the IC is thermally cycled. The ductility of lead materials such as aluminum will allow them to accommodate some stress, but if the stress is compressive, that accommodation will often be in the form of hillocks, which can punch through interlevel insulators and cause shorts. If the stress is tensile, as, for example, in going over steps, voids may result (90, 102).

10.9 DEPOSITION METHODS

The three usual methods of depositing lead material are evaporation, sputtering, and CVD. (For a general description of these processes and the kind of equipment used, see Chapter 4.) In some specialized cases, electrodeposition is used, and focused ion beams can be used for very slowly and directly depositing leads on the wafer in the desired pattern.

10.9.1 Surface Cleanup

Before deposition, the surface must be clean of oxide films that might prevent good contact, organic films that impair metal adhesion, and particulates that prevent good coverage or cause pinholes in overlaying interlevel insulators. (General surface cleanup procedures were discussed in Chapter 3.)

Contact windows are particularly difficult to clean since the oxide step tends to collect debris. When brushes are used for surface cleaning, if the water supply is not adequate or the brushes are too close to the surface, brush shavings will be left at the edge of the windows. Surface tension makes it difficult to satisfactorily wet etch very fine features, and plasma etching is often used instead. However, plasmas may not be able to remove some residues and often leave a thin oxide on silicon surfaces. If the contaminants are only organic, an ozone treatment (sometimes used for resist stripping) can be used for removal and not produce excessive oxide (103). In situ ion milling is also sometimes used as a final clean. More often, a quick dip in an weak HF etch, such as 1 HF to 100 H_2O, either buffered or unbuffered, followed by spin rinsing in DI water, is used just before insertion into the deposition chamber.

Rinse/dry equipment can reintroduce surface contamination if, for example, the drying gas (usually nitrogen) is contaminated with oil or particulates or if the rinse water is not completely removed (104). Particulates on the surface can also arise from broken wafers in spin-rinse/dryers and from particles that collect in the bottom of the deposition chamber and are then stirred up by gases entering during the venting operation. The latter occurs after the deposition, but often the metal film will be hot enough to cause the particles to adhere extraordinarily well. Spattering of an evaporation source due to excessive evaporation rates can cause globules of the evaporant to be attached to the film.

10.9.2 Metal Evaporation

While the vacuum equipment is essentially the same for evaporating any material, several kinds of evaporate sources may be used. Probably the oldest, and applicable only to metals, is the use of a filament heater comprised of several wires twisted together and heated by a high electric current. The metal to be evaporated is in wire form and wrapped around the filament. When the metal is heated, it melts, wets the filament surface, and evaporates. Heating materials must have a higher melting point than that of the evaporate and must not alloy appreciably with it. As a simple alternative, the filament may be a strip formed into a boat so that the evaporant can be contained, thus removing the restriction of wetability as well as allowing chunks or powders to be used. Both of these sources suffer from the inability to hold the large amounts of material necessary for rapid sequential evaporations while using loadlock systems. Further, filaments are prone to rapid burnout. To solve these problems, as well as filament/evaporant compatibility, large water-cooled crucibles with the center portion heated by an electron beam are often used (E-beam evaporation). Alternatively, a nonconductive, noncontaminating crucible can be used and a contained conductive evaporant heated inductively (IN-Source).[19] To provide yet more capacity, additional evaporant in the form of spooled wire can be housed inside the vacuum chamber and periodically reeled off into the crucible.

When multiple-component evaporations are desired–for example, for aluminum containing a few percent copper or silicon—control of the final composition is difficult because of the differences in vapor pressure. If the vapor pressures of the two constituents are not the same, then one will vaporize more rapidly than the other and gradually change the composition of the melt. Two independent sources can be used, with independently controlled rates to give the desired composition. Thin alternating layers can be deposited and then diffused together. Flash evaporation, in which small quantities of each component, such as a fine wire alloy of the correct composition, are continually fed to the evaporation chamber and immediately evaporated, is also sometimes used (105, 106). However, sputtering from a target of the desired composition is more common.

10.9.3 Sputtering

Sputtering is one common way of depositing metal alloys such as Al/Cu or Ti/W when sputtering targets are available. The problems in producing acceptable targets are threefold:

1. It is necessary to maintain a uniform composition throughout; otherwise, as the target is depleted, composition of the sputtered film will change.
2. The target must be physically strong enough to not break during usage and must be amenable to being attached to a backing plate.
3. The fabrication of the target, usually by powder metallurgy, must not introduce undesirable impurities. Pure or nearly pure metals such as platinum, gold, and doped aluminum seldom are a problem for any size target. However, brittle materials such as silicides are likely to have higher breakage rates.

---

[19]Trademark of Applied Materials, Inc.

When multiple layers are deposited, a series of sputter targets can be sequentially moved into place more easily than multiple evaporation sources. Sputtering also offers a distinct advantage for depositing low-vapor-pressure, high-melting-point materials like Mo, W, and Pt since source heaters of very high temperature are not required. Because deposition pressures are higher in sputtering than they are in evaporation, step coverage is slightly better. However, even though the pressure used is relatively high, provision must be made for pumping down to a hard vacuum in order to clean the system of oxygen and water vapor. In addition, the sputter gas, such as argon, must be very pure. Otherwise, the deposited films may not adhere well, will appear hazy, and will have higher than anticipated resistivity. It was such contamination that prevented early acceptance of sputtering for metallization, but after systems were improved to eliminate excessive residual gases, they were widely accepted.

10.9.4 Chemical Vapor Deposition

As has already been discussed, chemical vapor deposition provides better step coverage than sputtering or evaporation and is often used for interlevel insulator depositions. However, even though CVD of many metals is possible and has been studied for over 50 years, the application to semiconductor processing is quite recent. Metals germane to wafer fabrication that have been vapor phase deposited for other applications include Al, Au, Cr, Cu, Mo, Pt, Ta, Ti, and W (107). The fact that they can be deposited does not necessarily mean that the deposition temperature is low enough to be useful, that the film will adhere properly, that the electrical resistivity will be low enough, or that the surface quality will be satisfactory. The two metals that have been most studied for IC metallization are aluminum and tungsten. In both cases, low-pressure systems operating at a few hundred degrees C are being used.

Aluminum can be vapor deposited from $AlCl_3$ and from several aluminum-bearing organic compounds (107, 108). Tri-isobutyl aluminum thermal decomposition is used most (107–109) and is usually preceded by treating the wafer with $TiCl_4$ in order to promote nucleation. The overall reaction is

$$2Al(C_4H_9)_3 \rightarrow 2Al + 3H_2 + 6C_4H_8$$

and proceeds above about 220°C. Typical LPCVD pressures are in the 0.2–0.5 torr range (108). A rough and nonspecular surface, along with the lack of a process to co-deposit copper or some other material to improve the electromigration characteristics of aluminum, has thus far prevented commercial usage of the LPCVD process.

Tungsten can be deposited by the hydrogen reduction of WF_6, WCl_6, and WBr_6; by the silicon reduction of WF_6 ("displacement reaction"); and by the thermal decomposition of $W(CO)_6$ (tungsten carbonyl) (107, 108). It is relatively easy to produce good tungsten layers by depositing in the 250°C–500°C range, but the problems of integrating the procedure into an IC manufacturing flow have proven numerous. These kinds of problems are by no means unique to tungsten metallization and constitute a good case study of the way in which unexpected interactions can cause difficulties in the introduction of new processes.

WF_6 is the source commonly used for semiconductor application studies. Reactions include (a)

$$WF_6 + 3H_2 \rightarrow W + 6HF$$

and (b)

$$2WF_6 + 3Si \rightarrow 2W + 3SiF_4$$

The application most seriously considered is for via plugs. The deposition of a plug depends on selectively depositing tungsten only in the via openings. Thus, deposition conditions must be chosen so that nucleation does not occur on the interlevel oxide. In principle, reaction b just given is self-limiting at the thickness where appreciable WF_6 can no longer diffuse through the tungsten film to the silicon. However, for selectively depositing on silicon in contact openings, there is no covering of tungsten at the silicon–oxide interface, and considerable unwanted etching of the silicon at the window edges may occur. The effect can be suppressed but not entirely eliminated by the use of hydrogen (reaction a) since reaction b will proceed even with hydrogen present (108). One alternative is to use titanium silicide both as a barrier between the silicon and tungsten and as a means of making a very-low-resistance contact to the silicon. However, the reaction of WF_6 with the Ti in the $TiSi_2$ gives TiF_4, which causes high-resistance contacts (110, 111). When tungsten is used to contact aluminum rather than silicon, a problem similar to that of $TiSi_2$ occurs. That is, through the reaction

$$WF_6 + 2Al \rightarrow W + 2AlF_3$$

a small amount of aluminum fluoride is formed. Aluminum fluoride is a solid at deposition temperatures and forms a high-resistance interfacial layer. Higher temperatures and deposition rates minimize this effect (112).

Molybdenum and silicide CVD depositions for IC metallization applications have also been studied (113, 114). The silicides can, in principle, be deposited by the simultaneous reduction of a metal compound and a silicon compound. The thermal decomposition of silane (SiH_4) is a practical source of silicon since the reaction will proceed at temperatures of 400°C or less.

10.10 PROCESS CONTROL

The metallization parameters regularly monitored include the width of the leads (readily measured by commercial equipment) and the metal thickness and/or sheet resistance. In some cases, the metallization resistivity is itself of importance, but usually resistivity measurements are an alternative to more conventional thickness measurement. The resistivity ratio (defined in the next section), often used to check metal deposition processes as they are developed, also has merit as a process control tool.

Step height measurements using a stylus profilometer are easy when the step is more than a few thousand angstroms thick. Thus, after a typical metal leads layer has been patterned, its thickness can be readily determined. For very thin layers, however, such as platinum to be used for platinum silicide contacts, either optical transmissivity or sheet resistance measurements are more appropriate. Sheet resistance test patterns such as have been described for diffusion sheet resistance measurements can be used either on separate test wafers or on drop-in test chips (115). Optical transmissivity measurements require separate transparent wafers. Calibration curves relating the resistance or optical transmission to thickness are often not constructed. Rather, for process control, a range of the measured parameter that gives satisfactory results is specified.

When metallization composition is changed, bondability testing may also be required as part of the new metal qualification. The addition of silicon to aluminum metallization, for example, makes the bonding of wires to it considerably more difficult. The testing details will depend on the specific bonding procedure being used.

10.10.1 Resistivity Ratio

The resistivity ρ can be approximated by the sum of three terms:

$$\rho \cong \rho_1(T) + \rho_2(G) + \rho_3(I) \qquad 10.25$$

where ρ_1 is a function of temperature, ρ_2 depends on crystallographic disorder such as grain boundaries, and ρ_3 depends on the impurities present. For temperatures near absolute zero, ρ_1 becomes small, and ρ is dominated by ρ_2 and ρ_3. Normally, by room temperature, ρ_1 will be much larger. Only when disorder or impurities become excessive will ρ_2 and ρ_3 become noticeable. Thus, the resistivity ratio

$$RR = \frac{\rho_{\text{room}}}{\rho_{4.2\text{K}}} \cong \frac{\rho_1}{\rho_2 + \rho_3} \qquad 10.26$$

is a measure of the resistivity quality and can be used to evaluate film deposition methods (116). Typical ratios are around 30. Lower numbers indicate a more than normal amount of defects and/or impurities.

10.10.2 Current Capacity

Current-carrying capability testing is generally done on a continuing basis, with samples being periodically collected and subjected to accelerated electromigration testing. The need for routine checking arises because of the sensitivity of electromigration to various processing parameters such as film grain size and composition. Usually, electromigration testing is done by packaging test chips, but it can, in principle, be done in a limited fashion at the wafer level (117). Better control of the silicon temperature is possible because heat sinking is not through a package, subsequent examination and failure analysis are easier, and the time from wafer completion to test is much shorter. Unfortunately, the equipment cost per lead tested is much more expensive than that used in packaged testing.

The likelihood of burnout due to sudden current surges ("zotting") is not as amenable to routine testing and generally involves discharging a capacitor between probes on either side of suspected weak links. Such weak spots will usually be where the metal goes over steps, but some may be due to excessively narrow leads caused by lithography or etching problems. The capacitor size, the series resistance, and the capacitor voltage must be tailored to specific conditions. To simulate the electrostatic behavior of a person, a 150 pF capacitor in series with 2000 Ω is satisfactory (118). The voltage used may be as high as 10 kV, depending on just what the environmental conditions are. A much larger capacitor and a smaller resistance are needed when radiation-hardened circuits or those to be used in adverse industrial applications are tested.

10.10.3 Lead Adherence

On a much less frequent basis, film adhesion to its underlying substrate is measured. One of the oldest tests is the use of Scotch tape to see whether adhesion is better between tape and film or between film and substrate. Such a test is very subjective and can give only go/no-go results. The results depend on the adhesive qualities of the tape (which may vary from batch to batch), the peel rate, and the peel angle (119). A somewhat more scientific approach is to bond a wire to the surface with epoxy and then measure the force required to pull the film off the substrate (providing the glue–film interface does not fail first). The shear strength between film and substrate can, in principle, be determined quantitatively by dragging a sharp stylus across the film and increasing the force on the stylus until the film is scraped away at the bottom of the groove, leaving bare substrate. The shearing force F_s is given by (120)

$$F_s = \frac{a}{\sqrt{r^2 - a^2}} - P \qquad 10.27$$

where r is the radius of the stylus tip, P is the indentation hardness of the substrate. $a = \sqrt{w/\pi P}$ where w is the stylus force required for a clean scrape. The stylus must be very hard, and its tip very small in order to not require an excessive force. For aluminum, a minimum thickness of 1000 Å is recommended. The substrate should be a minimum of 1/8 inch thick to minimize deflection, and if the stylus is a diamond phonograph needle with a 0.7 mil radius, a force of less than 300 grams will be required (121).

10.10.4 Process Problems Requiring Test Patterns for Detection

The incidence of simple process-induced defects must be monitored in order to control the process and to assess the yield loss associated with the metallization steps. These defects primarily consist of open leads, shorts between adjacent leads, and high-resistivity metal-to-metal contacts at vias. For any practical process, such defects occur very infrequently, so any testing must involve long lengths of interconnect and many vias. Historically, patterns involving an appropriate number of potential failure sites have been used, with each pattern then being checked for failure. However, on a given pattern, the number of failures was not determined. As an alternative, by deliberately introducing a resistor between various segments of the pattern and then measuring the resistance after processing, the number of defects per pattern can be estimated (122).

SAFETY — CHAPTER 10

Safety precautions in metallization are primarily centered in making sure that the vacuum chambers do not collapse and that noxious fumes from CVD depositions are properly contained. At low pressure, approximately 15 pounds per square inch press the vacuum chamber inward. Metal chambers give little danger even if the pressure does cause the chamber to collapse. With the glass jars that used to be common, failure of the brittle glass caused the glass fragments to be propelled toward the center of the enclosure (an implosion). Their momentum then carried them on past the center and, unless an arresting metal safety shield was in place, out the other side, where extensive damage could occur. The CVD equipment sometimes used for metal deposition may use hydrogen as a carrier, in which case there is risk of fire or explosion. (These issues were discussed in Chapter 4.) In addition, many of the source materials and deposition by-products are corrosive and/or poisonous. Thus, appropriate storage and exhaust facilities must be provided.

KEY IDEAS — CHAPTER 10

- ☐ The lower the resistivity of the semiconductor, the lower the metal–semiconductor contact resistance. To make useful contacts, the doping level should generally be greater than 10^{19} atoms/cc.

- Aluminum is the most commonly used contact to silicon. To make good contact, it needs to be alloyed or sintered. These processes cause dissolution of the silicon and are crystallographic-plane sensitive.
- Many silicides form lower-resistance contacts to silicon than does aluminum.
- When either metals or silicides contact high-resistivity silicon or gallium arsenide, a Schottky barrier diode is formed. Schottky diodes are majority carrier devices and offer some circuit advantages over pn junction diodes.
- As lead geometries get smaller, lower-resistivity lead and gate materials are required.
- In many cases, it is desirable to subject leads to temperatures of several hundred degrees C after they are formed. This fact has lead to the use of polysilicon, silicides, or refractory metals for some interconnects.
- The high current densities found in IC leads ($>10^5$ A/cm^2) make them susceptible to electromigration failure. Copper- or silicon-doped aluminum is better than pure aluminum, and gold is better than doped aluminum.
- Plastic packages allow small amounts of moisture to penetrate to the chip. If the moisture contacts aluminum leads, electrolytic action will cause their oxidation and ultimate failure.

PROBLEMS CHAPTER 10

1. For the case of $\rho = 1$ Ω-cm and $a = 5$ μm, plot r_s versus ℓ of Eq. 10.3 over enough range of ℓ for r_s to essentially vary from the value given by Eq. 10.1 to that given by Eq. 10.4. Based on this curve, what ℓ/a ratio is required to approximate an infinite thickness?
2. What would the pits left in sintered aluminum contacts on a (110) silicon wafer be expected to look like? Explain your answer.
3. If a 1 μm thick aluminum pad is used for a contact to silicon, what thickness of silicon will be removed if the contact is sintered at 450°C and if the removal is uniform over the entire contact area? Which wafer orientation would be most likely to produce a flat aluminum–silicon interface?
4. If two adjacent silicon Schottky diodes of the same area are made, one of Al and one of PtSi, if n of each of them is 1.05, and if the forward voltage of the aluminum diode is 1 μA at 0.35 V, what will be the PtSi diode voltage when 100 μA are flowing?
5. What is the thickness change when $MoSi_2$ is made from a 3000 Å layer of Mo on a 10,000 Å thick layer of polysilicon? The density of $MoSi_2$ is ~6 grams/cm^3.
6. If gold leads are to be used, what are two potential problems?
7. How much would the MTF be expected to decrease if the current through an interconnect were increased from 10μA to 1 mA?
8. If a lead (interconnect) system is experimentally observed to have a 2000 hours mean time to failure at 215°C and an activation energy of 0.7 eV, what would be the mean time to failure if the temperature were dropped to 55°C?
9. If, because of moisture, a thick aluminum lead is partially anodized to Al_2O_3, would a polyimide or an SiO_2 overcoat be most likely to crack during anodization? Why? List a lead material less susceptible to moisture-induced failure.

CHAPTER REFERENCES 10

1. Paul Ho, "Integrated Circuit Metallization," in 1985 CEI Semiconductor Materials and Process Technologies course.
2. D.P. Kennedy and P.C. Murley, *IBM J. Res. Develop. 12*, pp. 242–250, 1968.
3. G. D'Andrea and H. Murrmann, "Correction Terms for Contacts to Diffused Resistors," *IEEE Trans. on Electron Dev. ED-17*, pp. 481–482, 1970.
4. H.H. Berger, "Models for Contacts to Planar Devices," *Solid-State Electronics 15*, pp. 145–158, 1971.
5. M.P. Lepselter and J.M. Andrews, "Ohmic Contacts to Silicon," pp. 159–186, in Bertram Schwartz, ed., *Ohmic Contacts to Semiconductors*, Electrochemical Society, New York, 1969.
6. C.Y. Chang, Y.K. Fang, and S.M. Sze, "Specific Contact Resistance of Metal–Semiconductor Barriers," *Solid-State Electronics 14*, pp. 541–550, 1971.
7. N. Braslau, "Alloyed Ohmic Contacts to GaAs," *J. Vac. Sci. Technol. 19*, pp. 803–807, 1981.
8. H.H. Berger, "Contact Resistance and Contact Resistivity," *J. Electrochem. Soc. 119*, pp. 507–514, 1972. (This paper includes a review of other methods of measuring contact resistance. See included references.)
9. R. Stall et al., "Ultra Low Resistance Ohmic Contacts to n-GaAs," *Electronics Letters 15*, pp. 800–801, 1979.
10. J.W. Faust et al., "Molten Metal Etches for the Orientation of Semiconductors by Optical Techniques," *J. Electrochem. Soc. 109*, pp. 824–828, 1962.
11. Tchang-II Chung, "Study of Aluminum Fusion into Silicon," *J. Electrochem. Soc. 109*, pp. 229–234, 1962.
12. P.A. Totta and R.P. Sopher, "SLT Device Metallurgy and Its Monolithic Extension," *IBM J. Res. Develop. 13*, pp. 226–238, 1969.
13. Harry Sello, "Ohmic Contacts and Integrated Circuits," pp. 277–298, in Bertram Schwartz, ed., *Ohmic Contacts to Semiconductors*, Electrochemical Society, New York, 1969.
14. R.J. Anstead and S.R. Floyd, "Thermal Effects on the Integrity of Aluminum to Silicon Contacts in Silicon Integrated Circuits," *IEEE Trans. on Electron Dev. ED-16*, pp. 381–386, 1969.
15. George L. Schnable and Ralph S. Keen, "Aluminum Metallization—Advantages and Limitations for Integrated Circuit Applications," *Proc. IEEE 57*, pp. 1570–1580, 1969.
16. J.M. McCarthy, "Failure of Aluminum Contacts to Silicon in Shallow Diffused Transistors," *Microelectronics and Reliability 9*, pp. 187–188, 1979.
17. Arthur J. Learn, "Evolution and Current Status of Aluminum Metallization," *J. Electrochem. Soc. 123*, pp. 894–906, 1976. (This review paper has over 250 references in it.)
18. L.A. Berthoud, "Aluminium Alloying in Silicon Integrated Circuits," *Thin Solid Films 43*, pp. 219–327, 1977.
19. R.A. Levy et al., "In-Source Al–0.5% Cu Metallization for CMOS Devices," *J. Electrochem. Soc. 132*, pp. 159–168, 1985.
20. B.L. Kuiper, U.S. Patent 3,382,568.
21. S.P. Bellier and L.B. Ehlert, "An Improved Metallization Process for Silicon Transistors," pp. 304–314, in Howard R. Huff and Ronald R. Burgess, eds., *Semiconductor Silicon/73*, Electrochemical Society, Princeton, N.J., 1973.
22. H.M. Naguib and L.H. Hobbs, "Al/Si and Al/Poly-Si Contact Resistance in Integrated Circuits," *J. Electrochem. Soc. 124*, pp. 573–577, 1977.
23. H.M. Naguib and L.H. Hobbs, "The Reduction of Poly-Si Dissolution and Contact Resistance at Al/n-Poly-Si Interfaces in Integrated Circuits," *J. Electrochem. Soc. 125*, pp. 169–171, 1978.
24. M. Eizenberg, "Applications of Thin Alloy Films in Silicon Contacts," pp. 348–360, in Kenneth E. Bean and George A. Rozgonyi, eds., *VLSI Science and Technology/84*, Electrochemical Society, Princeton, N.J., 1984.
25. A.S. Berezhnoi, *Silicon and Its Binary Systems*, Consultants Bureau, New York, 1960.
26. M.P. Lepselter, "Beam-Lead Technology," *Bell Syst. Tech. J. 45*, pp. 233–253, 1966.

27. S. Yanagisawa and T. Fukuyama, "Reaction of Mo Thin Films on Si (100) Surfaces," *J. Electrochem. Soc. 127*, pp. 1150–1156, 1980.
28. V.I. Rideout, "A Review of the Theory and Technology for Ohmic Contacts to Group III–V Compound Semiconductors," *Solid-State Electronics 18*, pp. 541–550, 1975.
29. R.H. Cox and H. Strach, "Ohmic Contacts for GaAs Devices," *Solid-State Electronics 10*, pp. 1213–1218, 1967.
30. N. Braslau, J.B. Gunn, and J.L. Staples, "Metal–Semiconductor Contacts for GaAs Bulk-Effect Devices," *Solid-State Electronics 10*, pp. 381–385, 1967.
31. Ajit Rode and J. Gordon Roper, "Gallium Arsenide Digital IC Processing—A Manufacturing Perspective," *Solid State Technology*, pp. 209–215, February 1985.
32. M. Heiblum et al., "Characteristics of AuGeNi Ohmic Contacts to GaAs," *Solid-State Electronics 25*, pp. 185–195, 1982.
33. P.A. Barnes et al., "Ohmic Contacts Produced by Laser Annealing Te-Implanted GaAs," *Appl. Phys. Lett. 33*, pp. 965–967, 1978.
34. P.A. Pianetta et al., "Non-Alloyed Ohmic Contacts to Electron-Beam Annealed Se-Ion-Implanted GaAs," *Appl. Phys. Lett. 36*, pp. 597–599, 1980.
35. Y.I. Nissim et al., "Non-Alloyed Contacts in n-GaAs by CW Laser Assisted Diffusion from a SnO_2/SiO_2 Source," *IEEE Trans. on Electron Dev. ED-28*, pp. 607–609, 1981.
36. H.R. Grinolds and G.V. Robinson, "Pd/Ge Contacts to n-Type GaAs," *Solid-State Electronics 23*, pp. 573–585, 1980.
37. E.D. Marshall et al., "Non-Alloyed Ohmic Contacts to n-GaAs by Solid Phase Epitaxy," *Appl. Phys. Lett. 47*, pp. 298–300, 1985.
38. P.A. Barnes and A.Y. Chou, "Non-Alloyed Ohmic Contacts to n-GaAs by Molecular Beam Epitaxy," *Appl. Phys. Lett. 33*, pp. 651–653, 1978.
39. W.T. Tsang, "In Situ Ohmic-Contact Formation to n- and p-GaAs by Molecular Beam Epitaxy," *Appl. Phys. Lett. 33*, pp. 1022–1025, 1978.
40. J.V. DiLorenzo et al., "Non-Alloyed and In Situ Ohmic Contacts to Highly Doped n-Type GaAs Grown by Molecular Beam Epitaxy for Field Effect Transistors," *J. Appl. Phys. 50*, pp. 951–954, 1979.
41. W.T. Anderson et al., "Development of Ohmic Contacts for GaAs Devices Using Epitaxial Germanium Films," *IEEE J. Solid-State Circuits SC-13*, pp. 430–435, 1978.
42. R.T. Galla et al., "Evaluation of the Interfacial Resistance of Thin Film Interconnections," *Microelectronics and Reliability 7*, pp. 185–212, 1968.
43. A.E. Michel et al., "Base Contacts for High-Speed Germanium Transistors," pp. 243–252, in Bertram Schwartz, ed., *Ohmic Contacts to Semiconductors*, Electrochemical Society, New York, 1969.
44. W. Shockley, "Research and Investigation of Inverse Epitaxial UHF Power Transistors," *Final Tech. Rpt. Al-TDR-64-207*, Air Force Avionics Laboratory, Air Force Systems Command, Wright-Patterson Air Force Base, Ohio, 1964.
45. G.K. Reeves, "Specific Contact Resistance Using a Circular Transmission Line Model," *Solid-State Electronics 23*, pp. 487–490, 1980.
46. Chung-Yu Ting and Charles Y. Chen, "A Study of the Contacts of a Diffused Resistor," *Solid-State Electronics 14*, pp. 433–438, 1971.
47. A.Y.C. Yu and E.H. Snow, "Surface Effects on Metal–Silicon Contacts," *J. Appl. Phys. 39*, pp. 3008–3016, 1968.
48. M.P. Lepselter and S.M. Sze, "Silicon Schottky Barrier Diode with Near-Ideal *I–V* Characteristics," *Bell Syst. Tech. J. 47*, pp. 195–208, 1968.
49. A.Y.C. Yu and C.A. Mead, "Characteristics of Aluminum–Silicon Schottky Barrier Diode," *Solid-State Electronics 13*, pp. 97–104, 1970.
50. S.M. Sze, *Physics of Semiconductor Devices*, 2d ed., John Wiley & Sons, New York, 1981.
51. Howard C. Card, "Aluminum–Silicon Schottky Barriers and Ohmic Contacts in Integrated Circuits," *IEEE Trans. on Electron Dev. ED-23*, pp. 538–544, 1976.
52. J.M. Shannon, "Control of Schottky Barrier Height Using Highly Doped Surface Layers," *Solid-State Electronics 19*, pp. 537–543, 1976.
53. P.B. Ghate, "Electromigration-Induced Failures in VLSI Interconnects," pp. 292–299, in *Proc. 20th Annual IEEE Reliability Physics Symposium,* March 1982.
54. R.W. Keyes, "The Evolution of Digital Electron-

ics toward VLSI," *IEEE Trans. on Electron Dev. ED-26*, pp. 271–278, 1979.
55. Krishna C. Saraswat, "Effect of Scaling of Interconnections on the Time Delay of VLSI Circuits," *IEEE Trans. on Electron Dev. ED-29*, pp. 645–650, 1982.
56. William E. Engeler and Dale M. Brown, "Performance of Refractory Metal Multilevel Interconnection System," *IEEE Trans. on Electron Dev. ED-19*, pp. 54–61, 1972.
57. N. Yamamoto et al., "Fabrication of Highly Reliable Tungsten Gate MOS VLSIs," *J. Electrochem. Soc. 133*, pp. 401–407, 1986.
58. L.L. Vadasz et al., "Silicon Gate Technology," *IEEE Spectrum 6*, pp. 28–35, 1969.
59. V. Ramakrishna et al., "Future Requirements for High-Speed VLSI Interconnections," pp. 27–32, in *Proc. IEEE VLSI Interconnection Conference*, Santa Clara, Calif., 1987.
60. Malcomb Beasley, "Superconductor Material Development for VLSI Interconnect," Multilevel Interconnection State-of-the-Art Seminar, Santa Clara, Calif., 1987.
61. Joseph W. Goodman et al., "Optical Interconnections for VLSI Systems," *Proc. IEEE 72*, pp. 850–866, 1984.
62. R. Selvaraj et al., "Optical Interconnections Using Integrated Waveguides in Polyimide for Wafer Scale Integration," pp. 306–313, in *Proc. IEEE VLSI Interconnection Conference*, Santa Clara, Calif., 1987.
63. P.B. Ghate et al., "Application of Ti:W Barrier Metallization for Integrated Circuits," *Thin Solid Films 53*, pp. 117–128, 1978.
64. P.B. Ghate, "Metallization for Very Large-Scale Integrated Circuits," *Thin Solid Films 93*, pp. 359–383, 1982.
65. M.A. Nicolet and M. Bartur, "Diffusion Barriers in Layered Contact Structures," *J. Vac. Sci. Technol. 19*, pp. 786–793 and references therein, 1981.
66. B. Lee et al., "Effect of Oxygen on the Diffusion Properties of TiN," pp. 344–350, in *Proc. IEEE VLSI Interconnection Conference*, Santa Clara, Calif., 1987.
67. M.-A. Nicolet, "Diffusion Barriers in Thin Films," *Thin Solid Films 52*, pp. 415–554, 1978.
68. Joel R. Shappirio, "Diffusion Barriers in Advanced Semiconductor Device Technology," *Solid State Technology*, pp. 161–166 and references therein, October 1985.
69. N. Kumar et al., "Fabrication of RF Reactively Sputtered TiN Thin Films," *Semiconductor International*, pp. 100–104, April 1987.
70. M. Whittmer et al., "Effect of Cu on the Kinetics and Microstructure of Al_3Ti Formation," *J. Electrochem. Soc. 132*, pp. 1450–1455, 1985.
71. I.A. Blech, "Evaporated Film Profiles over Steps in Substrates," *Thin Solid Films 6*, pp. 113–118, 1970.
72. W.G. Oldham et al., "A General Simulator for VLSI Lithography and Etching Processes: Part II—Application to Deposition and Etching," *IEEE Trans. on Electron Dev. ED-27*, pp. 1455–1459, 1980.
73. A.C. Adams, "Plasma Deposition of Inorganic Films," *Solid State Technology*, pp. 135–139, April 1983.
74. Yoshio Homma and Sukeyoshi Tsunekawa, "Planar Deposition of Aluminum by RF/DC Sputtering with RF Bias," *J. Electrochem. Soc. 132*, pp. 1466–1472, 1985.
75. C.Y. Ting, "Silicides for Contacts and Interconnects," Tech. Digest Int. Electron Dev. Meeting, San Francisco, December 1984.
76. Pieter Burggraaf, "Silicide Technology Spotlight," *Semiconductor International*, pp. 293–298, May 1985.
77. A.E. Morgan et al., "Characterization of a Self-Aligned Cobalt Silicide Process," *J. Electrochem. Soc. 134*, pp. 925–935 and references therein, 1987.
78. A.I. Akinwande and J.D. Plummer, "Process Dependence of the Metal Semiconductor Work Function Difference," *J. Electrochem. Soc. 134*, pp. 2297–2303 and references therein, 1987.
79. Papers in *Proc. IEEE VLSI Interconnection Conference*, Santa Clara, Calif., 1987.
80. C.Y. Ting et al., "Study of Planarized Sputter-Deposited SiO_2," *J. Vac. Sci. Technol. 15*, pp. 1105–1112, 1978.
81. Gregory C. Smith and Andrew J. Purdes, "Sidewall-Tapered Oxide by Plasma-Enhanced Chemical Vapor Deposition," *J. Electrochem. Soc. 132*, pp. 2721–2725, 1985.
82. J.S. Mercier et al., "Dry Etch-Back of Overthick PSG Films for Step-Coverage Improvement," *J. Electrochem. Soc. 132*, pp. 1219–1222, 1985.

83. A.C. Adams and C.P. Capio, "Planarization of Phosphorus-Doped Silicon Dioxide," *J. Electrochem. Soc. 128*, pp. 423–429, 1981.
84. O. Wadi et al., "Mask Preparation for Small Dimension Ion Milling by Two-Step Lift-Off Process," *J. Electrochem. Soc. 124*, pp. 959–960, 1977.
85. T. Sakurai and T. Serikawa, "Lift-Off Metallization of Sputtered Al Alloy Films," *J. Electrochem. Soc. 126*, pp. 1257–1260, 1979.
86. Moshe Oren and A.N.M. Masum Choudhury, "Interconnect Metallization Technique for GaAs Digital ICs," *J. Electrochem. Soc. 134*, pp. 750–752, 1987.
87. H.B. Huntington and A.R. Grone, "Current-Induced Marker Motion in Gold Wires," *J. Phys. Chem. Solids 20*, pp. 88–98, 1961.
88. I.A. Blech et al., "A Study of Failure Mechanisms in Silicon Planar Epitaxial Transistors," *Rome Air Development Center Tech. Rpt. TR 66–31*, December 1965.
89. James R. Black, "Electromigration Failure Modes in Aluminum Metallization for Semiconductor Devices," *Proc. IEEE 57*, pp. 1587–1594, 1969. (See also the included references for background on the work prior to 1969.)
90. P.B. Ghate, "Reliability of VLSI Interconnections," pp. 321–337, in *Proc. American Institute of Physics Conference*, no. 138, New York, 1986.
91. James R. Black, "RF Power Transistor Metallization Failure," *IEEE Trans. on Electron Dev. ED-17*, pp. 800–803, 1970.
92. Frederick H. Reynolds, "Thermally Accelerated Aging of Semiconductor Components," *Proc. IEEE 62*, pp. 212–222, 1974.
93. Robert B. Comizzoli, "Bulk and Surface Conduction in CVD SiO_2 and PSG Passivation Layers," *J. Electrochem. Soc. 123*, pp. 386–391, 1976.
94. Naoyuki Nagasima et al., "Interaction between Phosphosilicate Glass Films and Water," *J. Electrochem. Soc. 121*, pp. 434–438, 1974.
95. H. Koelmans, "Metallization Corrosion in Si Devices by Moisture Induced Electrolysis," pp. 168–171, in *Proc. 12th Annual IEEE Reliability Physics Symposium*, New York, 1974.
96. W.M. Paulson and R.W. Kirk, "The Effects of Phosphorus-Doped Passivation Glasses on the Corrosion of Aluminum," pp. 172–179, in *Proc. 12th Annual IEEE Reliability Physics Symposium*, New York, 1974.
97. E.P.G.T. van de Ven and H. Koelmans, "The Cathodic Corrosion of Aluminum," *J. Electrochem. Soc. 123*, pp. 143–144, 1976.
98. Nicholas Lycoudes, "The Reliability of Plastic Microcircuits in Moist Environments," *Solid State Technology*, pp. 53–62, October 1978.
99. T. Wada et al., "Relationship between Width and Spacing of Aluminum Electrodes and Aluminum Corrosion on Simulated Microelectronic Circuit Patterns," *J. Electrochem. Soc. 134*, pp. 649–653, 1987.
100. E.B. Hakim and J.R. Shappiro, "Failure Mechanisms in Gold Metallized Sealed Junction Devices," *Solid State Technology*, pp. 66–68, April 1975.
101. W.H. Becker and S.R. Pollack, "The Formation and Degradation of Ti–Ag and Ti–Pd–Ag Solar Cell Contacts," pp. 40–50, in *Proc. 8th IEEE Photovoltaic Specialists Conference*, Seattle, 1970.
102. S.K. Groothuis and W.H. Schroen, "Stress Related Failures Causing Open Metallization," pp. 1–8, in *Proc. 25th Annual IEEE International Reliability Physics Symposium*, 1987.
103. H. Norstrom et al., "Dry Cleaning of Contact Holes Using Ultraviolet (UV) Generated Ozone," *J. Electrochem. Soc. 132*, pp. 2285–2287, 1985.
104. Vance Hoffman, "Practical Troubleshooting of Vacuum Deposition Processes and Equipment for Aluminum Metallization," *Solid State Technology*, pp. 47–56, December 1978.
105. Arthur J. Learned, "Aluminum Alloy Film Deposition and Characterization," *Thin Solid Films 20*, pp. 261–279, 1974.
106. Tom Strahl, "Flash Evaporation, An Alternative to Magnetron Sputtering in the Production of High-Quality Aluminum Alloy Films," *Solid State Technology*, pp. 78–82, December 1978.
107. Carroll F. Powell, Chap. 10, "Chemically Deposited Metals," in Carroll F. Powell et al., eds., *Vapor Deposition*, John Wiley & Sons, New York, 1966.
108. R.A. Levy and M.L. Green, "Low Pressure Chemical Vapor Deposition of Tungsten and Aluminum for VLSI Applications," *J. Electro-*

chem. Soc. 134, pp. 37C–49C and references therein, 1987.

109. M.J. Cooke et al., "LPCVD of Aluminum and Al–Si Alloys for Semiconductor Metallization," *Solid State Technology*, pp. 62–65, December 1982.
110. E.K. Broadbent et al., "Growth of Selective Tungsten on Self-Aligned Ti and PtNi Silicides by Low Pressure Chemical Vapor Deposition," *J. Electrochem. Soc. 133*, pp. 1715–1721, 1986.
111. Gregory C. Smith et al., "Damage to $TiSi_2$ Clad Doped Silicon due to CVD Selective Tungsten Deposition," pp. 155–161, in *Proc. IEEE VLSI Interconnection Conference*, Santa Clara, Calif., 1987.
112. R. Chow and S. Kang, "Selective Chemical Vapor Deposition of Tungsten on Aluminum," pp. 208–215, in *Proc. IEEE VLSI Interconnection Conference*, Santa Clara, Calif., 1987.
113. Jim Crawford, "Refractory Metals Pace IC Complexity," *Semiconductor International*, pp. 84–86, March 1987.
114. Suresh Sachdev and Robert Castellano, "CVD Tungsten and Tungsten Silicide for VLSI Applications," *Semiconductor International*, pp. 306–310, May 1985.
115. Sheldon C.P. Lem and Doug Ridley, "An Overview of Thickness Measurement Techniques for Metallic Thin Films," *Solid State Technology*, pp. 99–103, February 1983.
116. C.R. Fuller and P.B. Ghate, "Magnetron-Sputtered Aluminum Films for Integrated Circuit Interconnections," *Thin Solid Films 64*, pp. 25–37, 1979.
117. Janet M. Towner, "Electromigration of Thin Films at the Wafer Level," *Solid State Technology*, pp. 197–200, October 1984.
118. T.M. Madzy and L.A. Price, "Module Electrostatic Discharge Simulator," pp. 36–40, in *Proc. Electrical Overstress/Electrostatic Discharge Symposium*, Rome Air Development Reliability Analysis Center, 1979.
119. D.W. Aubrey et al., *J. Appl. Polymer Sci. 13*, p. 2193, 1969.
120. P. Benjamin and C. Weaver, "Measurement of Adhesion of Thin Films," *Proc. Roy. Soc. A. 254*, pp. 163–176, 1960.
121. Murray Bloom, "Development of Improved Test Standards for Monolithic Circuits," *TRW Final Rpt. NAS8–24388*, 1970.
122. Richard Spencer, "Novel IC Metallization Test Structures for Drop-In Process Monitors," *Solid State Technology*, pp. 201–205, September 1983.

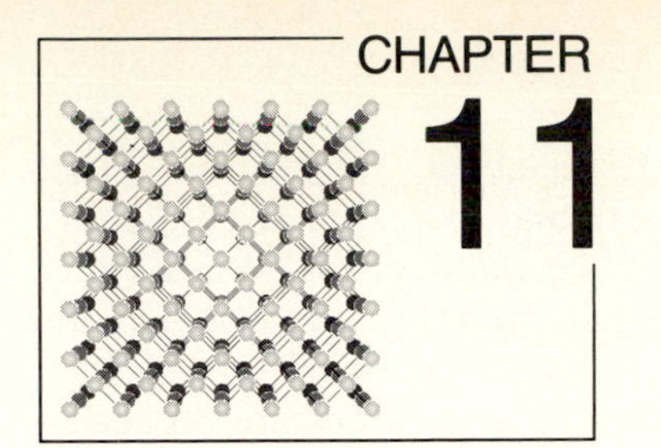

CHAPTER 11

Yields and Yield Analysis

11.1 INTRODUCTION

Three levels of wafer and circuit testing have evolved, two of which are performed in the wafer fabrication facility. The first level of testing consists of measuring a few parameters as soon as possible in order to provide quick feedback on process behavior. These measurements may be either wafer oriented, as are, for example, ion implant sheet resistances, or device oriented, as is the threshold voltage of a MOS test transistor. The results of these tests are generally reflected in the number of wafers passed at a given process inspection point (yield at that process step). The next level is the automatic DC testing (multiprobing) done on each chip after wafer processing is completed but before the wafer is broken into individual chips. The tests performed at this level are primarily designed to screen out chips that are likely to fail the more comprehensive AC/DC/temperature range third level of testing. However, the results are also an important measure of wafer-fab performance (multiprobe yield). Third-level testing is performed in the assembly/test area after packaging, and the primary goal is to ensure that the product meets customer specifications.

In addition to these test points, tests, and yields, there are also several materials-oriented inspection points and yields prior to a slice's entering the wafer fabrication facility. The more salient of these are included in the yield definitions of the next section.

11.2 YIELD DEFINITIONS

Crystal yield, the grams of crystal within specification divided by the grams of polysilicon required, is primarily a measure of how well the crystal resistivity, perfection, and oxygen content are kept within the required range. It is also a measure of how well the equipment is maintained (so that it does not fail during the 24 hours or so that it takes to pull a crystal) and of how wide a range of resistivities the manufacturer is able to sell. A typical value is about 50%. By recycling (remelting, redoping, and repulling) out-of-specification crystals, the silicon yield can be raised to perhaps 65%.

The next yield to be considered is that of converting from silicon crystal to slice. *Crystal-to-slice yield* may be expressed in grams/gram, grams/slice, or grams/cm^2 of slice and depends on how much silicon is lost when the crystal is ground from its as-grown diameter to the slice diameter, how thick the saw blade used for slicing is relative to the slice thickness, and how many slices are broken or rejected for edge chipping, poor polishing, and so on. A typical value is 50%. Thus, about 4 grams of polysilicon for each gram of slice is to be expected.

At this point, the slice enters the wafer processing area, where it becomes a wafer. One definition of *process yield* is the number of wafers to multiprobe produced in a specified time (often one month) divided by the number of starting slices used to produce those wafers. The overall process yield is generally broken into half a dozen or more intermediate yields relating to each of the major process steps. Typical overall process yield is 90%–95%. However, this yield is based only on those slices that are initially earmarked for finished salable units. Substantial quantities of incoming slices are also used for pilots, engineering runs, and so on. When these slices are considered, the overall wafer-fab area wafer out per slice in yield is closer to 70%. Thus, depending on whether process performance is being evaluated or the total number of slices required for some quantity of ICs is being estimated, two different yield numbers are required.

The individual chips on each wafer are checked electrically at multiprobe. The ratio of good chips to the total number of chips on a wafer is that wafer's *multiprobe yield*. The average value for a given circuit can range from a few percent to over 90%, depending on the complexity of the circuit and the proficiency of the wafer-fab area. There are a few pitfalls in determining this yield, and they stem from the determination of how many potential chips are on a wafer. One method of determining total chips is by means of the number of times the probe head comes down to make a measurement. However, this number will often include all of the partial chips on the periphery of the wafer as well as the whole ones over the rest of the wafer. Using this number for the potential chips will give a pessimistic yield number. Using the calculated number[1] of whole chips that can be placed on a wafer gives the most realistic yield number. Often, however, the chips in the outer few millimeters of the edge will be excluded from the potential count and thus give an inflated

---

[1]Simple programs can be written for either programmable calculators or desktop computers that will give this number. For large chips, their placement on the wafer may make a substantial difference in the number.

TABLE 11.1

Major Semiconductor Manufacturing Yield Points

Yield	*Definition*
Crystal yield*	Polycrystal to single-crystal conversion
Slice yield	Single-crystal boule to polished slice conversion
Process yield	Slice to finished wafer conversion
Multiprobe yield	Ratio of good chips per lot to total chips per lot
Assembly/test yield	Ratio of good finished units to good chips

*In the case of compound semiconductors, such as gallium arsenide, an additional loss point occurs prior to crystal yield. It is at the compounding step, where the constituents gallium and arsenic are combined to give polycrystalline gallium arsenide.

yield number. Multiprobe yield can be considered as the product of the functional yield and the parametric yield. A functional chip is one that "wiggles" (one that will roughly perform the function for which it was intended but that will not necessarily meet all of the specifications required to have a useful circuit). For example, a gate output might go from high to low as the appropriate inputs are changed and thus be functional. However, it might use too much power, switch too slowly, have too much leakage, or not have enough voltage swing and thus not meet the total set of specifications. Such a gate would then be a parametric failure. Ordinarily, the two yields will not be separated, but occasionally they will be for some yield analysis procedures.

After the wafers leave the wafer fabrication facility, they are broken into individual chips. The good chips are packaged and electrically tested to the finished-unit specifications. The *assembly/final test yield* is the number of units passing this final test in some time period divided by the number of chips required to produce those units. This final testing will check both DC and AC performance, usually at several temperatures. The assembly/final test yield will usually be in the 90%–95% range.

All of the yields that have been discussed, along with their definitions, are summarized in Table 11.1.

11.3 METHODS OF YIELD MEASUREMENT

Most wafer fabrication yield determinations are based on either visual examinations, special processing-oriented tests such as were described in the various processing chapters, curve tracer measurements, or multiprobe results. The main use of curve tracer examinations, however, is in follow-up defect analysis. By far, the largest body of yield data is taken by automatic multiprobe equipment.

11.3.1 The Curve Tracer

The basic elements of a curve tracer are shown in Fig. 11.1. The sweep voltage will usually be rectified 60 Hz, and the current or voltage step generation synchronized with it as shown in Fig. 11.1. Both the number of steps and the step amplitude may be varied. For diode junction *I–V* measurements, the diode is connected between terminals 1 and 3, and the step generator is not used. For bipolar transistor characteristics, a current step generator is connected to the base. For MOS transistor characteristics, a voltage step generator is connected to the gate.

To facilitate contacting the wafer, a probe station is used that has a chuck to hold the wafer and allow for x and y motion; two or three fine-wire probes with x, y, and z motion; and a stereoscopic microscope. A microscope and fine-probe wires with sharp points

FIGURE 11.1

Schematic of a typical curve tracer.

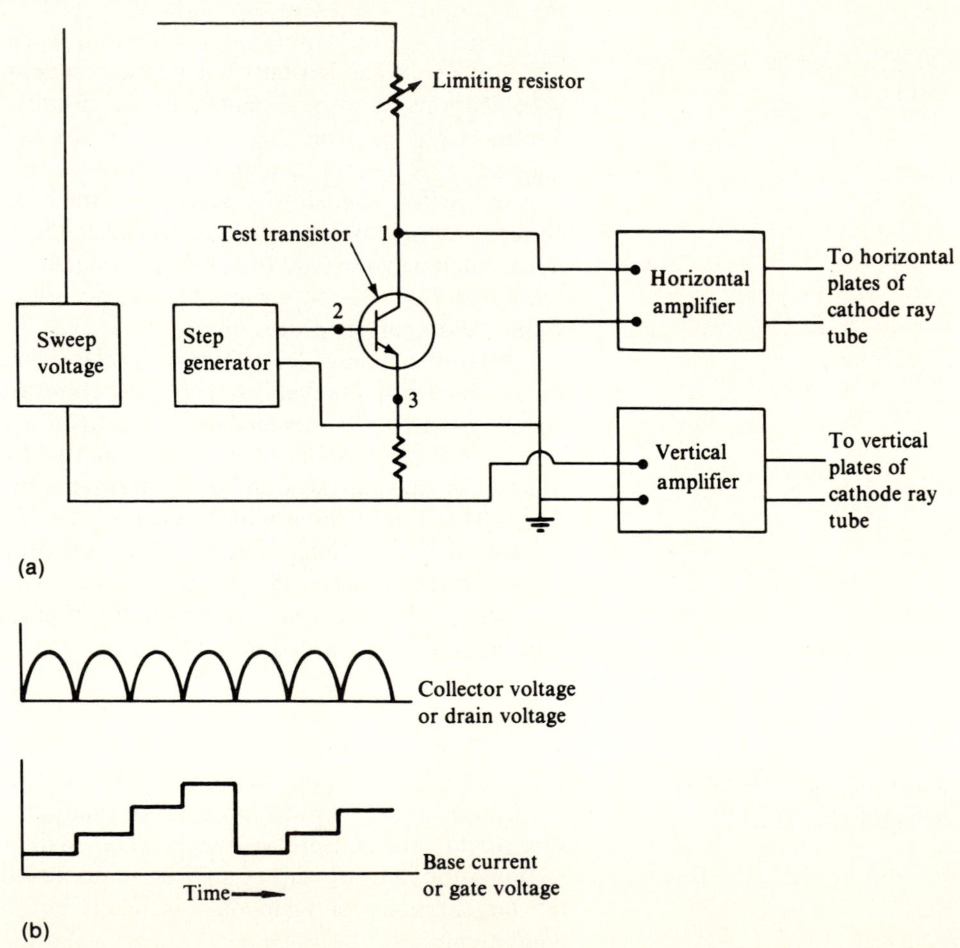

are required since leads may be only 1–3 μm across. When it is necessary to electrically isolate some component so that it can be individually examined, the appropriate interconnecting leads must be broken. A common method is to place two probes close together on a lead and use a capacitor discharge to vaporize the section of lead between the probes. An alternative method of breaking leads is to use an isolation mask (bipolar circuits) or a moat mask (MOS circuits) and etch through all interconnections.

11.3.2 Automatic Multiprobe Testing

Computer-controlled testers are required both to provide the speed necessary to process large volumes of chips and to direct the intricate tests of complex circuits. A block diagram of an automatic test system is shown in Fig. 11.2. The multiprobe station is similar to the

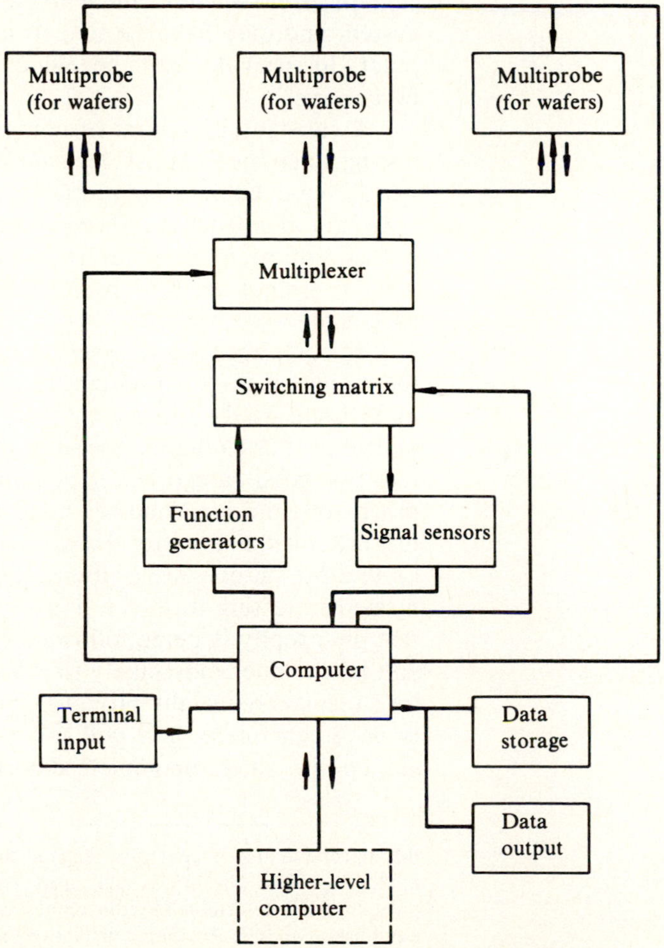

FIGURE 11.2

Block diagram of an automatic test system.

hand prober just described except that it has many more contacting probes, and the chucked wafer can be lowered away from the probes, shifted laterally, and raised again. Thus, the probes can be adjusted once to contact all of the appropriate pads on a chip, and then after each chip is tested, the wafer can automatically be moved to the next position. Initial positioning can either be manual or automatic through the use of pattern recognition technique. As indicated in Fig. 11.2, more than one multiprobe station may be connected to the test system via a multiplexer. Feeding into the multiplexer through a switching matrix is a multitude of function generators and signal sensors. The function generators include such items as DC power supplies and pulse generators.[2] The sensors are digital voltmeters, ammeters, comparators, and so on. The matrix directs the outputs of the function generators to the appropriate probes and connects the required sensors to proper probes. The local computer supplies timing and direction to all components of the system and may itself be tied to a higher-level management computer to provide test specifications and more complex data processing.

A substantial amount of computation may be required during testing since the fail/pass decision may be based on the outcome of various calculations. These may range from a simple calculation of transistor gain from measured currents to a regression analysis of a number of room-temperature measurements in order to predict whether or not the final packaged unit will pass high-temperature specifications.

As chips are tested, rejects are indicated by an ink dot placed in the middle of the chip. Often, two inkers with a different color ink in each are used so that a category in addition to good/bad can be indicated. For example, operational amplifier chips might be sorted into two good categories, depending on open loop gain. When the parameters are read and recorded for all chips on each wafer, the test order will not matter. However, to conserve test time, a "home-on-first-fail" procedure is often used so that valuable test time is not wasted in testing further a unit that has already failed. When this test philosophy is being followed, it is important to order the tests so that the ones most likely to fail are done first. Printouts of each lot's results are usually provided, but if a home-on-first-fail strategy is used, complete data will not be available. From a diagnostic standpoint, since incomplete data is a handicap, a full characteri-

---

[2]Multiprobe testing is generally referred to as "DC," but in the interest of minimizing test time, the inputs and outputs are really short pulses. The minimum pulse length is determined by the amount of time for the device to reach electrical equilibrium and the *RC* time constant of the connecting lines.

zation program will often be routinely run on a small percentage of wafers.

11.4 YIELD TRACKING

11.4.1 Process and Multiprobe Yield Tracking

Process yield or multiprobe yield can be plotted on a daily or a lot-by-lot basis. Either method gives a time series that can then be fitted with a trend line. Because of the rather substantial daily fluctuations, it is sometimes difficult to tell immediately when a break in the trend line occurs. One method of enhancing the breakpoint is to use the Cumulative Sum method (1, 2), in which not the yield Y_i but a sum S_i of the form

$$S = \sum_{i=1}^{n}(Y_i - Y_0) \qquad 11.1$$

is plotted. When the yields are constant, the curve is a straight line that will have a zero, positive, or negative slope, depending on the value chosen for Y_0. Ordinarily, a long-term yield average would be used for Y_0, although it may be slightly adjusted to give the desired initial slope. With this approach, if yields start drifting downward, there will almost immediately be a sharp break in the curve.

For wafer-fab areas that process many different device types, it may be impossible to maintain device continuity from week to week and thus difficult to adequately track multiprobe yield. One alternative is to track defect density, which is discussed in the next section. A defect density can be determined from multiprobe yield, IC area, and process complexity, which, in principle, is independent of the specific device being built.

11.4.2 Split Lot Testing

When process changes are considered, one of the more common ways of deciding whether such a change will in fact help yields (or costs) is to run a series of split lots. Split wafer lots all have common processing up to some particular step to be evaluated. At that point, the mother lot is split, and the various sublots are treated differently. Then, after the step or steps being evaluated are completed, the sublots are again merged for the remainder of the process. At the end, the various sublots are multiprobed separately, and decisions on the efficacy of the process changes are made on the basis of these yields. An improper interpretation could be very costly. Consequently, it is very important to accurately determine whether or not significant differences exist between various splits. What constitutes a significant difference is not defined in absolute terms, but rather only in how much confidence one must have in order to be satisfied that there is or is not a difference.

The distribution of multiprobe yield (MPY) by wafer for a full 50 wafer lot will ordinarily closely approach a normal distribution and have $1s$ limits of about 5 yield points.[3] The standard deviation of the multiprobe yield of the lot can be calculated from

$$s = \left[\frac{\Sigma(M - \overline{M})^2}{n}\right]^{1/2} \tag{11.2}$$

It is, however, more easily calculated from

$$s = \left[\frac{\Sigma M^2}{n} - \left(\frac{\Sigma M}{n}\right)^2\right]^{1/2} \tag{11.3}$$

where M is the multiprobe yield of each wafer, $\overline{M}$ is the arithmetic mean of the multiprobe yield of all of the wafers being considered, and n is the number of wafers. When the sample size is small enough for $n - 1$ to be appreciably less than n, $n - 1$ should be substituted for n in order to give an "unbiased" estimate of s.

If a random sample of N wafers is taken from a larger body of wafers with a normal distribution, it too should have a normal distribution. Its mean will not necessarily be the same as that of the whole lot. The dispersion[4] of the sample lot means will be given by

$$s_N = \frac{s}{\sqrt{N}} \tag{11.4}$$

For large N, the dispersion will be very low, and the sample mean will be essentially that of the lot. For $N = 1$, the mean of a lot of 1 is its MPY, and that mean's dispersion is s, or the same as that of the multiprobe yield of the lot. Thus, when the yield of a small split from a considerably larger lot is examined, the expected dispersion of the mean of a split, assuming that it is no different from the rest of the lot, can be calculated by Eq. 11.4. Then, the actual value of the mean of the split can be compared to that of the rest of the lot, and any difference is judged in light of the expected spread.

EXAMPLE ☐ Suppose that 9 wafers of a full 50 wafer lot had a different flow and that the multiprobe average of the other 41 wafers was 70% with a $1s$ value of 6 percentage points. What is the significance of a 9 slice MPY average of 74%?

---

[3]Remember that σ is the standard deviation of some property of a large population of objects and is defined so that $\pm 1\sigma$ from the arithmetic mean of a normal distribution will include 68.3% of the population, $\pm 2\sigma$ will include 95.5%, and $\pm 3\sigma$ will include 99.7%. In this case, the population is a collection of wafers, and the property is the multiprobe yield of each wafer. For samples from the large population, s rather than σ is used to denote standard deviation.

[4]One measure of dispersion is standard deviation. Relative dispersion is given by $s/\overline{M}$.

From Eq. 11.2, the expected dispersion s_N of the multiprobe average of lots of 9 wafers each would be

$$s_N = \frac{6}{\sqrt{9}} = 2 \text{ points}$$

The MPY of 74% is 4 points, or $2s$ away from the main lot yield of 70%. If these 9 slices were of the same population as the other 41, then there would have been only a 2.2% chance that their average yield would have been $2s$ away from the main average on the high side since 2.2% of the distribution is outside $2s$ on the high side. One can conclude that there is a 98% chance that these 9 wafers are not of the same population—that is, that the process change actually helped the yield. However, there is no implication that there is a 98% chance that the average yield of the new process will be *4 points* better than the old. The only implication is that there is a good chance that the new process is somewhat better than the old one.[5] □

Occasionally, a test lot will be completely split into small groups. There might, for example, be 5 lots of 9 wafers each. Again, the problem is one of determining whether or not any observed differences in multiprobe yield are significant. In this case, there is no major population to compare against, but it can first be postulated that all lots are random samples from the same population. If this is true, the standard deviation of the population from whence they came (pooled estimate) can be approximated by (3)

$$s_{\text{ap}}^2 = \frac{\Sigma M_{i1}^2 - [(\Sigma M_{i1})^2/N_1] + \Sigma M_{i2}^2 - [(\Sigma M_{i2})^2/N_2] + \ldots}{(N_1 - 1) + (N_2 - 1) + \ldots} \qquad 11.5$$

where M_{ij} is the multiprobe yield of the ith wafer of the jth split. The standard deviation s_{m1-m2} (dispersion) of the means of the multiprobe yields M_1 and M_2 of two samples picked from a sample with standard deviation of s is given by

$$s_{M1-M2} = s\left(\frac{1}{N_1} + \frac{1}{N_2}\right)^{1/2} \qquad 11.6$$

where N_1 and N_2 are the number of wafers in each lot. In this case, s is not available, but an approximate value s_{ap} can be calculated from Eq. 11.5 and used instead. The observed difference between

[5]For procedures on estimating the probable range of the improvement and for determining the size of sample needed for a given degree of confidence, the reader should see, for example, A.R. Alvarez et al., *Solid State Technology*, pp. 127–133, July 1983, or a standard statistics text.

TABLE 11.2

Area in One Tail of the Distribution Curve

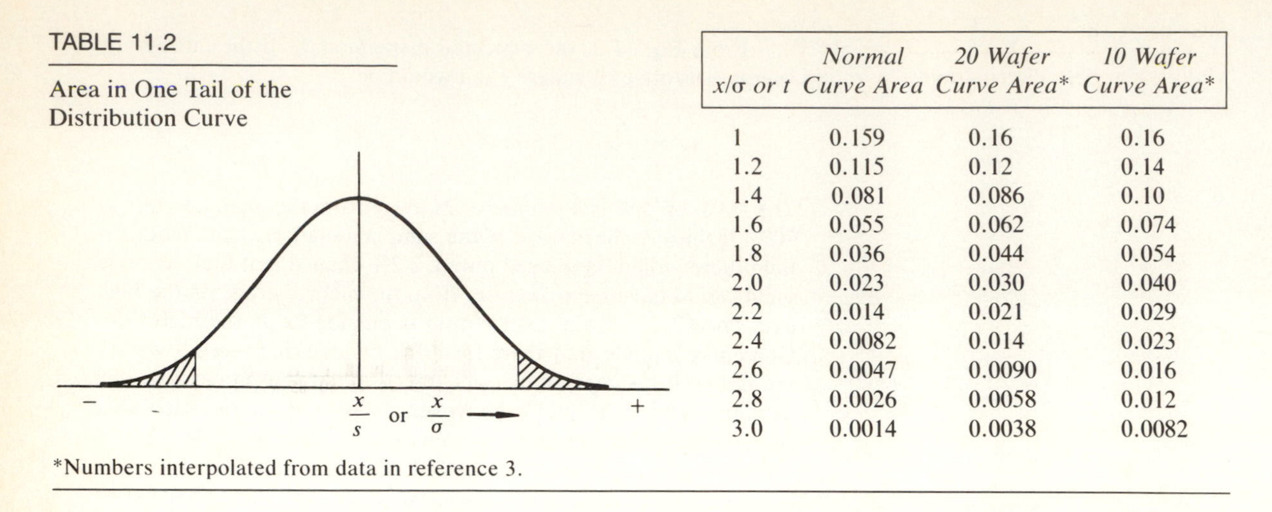

x/σ or t	Normal Curve Area	20 Wafer Curve Area*	10 Wafer Curve Area*
1	0.159	0.16	0.16
1.2	0.115	0.12	0.14
1.4	0.081	0.086	0.10
1.6	0.055	0.062	0.074
1.8	0.036	0.044	0.054
2.0	0.023	0.030	0.040
2.2	0.014	0.021	0.029
2.4	0.0082	0.014	0.023
2.6	0.0047	0.0090	0.016
2.8	0.0026	0.0058	0.012
3.0	0.0014	0.0038	0.0082

*Numbers interpolated from data in reference 3.

the two yields can then be compared to $s(M_1 - M_2)$ and a judgment as to its significance made. However, the normal probability curve data given in Table 11.2 do not, in principle, apply in this case. Instead, a "t" table, where t is the ratio of the yield difference to the standard error, is used with the approximate s values. The t table probabilities change with the number of wafers in the samples (N_1 and N_2 in Eq. 11.6). For demonstration purposes and because 5 and 10 slice splits are often used, Table 11.2 gives t only for $N_1 + N_2 = 20$ and $N_1 + N_2 = 10$. From a comparison of the t values with the x/s values in the table, it can be seen that, for a given difference in yield, the t table would predict a higher chance that the difference was insignificant and that the smaller the total number of wafers, the larger the yield difference required for significance. It can also be seen that if confidences in the 95% range are acceptable, the normal distribution curve can still be used.

11.5 DEFECT DENSITY

Two general kinds of multiprobe yield losses are observed in IC chips. One, which tends to extend over large areas of a wafer, occurs when one or more process parameters such as sheet resistance or diffusion depth drift out of range. The other occurs when localized defects cause either a single or a small cluster of failures. Examples are dirty contact windows and resist pinhole-induced diffusion defects. The defect density D is a measure of the density of localized defects and has proven quite useful in multiprobe yield prediction. Once a product is in production, it is usually possible to neglect the effect of broad-area process drift-induced defects because putting a process in production means matching the process variability to the required device parameter spread. Fig. 11.3 illus-

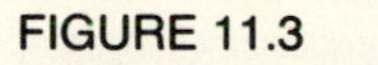
FIGURE 11.3

Multiprobe yield distribution for a large number of wafers.

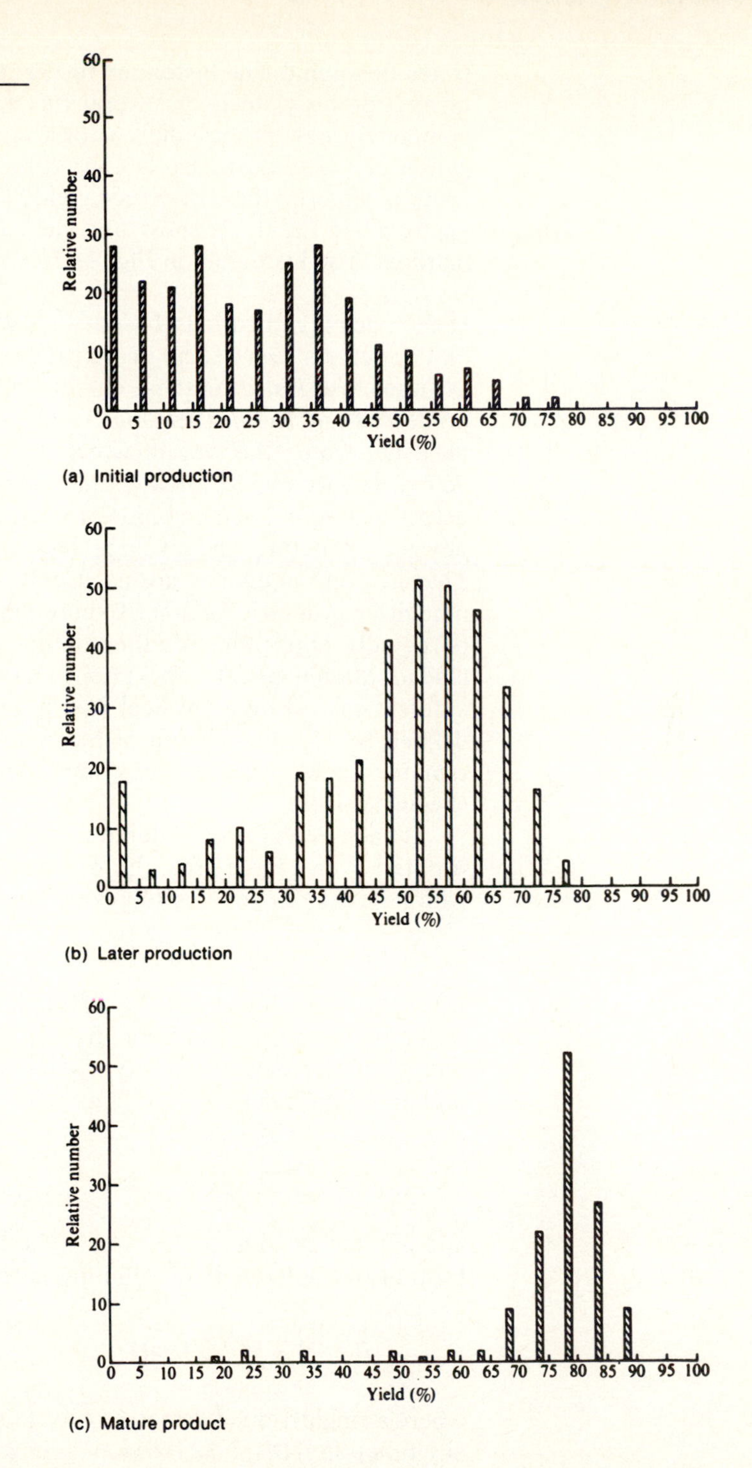

trates this point. The histogram in Fig. 11.3a shows the distribution of multiprobe yield over several lots when a new product with a completely new process flow is first introduced. The large congregation of yields near zero is caused by the actual process and the process required for high device yield not properly overlapping. The histogram in Fig. 11.3c shows the yield distribution after the product matures. The histogram in Fig. 11.3b will be discussed later.

11.5.1 Theory

Defect density theory is based on the existence of a number of "killer defects" very small in area and scattered over the wafer. The assumed distribution of these defects, $f(D)$, over the area of material processed and the mathematics used to convert this distribution to yield determine the way in which yield is predicted to vary with defect density and IC area. Some investigators restrict the use of defect-density-related equations to defects involving flaws in the IC physical structure and exclude flaws causing parametric failure. That is, yield equations are used only with functional failures and not with parametric failures. Experimentally, the equations seem to be equally applicable whether or not parametric failures are excluded. Mathematically, there appears to be little difference between a defect caused by a physical flaw in the IC structure and a defect defined by the intersection of the n-dimensional actual circuit parameter space with the envelope defined by the IC electrical specifications.

If each defect is considered to be identifiable and randomly distributed over a large area, the probability of finding a small area A with no defect (a good chip) is given by

$$P = e^{-AD} \qquad 11.7$$

where D is the density of the defects. This expression is generally referred to either as a Poisson distribution or as a Boltzmann distribution. If the defects are considered to be indistinguishable, the probability of finding a good chip in a small area A is given by

$$P = \frac{1}{1 + AD} \qquad 11.8$$

and is referred to as a Bose–Einstein distribution. A more general form of the Boltzmann distribution gives

$$P = \int e^{-D'A} f(D')dD' \qquad 11.9$$

where a single-valued defect density D is replaced with a defect density function $f(D')$, which has an average value D.

Early work assumed that the defect density was everywhere equal to D—that is, that $f(D') = 0$, except at $D' = D$ where it equaled 1. Then, the probability P of finding a chip area overlapping no defect (the yield Y of good chips) is given by (4, 5)

$$P = Y = e^{-DA} \tag{11.10}$$

or

$$Y(\%) = 100 \times e^{-DA}$$

Eq. 11.10 predicts that if a chip of area A has a yield Y_0, then a chip of size NA should have a yield Y of $(Y_0)^N$. This approach to yield projection was first proposed when the IC was just making its appearance. The yields of all chips were low at that time, and the projected yields of larger, more complex circuits looked hopeless. However, it soon became clear that the exponential gave a pessimistic view. Curves of yield versus area are illustrated in Fig. 11.4. Curve A is an Eq. 11.10 projection, based on an actual IC yield of 23% for a normalized chip area of 1. The data for curve B were obtained by determining the yield of a single circuit chip and then, on the same set of wafers, redetermining the yield of arrays of 2, 4, and 8 good adjacent chips. It can be argued that a multichip array is not a proper way to perform such an experiment since the yield of different parts of the same circuit can be quite different.[6] For example, the yield of the portion of a chip used for bonding pads should be very good and, for a small chip, is a much larger percentage of the total chip area than for a large chip. An array of adjacent chips would maintain the same high percentage of high-yielding bonding pad area and thus have artificially high yields at large areas. In most cases, however, this does not happen. As an example, curve C of Fig. 11.4 shows yields for a series of ICs that comprise a TTL gate family and that have the same yield–area trend as curve B. All of the ICs in the series have the same number of bonding pads but an increasing circuit complexity and area. The data were taken somewhat later in time than the data of curve B and thus explain why the yields are higher. For some classes of devices, a downward curvature has been observed (6), as shown in curve D. Thus, the argument is sometimes valid and must be considered.

If the defect density were uniform, the yield of all slices will be very closely grouped when the chip area is very small compared with the wafer area. That this fact is not observed can be seen from Fig. 11.3, and Murphy (7) in 1964 suggested that there could be a spread of defect densities. For example, the probability of finding a particular density might be highest for the value D and tail off on

---

[6]Actually, A is often defined not as the area of the chip, but as the portion that would be susceptible to defects. In some layouts, some substantial surface area has no active elements and is very insensitive to process-induced defects.

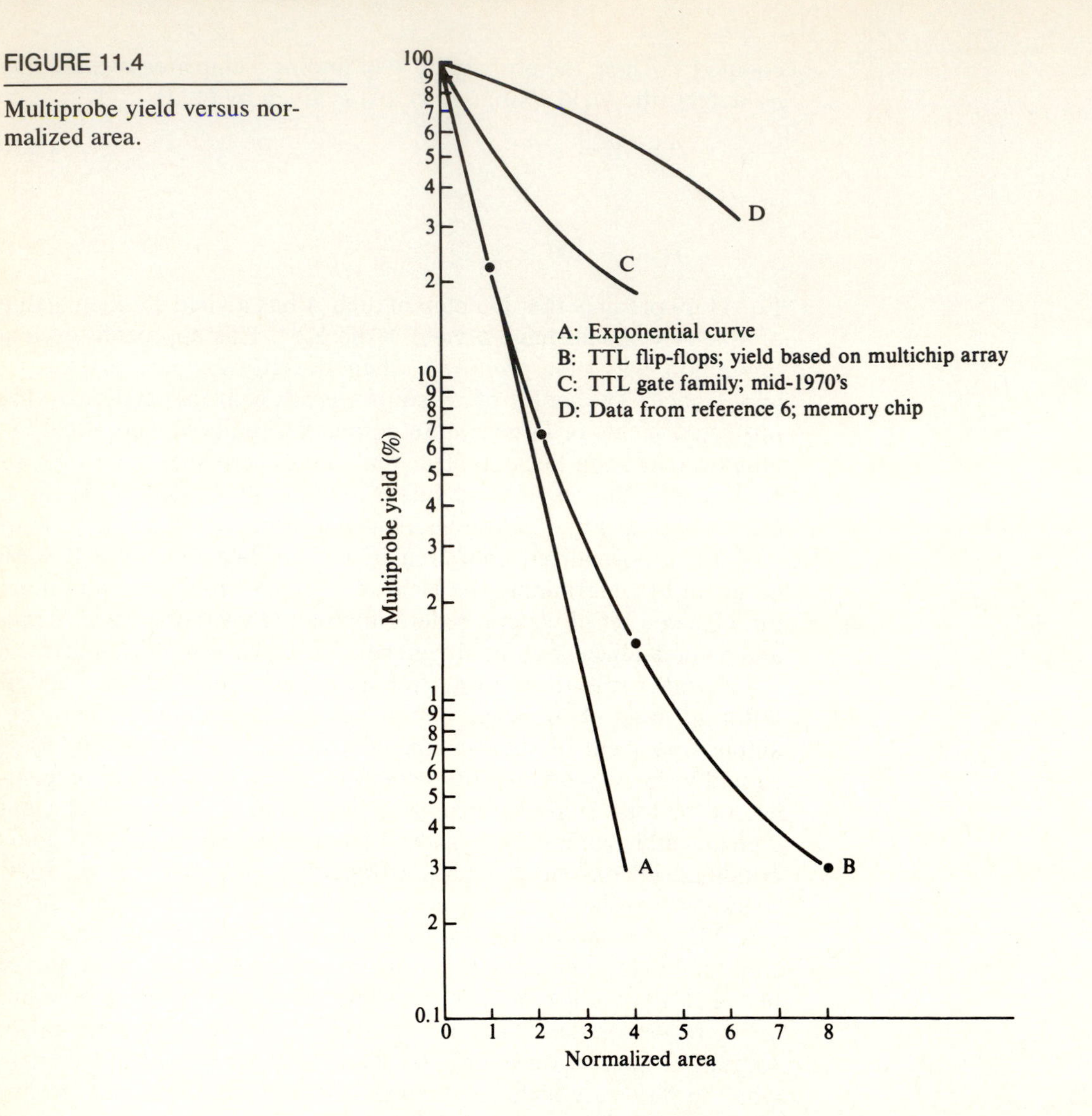

FIGURE 11.4

Multiprobe yield versus normalized area.

either side in the manner of a Gaussian curve. As approximations that made integration of Eq. 11.9 easier, a triangular distribution and a rectangular distribution, as shown in Fig. 11.5, were used in the calculations. The distribution smearing has the effect of changing the character of the curve from that of curve A in Fig. 11.4 to something more like that of curve B in Fig. 11.4. The triangular distribution, with the peak centered at the average defect density D and having a half width of D was the one most favored by Murphy and indeed is still in use. It results in

$$Y = \left(\frac{1 - e^{-DA}}{DA}\right)^2 \tag{11.11}$$

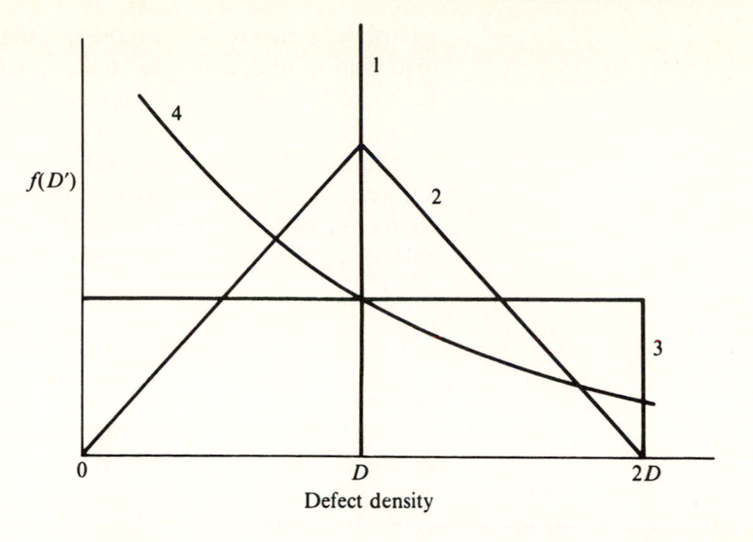

FIGURE 11.5

Defect distribution functions.

Use of the rectangular distribution of width twice the average defect density D gives

$$Y = \frac{1 - e^{-2DA}}{2DA} \tag{11.12}$$

A few years later, Seeds (8) suggested that there was a higher probability of a low defect density than a high one and used $(1/D)e^{-D'/D}$ for $f(D')$ (curve 4 of Fig. 11.5), which gives

$$Y = e^{-\sqrt{DA}} \tag{11.13}$$

and again results in a higher yield prediction at larger areas than does the exponential. Often, a noticeable reduction in yield occurs in going from the center to the edge of a wafer. Some approaches in handling this problem fold the radial variation in defect density into Eq. 11.10 in an attempt to provide more realistic yield projections (9, 10). In these cases, it is assumed that the average chip yield Y_w for a given wafer can be described by a series of yields associated with different areas of the wafer. That is,

$$Y_w = F_1 e^{-D1} + F_2 e^{-D2} + \cdots \tag{11.14}$$

where F_i is the fraction of the slice with an average defect density Di. A more generalized approach is to assume that $f(D')$ in Eq. 11.9 is a gamma distribution (11). A negative binomial distribution results, and the yield is given by

$$Y = (1 + \overline{D}AV)^{-1/V} \tag{11.15}$$

where V is the square of the coefficient of variance of the defect distribution and $\overline{D}$ is the defect density mean. The coefficient of variance is given by the standard deviation of the defect density divided by the mean $\overline{D}$.

Rather than choosing different density distribution functions, different statistics can also be used. Instead of the Boltzmann probability, Price (12), considering that the defects were indistinguishable, used Bose–Einstein statistics, which give for the yield

$$Y = \frac{1}{1 + AD} \qquad 11.16$$

This equation gives results similar to those of Eqs. 11.11 through 11.13. If several independent defect-producing mechanisms exist, as, for example, would be found in the different processing steps, then Eq. 11.16 can be generalized as follows:

$$Y = \frac{1}{(1 + AD_1)(1 + AD_2) \cdots} \qquad 11.17$$

The various equations just discussed are summarized in Table 11.3, and Fig. 11.6 shows a comparison of their projections. In most cases, it appears that while a particular equation may be most realistic for a given wafer-fab area it by no means has universal applicability. It should be noted that in all of these expressions, Y is a function of the DA product, not of D or A alone. Thus, the same differences among the curves showing up at the large areas of Fig.

TABLE 11.3

Defect Density–Yield Prediction Equations

Eq. No.	*Equation*	*Approach*	*Reference*
T1a	$Y = e^{-DA}$	Exponential (Poisson or Boltzmann)	(4, 5)
T1b	$Y = e^{-D1A}e^{-D2A} \cdots e^{-DnA}$		
T1c	$Y = e^{-NDA}$		
T2	$Y = \left(\frac{1 - e^{-DA}}{DA}\right)^2$	Murphy	(7)
T3	$Y = e^{-\sqrt{DA}}$	Seeds	(8)
T4a	$Y = \frac{1}{1 + DA}$	Price (Bose–Einstein)	(12)
T4b	$Y = \frac{1}{(1 + DA)^N}$		
T4c	$Y = \frac{1}{(1 + D_1A)(1 + D_2A) \cdots (1 + D_NA)}$		
T5	$Y = (1 + \overline{D}AV)^{-1/V}$	Negative binomial	(11)

FIGURE 11.6

Comparison of yield projection theories, with curves matched at yield of 50% and area of 0.23 cm². (The *D*'s required for this match are tabulated on graph.)

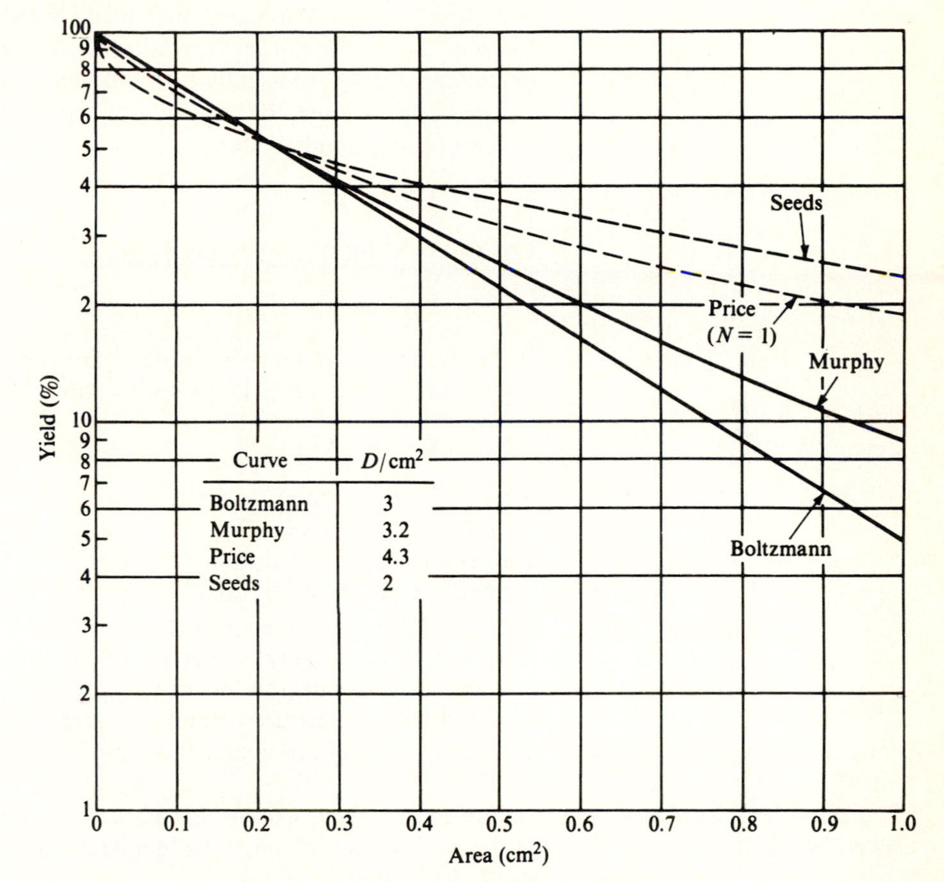

11.6 could just as easily occur for smaller areas and larger defect densities. If a given area and yield are used to calculate D through the various equations of Table 11.3, the D values may be quite different, as can be seen in the examples in Fig. 11.6. Despite the disparity in D values, if the yields are above perhaps 30%, the differences among the predictions of the various expressions are probably not as great as the daily experimental range of values that are observed. One practical approach to the choice of equations is to use regression analysis to see which equation best fits the data of a given wafer-fab area and then use that equation for that area.

Since defects are not all generated at the same time, it is helpful to have an expression that relates the number introduced at each process step to the final yield (a thing difficult to do with a simple polynomial fit of data). Generally, most of the defects will be generated at only a few of the many manufacturing steps. These steps will closely follow the major ones listed in the process flowcharts of Chapter 2 but may not include all of them. Those considered major defect generators are usually referred to as "critical" steps or levels. It can be postulated that the overall yield Y is the product of the yields of the individual steps. That is,

$$Y = Y_1 Y_2 \cdots Y_n \qquad 11.18$$

Eq. T1a of Table 11.3 then becomes

$$Y = e^{-D1A} e^{-D2A} \cdots \qquad 11.19$$

If the number of defects introduced are considered to be the same for each level, then Eq. 11.19 reduces to

$$Y = e^{-NDA} \qquad 11.20$$

where N is the number of critical levels. The derivation of Eq. T4a included this provision (Eq. 11.17) and gave rise to Eqs. T4b and T4c. It should be noted, however, that Eqs. T4b and T4c cannot be used interchangeably by making the D of T4a equal to the sum of the D's of T4c. It is also a property of Eqs. T4b and T4c that with the same defect density for each level, as N increases, the predicted Y more and more closely matches the Y of Eq. T1a. By the time $N = 8$ (not an unreasonable number of steps), the shape of the curves of Eqs. T1 and T4 are nearly indistinguishable.

11.5.2 Defect Density Determination

Before any projections can be made using an equation from Table 11.3, a value for D must be determined. Ordinarily, multiprobe yields for wafers made by the same processes but having chips of different sizes will be collected and a D determined from a best-fit curve to the data. Occasionally, however, monitor wafers with special test patterns whose yields can be correlated with a defect density for that step are processed together with production wafer lots. This method allows an update on the projected yield of the lots to be made after each process step that used the special monitor wafers. This aspect will be discussed separately.

The "area" in the various equations is usually taken to be that of the whole circuit, including bonding pads. In some cases, however, it may be more useful to consider an effective area more closely related to the specific area of the chip in which a killing defect could occur. The yield normally is the total multiprobe yield, although functional yield is sometimes used. The complexity factor N can be defined as 1 if all products to be examined are made by the same process and the goal of the study does not require defects by level. Since most defects are associated with the lithography steps, the number of patterning steps is often used for N. Sometimes, implant steps will each be counted as 1/2 since no etching is involved.

The usual presumption is that when a large number of devices are run by the same process (all of the same N value), a defect density applicable to the whole collection can be determined by best fitting all of the data points (yield and area of each individual IC type) with the chosen equation. Of the equations of Table 11.3, the exponential is, in principle, the easiest to fit. Standard expressions are available for an exponential best fit in which the differences among the observed ln Y and the best-fit predicted ln Y are minimized. Unfortunately, this approach does not constrain the yield to be 100% at zero area. The problem can be eliminated by deriving the best-fit expression based on the fitted equation $Y = e^{-bx}$ rather than $Y = ae^{-bx}$, as is normally done. In this case, the best-fit D is given by (13)[7]

$$D = \frac{-\Sigma \ln Y_i}{\Sigma N_i A_i} \qquad 11.21$$

where Y is used as a decimal fraction, not as a percentage. It may be tempting to calculate a D for each pair of points and average them. That is,

$$D_{\text{av}} = \frac{1}{N} \Sigma \frac{\ln Y_i}{N_i A_i} \qquad 11.22$$

Such a value will not match the best-fit D of Eq. 11.21, although it may be reasonably close.

Expression T2 of Table 11.3 is more difficult to use. An easy way to estimate a D for a series of data points is to plot the points onto paper with a family of curves already drawn on it as shown in Fig. 11.7. After connecting the points with a smooth curve, the D value can be estimated from the position of this curve relative to the

---

[7]What constitutes a "best fit" depends in part on the use to which it will be put. In the case just discussed, one could have, for example, decided on a procedure that would have minimized the sum of the deviations of the yields rather than deviations of the log of the yields.

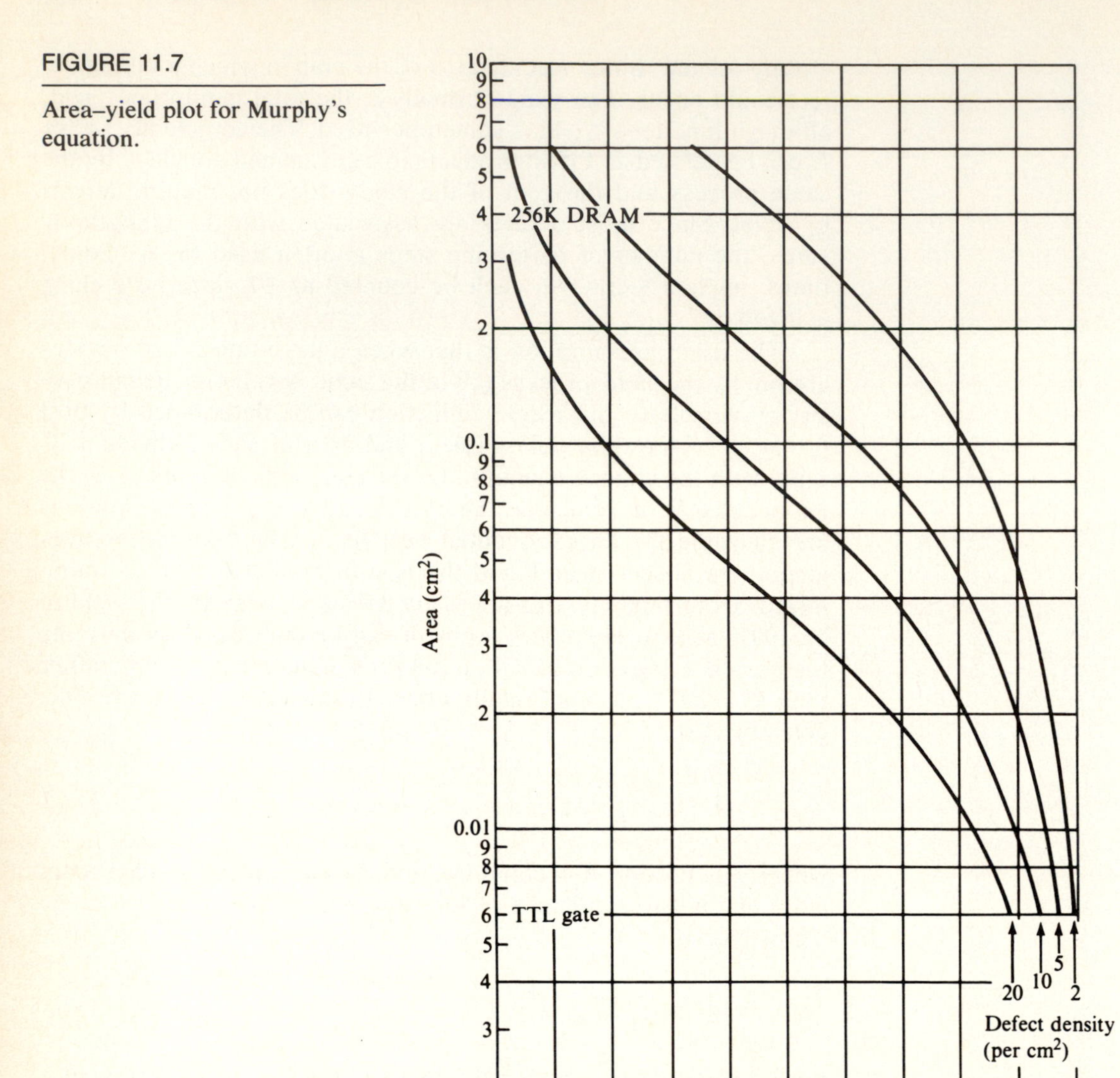

FIGURE 11.7

Area–yield plot for Murphy's equation.

location of the background curves. For calculating D from an individual Y–A value, a power series expansion can be used (14):

$$D = \frac{2Z + 1.33Z^2 + 1.11Z^3 + 1.007Z^4 + 0.953Z^5}{A} \tag{11.23}$$

where $Z = 1 - \sqrt{Y}$, and Y is used as a decimal fraction, not as a percentage.

In the case of Eq. T3, $Y = e^{-\sqrt{DA}}$, D for a single pair of Y–A values can be obtained from

$$D = \frac{(\ln Y)^2}{A} \tag{11.24}$$

For the D corresponding to a best-fit curve, the data points are plotted on semilog paper with yield along the log axis, and the square root of area (linear dimension of a square chip) is plotted on the other. The slope of the line will be $-\sqrt{D}$.

A best-fit D for expression T4b, $Y = 1/(1 + DA)^N$, can be obtained by rearranging as follows:

$$Y^{-1/N} = Z = 1 + DA \tag{11.25}$$

With this form, standard linear regression can be used for curve fitting. There is, however, one potential pitfall in that when Y is very small, Z becomes inordinately large and may weight the best fit in an undesirable manner. As in the case of Eq. T1c, yields associated with different N's but the same D value can all be combined into one curve in order to improve statistics.

11.6 CHIP YIELD PREDICTIONS BASED ON DEFECT DENSITY

Yield prediction is a necessary part of both IC design and the wafer fabrication operation. It can be used to estimate the yield of new devices and help ensure that the design will provide a part that will be economically successful. It can also be used to evaluate the yield of mature chips relative to others being run in the same wafer-fab shop.

11.6.1 New Devices

When a new part in which the only changes are chip size and circuit function is put into production, the use of historical data, plotted in the format of Fig. 11.6, can be used directly. A defect density is calculated, and from it and the new chip area, an expected yield can be determined. If the actual yield falls below the curve, it is almost always due to a layout flaw, such as a lead spacing inadvertently made too small. When a new "shrunk" device is introduced, it must be remembered that the yield is driven by two opposing forces. First, since it is a smaller part, there is less likelihood of a defect on it. Second, however, the tighter geometry makes the photolithography steps more susceptible to small-area defects.[8] Attempts have been made to fold the defect size distribution into the yield equation (15), but data on the distribution to use are difficult to surmise.

One approach to the problem is to assume that since most lithography defects are caused by particulates collecting on the wafer or mask surface, the defect size will be proportional to the particle size. Particles may be carried in the ambient air, held in suspension in the processing liquids and gases, and/or generated by moving parts of the slice-handling equipment. The airborne particulate density increases with decreasing particle size as shown in Fig. 11.8 as approximately $1/d^2$ where d is the diameter. The density of liquid-born particles follows the same general trend. A reasonable device sensitivity approximation is that a particle of diameter 1/2 the size of a feature (a line spacing or contact window opening) on the wafer surface can cause a killing photolithographic defect. A calculation of the defect density for a given level then gives a measure of defects greater than 1/2 the feature size ζ. Using the particulate size distribution of Fig. 11.8, the defect density can be calculated for some other feature size. Thus, all processing steps (levels) that have losses

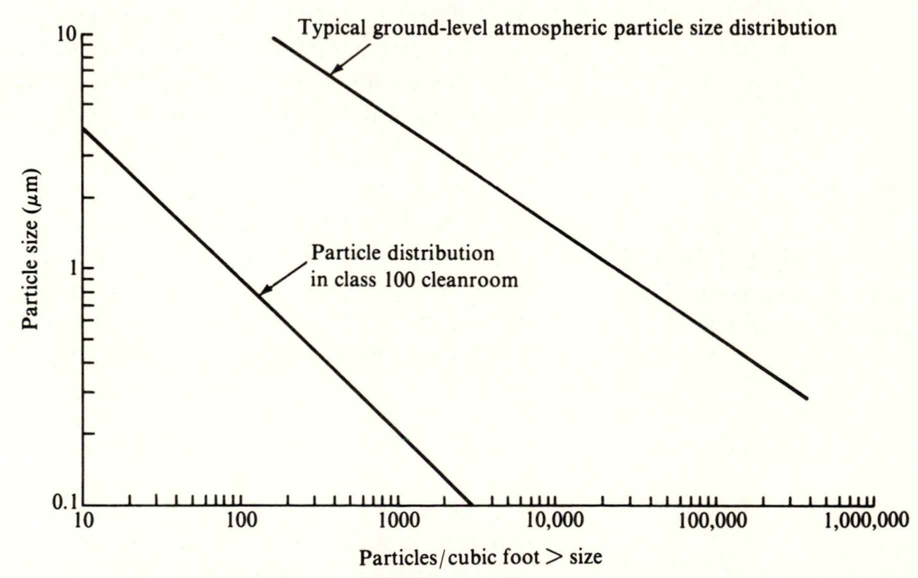

FIGURE 11.8

Particulate density above a given size versus size. (*Source:* Adapted from data in Philip W. Morrison, ed., *Environmental Control in Electronic Manufacturing,* Van Nostrand Reinhold Co., New York, 1973.)

[8]The defect density formulation discussed earlier assumed small-area "point" defects.

determined primarily by particulates, or photolithographic steps,[9] can be expected to show defect densities that will increase approximately[10] as $1/\zeta^2$. That is, for estimating purposes,

$$D(\zeta_2) \cong D(\zeta_1)\left(\frac{\zeta_1}{\zeta_2}\right)^2 \tag{11.26}$$

EXAMPLE ☐ Assume, in a 9 level process, that 6 levels are particle limited and that only one part type is run in the wafer-fab area. The chip has an area of 0.12 cm² and is running a 75% multiprobe yield. After a 15% linear shrink, what is the estimated new yield?

Since there is little extrapolation in area, choose the equation from Table 11.3 that is easiest to use. Further, assume that all levels have the same defect density. By using Eq. T1c, the initial D for *one level* is

$$D = \frac{\ln Y}{9 \times 0.12} = 0.27/\text{cm}^2$$

The new D_ζ for the shrunk lithographic levels is

$$D_\zeta = D\left(\frac{1}{0.85}\right)^2 = 0.27 \times 1.38 = 0.37$$

The new area is 0.087 cm², and

$$Y = (e^{-6AD_\zeta})(e^{-3AD}) = e^{-0.283} = 76\%$$

The projection in this case is that little change should occur in the multiprobe yield, but because of the smaller size, 38% more chips would be available. ☐

11.6.2 New IC Designs with New Processes

Projecting the yields of new designs with new processes is much more difficult than the case just discussed. There, the parametric yield was assumed to be very high and was neglected. However, in the case of new designs, it may well be the limiting factor. In any event,

$$Y = Y_P Y_D \tag{11.27}$$

---

[9]Assuming that the lithographic process is capable of printing the smaller geometries. If the process is already marginal or if the shrink is more than a few percent, yields may drop abruptly, not because of increased particles, but because of poor printing resolution.

[10]Allusion in reference 16 to unpublished work suggests that $1/\zeta^3$ might be more appropriate.

where Y_P is the parametric yield and Y_D is that due to defects. Devices with known high Y_P values can be run, and the wafer-fab defect density can be determined as before. Next, the process-induced variation of critical parameters important to the functioning of the new device is characterized in terms of a median value and standard deviation. For a MOS digital device, the channel length and width, the gate oxide thickness, and the flatband voltage are usually adequate. Then, the region in the space defined by those critical parameters that allow the IC to function properly is mapped out. The parametric yield can be estimated by integration of the probability distribution functions over this space (17). Such calculations are generally done in conjunction with SPICE or some similar circuit analysis program and will not be considered further in this book. The projected multiprobe yield (Eq. 11.27) is the product of this parametric yield and the previously determined defect-limited yield Y_D.

11.6.3 Devices in Production

Once an integrated circuit is in production, yield prediction is useful to see whether that circuit is running as well as it should. By examining a yield–area plot such as that of Fig. 11.9, low-yielding device types can be easily pinpointed and their expected yields projected. This particular figure shows the yields of 14 devices (□'s) that were just transferred into a wafer-fab area, along with 5 devices

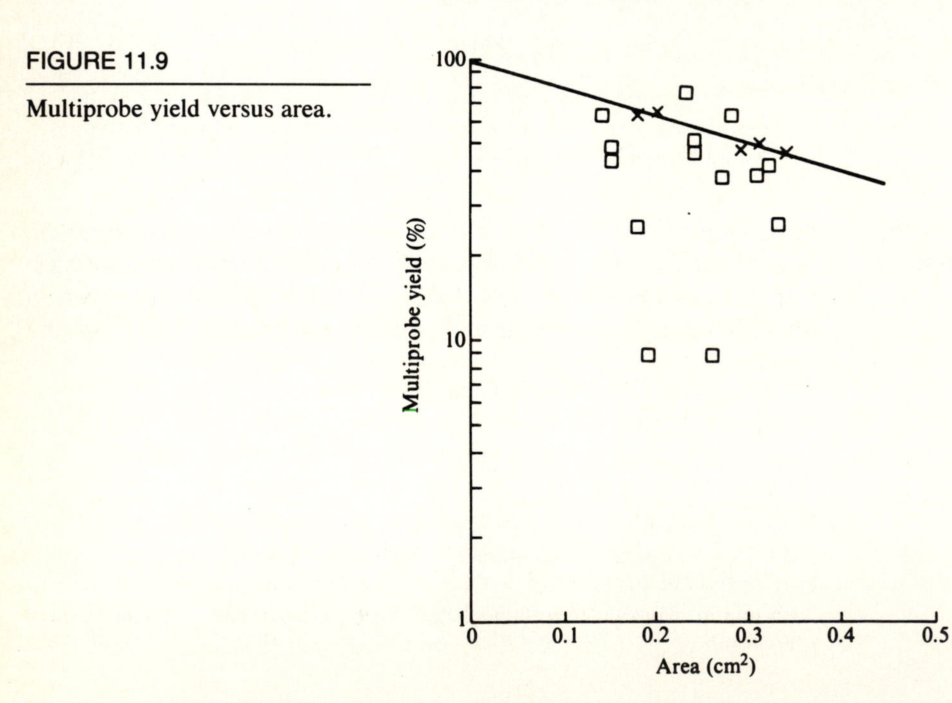

FIGURE 11.9

Multiprobe yield versus area.

($\times$'s) that were already running and whose yields were considered representative of the wafer-fab area. The 5 devices were used to draw the line used for projecting acceptable yields of the other 14 chips. Sometimes, it is necessary to switch production from one wafer-fab area to another, and it is desirable to predict the yield in the new area. Then, the determination of a defect density representative of each facility will allow the yield of transferred devices to be satisfactorily estimated.

11.6.4 Work in Progress

Since the cycle time for wafers in a fabrication area will generally be in the 3–4 week range, a means of projecting the multiprobe yield of lots in progress is sometimes helpful. Such a projection can afford a more accurate means of predicting output volume and can also give early warning of process problems. The projections can be made either on the basis of the yield of monitor wafers (pilot wafers) pulled at various stages of the process or from the results of regular quality control (QC) inspections for visual defects. The general procedure is to assume historical yields as lots begin and then, as the results come in for the intermediate yields, upgrade the projection. That is, when the lot begins,

$$Y_{\text{est}} = Y_{h1} \times Y_{h2} \times Y_{h3} \cdots \qquad 11.28$$

and, as time progresses,

$$Y_{\text{est}} = Y_{a1} \times Y_{h2} \times Y_{h3} \cdots$$

$$Y_{\text{est}} = Y_{a1} \times Y_{a2} \times Y_{h3} \cdots$$

where Y_{hi} is the historical yield of step i and Y_{ai} is the actual yield of step i.

11.6.5 When Yield Is Low

When the average multiprobe yield is low and the distribution by slice looks like that shown in Fig. 11.3a, yield projections based on defect density calculations are not very helpful since defects in the normal sense of the word are probably not limiting yield. The reason for the yield loss is more likely due either to the fact that the process is not matched to the device design or else to the fact that the process is in very poor control. However, the yields to be expected when the process is controlled are reasonably well predicted by the maximum yield observed, which is illustrated in the sequence of parts a, b, and c in Fig. 11.3. In part a, while the peak is at 35%–40%, there is some yield at 75%–80%. In part b, the peak has moved to 50%–55%, but there is still some yield to 75%–80%. In part c, the low-yielding slices have all disappeared, and the peak yield is now at 75%–80%.

11.7 DETERMINING CAUSES OF CHIP YIELD LOSS

Before one starts to search for the underlying physical reasons for the "mathematical" defects discussed up to this point, the yield Y of Eq. 11.27 needs to be defined in terms of the types of defects causing the yield loss. It can, for example, be written as

$$Y = Y_D Y_P = Y_{DV} Y_{DNV} Y_{PV} Y_{PP} \qquad 11.29$$

where $1 - Y_{DV}$ is the catastrophic yield loss due to visual defects, $1 - Y_{DNV}$ is the catastrophic yield loss due to nonvisual defects, $1 - Y_{PV}$ is the parametric failure due to visual defects, and $1 - Y_{PNV}$ is the parametric failure due to nonvisual defects.

The significance of the visual defects (the $Y_{DV} - Y_{PV}$ combination) can be judged from Table 11.4, which shows the percentage of total multiprobe failures that in three separate evaluations were ascribed to visual defects. Thus, it is apparent that with the current state-of-the-art processing, visual defects are a major indicator of

TABLE 11.4

Incidence of Visual Defects Deemed to Cause Failures

Evaluation	*Percentage of Total Failures*
Single-Level Metal Gate CMOS	
Aluminum shorts	40
Aluminum opens*	20
Scratches	10
Irregular oxide pattern	5
Alignment	5
Total	80
Single-Level Bipolar Logic	
Metal shorts	15
Metal opens*	15
Irregular oxide pattern	10
Oxide undercutting	25
Alignment	10
Total	75
Silicon Gate MOS Memory	
Metal shorts	30
Metal opens*	30
Polysilicon shorts	15
Oxide holes (not pinholes)	5
Total	80

*Metal opens due to lithography are easy to see. Those due to breaks in the metal where it goes over oxide steps may be difficult to see optically. SEM views may be required.

chip failures. However, the problem of relating a particular visual flaw to an electrical failure exists since usually more visual flaws occur than do electrical failures. It should be remembered that visual defects that do not cause multiprobe (electrical) failures may, in some product lines, still be cause for rejection during a subsequent visual inspection. Even though a visual defect, such as a lead partially etched away or poor step coverage, may not cause an immediate electrical failure, it can cause a reduction in the time to failure of an integrated circuit. Since visual defects cause such a large fraction of multiprobe failures, both optical and scanning electron microscope examinations must be an integral part of any study of yield losses.

Another approach to studying the reasons for chip failures is by electrical testing. Unlike visual examinations, where there is a problem of relating a particular visual defect to a specific electrical failure, this approach has the problem of relating the electrical failure to a specific processing error. Test masks with special patterns can be substituted for regular masks during some point in the process and used to produce electrical error signals easily identifiable when specific process problems arise. These test patterns may be placed on separate test wafers, incorporated into the main body of the chip, or, in some cases, put in the scribe line. Otherwise, it is often difficult to determine what defects are causing specific device failures. The design of these kinds of masks is dependent on the specific integrated circuit to be studied and generally must be prepared by a circuit design engineer. Performing straight electrical test diagnostics without benefit of test patterns is widely used and is a multiple-step process. First, the particular transistor, diode, resistor, capacitor, or lead whose failure is most likely to cause the observed integrated circuit failure must be determined. Specially written multiprobe programs are often used for this determination. These programs, in addition to the standard test program, are required since multiprobe tests are used primarily to ensure that the chip meets customer requirements rather than for diagnostic purposes. After the faulty element is isolated, curve tracer analysis is generally the most fruitful technique for determining why the component failed, and the next two sections of this chapter are devoted to that approach. Finally, based on the results of the curve tracer analysis, the processing errors can be determined.

11.8 JUNCTION TESTING

The pn and Schottky barrier junctions are basic building blocks of current semiconductor technology, and when they fail to perform properly, yield will suffer. The properties of a junction that are examined are the forward and reverse currents, the reverse breakdown voltage, and, occasionally, the series resistance. The current, forward or reverse, at a given voltage will almost never be less than projected and, if it is, will almost never be considered a reject. Higher than expected reverse voltage breakdown almost never occurs, and lower breakdown is usually because of excess reverse leakage current. Thus, junction analysis consists primarily of determining the origin of excess current.

11.8.1 Reverse-Biased pn Junctions

Current of a reverse-biased pn junction can be broken into three regions with quite separate behaviors as shown in Fig. 11.10. Before breakdown, which is where most junctions are intended to operate, the current is relatively low. This current is ordinarily referred to as "reverse current" and is discussed in more detail in the next paragraph. Above breakdown, the current (avalanche current) is quite large and is generally limited by external series resistance. In the region from near breakdown into breakdown, the current increases very rapidly (avalanche multiplication) and is proportional to $1/[1 - (V/V_B)^m]$ where V_B is the breakdown voltage and m is ~4 for n-type silicon and ~2 for p-type silicon.

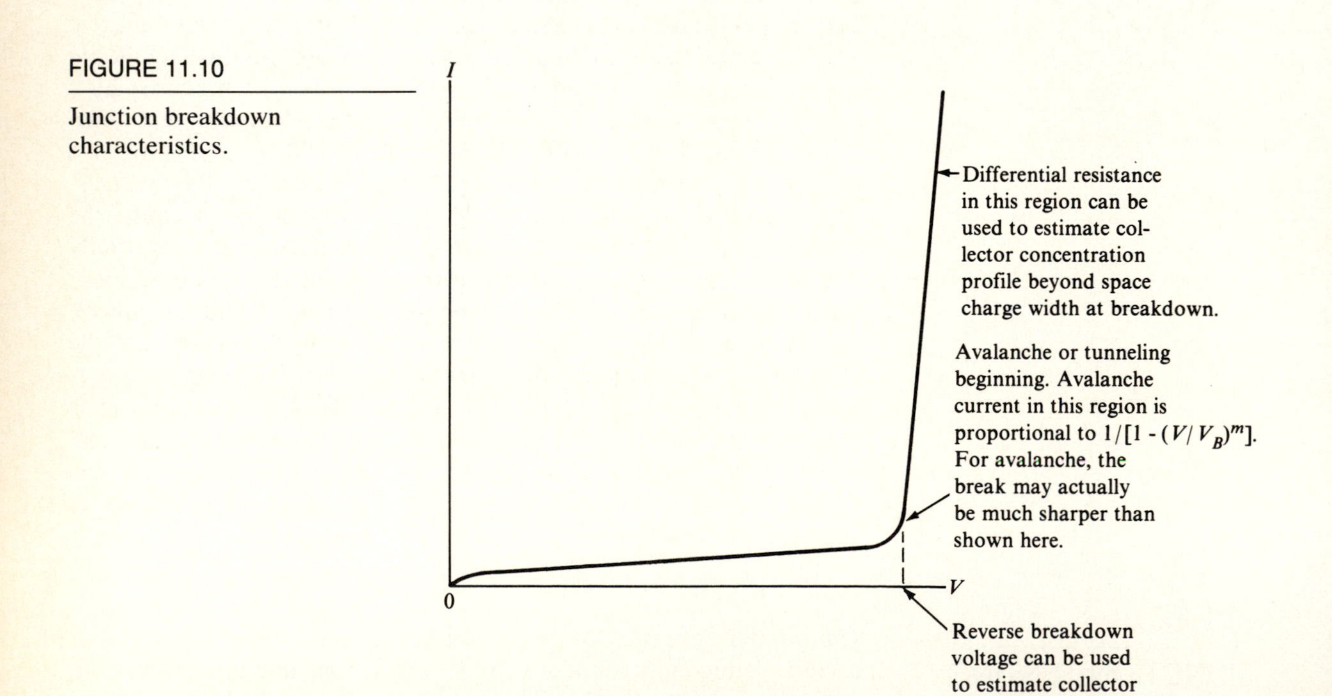
I
V
0
Differential resistance in this region can be used to estimate collector concentration profile beyond space charge width at breakdown.
Avalanche or tunneling beginning. Avalanche current in this region is proportional to $1/[1 - (V/V_B)^m]$. For avalanche, the break may actually be much sharper than shown here.
Reverse breakdown voltage can be used to estimate collector concentration.

FIGURE 11.10

Junction breakdown characteristics.

The reverse current of an ideal diode is diffusion current, given by

$$I_d = \left(\frac{n_i^2}{N_A}\right) qA \sqrt{\frac{D_n}{\tau_n}} \qquad \text{for n}^+\text{p-diodes} \qquad 11.30\text{a}$$

$$I_d = \left(\frac{n_i^2}{N_D}\right) qA \sqrt{\frac{D_p}{\tau_p}} \qquad \text{for p}^+\text{n-diodes} \qquad 11.30\text{b}$$

where q is the electronic charge, k is Boltzmann's constant, A is the area of the junction, n_i is the intrinsic carrier concentration, $D_{p,n}$ is the carrier diffusion constant, $\tau_{p,n}$ is the carrier lifetime outside the space charge region, and $N_{A,D}$ is the doping level of the lightly doped side of the diode. This current is independent of voltage. That is,

$$I_d :: V^0 \qquad 11.31$$

It is exceedingly small and seldom seen at room temperature, being overshadowed by the generation current, which is described next. The temperature dependence of the diffusion current is that of n_i^2 and is given by

$$I_d(T) = BT^3 e^{-1.21/kT} \qquad \text{for Si} \qquad 11.32\text{a}$$

$$I_d(T) = CT^3 e^{-1.52/kT} \qquad \text{for GaAs} \qquad 11.32\text{b}$$

where B and C are constants and T is the temperature in K.

When generation–recombination centers are present in the space charge region, there will be a generation current component, given by

$$I_g = \frac{qn_iwA}{\tau_g} \qquad 11.33$$

where w is the width of the space charge and τ_g is the carrier generation lifetime in it. The lifetime τ_g depends on the number of recombination centers in the space charge region and in some cases is deliberately reduced by the intentional addition of lifetime killers such as Au or Pt. When this situation occurs, the reverse current will go up commensurately. If I_g is excessive and there is no intentional lifetime reduction, then gettering may be needed (see Chapter 8). Since the lifetime is relatively insensitive to temperature, I_g increases as n_i and thus not nearly as fast as I_d. Even though generation current almost always dominates the reverse current of silicon diodes at room temperature, diffusion current will be dominant at the high end of the temperature operating range. The smaller the bandgap, the more likely diffusion current will be the most important term, and, in the case of germanium, generation current is seldom seen. A GaAs diode would need to be above perhaps 200°C before I_d becomes an appreciable part of the total current.

Eq. 11.33 does not directly indicate a dependence of I_g on the applied voltage V_a, but it does show that I_g increases as the volume wA of the space charge region increases. A remains constant, but w will generally increase with V_a. For the specific cases of abrupt and linearly graded junctions,

$$w = \left[\frac{2\varepsilon\varepsilon_0(V_a + V_b)}{qN_{A,D}}\right]^{1/2} \quad \text{for abrupt junctions} \qquad 11.34$$

$$w = \left[\frac{3\varepsilon\varepsilon_0(V_a + V_b)}{qa}\right]^{1/3} \quad \text{for graded junctions} \qquad 11.35$$

where ε is the relative dielectric constant, ε_0 is the permittivity of free space, V_b is the built-in voltage, and a is the impurity gradient on the lightly doped side. Since V_b is less than a volt, and V_a is generally several volts,

$$I_g :: (V_a)^{1/2} \quad \text{or} \quad (V_a)^{1/3} \qquad 11.36$$

In the event of a resistive shunt R_s across the junction, the current component I_s will be given by

$$I_s = \left(\frac{1}{R_s}\right) V_a^1 \qquad 11.37$$

This problem is external to the junction and could, for example, be due to an extraneous diffusion or extra circuit components not detached from the junction being examined. A thorough microscopic examination of the circuit should locate the shunt.

If metallic precipitates or other defects that can cause tunneling are present in the space charge region, the current component I_t will be present. I_t will be a much stronger function of applied voltage than any of the others, increasing as V_a^n where n can be in the 5–8 range (18):

$$I_t :: V_a^n \qquad 11.38$$

For such precipitates to occur, a combination of a high concentration of elements such as Ni, Cu, or Fe must be present, along with crystallographic defects and a slow cooldown.

These various currents are depicted in Figs. 11.11 and 11.12 and are plotted on both linear scales and log–log scales. Linear scales

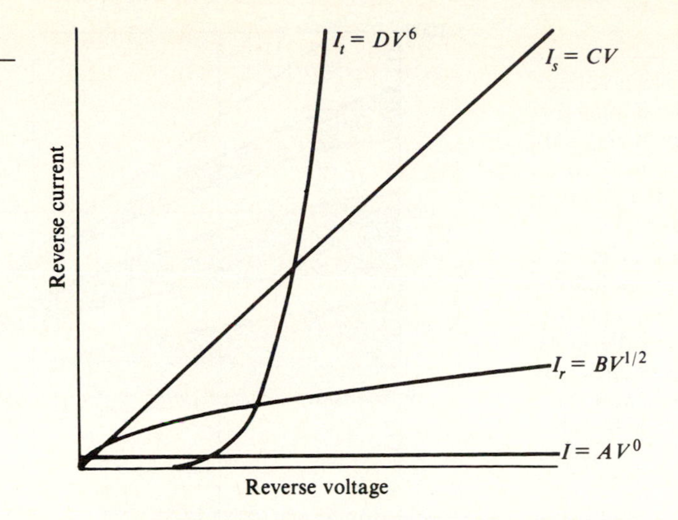

FIGURE 11.11

Curve tracer *I–V* plots of reverse pn junction current showing behavior of different current components (linear scales).

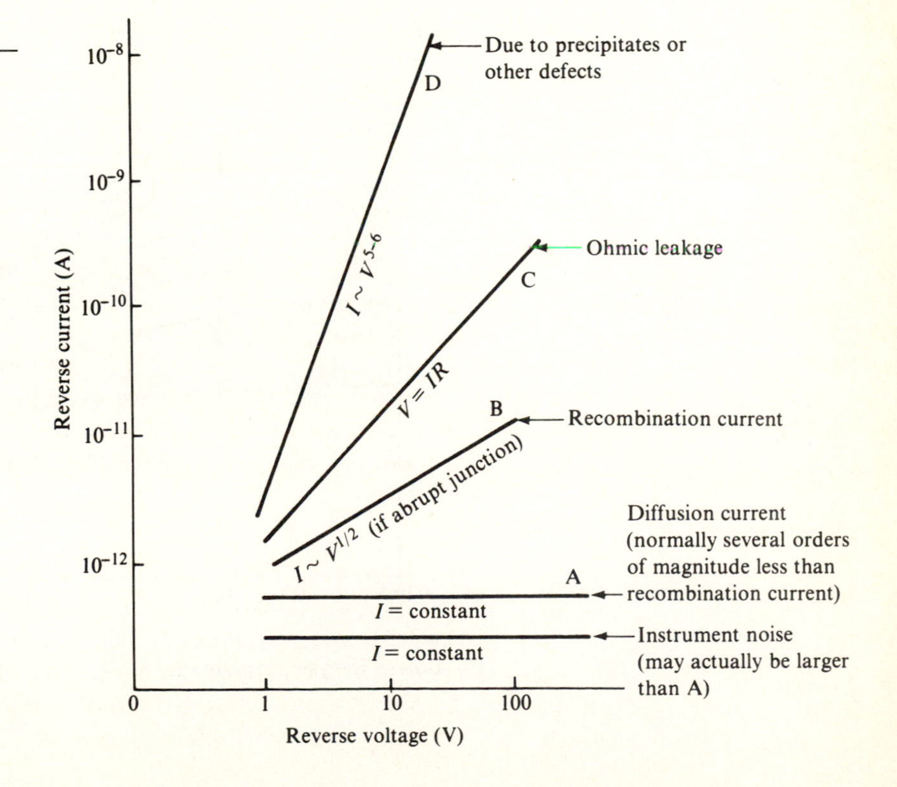

FIGURE 11.12

Functional relation between current and voltage for several modes of current generation in reverse-biased pn junction diodes.

FIGURE 11.13

Calculated breakdown voltage versus background doping for an abrupt junction having various radii of curvature. (The upper line is for a flat junction. The upper bound of each shaded area is for a cylindrical junction with radius as indicated. The lower bound is for a spherical junction of the same radius.) (*Source:* Reprinted with permission from *Solid State Electronics 9,* S.M. Sze and G. Gibbons, Copyright 1966, Pergamon Press plc.)

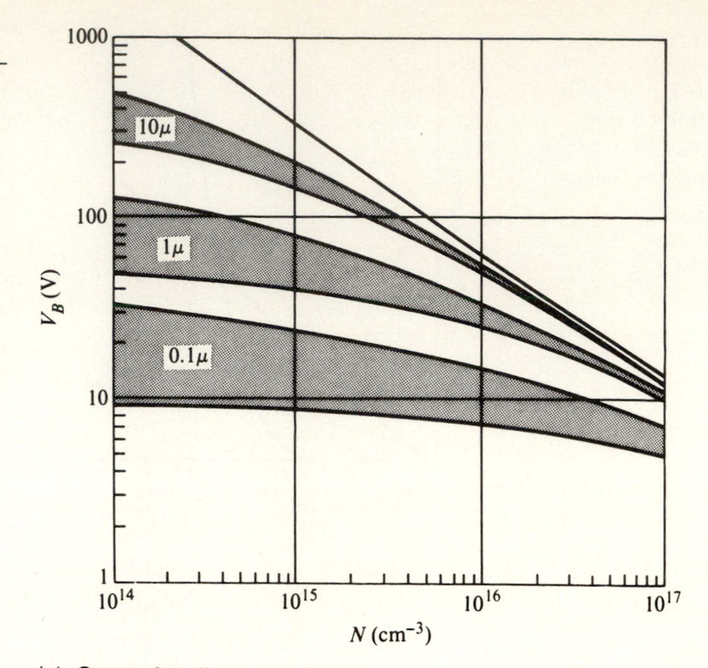

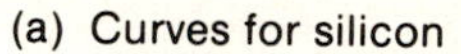

(a) Curves for silicon

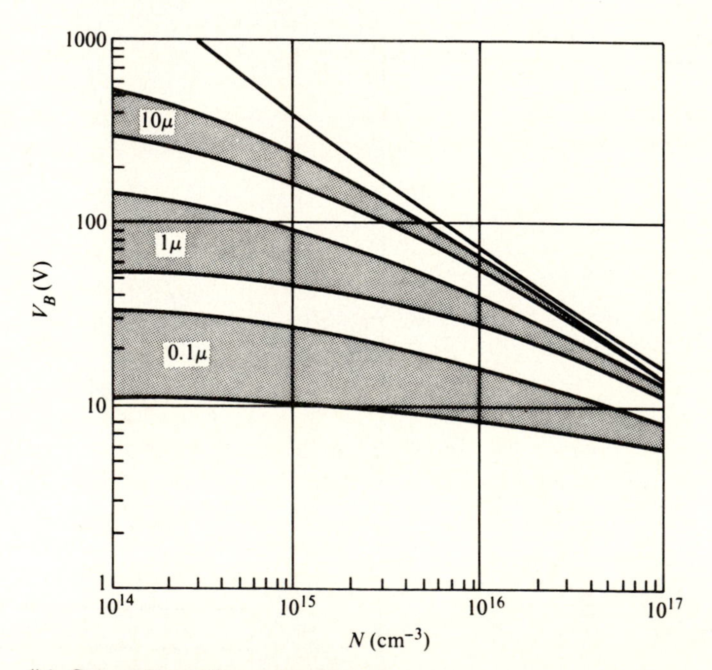

(b) Curves for gallium arsenide

are used on curve tracers and are appropriate for a quick overview of potential problems, but log–log scales are better for analytical interpretation. While the total current is the sum of all the components, if there is a problem of excess current, one will usually dominate. Then, since there is a different voltage dependency on each, the offending mechanism can generally be determined.

Voltage breakdown occurs when the voltage across the space charge region becomes large enough for carrier multiplication to occur (avalanche breakdown) or for tunneling to begin. The latter occurs only when both sides of the junction are heavily doped, which in ICs would only be because of a processing difficulty. If the junction is abrupt and one side is heavily doped, the breakdown voltage V_B depends on the impurity concentration of the lightly doped side as shown in each top curve of Fig. 11.13. If the space charge region moves out enough to touch a low-resistivity region (or an ohmic contact), breakdown will occur at that point, and V_B will depend on the width of the lightly doped side. If the junction is graded instead of abrupt, breakdown will depend on the gradient a and will be greater than for an abrupt junction. When a junction is physically curved, as, for example, at the periphery of a planar junction, the electric field increases as the radius of curvature decreases so that V_B depends on the curvature. This effect for varying degrees of curvature is also shown in Fig. 11.13. As junctions become shallower and their area smaller, curvature becomes more pronounced, and V_B for a given doping level decreases. It can also be seen from Fig. 11.13 that as the radius of curvature decreases, the V_B dependence on doping decreases.

11.8.2 Reverse Current Curve Tracer Diagnostics

Fig. 11.14 shows examples of the different kinds of curves that may occur when reverse-biased pn junctions (diodes) are examined with a curve tracer. Fig. 11.14a shows a good diode, assuming that the current gain is properly set. It also has the "textbook" shape. The other curves in Fig. 11.14 are all of good diodes. The loop in Fig. 11.14b is due to capacitance in the leads or junction and may be at least partially balanced out in most tracers. Stray pickup can cause the trace to be quite distorted. Fig. 11.14c shows the trace from drain to gate of a junction FET with the source floating. Pickup caused the loop, which occurs below 10 V. Above 10 V, the channel pinched off, and there was no longer any pickup. The trace in Fig. 11.14d is caused by charging of the emitter–base capacitance of a bipolar transistor (this curve was obtained by measuring between collector and emitter). These distorted traces of good junctions illustrate the point that care must be taken in interpreting curve tracer data.

FIGURE 11.14

Various I–V plots of good reverse-biased pn junctions as they appear on a curve tracer.

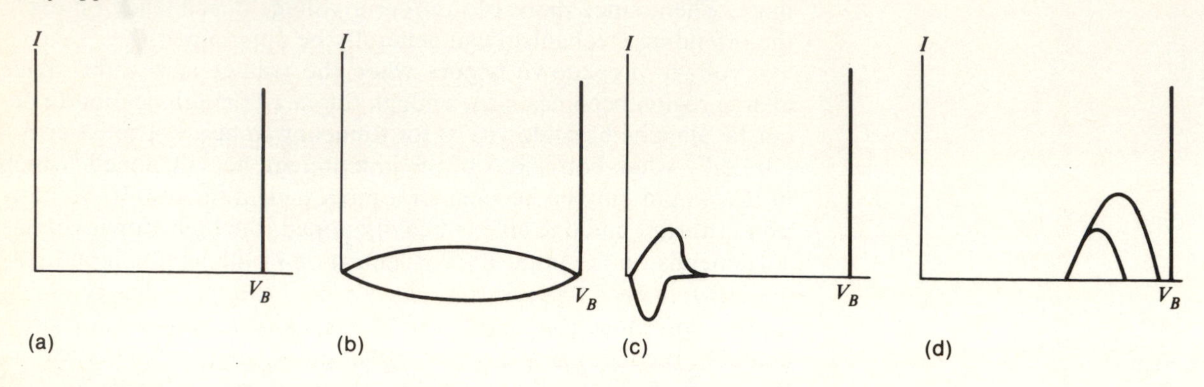

FIGURE 11.15

I–V curve tracer plots of reverse-biased pn junctions in parallel with various kinds of resistive paths.

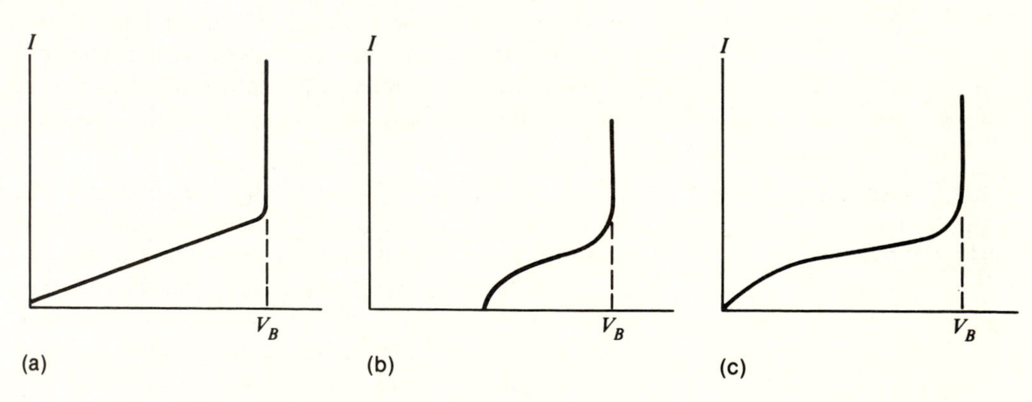

Fig. 11.15a has excessive leakage current that varies linearly with voltage and hence is due to a resistive shunt that does not pinch off.[11] Fig. 11.15b is characteristic of a low breakdown region connected to a higher breakdown region with a path that can pinch off.

[11]Pinchoff is the narrowing of the resistive path with increasing voltage and, as in a FET, occurs because of a space charge region moving into the path.

The most common example is a diffusion pipe connecting the collector and emitter of a bipolar transistor. This behavior is observed in looking at the collector–base junction. Fig. 11.15c has excessive current that appears to start at a zero voltage. Such behavior can be due either to generation current as shown in curve B of Fig. 11.12 or to a resistive path pinching off. Superficially, they appear to have much the same shape, but recombination current should continue to rise until breakdown with an approximate I–$\sim V^{0.5}$ dependency, whereas resistive pinchoff current will eventually saturate. Breakdown may occur before saturation, however. Larger-than-expected recombination current can occur because of excessive lifetime killers or because a surface inversion layer has extended the diode area.

Various inversion layer possibilities are shown in Fig. 11.16. If the inversion terminates as in Fig. 11.16a, excess current commensurate with the extra area will result. Depending on the value of the surface recombination velocity, the generation current in thin inversion layers can be substantial (19). Should the inversion layer extend

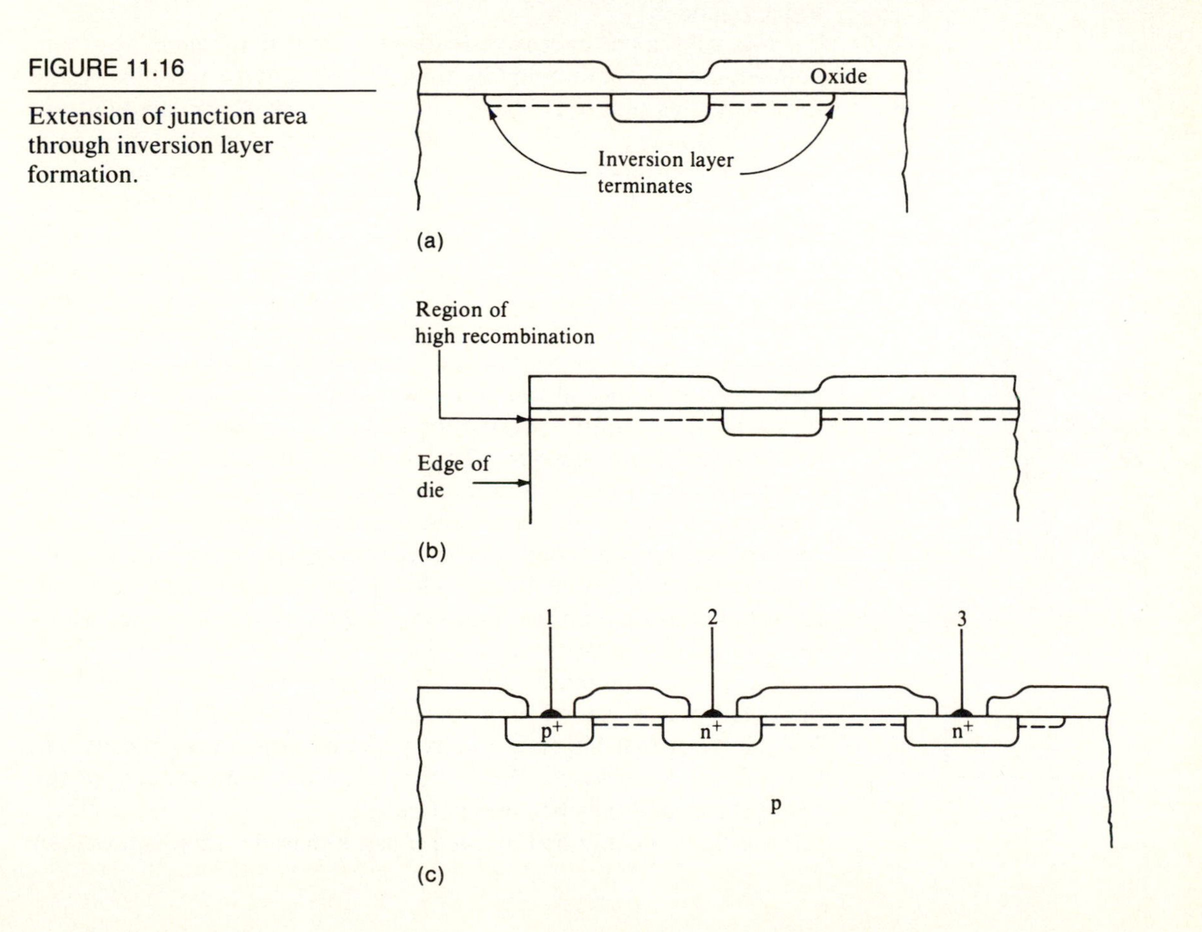

FIGURE 11.16

Extension of junction area through inversion layer formation.

FIGURE 11.17

I–V curve tracer plots of leaky reverse-biased pn junctions.

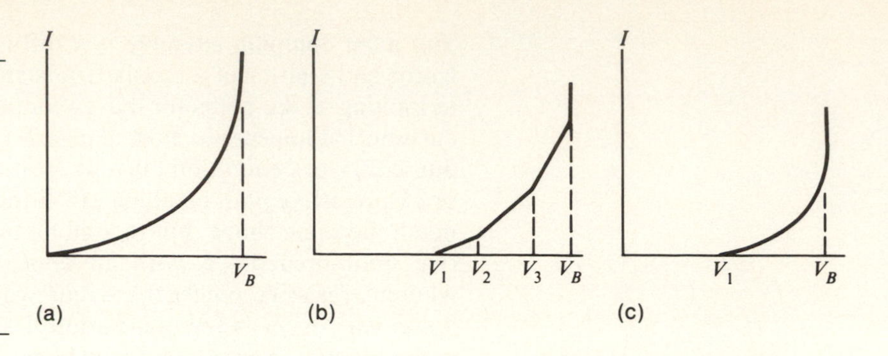

FIGURE 11.18

Miscellaneous *I–V* curve tracer plots of reverse-biased junctions.

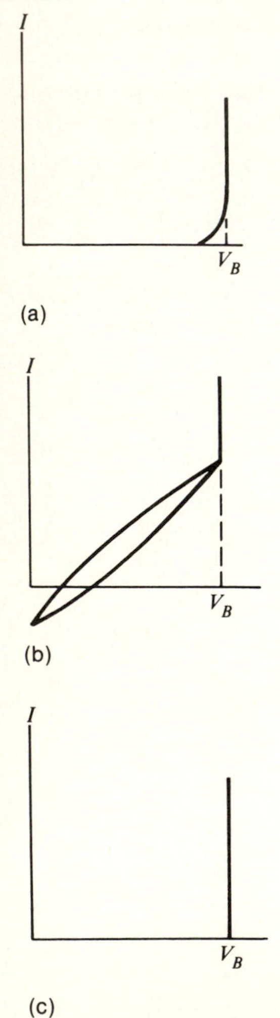

to a region of high recombination density such as the edge of a chip (Fig. 11.16b), orders of magnitude higher current can flow (20). If an inversion layer connects two n^+-regions as in Fig. 11.16c, a resistive path that will pinch off will be placed in parallel with the two back-to-back diodes.

Fig. 11.17a has excessive leakage current that begins near zero voltage, increases more rapidly than V, and hence is most likely due to precipitates (curve D of Fig 11.12). Because of the rapid current rise with voltage, such a curve could under some circumstances look like a forward diode ($I = Ae^{qV/kT}$). A log–log plot of I versus V over a few decades of current will allow separation. Fig. 11.17b is a case of premature breakdown. An initial low breakdown occurs in a small area of the junction at voltage V_1. Following it is a series of regions breaking down in succession at progressively higher voltages. When closely spaced, they give the illusion of the smooth curve of Fig. 11.17c. Possible causes are localized surface accumulation layers, localized high-concentration diffusion sources, or high fields due to high junction curvature at diffusion spikes. It is often possible to see light emitted from these small avalanching regions. The color will appear reddish if the light originates below the surface and white if it originates on the surface. If the voltage is held just above where the first breakdown shows on the curve tracer, one spot will usually be seen. Then, as the voltage is slowly raised, additional spots will light up in unison with the additional segments of current increases seen on the curve tracer. Since the spots are small and the light is dim, a microscope must be used either covered with a black cloth or placed in a darkened room. Another possibility for the curve of Fig. 11.17c is that the curve really looks like the one of Fig. 11.17a but only at point V_1 did the current exceed the noise level of the curve tracer and thus become noticeable.

Fig. 11.18a could be the trace of a good diode. If V_B is less than about 5 V, the trace could be that of a zener diode, which has a much softer breakdown (more gentle rounding of the knee of the curve) than avalanche breakdown. The trace might also be that of a Schottky diode. Schottky diode reverse current near breakdown appears excessive compared to that of a pn junction, and the break is not quite as sharp as expected. A log–log current voltage plot for Schottky diodes looks much different from that of a pn junction and is discussed in the next section. The trace in Fig. 11.18b is that of an otherwise good junction that has polarizable surface ions such as might be present in a wet package. While the trace in Fig. 11.18c looks like that of Fig. 11.18a, the breakdown, while sharp, is not at the correct voltage V_B. If the breakdown voltage is high, then almost certainly a material of higher-than-planned resistivity was used. If it is low, it could be because of a material with lower-than-planned resistivity; because the space charge reached through to a low-resistivity region and ceased widening before breakdown was reached; because excessive junction curvature produced high fields and premature breakdown; or because a surface accumulation layer[12] caused a narrowing of the space charge region at the surface as shown in Fig. 11.19. It is also possible that the final V_B is correct but that the low breakdown as observed on the curve tracer is because the junction is leaky and the current gain is set very high. In this case, a behavior like that shown in Fig. 11.20a will appear as in Fig. 11.20b.

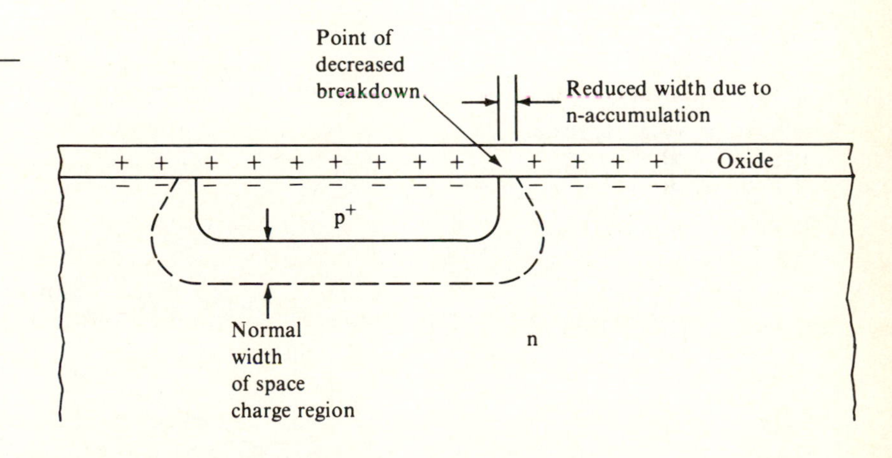

FIGURE 11.19

Method by which accumulation reduces breakdown voltage. (If p-type accumulation occurred rather than n, this effect would be observed in n^+p diodes.)

[12]For a discussion of the possible causes of accumulation and inversion layers, see Chapter 3.

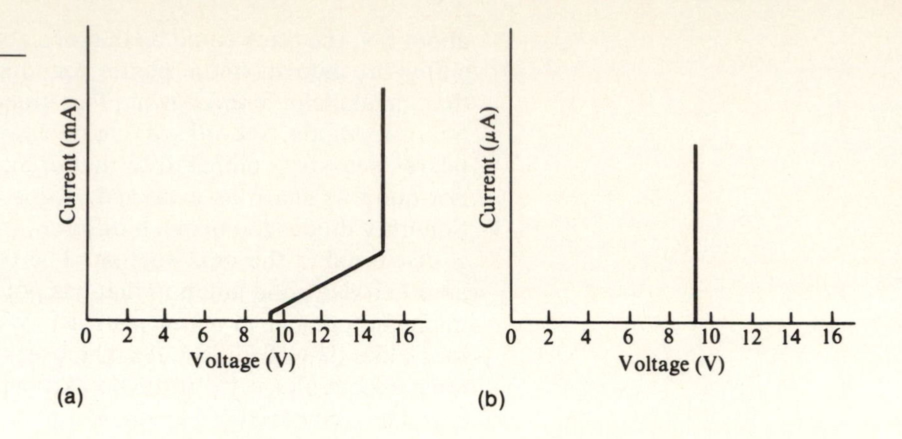

FIGURE 11.20

Effect of scale change on appearance of breakdown characteristics.

11.8.3 Reverse-Biased Schottky Junctions

The reverse current of an ideal Schottky diode is given by

$$I_s = AT^2A^{**}e^{-q\phi_B/kT} \qquad 11.39$$

where A is the diode area, A^{**} is the effective Richardson constant, and ϕ_B is the potential barrier. Since ϕ_B is voltage dependent, unlike a pn junction with only diffusion current, a theoretical Schottky diode reverse current never saturates but increases steadily with voltage as shown in Fig. 11.21. In addition to this current, there can also be (as in pn junctions) recombination current and tunnel cur-

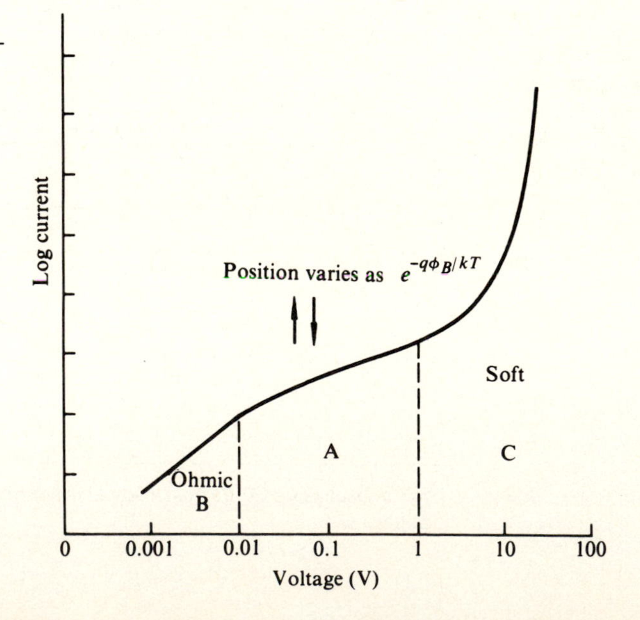

FIGURE 11.21

Schottky diode reverse behavior plotted on a log–log scale.

rents associated with the space charge region. Because of the high field associated with the very sharp curvature at the edge of the metal, additional potential barrier reduction occurs at the edge. Thus, Schottky diodes usually have either pn junction guard rings or field plates. Excess reverse current or low breakdowns can then be due to either pn junction or Schottky problems. The most likely cause is a lowering of the potential barrier due to a Schottky metal contamination. In this case, the current will be higher than expected, but the shape of the I–V curve should still be like that of Fig. 11.21. If a pn junction guard ring is suspected, the Schottky metal can be stripped away and the remaining current examined. To fully determine the various excess current sources, it may be necessary to use a special test structure with an isolated guard ring that can be independently biased. However, such a structure usually requires deviations from the standard process flow and is thus of limited usefulness.

11.8.4 Forward-Biased pn Junctions

Like reverse current, forward current I_f of a diode has several components whose relative magnitudes may change with design, processing deviations, and stress. For a nondegenerate pn junction, the components are the diffusion current I_{fd}, composed of current diffusing from n to p and from p to n; bulk recombination current I_{fr}, which arises because some of the injected minority carriers recombine in the space charge region; surface recombination current I_{fs}, from carriers that recombine at the surface; and a shunt component current I_{fsh}, which arises if a shunting resistor is present. Schottky diode forward current was discussed in Chapter 10. Note here that because Schottky diodes are primarily majority carrier devices, the only current component that they might have in common with pn junction diodes is I_{fsh}.

I_f is given by

$$I_f = I_{fd} + I_{fr} + I_{fs} + I_{fsh} \qquad 11.40$$

The forward diffusion current has the form

$$I_{fd} = I_D\,(e^{qV/kT} - 1) \qquad 11.41$$

where I_D is the sum of the two I_d values of Eq. 11.30. When V is greater than a few kT/q (0.026 V at room temperature), Eq. 11.41 becomes

$$I_{fd} = I_d e^{qV/kT} \qquad 11.42$$

The forward current due to recombination in the space charge region is given by

$$I_{fr} = I_r e^{qV/nkT} \qquad 11.43$$

where

$$I_r = \frac{qn_i wA}{\tau_g} \qquad 11.44$$

Eq. 11.44 is the same expression as that for generation current given in Eq. 11.33. n is an exponent that varies with the source of centers, but for space charge recombination it is usually about 2. If it originates in an inversion layer or channel,[13] it will be greater than 2 and normally less than 4 (21).

The surface recombination current has the form

$$I_{fs} = I_s e^{qV/2kT} \qquad 11.45$$

where I_s depends, not on the lifetime in the bulk, but on the surface recombination velocity s_0. That is,

$$I_s = \frac{n_i s_0}{2} \qquad 11.46$$

If a shunt resistance R_{sh} is present, then

$$I_{sh} = \frac{V}{R_{sh}} \qquad 11.47$$

These various currents are shown pictorially in Fig. 11.22. If a series resistance R_s is present, then the voltage across the junction is less than the applied voltage V_a by an amount $I_f R_s$. When the

FIGURE 11.22

Components of current in a forward-biased pn junction.

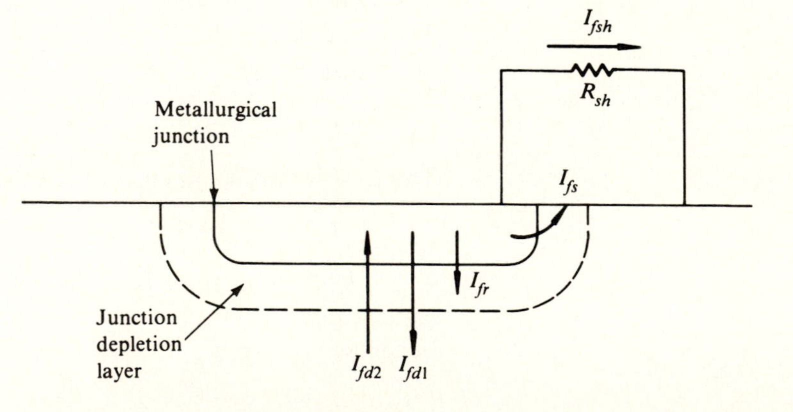

[13]In this context, channeling occurs when the surface potential on each side of the junction is the same. Channeling can also refer to an inversion layer connecting two otherwise isolated regions of the same type.

current is high enough for the injected carrier concentration to be comparable to the doping concentration, the form of the diffusion current of Eq. 11.41 changes to

$$I_{fd} = I_{d'} e^{qV/2kT} \qquad 11.48$$

and thus might be mistaken for recombination current. Note that $I_{d'}$ does not equal I_d of Eq. 11.30 and that V is the voltage across the junction and not V_a across the terminals. This is an important consideration since the effect only occurs at very high currents, and that is when a series resistance voltage drop is most noticeable. The onset of the $qV/2kT$ regime begins approximately when (22)

$$V = \left(\frac{2kT}{q}\right) \ln\left(\frac{2N}{n_i}\right) \qquad 11.49$$

where N is the doping concentration of the lightly doped side of the junction. Although reverse current I–V curves are most conveniently plotted on a log–log scale, forward characteristics are better displayed on log-I–linear-V coordinates. Fig. 11.23 shows a com-

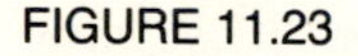

FIGURE 11.23

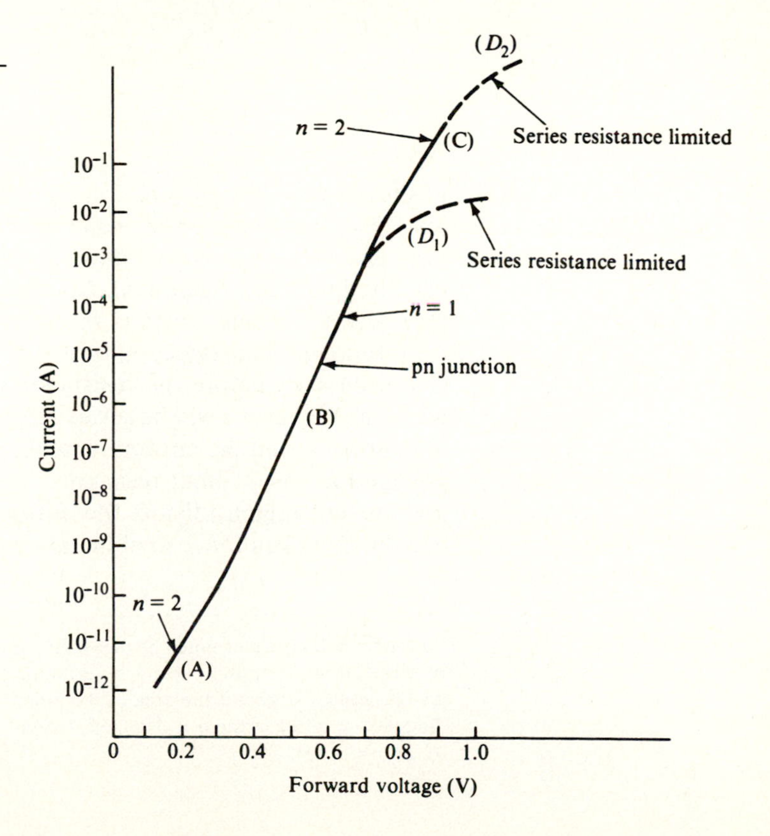

Forward pn junction I–V curve showing various components. (Values are typical for silicon.)

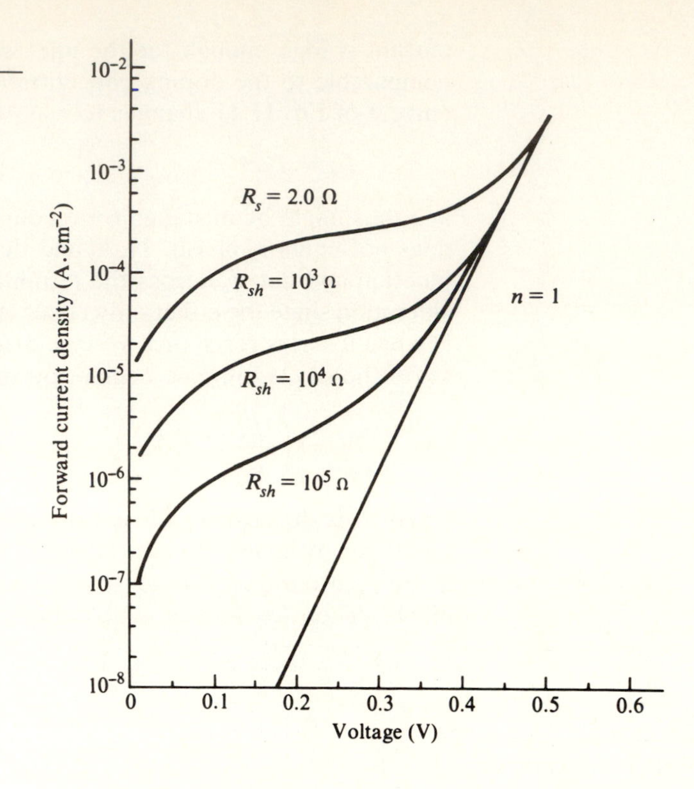

FIGURE 11.24

Calculated effect of shunt resistance on forward pn junction characteristics. (*Source:* Adapted from Richard J. Stirn, National Workshop on Low Cost Polycrystalline Silicon Solar Cells, sponsored by the National Science Foundation and the Energy Research and Development Administration, Southern Methodist University, Dallas, 1976.)

posite curve for a pn junction exhibiting most of the components just described and the regions in which they are ordinarily seen. At very low voltages, recombination current with a $2kT/q$ dependence usually dominates (region A). For higher voltages, diffusion current ($1kT/q$) will become larger as shown in region B. If series resistance is low enough, high injection level diffusion current ($2kT/q$) of region C will be seen before the resistance limit of D_2 is reached. Otherwise, the I–V curve will be series resistance limited sooner as shown in region D_1, and the current increase is directly proportional to the voltage increase. Shunt resistance will affect the I–V curve in the low current region, distort the curve as shown in Fig. 11.24, and may be mistaken for recombination current with a high n value.[14]

[14] n can be quickly determined by measuring the voltage differential ΔV in millivolts required to increase the current by a factor of 10. Under these conditions, $n = \Delta V/60$. Note also that if the area of the junction increases by a factor of 2 and the diode is operating in the $n = 1$ region, the voltage for a fixed current will decrease by 18 mV.

Not listed in Eq. 11.40 is tunnel current, which might be present if both sides of the junction were to be very heavily doped. It appears as an excess current superimposed on the normal current, peaks at about 100 mV for silicon, and will have virtually disappeared by the time V reaches 300 mV. It is also possible for adjacent junction interactions to cause enhanced forward current.

11.9 MOS TRANSISTOR TESTING

Fig. 11.25 shows cross sections, circuit diagrams, polarities, and terminology for MOS transistors. Both enhancement-mode and depletion-mode transistors are made, although the former are more common. The testing is somewhat different for the two types, but in either case, much of it can be done with a curve tracer. Three small-signal transistor parameters often measured are as follows:

1. Channel transconductance g_m

$$\left.\frac{\partial I_{DS}}{\partial V_{GS}}\right|_{V_{DS}=\text{constant}}$$

Usually, V_{DS} is made large enough for I_{DS} to be saturated, in which case $g_m = g_{m(\text{sat})}$.

2. Channel (drain) conductance g_D

$$\left.\frac{\partial I_{DS}}{\partial V_{DS}}\right|_{V_{GS}=\text{constant}}$$

3. Gain μ

$$\left.\frac{\partial V_{DS}}{\partial V_{GS}}\right|_{I_{DS}=\text{constant}}$$

11.9.1 Threshold Voltage

The most commonly checked parameter in diagnosing problems with enhancement[15] transistors is the threshold voltage V_T. The threshold voltage is the gate voltage required to just form the conductive channel between source and drain. V_T can be determined with the transistor operating either in the unsaturated mode or in the

[15]For zero gate voltage, there is no channel and, except for reverse current leakage, no current flow from source to drain of an enhancement transistor. Depletion transistors have a channel and current flow at zero gate voltage. That current can be either increased or decreased with a change of polarity of the gate voltage.

FIGURE 11.25

MOS transistor structure, polarities, and terminology.

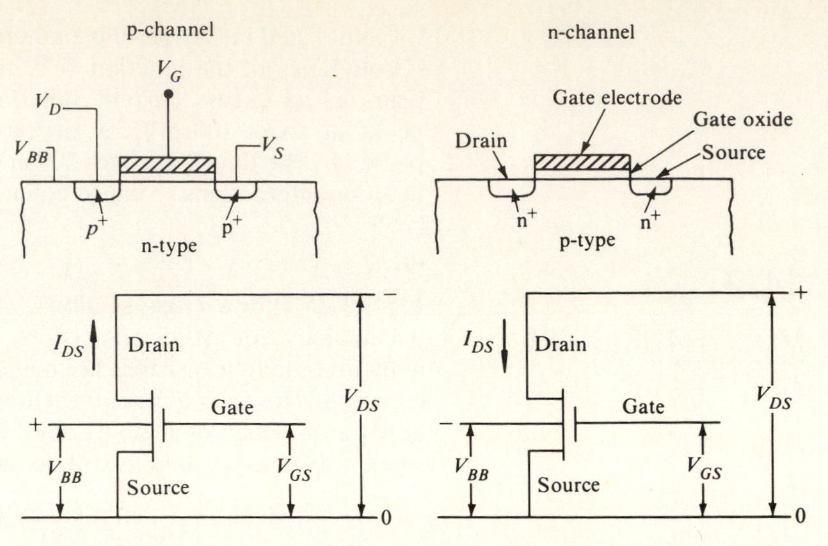

(a) Polarity: For depletion mode, V_{GS} and V_{DS} will be of opposite polarity. For enhancement mode, V_{GS} and V_{DS} will have the same polarity.

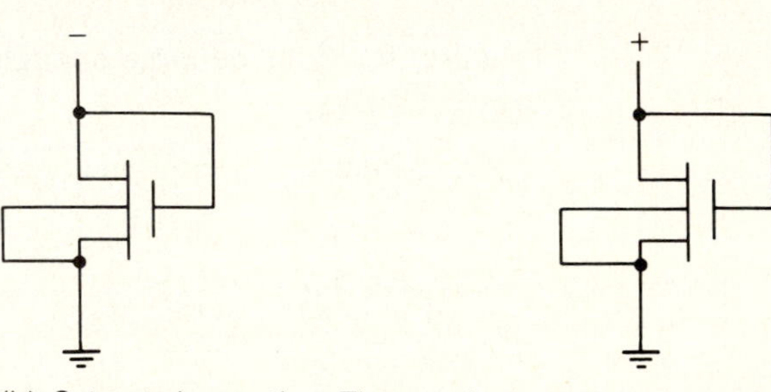

(b) Saturated operation: The gate is shorted to the drain, and the substrate is shorted to the source

Symbol	Meaning	Symbol	Meaning
V_D	Drain voltage	V_{GS}	Gate–source voltage
V_S	Source voltage	V_{BB}	Substrate back bias voltage (substrate to source)
V_G	Gate voltage		
V_{DS}	Drain–source voltage	I_{DS}	Drain–source current

(c) Terminology

saturated mode.[16] The saturation-mode method is somewhat simpler and is more often used. For unsaturation, when $V_{GS} - V_T >> V_{DS}$,

$$I_{DS} = k\left[V_{DS}(V_{GS} - V_T) - \frac{V_{DS}^2}{2}\right] \qquad 11.50$$

where k is the conduction factor. For a small fixed V_{DS}, an extrapolation to $I_{DS}/V_{DS} = 0$ of a plot of I_{DS}/V_{DS} versus V_{GS} will give V_T. For operation in the saturated mode, when $V_{GS} - V_T \leqslant V_{DS}$,

$$I_{DS} = \frac{k}{2}(V_{GS} - V_T)^2 \qquad 11.51$$

A plot of V_{GS} versus $\sqrt{I_{DS}}$ extrapolated to $I_{DS} = 0$ will give V_T. A procedure often used is to pick two points from a curve tracer plot of I_{DS} versus V_{GS} and, by using Eq. 11.51, calculate V_T. For example, if V_{GS1} is the voltage at one I_{DS} and V_{GS2} is the voltage for an I_{DS} of 10 times the first,

$$V_T = 1.46V_{GS1} - 0.46V_{GS2} \qquad 11.52$$

Sometimes, in order to simplify measurement, V_{TX} for a given transistor may be specified. It designates the gate voltage required to cause some preselected small current to flow and thus requires only one measurement. In V_T measurements, the source and substrate must be tied together; otherwise, V_T or V_{TX} will be displaced by approximately $(1/2)\sqrt{V_{BB}}$. Note that to keep from forward biasing the substrate source junction, the substrate must be positive with respect to the source for p-channel transistors and negative for n-channels. The gate and drain can be tied together to ensure saturation operation and also to give a two-terminal device that is easy to connect to a curve tracer. The V_T measured on a long-channel transistor will not be directly applicable to short-channel devices on the same wafer. In general, short-channel V_T's will be less.

When viewed on a curve tracer, a threshold of a few tenths of a volt might be confused with a forward-biased junction, and one above 5–6 V might be mistaken for a reverse-biased junction. To decide whether the trace belongs to a pn junction or a MOS transistor, plot V_{GS} versus $\sqrt{I_{DS}}$ for a range of voltages above where the break appears. If the trace is indeed for a MOS device, the curve will be a straight line. When the threshold voltage is not as expected,

[16]For more details of device behavior in these regions, see, for example, S.M. Sze, *Physics of Semiconductor Devices*, 2d ed., John Wiley & Sons, New York, 1981.

a change in trapped charge is generally the cause. However, a change in the interface charge density, oxide thickness, wafer doping level, or gate electrode work function could also be the trouble. Thresholds are adjusted via ion implantation, and it is those transistors whose V_T's are usually measured. However, it is also very helpful to know what the V_T would have been without adjustment. Therefore, one or more test transistors may have the threshold adjust omitted. Their V_T's are often referred to as "natural" thresholds.

If the MOS transistor to be studied consists of a lead, the field oxide, and two adjacent diffused regions, the V_T is referred to as the "thick-field turn-on." When this voltage is too low, surface inversion and unwanted shorting of components can occur when voltage is applied to the lead. The causes of a low thick-field turn-on are the same as those causing low V_T's of transistors.

11.9.2 Subthreshold Leakage Current

When the gate voltage is less than V_T, I_{DS} does not go to zero. Rather, a small current, the subthreshold leakage current, remains (23). It has the general character shown in Fig. 11.26. The lower current limit is the reverse current of the drain–substrate junction. The slope $\partial(\log I_{DS})\partial V_{GS}$ of the curve below threshold is proportional to the gate oxide thickness and the square root of substrate doping (24).

11.9.3 Pinchoff Voltage

For depletion-type transistors, the pinchoff voltage V_P is equivalent to V_T for enhancement transistors and is also given by Eqs. 11.50 and 11.51. For these transistors, saturation is not obtained by tying gate to drain. The drain voltage must be substantially greater than the gate voltage. I_{D0} is defined as the current at $V_{GS} = 0$. By measuring it and the current I_{D2} at some other gate voltage V_{GS2}, V_P from Eq. 11.51 is

$$V_P = \frac{V_{GS2}}{1 - \sqrt{I_{D2}/I_{D0}}} \qquad 11.53$$

and is somewhat analogous to Eq. 11.52 for enhancement-mode transistors. Alternatively, V_{GS} versus $\sqrt{I_{DS}}$ can be plotted, and the curve can be extrapolated back to $\sqrt{I_{DS}} = 0$. The depletion transistor counterpart to subthreshold current is subpinchoff current (25).

11.9.4 Conduction Factors *k* and *k'*

The conduction factor k is defined through Eq. 11.50 or Eq. 11.51. To remove the effect of channel length and width, k', defined as $k' = (l/w)k$, is sometimes used instead of k. w is the width of the transistor channel, and l is its length. From Eq. 11.51, when the transistor is operating in the saturated mode, a plot of $\sqrt{I_{DS}}$ versus

FIGURE 11.26

Subthreshold current as shown on (a) $\sqrt{I_{DS}}$ versus V_{GS} plot used to determine threshold voltage and (b) log I_{DS} versus log V_{GS} plot. (*Source:* Subthreshold data from W. Milton Gosney, *IEEE Trans. on Electron Dev. ED-19*, p. 213, 1972.)

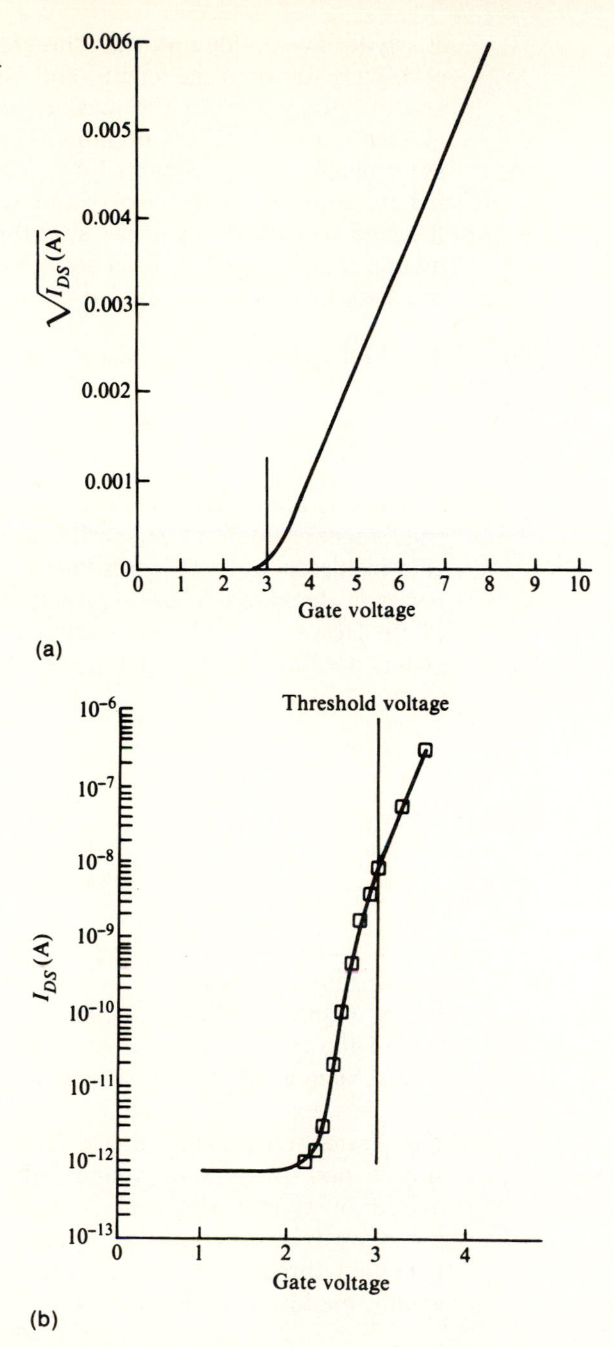

V_{GS} will have a slope of $\sqrt{k}$. The channel width is usually considerably greater than the length, and, as a consequence, the effect of lateral diffusion under the masking oxide on the width can be neglected in making k' calculations. Thus, the width can generally be determined closely enough by a measurement of the mask. The length, however, must be corrected by the amount of diffusion under the gate. For short-channel transistors, the length obtained in this manner may still be in substantial error, and special long-channel transistors are usually used.

11.9.5 Channel Mobility

To a first approximation,

$$k' = \frac{\mu \varepsilon \varepsilon_0}{2t} \qquad 11.54$$

where μ is the carrier mobility in the MOS channel, not the MOS small-signal gain discussed earlier. ε is the dielectric constant of the gate insulator, ε_0 is the permittivity of free space, and t is the gate insulator thickness. Thus, in principle, μ can be calculated from Eq. 11.54. However, for better accuracy, the drain conductance as $V_{DS} \rightarrow 0$ is normally used. For $V_{DS} << (V_{GS} - V_T)$,

$$I_{DS} = \left(\frac{w}{lt}\right)\mu \varepsilon \varepsilon_0 (V_{GS} - V_T) V_{DS} \qquad 11.55$$

Thus,

$$g_D = \frac{\partial I_{DS}}{\partial V_{DS}} = \left(\frac{w}{lt}\right)\mu \varepsilon \varepsilon_0 (V_{GS} - V_T) \qquad 11.56$$

Often, g_D near V_{DS} for several values of V_{GS} will be determined in order to obtain a better value. It should be remembered that μ is a tensor quantity and even in cubic crystals will depend on the direction of the current flow when the current is restricted to very thin layers, such as MOS inversion layers (26).

11.9.6 Drain Breakdown Voltage

The drain breakdown voltage BV_{DSS} is measured by tying gate, source, and substrate to ground and measuring the drain voltage required for some small preselected current. The limiting factor in this breakdown is generally the space charge region from the drain punching through to the source. Thus, the shorter the channel length, the lower the BV_{DSS}.

11.10 BIPOLAR TRANSISTOR DIAGNOSTICS

The bipolar transistor DC parameters that may require checking can be divided into the four categories of current gain, breakdown voltages, reverse bias currents, and resistance-dependent properties such as saturation voltage (the last of which is not discussed here).

11.10.1 Current Gain

Since the common emitter transistor configuration is most commonly used in bipolar ICs, h_{fe} is the current gain of most interest. Table 11.5 lists and defines it and several other gain parameters of interest. h_{fe} and α can be measured in a straightforward manner on a curve tracer. Fig. 11.27 shows typical traces when the curve tracer is connected to show common emitter and common base characteristics. In each case, the horizontal axis represents collector–emitter voltage, and the vertical axis represents collector current. In Fig. 11.27a, the family of curves comes from a discrete set of base currents (2, 4, 6, 8, and 10 μA). In Fig. 11.27b, the emitter current is increased in steps of 500 μA, and since α is very close to unity for this transistor, I_C very closely follows the input emitter current.

When h_{fe} is too high, a narrower-than-expected base is the probable problem. When it is too low over the whole current range, the problem is usually a base that has become too wide, thus reducing the base transport efficiency β. However, if the emitter doping level were to be substantially reduced, the emitter efficiency γ would decrease because of a decrease in the ratio of minority to majority carriers injected into the base. When the problem is only a reduction

TABLE 11.5

Current Gain Parameters

Symbol	*Meaning*	*Definition*
h_{fe}	Small-signal current gain (common emitter configuration)	$\partial I_C/\partial I_B$ where I_C = collector current and I_B = base current.
Inv. h_{fe}	Inverse small-signal current gain	Measured with collector and emitter terminals interchanged.
h_{FE}	Large-signal current gain (common emitter configuration)	I_C/I_B (see Fig. 11.27a to judge likely value difference from h_{fe}).
α	Small-signal current gain (common base configuration)	$\partial I_C/\partial I_E$ where I_E = emitter current; $h_{fe} = \alpha/(1 - \alpha)$ where $\alpha = \gamma\beta$.
γ	Emitter efficiency	Ratio of minority carrier current crossing into base to total emitter current; $\gamma = \gamma_1\gamma_2$.
γ_1		Ratio of injected emitter minority current to injected majority current.
γ_2		Fraction of injected emitter current lost to recombination in emitter–base space charge region.
β	Base transport efficiency	Fraction of minority current surviving passage across base.
β	Archaic symbol for h_{fe}	

FIGURE 11.27

Typical curve tracer display of (a) common emitter and (b) common base characteristics.

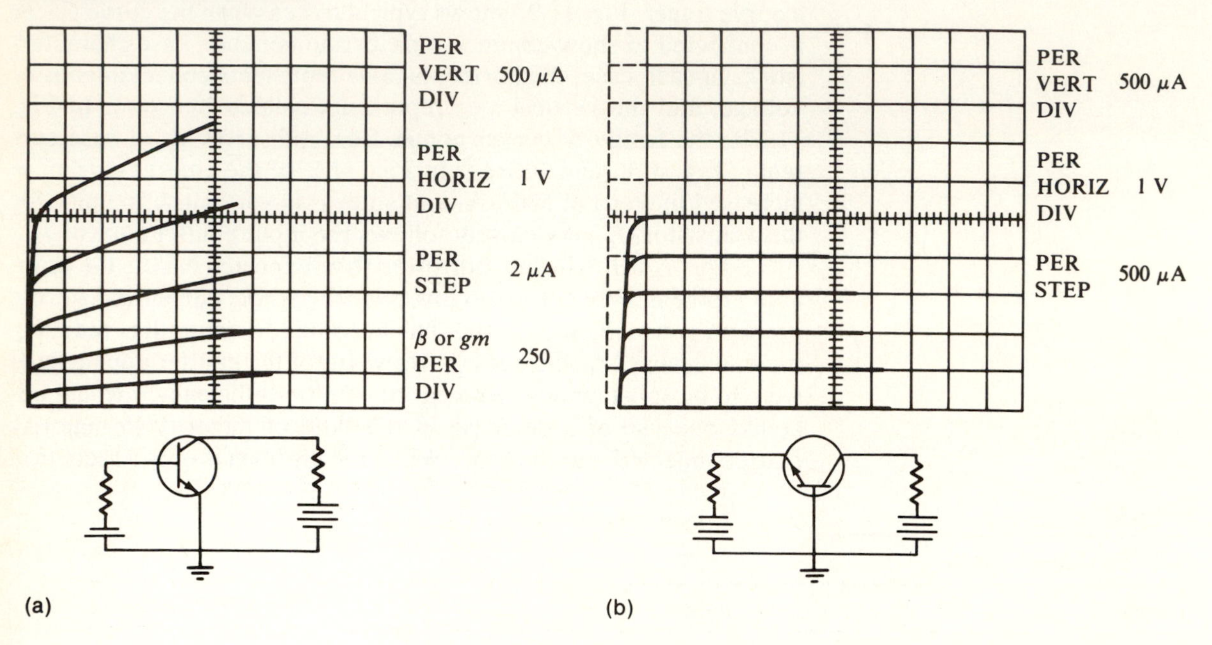

in low-current h_{fe}, then the difficulty is almost certainly emitter efficiency. However, the emitter efficiency reduction in this case is due to excessive carrier recombination in the emitter–base space charge region (bulk or surface).

To see whether β or the emitter efficiency γ is limiting h_{fe}, the common emitter output admittance h_{oe} can be examined. As normally defined,

$$h_{oe} = \left.\frac{\partial I_C}{\partial V_{CE}}\right|_{I_B} \quad 11.57$$

which is just the slope of a curve in Fig. 11.27a. Many curve tracers allow curves similar to those of Fig. 11.27a to be displayed, but with constant emitter–base voltage (V_{BE}) steps. Generally, either set of curves can be chosen by a simple switch change so that the two can be alternately observed on the screen. If the slope of the two curves are very close, then emitter efficiency determines current gain. If base transport is limiting h_{fe}, the slope of the constant I_B curve will be approximately twice that of the constant V_{BE} curve (27).

The emitter–base current is of the form $I_B = I_0 e^{qV_{BE}/nkT}$ where $n = 1$ implies predominately diffusion current and $n = 2$ implies predominately recombination current (see section 11.8.4) and a much lower h_{fe}. When $n = 1$, $\gamma_2 = 1$ and γ_1 determines γ. When $n = 2$, the reverse is true. The V_{BE} region where each mechanism dominates can be seen from either a log I_B–V_{BE} or a $1/h_{FE}$–log I_C plot. For each region, n can be determined directly from a log I_B–V_{BE} curve or from the slope S of the log $(1/h_{FE})$–log I_C curve through the relation $n = 1/(1 + S)$ (28). Fig. 11.28 shows log I_B–V_{BE} plots for two transistors. Unit 1 has very poor low-current h_{FE}, which is reduced to 1 for base current of only 0.2 μA. Unit 2 is considered normal, and although the curve does not extend that low, it can be seen that base current will be less than 0.001 μA before h_{FE} is reduced to 1. Fig. 11.29 shows a $1/h_{FE}$ plot for unit 1. While not shown on either of the figures, at somewhat higher currents, h_{FE} will start decreasing with increasing current because of high current density effects. Should it be desired to determine whether bulk or surface recombination is dominant, an I_B–T plot may allow a determination. Bulk recombination processes usually have an activation energy of about $E_G/2$, where E_G is the bandgap. Surface recombination tends to be nearly temperature independent (29).

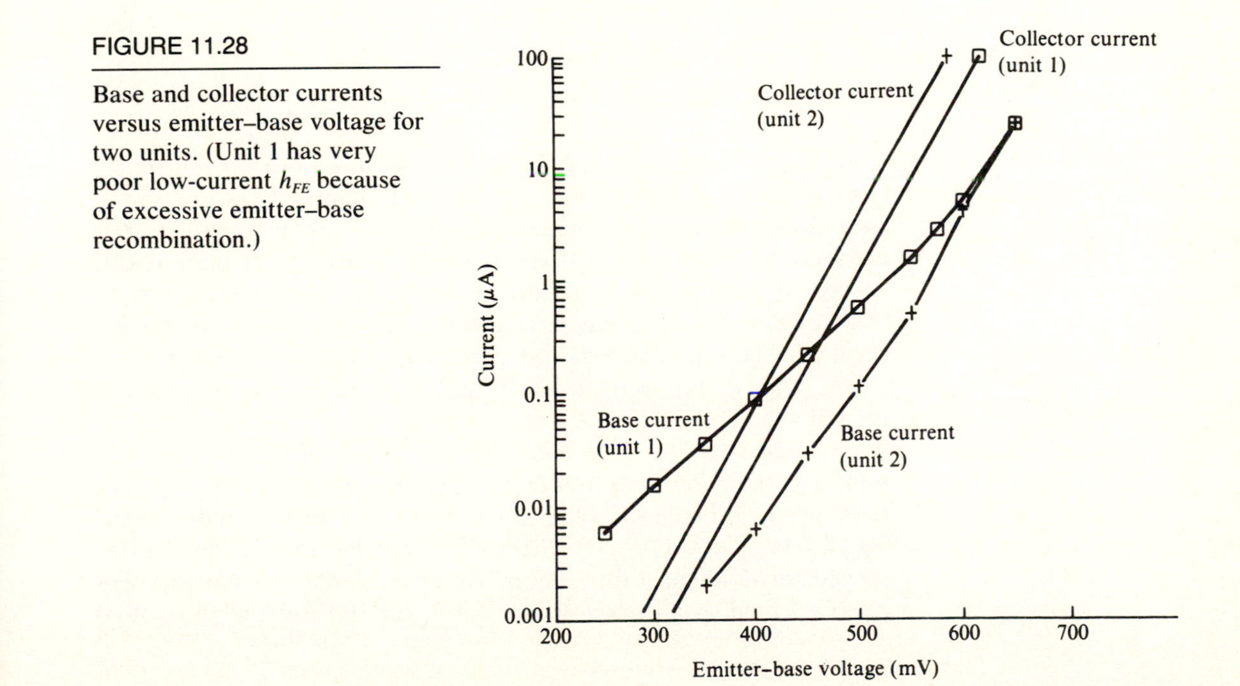

FIGURE 11.28

Base and collector currents versus emitter–base voltage for two units. (Unit 1 has very poor low-current h_{FE} because of excessive emitter–base recombination.)

FIGURE 11.29

Plot of $1/h_{FE}$ versus current showing a region of high emitter–base recombination for unit 1.

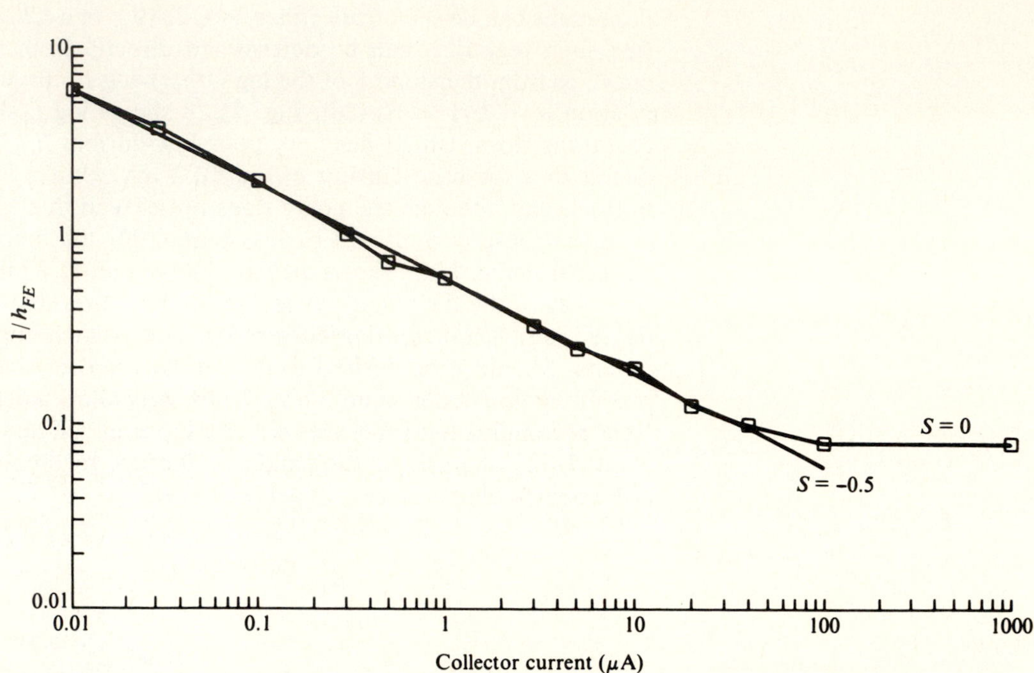

11.10.2 Reverse Breakdown Voltages

When a device with multiple junctions, such as a bipolar transistor, is examined, a variety of interactions can change breakdown voltages. Depending on the manner in which the junctions are electrically connected, several different breakdown voltages may be observed. By studying their relative magnitudes, considerable insight into some varieties of processing problems can be gained. The bipolar transistor breakdown voltages are listed in Table 11.6.

$V_{(BR)CEO}$ is the voltage most different from isolated junction breakdown. The difference arises because when a bipolar transistor is operated with the base floating, avalanche will occur when the multiplication factor (section 11.8.1) approaches $h_{FE}/(1 + h_{FE})$ rather than when it becomes very large. Simple theory[17] predicts that $V_{(BR)CEO}/V_{(BR)CBO} \cong (1/h_{FE})^{1/P}$ where P is a number experimentally observed to be about 4 for silicon. As an example of what this will predict, consider a transistor with an h_{FE} of 100 and a $V_{(BR)CBO}$ of 60

[17]For a discussion of this effect, see any standard text on semiconductor devices.

TABLE 11.6

Summary of Active Bipolar Transistor Device Breakdown Voltages and Their Abbreviations

IEEE Symbol	*Common Symbol*	*Definition*
$V_{(BR)CBO}$	BV_{CBO}	Reverse breakdown voltage, collector to base, with emitter open: This voltage is the same as that which would be observed if there were no emitter–base junction—that is, if $V_{(BR)CBO} = V_{(BR)}$.
$V_{(BR)CEO}$	BV_{CEO}	Reverse breakdown voltage, collector to emitter, with base open: This voltage is ordinarily much less than $V_{(BR)CBO}$ and is dependent on the current gain of the transistor.
$V_{(BR)CES}$	BV_{CES}	Reverse breakdown voltage, collector to emitter, with base shorted to emitter: This voltage lies between $V_{(BR)CEO}$ and $V_{(BR)CBO}$ and should be close to $V_{(BR)CBO}$.
$V_{(BR)CER}$	BV_{CER}	Reverse breakdown voltage, collector to emitter, with a specified resistance between base and emitter: For $R = \infty$, $V_{(BR)CER} = V_{(BR)CEO}$, and for $R = 0$, $V_{(BR)CER} \equiv V_{(BR)CES}$. Thus, $V_{(BR)CER}$ will always lie between $V_{(BR)CES}$ and $V_{(BR)CEO}$.
$V_{(BR)EBO}$	BV_{EBO}	Reverse breakdown voltage, emitter to base, with collector open.

V. Then, $(1/100)^{1/4} = 0.32$, so the collector–emitter breakdown will be just over 20 V. If a resistor is connected from base to emitter, the collector–emitter I–V trace will snap down from $V_{(BR)CBO}$ to $V_{(BR)CEO}$ or some intermediate value, depending on the resistor value, as the current is increased. In the case of silicon, the transistor itself behaves as though it has a built-in resistor. Germanium, however, does not show the effect.

When the observed CEO/CBO ratio is larger than expected, it is generally because the $V_{(BR)CBO}$ value being used is not the true value, but one reduced because of, for example, a collector region that is too thin. If the ratio is smaller than predicted, punchthrough may be limiting $V_{(BR)CEO}$. Punchthrough occurs when the collector–base junction space charge region widens into the base enough to reach the emitter–base junction. $V_{(BR)CEO}$ will be reduced, but $V_{(BR)CBO}$ will be unaffected. It can be detected by noting when (or if) the emitter floating potential[18] begins to rapidly increase as V_{CB} is

[18]Emitter floating potential is the voltage between emitter and ground with the emitter floating and a reverse voltage applied from collector to base. With no punchthrough, the value for silicon will be about 200 mV.

varied over the operating range. Punchthrough[19] can be caused either by a thinner-than-expected base (perhaps only locally due to a nonuniform diffusion front) or by a higher-than-normal base resistivity.

The ratio $V_{(BR)CES}/V_{(BR)CBO}$ is normally close to 1. The exact value depends on inverse current gain, but it should be greater than 0.8

FIGURE 11.30

Examples of pipe formation in pn structures.

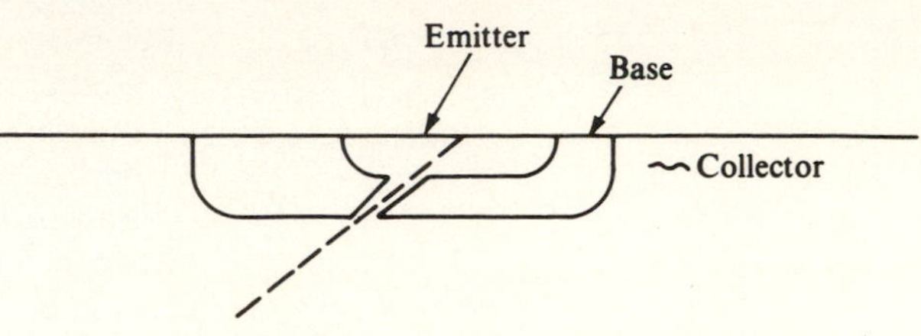

(a) Pipe formed by enhanced diffusion from emitter along a crystallographic defect

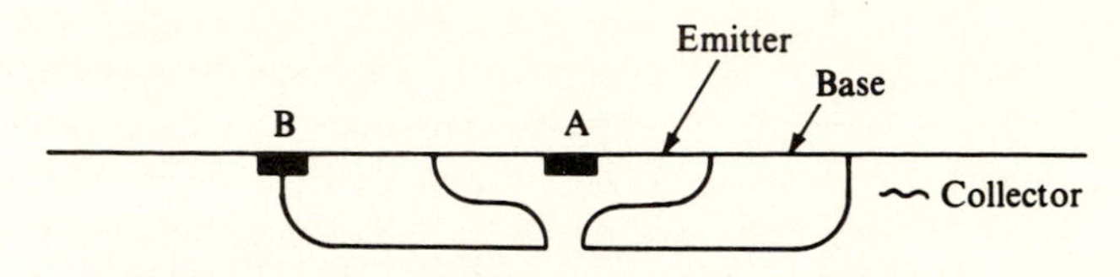

(b) Pipe formed by a high concentration of local emitter type source at A. Note that if the source were moved from A to B, the pipe would disappear, but a low collector–base breakdown would replace it.

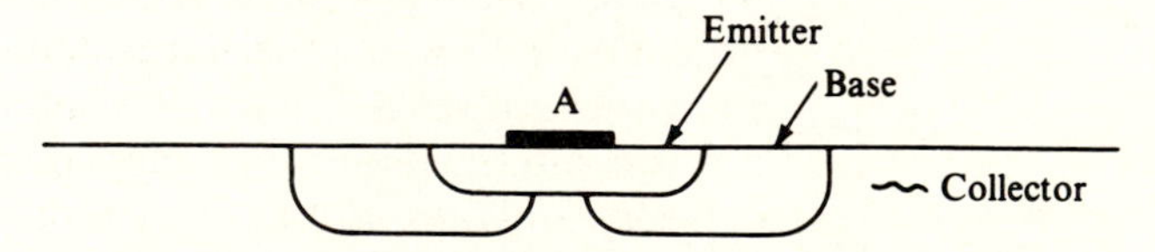

(c) Pipe formed by local masking at A during base diffusion

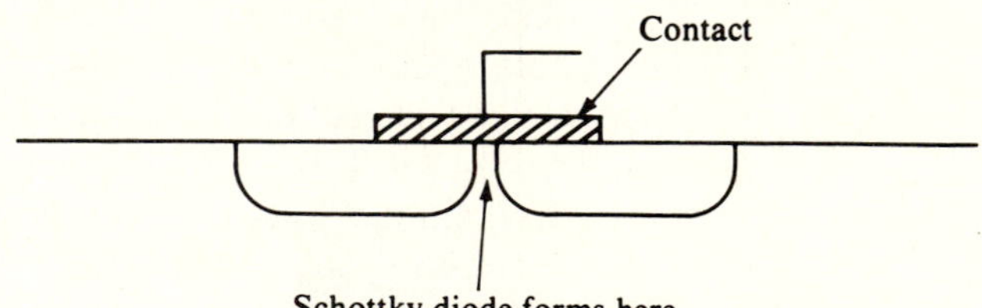

(d) Pipe similar to the one in part c, but the diode diffusion was partially blocked in a region later covered by a contact

[19]Depending on the specific design, punchthrough may occur even on good transistors, but outside the normal operating range.

for an inverse gain of less than 0.5. If it is appreciably less, punch-through is probably limiting it.

When pipes are present, they will radically change the relations among the various breakdown voltages. Pipes are small cross-sectional electrical paths formed by diffusion flaws and are more likely to be a problem in bipolar transistors. Examples of pipes and some of their causes are shown in Fig. 11.30. If a pipe like the one of Fig. 11.30a, 11.30b, or 11.30c occurs, the emitter–base breakdown voltage $V_{(BR)EBO}$ will appear normal. However, the collector–base breakdown will also initially equal $V_{(BR)EBO}$ since the emitter and collector are tied together by the pipe. However, as the voltage is increased, the current will not increase as rapidly as normally observed during avalanche since the resistance of the pipe will act as a current limiter. The space charge movement into the pipe will eventually pinch off additional current flow so that the classical pipe *I–V* curve looks like the curve of Fig. 11.15b. (In the rare event that the pipe is caused by a metallic path, or if the pipe has a very large cross section, the path will appear resistive and will not pinch off.) When a pipe is present, the application of an emitter–collector voltage will cause excess current immediately. If the pipe is small enough, it will pinch off and allow $V_{(BR)CEO}$ to be observed and will look like the curve of Fig. 11.15c. Inversion layers sometimes form paths across device surfaces, and their effect may superficially resemble pipe behavior. With voltage between collector and emitter, both pipe and surface inversion will have the same characteristics. However, in looking from collector to base, only an inversion layer will cause current flow to begin immediately. Fig. 11.30d shows a pipe-like defect that will cause a Schottky diode to form over a small area and be in parallel with the pn junction. Since no care will have been exercised in forming the Schottky diode, it will be very leaky in the reverse direction and hence cause the whole junction to appear leaky. This defect can be found in MOS as well as in bipolar structures.

11.10.3 Reverse and Forward Currents

The bipolar transistor reverse current definitions are given in Table 11.7. When behavior is normal, I_{CBO} is just that of a reverse-biased pn junction, and the others can be calculated[20] from it and forward and reverse h_{FE}. For most cases, $I_{CBO} < I_{CES} < I_{CEO}$. Note that while these currents have been defined as "saturation currents," they are seldom completely independent of voltage. Thus, when a numerical value for I_{XYZ} is given, the voltage at which it was measured must

[20]See a standard text on transistor theory.

TABLE 11.7

Bipolar Transistor Reverse Current Definitions

Symbol	*Meaning*
I_{EBO}	Emitter saturation cuirrent, with collector open
I_{ECO}	Emitter saturation current, with base open
I_{CBO}	Collector saturation current, with emitter open
I_{CEO}	Collector saturation current, with collector open
I_{CES}	Collector saturation current, with base shorted to emitter
I_{CER}	Collector saturation current, with base connected to emitter through a resistor R

also be given. Excess reverse currents that may flow in the various configurations were discussed in the previous section.

Forward transistor junction currents also depend on the manner in which the junctions are connected. Since IC diodes are often made by shorting out one junction, the resulting forward current–voltage characteristics may differ from those expected from an isolated junction (30). In particular, the emitter–base current with the collector–base junction shorted is substantially different, as shown in Fig. 11.31. When such a curve is encountered unexpectedly, at lower current levels it might be identified as a Schottky forward, but as the current and voltage are increased (as shown in the figure), it will switch back toward normal isolated pn junction behavior. The point of switchback is dependent on the transistor characteristics and collector resistivity and occurs when enough current flows to saturate the transistor.

11.11 YIELD ECONOMICS

Since the processes, yields, and economics of a semiconductor operation are interrelated, it is necessary to understand the impact of the various yields and processes on overall IC manufacturing costs in order to plan efficient yield improvement programs. As will be seen in the next section, for the case of silicon ICs, the cost of the silicon itself is generally a very small part of the overall cost. The cost of manufacturing the chip will range from about 1/5 the assembled unit cost for a very small device such as a TTL gate to perhaps 9/10 the assembled unit cost for a large chip device like a 1 megabit DRAM. Assembly/test yields are generally above 95%, while multiprobe yields probably average no more than 50% (although some chip yields may exceed 90%). Thus, because of the relatively low chip yields and the relatively high percentage of the total finished-unit cost represented by chips, most IC manufacturing emphasis is on chip yield improvement.

FIGURE 11.31

Effect of a shorted collector–base junction on emitter–base current.

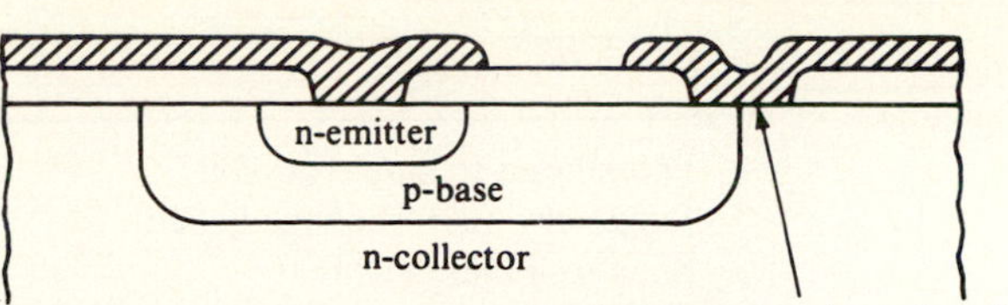

(a)

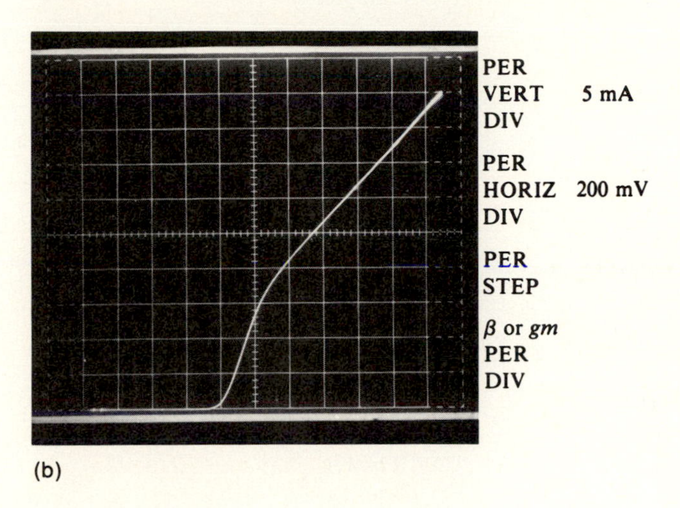

(b)

11.11.1 Silicon Usage

The overall yield for the silicon semiconductor manufacturing operation, from polycrystalline semiconductor-grade silicon to the outgoing packaged unit (IC or discrete transistor), is sometimes calculated in terms of grams of polysilicon per device. Such yields are useful in determining the industry requirements for polysilicon, given a device unit market projection. These requirements, in turn, are examined by IC manufacturers interested in knowing whether or not future polysilicon shortages that could limit their growth are likely; by polysilicon manufacturers concerned about whether they are over- or underbuilding poly plant capacity; and by government agencies concerned about the health of the semiconductor industry in their respective countries.

The actual quantity of single crystal in a packaged unit is quite small. A silicon 256K DRAM has about 0.05 gram of silicon per unit; a TTL gate IC, about 0.5 milligram. Table 11.8 summarizes the silicon losses at the yield points discussed earlier and shows that only about 10% of the starting polysilicon actually finds its way into a finished IC. Thus, if 0.05 gram of silicon is in a package, then no

TABLE 11.8

Overall Silicon Yield

Step	Yield (% by weight)
Polysilicon to single crystal	50
Single crystal to polished slice	50
Slice to finished wafer	70
Finished wafer to chip*	40–80
Chip to finished IC	93
Overall yield	6.5–13

*If the wafer is thinned before being broken into individual chips, the yield at this point will be commensurately less.

TABLE 11.9

Elements of Cost for Producing a Finished Good Wafer

Item	Normalized Cost*
Starting slice	20
Other direct material	12
Masks/slice	
Chemicals/slice	
Labor	12
Labor-associated overhead	8
Overhead materials	8
DI water	
Gases	
Other	
Process control slice	
Depreciation	40
Facility	
Equipment	
Total	100

*Estimated values. Actual numbers will vary substantially, depending on size and age of facility, capacity utilization, kind of product being built, and so on.

more than 0.5 gram of polysilicon would have been used, and at a price of $60/kilogram, the cost of the polysilicon would contribute only 3¢ to the cost of the chip.

11.11.2 Wafer Cost

The actual cost of producing a wafer in a given IC factory is generally considered to be a trade secret, but elements that make up the cost are given in Table 11.9. Tables such as this one can be used to help estimate whether, for example, the yield or productivity improvement to be derived from a new piece of equipment will more than offset the added depreciation.

11.11.3 Wafer Diameter

The purpose of the original drive to change from 1.5 inch to 2 inch diameter slices was to reduce labor. The reasoning was that no more effort should be involved in handling 2 inch wafers than in handling 1.5 inch wafers. However, the move to larger slices also enabled more wafer area to be processed in the same floor space and thus could delay the construction of new facilities. In many cases, larger wafers could be processed in the same equipment and thus give increased capacity with minimal capital expenditure.

Multiprobe yield has been observed to increase along with diameter. This increase is due to two related effects. The first is that as diameter increases, the area on the periphery containing partial chips becomes a smaller part of the total wafer area. The second is that the outer ring of whole chips is generally yield depressed because of such things as wafer–edge rounding, resist-edge beading, shadowing of the periphery by reference pins of front-side referencing equipment, and damage from wafer-handling equipment. Both of these effects are shown in Fig. 11.32. The larger the chip is with respect to the wafer, the more pronounced is the improvement in going to larger diameters.

As the wafer diameter has increased in 25 mm increments from 50 mm to 150 mm, the chip cost has decreased. The maximum economical diameter at a given time has been dictated by various economic or technological factors such as labor costs or the availability of suitable equipment. However, at some point, a limit is almost certain to be set by the semiconductor's high-temperature yield point and thermal conductivity.

FIGURE 11.32

Effect of wafer diameter on multiprobe yield. (The example shown would scale to $1 \times 1\ \text{cm}^2$ chips on a 150 mm wafer, and 23% of the total chips are in this 150 mm row. Thus, if they have a reduced yield, the yield of the whole wafer is severely impacted.)

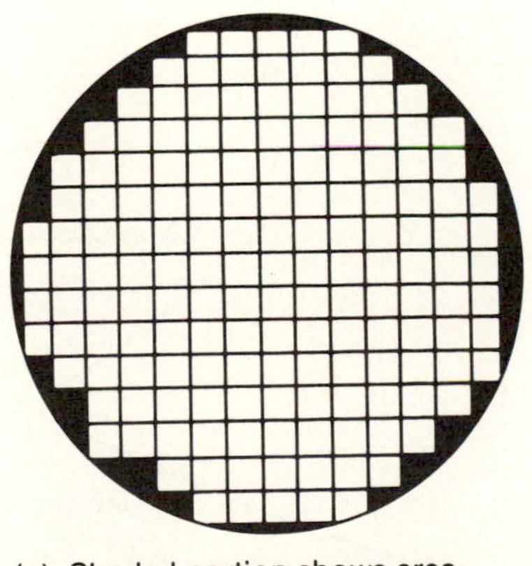

(a) Shaded portion shows area lost due to incomplete chips

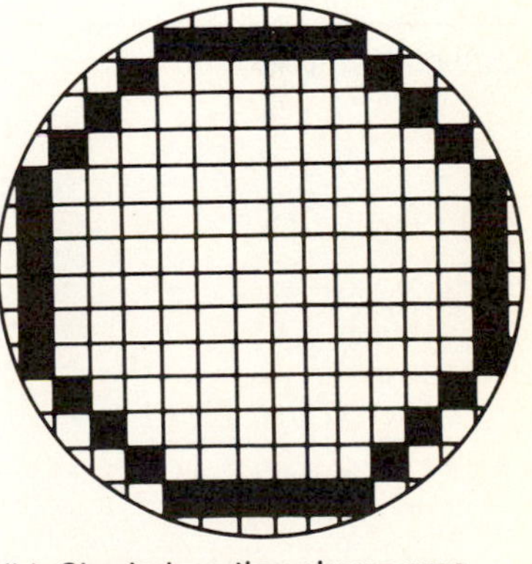

(b) Shaded portion shows area occupied by outer row of whole chips

11.11.4 Effect of Chip Size on Yield and Cost

Since for a fixed process, the chip multiprobe yield decreases as the chip size increases, the number of good chips per wafer decreases with chip size both because of decreasing yield and because of fewer potential chips per wafer. The cost of producing a wafer is essentially independent of chip size as long as the process remains the same, so the cost per chip will increase with increasing chip size in the manner shown in Fig. 11.33. The designer must therefore be as frugal with area as possible in the initial design. After a new chip is introduced and the manufacturing and design kinks are eliminated, a size reduction (shrinkage) of the chip can be undertaken in order to move down the chip cost curve. Initially, the mask image size is reduced by a small amount (perhaps 10% in linear dimensions). Such a reduction reduces all geometries and may make the yield drop instead of increase. However, experience has shown that generally some reduction is possible before this happens. Even when a marked yield drop does occur, it is likely that there are a few localized spacing problems that can be corrected with a simple mask revision.

As just mentioned, the simplest method of reducing chip size (other than by clever designs) is by geometry reduction. However, as complexity passes some point, the area required for interconnections becomes a limitation. Then, a move is generally made to multilevel metallization, which can only be done if the extra steps required for the additional levels of metallization do not reduce the yield as much as the chip area shrinkage helps it. Indeed, not until

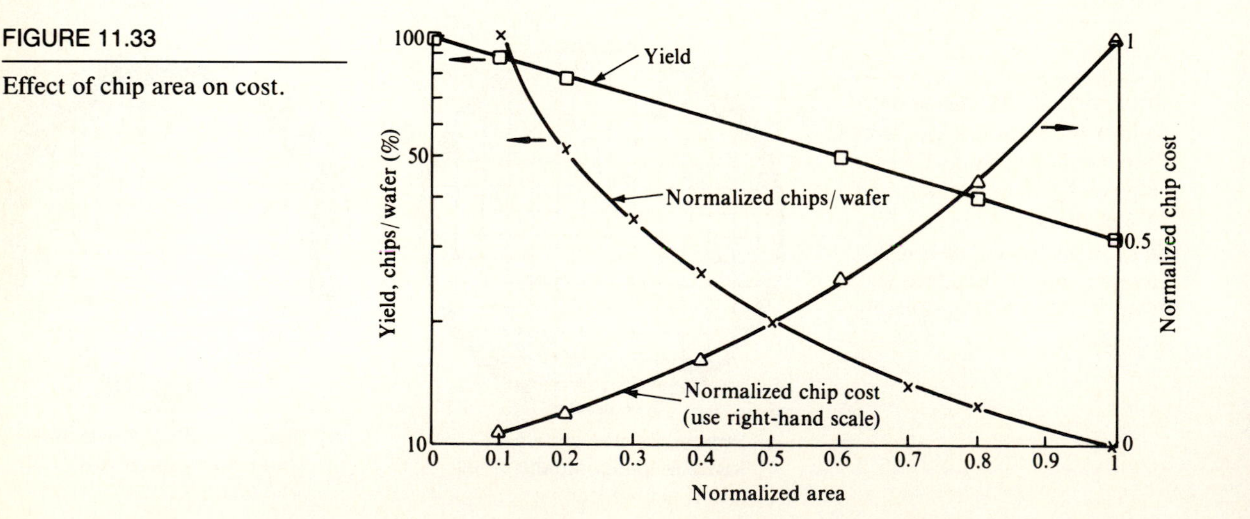

FIGURE 11.33

Effect of chip area on cost.

the late 1970s was the process improved to the point that a yield crossover could be achieved.

11.11.5 Revenue from Wafers

The discussion thus far has been related to the costs accrued in producing a wafer. In a competitive world, these costs must be covered by the revenue derived from the wafer.[21] Such revenue is generally described as the net sales billed per wafer, or NSB/wafer. The NSB depends on what value the market places on the specific circuits built on the wafer and on the number of good chips on the wafer. A wafer containing a very large number of small simple devices—a TTL gate, for example—can have a higher value than a mature product requiring a much larger chip, such as a 64K MOS DRAM. The reason is that the selling price of small-chip devices is more influenced by packaging costs than by chip complexity. At the other end of the spectrum, the relatively new, more complex, large-chip circuits command enough higher price to more than offset the reduced number of chips and thus cause the NSB/wafer to again rise. Table 11.10 gives details of the makeup of NSB/wafer for some representative ICs, and Fig. 11.34 shows the general trend of NSB/wafer versus chip area. The values for the low and middle portions of the curve are relatively stable with time (neglecting inflation) since the products and packaging techniques are quite mature. The upper end of the curve keeps shifting as new products move down the experience curve.[22]

TABLE 11.10

Elements Needed To Calculate NSB/Wafer

Product	*Chip Size (cm^2)*	*Number of Chips/Wafer*	*Overall Yield (%)*	*Price/Unit ($)*	*NSB ($)*
TTL gate	0.0064	18,875	75	0.12	1700
64K DRAM	0.15	776	66	0.75	384
256K DRAM	0.39	276	60	2.00	331
Microprocessor	0.5	220	40	20.00	1760

Note: Estimated numbers are for a 125 mm wafer diameter for the mid-1980s. Actual yields and costs will vary from manufacturer to manufacturer and are generally considered to be trade secrets.

[21] A "merchant" semiconductor operation is assumed. Captive wafer-fab areas may have more leeway since their products are typically custom ICs with high leverage at the system level.

[22] The experience curve is an expansion of the "learning curve" concept that was apparently introduced by the aircraft industry during World War II.

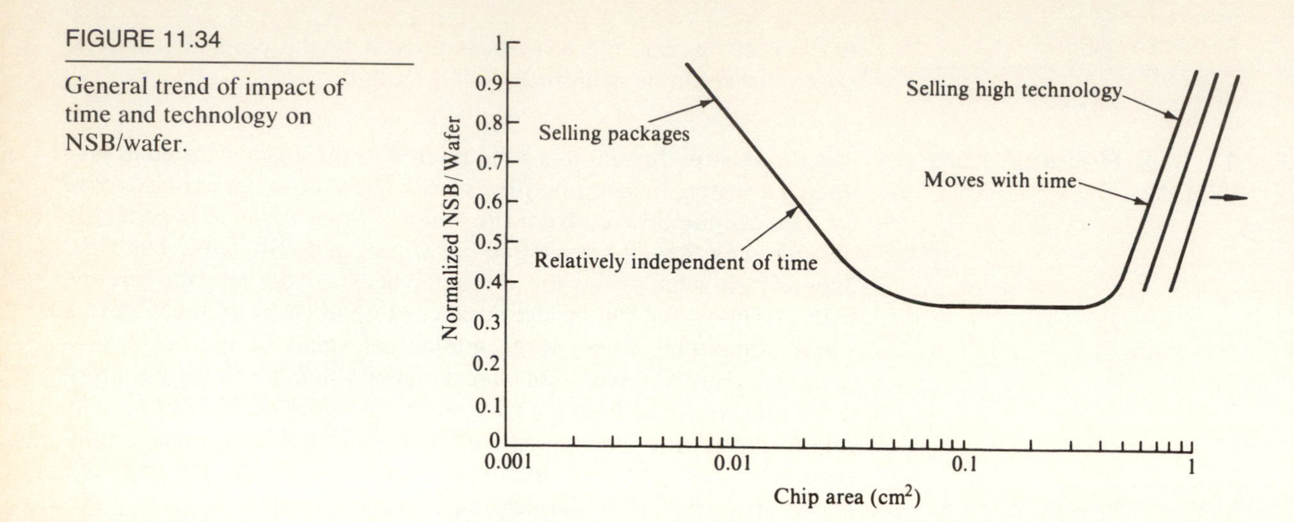

FIGURE 11.34

General trend of impact of time and technology on NSB/wafer.

11.12 FACTORS AFFECTING YIELD AND YIELD IMPROVEMENT

The discussion thus far has been directed primarily at ways of evaluating and tracking yield, with some discussion of how various yields affect the final economic performance of a wafer-fab facility. In order to improve the yield, it is of course first necessary to determine the specific physical defect causing a problem, as was briefly discussed in the previous section. However, to determine the root cause and to effect a permanent cure may be a much more difficult task. Fig. 11.35 depicts factors, many rather obscure, that can affect yields. Despite the complexity of this figure, the factors can be broadly grouped as follows:

People
Equipment
Incoming materials
Device design
Semiconductor processing

Only the last item, semiconductor processing, is dependent on items historically considered as the domain of process engineering. It is clear, however, that the whole span of possibilities, and not just the technical aspects of semiconductor processing, must be considered before instituting changes designed to permanently improve yields. To this end, Fig. 11.35 can be used as a checklist as each factor is considered in the context of the specific manufacturing environment.

FIGURE 11.35

Factors affecting yields.

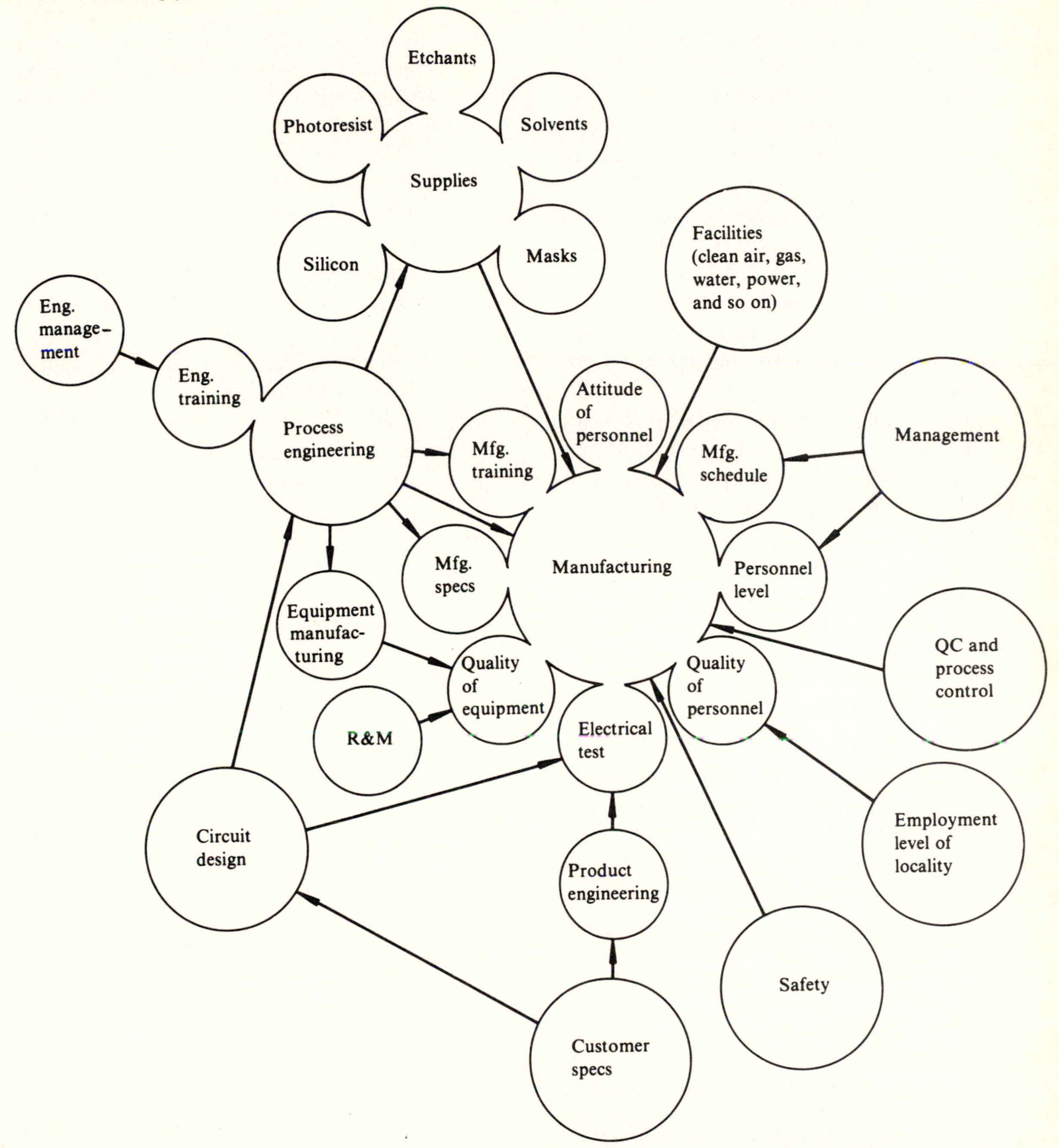

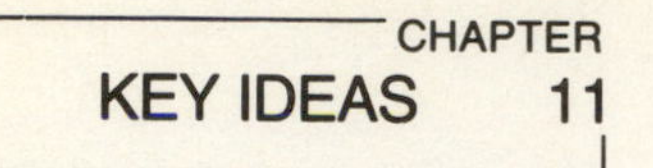

- □ Simple theory predicts that device yield Y versus area A should vary as $Y = e^{-DA}$ where D is the defect density.
- □ The simple yield expression $Y = e^{-DA}$ gives a pessimistic prediction for large DA products.
- □ Multiprobe tests designed to test chips to customer specifications are usually not satisfactory for yield diagnostics.
- □ Unless special analytical procedures are used, long-term yield trend lines are generally too insensitive to changes to be very useful.
- □ Because of the normal variability in wafer yield, split lot results must be carefully analyzed to determine whether or not any observed yield differences are significant.
- □ Before an electrical study of a malfunctioning circuit is begun, the circuit should be given a careful visual (microscopic) examination.
- □ In both MOS and bipolar circuits, the study of pn junction leakage current and breakdown voltage is a key part of low-yield analysis.
- □ The dominant component of forward or reverse junction current can generally be determined from, respectively, a log I–V or a log-I–log-V plot.
- □ A common cause of junction failure is an excessive concentration of heavy metals in or near the junction space charge regions.
- □ Low threshold voltage is a common cause of MOS device failure.
- □ In most cases, the cost of the silicon in a silicon IC is a small part of the total IC cost.
- □ In addition to the technical issues involved in yield improvement, a large number of economic and people-related factors must also be considered.

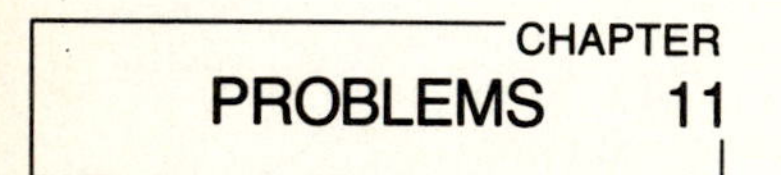

1. If the multiprobe yield of a group of similar devices with 5 significant levels is as follows (yield in percent/area in mm²), determine their defect density. Which equation was used? Why?

55	6
39	9.4
38	10
30	12
21	16
15	19

2. If a chip of area 1 mm² has a yield of 83%, what would the expected yield be for a chip of area 1.5 mm²? Explain your choice of equations.

3. If an incoming 125 mm slice has 8 randomly spaced point defects on it, how much will the yield of a 0.5 cm² chip be reduced by these defects if there are 8 processing levels, each of which introduces 0.1 defect/cm²?

4. What, if anything, is wrong with the diode whose reverse I–V characteristics are as follows? (Units are A/V.)

4×10^{-12}	0.5
7×10^{-12}	1
2×10^{-10}	4
10^{-8}	20
10^{-7}	20.5
10^{-3}	23

5. What device characteristic is most likely being measured when the following *I–V* data are obtained? (Units are A/V.) Plot the data in a manner that demonstrates your answer.

10^{-12}	0.5
1.2×10^{-12}	2
2×10^{-11}	2.5
9×10^{-9}	3
1.4×10^{-6}	4
2.5×10^{-5}	7

6. Sketch the impurity profile for a diode most likely to have the following *I–V* curve. (Units are A/V.)

10^{-12}	− 1
1.25×10^{-12}	2
1.6×10^{-12}	5
2×10^{-12}	10

7. When a large number of 1 cm^2 silicon IC chips is produced, estimate the amount of semiconductor-grade polysilicon required for each one if the multiprobe yield is 55%. Assume that the saw used in slicing the crystal into slices cuts a 500 μm wide slot. List all other assumptions made.

CHAPTER 11 REFERENCES

1. William H. Woodall, "The Design of CUMSUM Quality Control Charts," *J. Quality Technol. 18*, pp. 99–102, 1986. (For older references, see H. Mack Truax, "Cumulative Sum Charts," *Industrial Quality Control*, pp. 18–25 and references therein, December 1961.
2. The application of the Cumulative Sum Approach to yield tracking was suggested by John Conroy of Texas Instruments Incorporated.
3. Frederick E. Croxton and Dudley J. Cowden, *Applied General Statistics*, Prentice-Hall, Englewood Cliffs, N.J., 1955.
4. J.T. Wallmark, "Design Considerations for Integrated Electronic Devices," *Proc. IRE 48*, pp. 293–300, 1960.
5. S.R. Hofstein and F.P. Heiman, "The Silicon Insulated-Gate Field-Effect Transistor," *Proc. IEEE 51*, pp. 1190–1202, 1963.
6. C.H. Stapper, "Comments on 'Some Considerations in the Formulation of IC Yield Statistics'," *Solid-State Electronics 24*, pp. 127–132, 1981.
7. B.T. Murphy, "Cost–Size Optima of Monolithic Integrated Circuits," *Proc. IEEE 52*, pp. 1537–1545, 1964.
8. R.B. Seeds, "Yield and Cost Analysis of Bipolar LSI," *Proc. IEEE Int. Elec. Dev. Meeting*, October 1967.
9. Anil Gupta and Jay W. Lathrop, "Yield Analysis of Large Integrated-Circuit Chips," *IEEE J. Solid-State Circuits*, pp. 389–395, 1972.
10. R.M. Warner, Jr., "Applying a Composite Model to the IC Yield Problem," *IEEE J. Solid-State Circuits*, pp. 86–95, 1974.
11. Charles H. Stapper, "On a Composite Model to the IC Yield Problem," *IEEE J. Solid-State Circuits SC-10*, pp. 537–539, 1975.
12. John E. Price, "A New Look at Yield of Integrated Circuits," *Proc. IEEE 58*, pp. 1290–1291, 1970.
13. This derivation is courtesy of Fred Strieter.
14. This derivation is courtesy of Steve Cowdrey.
15. Albert V. Ferris-Prabhu, "Role of Defect Size Distribution in Yield Modeling," *IEEE Trans. on Electron Dev. ED-32*, pp. 1727–1736, 1985.
16. C.H. Stapper, A.N. McLaren, and M. Dreckmann, "Yield Model for Productivity Optimization of VLSI Memory Chips with Redundancy and Partially Good Product," *IBM J. Res. Develop. 24*, pp. 398–409, 1980.
17. Paul Cox et al, "Statistical Device Characterization and Parametric Yield Estimation," *Solid State Technology,* pp. 154–160, and references therein, August 1985.
18. William Shockley, "Problems Related to p-n Junctions in Silicon," *Solid-State Electronics 2*, pp. 35–67, 1961.
19. J.C. Inkson, "An Investigation of Inversion Layer Induced Leakage Current in Abrupt p-n Junctions," *Solid-State Electronics 13*, pp. 1167–1174, 1970.

20. D.J. Fitzgerald and A.S. Grove, "Mechanism of Channel Current Formations in Silicon p-n Junctions," in M.E. Goldberg and Joseph Vaccaro, eds., *Physics of Failure in Electronics*, Vol. 4, Rome Air Development Center, 1966.
21. Chih-Tang Sah, "Effect of Surface Recombination and Channel on p-n Junction and Transistor Characteristics," *IRE Trans. on Electron Dev. ED-9*, pp. 94–108, 1962.
22. Sorab K. Ghandhi, *The Theory and Practice of Microelectronics*, p. 282, John Wiley & Sons, New York, 1968.
23. W. Milton Gosney, "Subthreshold Drain Leakage Currents in MOS Field Effect Transistors," *IEEE Trans. on Electron Dev. ED-19*, pp. 213–219, 1972.
24. R.R. Troutman, "Subthreshold Slope for Insulated Gate Field-Effect Transistors," *IEEE Trans. on Electron Dev. ED-22*, pp. 1049–1051, 1975.
25. Thomas E. Hendrickson, "Subpinchoff Conduction in Depletion-Mode IGFETS," *IEEE Trans. on Electron Dev. ED-25*, pp. 425–431, 1978.
26. D. Coleman, R.T. Bate, and J.P. Mize, "Mobility, Anisotropy and Piezoresistance in Silicon p-Type Inversion Layers," *J. Appl. Phys. 39*, pp. 1923–1931, 1968.
27. Lowell E. Clark, "High Current-Density Beta Diminution," *IEEE Trans. on Electron Dev. ED-17*, pp. 661–666, 1970.
28. W.H. Schroen, J.G. Aiken, and G.A. Brown, "Reliability Improvement by Process Control," *Proc. 10th Annual Reliability Physics Symposium*, pp. 42–48, 1972.
29. A.A. Bergh and C.Y. Bartholomew, "The Effect of Heat-Treatment on Transistor Low Current Gain with Various Ambients and Contamination," *J. Electrochem. Soc. 115*, pp. 1282–1286, 1968.
30. David K. Lynn et al., eds., *Analysis and Design of Integrated Circuits*, p. 253, McGraw-Hill Book Co., New York, 1967.

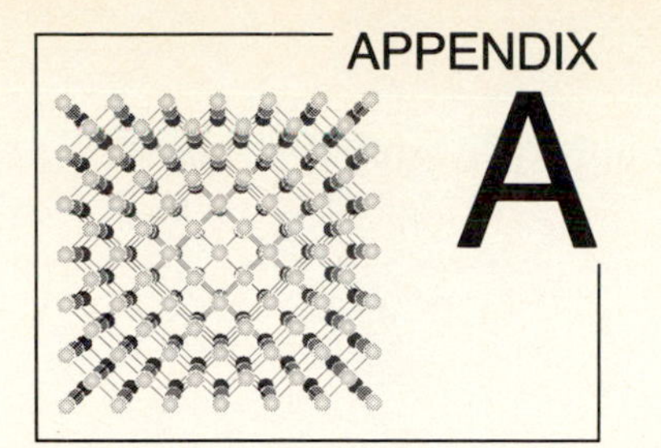

APPENDIX A

Crystallography

A.1 Silicon and Gallium Arsenide Structure

At atmospheric pressure, silicon (along with diamond, germanium, and gray tin) has a diamond cubic structure. Its point group is m3m, and its space group is Fd3m. Gallium arsenide and other III–V compounds have a zinc blende structure. The diamond lattice can be described as two interpenetrating face-centered cubes displaced $(a_0/4, a_0/4, a_0/4)$ from each other along the right-handed x–y–z coordinate system as shown in Figs. A.1a and A.2. a_0 is the lattice spacing (the length of the unit cell). Ordinarily, distances are normalized to a_0 so that the position is written as (1/4,1/4,1/4). When one of the face-centered cubes is populated solely by one element and the other cube is populated by a different element, the structure is zinc blende. The two cubes are shown isometrically in Fig. A.1a, while Fig. A.1b shows a top view of the two cubes plus an additional cube. Fig. A.2 shows the unit cell, which is made up of atoms from each of the two face-centered arrays. Shown in this figure are 18 atoms, but only 8 of them belong to a single unit cell. The others are parts of adjacent cells that are not completely shown.

The atoms are bonded together as shown in Fig. A.3. Each atom is surrounded by four others in tetrahedral fashion. In the case of the zinc blende structure, the four bonds from one constituent atom all go to atoms of the other constituent. The bond length is twice the covalent radius for silicon and is the sum of the covalent radii of the two constituents for zinc blende. The angle between any two adjacent bonds is 109° 28′.

When they are viewed in certain directions, the atoms and their bonds appear to lie in sheets. To describe the position of these sheets (planes) and the direction of the bonds, it is convenient to use Miller indices,[1] which were originally developed to map the external

[1]Named after William H. Miller, professor of mineralogy at the University of Cambridge from 1832 to 1880.

FIGURE A.1

(a) View of two interpenetrating face-centered cubes that comprise diamond lattice. (b) Top view of position of atoms with respect to plane passing through atoms 1, 2, 3, and 4.

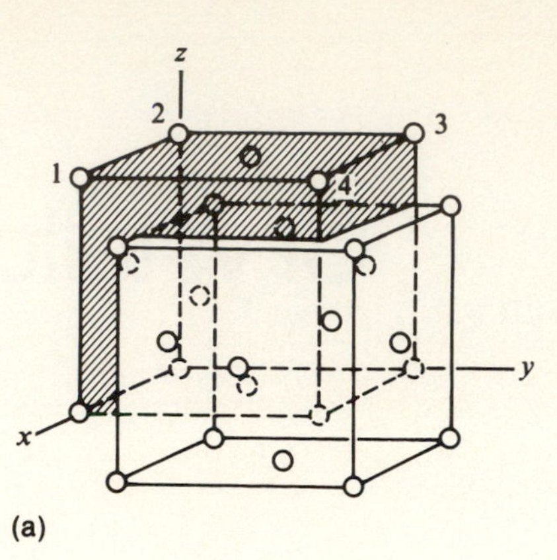

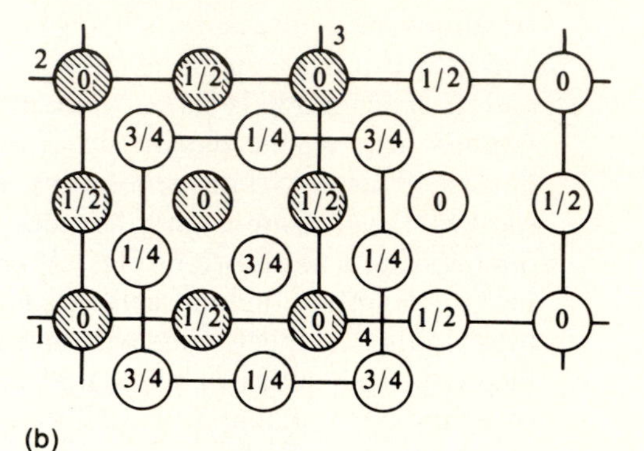

FIGURE A.2

Diamond lattice. [The atoms actually belonging to the unit cell are those with coordinates (0,0,0); (0, ½, ½); (½, 0, ½); (½, ½, 0); (¼, ¼, ¼); (¼, ¾, ¾); (¾, ¼, ¾); and (¾, ¾, ¼).] (*Source:* Adapted from R.W.G. Wyckoff, *Crystal Structures,* Interscience Publishers, New York, 1960.)

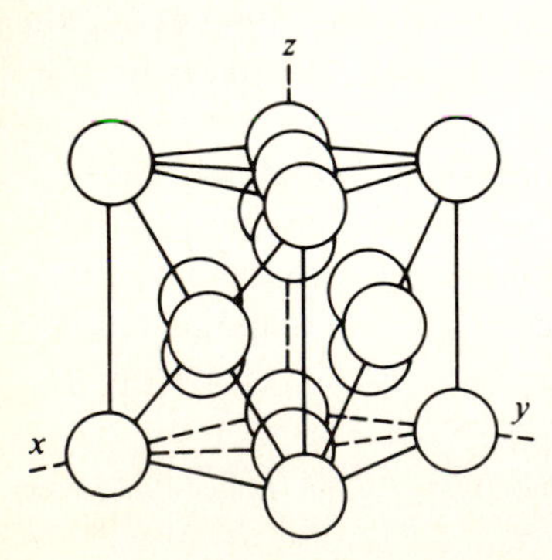

faces of crystals. The Miller indices of a plane intersecting the x–y–z axes at distances a, b, and c, respectively, from the origin are the smallest set of integers in the ratio of $1/a$ to $1/b$ to $1/c$. For describing the plane containing a specific set of atoms within a unit cell, the reciprocals are used directly. Miller indices are written as $(hk\ell)$. Negative numbers are written with a bar over the number, as, for example, $(1\bar{1}1)$, which is read as "one bar-one one." Fig. A.4 shows the low indices planes (111), (110), and (010). Note that when a plane is parallel to an axis, it intersects it at infinity, and the reciprocal is zero. Thus, a (010) plane is parallel with both the x and z axes. There are eight planes that make up the 111 family: (111), $(\bar{1}\bar{1}\bar{1})$, $(\bar{1}11)$, $(1\bar{1}\bar{1})$, $(\bar{1}\bar{1}1)$, $(11\bar{1})$, $(1\bar{1}1)$, and $(\bar{1}1\bar{1})$. They are referred to collectively by using braces instead of parentheses—for example, {111}.

A crystallographic direction is written as $[hk\ell]$. In the cubic system, the $[hk\ell]$ direction is perpendicular to the $(hk\ell)$ plane. A whole family of $hk\ell$ directions is designated as $<hk\ell>$. A direction is a vector, and if it goes from the origin to $x = u$, $y = v$, $z = w$, it has the direction coordinates $[uvw]$. The direction indices $[hk\ell]$ are the

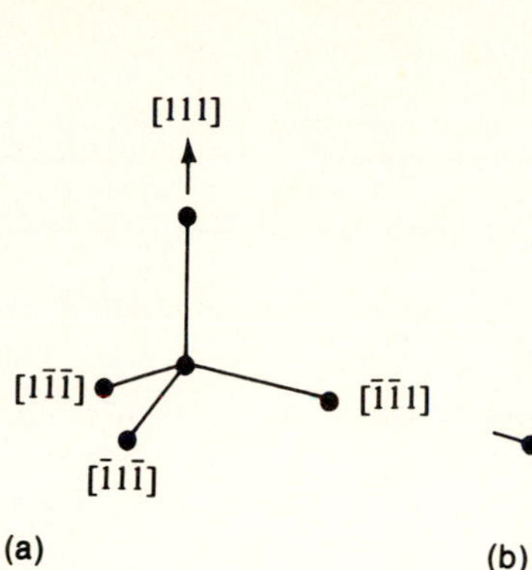

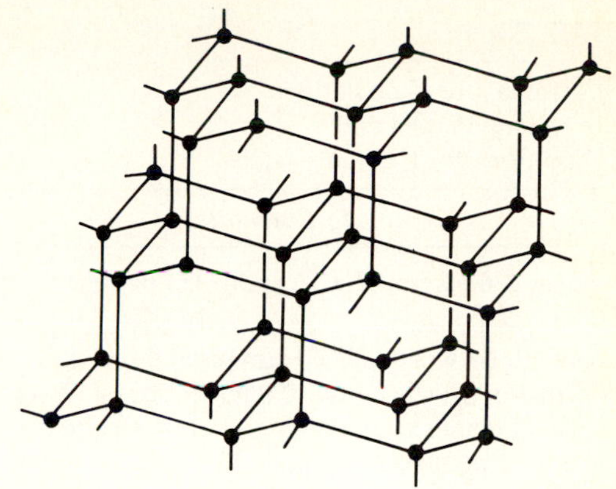

FIGURE A.3

(a) Single atom tetrahedrally bonded to four other atoms. (b) Several atoms joined together to form diamond lattice. (*Source:* Adapted from J. Hornstra, *J. Phys. Chem. Solids 5*, pp. 129–141, 1958.)

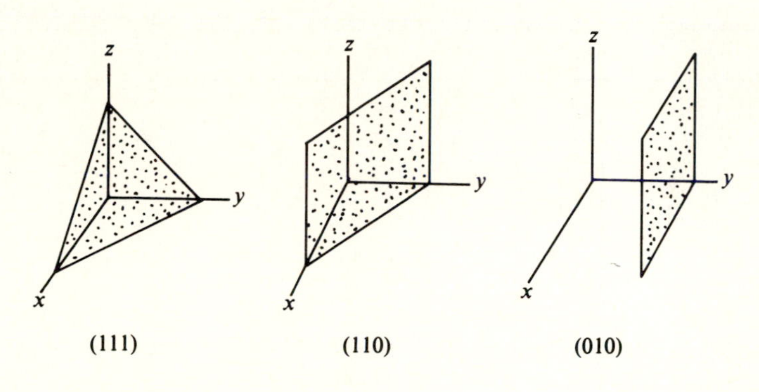

FIGURE A.4

Examples of low indices planes.

smallest set of integers having the ratio of u:v:w. Just as all parallel planes have the same Miller indices, all parallel directions have the same direction indices. Table A.1 provides helpful formulas to use in working with planes and directions and also gives a short table of angles.

All of the planes that intersect at a common line (an infinite number) are said to belong to the same zone, and the common line is called the zone axis. Two planes (HKL) and ($hk\ell$) will have the same zone axis [uvw] if [uvw] is perpendicular to the normal of each plane—that is, when

$$Hu + Kv + Lw = hu + kv + \ell w = 0$$

Fig. A.5 shows a series of planes belonging to the [110] zone. One way of looking at this series of planes is to consider that they resulted from rotating the (001) plane about a line in the [110] direction. Another is that the (001) plane was tilted a given number of

TABLE A.1

Formulas for Calculations Involving Planes and Directions for Cubic Crystals

To Calculate:	*Use the Formula:*
Angle ϕ between (HKL) and $(hk\ell)$ planes	$\cos\phi = \dfrac{Hh + Kk + L\ell}{[(H^2 + K^2 + L^2)(h^2 + k^2 + \ell^2)]^{1/2}}$
Angle θ between $[HKL]$ and $[hk\ell]$ directions	Use expression above for $\cos\phi$.
Direction $[uvw]$ of line of intersection of (HKL) and $(hk\ell)$ planes (this line is sometimes called trace of one plane on the other)	$u = K\ell - kL$ $v = Lh - H\ell$ $w = Hk - hK$
Spacing between adjacent $(hk\ell)$ planes	$d = \dfrac{a_0}{(h^2 + k^2 + \ell^2)^{1/2}}$

Angles between Planes or Directions for Some Selected Low Indices Planes

HKL	*hkℓ*	*Angle*			
100	100	90			
	110	45	90		
	111	54.74			
	311	25.24	72.45		
110	110	60	90		
	111	35.26	90		
	311	31.48	64.76	90	
111	111	70.53			
	311	29.50	58.52	79.98	
311	311	35.10	50.48	62.96	84.78

For a more extensive table (up to 554/554), see R.J. Peavler and J.L. Lenusky, "Angles between Planes in Cubic Crystals," *IMD Spec. Rpt. No. 8,* American Institute of Mining, Metallurgical and Petroleum Engineers.

degrees toward the nearest (111) plane. This latter terminology is often used in describing the manner in which wafers are slightly misoriented from a low index plane and is best understood by looking at a stereographic projection.[2]

Examples of stereographic projections for the three common low indices planes are shown in Figs. A.6 and A.7. Dots on the periphery represent planes that are perpendicular to the projection plane (plane of the paper) and oriented so that they are tangent to the peripheral circle at the point of the dot (Fig. A.6b). An interior dot represents a plane making an angle other than 90° with the pro-

[2]For a complete discussion of the mechanics of stereographic projections, see a standard textbook on X-ray crystallography.

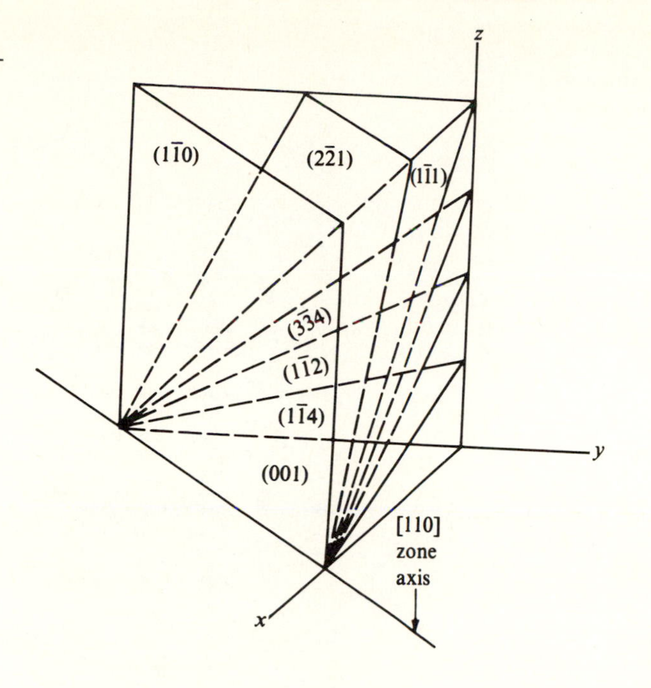

FIGURE A.5

Examples of planes belonging to [110] zone.

jection plane. The angle between the traces of any two planes intersecting the stereographic plane, as, for example, angle θ in Fig. A.6c, is the same as the angle between the two lines joining the dots with the center of the projection (angle θ' in Fig. A.6c). Angles between planes represented by interior dots and the projection plane can be determined graphically with a transparent overlay stereographic net, such as a Wulff net. Such a net is not shown, but the closer the dot is to the center of the projection, the smaller the angle between that plane and the projection plane. Thus, using the terminology of tilting toward the nearest {111} plane, it can be seen from Fig. A.6 that any of the four {111} planes are equally close and that to tilt to any of them requires rotating about a <110> zone axis. If the intent was to tilt away from the (111) plane toward the nearest {110}, then reference to Fig. A.7a shows that the zone axis should be a <110> so that the first {110} encountered would be either a (101), a (011), or a (110). Rotation about a <121> zone axis would lead to the (110) planes on the periphery being the first ones reached and would require a greater tilt angle (90°).

Fig. A.8 shows how the atoms and their bonds appear when projected onto (100), (110), and (111) planes—that is, how they appear if viewed by looking into the lattice along [100], [110], and [111] directions, respectively. The scale is the same for all three views, so it is clear why the [110] direction is referred to as the "open" direc-

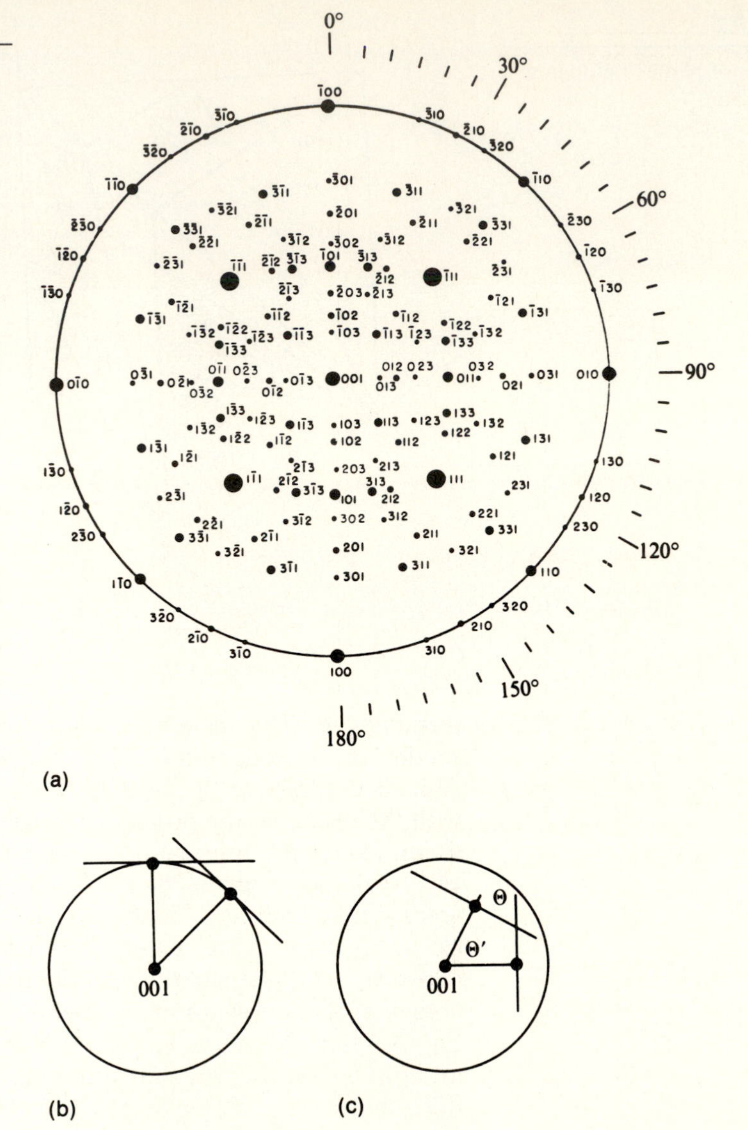

FIGURE A.6

(a) 001 stereographic projection. (b) Planes represented by peripheral dots are perpendicular to projection plane. (c) Planes represented by interior dots intersect projection plane at an angle other than 90°.

tion. Fig. A.8a shows that three atomic layers exist between the (010) planes at $x = 0$ and at $x = a_0$ [there are also three between the (100)'s and (001)'s]. These are often referred to as (040) planes. Similarly, the spacing for the (110)'s is such that to properly locate them with respect to the individual atoms requires that they be referred to as (220)'s. Table A.2 lists a_0 for several common semiconductors as well as the separation of low indices planes for silicon and gallium arsenide.

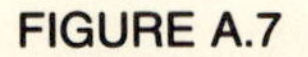

FIGURE A.7

111 and 110 stereographic projections.

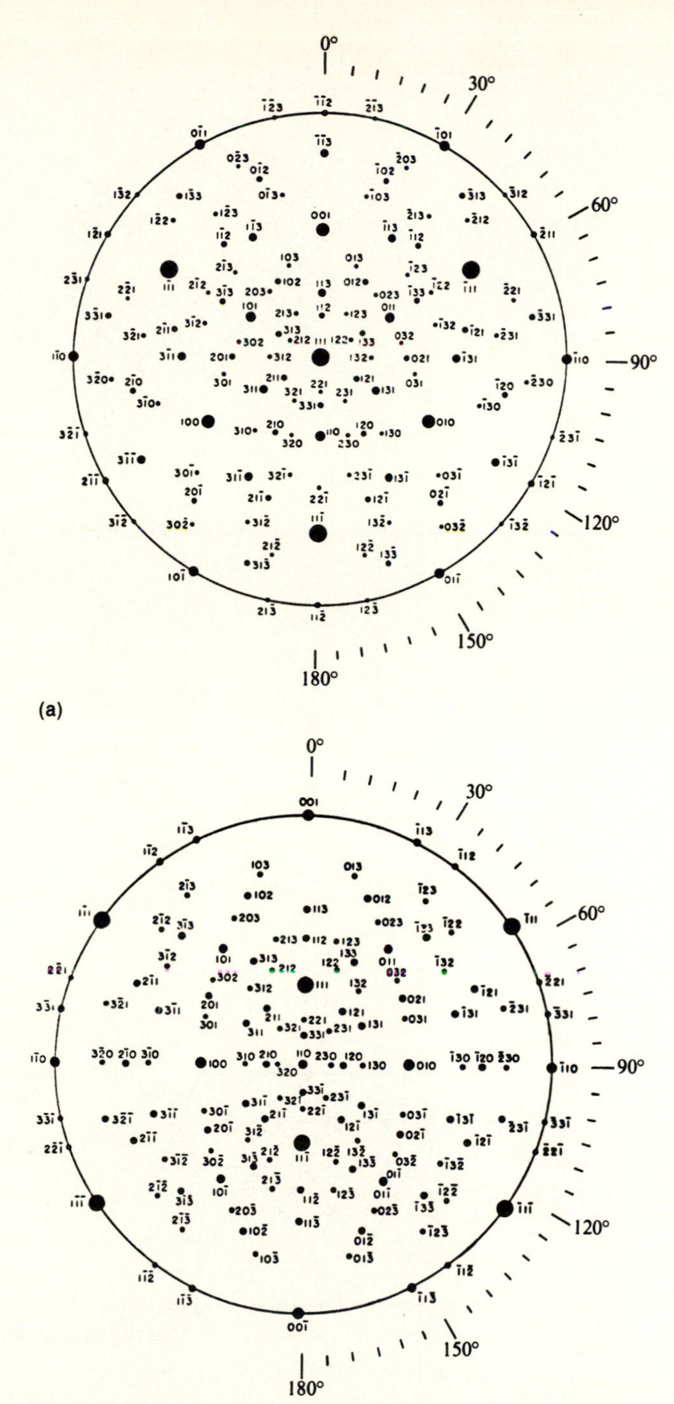

FIGURE A.8

Views of atoms and bonds when sighting into crystal in [100], [110], and [111] directions, respectively.

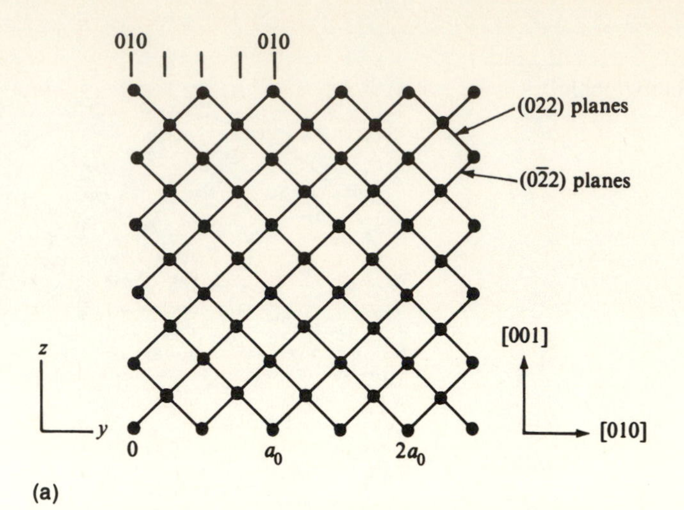

(a)

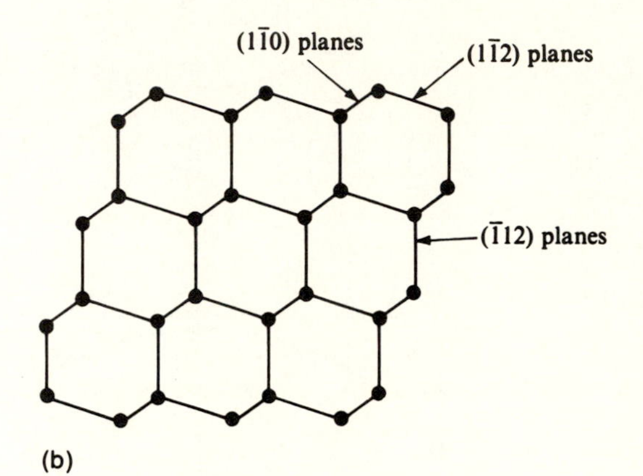

(b)

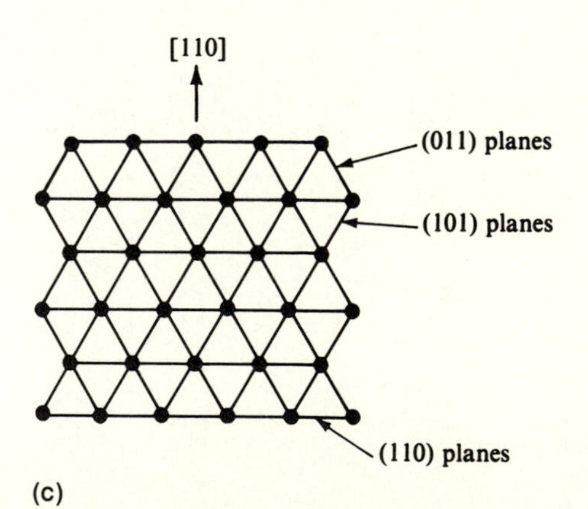

(c)

TABLE A.2

Lattice and Plane spacings (Å)

Material	a_0	*Plane*	*Spacing* *Si*	*GaAs*
Diamond	3.668	100	5.43	5.65
SiC	4.3596	400	1.36	1.41
Si	5.4307	110	3.84	4.00
GaP	5.4504	220	1.92	2.00
GaAs	5.6533	111	3.14	3.26
Ge	5.6575	444	0.784	0.815

Fig. A.8b is reproduced in Fig. A.9a, but dashed lines have been added to indicate the traces of the parallel sheets of atoms that make up (111) planes. The two closely spaced layers are separated substantially from the next set of double layers. Normally, "the (111) plane" is considered to include both sheets, and the separation of (111) planes as calculated from Table A.1 gives the distance from the middle of one double layer to the middle of the next one. From Fig. A.9, it can be seen that the distance is also given by the length of the bond plus the separation of the close spaced layers. This separation is 1/4 the (111) spacing. Fig. A.9b is the same as Fig A.3b except that parts of two planes have been shaded to show the double sheet just described. These two sheets are multiply bonded together, with three of the four bonds of each atom being used for that purpose. The fourth bond goes to the next set of double layers. It is this paucity of bonds that allows slip and cleavage to rather easily occur between adjacent silicon (111) planes. Table A.3 gives the number

FIGURE A.9

(a) A 110 section of diamond lattice showing location of two sheets of atoms comprising a (111) plane. (b) A perspective view with a portion of two pairs of close-spaced (111) planes shaded. [These show two of the four orientations of (111) planes of the lattice; the other two are perpendicular to the $[\bar{1}\bar{1}1]$ and $[\bar{1}\bar{1}1]$ directions.]

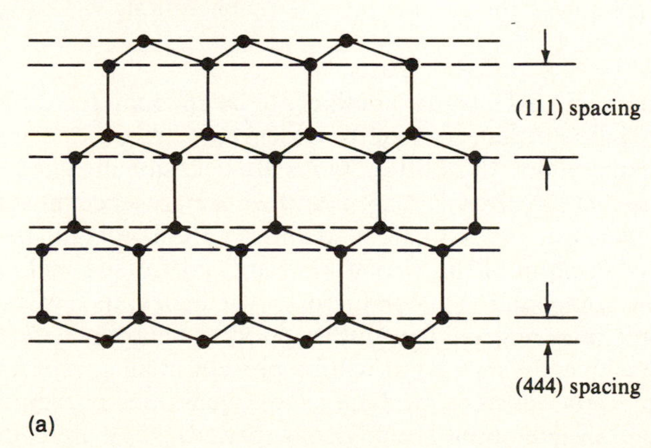

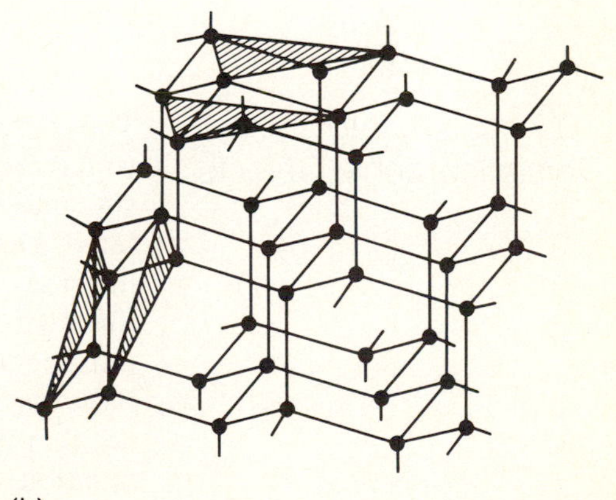

TABLE A.3

Plane-to-Plane Bond Density for Diamond and Zinc Blende Structures

Plane	*Bonds/Atom*	*Bonds/a_0^2*	*Bonds/cm²*	*Relative Bonds/ Unit Area*	*Bonds/cm² (Si)*
(111)*	1	$4/\sqrt{3}$	$4/(d^2\sqrt{3})$	1	7.9×10^{14}
(110)	1	$2\sqrt{2}$	$2\sqrt{2}/d^2$	1.22	9.6×10^{14}
(100)	2	4	$4/d^2$	1.73	1.4×10^{15}
(111)†	3	$12/\sqrt{3}$	$12/(d^2\sqrt{3})$	3	2.4×10^{15}

*Wide spacing.
†Close spacing.

Source: J.H. Brunton, "Fracture Propagation in Diamond," *Proc. 1st Int. Cong. on Diamonds in Industry,* Paris, 1962.

of bonds/atom that go from one plane to the next, as well as the number of bonds/cm².

The gallium arsenide double layer has one layer totally of gallium and the other totally of arsenic. Gallium arsenide shows some ionic bonding characteristics, and the attraction between the two layers provides enough strength to offset the low bond density and prevent easy cleavage between (111) planes. There is also only one bond per atom between (110) planes, in which case, with the adjacent planes each having equal numbers of gallium and arsenic atoms, ionic bonding is less, and cleavage occurs. The difference in composition of each layer of a gallium arsenide (111) double layer causes several properties to appear differently when this layer is approached from the [111] or the $[\bar{1}\bar{1}\bar{1}]$ direction. One example is etching, in which a (111) face behaves differently depending on whether the gallium or the arsenic layer is reached first.

A.2 Crystallographic Defects

The crystallographic defects that are likely to be encountered during wafer processing and that are readily observable are dislocations, slip, and stacking faults. In addition, twins are occasionally encountered. Defects that are not so apparent and whose presence must be inferred are vacancies, interstitials, and antistructure. An antistructural defect is an atom of the wrong kind at a lattice site and can only occur in compound semiconductors. Vacancies and interstitials, also sometimes referred to as either point or Schottky defects, are thermodynamically stable and will be present in all crystals. In addition, some process steps produce excess quantities of them. If a vacancy and an interstitial are adjacent (paired), they are called

Frenkel defects. Vacancies also sometimes pair with impurities in the crystal to form various complexes. Some of these will occur during high-temperature operations such as diffusion, but they are particularly likely to form during irradiation by high-energy neutrons, electrons, or protons.

A vacancy is an atom missing from a lattice site. Since the bonding is disrupted, vacancies can have a charge and, in silicon, are thought to be able to exist as neutrals (V^0), singly charged donors (V^+), singly charged acceptors (V^-), and doubly charged acceptors ($V^=$) (1). The relative amounts of each type will depend on the temperature and doping level. For silicon, the equation for the neutral equilibrium concentration at a temperature T (K) is of the form

$$N_V^0 = Ne^{-E_V^0/kT}$$

where E_V^0 is ~2.3 eV and N is the number of silicon atoms per unit volume (2). In the case of compound semiconductors, the number of vacancies for each constituent may be different. For gallium arsenide, the calculated concentrations, assuming equilibrium vapor pressure, are (3)

$$N_V^0 = (3.3 \times 10^{18})e^{-0.4/kT} \qquad \text{for gallium}$$

$$N_V^0 = (2.2 \times 10^{20})e^{-0.7/kT} \qquad \text{for arsenic}$$

The concentration of charged vacancies is related to the neutral concentration through the difference between the location of the charged vacancy energy level E_V and the Fermi level E_F in the bandgap. For silicon (4),

$$N_{V+} = N_V^0 e^{(E_{V+} - E_F)/kT}$$

$$N_{V-} = N_V^0 e^{(E_F - E_{V-})/kT}$$

$$N_{V=} = N_V^0 e^{(2E_F - E_{V-} - E_{V=})/kT}$$

An interstitial is an extra atom residing in the space between the normal lattice sites. In the diamond and zinc blende lattices, there are five spaces in the interior of each unit cell and three centered on the unit cell walls. In each case, the location is vertically down half a space from the atoms shown in the plan (top) view of Fig. A.1. The interior interstitials are referred to as hexagonal sites; the other three, as tetragonal. Energy considerations make it appear that only the hexagonal sites are stable locations (5). In semiconductors with zinc blende structure, there are two further splits of interstitial sites, those whose nearest atomic neighbors are either group III elements or group V elements.

The most commonly discussed dislocations are screw and edge.

FIGURE A.10

(a) Screw dislocation. (b) Edge dislocation.

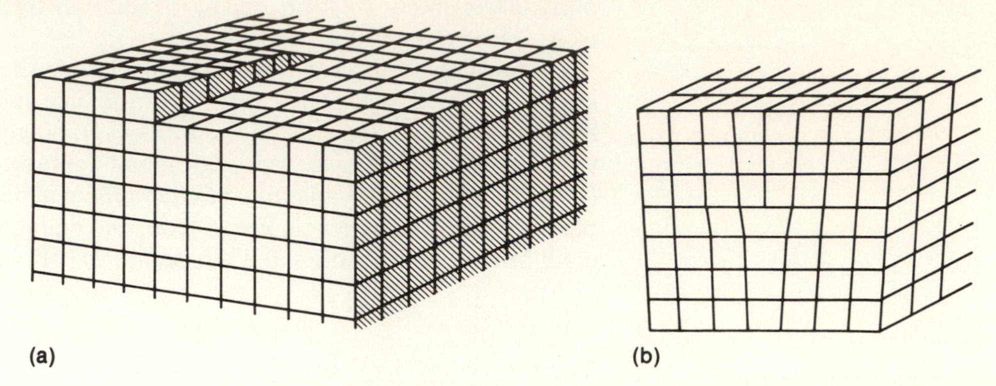

Fig. A.10a shows an idealized structure containing a screw dislocation. Edge dislocations are generally depicted as the disorder caused by an extra layer of atoms introduced into the crystal as shown in Fig. A.10b. This disorder will be at the edge of the extra plane, and crystal symmetry will dictate what orientation "extra" planes actually occur and the direction in which the edge of the plane will lie. In the case of silicon, various dislocation configurations are possible (6), some of which have both edge and screw characteristics. The one usually observed is the 60° dislocation. It is shown in Fig. A.11, is located at the edge of an extra (111) double layer, and runs in a [110] direction. Dislocations in semiconductors are generally formed, not by the physical introduction of an additional layer, but by a shearing deformation causing part of the lattice to slip relative to the rest, as indicated in Fig. A.12.

Misfit dislocations occur when two lattices that are not exactly the same size meet. Where only a small mismatch occurs, the bonds will distort and join; when the cumulative mismatch becomes on the order of the spacing of the planes, a whole plane on one side will fail to bond and will give the appearance of an extra half-plane as shown in Fig. A.13. The net result is a configuration that looks just like that of Fig. A.12b. The dislocations run in [110] directions, and it is (111) planes that appear to have been added, but the dislocations themselves are all confined to lie in the interface between the two lattices. Such dislocations are most commonly observed when a material is epitaxially overgrown onto a substrate with differing lattice constants. This occurs, for example, when high-resistivity silicon is grown onto a heavily doped substrate or when gallium arsenide is grown onto silicon.

FIGURE A.11

A 60° dislocation in a diamond lattice. [The extra half-plane is a (111) double layer. The dislocation at the end of the plane, indicated by the arrow, travels in a [110] direction.]
(*Source:* Reprinted with permission from R.G. Rhodes, *Imperfections and Active Centres in Semiconductors*, Copyright 1964, Pergamon Press plc.)

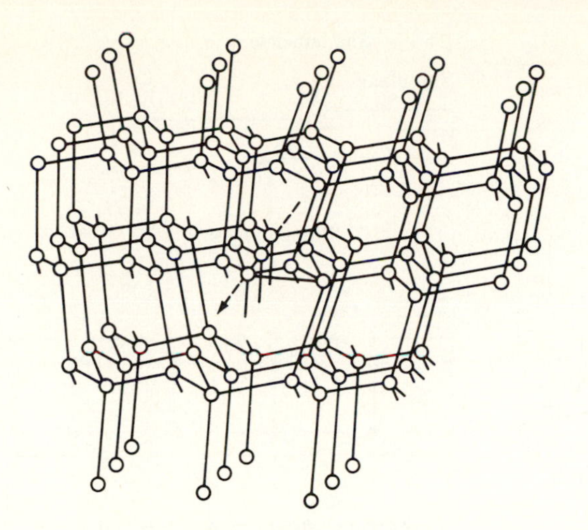

FIGURE A.12

Formation of an edge dislocation by mechanical deformation. (Application of stress causes bonds A–2, B–3, C–4, and so on to break and change to A–1, B–2, C–3, and so on.)

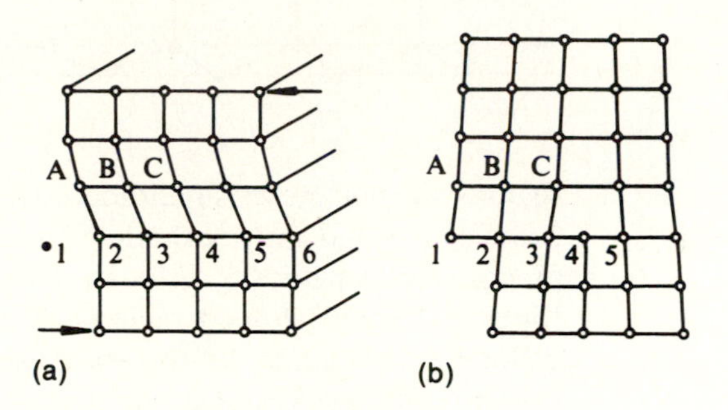

FIGURE A.13

Misfit dislocations caused by matching of two lattices with differing lattice spacings a_1 and a_2. (In an actual situation, the lattice spacings would be very close together, and the dislocations would be widely separated.)

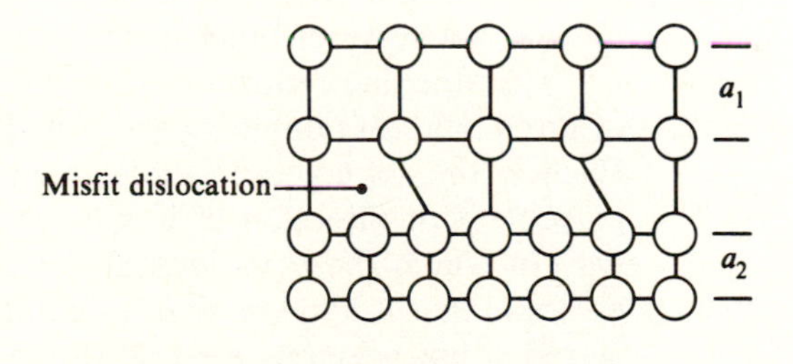

Slip is the plastic deformation of a crystal by one part slipping across the other as shown in Fig. A.14. The plane along which the shear motion occurs is the slip plane and, for silicon, is the {111} family. If slip over the whole plane were to occur simultaneously and all of the motion were to be in exact atomic units, after slip, there would be no crystal defects. Unfortunately, this is not true; all

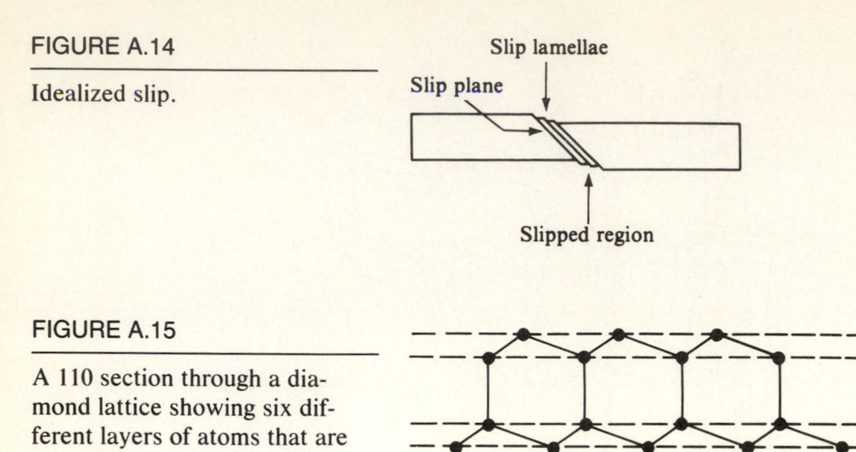

FIGURE A.14

Idealized slip.

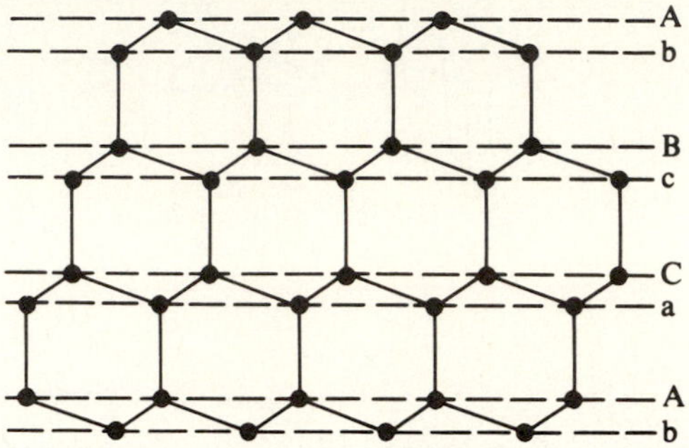

FIGURE A.15

A 110 section through a diamond lattice showing six different layers of atoms that are encountered in moving in a <111> direction.

regions do not move simultaneously. The result is deformation in some places and slip in others. Thus, it appears that numerous extra planes, along with the dislocations at their boundaries, have been formed. Hence, slipped regions in a wafer usually cause rows of dislocations to intersect the wafer surface. It is the etch pits associated with these dislocations that are generally referred to as "slip lines." In some cases, slip steps can be detected at the wafer surface by microscopic observation with interference contrast or by scanning with a very sensitive surface profilometer.

The stacking sequence of layers in a <111> direction consists of three pairs of double layers (Ab, Bc, Ca) as shown in Fig. A.15. Each of the six layers in the sequence is different in either atomic placement or direction of bonds. Stacking faults are errors in the way in which these various layers are stacked. The result of the errors is that it appears that an additional layer either has been inserted or has been removed. However, unlike the dislocation, which has an additional double layer, a stacking fault involves the addition or removal of the bottom of one double layer and the top of the double layer below it—that is, bB, cC, or aA—which is necessary since these are the pairs that have atoms in the same positions. Fig. A.16a shows two planes marked for removal, and Fig. A.16b shows the result when the lattice collapses and closes the gap. One double layer is introduced with bonding that is different from the remainder of the crystal. The result is referred to as an intrinsic stacking fault. One way in which such a fault could occur is by the aggregation of

FIGURE A.16

A (110) section through a diamond lattice showing formation of an intrinsic stacking fault by removal of two adjacent layers of atoms.

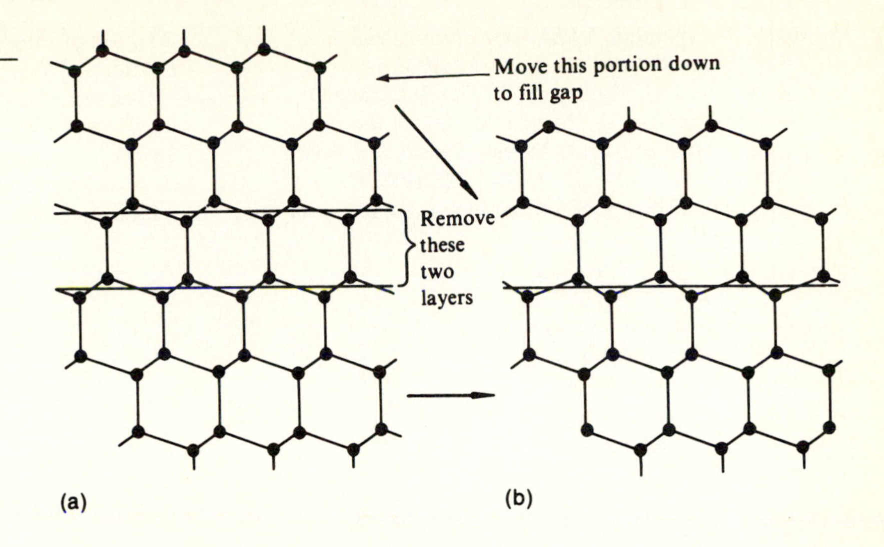
Move this portion down to fill gap
Remove these two layers
(a)
(b)

a large number of vacancies and the subsequent collapse of the lattice. If an extra layer is inserted into the lattice, then two double layers are different, and the fault is extrinsic. They can grow by the capture of self-interstitials. When the extra or missing layers terminate within the crystal, dislocations will occur at the edges of the fault.

Twinning is a crystal defect characterized by a twin plane separating two crystal regions of different orientation. The atomic positions on each side of the plane are mirror images, and the nearest neighbors across the boundary are in the correct position. Crystal symmetry requirements restrict the possible twin-plane orientations as well as the orientation of the material on each side of the twin. In the case of a diamond lattice, (111) and (112) planes are allowed, but when detailed energy requirements are considered, only the (111) is expected (7) and is the only one that has been observed. Zinc blende, with its two components, can, in principle, have a twin with the nearest neighbors in the correct positions and the proper constituents at the locations, or else, while the positions are correct, the constituents can be reversed. Twinning most often occurs during crystal growth and will generally be screened out before reaching a wafer-fab facility.

APPENDIX A: REFERENCES

1. G.D. Watkins, *Radiation Damage and Defects in Semiconductors*, p. 228, Institute of Physics, London, 1973.
2. R.G. Rhodes, *Imperfections and Active Centres in Semiconductors*, Pergamon Press, New York, 1964.

3. Sorab K. Ghandhi, *VLSI Fabrication Principles*, John Wiley & Sons, New York, 1983.
4. C.P. Ho and J.D. Plummer, "Si/SiO_2 Interface Oxidation Kinetics: A Physical Model for the Influence of High Substrate Doping Levels," *J. Electrochem. Soc. 126*, pp. 1516–1522, 1979.
5. K. Weiser, "Theory of Diffusion and Equilibrium Position of Interstitial Impurities in the Diamond Lattice," *Phys. Rev. 126*, pp. 1427–1436, 1962.
6. J. Hornstra, "Dislocations in the Diamond Lattice," *J. Phys. Chem. Solids 5*, pp. 129–141, 1958.
7. E.I. Salkouitz and F.W. Von Batchelder, "Twinning in Silicon," *J. Metals 4*, p. 165, 1952.

APPENDIX B

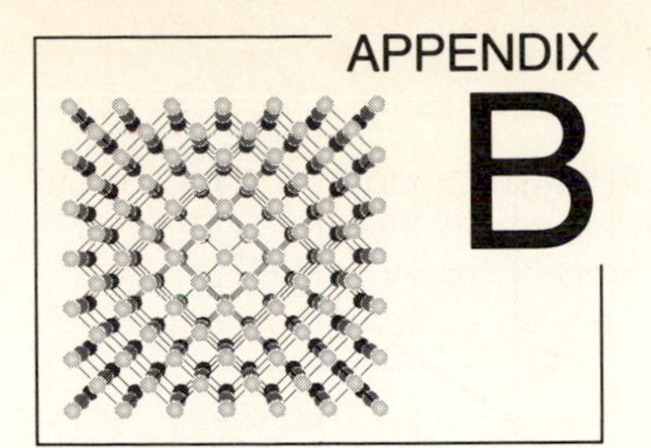

Phase Diagrams

B.1 BINARY PHASE DIAGRAMS

Phase diagrams are very helpful in compactly showing the melting and freezing behavior of mixtures of two or more components. They are a series of curves, generally at constant pressure, that map out the solid and liquid regions as a function of temperature and composition.[1] All of the phase diagrams must be experimentally determined, and the details for a large portion of the possible two-element combinations have now been published (1). An example of a binary phase diagram is shown in Fig. B.1. It is for the Al–Si mixture and shows composition along the x axis, ranging from 100% Al on the left to 100% Si on the right. Temperature is on the y axis, and for this diagram (and most others), the pressure is 1 atm. This curve is one of the relatively simple binary eutectic diagrams, which include combinations such as Sb–Si, Be–Si, Ga–Si, Au–Si, and Ag–Si. The three main classes of binary phase diagrams are isomorphous, eutectic, and peritectic.

The simplest of the binary systems is the isomorphous, which only occurs when the two components A and B are mutually soluble in all proportions, both as a liquid and as a solid. Not many combinations meet these requirements, but Ge–Si and Se–Te are two examples that do. The isomorphous curve of components A and B is shown in Fig. B.2. Above the top line (liquidus line), all of the material is melted. Below the bottom line (solidus line), all of the material is frozen. In the region between the two curves, there will be both liquid and solid material. The behavior of a frozen (solid) isomorphous alloy of composition X is as follows: As the temperature is increased, melting will begin when the temperature reaches T_1.

[1]Despite the fact that the data must be obtained experimentally, the phase diagram concept is based on thermodynamical principles, and some general rules are available to help in shaping the curves. For more details, the reader should consult a standard text on phase diagrams.

FIGURE B.1

Al–Si phase diagram. (*Source:* Max Hansen and Kurt Anderko, *Constitution of Binary Alloys*, © 1958 by the McGraw-Hill Book Co., New York. Used with permission.)

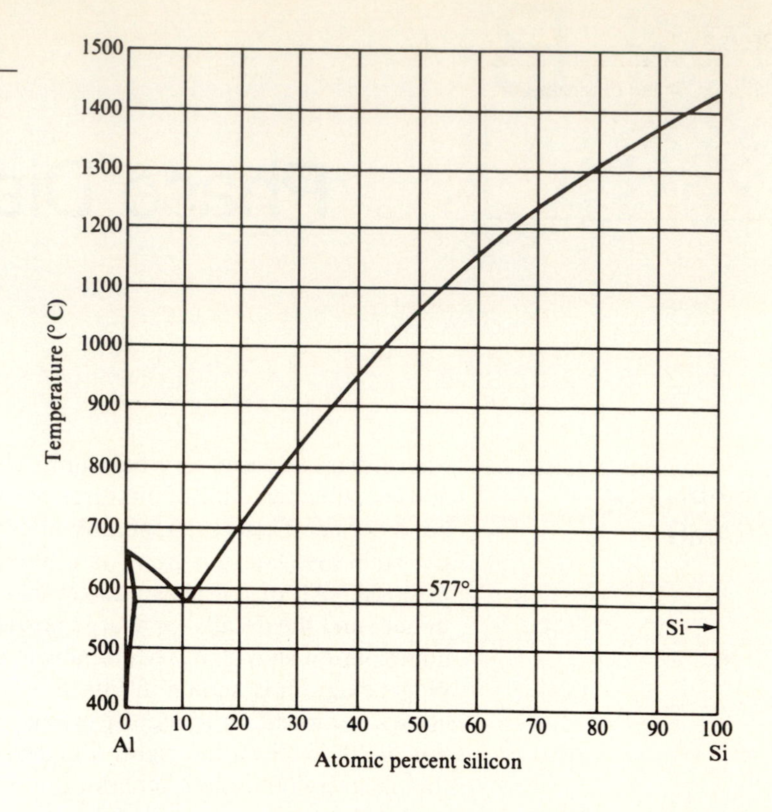

The composition of the melt (composition Z) is given by the point at which the horizontal constant temperature line (tie line) intersects the liquidus curve. (The tie line gives the composition of the two phases that can co-exist in equilibrium at a given temperature and pressure.) As the temperature is raised the tie line moves upward, indicating a change in the composition of both the melt and the remaining solid. Finally, at T_3, the solid disappears, and the alloy is completely melted. In cooling, the reverse occurs. All of the material remains molten until temperature T_3, at which point material with composition Y begins to freeze out. As the temperature decreases, the composition of the solid gradually changes[2] until temperature T_1 and composition X is reached, at which point the melt is gone, and the material is completely frozen.

Fig. B.3 shows the classical eutectic phase diagram. The region

[2]It must be noted that under equilibrium conditions, it is not just the material melting or just frozen that has a composition determined by the intersection of the tie line with the solidus curve. *All* of the solid material must have that composition. That is, the melting or freezing process must proceed slowly enough for solid-state diffusion to maintain the equilibrium composition of the solid.

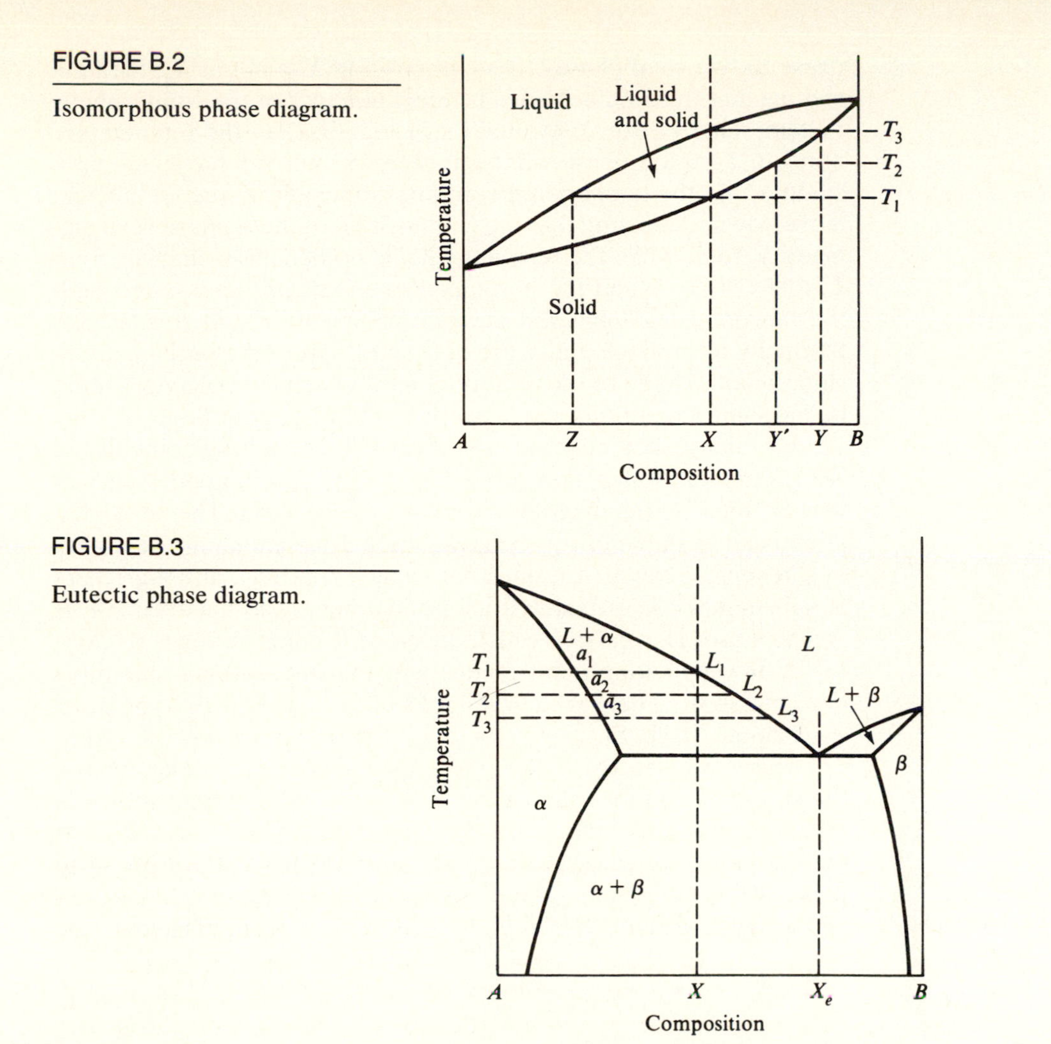

FIGURE B.2

Isomorphous phase diagram.

FIGURE B.3

Eutectic phase diagram.

labeled α encloses the α solid phase, which is comprised of component A saturated with component B. The maximum amount of B at any temperature that can be in the α phase is read from the curve that separates the α region from the liquid or the α + β region. Similarly, the β region is composed of component B saturated with component A. If the temperature for a given composition is above the top curve, all material will be molten. The minimum in the liquidus line is the eutectic point, and the composition X_e will have the lowest melting point for the alloy A–B. Further, all material of this composition will melt and freeze isothermally at the eutectic temperature, producing a mixture of α and β. Thus, at the eutectic point, as the temperature is lowered, there is a change from a liquid phase to two solid phases. If a composition X is not at X_e, then the melting and freezing behavior is much like that of the isomorphous system. That is, for X located as in Fig. B.3, as the temperature drops to T_1, the α phase will begin freezing and will have a composition α_1. As the temperature continues to decrease, the tie line will move down, indicating that the composition of the α phase is changing to α_2 to α_3 while the remaining liquid composition changes from L_1 to L_2 to L_3. When the horizontal line (eutectic line) is reached, the remaining melt has the eutectic composition and freezes isothermally to produce a mixture of α and β. Were the initial composition point X to lie to the right of point X_e, then the behavior would be the same except that the β phase would freeze out first.

For many alloy constituents, there will be so little solubility of A in B and/or B in A that the α and β regions will appear only as vertical lines on the diagram, as shown in Fig. B.4a. The Al–Si diagram is an example of this behavior in that the aluminum solubility in silicon is so low that it does not show[3] (Fig. B.1), although there is substantial silicon solubility in aluminum. Another example is Au–Si. Cases also exist in which the eutectic composition is so close to the lower melting point component that the eutectic minimum does not show. Then, the curve looks like either Fig. B.4b or B.4c. An example is Sb–Si.

There are also two other phase diagrams in the eutectic class, the eutectoid and the monotectic. The eutectoid diagram, shown in Fig. B.5a, looks like the eutectic except that all phases are solid. At higher temperatures, some sort of transition from the γ phase to liquid will occur. There do not appear to be any examples with silicon as a constituent. Cr–Ni is thought to show such behavior. The

FIGURE B.4

Effect of terminal solid solubility range on eutectic phase diagram.

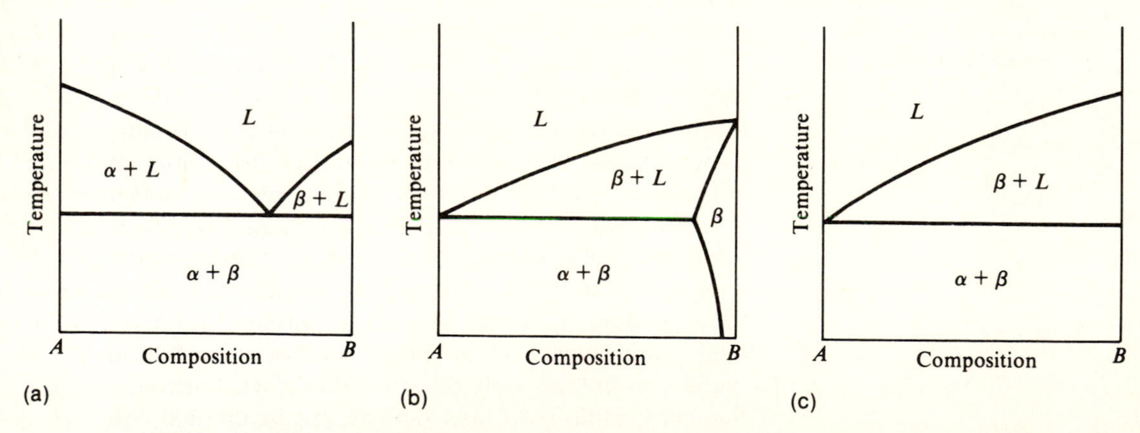

[3]Curves for the solubility of various elements in silicon are given in Chapter 8.

FIGURE B.5

(A) Eutectoid and (b) monotectic phase diagrams.

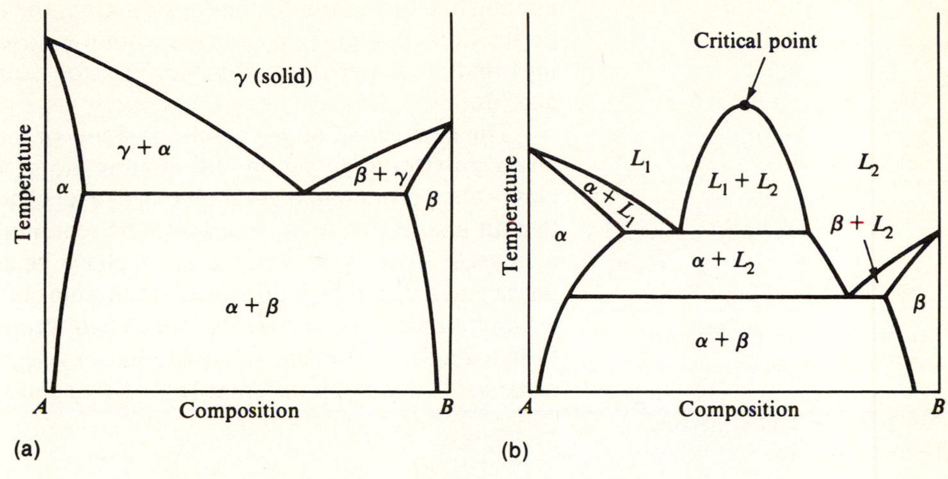

monotectic diagram is shown in Fig. B.5b. At the monotectic composition, if the temperature is high enough, one liquid phase exists, but as the temperature is decreased, it separates into two liquid phases. As the temperature is dropped further, a solid phase forms, and one liquid phase remains. Further cooling gives a mixture of two solid phases. Bi–Si and Pb–Si behave in this manner.

A peritectic transformation occurs if, as the temperature rises, a solid phase transforms into a liquid and a new solid phase. Fig. B.6a shows a peritectic phase diagram. If the composition is at the peritectic value, decreasing temperature causes the α phase to freeze out until the peritectic line is reached (T_p), at which time the liquid and α combine to form the β phase. Such reactions are generally extremely slow since the first β to form will create a diffusion barrier that partially isolates the α phase from the remaining liquid. Often, the phase formed during the peritectic reaction will not involve one of the terminal phases but rather a compound of the two terminal phases. In this case, the curve might appear as in Fig. B.6b. This figure also illustrates a feature exhibited by many phase diagrams, which is that the overall phase diagram may be a composite incorporating different reactions in different compositional ranges. In this particular case, for compositions with B less than Y, the system behaves as peritectic. However, if there is more B than Y, the system is eutectic. X and Z are, respectively, the peritectic and the eutectic compositions. In the event that there is little compositional range for each solid phase (α, β, γ), the diagram reduces to that of

Fig. B.6c, which is more typical of those with silicon as a constituent. If the peritectic reaction occurs very close to one of the terminal phases, the curve can look like that of Fig. B.4b or B.4c. In this circumstance, without further information, the difference between a peritectic and a eutectic diagram cannot be ascertained. Silicon alloys that show peritectic behavior are, for example, As–Si, Ca–Si, and Mo–Si.

The peritectoid and syntectic systems of the peritectic class are analogous to the eutectoid and monotectic systems of the eutectic class. The peritectoid diagram looks like a peritectic diagram except that all phases are solid. In a syntectic reaction, as the temperature increases, instead of a single solid phase separating into a liquid phase and a different solid phase, it separates into two liquid phases.

Sometimes, a combination of the two terminal components will form a single solid phase that will melt to give only one phase. This behavior is distinctly different from either the eutectic system (two

FIGURE B.6

(a) Peritectic phase diagram. (b) Phase diagram showing both peritectic and eutectic behavior. (c) Effect of narrow terminal phases.

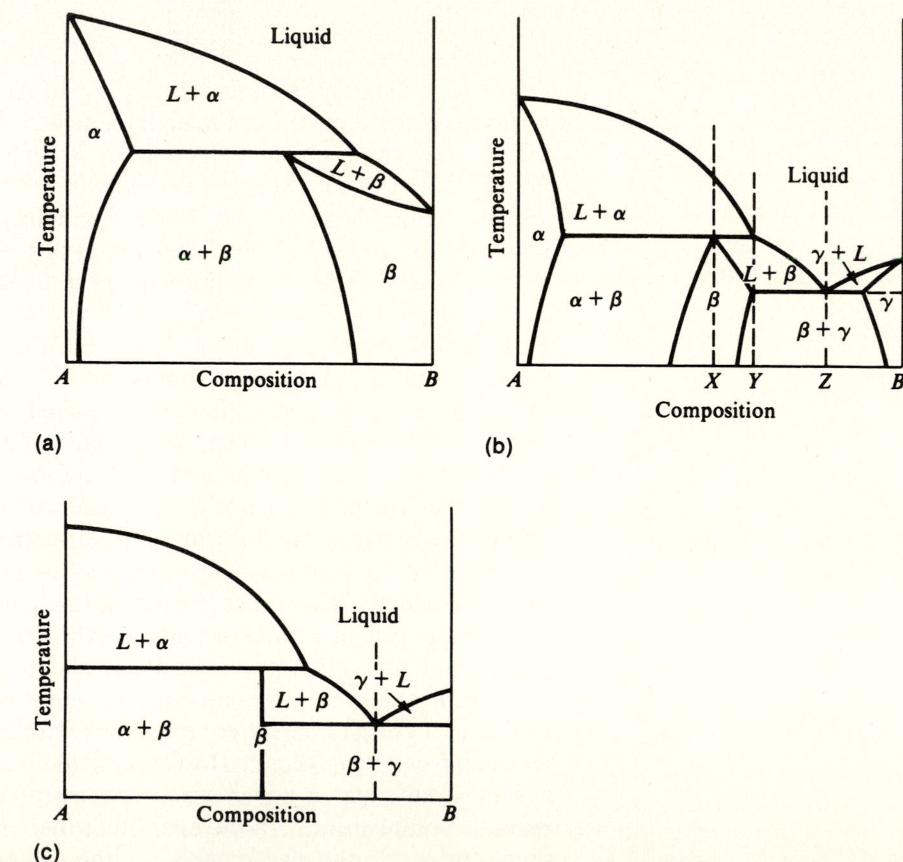

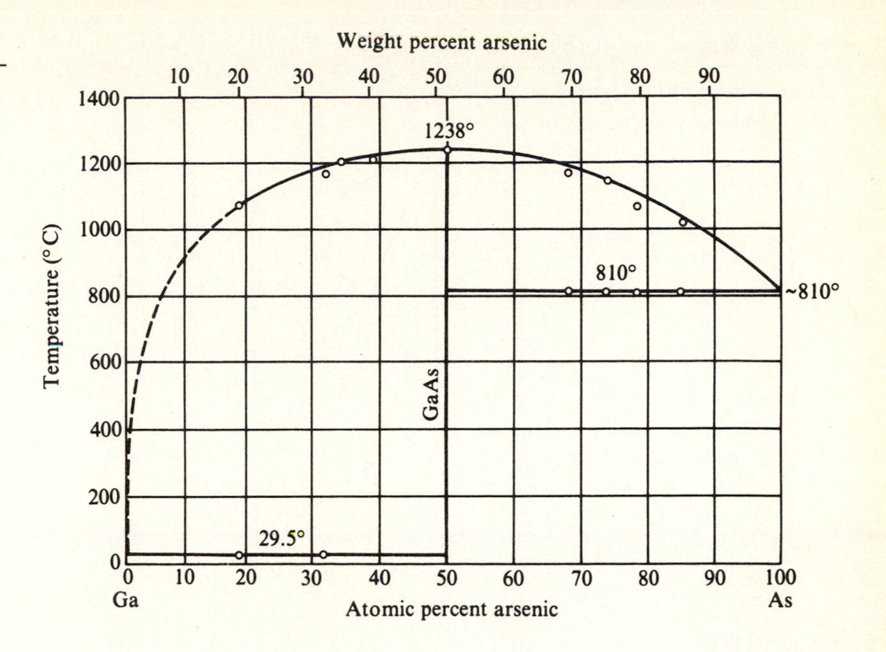

FIGURE B.7

Ga–As phase diagram. (*Source:* Max Hansen and Kurt Anderko, *Constitution of Binary Alloys,* © 1958 by the McGraw-Hill Book Co., New York. Used with permission.)

solid phases melt to give a liquid) or the peritectic system (one solid phase melts to give another solid phase plus a liquid) and is referred to as congruent melting. Such phases are sometimes called intermetallics and include compound semiconductors such as gallium arsenide, gallium phosphide, and indium antimonide. Also included are many of the silicides discussed in Chapter 10, although some of them apparently undergo peritectic (incongruent) melting, which includes, for example, Pt_5Si_2. When a congruently melting phase is formed, it divides the phase diagram into independent sections as is shown, for example, in the As–Ga diagram of Fig. B.7. The diagram is divided into two sections, each of which looks like Fig. B.4c.

B.2 TERNARY PHASE DIAGRAMS

When three elements are involved, such as gallium, arsenic, and phosphorus, the phase diagrams are referred to as ternary and become much more complex. An equilateral triangle is used as the base of a three-dimensional composition–temperature diagram as shown in Fig. B.8. The three corners of the triangle (Gibbs triangle) are labeled, respectively, as component *A*, component *B*, and component *C*. Along each leg, the percent composition scale goes from 0 to 100%. Each of the three sides of the triangular cross-sectional prism of Fig. B.8 depicts a conventional binary phase diagram of the components *A*–*B*, *B*–*C*, and *C*–*A*. In this example, each pair of components forms a simple eutectic of the form shown in Fig. B.4a. By taking a section parallel to the base at a given temperature, an iso-

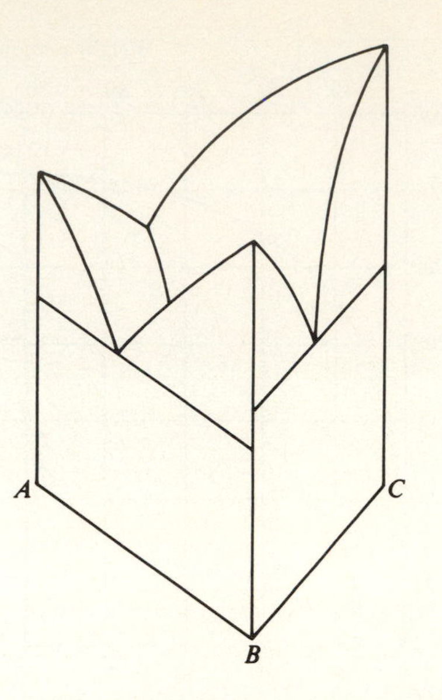

FIGURE B.8

Three-dimensional view of a ternary phase diagram.

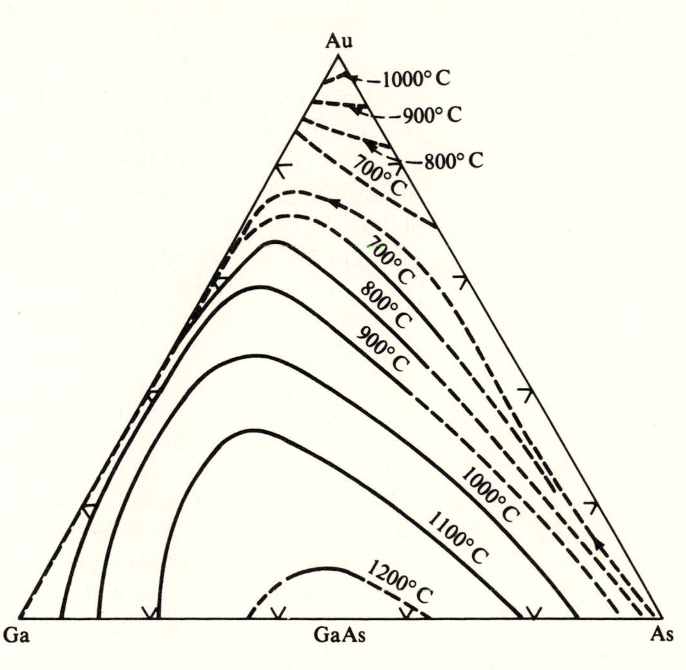

FIGURE B.9

Isothermal section of Ga–As–Au ternary phase diagram. (Because this diagram is a particularly simple one, there are only one or two curves per temperature section. Thus, curves for several temperatures can be combined on one section.) (*Source:* Adapted from M.B. Panish, *J. Electrochem. Soc. 114,* p. 516, 1967.)

thermal section involving all three components is obtained as shown in Fig. B.9. This particular section is not from the diagram of Fig. B.8, but rather is of Ga–As–Au (2). To follow the behavior as a function of temperature, it is then necessary to have a series of sections covering the temperature range of interest. In this figure, the intersection of the solidus lines with the plane is shown for the plane

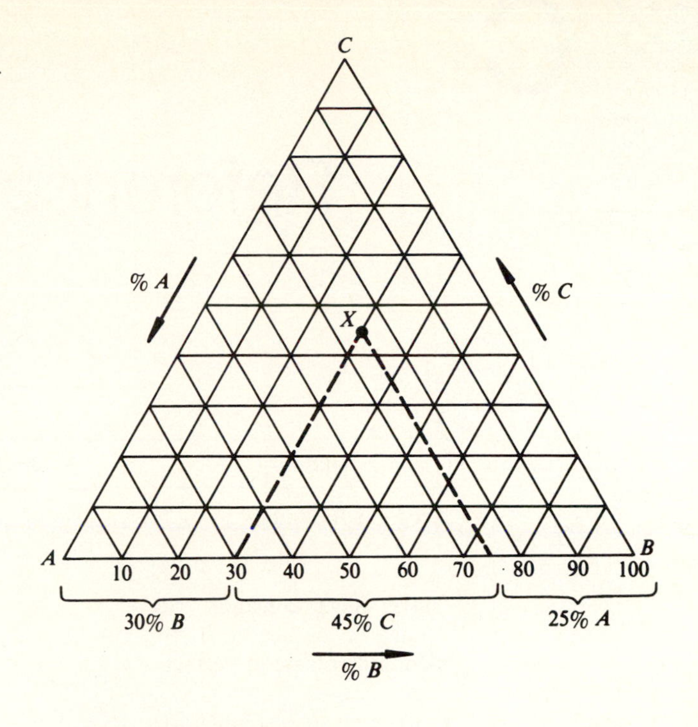

FIGURE B.10

Determination of composition from triangular scale. (Normally, the interior lines will not be drawn on the diagram. This scale is shown moving counterclockwise. Some diagrams use a clockwise scale.)

located at several different temperatures. Thus, one figure suffices to show several different temperature sections. The composition of the material at any point X in the triangle is read from the scales as shown in Fig. B.10. Draw a line through X parallel to the triangle leg opposite apex A. The intersection of the line with either of the other two legs gives the amount of A. Use a similar construction to find the amount of B. C can either be read as shown in the figure or calculated from $C = 100 - A - B$.

Since, as was discussed earlier, a congruently melting compound can be used as a terminal phase of a phase diagram, it is not unusual, particularly in the case of oxides, to find, for example, an SiO_2–P_2O_5 diagram. This concept can also be extended to ternary phase diagrams; SiO_2–P_2O_5–B_2O_3 is an example (3). The result is that some four-component, such as Si–P–B–O, systems can be described with the procedures normally used for ternary materials.

APPENDIX
REFERENCES B

1. Max Hansen and Kurt Anderko, *Constitution of Binary Alloys*, 2d ed., McGraw-Hill Book Co., New York, 1958.
2. Ernest M. Levin, Carl R. Robbins, and Howard F. McMurdie, *Phase Diagrams for Ceramists*, The American Ceramic Society, Columbus, Ohio, 1964.
3. M.B. Panish, "Condensed Phase Systems of Gallium and Arsenic with Group IB Elements," *J. Electrochem Soc. 114*, p. 516, 1967.

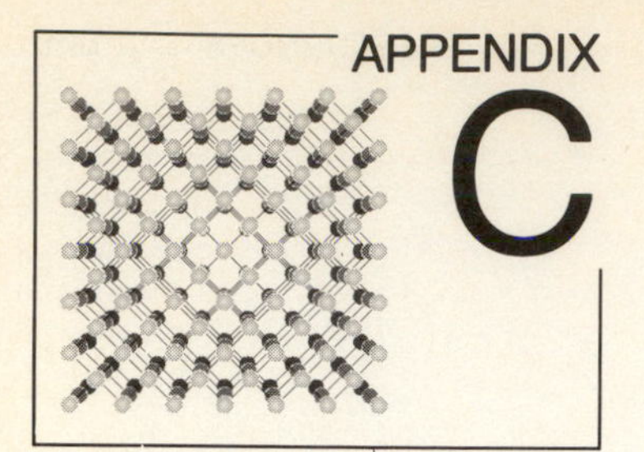

APPENDIX C

Reference Tables

TABLE C.1

Numerical Values of Useful Constants

Quantity	*Symbol*	*Value*
Avogadro's number	n	6.023×10^{23} atoms per gram molecular weight
Universal gas constant	R	8.31×10^{7} ergs/K·mol 1.99×10^{-3} kcal/K·mol
Standard volume of perfect gas	V	22.41 liters/mol
Boltzmann's constant (R/n)	k	8.62×10^{-5} eV/K 1.38×10^{-16} ergs/K 1.38×10^{-23} J/K
Electron charge	q	1.6×10^{-19} C
Electron rest mass	m_o	9.108×10^{-28} gram
Planck's constant	h	4.14×10^{-15} eV·s 6.62×10^{-27} erg·s
Permittivity of free space	ε_0	8.86×10^{-14} F/cm
Permeability of free space	μ_0	1.26×10^{-8} H/cm
Velocity of light	c	3×10^{10} cm/s

TABLE C.2

Length, Energy, and Pressure Conversions

Quantity		*Value*
Length		
1 Å	=	10^{-8} cm
1 nm	=	10^{-7} cm
1 μm (micron)	=	10^{-4} cm
Energy		
1 eV	=	1.602×10^{-19} J
	=	1.602×10^{-12} ergs
	=	3.832×10^{-20} cal

TABLE C.2 (*continued*)

Length, Energy, and Pressure Conversions

Quantity		*Value*
1 e*V*	=	23.05 kcal/mol
1 J	=	10^7 ergs
1 gram-cal	=	4.181 J (at 20°C)
1 W·s	=	1 J = 0.24 cal
Pressure		
1 atm	=	1.013×10^6 dynes/cm^2
	=	14.7 lb/in.^{2}
	=	760 mm Hg
1 mm Hg	=	1333 dynes/cm^2
	=	133.3 Pa
1 dyne/cm^2	=	7.502×10^{-4} mm Hg
1 Pa (pascal)	=	10 dynes/cm^2
	=	1 N/m^2
	=	0.0075 mm Hg
	=	0.0075 torr
1 torr	=	1 mm Hg
	=	133.3 Pa
1 bar	=	10^6 dynes/cm^2
1 kPa	=	10^4 dynes/cm^2
1 MPa	=	10^7 dynes/cm^2
1 GPa	=	10^{10} dynes/cm^2
1 kg/mm^2	=	9.8×10^7 dynes/cm^2
1 lb/in.^{2}	=	6.89×10^4 dynes/cm^2

TABLE C.3

Greek Alphabet

	Lowercase Letter	*Uppercase Letter*		*Lowercase Letter*	*Uppercase Letter*
Alpha	α	Α	Nu	ν	Ν
Beta	β	Β	Xi	ξ	Ξ
Gamma	γ	Γ	Omicron	ο	Ο
Delta	δ	Δ	Pi	π	Π
Epsilon	ε	Ε	Rho	ρ	Ρ
Zeta	ζ	Ζ	Sigma	σ	Σ
Eta	η	Η	Tau	τ	Τ
Theta	θ	Θ	Upsilon	υ	Υ
Iota	ι	Ι	Phi	φ	Φ
Kappa	κ	Κ	Chi	χ	Χ
Lambda	λ	Λ	Psi	ψ	Ψ
Mu	μ	Μ	Omega	ω	Ω

TABLE C.4

International System of Units

Quantity	*Unit*	*Symbol*
Length	meter	m
Mass	kilogram	kg
Time	second	s
Temperature	kelvin	K
Current	ampere	A
Frequency	hertz	Hz (l/s)
Force	newton	N ($kg \cdot m/s^2$)
Pressure	pascal	Pa (N/m^2)
Energy	joule	J ($N \cdot m$)
Power	watt	W (J/s)
Electric charge	coulomb	C ($A \cdot s$)
Potential	volt	V (J/C)
Conductance	siemens	S (A/V)
Resistance	ohm	Ω (V/A)
Capacitance	farad	F (C/V)
Magnetic flux	weber	Wb ($V \cdot s$)
Magnetic induction	tesla	T (Wb/cm^2)
Inductance	henry	H (Wb/A)

Index

SEMICONDUCTOR-RELATED ACRONYMS

Acronym	Meaning
MPY	Multiprobe yield
MR	Metal removal (after a resist patterning step)
MSI	Medium-scale integration
MTF	Modulation transfer function or median time to failure
NA N.A.	Numerical aperture
NAA	Neutron activation analysis
NMOS	n-channel MOS transistor
NSB	Net sales billed
NSE	Net sales entered
NTV	Nonlinear thickness variation
ODE	Orientation-dependent etch
OISF	Oxidation-induced stacking faults
OR	Oxide removal (after a resist patterning step)
OSF	Oxidation-induced stacking faults
PAL	Programmable logic array
PECVD	Plasma-enhanced chemical vapor deposition
PEL	Permissible exposure limit
PGMA	Poly–(glycidyl methacrylate), an E-beam or X-ray resist
PLA	Programmable logic array
PMMA	Poly–(methyl methacrylate), an E-beam or X-ray resist
PMOS	p-channel MOS transistor
PO	Protective overcoat
ppma	Parts per million atomic
PROM	Programmable read-only memory
PSG	Phosphosilicate glass
QA	Quality assurance
QRA	Quality and reliability assurance
RAM	Random access memory
RBS	Rutherford backscattering
RF	Radio frequency
RGA	Residual gas analyzer
RIBE	Reactive ion beam etching
RIE	Reactive ion etching
RISC	Reduced instruction set computing
RO	Reverse osmosis (used in water purification)
ROM	Read-only memory
RTA	Rapid thermal anneal
RTP	Rapid thermal processing

Note: Many of these acronyms are not used in the text; some that are related to devices are included here for the sake of completeness.